Aircraft Stability and Control for Pilots and Engineers

AF740408

Pitman's Aeronautical Engineering Series

The books in this series are intended primarily as textbooks for students at universities and technical colleges who are studying aeronautical engineering in one or more of its various branches with the aim of obtaining degrees or diplomas in aeronautical engineering or allied subjects. Their contents, however, are not confined to the syllabus of any particular university or college, and they may profitably be read by anyone engaged in aircraft engineering, or interested in aeronautical problems, who wishes to further his acquaintance with the subjects covered by the series.

A knowledge of mathematics and physics to approximately the A-level standard of the General Certificate of Education is assumed, but an intelligent reader who has not reached that standard should have no difficulty in following the main parts of the text. The majority of books in the series contain worked examples, and other questions for the student to work out for himself, and also references for further study—thus the books should be as helpful to those who wish to teach themselves as to those engaged on recognized courses.

It is unfortunate that at the present time, when co-operative efforts are being made between the aircraft industries of the United States, Great Britain and the Continent of Europe, there still exist differences in the systems of units and symbols and even in the aeronautical technical terms, used in these countries; but, while the units and terms usually accepted in aeronautical circles in Great Britain have so far been used in this series, the problem has been borne in mind with a view to making later editions of the books more international in character.

It is intended that the series will eventually cover all subjects which can be classified under the general heading of Aeronautical Engineering, e.g. aerodynamics, stability and control, aircraft propulsion, aircraft materials and structural design, performance, rockets and space flight, and even hovercraft.

Air Vice-Marshal A. C. KERMODE, *R.A.F.* (*Ret'd.*)
Advisory Editor

Aircraft Stability and Control for Pilots and Engineers

Sqn. Ldr. B. Dickinson
MBE, MA, DCAe, AFRAeS, RAF
Formerly Chief Ground Instructor,
Empire Test Pilots' School,
Farnborough, England.
Exchange Professor, United States Air Force
Institute of Technology, Wright-Patterson
Air Force Base, Ohio, U.S.A.

London
SIR ISAAC PITMAN & SONS LTD

First published 1968

SIR ISAAC PITMAN & SONS LTD
Pitman House, Parker Street, Kingsway, London, WC2
Pitman House, Bouverie Street, Carlton, Victoria 3053, Australia
P.O. Box 7721, Johannesburg, Transvaal
P.O. Box 6038, Portal Street, Nairobi, Kenya

PITMAN PUBLISHING CORPORATION
20 East 46th Street, New York, N.Y. 10017

SIR ISAAC PITMAN & SONS (CANADA) LTD
Pitman House, 381–383 Church Street, Toronto

DEDICATION
To My Mother

SBN: 273 43910 3

Made in Great Britain at the Pitman Press, Bath

Preface

The material incorporated in this book is based on the lecture notes I have prepared for the stability and control section of the course at the Empire Test Pilots' School, Farnborough, England. Although initially intended for student test pilots, it is hoped that this material will be of value to all aeronautical engineering students, and to scientists working in flight test departments. The aim throughout has been to give a clear physical explanation of the problems involved; the amount of mathematics has been kept to a minimum and no numerical design data has been included. It is thought that the book will provide a useful introduction to the more advanced textbooks and reports referred to in the text.

In addition to dealing with the various aspects of stability and control of conventional aircraft designs, the problems associated with canard aircraft, variable sweep aircraft, slender delta wing aircraft, helicopters and other types of V.T.O.L. aircraft have also been considered. A number of topics have been included (e.g. power controls, aeroelasticity and automatic control systems) which, although important in any considerations of stability and control, are themselves very specialized fields, and I would like to express my appreciation to Mr. J. Lang (Chapter 12), Mr. J. C. A. Baldock (Chapter 14) and Mr. G. C. Howell (Chapter 19) for reading through these sections of the manuscript and making helpful criticisms.

I am grateful to the Ministry of Aviation for granting me permission to publish the material in this book, and to Group Captain R. A. Watts, A.F.C., R.A.F., the Commandant of the Empire Test Pilots' School, for his encouragement during its preparation.

Acknowledgement must also be made to the Controller of Her Majesty's Stationery Office, the Advisory Group for Aerospace Research and Development (AGARD), Macmillan Co. Ltd., Prentice-Hall International Inc., the American Institute of Aeronautics and

Astronautics and Sir Isaac Pitman & Sons Ltd. for permission to reproduce figures from some of their publications.

Finally, I would like to acknowledge the support of my wife and to thank her for her encouragement during the time spent in writing this book.

Farnborough, Hants.
England

B. DICKINSON

Contents

List of Symbols

The following is a list of the more important symbols used in this book. However, since other authors often use different symbols and different definitions, it is essential when reading aeronautical literature to make sure how a particular symbol is being used and defined.

Symbol	*Meaning and definition*
A	Aspect ratio: $A = b^2/S$.
A (or I_x)	Moment of inertia about the rolling ($0x$) axis.
A	Rate of change of wing lift coefficient with incidence: $A = \dfrac{\partial C_L}{\partial \alpha}$
A_1, A_2, A_3	Rate of change of tailplane lift coefficient with tailplane incidence, elevator angle and elevator tab angle, respectively: $A_1 = \dfrac{\partial C_L'}{\partial \alpha'}, A_2 = \dfrac{\partial C_L'}{\partial \eta}, A_3 = \dfrac{\partial C_L'}{\partial \beta}$
A_1, A_2	Coefficients of the λ^4 terms in the longitudinal and lateral stability quartic equations, respectively.
a	Speed of sound.
a, a_1, a_2, a_3	Values of the wing lift and tailplane lift equation coefficients A, A_1, A_2, and A_3 when they are independent of forward speed.
a_0	Coning angle of a helicopter rotor blade.
$a_1, a_2 \ldots$	Coefficients in the helicopter rotor blade flapping equation.
B (or I_y)	Moment of inertia about the pitching ($0y$) axis.
B_0	Value of elevator hinge moment coefficient when $\alpha' = \eta = \beta = 0$.

B_1, B_2, B_3	Rate of change of elevator hinge moment coefficient with tailplane incidence, elevator angle and elevator tab angle, respectively: $$B_1 = \frac{\partial C_H}{\partial \alpha'}, B_2 = \frac{\partial C_H}{\partial \eta}, B_3 = \frac{\partial C_H}{\partial \beta}$$
B_1	Helicopter longitudinal cyclic pitch control angle.
B_1, B_2	Coefficients of the λ^3 terms in the longitudinal and lateral stability quartic equations, respectively.
b_0, b_1, b_2, b_3	Values of the elevator hinge moment equation coefficients B_0, B_1, B_2 and B_3 when they are independent of forward speed.
b	Wing span.
b	Damping coefficient.
$b_1, b_2 \ldots$	Coefficients in the helicopter rotor blade flapping equation.
C (or I_z)	Moment of inertia about the yawing ($0z$) axis.
C_1, C_2	Coefficients of the λ^2 terms in the longitudinal and lateral stability quartic equations, respectively.
C_L	Lift coefficient: $C_L = \frac{L}{\frac{1}{2}\rho V^2 S}$ (L = Lift).
C_L'	Tailplane lift coefficient: $C_L' = \frac{L'}{\frac{1}{2}\rho V^2 S}$.
C_D	Drag coefficient: $C_D = \frac{D}{\frac{1}{2}\rho V^2 S}$.
C_M (or C_m)	Pitching moment coefficient: $C_M = \frac{M}{\frac{1}{2}\rho V^2 S\bar{c}}$.
C_{M0} (or C_{m0})	Zero-lift pitching moment coefficient.
C_H	Hinge moment coefficient: $C_H = \frac{H}{\frac{1}{2}\rho V^2 S_\eta c_\eta}$.
C_R	Resultant aerodynamic force coefficient.
C_l	Rolling moment coefficient: $C_l = \frac{L}{\rho V^2 S(b/2)}$. ($L$ = Rolling moment).
C_n	Yawing moment coefficient: $C_n = \frac{N}{\rho V^2 S(b/2)}$.
$C_{1/2}$ (or $N_{1/2}$)	Number of cycles to half amplitude.
c	Wing chord.

$\bar{c}$ — Standard mean chord: $\bar{c} = S/b$ (also used in this book for the mean aerodynamic chord, see Chapter 1, para. 3.2).

$\bar{\bar{c}}$ — Mean aerodynamic chord.

c_ξ, c_η, c_ζ — Chord of the control aft of the hinge line for the aileron, elevator and rudder, respectively.

D (or I_{yz}) — Product of inertia about $0y$ and $0z$.

D — Drag.

D — Differential operator: $D \equiv \frac{d}{dt}$ or $\frac{d}{d\tau}$.

D_1, D_2 — Coefficients of the λ terms in the longitudinal and lateral stability quartic equations, respectively.

E (or I_{xz}) — Product of inertia about $0x$ and $0z$.

E_1, E_2 — Terms independent of λ in the longitudinal and lateral stability quartic equations, respectively.

F (or I_{xy}) — Product of inertia about $0x$ and $0y$.

g — Acceleration due to gravity (32·2 ft/sec^2).

H — Hinge moment.

H — Helicopter rotor H-force.

H_m — Stick-fixed manoeuvre margin: $H_m = h_m - h$.

H_m' — Stick-free manoeuvre margin: $H_m' = h_m' - h$.

H_n — Stick-fixed static margin (Simple theory).
Stick-fixed C.G. margin: $H_n = h_n - h$.

H_n' — Stick-free static margin (Simple theory).
Stick-free C.G. margin $H_n' = h_n' - h$.

h — Non-dimensional distance of the C.G. aft of the leading edge (of the mean aerodynamic chord) of the wing.

h_m — Non-dimensonal distance of the stick-fixed manoeuvre point aft of the leading edge of the wing.

h_m' — Non-dimensional distance of the stick-free manoeuvre point aft of the leading edge of the wing.

h_n — Non-dimensional distance of the stick-fixed neutral point aft of the leading edge of the wing.

h_n' — Non-dimensional distance of the stick-free neutral point aft of the leading edge of the wing.

h_0	Non-dimensional distance of the wing aerodynamic centre aft of the leading edge of the wing.
I_1	Moment of inertia of helicopter rotor blade about the flapping hinge.
i_A, i_B, i_C	Non-dimensional moment of inertia coefficients about the rolling, pitching, and yawing axes, respectively.
i_E	Non-dimensional product of inertia coefficient.
i	Longitudinal dihedral; the tailplane setting relative to the wing.
K	Response factor.
K_m	Restoring margin.
K_n	Stick-fixed static margin: $K_n = \left(\frac{-dC_M}{dC_R}\right)_{fixed}$; on simple theory $K_n = H_n$.
K_n'	Stick-free static margin: $K_n' = \left(\frac{-dC_M}{dC_R}\right)_{free}$; on simple theory $K_n' = H_n'$.
k	Spring stiffness coefficient.
L	Lift.
L	Rolling moment.
L' (or L_T)	Tailplane lift.
l' (or l_T)	Tail arm.
l_F	Fin arm.
l_p	Rate of change of rolling moment coefficient with rate of roll: $l_p = \frac{\partial C_l}{\partial(pb/2V)} = \frac{L_p}{\rho VSs^2}$.
l_r	Rate of change of rolling moment coefficient with rate of yaw: $l_r = \frac{\partial C_l}{\partial(rb/2V)} = \frac{L_r}{\rho VSs^2}$.
l_v (or l_β)	Rate of change of rolling moment coefficient with sideslip velocity: $l_v = l_\beta = \frac{\partial C_l}{\partial \beta} = \frac{L_\beta}{\rho V^2 Ss} = \frac{L_v}{\rho VSs}$.
l_ξ	Rate of change of rolling moment coefficient with aileron deflection: $l_\xi = \frac{\partial C_l}{\partial \xi} = \frac{L_\xi}{\rho V^2 Ss}$.
l_ζ	Rate of change of rolling moment coefficient with rudder deflection: $l_\zeta = \frac{\partial C_l}{\partial \zeta} = \frac{L_\zeta}{\rho V^2 Ss}$.

M	Mach number: $M = \frac{V}{a}$.
M	Pitching moment.
m	Mass of aircraft: $m = \frac{W}{g}$.
m_e	Gearing between the stick and the control surface.
m_q	Rate of change of pitching moment coefficient with rate of pitch.
m_u	Rate of change of pitching moment coefficient with velocity increment along $0x$ axis.
m_w	Rate of change of pitching moment coefficient with velocity increment along $0z$ axis.
$m_{\dot{w}}$	"Downwash-lag" derivative.
m_η	Rate of change of pitching moment coefficient with elevator deflection.
N	Normal force.
N	Yawing moment.
$N_{1/2}$ (or $C_{1/2}$)	Number of cycles to half amplitude.
n	Total (or indicated) g.
n_a	Applied g. In a pull-out $n_a = (n - 1)$.
n_p	Rate of change of yawing moment coefficient with rate of roll: $n_p = \frac{\partial C_n}{\partial(pb/2V)} = \frac{N_p}{\rho V S s^2}$.
n_r	Rate of change of yawing moment coefficient with rate of yaw: $n_r = \frac{\partial C_n}{\partial(rb/2V)} = \frac{N_r}{\rho V S s^2}$.
n_v (or n_β)	Rate of change of yawing moment coefficient with sideslip velocity: $n_v = n_\beta = \frac{\partial C_n}{\partial \beta} = \frac{N_\beta}{\rho V^2 S s} = \frac{N_v}{\rho V S s}$.
n_ζ	Rate of change of yawing moment coefficient with rudder deflection: $n_\zeta = \frac{\partial C_n}{\partial \zeta} = \frac{N_\zeta}{\rho V^2 S s}$.
n_ξ	Rate of change of yawing moment coefficient with aileron deflection: $n_\xi = \frac{\partial C_n}{\partial \xi} = \frac{N_\xi}{\rho V^2 S s}$.

P (or T)	Periodic time of oscillation.
P	Stick force.
P	Atmospheric pressure.
P_0	Standard atmospheric pressure at sea-level.
p	Relative atmospheric pressure: $p = \frac{P}{P_0}$.
p	Rate of roll.
$\bar{p}$	Non-dimensional rate of roll: $\bar{p} = p\hat{t}$.
q	Dynamic pressure: $q = \frac{1}{2}\rho V^2 = \frac{1}{2}\rho_0 V_i^2$.
q	Rate of pitch.
$\bar{q}$	Non-dimensional rate of pitch: $\bar{q} = q\hat{t}$.
R	Resultant aerodynamic force.
R	Helicopter rotor disc radius.
R	Routh's Discriminant.
Re	Reynolds Number.
r	Radius of turn.
r	Rate of yaw.
$\bar{r}$	Non-dimensional rate of yaw: $\bar{r} = r\hat{t}$.
S	Wing area.
S_F	Fin area.
S' (or S_T)	Tailplane area.
S_ξ, S_η, S_ζ	Area behind the hinge of the aileron, elevator and rudder control surfaces, respectively.
s	Wing semi-span: $s = \frac{1}{2}b$.
T (or P)	Periodic time of oscillation.
T	Thrust force.
T	Absolute temperature.
T_0	Standard absolute temperature at sea-level.
t	Time.
$\hat{t}$	Unit of aerodynamic time: $\hat{t} = \frac{W}{g\rho SV} = \frac{m}{\rho SV} = \frac{VC_L}{2g}$.
$t_{1/2}$ (or $T_{1/2}$)	Time to one-half of the initial amplitude.
$t_{1/e}$ (or $T_{1/e}$)	Time to $1/e$ of the initial amplitude, i.e. the time constant.
t_2 (or T_2)	Time to double the initial amplitude.

U	Component of the velocity of the C.G. along $0x$ in steady flight. [*Note:* When wind axes are being used this is the same as the forward speed of the aircraft and $U \equiv V$, where V represents the forward speed].
u	Component of the velocity disturbance along $0x$.
$\bar{u}$	Non-dimensional value of u: $\bar{u} = \dfrac{u}{V}$.
V	Component of the velocity of the C.G. along $0y$ in steady flight. [*Note:* When wind axes are being used $V = 0$].
V	Forward speed of the aircraft relative to the air. True air speed, T.A.S.
V_i (or V_e)	Equivalent airspeed: E.A.S. $= V\sqrt{\sigma}$.
$\bar{V}$	Tail volume ratio: $\bar{V} = \dfrac{S'l'}{S\bar{c}}$.
$\bar{V}_F$	Fin volume ratio: $\bar{V}_F = \dfrac{S_F l_F}{Sb}$.
v	Component of the velocity disturbance along $0y$. Sideslip velocity.
v_e	Equivalent sideslip velocity: $v_e = v\sqrt{\sigma}$.
$\bar{v}$ (or β)	Non-dimensional sideslip velocity: $\bar{v} = \dfrac{v}{V}$. For small sideslip angles $\bar{v} = \beta$.
W	Weight of aircraft.
W	Component of the velocity of the C.G. along $0z$ in steady flight. [*Note:* When wind axes are being used $W = 0$].
w	Component of the velocity disturbance along $0z$.
$\bar{w}$	Non-dimensional value of w: $\bar{w} = \dfrac{w}{V}$. For small values of w, $\bar{w} = \Delta\alpha$.
w_g	Vertical gust velocity.
X	Component of the force along $0x$.
Y	Component of the force along $0y$.
y_v (or y_β)	Rate of change of lateral force coefficient with sideslip velocity: $y_v = y_\beta = \frac{1}{2}\dfrac{\partial C_Y}{\partial \beta} = \dfrac{Y_\beta}{\rho V^2 S} = \dfrac{Y_v}{\rho V S}$.

Z Component of the force along 0*z*.

z_w Rate of change of the *Z*-force coefficient with velocity disturbance along 0*z*.

z_η Rate of change of the *Z*-force coefficient with elevator deflection.

Greek Symbols

α	alpha	Angle of incidence. Angle of attack.
α'		Tailplane incidence.
α_0		No-lift angle of attack.
β	beta	Tab angle.
β		Angle of sideslip: $\beta \simeq \frac{v}{V}$ (where $V =$ forward speed).
β		Helicopter rotor blade flapping angle.
Γ	gamma	Dihedral angle.
γ	gamma	Lock number or the blade mass factor of a helicopter: $\gamma = \frac{c\rho a R^4}{I_1}$.
γ		Ratio of the specific heats of a gas: $\gamma = \frac{c_p}{c_v}$.
Δ	delta	Increment in . . .
δ	delta	Helicopter rotor lag angle.
δ		Logarithmic decrement (i.e. log. dec.).
$\delta_a, \delta_e, \delta_r$ (or ξ, η, ζ)		Aileron, elevator and rudder control surface deflections, respectively.
ε	epsilon	Downwash angle.
ε		Angle of inclination of the principal inertia axis.
ε_D		Damping angle.
ζ	zeta	Rudder control surface deflection.
ζ		Relative damping.
η	eta	Elevator control surface deflection.
η_T		All-moving tailplane deflection.
θ	theta	Angle of pitch.

θ		Relative temperature: $\theta = \frac{T}{T_0}$.
Λ	lambda	Sweepback angle.
λ	lambda	Root of stability quartic: $\lambda = -\mu \pm i\omega$.
μ	mu	Coefficient of viscosity.
μ		Damping factor. Real part of a complex λ root.
μ		Helicopter tip speed ratio: $\mu \simeq \frac{V}{\Omega R}$.
μ_1		Longitudinal aircraft relative density: $\mu_1 = \frac{W}{g\rho Sl'} = \frac{m}{\rho Sl'}$.
μ_2		Lateral aircraft relative density: $\mu_2 = \frac{W}{g\rho Ss} = \frac{m}{\rho Ss}$.
ξ	xi	Aileron control surface deflection.
π	pi	Constant = 3·1416.
ρ	rho	Air density.
ρ_0		Standard air density at sea-level.
σ	sigma	Relative air density: $\sigma = \frac{\rho}{\rho_0}$.
τ	tau	Non-dimensional time: $\tau = \frac{t}{\hat{t}}$ airsecs.
ϕ	phi	Angle of bank.
χ	chi	Constant of proportionality in eqn. (2.65).
ψ	psi	Constant of proportionality in eqn. (2.64).
ψ		Angle of yaw.
ψ		Azimuth angle of helicopter rotor blade measured from downwind position in the direction of rotation.
Ω	omega	Rate of spin.
Ω		Helicopter rotor r.p.m.
ω	omega	Frequency (rads/sec). Imaginary part of a complex λ root.
ω_n		Natural undamped frequency.

American Symbols and Sign Conventions

When reading any book or report on the subject of stability and control, it is essential to note very carefully the definitions of the symbols being used and the particular sign convention employed (see Chapter 1). The American systems are similar to the British; the system of axes used is the same, as are many of the symbols (u, v, w, p, q, r, etc.) and the definitions of some of the basic coefficients C_L, C_D, C_m, etc. There are, however, some important differences that should be noted—

(i) The American system does not use many of our abbreviations, e.g. $C_{L\alpha}$ is used, instead of a, for the lift curve slope; I_{xx}, I_{yy} and I_{zz}, instead of A, B and C, for the moments of inertia; C_{H_α}, $C_{H_{\delta e}}$, $C_{H_{\delta t}}$, instead of b_1, b_2 and b_3, for the hinge moment characteristics of a control surface, etc.

(ii) The American system uses δ as the symbol for control surface deflection and denotes the particular control surface concerned by the appropriate suffix. Thus δ_a, δ_e and δ_r are used instead of ξ, η and ζ to denote aileron, elevator and rudder deflections, respectively. Particular care must be taken over the sign convention used for control deflections; often the definition of "positive control deflections produce positive control moments" is used, which is the opposite of ours.

(iii) On the American system the unit of aerodynamic time is often defined in the same way as ours, though the symbol τ is used, instead of $\hat{t}$. Sometimes an alternative definition $\tau = l/V$ (or $\tau = b/2V$ for lateral work) is used, where l is a longitudinal linear dimension of the aircraft, b is the wing span and V is the undisturbed true airspeed. (It should be noted that V is being used here for the forward speed of the aircraft; it is sometimes replaced by V_0, or, when wind axes are being used, by U.) The American relative density parameters are denoted by the symbol μ and usually defined in the longitudinal case by $m/\rho S\bar{c}$ and in the lateral case by $m/\rho Sb$. The corresponding British parameters are $\mu_1 = m/\rho Sl'$ and $\mu_2 = m/[\rho S(\frac{1}{2}b)]$ and by comparing these definitions it can be seen that $\mu_1 = \mu\bar{c}/l'$ and $\mu_2 = 2\mu$.

(iv) The American notation for the stability and the control derivatives is C_{mq}, C_{lr}, C_{nr}, C_{mu}, etc., instead of the British m_q, l_r, n_r, m_u, etc. These derivatives have the same physical interpretations (e.g. C_{m_q} is the damping-in-pitch derivative), but they are defined in different ways. The American and British symbols can be related to one another; taking C_{m_q} and m_q as an example, the respective definitions are given by

$$C_{m_q} = \frac{\partial C_m}{\partial(q\bar{c}/2V)} \text{ and } m_q = \frac{\partial M/\partial q}{\rho VSl'^2}$$

In both systems $C_m = \dfrac{M}{\frac{1}{2}\rho V^2 S\bar{c}}$ and thus C_{m_q} can be written

$$C_{m_q} = \frac{\partial\left(\dfrac{M}{\frac{1}{2}\rho V^2 S\bar{c}}\right)}{\partial(q\bar{c}/2V)} = \frac{\partial M/\partial q}{\frac{1}{2}\rho V^2 S\bar{c}(\bar{c}/2V)}$$

Substituting $m_q\rho VSl'^2$ for $\partial M/\partial q$, we have

$$C_{m_q} = \frac{m_q\rho VSl'^2}{\frac{1}{2}\rho V^2 S\bar{c}(\bar{c}/2V)}$$

hence

$$C_{m_q} = \frac{4l'^2}{\bar{c}^2} m_q$$

Similar relationships can be derived between the other derivatives and in some cases the British and American derivatives are identical to one another (Ref. 1, pages 118 and 373).

References

1. Babister, A. W.; *Aircraft Stability and Control* (Pergamon Press, 1961).
2. Seckel, E.; *Stability and Control of Airplanes and Helicopters* (Academic Press, New York, 1964).
3. NACA and NASA Reports.
4. *Cornell Aeronautical Reports*, Cornell University.
5. *USAF Stability and Control Datcom* (Oct. 1960, revised Nov. 1965), Wright-Patterson Air Force Base, Ohio.

Author's notes

(i) In this book the pitching moment coefficient is denoted by C_M, though C_m is more usually used. C_l and C_n are used to denote the rolling and yawing moment coefficients, to prevent confusion with C_L and C_N, the lift and normal force coefficients, respectively.

(ii) For printing convenience the quantities C_{L_α}, C_{M_0} . . . have been printed as $C_{L\alpha}$, C_{M0} . . ..

(iii) After this book had been written a report was issued (*RAE Technical Report* 66200, Parts 1 to 5, June 1966) which proposed a new system of notation and nomenclature applicable to the dynamics of aircraft and missiles. Though basically similar to the old system there are some important changes in the notation used and the definitions employed. The report contains conversion tables showing the relationships between the old and the new symbols. The new system has been adopted by the Royal Aeronautical Society and it is incorporated in their new Dynamics Data Sheets.

2—(A.203)

Dimensions, Units and Non-dimensional Quantities

Quantities arising in Mechanics and Aeronautics can be expressed in terms of three fundamental physical quantities—mass M, length L and time T. The units commonly used for these quantities are the SLUG for the unit of mass, the FOOT for the unit of length and the SECOND for the unit of time. The dimensions and units of some important quantities are given in Table I. The force is a derived quantity, since it can be expressed as the product of a mass and an acceleration and it has the dimensions of MLT^{-2}. The unit of force is the POUND and the slug is defined as that mass which acted upon by a force of 1 lb accelerates at a rate of 1 ft/sec² (i.e. 1 slug $\equiv$ 32·2 lb mass). It is important to realize that though the actual units of a particular quantity vary from one different system of units to another, the physical dimensions of the quantity remain the same.

The analysis of the stability and control of aircraft is usually carried out in terms of non-dimensional groups. Any quantity can be made non-dimensional by dividing by some other quantity with the same units and in order to standardize the work it is necessary to define the quantities that are to be used as the divisors. The lift coefficient C_L, for example, is widely used to represent the lift characteristics of an aircraft; it is non-dimensional, and defined by

$$C_L = \frac{L}{\frac{1}{2}\rho V^2 S}$$

It is advantageous to work in terms of these non-dimensional coefficients, because they are defined in a way that makes them independent of the size of the aircraft and the speed and height at which it is flying (though they are subject to scale effects and compressibility effects). The values of the coefficients are also independent of the particular system of units being employed. Table II shows the standard divisors used for deriving the non-dimensional stability and control derivatives and the non-dimensional inertia coefficients.

Table I. The Physical Dimensions and Units of Various Quantities

QUANTITY	PHYSICAL DIMENSIONS	UNITS
Area A	L^2	ft^2
Velocity V	$L\,T^{-1}$	ft/sec
Acceleration a	$L\,T^{-2}$	ft/sec^2
Force or Weight W	$M\,L\,T^{-2}$	slugs ft/sec^2 or lb
Pressure P	$M\,L^{-1}\,T^{-2}$	lb/ft^2
Density ρ	$M\,L^{-3}$	$slugs/ft^3$
Viscosity μ	$M\,L^{-1}\,T^{-1}$	slugs/ft sec or lb sec/ft^2
Moment	$M\,L^2\,T^{-2}$	lb ft
Energy or Work	$M\,L^2\,T^{-2}$	ft lb
Momentum	$M\,L\,T^{-1}$	slugs ft/sec
Power	$M\,L^2\,T^{-3}$	ft lb/sec
Moment of Inertia	$M\,L^2$	slugs ft^2
Angle θ	— (i.e. $M^0\,L^0\,T^0$)	radians
Angular Velocity $\dot{\theta}$	T^{-1}	rads/sec
Angular Acceleration $\ddot{\theta}$	T^{-2}	$rads/sec^2$

Table II. The Non-dimensional Inertia Coefficients and Stability and Control Derivatives

DIMENSIONAL QUANTITIES	UNITS	DIVISORS	NON-DIMENSIONAL QUANTITIES	DESCRIPTION
$X\ Y\ Z$	lb	$\frac{1}{2}\rho V^2 S$	$C_x\ C_y\ C_z$	Force coefficients
$L\ M\ N$	lb ft	$\frac{1}{2}\rho V^2 S l_1$	$C_l\ C_m\ C_n$	Moment coefficients
X_u Z_u Y_v X_w Z_w	$\frac{\text{lb}}{\text{ft/sec}}$	$S\rho V$	x_u z_u y_v x_w z_w	Force-velocity derivatives
Y_p X_q Z_q Y_r	$\frac{\text{lb}}{\text{rad/sec}}$	$S\rho V l_2$	y_p x_q z_q y_r	Force-angular velocity derivatives
M_u L_v N_v M_w	$\frac{\text{lb ft}}{\text{ft/sec}}$	$S\rho V l_2$	m_u l_v n_v m_w	Moment-velocity derivatives
L_p N_p M_q L_r N_r	$\frac{\text{lb ft}}{\text{rad/sec}}$	$S\rho V l_2^2$	l_p n_p m_q l_r n_r	Moment-angular velocity derivatives
Y_ξ X_η Z_η Y_ζ	$\frac{\text{lb}}{\text{rad}}$	$\rho V^2 S$	y_ξ x_η z_η y_ζ	Force-control movement derivatives
L_ξ N_ξ M_η L_ζ N_ζ	$\frac{\text{lb ft}}{\text{rad}}$	$\rho V^2 S l_2$	l_ξ n_ξ m_η l_ζ n_ζ	Moment-control movement derivatives
$M_{\dot{w}}$	$\frac{\text{lb ft}}{\text{ft/sec}^2}$	$\frac{Wl_2}{g}$	$m_{\dot{w}}$	Moment-downwash lag derivative
$A\ B\ C$ E	slugs ft²	$\frac{Wl_2^2}{g}$	$i_A\ i_B\ i_C$ i_E	Inertia coefficients

where l_1 and l_2 are defined below.

	LONGITUDINAL STABILITY GROUP	LATERAL STABILITY GROUP
l_1	$\bar{c}$ = mean chord	b = wing span
l_2	l' = tail arm	$s = \frac{1}{2}b$ = semi-span

1 Introduction to Stability and Control

1. Introduction

Though an aircraft is designed mainly from performance considerations, it must also possess acceptable handling characteristics. However, if necessary, these can be achieved by artificial methods. An aircraft is completely free to move in space, and its motion can be represented with respect to a set of three mutually perpendicular axes. It is desirable that the aircraft should be stable about each of these three axes. The aircraft must also be controllable because the pilot needs to be able to trim the aircraft in flight, and, when necessary, to manoeuvre the aircraft about each of the axes.

Stability and control are closely inter-related since the amount of stability possessed by an aircraft determines the control forces and movements that are required to trim and manoeuvre the aircraft. The problems are made more complicated by the effects of compressibility on the aerodynamic characteristics and by the fact that the aircraft is not a rigid structure but suffers from aeroelastic distortion.

Various stability and control criteria have been laid down for aircraft. Some of these criteria can be expressed in a quantitative form, but the latest tendency is to rely more on the pilot's qualitative assessments. The assessment of the handling qualities of an aircraft is dependent upon the combined effect of the stability and control characteristics. The physical limitations of a human pilot may be an important factor in the design of a control system, though on most high-performance aircraft the pilot's force limitation is eliminated by the use of power-operated controls.

2. The Aerodynamics of Aircraft Design

2.1. Design for Performance

The overall design of any aircraft is always a compromise between the various aerodynamic, structural, propulsion, economic, and operational requirements. The amount of emphasis placed on each of these particular factors depends on the type of aircraft being considered. In the case of a military aircraft, the operational requirements are the predominant ones and economic considerations are only of secondary importance. On a civilian transport aircraft, however, the economy of operation is one of the major considerations in the design.

There is no one single design that is the best solution for any given specification. Different designers will arrive at different final solutions, by compromising in different ways. This explains why a particular specification can be satisfied more or less equally well by a number of designs which may vary widely in their external configuration. As the top speed requirement of aircraft has increased, considerable changes have had to be made to the design shapes of aircraft because, unfortunately, a configuration which is the optimum for one speed range is not the optimum for a different speed range.

The performance of an aircraft is mainly determined by its thrust/drag characteristics. For a given amount of installed thrust, the level flight top speed of an aircraft can be increased by reducing the overall drag. The climbing and manoeuvring performance of the aircraft depends on the excess thrust available, and the range performance is determined by the cruise drag of the aircraft and the specific fuel consumption of the engine. The low speed performance, on the other hand, is usually determined by the value of the maximum lift coefficient that the wing can produce, although pilot view, passenger comfort, and certain stability and control characteristics (e.g. pitch-up), may also be important factors.

The engines can be mounted in a number of different places on the aircraft depending on the type and number of engines that are being installed. They are usually buried in the fuselage or in the wing roots, mounted in pods under the wing, or in nacelles on the wing. If jet engines are used, then the designs of the intake and the exhaust nozzle are particularly important in determining the efficiency of the installation. On aircraft designed for high supersonic speeds, it is essential that the intake and exhaust nozzle are of variable geometry design, in order to maintain a high efficiency throughout the speed range of the aircraft.

During the last few years, the rear-fuselage engine layout has been used quite a lot on civilian transport designs. A number of advantages are claimed for this configuration:

(i) The wing is "cleaner" and more efficient; the drag is reduced and high lift devices (e.g. trailing-edge flaps and leading-edge flaps and slots) are more effective because of the absence of cut-outs.

(ii) The asymmetric yawing moment resulting from an engine failure is relatively small.
(iii) The noise in the passenger cabin is reduced.

Even this design is not an ideal one however and there are a number of disadvantages associated with it:

(i) There may be an increase in drag due to interference between the flow round the engine nacelles and the rear fuselage.
(ii) The wing has to be made stronger to carry the aerodynamic loads in flight because when the engines are positioned on the wings they tend to relieve these loads to some extent. The fuselage also has to be made stronger to support the engines. This results in an appreciable increase in all-up weight.
(iii) There may be possible engine intake problems, particularly at the higher angles of attack.
(iv) Because the engines are attached to the fuselage where the tailplane is normally positioned, this has to be moved. Most rear-engined aircraft have a T-shaped tail, with the horizontal tailplane on the top of the vertical fin. This particular configuration may lead to very serious stability and control problems.

Some project design studies were made recently for the Boeing 737 twin turbofan aircraft in which both wing-mounted and fuselage-mounted engine layouts were investigated. Boeing's concluded that, in fact, the rear-engine configuration had few advantages and that a large saving in weight could be achieved by reverting to the more conventional design with wing-mounted nacelles. They found that the rear-engine configuration did not necessarily have better C_{Lmax} or lower drag characteristics, and that there was no appreciable noise reduction during the cruising phase of the flight. It would appear that the popularity of the rear-engine layout has been based more on aesthetic grounds than on technical ones.

On older aircraft designs the airframe and engine were usually designed separately. Within recent years, however, the airframe and the engine have been more closely integrated in the design of aircraft and "thrust" and "lift" are no longer distinct quantities (Ref. 1). The exhaust of the engine has been used both directly (e.g. as in the Hawker 1127 Kestrel) and indirectly (e.g. in boundary layer control and jet-flap designs) to produce or increase the lift of the aircraft.

The total drag of an aircraft is made up of a number of separate parts and the relative importance of these drag components varies throughout the speed range of the aircraft. The total drag can be broken down in various ways and different names are used for the component parts. At subsonic speeds the drag is usually considered to be made up of form drag, skin friction drag and vortex (or induced) drag. The form drag and skin friction are sometimes added together and referred to as profile or "lift-independent drag," whereas vortex drag is referred to as "lift-dependent drag."

Form drag can be reduced by conventional "streamlining" techniques and skin friction drag by maintaining smooth surfaces and by the application of boundary layer control methods. Vortex drag is associated with the wing tip vortices of the aircraft, and it can be reduced by increasing the span (i.e. increasing the aspect ratio) of the wing. Form drag and skin friction drag both increase with increase in forward speed ($\propto V^2$), whereas vortex drag decreases with increasing speed ($\propto 1/V^2$). Thus vortex drag is the dominant factor in the design of a low speed aircraft and the optimum shape consequently consists of the "classical" configuration of a streamline-shaped load-carrying fuselage supported by a high aspect ratio unswept rectangular or slightly tapered wing.

At high subsonic speeds additional forms of drag appear; these are associated with the shock wave pattern which is produced round the aircraft at these speeds. The onset of the compressibility drag rise on an aircraft can be delayed by the use of sweepback. Sweepforward would be equally effective in this respect but as we shall see later it does lead to serious aeroelastic problems. For this reason large amounts of forward sweep are not used, though various studies incorporating M- and W-shaped wings have been made for low supersonic speed aircraft designs and some aircraft have been built with a limited amount of sweepforward. Though sweepback is a very useful design feature for reducing the drag of the aircraft, it does introduce some unfavourable handling characteristics (e.g. "pitch-up"). The optimum configuration for this speed range consists either of

(i) a sweptback wing planform with a separate fuselage, or
(ii) an integrated, or partially integrated, delta-wing configuration.

The wing span and aspect ratio values are lower than those of the optimum subsonic design. If the horizontal tailplane is eliminated to reduce the overall drag, then the aircraft may suffer from certain stability and control deficiencies (e.g. poor manoeuvrability and spin recovery characteristics, lack of longitudinal damping, etc.). To keep the overall drag to as low a value as possible, the aircraft has to be designed in accordance with the transonic area rule. To satisfy this rule the cross-sectional area of the aircraft, measured at right angles to the longitudinal axis, must vary smoothly along the length of the aircraft from the nose to the tail. On some aircraft this has resulted in a "waisting" of the fuselage, but this is not always necessary. In other cases, extra cross-sectional area has had to be provided by the addition of bulges on the fuselage or nacelle-like bodies on the wings. These bodies are sometimes referred to as "Whitcomb bodies" or "Küchemann carrots."

At supersonic speeds the wave drag becomes predominant. It is made up of two components, the wave drag due to lift and the wave drag due to volume. The wave drag due to volume can be reduced by increasing the length of the aircraft (i.e. by increasing the fineness ratio), and the wave drag due to lift can be reduced by spreading the lift out over the length of the aircraft, and by designing the wing so that the chordwise distribution of lift is elliptically shaped. (This should be compared with the elliptical spanwise lift distribution that is required to minimize the induced drag.) The optimum configuration for speeds between about $M = 2$ and $M = 3$ can be shown to be that of the integrated slender-delta wing layout, with a span/length ratio of between 0·4 and 0·5 and an aspect ratio of about unity.

A detailed discussion of the determination of the various optimum configurations throughout the speed range from $M = 0$ to $M = 5{\cdot}0$ is given in Ref. 2.

2.2. Design for Stability and Control

Having determined the basic design of the aircraft on the basis of the performance requirements, it is then necessary to ensure that the stability and control characteristics are satisfactory throughout its speed and height range. Aerodynamic modifications can be made to the design of the aircraft to achieve the desired handling qualities, as long as these modifications do not reduce the performance of the aircraft below the required specification.

The philosophy of the design of an aircraft from stability and control considerations is summarized in Ref. 3. To begin with, as long as aircraft were manually operated by the pilot, the aim of the designer was to produce an aircraft which possessed inherent stability in all conditions of flight. As the forward speeds have increased, the design configuration of the aircraft has changed and aeroelastic and compressibility effects have become more important. The design changes have resulted in important changes in both the aerodynamic and the inertia characteristics of aircraft. On low speed designs the mass of the aircraft is usually fairly evenly distributed about the roll and pitch axes and the ratio of the pitching inertia B to the rolling inertia A is of the order of unity. On high speed designs however the mass is mainly concentrated in the fuselage and the B/A ratio is very much greater. As a result of these changes, it has become very much more difficult to design an aircraft which possesses satisfactory stability and control characteristics throughout the complete speed and height range.

Many modern high-performance aircraft suffer from certain stability and control deficiencies within their flight envelope. It is often impossible to remedy these by natural aerodynamic means and various forms of stability augmentation systems (e.g. autostabilizers, auto-throttles, auto-Mach trimmers, etc.) have to be incorporated into the design in order to achieve acceptable handling qualities. The use of these devices gives the aircraft designer very much more freedom, since he is able to design the aircraft more from considerations of performance and operational efficiency, without having to compromise the design in order to obtain the required handling characteristics. The basic stability of the aircraft may be allowed to vary over a wide range, even, possibly, becoming negative, the use of the stability augmentation systems ensuring satisfactory handling characteristics. However, although these augmentation systems can be duplicated or triplicated to provide the necessary safety and reliability, it is desirable to retain as much natural stability as possible. As Ref. 3 points out, if an aerodynamic solution can be found, then, once obtained, it requires no maintenance, whereas the various artificial stabilization systems do.

In the case of simple non-powered control systems, the various control characteristics of an aircraft (e.g. stick forces, stick movements, aircraft responses, etc.) are very closely related to the corresponding stability characteristics. The human pilot can be considered as "closing" the loop of the airframe/control system and his physical capabilities and limitations are important factors in the design of the control system of the aircraft. Within recent years it has been necessary to incorporate power-operated controls and various types of automatic feedback control systems (e.g. manoeuvre demand and adaptive systems). Up to now, nearly all control systems, both powered and non-powered ones, have used mechanical signalling (e.g. rods, cables, etc.) between the pilot and the control surfaces. These mechanical systems suffer from backlash, cable stretch and problems associated with temperature effects and airframe structural distortion. Flight tests have been carried out using electric signalling controls and these will be used on some future aircraft (e.g. the B.A.C./Sud Concorde). The use of electric signalling systems avoids most of the problems associated with mechanical systems. They are also much more accurate, and they enable the signals from the autopilot and the autostabilizers to be mixed in more efficiently.

3. Basic Aerodynamics

3.1. Flow Patterns and Pressure Distributions

The overall performance and stability and control characteristics of the final design are determined by the basic aerodynamic characteristics of the various component parts of the aircraft and the way in which they vary under different conditions of flight. When a body moves through the air, the pattern of the airflow round the body is dependent on a large number of factors, i.e. the shape and size of the body and its attitude relative to the airflow, the density (ρ) and viscosity (μ) of the air and the relative speed of motion (V). By using dimensional analysis, it can be shown that a number of these variables can be grouped together to give two dimensionless parameters, the Reynolds Number ($Re = \rho Vl/\mu$), and the Mach number ($M = V/a$), where l is a convenient dimension of the body representing its size and a is the speed of sound. The density ρ is usually expressed in slugs/ft^3, μ in slugs/ft sec, l in ft, and V and a in ft/sec.

At low forward speeds, the airflow pattern round a body is dependent on three main factors

(i) the shape of the body,
(ii) the attitude of the body,
(iii) the Reynolds Number.

The aerodynamic characteristics of a body usually vary with Reynolds Number, and this variation is known as Scale Effect. It should be carefully noted however that Scale Effect is not just a question of "size," it also involves air density, air viscosity, and the relative airflow velocity. Thus the value of *Re* in wind tunnel experiments can be altered by changing any of these parameters (e.g. a reduction in *Re* due to the small size of a model can be made up for by increasing the air density, and compressed-air wind tunnels are used for this purpose).

As the forward speed is increased, shock waves are produced and these shock waves modify the flow pattern round the body. At supersonic speeds the changes in the flow characteristics (i.e. pressure, flow direction, etc.) are restricted to the shock wave and expansion regions.

As the air flows round a body, there is a change in the local relative airflow velocity and associated with these speed changes there are corresponding pressure changes. As a consequence of the law of conservation of energy, an increase of velocity results in a decrease of pressure and a decrease of velocity results in an increase of pressure. Fig. 1.1*a* shows a typical pressure distribution for a conventional aerofoil section at a low speed and moderate angle of attack and Fig. 1.1*b* shows the theoretical pressure distribution on a double-wedge aerofoil section at supersonic speeds. The pressure distribution can be shown diagrammatically either by representing the pressure changes perpendicular to the surface of the aerofoil (as in Fig. 1.1*a*), or perpendicular to the chord line (as in Fig. 1.1*b*). On a finite wing, the pressure distribution round the wing section varies along the span, due to the three-dimensional

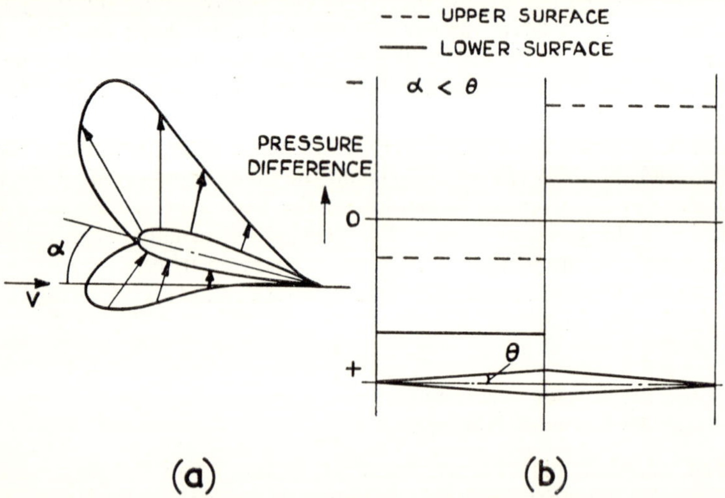

Fig. 1.1. (*a*) Typical subsonic pressure distribution
(*b*) Typical supersonic pressure distribution

effects occurring in the flow. At subsonic speeds, the wing tip effects theoretically extend across the whole span, but at supersonic speeds, on a wing of moderate aspect ratio, the three-dimensional effects are confined to the region of the Mach cones at the wing tips.

3.2. Lift, Drag and Pitching Moments

The overall effect of all the small forces acting on a wing section can be represented in various ways and the actual method used depends on which one is the most convenient for the particular problem being considered. The simplest way (but not, as we shall see later, the most convenient one for stability work) is to represent the overall effect by a single aerodynamic resultant force (R) acting at a particular point on the chord line known as the CENTRE OF PRESSURE (C.P.). The C.P. is not a fixed point, and it usually moves forward with increase of incidence. The resultant force is usually resolved into two directions at right angles to one another, either perpendicular and parallel to the relative airflow, or perpendicular and parallel to the chord line. Resolving with respect to the relative airflow, the components of the resultant force are referred to as "lift" and "drag" (Fig. 1.2)—

Lift L—the component of the resultant aerodynamic force perpendicular to the relative airflow.

Drag D—the component of the resultant aerodynamic force parallel to the relative airflow.

The sectional lift and drag characteristics vary across the span and they can be integrated to give the overall lift and drag of the wing. These aerodynamic forces are usually represented in terms of the corresponding non-dimensional coefficients. Since the lift and drag are forces (with units of lb) they can be made non-dimensional by dividing by any group of quantities which also has the dimensions of a force. In order to standardize the definitions, well-known standard quantities are used as the divisor. The quantity $\frac{1}{2}\rho V^2$ often occurs in our work on aerodynamics; at low speeds this quantity represents the dynamic pressure of the airflow and it is often denoted by the symbol q. It has the units of a pressure (i.e. lb/ft^2). To obtain a force, this standard pressure has to be multiplied by a standard area. The standard area usually chosen for aircraft work is the gross wing area (S ft^2) which is usually obtained by cutting off any overhanging nacelles, etc., and extending the leading and trailing edges of the wing to the centre line of the fuselage (Fig. 1.3). The non-dimensional lift and drag coefficients are then defined by

$$\left.\begin{aligned}\text{Lift coefficient } C_L &= \frac{L}{\frac{1}{2}\rho V^2 S}\\ \text{Drag coefficient } C_D &= \frac{D}{\frac{1}{2}\rho V^2 S}\end{aligned}\right\} \tag{1.1}$$

The forces acting on the aerofoil section produce a pitching moment M about all points on the wing, other than the centre of pressure itself. Because the value of the pitching moment depends on the point about which it is measured, it is essential to specify the particular datum point. In practice, the pitching moments are usually measured about either the leading edge or the $\frac{1}{4}$-chord point of the wing. The non-dimensional pitching moment coefficient C_M is defined by

$$C_M = \frac{M}{\frac{1}{2}\rho V^2 S\bar{c}} \tag{1.2}$$

The pitching moment M has units of lb ft, and therefore, to make the coefficient non-dimensional, it must be divided by a quantity which also has the dimensions of force × length. The quantity $\frac{1}{2}\rho V^2 S$ has been seen to have the dimensions of force and therefore an extra standard length is required in the denominator. The value usually

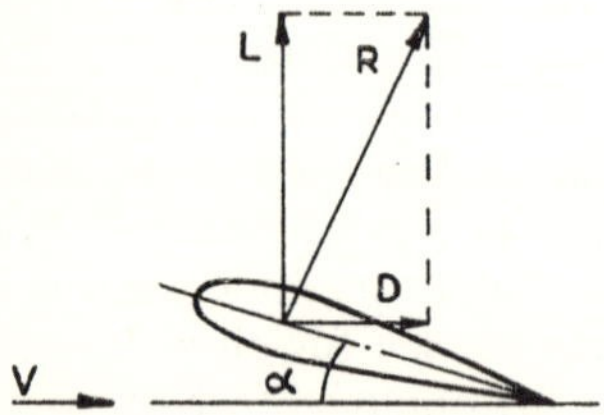

Fig. 1.2. Forces acting on an aerofoil

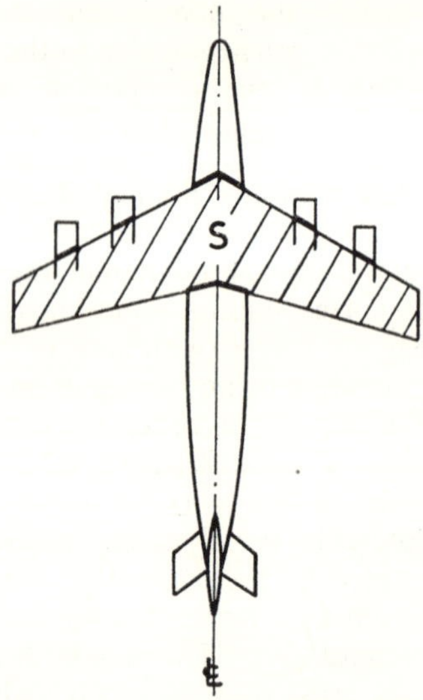

Fig. 1.3. Definition of wing area

used for the standard length is $\bar{c}$, the mean chord of the wing. The mean chord can be defined in two different ways as follows.

(*a*) *Mean Geometric Chord or Standard Mean Chord* (S.M.C.)

The standard mean chord is defined very simply as the ratio of the gross wing area S, divided by the wing span b

$$\bar{c} = \frac{S}{b} \tag{1.3}$$

Thus it merely represents the chord length of the equivalent rectangular wing which has the same gross area and span as the wing being considered. If the semi-span of the wing $\frac{1}{2}b$ is represented by the symbol s, then the area of the wing can be expressed by the equation

$$S = \int_{-s}^{s} c\, dy \tag{1.4}$$

where c is the local chord length of the wing section (Fig. 1.4). Similarly, the wing span can be expressed in the form

$$b = \int_{-s}^{s} dy \tag{1.5}$$

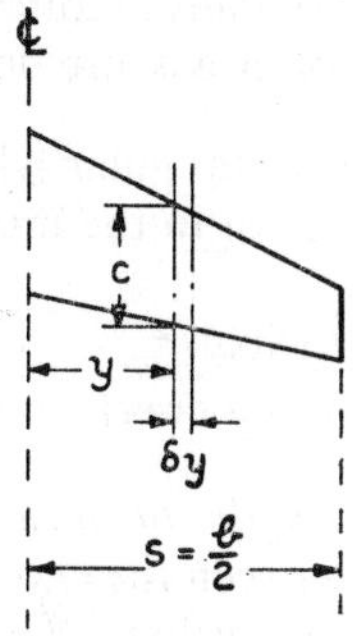

Fig. 1.4. Wing geometry notation

Thus eqn. (1.3) can be rewritten as

$$\bar{c} = \frac{\int_{-s}^{s} c\, dy}{\int_{-s}^{s} dy} \tag{1.6}$$

(*b*) *Mean Aerodynamic Chord* (M.A.C.)

The mean aerodynamic chord is defined by

$$\bar{\bar{c}} = \frac{\int_{-s}^{s} c^2\, dy}{\int_{-s}^{s} c\, dy} \tag{1.7}$$

or alternatively, since $\int_{-s}^{s} c\, dy = S$, by

$$\bar{\bar{c}} = \frac{\int_{-s}^{s} c^2\, dy}{S} \tag{1.8}$$

The mean aerodynamic chord is generally used in the definitions involved in stability and control. However, in this book, we will use the symbol $\bar{c}$ for convenience, simply referring to it as the mean chord of the wing.

3.3. Variation of the Lift and Pitching Moment Characteristics

The values of the non-dimensional coefficients of a wing depend on a large number of factors, e.g. the sectional shape and the way it varies along the span, the sweepback, the aspect ratio, the taper ratio and the geometric twisting of the wing, the angle of attack of the wing and the values of the Reynolds Number and the Mach Number. The aspect ratio and angle of attack are defined as follows—

(i) *Aspect ratio A* On a rectangular wing it is defined simply as the ratio (span/chord), but in the more general case it is given by

$$\text{Aspect ratio} = \frac{(\text{Span})^2}{(\text{Wing area})} = \frac{b^2}{S} \tag{1.9}$$

(ii) *Angle of attack or angle of incidence* α This is usually defined as the angle between the chord line of the wing and the direction of the relative airflow. The terms "angle of attack" and "angle of incidence" are often used synonymously with one another; it is not often necessary to refer to the rigger's angle of incidence, i.e. the angle that a lifting surface is attached to the fuselage of the aircraft.

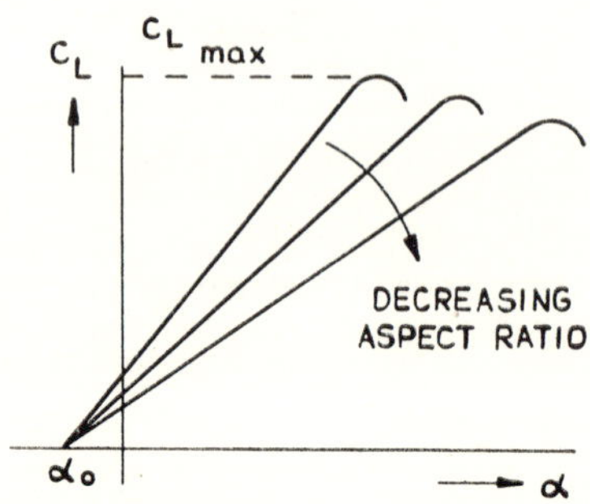

Fig. 1.5. Effect of aspect ratio on the C_L/α curve

Some typical C_L/α curves for a conventional cambered subsonic wing are shown in Fig. 1.5. The lift coefficient is zero at a small negative angle of attack (i.e. the no-lift angle, α_0), it increases linearly with increase of angle of attack up to a value of $C_{L max}$ at the stalling angle, and then the value of the lift coefficient decreases. A reduction in aspect ratio does not change the no-lift angle (α_0), and has little effect on the value of $C_{L max}$, but it decreases the value of the lift curve slope $dC_L/d\alpha$ and increases the value of the stalling angle.

The angle of no-lift is determined by the amount of camber possessed by the wing. Though the overall lift produced by the wing in this condition is zero, small pressure changes still occur round the wing and a typical zero-lift pressure distribution is shown diagrammatically in Fig. 1.6.

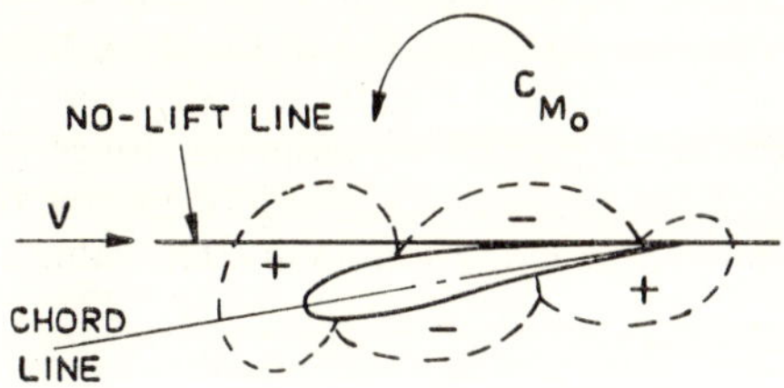

Fig. 1.6. Zero-lift pressure distribution

Since the C_L/α curve is linear over most of its range, the value of C_L can be expressed in the form

$$C_L = \frac{dC_L}{d\alpha}(\alpha - \alpha_0) \tag{1.10}$$

Substituting the symbol a for $dC_L/d\alpha$, this becomes

$$C_L = a(\alpha - \alpha_0) = a_0 + a\alpha \tag{1.11}$$

where $a_0(= -a\alpha_0)$ is the value of the lift coefficient at zero angle of attack. If the angle of attack of the wing is measured with respect to the no-lift line ($\alpha_{N.L.}$) instead of with respect to the chord line (Fig. 1.6), this equation becomes simpler—

$$C_L = a\,\alpha_{N.L.} \tag{1.12}$$

For this reason, in stability work, the angles of attack of the wing and the tailplane are sometimes referred to the no-lift lines.

The value of C_{Lmax} determines the stalling speed and the shape of the C_L/α curve around the stall is important in determining the stalling characteristics. The value of C_{Lmax} is very dependent on Scale Effect and it can be increased by using the various high-lift devices that are available.

The curve of C_M against C_L for a wing section is also usually linear up to the stalling angle. Typical curves for three different reference points are shown in Fig. 1.7, nose-up moments being taken conventially as positive moments. Since the graphs are straight lines, the pitching moment coefficient C_M can be expressed in the form

$$C_M = C_{M0} + \frac{dC_M}{dC_L}C_L \tag{1.13}$$

where C_{M0} is the intercept on the vertical axis (i.e. the pitching moment coefficient at zero lift) and dC_M/dC_L is the slope of the line. It can be seen that the value of C_{M0} is independent of the position of the point about which the moments are measured and that for a conventional cambered aerofoil it usually has a negative (i.e. nose-down) value. This zero-lift moment is present as a result of the small pressure changes occurring round the wing when the overall lift is zero. It can be seen from Fig. 1.6 that the pressure distribution is such as to produce a nose-down moment. Since the total lift is zero, the effects of the pressure distribution can not be represented by a single resultant force acting at a point. The value of the moment

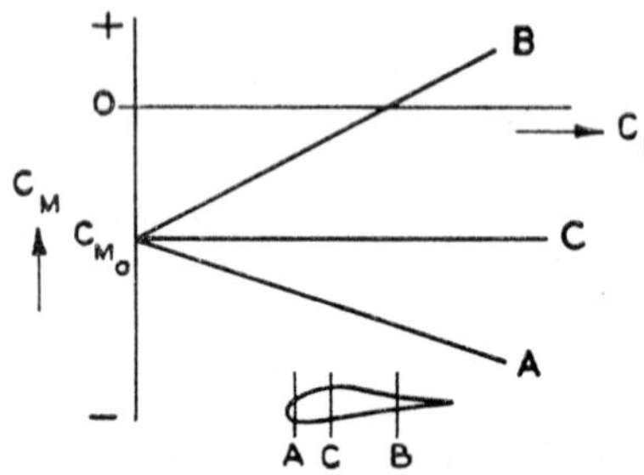

Fig. 1.7. C_M/C_L curves of an aerofoil for different datum points

can be represented by a couple acting on the aerofoil and it is a basic fact that the moment of a couple is independent of the point about which the moments are taken. (It is worthwhile pointing out here that, although the zero-lift condition is not a practical one that will occur in normal level flight, it does provide a very useful reference point for use in the mathematical analysis of stability and control.)

When the reference point is near the leading edge of the aerofoil, the C_M/C_L curve has a negative slope, the pitching moment becoming more nose-down with increase of incidence. As we shall see later, this is the condition required for positive longitudinal static stability. On the other hand, when the reference point is near the trailing edge of the aerofoil, the C_M/C_L curve has a positive slope and the pitching moment becomes less nose-down and, eventually, more nose-up as the incidence is increased. For one particular point, somewhere between the leading and trailing edges of the wing, the slope of the line will be zero and the pitching moment will remain constant with increase of incidence.

However if there is a point about which the pitching moment remains constant while the incidence (and hence the lift) increases, the changes in lift must be acting through this point (otherwise they would produce a change in the pitching moment). This point is known as the AERODYNAMIC CENTRE (A.C.), and it can thus be defined either as

(i) that point about which the pitching moment remains constant with increase of incidence, equal to the value of the zero-lift moment, i.e. $C_M = C_{M0} =$ constant; $dC_M/dC_L = 0$, or
(ii) that point at which the incremental changes in lift produced by changes of incidence can be considered to act (i.e. the pressure changes acting on the wing, as a result of a change in incidence from the no-lift value, can be represented by a resultant force acting at the A.C.).

From eqn. (1.13) it can be seen that the overall effect of the pressure distribution round an aerofoil on the pitching moment characteristics can be considered to be made up of two parts

(i) the zero-lift pressure distribution, which produces the zero-lift moment C_{M0};
(ii) the pressure distribution due to incidence (i.e. lift), which has a resultant going through a fixed point (the Aerodynamic Centre) and which produces the term proportional to C_L.

We thus have an alternative way of representing the overall lift and

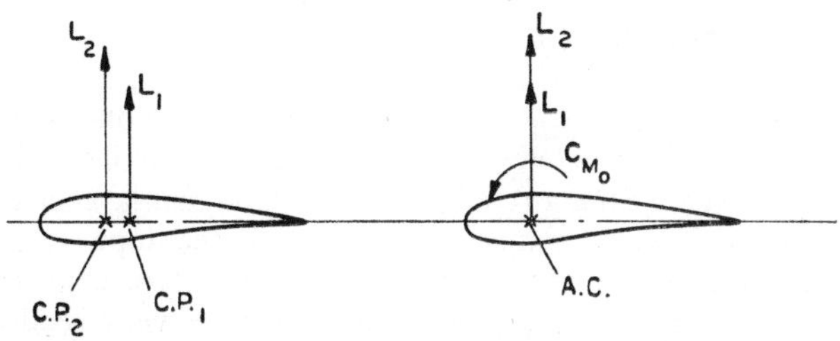

Fig. 1.8. Centre of pressure and aerodynamic centre representations

pitching moment effects of the pressure changes acting round a wing section. The two methods are compared in Fig. 1.8. In the one case the effects are represented by a variable force acting at a variable point (i.e. the Centre of Pressure), and in the other case, by a variable force acting at a fixed point (i.e. the Aerodynamic Centre), together with the constant component of the pitching moment coefficient (i.e. C_{M0}). Though this second method appears to be a more complicated representation, it will be found to be more convenient for use in stability analysis, the important point being that only one of the quantities involved (i.e. the lift coefficient C_L) is varying, the others (i.e. the A.C. position and C_{M0}) are constant.

The pitching moment coefficient equation can be expressed in the form

$$C_M = C_{M0} + C_L\left(\frac{x}{\bar{c}} - \frac{x_{A.C.}}{\bar{c}}\right) \tag{1.14}$$

where x is the distance of the reference point and $x_{A.C.}$ the distance of the aerodynamic centre from, say, the leading edge of the wing. By comparing eqns. (1.13) and (1.14), it can be seen that

$$\frac{dC_M}{dC_L} = \frac{x}{\bar{c}} - \frac{x_{A.C.}}{\bar{c}} \tag{1.15}$$

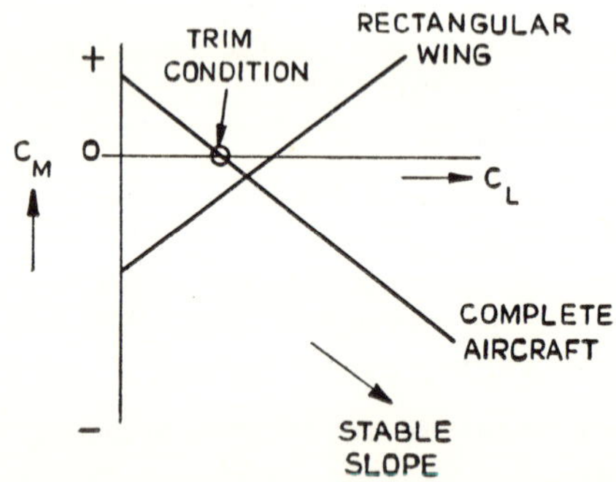

Fig. 1.9. C_M/C_L curve for the complete aircraft

Thus the slope of the C_M/C_L curve is equal to the non-dimensional distance between the reference point (which in practice will be the centre of gravity of the aircraft) and the aerodynamic centre of the wing; if the reference point is ahead of the A.C. the slope is a negative one, if the reference point is behind the A.C. the slope is positive.

At subsonic speeds the aerodynamic centre of wing sections lies at, or near to, the $\frac{1}{4}$-chord point, irrespective of the sectional shape. At supersonic speeds symmetrical sections are often used and the centre of pressure and the aerodynamic centre coincide with one another at about the $\frac{1}{2}$-chord point.

A rectangular finite wing planform will usually have a negative value of C_{M0} and a positive value of dC_M/dC_L (since the C.G. will normally lie farther back than the $\frac{1}{4}$-chord point). As we shall show in Chapter 2, to produce an aircraft that is stable longitudinally and can be trimmed (i.e. $C_M = 0$), the signs of both the quantities C_{M0} and dC_M/dC_L have to be reversed (Fig. 1.9). The zero-lift moment of a wing alone can be made more positive by using aerofoil sections with reflex curvature, or by incorporating sweepback and washout on the wing. In the latter case, when the overall lift of the wing is zero, a download is produced at the tips and an upload at the root and this produces a nose-up pitching moment (Fig. 1.10). The sweeping back of the wings also moves the A.C. of the wing rearwards by a correspondingly greater amount than the C.G. of the aircraft (since most of the weight is in the fuselage) and when the A.C. moves behind the C.G. the aircraft becomes longitudinally stable. The more usual way of achieving these results however is by means of fitting a horizontal tailplane, fixed on the fuselage at a negative angle with respect to the wing. This "longitudinal dihedral" as it is called produces the positive C_{M0} that is required, because

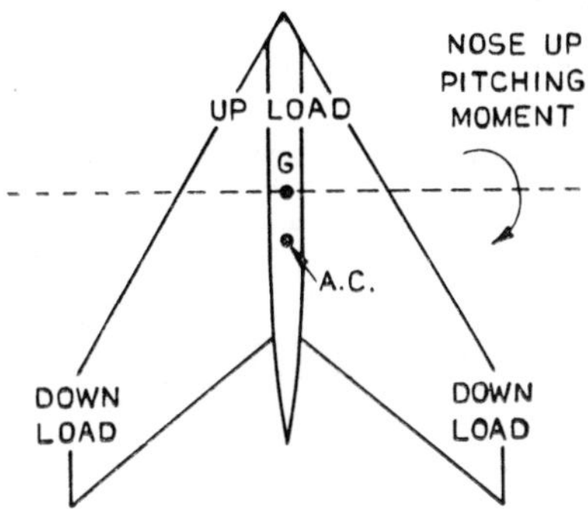

Fig. 1.10. Zero-lift load distribution on a sweptback wing

when the wing is producing no lift the tailplane has a download acting on it which produces a nose-up pitching moment. The presence of the horizontal tailplane also ensures positive longitudinal stability.

3.4. Control Surface Characteristics

The majority of aircraft control systems use the conventional flap-type control, though a number of modern aircraft are using spoilers and all-moving surfaces. Though control systems are dealt with in more detail in Chapters 10 and 12 a few of the basic principles of low speed controls will be introduced here since some of them are referred to in the earlier work on stability. When a flap is deflected relative to the fixed part of an aerofoil, the pressure distributions round both the flap and the aerofoil are altered. The overall change in the aerodynamic force acting on the aerofoil-control surface combination produces a moment about the appropriate aircraft axis (i.e. a rolling, pitching, or yawing moment) whereas the change in the aerodynamic force acting only on the control surface produces a moment acting about the hinge line of the control. These aerodynamic hinge moments usually oppose the motion of the control and in the case of a non-powered control system the pilot has to apply a force to his stick or rudder pedals to overcome these moments.

3.4.1. LIFT CHARACTERISTICS

Let us consider the horizontal tailplane-elevator-elevator trim tab combination. Using eqn. (1.11) we can write down the total lift coefficient C_L' on the whole unit when the tailplane is at an angle of attack of α' and the elevator and elevator trim tab are in their neutral positions, by

$$C_L' = a_0 + a_1\alpha' \tag{1.16}$$

where $a_1 = dC_L'/d\alpha'$, the tailplane lift curve slope. The symbol a_0 represents the tailplane lift at zero angle of attack; this term

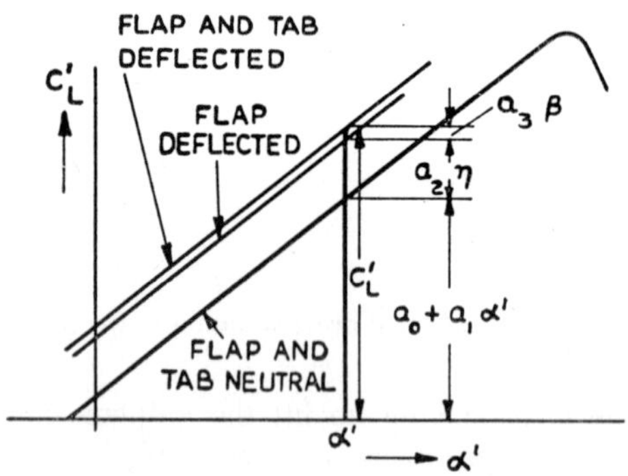

Fig. 1.11. Effect of flap and tab deflection on the C_L'/α' curve

disappears from the equation if the tailplane has a symmetrical section or if the incidence is measured with respect to the tailplane no-lift line.

When the elevator is deflected through an angle η an additional tailplane lift is produced (Fig. 1.11). This additional lift increases linearly with increase of flap deflection and it can be represented by an additional term in the equation equal to $a_2\eta$. Similarly, deflection of the elevator trim tab produces more lift (Fig. 1.11) which can be represented by an additional term $a_3\beta$. If it is assumed that all these lift contributions are independent of one another over the range of incidences and deflections which we are normally interested in, then the total tailplane lift coefficient is given by

$$C_L' = a_0 + a_1\alpha' + a_2\eta + a_3\beta \tag{1.17}$$

where $a_1(=\partial C_L'/\partial\alpha')$ is the rate of change of tailplane lift coefficient with tailplane angle of attack at constant flap and tab deflection,

$a_2(=\partial C_L'/\partial\eta)$ is the rate of change of tailplane lift coefficient with flap deflection at constant tailplane angle of attack and tab deflection, and

$a_3(=\partial C_L'/\partial\beta)$ is the rate of change of tailplane lift coefficient with tab deflection at constant tailplane angle of attack and flap deflection. The values of a_1, a_2 and a_3 can be expressed either in units of per radian or per degree; in eqn. (1.17) the values of α', η and β must be expressed in radians or degrees as appropriate.

The values of a_1, a_2 and a_3 are normally positive, a positive change of incidence and positive (i.e. down) flap and tab deflections producing an increase in tailplane lift. The numerical values are different

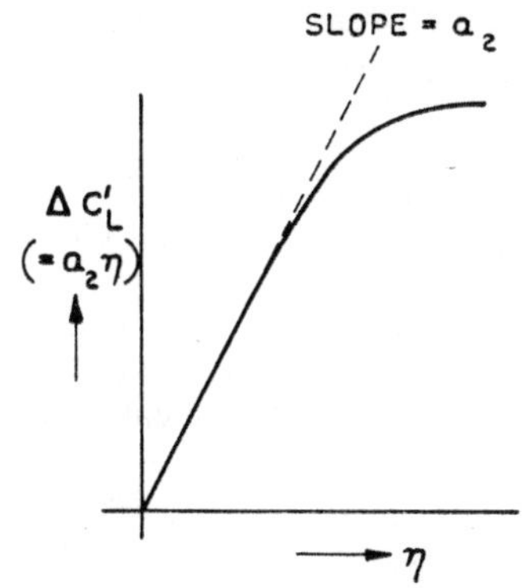

Fig. 1.12. Lift increment due to flap deflection

however, a_1 being greater than a_2 while a_3 is usually small enough to be neglected (i.e. the small tab does not produce an appreciable change in the overall tailplane lift coefficient compared to the effects of the tailplane and the elevator control). In practice the tab is normally deflected in the opposite direction to the elevator and in this case the small $a_3\beta$ term would have the opposite sign to that of the $a_2\eta$ term.

At the higher incidences and flap deflections the characteristics become non-linear (Figs. 1.11 and 1.12) and the values of a_1, a_2 and a_3 are no longer constants. In practice their values are also modified by the effects of compressibility and in the general analysis of stability and control the symbols A_1, A_2 and A_3 are used.

3.4.2. PITCHING MOMENT CHARACTERISTICS

The tailplane lift is given by

$$L' = C_L'(\tfrac{1}{2}\rho V^2)'S' \tag{1.18}$$

where $(\frac{1}{2}\rho V^2)'$ and S' are the values appropriate to the tailplane. The pitching moment produced about the C.G. of the aircraft by this lift force is obtained by multiplying by the tailarm, the distance between the C.G. and the C.P. of the tailplane (l'). The pitching moment of the aircraft is made non-dimensional by dividing by the value of $\frac{1}{2}\rho V^2 S\bar{c}$ appropriate to the wing and hence the tailplane contribution to the overall pitching moment coefficient is given by

$$\text{Tailplane } C_M = \frac{(\frac{1}{2}\rho V^2)'}{(\frac{1}{2}\rho V^2)} \frac{S'l'}{S\bar{c}} C_L'$$

Assuming that $(\frac{1}{2}\rho V^2)' = \frac{1}{2}\rho V^2$ and replacing $\frac{S'l'}{S\bar{c}}$ by $\bar{V}$, the tail volume coefficient,

$$\text{Tailplane } C_M = \bar{V}C_L' = \bar{V}(a_1\alpha' + a_2\eta + a_3\beta) \qquad (1.19)$$

when $a_0 = 0$.

3.4.3. HINGE MOMENT CHARACTERISTICS

The total hinge moment coefficient can be expressed in a similar way by

$$C_H = b_0 + b_1\alpha' + b_2\eta + b_3\beta \qquad (1.20)$$

where b_0 is the value of the hinge moment coefficient when

$$\alpha' = \eta = \beta = 0, \quad b_1 = \frac{\partial C_H}{\partial \alpha'}, \; b_2 = \frac{\partial C_H}{\partial \eta} \text{ and } b_3 = \frac{\partial C_H}{\partial \beta}$$

The values of b_1, b_2 and b_3 can also be expressed in units of per radian or per degree. The values are normally negative, an increase of tailplane incidence and positive flap and tab deflections producing, using our sign convention, a negative hinge moment.

The coefficient b_2 can be described as the control "heaviness" parameter since it is the major factor determining the magnitude of the pilot's control forces. It is desirable that b_2 should always be negative (i.e. the hinge moment acting to oppose the motion of the control) otherwise a control deflection would produce a hinge moment tending to increase the deflection and the control would be "overbalanced" (Fig. 1.13).

Though the effect of the trim tab on the tailplane lift is small it has a very important effect on the hinge moment characteristics. It is moved in the opposite direction to the control surface in order that the $b_3\beta$ term in eqn. (1.20) will oppose the $b_2\eta$ term. When the hinge moment coefficient is zero, the control forces are said to be trimmed out. It should be noted that the word "trim" is used in two different senses, in one case to indicate a balance of the pitching moments about the C.G. (i.e. the aircraft is trimmed out) and in the other case to indicate a balance of the moments about the control hinge line (i.e. the control forces are trimmed out).

The coefficient b_1 can be described as the control surface "floating" parameter. It is usually negative, but it can be made positive by

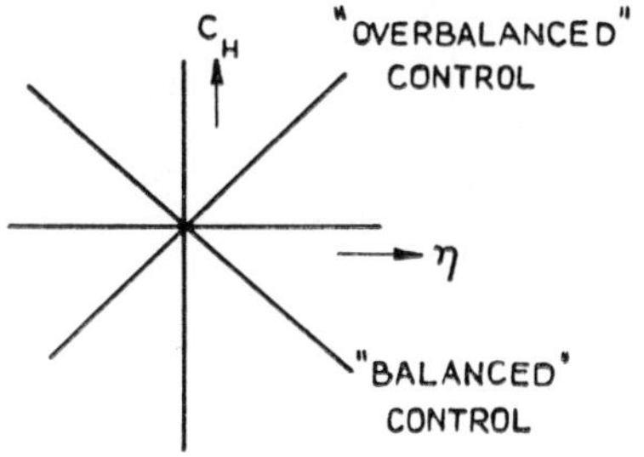

Fig. 1.13. Balanced and overbalanced control surfaces

suitable design and the sign of b_1 determines whether the control surface floats "with" (b_1 negative) or "against" (b_1 positive) the wind when the angle of attack of the tailplane is increased (Fig. 1.14). It is sometimes found to be beneficial to make b_1 positive while still retaining b_2 negative.

The coefficient b_1 represents the effect on the hinge moments of the control of changing tailplane incidence and b_2 the effect of

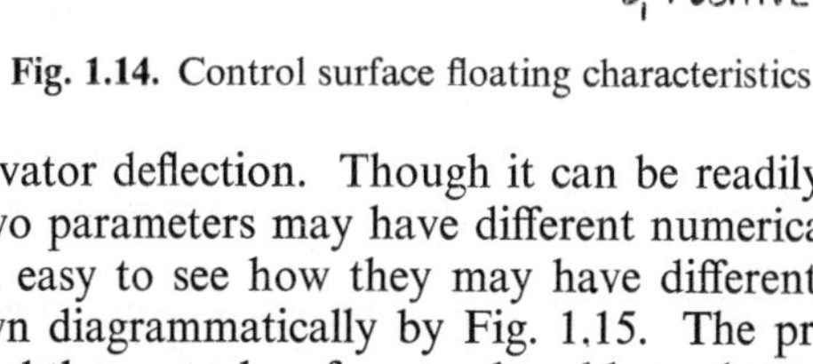

Fig. 1.14. Control surface floating characteristics

changing elevator deflection. Though it can be readily appreciated that these two parameters may have different numerical values it is not quite so easy to see how they may have different signs. This can be shown diagrammatically by Fig. 1.15. The pressure distribution around the control surface produced by a change of tailplane incidence will not be the same as that produced by a change of flap deflection. Thus the resultant force will act in a different position along the wing chord. Let us assume that the resultant force produced by changing flap deflection (R_2) lies behind that produced by changing tailplane incidence (R_1). When the hinge line is at the leading edge of the control surface both b_1 and b_2 will have negative values. As the hinge line is moved back however it will eventually reach a position at which it is behind R_1 and in front of R_2. In this

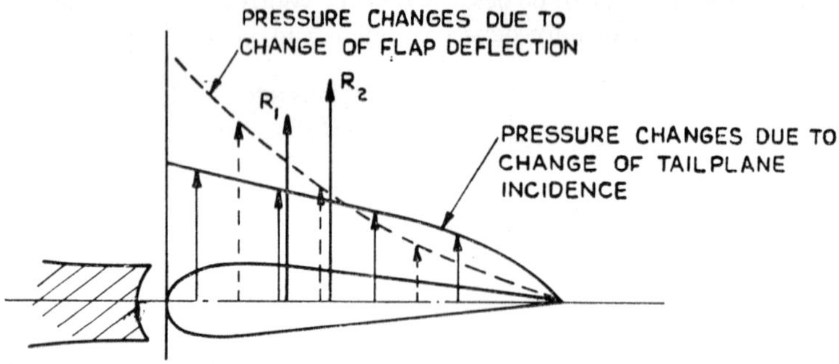

Fig. 1.15. Pressure distribution changes due to tailplane incidence and flap deflection

condition b_1 will have a positive value and b_2 will have a negative value.

4. The Analysis of Stability Characteristics

4.1. Basic Definitions

Before discussing the way in which the stability characteristics of an aircraft can be determined, a number of important terms will be defined.

(*a*) *Equilibrium* A body is in equilibrium when it is in a state of rest or uniform motion in a straight line and the forces acting on it are balanced out. The definition can be extended to cover those bodies in uniform motion in a curved path. There is, in these cases, a resultant force and an acceleration towards the centre of the curved path, but they can be considered as cases of dynamic equilibrium. Stability is a property of the equilibrium state and there are two types of stability to consider, static stability and dynamic stability.

(*b*) *Static Stability* This is concerned with the forces and moments produced by a small disturbance from the condition of equilibrium. It determines whether or not the body will *initially* tend to return, of its own accord, towards the equilibrium condition, once the disturbance is removed. There are three possibilities

(i) a body is statically *stable* when it tends to return to the equilibrium position (Fig. 1.16*a*).

(ii) a body is statically *unstable* when it tends to diverge further away from the equilibrium position (Fig. 1.16*b*).
(iii) a body possesses *neutral* static stability when it remains in the disturbed position (Fig. 1.16*c*).

The direction of the applied disturbance may be important. Fig. 1.16*d* shows a particular system which is stable to a disturbance in

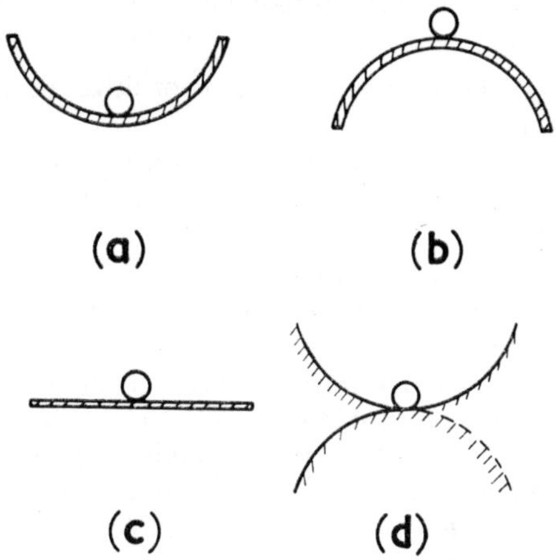

Fig. 1.16. Diagrammatic representation of static stability
(*a*) Static stability (*b*) Static instability
(*c*) Neutral stability (*d*) Mixed stability

one direction (i.e. in the plane of the page) and unstable to a disturbance in the other direction (i.e. at right angles to the plane of the page). An aircraft is a system which has a number of degrees of freedom, and it is possible for an aircraft to be statically stable with respect to some disturbances and statically unstable with respect to others.

The degree of static stability possessed by a body is expressed by the ratio

$$\frac{\text{Restoring effect produced as a result of the disturbance}}{\text{Magnitude of the disturbance}}$$

By convention, a body possessing static stability is said to have positive stability, static instability is negative stability and neutral stability is zero stability.

(*c*) *Dynamic Stability* This is concerned with the subsequent behaviour of a body which possesses static stability. The motion usually consists of oscillations about the equilibrium position and there are once again three possibilities

(i) a body is dynamically *stable* if the oscillations eventually damp out (i.e. converge).
(ii) a body is dynamically *unstable* if the oscillations are undamped or divergent, increasing in amplitude.
(iii) a body possesses *neutral* dynamic stability if the oscillations are undamped and maintain a constant amplitude.

In a similar way, these are referred to as positive, negative and zero stability, respectively.

By this definition, a body which is statically unstable must also be dynamically unstable, because the oscillations, even if they are damped oscillations, cannot be damped out about the original equilibrium position. It is possible to extend the definition of positive dynamic stability to cover the case in which the oscillations die away about a mean line which is slowly diverging away from the original equilibrium condition. This type of motion sometimes occurs when considering the lateral disturbance of an aircraft. The overall motion is made up of three separate modes two of which are the spiral mode and the Dutch Roll oscillation. If the spiral mode is unstable and the Dutch Roll oscillation stable, then the oscillations will be found to damp out about the diverging spiral motion. In this case it is better to consider each mode separately.

The time histories of a body possessing different types of stability are shown diagrammatically in Fig. 1.17. Fig. 1.17*a* represents a limiting form of dynamic stability referred to as a "Subsidence" or "Dead-beat" motion.

(*d*) *Stick-fixed Stability* (Also known as Control-fixed stability or Control-position stability.) This is the stability of the aircraft (both static and dynamic) with the stick (and hence it is assumed the control surfaces) held fixed. It is the amount of basic stability which is inherent in the aircraft. When the above assumption breaks down it is necessary to distinguish carefully between stick-fixed and control-fixed stability. It is sometimes referred to as control-position stability because, as we shall show in later work, it is this stability which determines the various control positions (i.e. stick movements) that are required to trim and manoeuvre the aircraft.

(*e*) *Stick-free Stability* (Also known as Control-free stability or Control-force stability.) This is the stability of the aircraft (both static and dynamic) with the stick left free and the control surfaces taking up a position of equilibrium (i.e. "floating") under the action of their hinge moments. It is the basic stability of the aircraft modified by the effects of the control movements. The difference between the stick-free and stick-fixed stabilities is very dependent

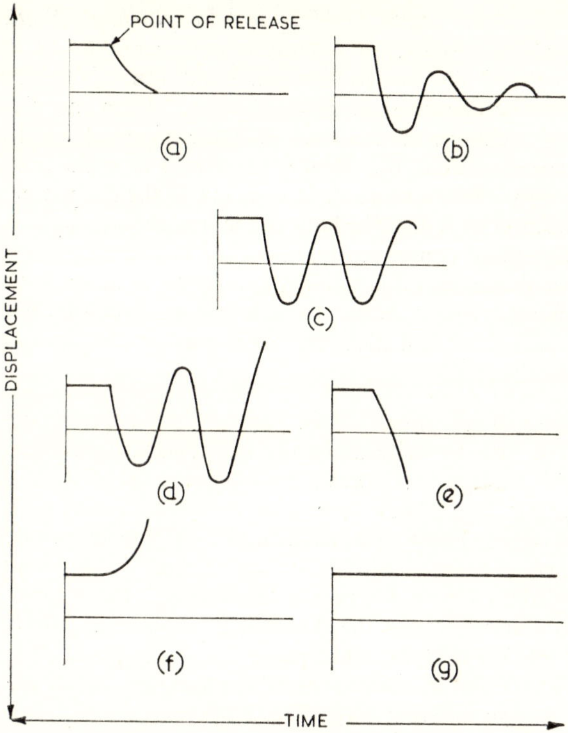

Fig. 1.17. Time histories of various responses to a small disturbance
(*a*) and (*b*) Static and dynamic stability
(*c*) Neutral dynamic stability
(*d*) and (*e*) Dynamic instability
(*f*) Static instability
(*g*) Neutral static stability

on the characteristics of the control surfaces (i.e. b_1 and b_2) and the control circuits.

Stick-free stability is sometimes referred to as control-force stability because, as we again show later, it is this stability which determines the various control forces that are required to trim and manoeuvre an aircraft fitted with non-powered control systems.

In the case of an aircraft fitted with a perfectly irreversible power control system, the control surfaces will not float even if the stick is left free. In these circumstances there will be no difference between the stick-fixed and stick-free stabilities and the stick-fixed stabilities will determine both the stick movements and the stick forces involved.

(*f*) *Longitudinal Static Stability and Longitudinal* (*Static*) *Manoeuvre Stability* When an aircraft is considered to be disturbed longitudinally from the equilibrium condition of trimmed level flight it can be disturbed in different ways. When the disturbance consists of a change of incidence or angle of attack (and possibly forward speed as well), the stability is referred to as the longitudinal static stability. When however in addition to the change in angle of attack, the aircraft is disturbed out of the straight line flight path into a curved one and consequently, disturbed in rate of pitch, then the stability is altered and this stability is referred to as the longitudinal (static) manoeuvring stability. In each case it is necessary to consider both the stick-fixed and stick-free conditions.

4.2. Systems of Axes

The motion of an aircraft is in general very complicated and in order to be able to write down the mathematical equations representing this motion it is necessary to define a set of three mutually perpendicular axes. These three axes have their origin at the centre of gravity of the aircraft and usually move with the aircraft during the disturbance. For this reason they are referred to as "body" axes. The longitudinal axis $0x$ points "forwards," the lateral axis $0y$ points to starboard and the directional or normal axis $0z$ points "downwards" (Fig. 1.18). The plane $0xz$ is the plane of symmetry of the aircraft and the directions of the axes $0x$ and $0z$ are defined precisely in two different ways (Fig. 1.19)

(i) *Wind* (*or Stability*) *Axes System* In this system of axes the $0x$-axis is made to point in the direction of the initial undisturbed motion (i.e. in the opposite direction to that of the undisturbed relative "wind"). These axes usually remain fixed in the aircraft during the motion following the disturbance, but they are orientated differently in the aircraft for different initial equilibrium conditions.

(ii) *Principal Inertia Axes System* In this system of axes the $0x$-axis is made to point in the direction of the principal inertia axis of the aircraft, and the corresponding system of axes is known as the Principal Inertia Axes System. These axes remain fixed in the aircraft during the motion following a disturbance and are always the same irrespective of the initial conditions of flight.

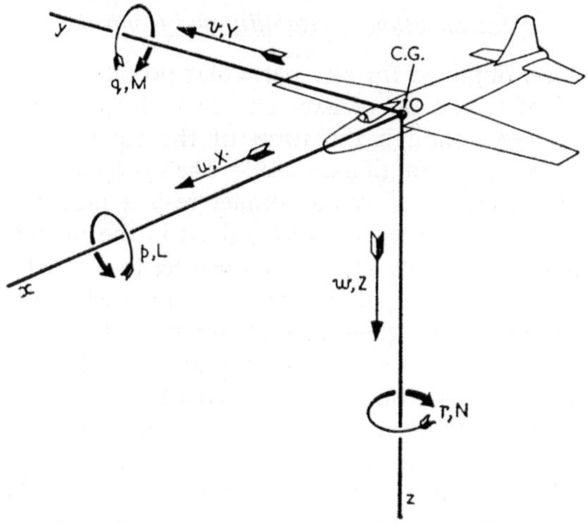

Fig. 1.18. Notation for velocities, forces and moments

AXIS	VELOCITY ALONG AXIS	FORCE ALONG AXIS	ANGULAR VELOCITY ABOUT AXIS	MOMENT ABOUT AXIS	MOMENT OF INERTIA ABOUT AXIS
0x	u	X	p	L	A
0y	v	Y	q	M	B
0z	w	Z	r	N	C

Note—Arrows denote the positive sense of the forces, velocities and moments

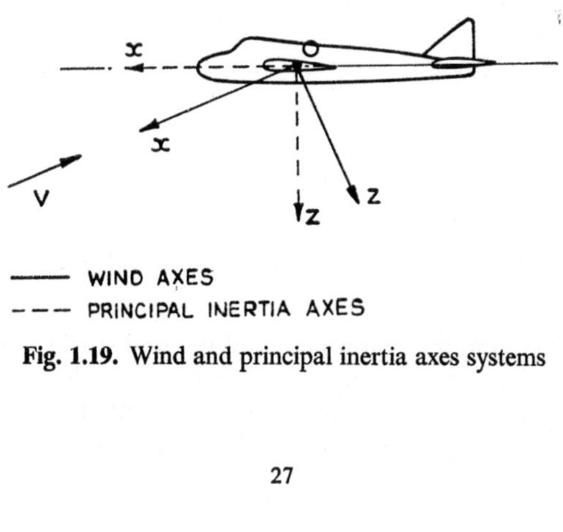

Fig. 1.19. Wind and principal inertia axes systems

The solution obtained for any particular problem must of course be independent of the set of axes chosen to formulate the mathematical equations, though the form of the equations themselves will depend on the system of axes used. The system chosen is simply a matter of convenience. Wind tunnel results may be measured with respect to wind axes, or to axes fixed in the model, measurements in flight are usually made with respect to a fixed set of axes in the aircraft. If principal inertia axes are used the product of inertia terms are zero, if wind axes are used $W = 0$. Standard formulae can be found for converting quantities referred to one set of axes to the corresponding values referred to the other set of axes. Up to now the stability axes system has generally been used for aircraft stability and control problems. The principal inertia system may be preferable for missiles and for very slender delta aircraft configurations (Ref. 4) and for the analysis of aircraft motions when the inertia effects are large compared with the aerodynamic effects (e.g. spinning).

4.3. Notation and Sign Convention

In addition to a standard set of axes, it is also necessary to use a standard notation and a standard sign convention. Various systems are in use and when reading any book or report on this subject it is essential to note very carefully the definitions of the symbols and the particular sign convention being used.

The system of notation used for the linear and angular velocities and the aerodynamic forces and moments is shown in Fig. 1.18. The symbols used for the corresponding angles of rotation and control surface deflections are shown in the table below.

AXIS	$0x$	$0y$	$0z$
Angles of rotation	ϕ (bank)	θ (pitch)	ψ (yaw)
Control surface deflections	ξ (aileron)	η (elevator)	ζ (rudder)

There are two particular points which might give rise to some confusion.

(i) In this system the components of the undisturbed speed of the aircraft are represented by U, V and W and the corresponding disturbances by u, v and w. Thus the symbol V should strictly be retained for the initial sideslip velocity, whereas in fact it is often used for the forward speed of the aircraft which should in the wind axes system be U. This does not usually create any problem since the initial sideslip velocity is usually zero.

(ii) The symbol L is used for rolling moment whereas in general aerodynamic work it is also used to represent the lift force. In stability notation a force acting in the plane of symmetry in a direction perpendicular to the relative airflow is a Z-force. However the concept of lift and lift coefficient C_L has become so familiar that it is still used in the stability analysis. Here again confusion does not usually arise since longitudinal and lateral motions are generally dealt with separately. Lower case letters are often used as the suffixes of the moment coefficients, there is then no confusion between C_l, the rolling moment coefficient, and C_L, the lift coefficient.

The sign convention used is as follows—

(*a*) Directions, linear velocities and forces are positive when forwards, to starboard and downwards.
(*b*) Angular velocities, moments and control deflections are positive when the direction of rotation is clockwise about the appropriate axis, when viewed from the C.G. in the positive direction of that axis.

The arrows on Fig. 1.18 indicate the positive directions of the various quantities. One point needs comment and that is the definition of positive control movements. Clockwise deflections (i.e. starboard aileron down, elevator down, and rudder to port) are positive ones and this means that all positive control movements produce negative control moments (i.e. negative rolling, pitching and yawing moments about the C.G. of the aircraft). The other alternative is to define control deflections as positive when they produce positive control moments but in that case there is no uniformity in the various "positive" directions of rotation.

4.4. Equations of Motion

The mathematical formulation of the motion of an aircraft involves the equations of motion of the aircraft and, if applicable, its control surfaces, allowance being made for aeroelastic distortion and compressibility effects as necessary. The equations are basically very simple, being a form of Newton's law relating the force and the linear acceleration in linear motion (force = mass × acceleration) and the moment and the angular acceleration in angular motion (moment = moment of inertia × angular acceleration).

In the case of a rigid aircraft with moving control surfaces, there are three sets of equations, each set consisting of three separate equations

(*a*) The *force equations* which relate the forces X, Y and Z to the motion along the respective axes.
(*b*) The *moment equations* which relate the moments L, M and N to the motion about the respective axes.
(*c*) The *control hinge moment equations* which relate the moments about the hinge lines of the controls to the rotation of the control surfaces.

The expressions for the forces and moments include terms arising from gravitational, aerodynamic and inertia effects. The inertia

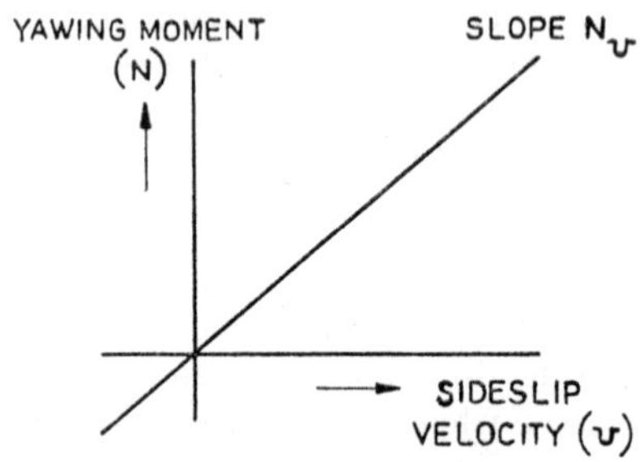

Fig. 1.20. Variation of yawing moment with sideslip velocity

terms include, in general, the direct inertia terms, the product of inertia terms and the inertia coupling terms. The formulation of these equations is simplified by the use of the Small Disturbance Theory in which all the deviations from the equilibrium position are assumed to be small. This allows the changes in the forces and moments to be expressed as linear functions in terms of the changes in linear velocities, angular velocities, etc. As an example of this, the total aerodynamic yawing moment acting on the aircraft in a disturbance can be considered as the sum of the separate yawing moments due to sideslip, rate of yaw, rudder deflection and so on. Fig. 1.20 shows how the yawing moment N may vary with sideslip velocity v. Over the linear range of this curve the total change in yawing moment produced as a result of a sideslipping motion can be represented by

$$\Delta N = \frac{\partial N}{\partial v} v \tag{1.21}$$

where $\dfrac{\partial N}{\partial v}$ is the slope of the N/v curve.

Similarly, the change in yawing moment due to the rate of yaw can be expressed as $\frac{\partial N}{\partial r} r$ and that due to rudder deflection as $\frac{\partial N}{\partial \zeta} \zeta$. The total change in yawing moment is then given by

$$\Delta N = \frac{\partial N}{\partial v} v + \frac{\partial N}{\partial r} r + \frac{\partial N}{\partial \zeta} \zeta + \ldots \tag{1.22}$$

The partial differentials $\partial N/\partial v$, $\partial N/\partial r$, $\partial N/\partial \zeta \ldots$ can be written in the form N_v, N_r, $N_\zeta \ldots$; they are examples of quantities that are referred to as Aerodynamic Derivatives. In some problems however this assumption of linearity may break down and it may be necessary to allow for a non-linear variation between the quantities involved.

The equations of motion may be left in their dimensional form or they may be made completely non-dimensional. Most standard text books use the non-dimensional form of the equations. This method has the advantage that the terms in the equation are independent of the size of the aircraft and the speed and height at which it is flying and thus the results are more generalized. This enables comparisons of the stability characteristics of different aircraft to be more easily appreciated. (This argument is similar to the one used to explain the advantages of working with non-dimensional lift and drag coefficients [i.e. C_L and C_D] instead of the actual lift and drag forces.) The disadvantages are

(i) that the form of the equations varies depending on the quantities used to make the different terms non-dimensional. It has already been emphasized that when reading any literature on the subject, it is extremely important to note the definitions and notation being used.

(ii) that even the weight of the aircraft and the time have to be made non-dimensional. The unit of time can be defined in different ways but on the British system it is denoted by $\hat{t}$ (i.e. t hat) and given by

$$\hat{t} = \frac{W}{g\rho SV} \text{ secs} \tag{1.23}$$

and the non-dimensional time is then expressed in units of airsecs given by

$$\tau = \frac{t}{\hat{t}} \text{ airsecs} \tag{1.24}$$

In some books (Refs. 5 and 6) the equations are not made non-dimensional and this has various advantages, one of which is that the natural time scale is retained (Ref. 6, page 220). Alternative forms of the equations of motion are given in Appendix F.

Solutions of these equations are usually made by making simplifying assumptions. Depending on the particular case being considered, some of the aerodynamic derivatives may be negligibly small and the corresponding terms can then be omitted from the equations. The inertia coupling terms can usually be omitted (unless spinning or roll coupling problems are being investigated) and in these circumstances the equations of motion can be divided into two separate groups, the longitudinal and lateral groups of equations. In the stick-fixed considerations, the longitudinal characteristics can be determined by solving the X, Z and M set of equations, while the corresponding lateral characteristics can be obtained from the Y, L and N set of equations. These equations are not derived in this book, but one form of the non-dimensional equations of motion is given for reference in Appendix B.

The equations are linear differential equations of the second order with constant coefficients and solutions are obtained by assuming that the solutions of these equations are exponential in form (i.e. $x = x_0 \, e^{\lambda t}$). In the analysis of stick-fixed stability, the problem is then reduced to one of solving an algebraic equation of the fourth power (i.e. a quartic equation for λ) of the form

$$A\lambda^4 + B\lambda^3 + C\lambda^2 + D\lambda + E = 0 \qquad (1.25)$$

where A, B, C, D and E are dependent on the values of the stability derivatives, the inertia coefficients, etc., and have different values for the longitudinal and lateral motions. The solutions of this equation determine the types of motion involved, i.e. whether the aircraft possesses static and dynamic stability, the amount of damping, the periodic time of the oscillations, etc. The root λ has four possible values; these may consist of four real roots, two real roots and one complex pair of roots, or two complex pairs of roots. If λ is real, the sign of λ determines whether the motion is convergent (λ negative, i.e. exponential decay) or divergent (λ positive, i.e. exponential growth). If λ is complex the motion is an oscillatory one and the sign of the real part of the root again indicates whether the motion is a convergent or a divergent one. The numerical value of the imaginary part of the root, on the other hand, determines the frequency (and hence the periodic time) of the oscillation.

5. Pilot's Assessment of Flying Qualities

5.1. Characteristics of the Human Pilot

As we have seen in the previous paragraph, it is sometimes possible to lay down specific numerical requirements for the stability and control characteristics of aircraft. In many cases, however, the numerical data does not give a true indication of the handling qualities of an aircraft and it is better to rely on the qualitative assessment of the test pilot. Thus the trend in official requirements is to make more general statements of what is required, giving quantitative information, where practicable, merely as a guide (Ref. 7). Desirable handling qualities are discussed in more detail at appropriate places in the text. Ref. 15 gives a detailed review of current handling criteria, discusses possible new forms of criteria for slender wing and V/STOL aircraft, and outlines current developments, both experimental and theoretical, in the study of handling and associated criteria.

The pilot's assessment of the handling qualities of an aircraft is dependent on a large number of factors, e.g. the stability and control

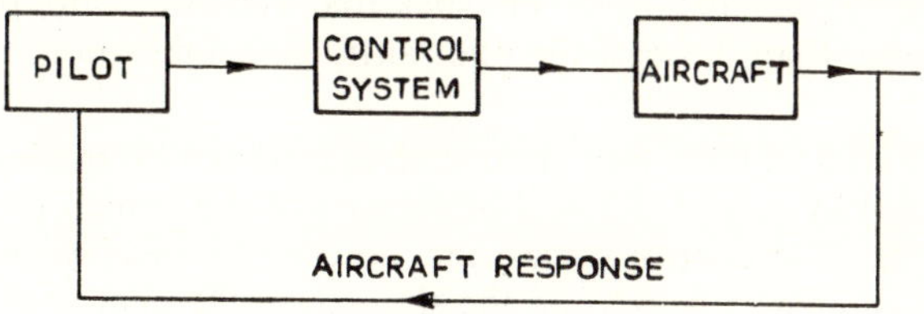

Fig. 1.21. The aircraft/pilot closed-loop system

characteristics of the aircraft, the cockpit layout, the weather conditions, the operational task to be performed, and so on. It is important here to emphasize the close connection between stability and control, since the stability characteristics of an aircraft determine the way in which it responds to control movements and atmospheric gusts. Though stability is dealt with under the headings of "stick-fixed" and "stick-free" stability, the pilot does not usually fly the aircraft this way. Except in certain special cases, he is not concerned with stick-fixed and stick-free motions as such. However, we shall in fact show that these particular stability characteristics are important in determining the control movements and forces involved and hence are important in helping to give the pilot the right "feel." To fly the aircraft the pilot is continually moving the controls around. He is in effect part of a "closed-loop" system, consisting of the pilot, the control system, and the airframe (Fig. 1.21), and he continually measures the response of the aircraft from the various cues that he is provided with (i.e. external visual cues, instrument display, sensation of g, etc.) and puts in the appropriate control inputs. With a pilot in the loop, the basic stability characteristics of the aircraft may be considerably modified. Thus for example, an aircraft which possesses stick-fixed spiral instability or Dutch Roll instability, may be quite satisfactory when the pilot is in the control loop using his controls (Ref. 8).

There are some limitations to this however, as follows—

(i) If the pilot has to correct all the time for a lack of basic stability in the aircraft, then the aircraft becomes much more difficult to fly and the pilot becomes tired more quickly.

(ii) If the periodic time of an oscillatory motion is small, the pilot may get his control inputs "out of phase" with the motion of the aircraft and the oscillation may be made more divergent. Longitudinal Pilot Induced Oscillations (P.I.O.S) are an example of this and in this case it will be better for the pilot either to keep the controls fixed or to let them go free since the basic stick-fixed or stick-free stabilities are probably better than those of the pilot/airframe combination.

The overall stability and control characteristics of an aircraft are determined by certain physical limitations of the human pilot (Ref. 9, page 2). The control forces and movements involved must not be large enough to be beyond the strength of the pilot or to be too fatiguing and they must, as far as possible, be in the conventional directions in order to give the pilot the correct "feel." Unconventional characteristics are often eliminated by artificial means (e.g. an auto-Mach trimmer) if they cannot be eliminated by aerodynamic methods.

In recent years a lot of work has been done in attempting to formulate a mathematical expression for the behaviour of the human pilot in order that the pilot can be included in the overall servo analysis of the complete aircraft system. The transfer function (see Chapter 19) of a pilot is, however, rather a complicated one and it does not remain constant. The behaviour of the pilot depends on the flying conditions (e.g. weather, day or night, etc.), on certain personal factors (e.g. training and experience) and on various other factors (e.g. noise, vibration, etc.). However it has been possible to specify numerical values for certain human characteristics (e.g. the average response time lag, for example, is usually taken as 0·25 secs), and on this basis it has been possible to formulate an approximate transfer function to represent the human pilot.

5.2. Representation of Pilot Opinion

The pilot's assessment of the handling characteristics of an aircraft can be presented in a number of different ways (*a*) *Numerical Rating Scales* A number of different rating scales are available which can be used to assess the handling characteristics. The most common of these scales is the Cooper Scale originally proposed in Ref. 10. This method suggests that the handling qualities being assessed should be assigned a number between 1 and 10, the scale being defined as follows—

1	Excellent
2	Good, pleasant to fly
3	Satisfactory, but with some mildly unpleasant characteristics
4	Acceptable, but with unpleasant characteristics
5	Unacceptable for normal operation
6	Acceptable for emergency condition only
7	Unacceptable even for emergency condition
8	Unacceptable—dangerous
9	Unacceptable—uncontrollable
10	Catastrophic

Though this rating scale is not an ideal one, it has been used quite a lot in correlating the results of flight tests and simulator experiments (Refs. 11 and 12). Ratings 1, 2 and 3 are considered satisfactory and under these conditions an aircraft could complete its primary mission. Ratings 4, 5 and 6 are acceptable for emergency operation; the aircraft could still be landed but could not carry out its mission. Ratings 7 to 10 are unacceptable.

Various modified forms of this rating scale exist (Refs. 13 and 14).

(*b*) *Iso-Opinion Plots* The values on the Cooper scale allocated by the pilot to a particular configuration can be marked on a graph at the appropriate point. By varying the particular parameters of the aircraft over a wide range of values it is possible to draw in the boundaries separating the combinations which have been assessed as "satisfactory" (i.e. Cooper ratings $<3\frac{1}{2}$) from those which are "unsatisfactory," "acceptable for emergency only," or "unacceptable" (i.e. Cooper rating $>6\frac{1}{2}$).

This method of representing the handling characteristics of an aircraft has become particularly useful since the introduction of ground-based simulators (Refs. 11 and 13) and variable-stability aircraft (Refs. 12 and 13) because these devices have allowed the stability parameters to be varied over a wide range.

The characteristics of the Dutch Roll oscillation are dependent on the values of the damping and the roll-yaw ratio. Fig. 1.22 shows a typical iso-opinion plot. Other examples are shown in Refs. 5, 12 and 14. The iso-opinion plots in Ref. 5 show the boundaries separating the regions with different assessments and describes in

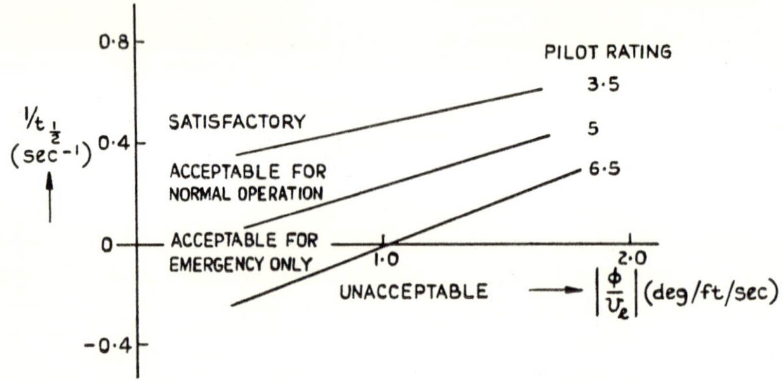

Fig. 1.22. Iso-opinion plot of Dutch Roll characteristics

words the reasons for the particular assessments allocated by the pilots.

References

1. NICHOLSON, L. F.; "Engine-airframe Integration," *Journal of the Royal Aeronautical Society*, Vol. 61, Nov. 1957, pages 711–720.
2. KÜCHEMANN, D.; *Some Considerations of Aircraft Shapes and Their Aerodynamics for Flight at Supersonic Speeds*, Unpublished M.O.A. Report.
3. LEE, G. H.; *The Aeroplane Designer's Approach to Stability and Control*, Agard Report No. 334, 1961.
4. PINSKER, W. J. G.; *The Lateral Motion of Aircraft and in Particular of Inertially Slender Configurations*, R.A.E. Report Aero 2656, 1961.
5. KOLK, W. R.; *Modern Flight Dynamics* (Prentice Hall International, London, 1961).
6. SECKEL, E.; *Stability and Control of Airplanes and Helicopters* (Academic Press, New York, 1964).
7. ALLWRIGHT, H. J.; *Design Aims for Stability and Control of Piloted Aircraft*, Agard Report No. 337, 1961.
8. HENDERSON, J. M.; "Low Speed Handling of a Slender Delta (H.P. 115)," *Journal of the Royal Aeronautical Society*, May 1965, pages 311–324.
9. DUNCAN, W. J.; *Control and Stability of Aircraft* (Cambridge University Press, 1952).
10. COOPER, G. E.; "Understanding and Interpreting Pilot Opinion," *Aero. Eng. Review*, Vol. 16, No. 3, March 1957, pages 47–51.
11. RATHERT, G. A., CREER, B. Y. and SADOFF, M.; *The Use of Piloted Flight Simulators in General Research*, Agard Report No. 365, 1961.

12. KIDD, E. A., BULL, G. and HARPER, R. P.; *In-flight Simulation—Theory and Application*, Agard Report No. 368, 1961.
13. WESTBROOK, C. B.; *Simulation in Modern Aero-Space Vehicle Design*, Agard Report No. 366, 1961.
14. KOVEN, W. and WASICKO, R.; *Flying Quality Requirements for United States Navy and Air Force Aircraft*, Agard Report No. 336, 1961.
15. O'HARA, F.; "Handling Criteria," *Journal of the Royal Aeronautical Society*, April 1967, page 271.

2 Subsonic Longitudinal Static Stability

1. Introduction

In Chapter 1 static stability was defined as the tendency of the aircraft to return to the equilibrium condition of flight, following a disturbance. In this chapter we will consider the static stability of the longitudinal (i.e. pitching) motion of the aircraft. The problem can be investigated using the full longitudinal equations of motion of the aircraft, taking into account all the possible variations of forward speed, vertical speed (or angle of attack) and rate of pitch. This is, however, not necessary and the problem can be analysed initially by a simple mathematical theory which uses only the pitching moment equation and allows only for angle of attack disturbances.

The simple theory produces some useful results and enables a good physical appreciation of the problem to be obtained. A large number of assumptions are inherent in this theory, however, and a better analysis can be made by a more general mathematical theory. This allows for both angle of attack and speed disturbances and enables the effects of compressibility and aeroelasticity to be allowed for.

It will be shown that, on an aircraft fitted with non-powered controls, the stick-fixed static stability possessed by the aircraft determines the magnitude of the stick and elevator movements required to change the forward speed of the aircraft from an initially selected trimmed value, whereas the amount of stick-free static stability determines the magnitude of the corresponding stick forces (or elevator trim tab deflections) required. If the aircraft is fitted with irreversible power-operated controls, then only the question of stick-fixed stability need be considered, since this, together with the particular feel system employed, will determine both the stick movements and forces involved. This chapter has been written from the point of view of a non-powered control system. Powered controls will be considered in a later chapter, though, obviously, some of the results obtained here will be directly applicable.

It is desirable that an aircraft should possess both stick-fixed and stick-free stability and the requirements for any aircraft are laid down in the appropriate regulations. This ensures that the pilot's stick movements and forces to trim will be in the conventional directions. Flight tests are performed to measure the trim curves, from which the values of the static stabilities can be determined. These tests are generally repeated for different C.G. positions, with different configurations and for various conditions of flight.

2. Simple Mathematical Theory of Static Stability

2.1. Basic Assumptions

To determine the factors which determine the stick-fixed and stick-free static stability of an aircraft, it is necessary to analyse the changes occurring in the forces and moments acting on the aircraft when it is disturbed in pitch. To begin with, we will consider only the simple theory and will assume an idealized state of affairs in which

(i) Propeller and slipstream effects can be neglected.
(ii) The structure does not distort.
(iii) Compressibility effects can be ignored.
(iv) The tailplane coefficients a_1, a_2, a_3, b_1, b_2, b_3 are *constants* independent of forward speed, etc.
(v) There is no vertical displacement of the C.G. from the chord line of the wing.
(vi) The tailplane lift is small in comparison with the wing lift and the movement of the centre of pressure on the tailplane can be ignored.
(vii) The flight path is approximately horizontal.

Only the effects of varying incidence will be allowed for, since, if all the above assumptions are true, any variation in forward speed will have no effect on the static stability.

These assumptions are approximately true only for a low speed aircraft in a shallow power-off glide.

2.2. The Equilibrium Equation

When an aircraft is in equilibrium, all the forces and moments acting on it must be balanced out, any nose-up pitching moments being balanced by exactly equal nose-down moments. The aircraft can be represented very simply by the system of forces and moments shown in Fig. 2.1. The Aerodynamic Centre method of representation is generally used and the effect of the wing is represented by a lift force L acting at this point, together with a zero-lift pitching moment

M_0. In Fig. 2.1 the zero-lift moment is shown acting in the conventionally positive direction; in practice, it is normally a negative (i.e. nose-down) moment. The equations deduced are correct, as long as the appropriate sign is used. The weight W acts vertically through the C.G., a distance x behind the A.C. and the tailplane lift L' acts at the aerodynamic centre of the tailplane, at a distance l' behind the C.G.* At low forward speeds, the tailload is usually upwards and at high forward speeds it is usually downwards; the

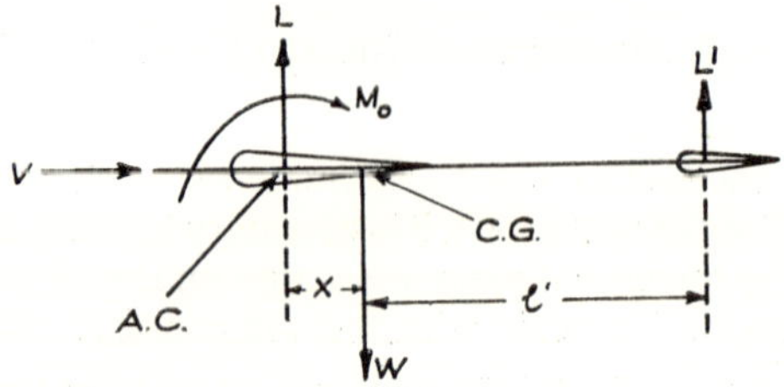

Fig. 2.1. Forces and moments acting on an aircraft

actual direction of the load does not, however, affect the following analysis.

The total pitching moment acting on the aircraft about its centre of gravity is given by

$$M = M_0 + (L \times x) - (L' \times l') \qquad (2.1)$$

This equation can be made non-dimensional by dividing throughout by $\frac{1}{2}\rho V^2 S\bar{c}$ where S is the wing area, $\bar{c}$ the wing mean chord and V the forward speed. The tailplane lift can be expressed in the form $L' = C_L'(\frac{1}{2}\rho V^2)'S'$, where C_L' is the tailplane lift coefficient, S' the tailplane area, and $(\frac{1}{2}\rho V^2)'$ the dynamic pressure at the tailplane. Assuming that the dynamic pressure is unchanged between the wing and the tailplane [i.e. $\frac{1}{2}\rho V^2 = (\frac{1}{2}\rho V^2)'$] equation (2.1) reduces to

$$C_M = C_{M0} + C_L \frac{x}{\bar{c}} - C_L' \frac{S'l'}{S\bar{c}} \qquad (2.2)$$

The distance of the C.G. behind the leading edge can be expressed as a fraction of the mean chord and denoted by h. If the A.C. of the wing is at a distance $h_0\bar{c}$ behind the leading edge, then $x/\bar{c} = h - h_0$. The quantity $S'l'/S\bar{c}$ is an important one to the designer, and is

* The zero-lift moment of the tailplane is assumed to be zero, because symmetrical sections are usually used.

generally known as the tail volume coefficient and denoted by $\bar{V}$. The overall pitching moment coefficient thus becomes

$$C_M = C_{M0} + (h - h_0)C_L - \bar{V}C_L' \tag{2.3}$$

We have seen in Chapter 1 that the tailplane lift coefficient can be expressed as

$$C_L' = a_1\alpha' + a_2\eta + a_3\beta \tag{2.4}$$

where a_1, a_2, and a_3 are non-dimensional coefficients and the control angles are positive for downward deflections. [*Note.* The symbol β is used to represent the deflection of the trim tab; if a geared tab is fitted it can be regarded as modifying the basic characteristics (e.g. a_2 and b_2) of the control surface.] The tailplane incidence α' is that relative to the local airflow at the tail and is given by

$$\alpha' = \alpha - i - \varepsilon \tag{2.5}$$

where α is the wing incidence, i the tailplane setting below the wing and ε the downwash angle (Fig. 2.2). The values of α and i must be

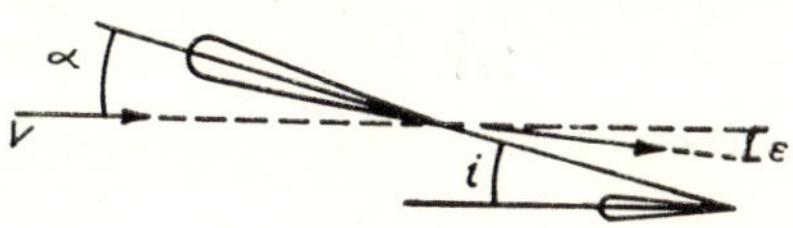

Fig. 2.2. Longitudinal dihedral and downwash

measured with respect to the same datum line, the wing chord line is often used, but the wing no-lift line simplifies the analysis.

Substituting in eqn. (2.3) from eqns. (2.4) and (2.5)

$$C_M = C_{M0} + (h - h_0)C_L - \bar{V}(a_1\alpha' + a_2\eta + a_3\beta)$$

or

$$\underline{\underline{C_M = C_{M0} + (h - h_0)C_L - \bar{V}[a_1(\alpha - i - \varepsilon) + a_2\eta + a_3\beta]}} \tag{2.6}$$

When the aircraft is in trim, the overall value of the pitching moment is zero and the right hand side of eqn. (2.6) can be equated to zero. If the values of h, $\bar{V}$, C_{M0}, etc., are known, they can be substituted in eqn. (2.6) and the necessary elevator angle required to trim calculated for various selected values of C_L (i.e. forward speed). Typical curves of elevator angle to trim plotted against C_L and forward speed are shown in Figs. 2.5 and 2.6; they will be considered in more detail later. We will, however, first discuss static stability, because there is a very close relationship between the elevator angle required to trim and the stick-fixed static stability.

2.3. Measurement of Static Stability

Consider the aircraft disturbed from its state of equilibrium. If, for convenience, we consider the disturbance to consist of an increase in the angle of attack of the aircraft, then, as long as the aircraft is not near the stall, the disturbance will result in an increase of lift on the wing and the tailplane (Fig. 2.3). These changes in the lift forces produce corresponding changes in the pitching moments about the C.G. of the aircraft. If the overall change in the pitching moment is a nose-down one, then the aircraft possesses static

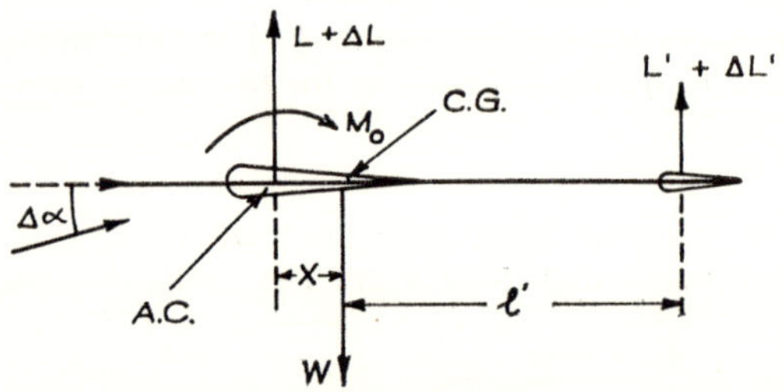

Fig. 2.3. Effect of angle of attack disturbance on the lift forces

stability. If, on the other hand, the change in pitching moment is a nose-up one, the aircraft is statically unstable.

The static stability can be measured by the ratio of the restoring moment produced by the disturbance to the magnitude of the disturbance. The restoring moment can be represented by ΔM and the corresponding coefficient by ΔC_M and the disturbance can be considered as one of incidence $\Delta\alpha$, or, since they are proportional to one another, one of lift coefficient ΔC_L. Thus the static stability can be measured in terms of $\Delta C_M/\Delta C_L$ or dC_M/dC_L.

Differentiating eqn. (2.6) we have

$$\frac{dC_M}{dC_L} = (h - h_0) - \bar{V}\left[a_1\left(\frac{d\alpha}{dC_L} - \frac{d\varepsilon}{dC_L}\right) + a_2\frac{d\eta}{dC_L} + a_3\frac{d\beta}{dC_L}\right] \tag{2.7}$$

since C_{M0} and i are constants. Substituting the symbol a for $dC_L/d\alpha$, this can be written

$$\frac{dC_M}{dC_L} = (h - h_0) - \bar{V}\left[\frac{a_1}{a}\left(1 - \frac{d\varepsilon}{d\alpha}\right) + a_2\frac{d\eta}{dC_L} + a_3\frac{d\beta}{dC_L}\right] \tag{2.8}$$

Static stability is positive when dC_M/dC_L is negative, since for stability a positive change in ΔC_L has to produce a negative change in ΔC_M and vice versa. Hence, reversing the signs in (2.8)

$$-\frac{dC_M}{dC_L} = -(h - h_0) + \bar{V}\left[\frac{a_1}{a}\left(1 - \frac{d\varepsilon}{d\alpha}\right) + a_2\frac{d\eta}{dC_L} + a_3\frac{d\beta}{dC_L}\right] \tag{2.9}$$

and this must be positive for stability.

2.4. Stick-fixed Static Stability

If the stick is held fixed during the disturbance, then $d\eta/dC_L$ and $d\beta/dC_L$ are both zero in equation (2.9) and the stick-fixed static stability becomes

$$\left(-\frac{dC_M}{dC_L}\right)_{fixed} = -(h - h_0) + \bar{V}\frac{a_1}{a}\left(1 - \frac{d\varepsilon}{d\alpha}\right) \tag{2.10}$$

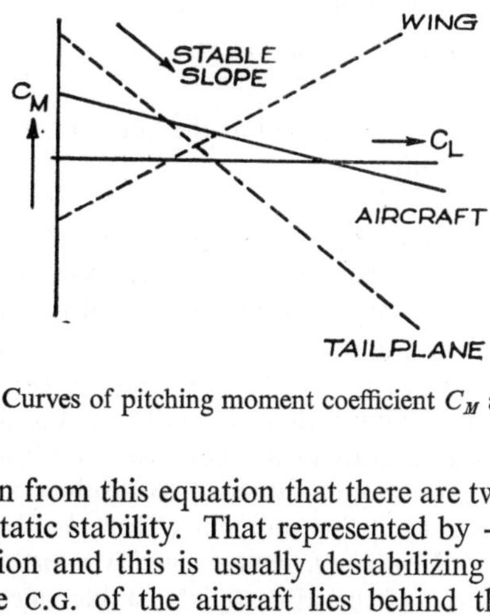

Fig. 2.4. Curves of pitching moment coefficient C_M against C_L

It can be seen from this equation that there are two contributions to the overall static stability. That represented by $-(h - h_0)$ is the wing contribution and this is usually destabilizing (Fig. 2.4) since $h > h_0$ (i.e. the C.G. of the aircraft lies behind the aerodynamic centre of the wing). The tailplane contribution is represented by $\bar{V}(a_1/a)(1 - d\varepsilon/d\alpha)$; this is stabilizing (Fig. 2.4) and obviously depends on such factors as tailplane area, tail-arm, tailplane lift curve slope and the downwash factor $(1 - d\varepsilon/d\alpha)$.

2.4.1. STICK-FIXED NEUTRAL POINT

All the quantities in eqn. (2.10) are fixed by the external configuration of the aircraft, except h, the position of the aircraft's C.G. By varying h the stability can be made positive, zero, or negative. Rearward movement of the C.G. increases $h - h_0$, and hence is destabilizing.

The particular position of the C.G. h_n which gives neutral stability is called the stick-fixed neutral point N. Substituting in eqn. (2.10)

$$0 = -(h_n - h_0) + \bar{V}\frac{a_1}{a}\left(1 - \frac{d\varepsilon}{d\alpha}\right) \qquad (2.11)$$

Subtracting (2.11) from (2.10), we have for any other C.G. position h

$$-\frac{dC_M}{dC_L} = h_n - h \qquad (2.12)$$

i.e. the static stability is equal to the non-dimensional distance between the C.G. and the neutral point. When the C.G. is ahead of the neutral point the aircraft is stable; when the C.G. is behind the neutral point the aircraft is unstable. The distance of the C.G. from the stick-fixed neutral point is called the *Stick-fixed* C.G. *Margin* and is denoted by H_n. The quantity $-dC_M/dC_L$ is called the *Static Margin* and so, on this simple theory

$$H_n = \text{C.G. margin} = \text{Static margin} \qquad (2.13)$$

The neutral point can be regarded as the overall aerodynamic centre of the complete aircraft.

2.4.2. ELEVATOR ANGLE REQUIRED TO TRIM

If the aircraft is flying in trim and it is desired to change the trimmed speed, then it is necessary to apply the elevator control in such a way as to rotate the aircraft to its new required incidence. If the control is then returned to its original position, the aircraft would, because of its static stability, tend to return to the initial trim condition. To *hold* the aircraft in the new trim condition the elevator must be deflected an amount sufficient to produce a moment equal and opposite to the stabilizing moment. Thus *the change in elevator angle required to hold a new trimmed speed is a measure of its stick-fixed static stability*. If the aircraft is a stable one, the change in elevator angle to hold the new trimmed speed will be in the same direction as the movement required to initiate the change.

A change in elevator angle $\Delta\eta$ produces a pitching moment coefficient change of $-\bar{V}a_2\Delta\eta$ (from eqn. (2.6)) and since the stabilizing moment coefficient ΔC_M is equal and opposite to this we have

$$\Delta C_M = \bar{V}a_2\Delta\eta$$

$$\therefore \quad -\frac{dC_M}{dC_L} = -\bar{V}a_2\frac{d\eta}{dC_L} \qquad (2.14)$$

Since, on this theory, the stick-fixed static stability is independent of C_L, eqn. (2.14) indicates that the curve of elevator angle to trim plotted against C_L is a straight line, with a slope given by

$$\frac{d\eta}{dC_L} = -\frac{H_n}{\bar{V}a_2} \tag{2.15}$$

If the aircraft is statically stable, the elevator angle curve has a negative slope, since both $\bar{V}$ and a_2 are positive quantities ($\bar{V}a_2$ can be regarded as the elevator pitching moment effectiveness). Rearward movement of the C.G. reduces the stability and the numerical

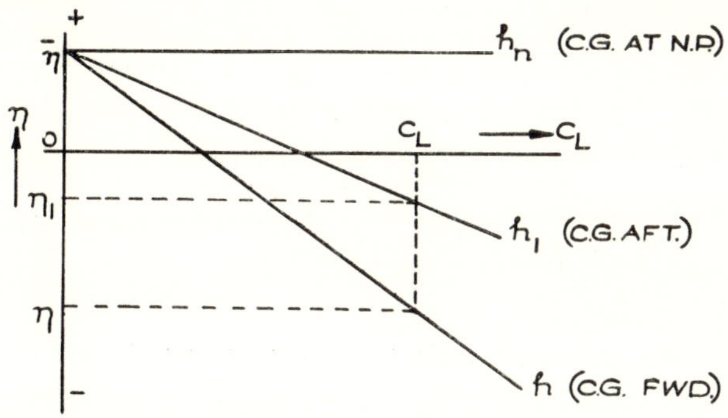

Fig. 2.5. Elevator angles to trim against C_L

value of the slope decreases (Fig. 2.5). If the values of $\bar{V}$ and a_2 (or $\bar{V}a_2$) are known, then the stick-fixed static stability can be determined by measuring the slope of the η/C_L curve and substituting the values in eqn. (2.14).

Equation (2.15) only determines the slope of the η/C_L curve, it does not determine the intercept on the vertical axis. The elevator angle to trim curve can be obtained from the basic trim equation (2.6). By comparing eqn. (2.15) and eqn. (2.6) it can be seen that, although the stick-fixed static stability determines the slope of the η/C_L line, the value of η at $C_L = 0$ is determined, for zero tab angle, by the values of C_{M0} and i. If the angles are measured with respect to the wing no-lift line, then, from eqn. (2.6) we can deduce that

$$\bar{\eta} = \frac{(C_{M0} + \bar{V}a_1 i)}{\bar{V}a_2} \tag{2.16}$$

The value of $(C_{M0} + \bar{V}a_1 i)$ can be regarded as the overall zero-lift moment coefficient of the aircraft. It should be remembered that C_{M0} is usually negative; the tailplane is set at an angle i below the wing (the so-called longitudinal dihedral), in order to make the value of $(C_{M0} + \bar{V}a_1 i)$, and hence the value of $\bar{\eta}$, positive. This enables the complete aircraft to satisfy the necessary trim and stability requirements at the same time.

From eqn. (2.16) it can be seen that the value of $\bar{\eta}$ is independent of the C.G. position and the stick-fixed static stability. The intercept is thus the same for all C.G. positions (Fig. 2.5).

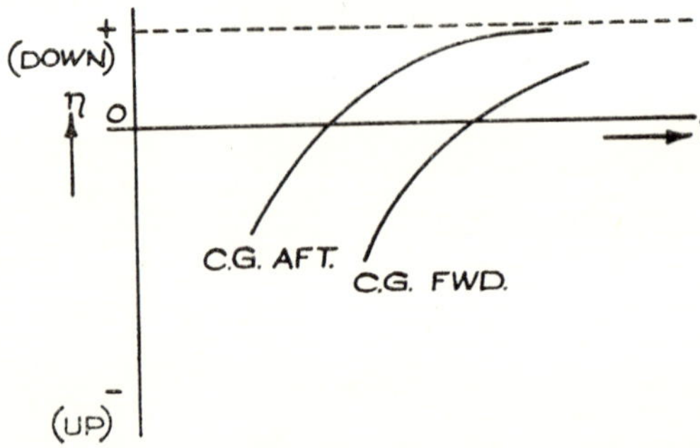

Fig. 2.6 Elevator angles to trim against forward speed

Equation (2.14) can be used to determine the elevator movement required to change the lift coefficient by a small amount ΔC_L from its trimmed value—

$$\Delta\eta = \frac{dC_M}{dC_L}\frac{\Delta C_L}{\bar{V}a_2}$$

$$= -\frac{H_n}{\bar{V}a_2}\Delta C_L \tag{2.17}$$

Since $C_L V^2 =$ constant, in level flight, the change in C_L and the corresponding change in speed are related by

$$\frac{\Delta C_L}{C_L} = -2\frac{\Delta V}{V}$$

If a_2 is expressed in radians, then substituting for ΔC_L in eqn. (2.17) gives

Elevator angle to change the trim speed

$$= 2C_L \frac{H_n}{\bar{V}a_2}\frac{\Delta V}{V} \text{ rads} \tag{2.18}$$

and the corresponding stick movement is then given by

Stick movement to change the trim speed

$$= \frac{2}{m_e} C_L \frac{H_n}{\bar{V}a_2} \frac{\Delta V}{V} \text{ ft} \tag{2.19}$$

where m_e is the stick gearing (radians/ft). It can be seen that the elevator and stick movements required to change the trimmed speed by a small amount ΔV are directly proportional to the stick-fixed static margin and that they decrease with increase of the trimmed speed.

2.4.3. FLIGHT TEST MEASUREMENTS

The stick-fixed static stability can be determined from flight tests by measuring the elevator angle required to trim over the C_L range, for any selected C.G. position h. The value of $\bar{V}a_2$ may not be known and an additional test is then required to determine this value. The aircraft is flown with a further aft C.G. position h_1 and the new elevator angle to trim curve is obtained (Fig. 2.5). The initial trim equation can be written

$$0 = C_M = C_{M0} + (h - h_0)C_L - \bar{V}[a_1\alpha' + a_2\eta + a_3\beta] \tag{2.20}$$

and the new trim equation, at the same value of C_L, as

$$0 = C_M = C_{M0} + (h_1 - h_0)C_L - \bar{V}[a_1\alpha' + a_2\eta_1 + a_3\beta] \tag{2.21}$$

Subtracting (2.21) from (2.20) and rearranging

$$(h_1 - h)C_L = \bar{V}a_2(\eta_1 - \eta)$$

and

$$\bar{V}a_2 = \frac{h_1 - h}{\eta_1 - \eta} C_L \tag{2.22}$$

Substituting for $\bar{V}a_2$ in eqn. (2.14), we obtain the stick-fixed static margin for the C.G. position h

$$H_n = -\frac{h_1 - h}{\eta_1 - \eta} C_L \frac{d\eta}{dC_L} \tag{2.23}$$

The static margin for the C.G. position h_1 is obtained from the same formula except that $(d\eta/dC_L)$ now corresponds to the slope of the h_1 curve. In practice $(d\eta/dC_L)$ for the C.G. position h can be expressed as $-(\bar{\eta} - \eta)/C_L$, where $\bar{\eta}$ is the zero-lift elevator angle (Fig. 2.5) and hence we can write

$$\left.\begin{aligned} \underline{\underline{H_n = \frac{h_1 - h}{\eta_1 - \eta}(\bar{\eta} - \eta)}} \\ \text{and} \quad \underline{\underline{H_{n1} = \frac{h_1 - h}{\eta_1 - \eta}(\bar{\eta} - \eta_1)}} \end{aligned}\right\} \qquad (2.24)$$

By determining the static margin at two different C.G. positions and plotting, as in Fig. 2.7, it is possible to extrapolate and determine the position of the stick-fixed neutral point.

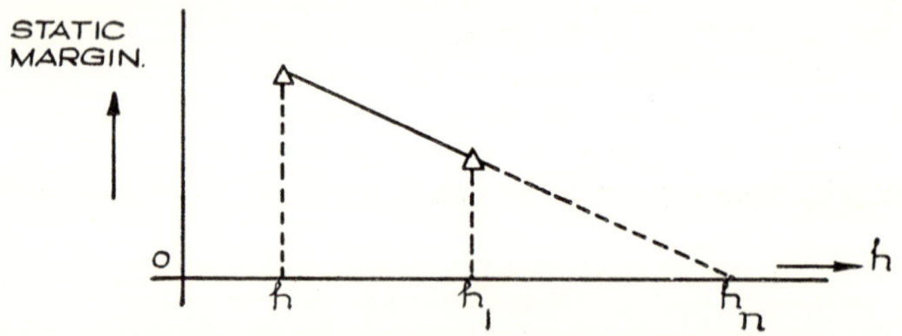

Fig. 2.7. Determination of the stick-fixed neutral point

2.4.4. SUMMARY

From eqn. (2.10) it can be seen that the factors affecting stick-fixed static stability are

(*a*) C.G. position
(*b*) tailplane area } i.e. tail volume coefficient
(*c*) tail arm }
(*d*) tailplane lift curve slope
(*e*) rate of change of downwash with incidence.

2.5. Stick-free Static Stability

If the aircraft is in trim and the stick left free, then during a disturbance the elevator will move and this will affect the stability characteristics. The elevator will move under the action of the control surface hinge moment and will take up an equilibrium position such that $C_H = 0$. For the initial trimmed state we can write

$$C_H = 0 = b_1\alpha' + b_2\eta + b_3\beta \qquad (2.25)$$

assuming $b_0 = 0$. If the incidence is changed in the disturbance to $(\alpha' + \Delta\alpha')$, then the elevator angle changes to $(\eta + \Delta\eta)$ (the trim tab setting remaining unchanged), such that

$$C_H = 0 = b_1(\alpha' + \Delta\alpha') + b_2(\eta + \Delta\eta) + b_3\beta \tag{2.26}$$

By subtracting (2.25) from (2.26), we have

$$b_1\Delta\alpha' + b_2\Delta\eta = 0$$

and the floating movement of the elevator is given by

$$\Delta\eta = -\frac{b_1}{b_2}\Delta\alpha' \tag{2.27}$$

Hence

$$\frac{d\eta}{dC_L} = -\frac{b_1}{b_2}\frac{d\alpha'}{dC_L}$$

Differentiating eqn. (2.5) with respect to C_L

$$\frac{d\alpha'}{dC_L} = \frac{d\alpha}{dC_L}\left(1 - \frac{d\varepsilon}{d\alpha}\right) = \frac{1}{a}\left(1 - \frac{d\varepsilon}{d\alpha}\right) \tag{2.28}$$

and hence

$$\underline{\underline{\frac{d\eta}{dC_L} = -\frac{b_1}{b_2}\frac{1}{a}\left(1 - \frac{d\varepsilon}{d\alpha}\right)}} \tag{2.29}$$

Substituting in eqn. (2.9) and putting $d\beta/dC_L = 0$, we have

$$-\frac{dC_M}{dC_L} = -(h - h_0) + \bar{V}\left[\frac{a_1}{a}\left(1 - \frac{d\varepsilon}{d\alpha}\right) - \frac{a_2 b_1}{a b_2}\left(1 - \frac{d\varepsilon}{d\alpha}\right)\right]$$

$$\underline{\underline{\left(-\frac{dC_M}{dC_L}\right)_{free} = -(h - h_0) + \bar{V}\frac{a_1}{a}\left(\left(1 - \frac{d\varepsilon}{d\alpha}\right)\left(1 - \frac{a_2 b_1}{a_1 b_2}\right)\right.}} \tag{2.30}$$

Hence the stick-free stability is identical to the stick-fixed stability, except that the tailplane contribution is modified by the factor $1 - (a_2b_1/a_1b_2)$ (compare eqns. (2.10) and (2.30)) and thus the stability is increased or decreased depending on whether this factor is greater, or less than, unity. The quantities a_1 and a_2 are always positive, b_2 should be negative for a normally balanced elevator and hence b_1 is the important variable. There are two possibilities

(i) b_1 is positive if the control is fitted with a large horn balance; the control then trails against the wind and the static stability is increased (i.e. the stick-free stability is greater than the stick-fixed).

(ii) if b_1 is negative the control trails with the wind and the stick-free stability is less than the stick-fixed.

2.5.1. STICK-FREE NEUTRAL POINT

By varying the C.G. position h, the stability can be made positive, zero, or negative; rearward movement of the C.G. being destabilizing. The particular position of the C.G. h_n' which gives neutral stability is called the stick-free neutral point N'. Substituting in (2.30)

$$0 = -(h_n' - h_0) + \bar{V}\frac{a_1}{a}\left(1 - \frac{d\varepsilon}{d\alpha}\right)\left(1 - \frac{a_2 b_1}{a_1 b_2}\right) \tag{2.31}$$

Subtracting (2.31) from (2.30), we have for any other C.G. position h

$$-\frac{dC_M}{dC_L} = h_n' - h \tag{2.32}$$

i.e. the stick-free static stability is equal to the non-dimensional distance between the C.G. and the stick-free neutral point (cf. para. 2.4.1 for the stick-fixed stability). The distance of the C.G. from the stick-free neutral point is called the *Stick-free* C.G. *Margin* and denoted by H_n'.

2.5.2. ELEVATOR TAB ANGLE AND STICK FORCE REQUIRED TO TRIM

If the aircraft is flying in trim and it is desired to change the trimmed speed, then it is necessary to apply the elevator control in such a way as to rotate the aircraft. A measure of the stick-free static stability is given by the difference between the new trim elevator angle required and *the angle that the elevator would have taken up, following the same incidence change, if the stick had been left free.* (Note the difference between this and para. 2.4.2 where the change in elevator angle from the initial trim position has to be measured.) As the stick-free elevator position is not known, it is necessary to use some other quantity to measure the stability and since the difference between the actual elevator position and its corresponding stick-free position must be directly related to the hinge moment, it is possible to use either (*a*) the elevator trim tab angle, or (*b*) the stick force required to trim.

(a) Elevator Tab Angle to Trim

If a change in tab angle of $\Delta\beta$ produces an additional change in elevator deflection $\Delta\eta$ (i.e. $\Delta\eta$ is the difference between the floating and trim positions of the elevator), then the corresponding change in pitching moment coefficient will be given by $-(\bar{V}a_2\Delta\eta + \bar{V}a_3\Delta\beta)$

[from eqn. (2.6)] and since the stabilizing moment coefficient (ΔC_M) is equal and opposite to this we have

$$\Delta C_M = \bar{V}a_2\Delta\eta + \bar{V}a_3\Delta\beta$$

and

$$-\frac{dC_M}{dC_L} = -\bar{V}\left(a_2\frac{d\eta}{dC_L} + a_3\frac{d\beta}{dC_L}\right) \tag{2.33}$$

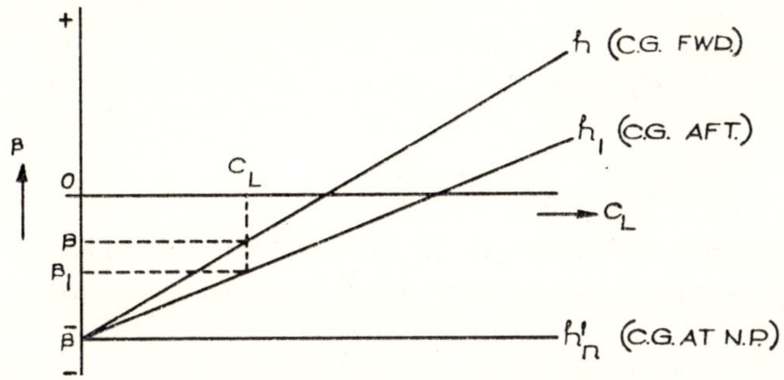

Fig. 2.8. Elevator tab angles to trim against C_L

To eliminate the change in the elevator angle from this equation we use the fact that the control hinge moments remain trimmed out ($\Delta C_H = 0$) and hence

$$b_2\Delta\eta + b_3\Delta\beta = 0 \tag{2.34}$$

$$\Delta\eta = -\frac{b_3}{b_2}\Delta\beta$$

and

$$\frac{d\eta}{dC_L} = -\frac{b_3}{b_2}\frac{d\beta}{dC_L} \tag{2.35}$$

Substituting in (2.33) we get

$$-\frac{dC_M}{dC_L} = -\bar{V}\left(-\frac{a_2b_3}{b_2} + a_3\right)\frac{d\beta}{dC_L}$$

$$-\frac{dC_M}{dC_L} = -\bar{V}a_3\left(1 - \frac{a_2b_3}{a_3b_2}\right)\frac{d\beta}{dC_L} \tag{2.36}$$

The curves of tab angle to trim against C_L are thus straight lines (Fig. 2.8) and providing we know $\bar{V}a_3$ $(1 - a_2b_3/a_3b_2)$, then we can determine the stick-free static stability by measuring the slope of the β/C_L curve and substituting in eqn. (2.36) above.

For normal balanced controls a_2b_3/a_3b_2 is usually greater than unity (because $a_2 \gg a_3$), hence $1 - (a_2b_3/a_3b_2)$ is negative and since $\bar{V}$ and a_3 are positive quantities, the slope of the β/C_L curve must be positive to give positive stability.

Equation (2.36) can be used to determine the elevator tab angle required to change the lift coefficient by a small amount ΔC_L from its trimmed value

$$\Delta\beta = \frac{dC_M}{dC_L}\frac{\Delta C_L}{[\bar{V}a_3(1 - a_2b_3/a_3b_2)]}$$

$$= \frac{-H_n'}{\bar{V}a_3(1 - a_2b_3/a_3b_2)}\Delta C_L \qquad (2.37)$$

Since $C_LV^2 = \text{constant}$, $\Delta C_L/C_L = -2\,\Delta V/V$ and expressing eqn. (2.37) in terms of ΔV, we have (if a_3 is expressed in radians)

$$\Delta\beta = -\frac{H_n'}{\bar{V}a_3\,(1 - a_2b_3/a_3b_2)}\left(-2C_L\frac{\Delta V}{V}\right)\text{ rads} \qquad (2.38)$$

Elevator tab angle to change the trim speed

$$= 2C_L\frac{H_n'}{\bar{V}a_3(1 - a_2b_3/a_3b_2)}\frac{\Delta V}{V}\text{ rads} \qquad (2.39)$$

(b) Stick Force to Trim

The stick force experienced by the pilot is given by

$$P = m_eC_H\tfrac{1}{2}\rho V^2S_\eta c_\eta\text{ lb} \qquad (2.40)$$

where m_e is the stick gearing (rads/ft) and C_H is the hinge moment coefficient. In terms of the elevator and tab deflections, we have (assuming $b_0 = 0$) that

$$C_H = b_1\alpha' + b_2\eta + b_3\beta \qquad (2.41)$$

For zero tab angle, eqn. (2.40) becomes

$$P = m_e[b_1\alpha' + b_2\eta]\tfrac{1}{2}\rho V^2S_\eta c_\eta$$

or, since $\alpha' = (\alpha - i - \varepsilon)$,

$$P = m_e[b_1(\alpha - i - \varepsilon) + b_2\eta]\,\tfrac{1}{2}\rho V^2S_\eta c_\eta$$

If all the angles are measured with respect to the wing no-lift line, α can be written as $\dfrac{C_L}{a}$ and $\varepsilon = \dfrac{d\varepsilon}{d\alpha}\alpha = \dfrac{d\varepsilon}{d\alpha}\dfrac{C_L}{a}$, hence

$$P = m_e\left[b_1\left(\frac{C_L}{a} - i - \frac{d\varepsilon}{d\alpha}\frac{C_L}{a}\right) + b_2\eta\right]\tfrac{1}{2}\rho V^2 S_\eta c_\eta \tag{2.42}$$

Using eqn. (2.15), the elevator angle to trim can be expressed as

$$\eta = \bar{\eta} - \frac{H_n}{\bar{V}a_2}C_L \tag{2.43}$$

and, substituting in eqn. (2.42), we get

$$P = m_e\left[b_1\left(\frac{C_L}{a} - i - \frac{d\varepsilon}{d\alpha}\frac{C_L}{a}\right) + b_2\left(\bar{\eta} - \frac{H_n}{\bar{V}a_2}C_L\right)\right]\tfrac{1}{2}\rho V^2 S_\eta c_\eta$$

Separating out the terms in C_L

$$P = m_e\left[\left(\frac{b_1}{a} - \frac{b_1}{a}\frac{d\varepsilon}{d\alpha} - \frac{b_2 H_n}{\bar{V}a_2}\right)C_L + (b_2\bar{\eta} - b_1 i)\right]\tfrac{1}{2}\rho V^2 S_\eta c_\eta$$

Multiplying out and replacing $C_L\tfrac{1}{2}\rho V^2$ by W/S

$$P = m_e\frac{W}{S}S_\eta c_\eta\left[\frac{b_1}{a}\left(1 - \frac{d\varepsilon}{d\alpha}\right) - \frac{b_2 H_n}{\bar{V}a_2}\right]$$
$$+ m_e S_\eta c_\eta(b_2\bar{\eta} - b_1 i)\tfrac{1}{2}\rho V^2$$

Rearranging

$$P = -m_e\frac{W}{S}S_\eta c_\eta\left\{\frac{b_2}{\bar{V}a_2}\left[H_n - \bar{V}\frac{a_1}{a}\frac{a_2 b_1}{a_1 b_2}\left(1 - \frac{d\varepsilon}{d\alpha}\right)\right]\right\}$$
$$+ m_e S_\eta c_\eta(b_2\bar{\eta} - b_1 i)\tfrac{1}{2}\rho V^2$$

and since

$$H_n - \bar{V}\frac{a_1}{a}\frac{a_2 b_1}{a_1 b_2}\left(1 - \frac{d\varepsilon}{d\alpha}\right) = H_n'$$

this becomes

$$P = m_e S_\eta c_\eta(b_2\bar{\eta} - b_1 i)\tfrac{1}{2}\rho V^2 - m_e\frac{W}{S}S_\eta c_\eta\frac{b_2 H_n'}{\bar{V}a_2} \tag{2.44}$$

The stick force thus consists of two parts, one which depends on the stick-free static stability H_n' and is independent of speed and the other which is proportional to the square of the speed and depends on such factors as C_{M0} and i. Eqn. (2.44) can be written in the form

$$P = A + BV^2 \tag{2.45}$$

where $A = -m_e \dfrac{W}{S} S_\eta c_\eta \dfrac{b_2 H_n'}{\bar{V} a_2}$ and $B = m_e S_\eta c_\eta (b_2 \bar{\eta} - b_1 i)\frac{1}{2}\rho$, and the variation is shown graphically in Fig. 2.9. The intercept on the vertical axis at $V = 0$ is A and this is dependent on the value of H_n'.

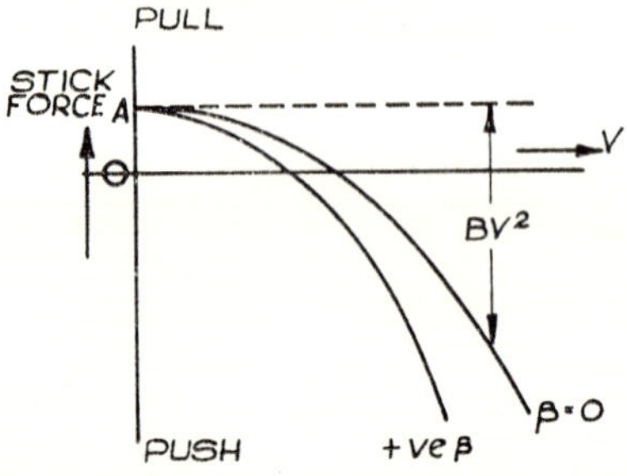

Fig. 2.9. Stick force against speed for different tab angles

If the aircraft possesses stick-free static stability this intercept is a positive one (i.e. a pull force).

If V_{trim} is the speed at which the stick force is zero, then from eqn. (2.44)

$$m_e S_\eta c_\eta (b_2 \bar{\eta} - b_1 i)\tfrac{1}{2}\rho V^2_{trim} = m_e \frac{W}{S} S_\eta c_\eta \frac{b_2 H_n'}{\bar{V} a_2}$$

Substituting $m_e \dfrac{W}{S} S_\eta c_\eta \dfrac{b_2 H_n'}{\bar{V} a_2} \dfrac{1}{V^2_{trim}}$ for $m_e S_\eta c_\eta (b_2 \bar{\eta} - b_1 i)\frac{1}{2}\rho$ in eqn. (2.44) we get

$$P = -m_e \frac{W}{S} S_\eta c_\eta \frac{b_2 H_n'}{\bar{V} a_2}\left(1 - \frac{V^2}{V^2_{trim}}\right) \tag{2.46}$$

Differentiating this equation

$$\frac{dP}{dV} = 2m_e \frac{W}{S} S_\eta c_\eta \frac{b_2 H_n'}{\bar{V} a_2} \frac{V}{V^2_{trim}}$$

and, at the trimmed speed,

$$\frac{dP}{dV} = 2m_e \frac{W}{S} S_\eta c_\eta \frac{b_2 H_n'}{\bar{V} a_2} \frac{1}{V_{trim}} \tag{2.47}$$

Thus the slope of the curve of stick force plotted against forward speed, at the trimmed speed, is a measure of the stick-free static stability.

From eqn. (2.47), the change in stick force ΔP required for a small change in the forward speed ΔV, from the initial trimmed value V_{trim} is given by

$$\Delta P = 2m_e \frac{W}{S} S_\eta c_\eta \frac{b_2 H_n'}{\bar{V} a_2} \frac{1}{V_{trim}} \Delta V \tag{2.48}$$

From a comparison with eqn. (2.45), it can be seen that the stick-force gradient, at the trimmed speed, can be expressed in the form of

$$\frac{dP}{dV} = -2 \frac{A}{V_{trim}} \tag{2.49}$$

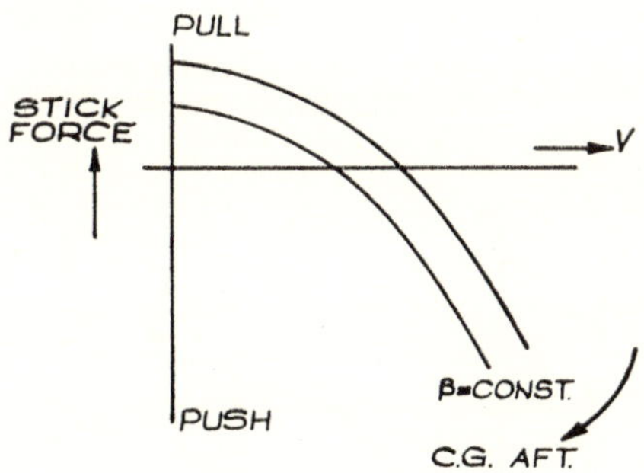

Fig. 2.10. Stick force against speed for different C.G. positions

The analysis has been performed for one tab position; any variation in the tab position changes the shape of the curve and the trimmed speed, but does not vary the value of the intercept on the vertical axis (i.e. A). From eqn. (2.47) it can be seen that, for a constant value of H_n', the gradient of the stick force curve is inversely proportional to the trim speed.

Rearward movement of the C.G. reduces the stick-free static stability, and thus reduces the value of the intercept A. It does not, however, alter the shape of the curve (Fig. 2.10) and the slopes of these curves are not indicative of the stick-free stability. However, if the trimmed speed is made identical for the two C.G. positions considered, by adjusting the trim tab, then the gradients of the stick force curves, at the trimmed speed, are directly proportional to the corresponding stick-free stabilities (Fig. 2.11).

The method so far described has the practical disadvantage that the gradient of the stick force curve is dependent on the particular value of the trimmed speed, as well as on the stick-free static stability. Unfortunately, if there is a significant amount of friction in the control system, it is extremely difficult for the pilot to trim the aircraft at precisely the speed required for each of the different C.G. positions. To reduce the effects of friction, the previous analysis

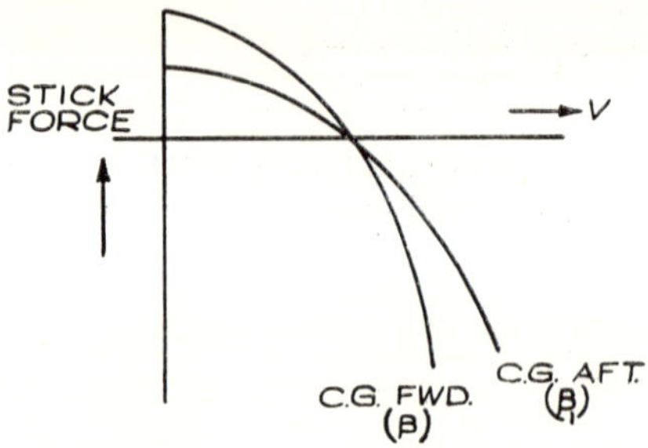

Fig. 2.11. Stick force against speed for different C.G. positions and tab angles

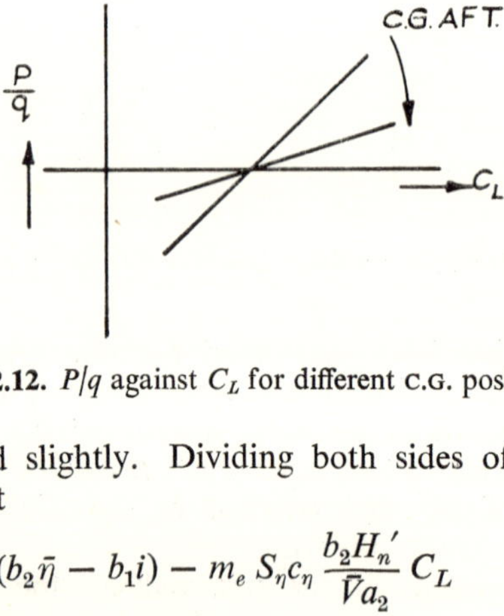

Fig. 2.12. P/q against C_L for different C.G. positions

can be modified slightly. Dividing both sides of eqn. (2.44) by $\frac{1}{2}\rho V^2$ or q we get

$$\frac{P}{q} = m_e S_\eta c_\eta (b_2 \bar{\eta} - b_1 i) - m_e S_\eta c_\eta \frac{b_2 H_n'}{\bar{V} a_2} C_L \tag{2.50}$$

and differentiating with respect to C_L

$$\frac{d}{dC_L}\left(\frac{P}{q}\right) = -m_e S_\eta c_\eta \frac{b_2 H_n'}{\bar{V} a_2} \tag{2.51}$$

Thus the curves of P/q plotted against C_L should be straight lines, the slope of the line being determined by H_n', the stick-free static stability (Fig. 2.12).

2.5.3. FLIGHT TEST MEASUREMENTS

In flight test work the stick-free static stability can be determined by measuring either the elevator tab angles to trim, or the corresponding stick forces. If the tab angles are measured, then an additional test is required to determine the value of $\bar{V}a_3(1 - a_2b_3/a_3b_2)$. The aircraft is flown with a further aft C.G. position h_1 and the new elevator tab angle to trim curve is obtained (Fig. 2.8). Then, by a comparison with eqn. (2.22)

$$\bar{V}a_3\left(1 - \frac{a_2b_3}{a_3b_2}\right) = \frac{h_1 - h}{\beta_1 - \beta} C_L \tag{2.52}$$

Substituting in eqn. (2.36), the stick-free static margin for the C.G. position h is given by

$$H_{n1}' = -\frac{h_1 - h}{\beta_1 - \beta} C_L \frac{d\beta}{dC_L} \tag{2.53}$$

The static margin for the C.G. position h_1 is obtained from the same formula except that $(d\beta/dC_L)$ now represents the slope of the h_1 curve. In practice $(d\beta/dC_L)$ can be expressed as $-(\bar{\beta} - \beta)/C_L$ where $\bar{\beta}$ is the zero-lift value (Fig. 2.8) and hence we can write

$$\left.\begin{aligned} H_n' &= \frac{h_1 - h}{\beta_1 - \beta}(\bar{\beta} - \beta) \\ \text{and}\quad & \\ H_{n1}' &= \frac{h_1 - h}{\beta_1 - \beta}(\bar{\beta} - \beta_1) \end{aligned}\right\} \tag{2.54}$$

By measuring the stick-free static margin at two different C.G. positions it is possible to determine the position of the stick-free neutral point, by linearly extrapolating the curve of static margin plotted against C.G. position (cf. Fig. 2.7 for the stick-fixed case).

If the method of measuring the stick forces is used, then the test is flown at two C.G. positions and the results are plotted in the form P/q against C_L. The pilot attempts to trim the aircraft at the same speed in each case, but, if he is not successful in this, the method of plotting eliminates the errors involved. By measuring the slopes of these curves at the appropriate trim speed, the position of the stick-free neutral point can be determined by linear extrapolation as before. Fig. 2.13 shows some test results for the Canberra Mk. 2, taken from Ref. 3, for different tailplane settings.

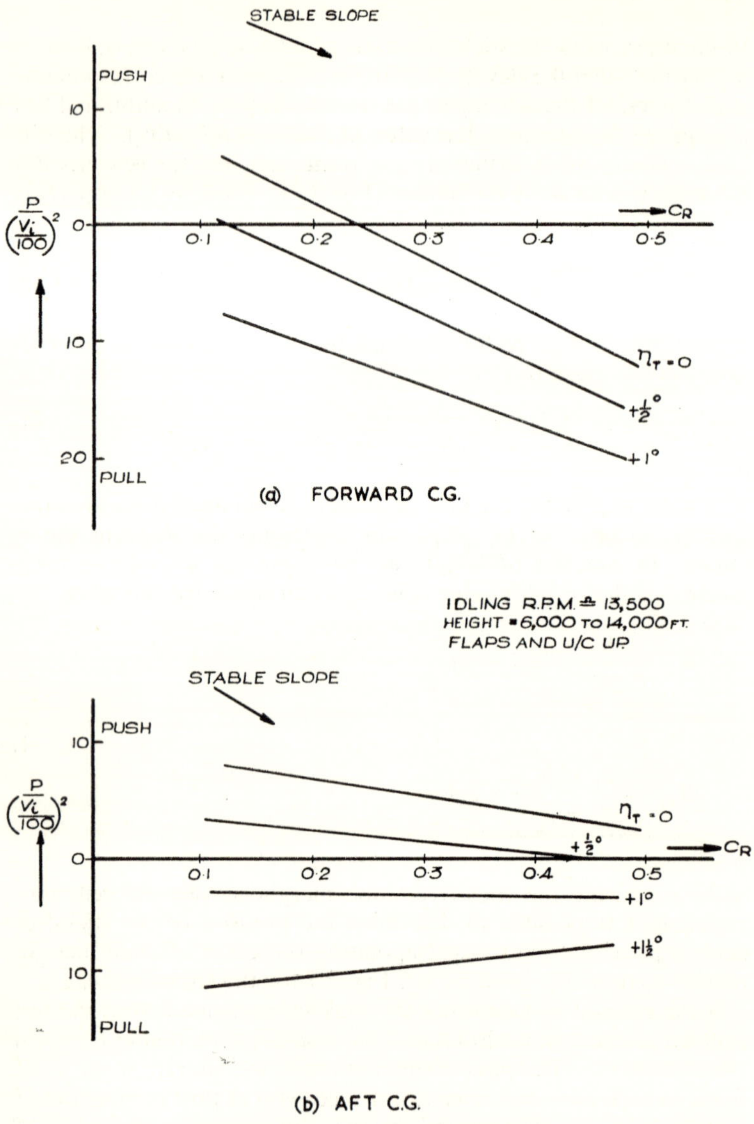

Fig. 2.13. Stick forces to trim [From Ref. 3]

2.5.4. SUMMARY

Additional factors to those already given in para. 2.4.4 can be seen from eqn. (2.30) to be

(*a*) the rate of change of tailplane lift coefficient with elevator deflection a_2;

(*b*) the rate of change of hinge moment coefficient with tailplane incidence b_1;

(*c*) the rate of change of hinge moment coefficient with elevator deflection b_2.

3. General Treatment of Static Stability

3.1. Complicating Factors

The preceding simplified mathematical analysis is based on the assumptions listed in para. 2.1. If these assumptions are valid, then the trim curves of elevator angle and elevator tab angle plotted against C_L are straight lines. The slopes of these lines are dependent on the position of the C.G., but they intersect the vertical axis (i.e. $C_L = 0$) at a common point. The C.G. margin and the static margin are numerically equal to one another, and the non-dimensional distance of the C.G. from the neutral point is a direct measure of the static stability. The stability values can be calculated, at any selected value of C_L, by substituting the appropriate numerical values in either eqns. (2.24) or (2.54) as required; they are independent of the C_L value selected and hence do not vary with forward speed.

It is, however, most unlikely that all these assumptions will be valid and the trim curves will usually depart appreciably from the straight lines of the simple theory. The stability will then vary throughout the speed range and the position of the neutral point will vary with C_L. The C.G. margin and the static margin will no longer be identical to one another and C.G. movement will not necessarily have the same effect on the static stability as the simple theory predicts. These complications tend to make the concept of a static margin less useful than before. However, in many cases, the trim curves do not depart very far from straight lines and under these conditions the results can be analysed in a similar way, and the corresponding static margins determined.

Non-linearities in the trim curves will arise as a result of the breakdown in any of the assumptions made in para. 2.1. Before trying to analyse these effects, however, it is essential to realize that, while the trim measurements are being made, changes occur in both the forward speed and the incidence of the aircraft. In the simple theory of para. 2, only incidence effects were allowed for; it was assumed that changes in forward speed did not affect the stability, because all the pitching moments were changed in the same proportion.

This assumption is satisfactory at low speeds, but breaks down when aeroelastic and compressibility effects are involved. In the more general theory of static stability, both incidence and speed effects have to be allowed for.

Other assumptions, which are inherent in the simple theory, are that the stability is not influenced in any way by the power unit and the effects of the slipstream and that the vertical position of the wing is unimportant. These effects will now be discussed.

3.1.1. POWER EFFECTS

The power effects are very important, they modify the static stability characteristics and determine the method of performing the actual flight tests. There are two possible ways of measuring the trim curves

(*a*) in level flight, adjusting the power or thrust of the engines to maintain level flight equilibrium,

(*b*) at a constant engine setting, adjusting the rate of climb, or descent, to maintain equilibrium flight.

In the first of these two methods, the changes of power or thrust will usually involve trim changes. These trim changes will be included in the trim curves and hence in the calculated values of the static margins. This does not give a true measure of the inherent static stability of the aircraft and the effects are usually eliminated by performing the test in accordance with method (*b*).

Propeller Aircraft

In the case of a propeller driven aircraft, the power-on stability is usually less than the corresponding power-off stability. The effects on the static stability can be split up into

(*a*) the direct effects, which consist of a small destabilizing contribution from the propeller acting as a small horizontal surface ahead of the C.G. (Fig. 2.14) and a contribution which depends on the vertical position of the thrust line relative to the C.G. A high thrust line is stabilizing, and a low thrust line destabilizing (Fig. 2.15),

(*b*) the indirect effects from the slipstream, which consist of a destabilizing contribution arising from the modification of the downwash over the tailplane (i.e. an increase in the value of $d\varepsilon/d\alpha$) and a contribution arising from the change in the dynamic pressure at the tail. This increases with increase of C_L

(Ref. 4, page 239) and the resulting effect on the stability depends on the loading of the tailplane. When the tailplane has an upload on it, an extra stabilizing contribution is produced, whereas if the tailplane has a download, the effect is a destabilizing one.

It should be noted, however, that some of the above effects are *incidence* disturbance effects and that the others are *speed* disturbance

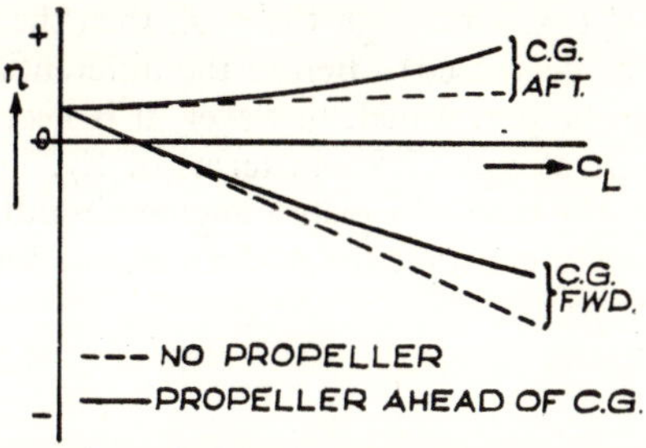

Fig. 2.14. Effect of propeller on the elevator angle to trim curve

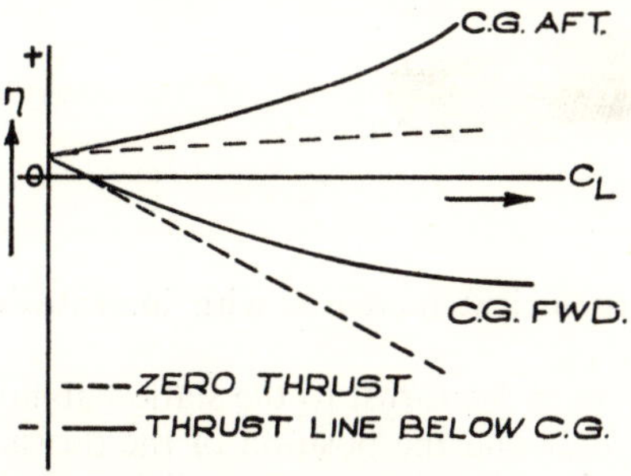

Fig. 2.15. Effect of position of the thrust line on the elevator angle to trim curve

effects. The speed disturbance effects are those involving the position of the thrust line and the changes in the dynamic pressure. It is not immediately obvious how these effects arise and the contribution from the vertical position of the thrust line is discussed more fully below.

On page 232 of Ref. 4, it is shown that the pitching moment coefficient equation has to be modified, due to the presence of the propeller, by the addition of two terms, one arising from the thrust force T and one from the normal force N which we will not consider further. The equations are usually written in terms of the pitching moment *coefficients* and a thrust coefficient is defined (in a similar way to the lift coefficient) as

$$T_c = \frac{T}{\rho V^2 D^2} \tag{2.55}$$

where T is the thrust and D a representative length. The effect on the static stability dC_M/dC_L is determined by whether or not T_c varies with C_L (i.e. if dT_c/dC_L has a value other than zero and if the thrust line does not pass through the C.G., then the stability will be either increased or decreased). Before the differentiation of T_c with respect to C_L can be performed, however, it is necessary to rewrite the right hand side of eqn. (2.55) in terms of C_L. Now T is itself a variable, since, in the case of a piston engine aircraft, it is the power (i.e. $T \times V$) which is constant for a given engine setting. Thus, we can write

$$\left.\begin{aligned} &T \propto 1/V \\ \text{and}\quad & \\ &T_c \propto 1/V^3 \end{aligned}\right\} \tag{2.56}$$

In terms of C_L, since $C_L \propto 1/\sqrt{V}$, we have

$$T_c \propto C_L{}^{3/2} \tag{2.57}$$

hence

$$\frac{dT_c}{dC_L} \propto C_L{}^{1/2} \tag{2.58}$$

and the thrust coefficient increases with increase in C_L (i.e. with decreasing speed).

The contribution of the thrust to the static stability depends upon the value of dT_c/dC_L and the position of the thrust line relative to the C.G. If the thrust line is a vertical distance z below the C.G., then the contribution to dC_M/dC_L is given by

$$\underline{\underline{\frac{dC_M}{dC_L} = \text{constant} \times z \times \frac{dT_c}{dC_L}}} \tag{2.59}$$

and since both z and dT_c/dC_L are positive in this case, the contribution to dC_M/dC_L is positive. Thus an increase in C_L results in a positive (i.e. nose up) contribution to C_M, tending to increase the incidence still further and a low thrust line is destabilizing. If the thrust line is above the C.G., then z is negative and because dT_c/dC_L is still positive, the contribution to dC_M/dC_L is a negative one. Thus a high thrust line is stabilizing.

Jet Aircraft

In the case of a jet aircraft, similar effects are present.

(*a*) the direct effects, which consist of a destabilizing component from the intake normal force (Ref. 4, page 245) and one depending on the position of the thrust line,

(*b*) the indirect effect of the extra downwash induced by the jet efflux, which may be destabilizing.

The contribution from the vertical position of the thrust line can be analysed in a similar way to that for a piston-engine aircraft. In this case, however, it is the thrust of the engine that is constant and hence

$$T_c \propto 1/V^2 \tag{2.60}$$

Equation (2.57) is now replaced by

$$T_c \propto C_L \tag{2.61}$$

and equation (2.58) by

$$\frac{dT_c}{dC_L} = \text{constant} \tag{2.62}$$

Thus T_c increases linearly with increase of C_L (i.e. increases with decreasing speed). In the case of a piston engine aircraft, T_c increases non-linearly with C_L, the rate increasing with increasing C_L (as shown in Fig. 5.20 on page 233 of Ref. 4).

Apart from this difference, however, the two cases are identical and so is the final result, that

(*a*) a high thrust line is stabilizing,

(*b*) a low thrust line is destabilizing.

3.1.2. WING POSITION EFFECTS

Assumption 2.1 (v) is no longer applicable when the wing is set high up or low down on the fuselage. A low wing position is destabilizing, and a high position stabilizing, the magnitude of the effect increasing with increasing C_L (i.e. with decreasing speed) [Fig. 2.16].

3.2. Generalized Theory

In the more general mathematical theory of static stability, a number of the assumptions inherent in the simple theory are no longer made and the effects due to the variations of forward speed have to be allowed for. The *Static margin* then includes both speed and incidence effects; the term *Restoring margin* is sometimes used for the incidence contribution to the static margin.

The analysis is still made for power-off gliding flight; the equilibrium condition is given by

$$W = R = C_R \tfrac{1}{2}\rho V^2 S \tag{2.63}$$

where $R[= \sqrt{(L^2 + D^2)}]$ is the resultant aerodynamic force and the aircraft is flown such that $C_R V^2$ remains constant. The theory is

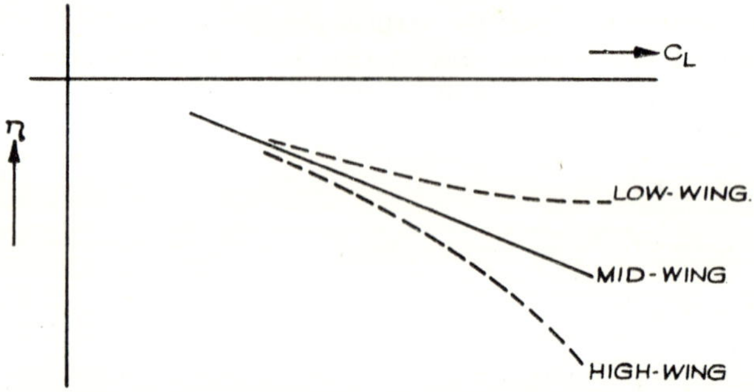

Fig. 2.16. Effect of vertical position of wing on the elevator angle to trim curve

now applicable for any value of the gliding angle; when this is small $C_R \simeq C_L$.

To allow for the speed effects, all the coefficients a_1, a_2, a_3, b_1, etc., are no longer regarded as constants, but as functions of speed (i.e. of C_R) and are now represented by A_1, A_2, A_3, B_1, etc.

The results obtained from the more general theory are very similar to those of the simple theory (except that C_R replaces C_L, A_2 replaces a_2, etc.) and because of this, the analysis will not be repeated here. (It can be found, if required, in Refs. 1 and 2.) There is, however, one very important and essential difference and that is that the static margins are no longer numerically *equal* to the corresponding C.G. margins, but are only *proportional* to them, the ratio varying with forward speed.

The stick-fixed static margin is now represented by K_n, to distinguish it from H_n, the stick-fixed C.G. margin and the relationship between these two quantities can be written

$$K_n = \psi H_n \tag{2.64}$$

where ψ is the constant of proportionality, which is a function of speed.

In a similar way, the stick-free static margin K_n' is related to the corresponding stick-free C.G. margin H_n' by

$$\underline{\underline{K_n' = \chi H_n'}} \tag{2.65}$$

where χ is another constant of proportionality, which is also a function of speed.

A measure of the static stabilities can still be obtained from the trim curves, though separate calculations have to be made at each selected value of C_R. The equations deduced for K_n and K_n' are similar to those [eqns. (2.24) and (2.54)] obtained from the simple theory. They are given by

$$\left.\begin{aligned} \underline{\underline{K_n = \frac{h_1 - h}{\eta_1 - \eta}(\bar{\eta} - \eta)}} \\ \underline{\underline{K_{n1} = \frac{h_1 - h}{\eta_1 - \eta}(\bar{\eta}_1 - \eta_1)}} \end{aligned}\right\} \tag{2.66}$$

and

$$\left.\begin{aligned} \underline{\underline{K_n' = \frac{h_1 - h}{\beta_1 - \beta}(\bar{\beta} - \beta)}} \\ \underline{\underline{K_{n1}' = \frac{h_1 - h}{\beta_1 - \beta}(\bar{\beta}_1 - \beta_1)}} \end{aligned}\right\} \tag{2.67}$$

The corresponding values of ψ and χ can also be obtained from

$$\psi = 1 - \left(\frac{\bar{\eta} - \bar{\eta}_1}{\eta - \eta_1}\right) \tag{2.68}$$

$$\chi = 1 - \left(\frac{\bar{\beta} - \bar{\beta}_1}{\beta - \beta_1}\right) \tag{2.69}$$

On the simple theory, these values both reduce to unity, since $\bar{\eta} = \bar{\eta}_1$ and $\bar{\beta} = \bar{\beta}_1$ (i.e. the intercepts on the vertical axes are independent of the C.G. position).

It should be carefully noted that the *static margin* gives a measure of the static stability of the aircraft for a given C.G. position, whereas the *C.G. margin* indicates how far aft the C.G. can be moved before the aircraft becomes unstable.

If the tangents drawn to the trim curves at any selected C_R value do not intersect on the vertical axis, there are four possibilities. These possibilities are illustrated by the elevator angle curves in

Fig. 2.17, though the following discussion is applicable to the tab angle curves also, with any necessary modifications. The tangents may either converge at the low values of C_L, intersecting one another to the right (Fig. 2.17*a*), or to the left (Fig. 2.17*b*), of the vertical axis, or they may be parallel to one another (Fig. 2.17*c*), or they may diverge (Fig. 2.17*d*). The value of ψ decreases from a value greater

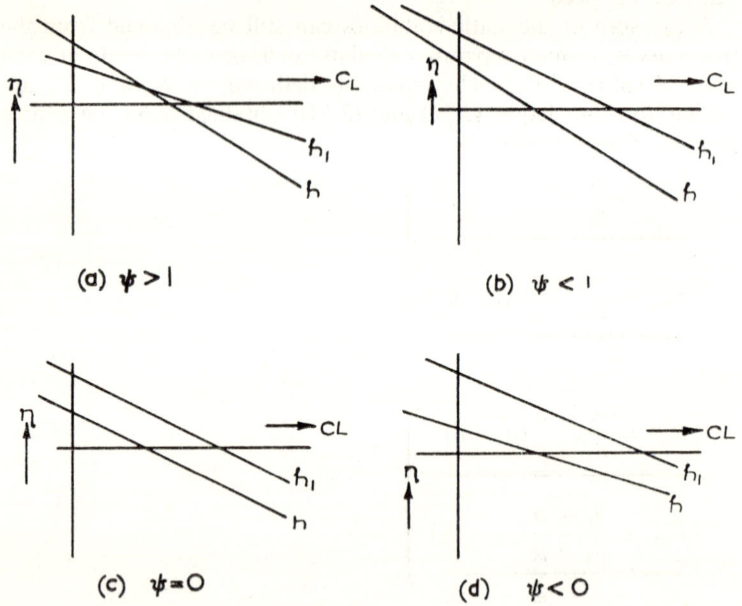

Fig. 2.17. The effect of ψ on the slopes of the tangents to the elevator trim curves

than unity (Fig. 2.17*a*), to a value less than unity (Fig. 2.17*b*), to zero (Fig. 2.17*c*) and finally in Fig. 2.17*d*, becomes negative. Interpreted in terms of C.G. movement, this means that rearward movement of the C.G. has the following effects

Fig. 2.17*a*—reduces the static stability at a faster rate than the C.G. movement itself.

Fig. 2.17*b*—reduces the static stability, but at a slower rate than the C.G. movement itself.

Fig. 2.17*c*—has no effect on static stability.

Fig. 2.17*d*—increases the static stability. This effect is, of course, exactly the opposite of that normally experienced and in this case, the stability is positive with a negative C.G. margin.

These results are presented diagrammatically in Fig. 2.18. Some actual values of ψ obtained from flight tests are shown in Fig. 14.11.

From this, it should be obvious that the static margin is dependent not only on the slope of the trim curves, but also on the separation between them. This separation is given by

$$(\eta_1 - \eta) = \frac{(h_1 - h)C_L}{\bar{V}a_2} \tag{2.70}$$

and is thus determined by $\bar{V}a_2$, the elevator pitching moment effectiveness. If $\bar{V}a_2$ is constant, then $(\eta_1 - \eta)$ decreases with decrease in C_L, becoming zero at $C_L = 0$. If the curves are parallel to one

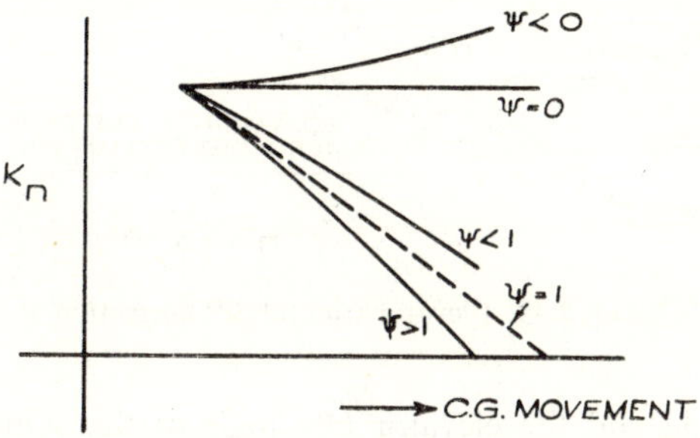

Fig. 2.18. Effect of C.G. movement on static stability

another (i.e. $\eta_1 - \eta$ = constant) then, since the numerator on the R.H.S. of eqn. (2.70) is proportional to C_L, the denominator is also proportional to C_L (i.e. the elevator effectiveness $\bar{V}a_2$ is proportional to C_L and decreases with increasing speed).

4. Flight Test Measurements

4.1. Static Stability Requirements

It is desirable that an aircraft should possess both stick-fixed and stick-free static stability, since they determine the stick movements and stick forces, respectively, to trim the aircraft at different forward speeds.

Stick-fixed stability is not of such primary importance, because the pilot is not normally concerned with the movements involved, unless they are large, or in the unexpected direction. He is usually more interested in the stick forces and these are determined by the stick-free stability. However, as we have seen, the two stabilities are closely related to one another and for a given value of stick-fixed stability, the stick-free stability can only be varied over fairly narrow limits, if undesirable handling characteristics are to be avoided.

The requirements are laid down in Av.P.970 Vol. 1, Chapter 601, paras. 9.1.1 and 9.1.2. The aircraft should possess stick-free static stability and the stick-fixed static margin should not fall below the stick-free margin by more than 0·05.

4.2. Flight Test Techniques

The test is usually flown with a fixed engine setting, climbing or diving the aircraft as necessary to maintain equilibrium flight.

Strictly, the elevator angles to trim should be measured with the trim tab kept fixed. In practice, it is easier to measure both the

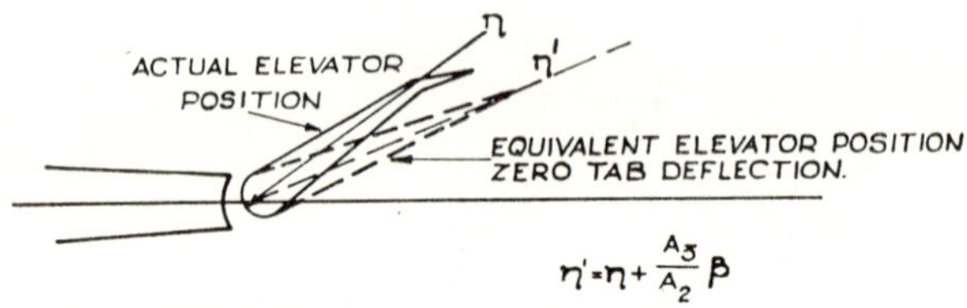

Fig. 2.19. Elevator trim tab lift correction

elevator angle and the elevator tab angle at the same time. This necessitates making a correction to the measured elevator angles to allow for the effects of the trim tab movement on the tailplane lift. The corrected elevator angle η' is given by

$$A_2\eta' = A_2\eta + A_3\beta$$

where η and β are the measured values. Hence

$$\eta' = \eta + \frac{A_3}{A_2}\beta \tag{2.71}$$

The corrected angle is usually slightly less (numerically) than the uncorrected one, since the ratio A_3/A_2 is small (about 0·1) and β is opposite in sign to η (Fig. 2.19).

Before the corrected elevator angles to trim can be calculated it is necessary to know, or assume, a value for A_3/A_2. An additional test can be carried out to determine this value. Suppose the aircraft is in trim at a particular speed with the stick force also trimmed out, the elevator deflection being η and the tab deflection β. If a small push force is now applied to the stick, and the trim tab moved so as to keep the aircraft in trim at the same speed, the deflections become, say, η_{i} and β_{i}. The tailplane lift will be the same in each case, hence

$$C_L' = A_1\alpha' + A_2\eta + A_3\beta = A_1\alpha' + A_2\eta_{\text{i}} + A_3\beta_{\text{i}}$$

Rearranging,

$$-A_2(\eta_{\text{i}} - \eta) = A_3(\beta_{\text{i}} - \beta)$$

and

$$\frac{A_3}{A_2} = -\frac{\eta_{\text{i}} - \eta}{\beta_{\text{i}} - \beta} \tag{2.72}$$

The ratio A_3/A_2 is thus given by the ratio of the change in elevator angle to the corresponding change in tab angle (the minus sign being required because if $\eta_{\text{i}} > \eta$ then $\beta_1 < \beta$).

It is advisable to repeat the test for a pull force. If the corresponding deflections are η_{ii} and β_{ii}, then, using the total surface movements

$$\frac{A_3}{A_2} = -\frac{\eta_{\text{i}} - \eta_{\text{ii}}}{\beta_{\text{i}} - \beta_{\text{ii}}} \tag{2.73}$$

The out-of-trim forces must be kept small, otherwise they may cause aero-elastic distortion of the tailplane and give incorrect results. It is necessary to determine the value of A_3/A_2 at a number of speeds throughout the speed range.

4.3. Flight Test Results

From the curves of corrected elevator angle to trim and elevator tab angle to trim, plotted against C_R, the static stabilities can be determined using eqns. (2.66) and (2.67). Some actual results are shown in Fig. 2.20.

By measuring the static margins for two different C.G. positions, the corresponding value of the neutral point can be determined by the method of linear extrapolation, similar to that shown in Fig. 2.7. In the more general case, however, separate curves have to be drawn for each selected value of C_R, since the positions of the neutral points are no longer independent of C_R, as they were on the simple theory (Fig. 2.21). It is also advisable to determine the trim curves at more than two C.G. positions, since the variation of stability with C.G. position will no longer be a linear one.

If, for any reason, it is decided not to analyse the trim curves in the way described above, then a qualitative assessment may be made. In this case, the following points should be noted

(i) the slopes of the curves always indicate whether the aircraft is stable, unstable, or neutrally stable.

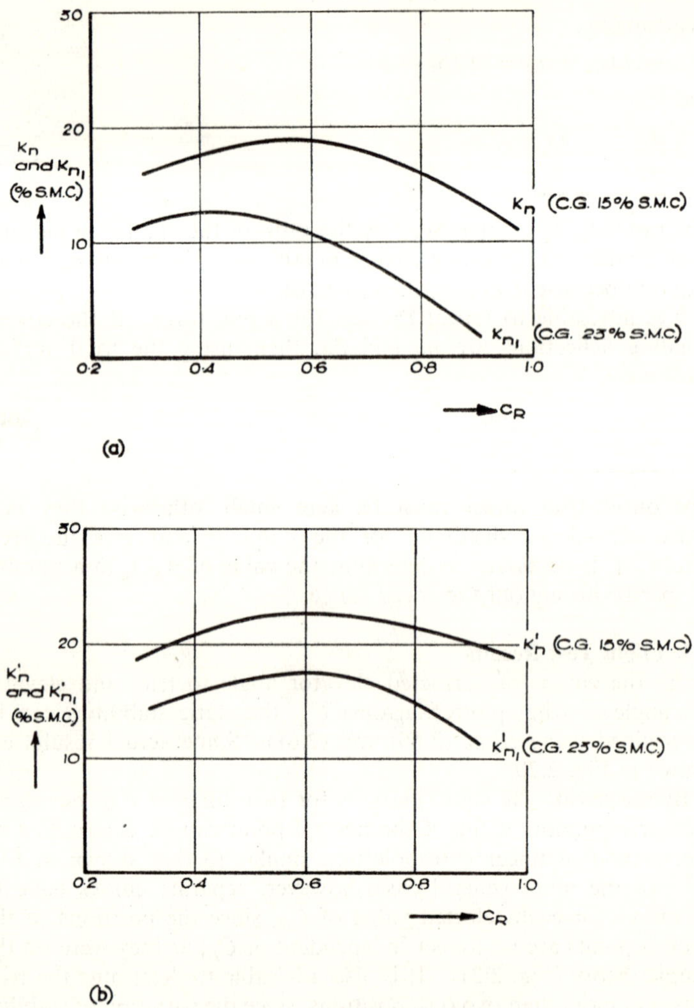

Fig. 2.20. (*a*) Stick-fixed static margins
(*b*) Stick-free static margins

(ii) the slopes of the curves can not be taken as proportional to the static stability, except when comparing the stability for two C.G. positions at a given value of C_R.

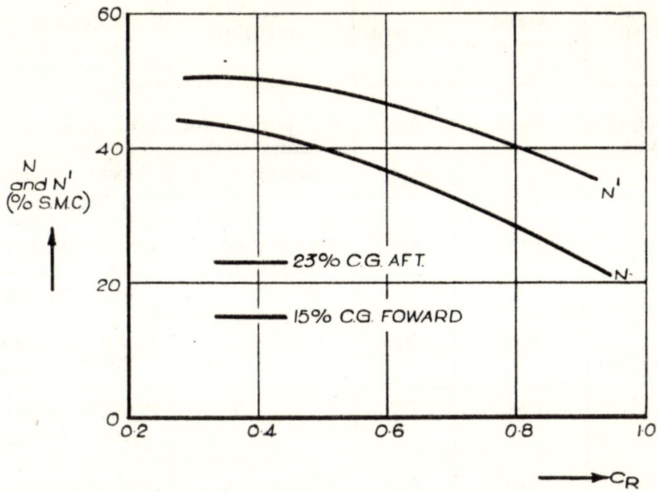

Fig. 2.21. Stick-fixed and free neutral points

(iii) any discussion of static stability over the speed range must be based on the values of K_n and K_n' and not on the slopes of the trim curves.

5. Summary

The work of this chapter can be summarized in the following tables—

Table 1. Definitions and Notations

NAME	STICK CONDITION	SYMBOL	MEANING	USED TO DETERMINE
Neutral point	fixed	h_n	C.G. position for neutral static stability	C.G. position for zero stick movement to change speed
	free	h_n'	C.G. position for neutral static stability	C.G. position for zero stick force to change speed
C.G. margin	fixed	H_n	$h_n - h$	Guide to permissible C.G. range
	free	H_n'	$h_n' - h$	Guide to permissible C.G. range
Static margin	fixed	K_n	Measures static stability stick fixed	Stick movement to change speed
	free	K_n'	Measures static stability stick free	Stick force to change speed

Table 2. Methods of Measurement

ITEM	STICK FIXED	STICK FREE
STATIC STABILITY 1. Neutral point	h_n	h_n'
2. C.G. margin Basic equation	$H_n = h_n - h = -\frac{dC_M}{dC_L}$ $= -(h - h_0) + \bar{V}\frac{a_1}{a}\left(1 - \frac{d\varepsilon}{d\alpha}\right)$	$H'_n = h_n' - h = -\frac{dC_M}{dC_L}$ $= -(h - h_0)$ $+\bar{V}\frac{a_1}{a}\left(1 - \frac{d\varepsilon}{d\alpha}\right)\left[1 - \frac{a_2 b_1}{a_1 b_2}\right]$
Method of measurement	$= -\bar{V} a_2 \frac{d\eta}{dC_L}$ $= -\frac{h_1 - h}{\eta_1 - \eta} C_L \frac{d\eta}{dC_L}$	$= -\bar{V} a_3 \left[1 - \frac{a_2 b_3}{a_3 b_2}\right] \frac{d\beta}{dC_L}$ $= -\frac{h_1 - h}{\beta_1 - \beta} C_L \frac{d\beta}{dC_L}$
3. Static margin Method of measurement	$K_n = \psi H_n$ $= -\bar{V} A_2 \frac{d\eta}{dC_R}$ $= -\frac{h_1 - h}{\eta_1 - \eta} C_R \frac{d\eta}{dC_R}$	$K_n' = \chi H_n'$ $= -\bar{V} A_3 \left[1 - \frac{A_2 B_3}{A_3 B_2}\right] \frac{d\beta}{dC_R}$ $= -\frac{h_1 - h}{\beta_1 - \beta} C_R \frac{d\beta}{dC_R}$

References

1. GATES, S. B. and LYON, H. M.; *A continuation of longitudinal stability and control analysis*, Part 1—General Theory, R. & M. 2027, 1944.
2. BABISTER, A. W.; *Aircraft Stability and Control* (Pergamon Press, Oxford, 1961).
3. *Canberra B. Mk. 2—Longitudinal Stability and Manoeuvrability Tests*, Unpublished M.O.A. Report.
4. PERKINS, C. D. and HAGE, R. E.; *Airplane Performance Stability and Control* (John Wiley, New York, 1949).

3 Subsonic Longitudinal Manoeuvring Stability

1. Introduction

In the previous chapter we have considered the stability of an aircraft when it is disturbed in incidence and speed from the condition of trimmed level flight. In the case of an aircraft fitted with non-powered controls, it was shown that the stick movements and forces required to trim the aircraft at different speeds in level flight are related to the stick-fixed and stick-free static stabilities, respectively. In addition to the ability to trim an aircraft in level flight, a pilot also needs to be able to manoeuvre the aircraft (i.e. to increase the normal acceleration or g value), and he is very interested in the associated stick movements and forces that are required. These characteristics can be determined by considering the stability of an aircraft when it is disturbed in incidence and normal acceleration (but not speed) from the condition of trimmed level flight. This stability is referred to as the Manoeuvring Static Stability of an aircraft and the stick movement per applied g and the stick force per applied g can be shown to be related to the stick-fixed and stick-free manoeuvre stabilities, respectively. It should be carefully noted that the initial condition is, as before, steady level flight, and that the difference between static and manoeuvre stability is that manoeuvre stability deals with a disturbance in incidence and g occurring at constant speed, whereas static stability deals with a disturbance in incidence and speed occurring at constant (i.e. $n = 1$) g.

When the normal acceleration is increased from the level flight value of $n = 1$, the aircraft departs from the straight line flight path. As a result of this, the manoeuvre stability is made up of two contributions, one resulting from the change in incidence involved (this is exactly the same as the incidence contribution which occurs in the analysis of basic static stability) and one resulting from the pitching motion of the aircraft as it moves round its curved flight path. The rotation of the aircraft about its lateral axis produces a pitching moment which tends to oppose the rotation and this results is an additional stabilizing contribution. The overall stability of an aircraft can be represented by dC_M/dC_L and in the case of manoeuvre stability this can be expressed in the form

$$\frac{dC_M}{dC_L} = \frac{\partial C_M}{\partial \alpha} \times \frac{d\alpha}{dC_L} + \frac{\partial C_M}{\partial q} \times \frac{dq}{dC_L} \tag{3.1}$$

The values of the stick movement per g and the stick force per g are very important in the assessment of the handling qualities of an aircraft and they should lie within the limits laid down in the appropriate requirements (Refs. 2, 3 and 4). It is desirable that an aircraft should be designed to have positive stick-fixed and stick-free manoeuvre stability, in order that the stick movements and forces required for manoeuvring flight are in the conventional directions. The manoeuvre stability is also important in determining the longitudinal short period oscillatory response of an aircraft to elevator control movements or atmospheric gusts, because, as we shall see in Chapter 8, the short period oscillation can be considered as one involving incidence and g changes at constant forward speed.

The simplest longitudinal manoeuvre to analyse is a pull-up from a dive at a steady speed, but the analysis can be extended to steady or wind-up turns and all three of these methods are used in flight test work. The results obtained from the simple theory are modified in practice, as a result of a breakdown in the various simplifying assumptions that are made. In this chapter, compressibility and aeroelastic effects are not allowed for; these effects are considered in Chapters 14 and 15.

2. Simple Mathematical Theory of Manoeuvre Stability

The analysis is similar to that for static stability discussed in Chapter 2 and many of the simplifying assumptions made then are also applicable here. In order to determine the additional contribution arising from the rotation of the aircraft, however, it is first of all necessary to obtain the basic mathematical relationships involved.

2.1. Basic Mechanics of the Pull-out Manoeuvre

Consider an aircraft which is initially in straight and level trimmed flight. If the aircraft is climbed and then dived and pulled out of the dive so that at the bottom of the pull-out it is at its original trimmed values of speed and height, then the aircraft can be considered (at least momentarily) as moving round the arc of a circle at a constant speed. If the conditions at the bottom of the pull-out are compared with the corresponding level flight conditions, the aircraft can be considered to have been "disturbed" in two ways (Fig. 3.1)

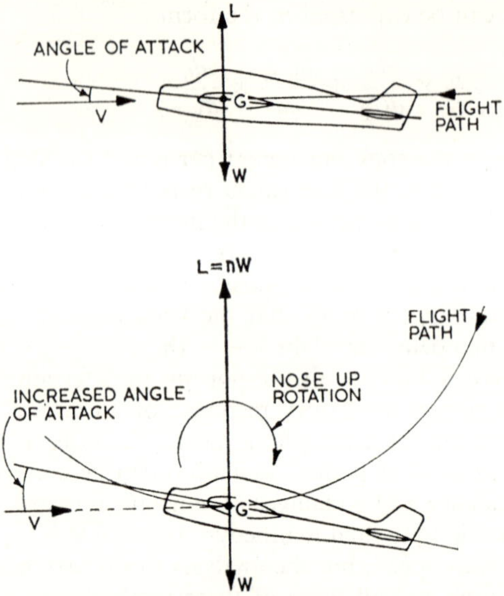

Fig. 3.1. Comparison of conditions in level flight and in a pull-out (*a*) Level flight. (*b*) Pull-out

(i) to a higher angle of attack, in order that the wing of the aircraft can develop the extra lift that is required to keep the aircraft in the curved flight path;
(ii) by having developed a steady nose-up rate of rotation about the centre of gravity, equal in magnitude to the rate of rotation of the aircraft about the centre of the pull-out.

These disturbances result in a pitching moment acting on the aircraft (if it is a stable one) in such a way as to tend to bring the aircraft out of the manoeuvre. In order to hold the aircraft in the pull-out, the pilot has to pull back on the stick and apply up elevator.

(i) *Change in Wing Angle of Attack*
In steady level flight at a speed V ft/sec (T.A.S.), the wing lift is equal to the weight of the aircraft and is given by

$$L = W = C_L \tfrac{1}{2}\rho V^2 S \tag{3.2}$$

In a pull-up manoeuvre at the same speed, the angle of attack of the wing has to be increased in order to increase the lift to a value of $(L + \Delta L)$ given by

$$L + \Delta L = nW = (C_L + \Delta C_L)\tfrac{1}{2}\rho V^2 S \tag{3.3}$$

where n is the total or "indicated" value of g (i.e. the reading on the accelerometer; $n = 1$ in level flight) and ΔC_L is the associated increase required in the value of the wing lift coefficient. By subtracting eqn. (3.2) from eqn. (3.3) we get

$$\Delta L = (n - 1)W = \Delta C_L \tfrac{1}{2}\rho V^2 S \tag{3.4}$$

Substituting n_a for $(n - 1)$, C_L for $W/(\tfrac{1}{2}\rho V^2 S)$ and rearranging, we get

$$n_a C_L = \Delta C_L \tag{3.5}$$

where n_a is the "applied" acceleration. To produce this increase of lift, the incidence of the wing has to be increased by an amount $\Delta\alpha$ given by

$$\Delta\alpha = \frac{\Delta C_L}{a} = \frac{n_a C_L}{a} \text{ rads} \tag{3.6}$$

where a is the lift curve slope of the wing, expressed in radians (i.e. per radian).

(ii) *Determination of the Rate of Rotation*

If the linear velocity of the aircraft at the bottom of the pull-out is V ft/sec (T.A.S.), the angular velocity in pitch q rads/sec and the radius of the circular flight path r ft, then the normal acceleration (i.e. the acceleration towards the centre of the circle) is given by

$$\frac{V^2}{r} = qV = (n - 1)g = n_a g \text{ ft/sec}^2$$

where g is the acceleration due to gravity (i.e. approx. 32 ft/sec^2). Thus the pitching velocity q is given by

$$q = \frac{(n - 1)g}{V} = \frac{n_a g}{V} \text{ rads/sec} \tag{3.7}$$

From eqn. (3.7) it can be seen that the rate of pitch associated with a particular value of applied g decreases with increase in forward speed.

2.2. Change in the Tailplane Angle of Attack

The change in attitude and the rotation of the aircraft both produce a change in the angle of attack of the horizontal tailplane to the local relative airflow. As a result of the increase in the wing angle of attack the tailplane angle of attack will be increased by the same amount, less a small correction due to the downwash effects

$$\Delta\alpha'_1 = \Delta\alpha\left(1 - \frac{d\varepsilon}{d\alpha}\right) \text{ rads} \tag{3.8}$$

Because the aircraft is rotating at a rate q, the tailplane is moving downwards with a linear velocity of ql' ft/sec. The relative airflow can be considered to be coming upwards with the same velocity and

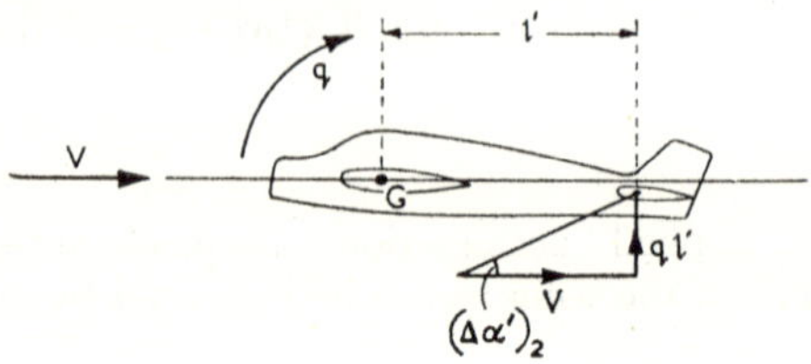

Fig. 3.2. Change in relative airflow at the tailplane due to rotation

the resulting change in angle of attack is given approximately by (Fig. 3.2)

$$\Delta\alpha'_2 = \frac{ql'}{V} \text{ rads} \tag{3.9}$$

The total change in tailplane angle of attack is given by

$$\Delta\alpha' = \Delta\alpha\left(1 - \frac{d\varepsilon}{d\alpha}\right) + \frac{ql'}{V} \text{ rads} \tag{3.10}$$

Substituting $q = n_a g/V$ from eqn. (3.7),

$$\Delta\alpha' = \Delta\alpha\left(1 - \frac{d\varepsilon}{d\alpha}\right) + \frac{n_a g l'}{V^2}$$

and replacing V^2 by $2W/(C_L \rho S)$

$$\Delta\alpha' = \Delta\alpha\left(1 - \frac{d\varepsilon}{d\alpha}\right) + \frac{g\rho S l'}{W}\frac{n_a C_L}{2} \tag{3.11}$$

The quantity $W/g\rho S l'$ is called the Relative Aircraft Density Parameter and it is denoted by the symbol μ_1. It is non-dimensional and increases with an increase in wing loading and height. Eqn. (3.11) can be rewritten

$$\Delta\alpha' = \Delta\alpha\left(1 - \frac{d\varepsilon}{d\alpha}\right) + \frac{n_a C_L}{2\mu_1} \tag{3.12}$$

Substituting ΔC_L for $n_a C_L$ from eqn. (3.5),

$$\Delta\alpha' = \Delta\alpha\left(1 - \frac{d\varepsilon}{d\alpha}\right) + \frac{\Delta C_L}{2\mu_1} \quad (3.13)$$

Dividing both sides of this equation by ΔC_L and replacing $d\alpha/dC_L$ by $1/a$ gives

$$\frac{d\alpha'}{dC_L} = \frac{1}{a}\left(1 - \frac{d\varepsilon}{d\alpha}\right) + \frac{1}{2\mu_1} \quad (3.14)$$

2.3. Stick-fixed Manoeuvre Stability

From the work of the previous chapter, the stick-fixed stability can be expressed in the form

$$\frac{-dC_M}{dC_L} = -(h - h_0) + \bar{V}a_1\frac{d\alpha'}{dC_L} \quad (3.15)$$

where $d\alpha'/dC_L$ is the rate of change of tailplane angle of attack with wing lift coefficient in the particular case being considered. By substituting the expression in eqn. (3.14), the stick-fixed manoeuvre static stability is given by

$$\frac{-dC_M}{dC_L} = -(h - h_0) + \bar{V}a_1\left[\frac{1}{a}\left(1 - \frac{d\varepsilon}{d\alpha}\right) + \frac{1}{2\mu_1}\right]$$

$$\frac{-dC_M}{dC_L} = -(h - h_0) + \bar{V}\frac{a_1}{a}\left(1 - \frac{d\varepsilon}{d\alpha}\right) + \frac{\bar{V}a_1}{2\mu_1} \quad (3.16)$$

By comparing this result with eqn. (2.10), it can be seen that the stick-fixed manoeuvre stability is greater than the corresponding static stability by an amount $\bar{V}a_1/2\mu_1$, the term resulting from the rotation of the aircraft during the pull-out manoeuvre. This additional stabilizing tailplane contribution decreases with increase of height, since the value of μ_1 increases. A physical explanation for this loss of stability can be given using Fig. 3.3. If an aircraft is considered to be flying at two different heights at the same V_i (i.e. the same C_L value) and the same g is applied in the two cases, then, from eqns. (3.6) and (3.8), the value of $\Delta\alpha'_1$ will be the same (assuming that the value of a is unchanged). Eqn. (3.7) however, shows that, since the T.A.S. increases with height, the rate of pitch q of the aircraft decreases. By substituting from eqn. (3.7) into eqn. (3.9), the corresponding change in tailplane angle of attack is given by

$$\Delta\alpha'_2 = n_a g l'/V^2 \quad (3.17)$$

Thus $\Delta\alpha'_2$ is inversely proportional to V^2 (or, alternatively, directly proportional to air density at the same E.A.S.), and since the T.A.S. corresponding to a given E.A.S. increases with height, the change in tailplane angle of attack associated with the rotation of the aircraft,

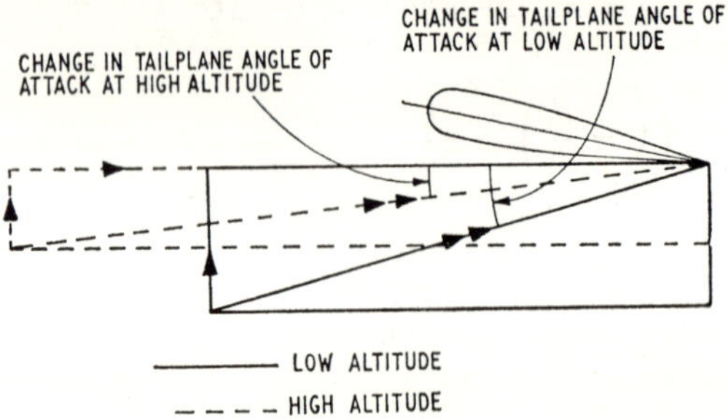

Fig. 3.3. Effect of altitude on the tailplane contribution

and, hence, the corresponding stabilizing contribution, decreases with height.

2.3.1. STICK-FIXED MANOEUVRE POINT

By varying h, the position of the C.G., the manoeuvre stability can be made positive, zero, or negative, rearward movement of the C.G. being destabilizing. The particular position of the C.G. h_m which gives neutral stability is called the stick-fixed manoeuvre point N_m. Substituting in eqn. (3.16)

$$0 = -(h_m - h_0) + \bar{V}\frac{a_1}{a}\left(1 - \frac{d\varepsilon}{d\alpha}\right) + \frac{\bar{V}a_1}{2\mu_1} \tag{3.18}$$

and by subtracting eqn. (3.18) from (3.16), we have for any other C.G. position

$$\left(\frac{-dC_M}{dC_L}\right)_{fixed} = h_m - h \tag{3.19}$$

Thus the stick-fixed manoeuvre stability is equal to the non-dimensional distance between the C.G. and the stick-fixed manoeuvre point. When the C.G. is ahead of the manoeuvre point the aircraft is stable, when the C.G. is behind the manoeuvre point the aircraft is unstable. The distance between the C.G. and the stick-fixed manoeuvre point is called the Stick-fixed Manoeuvre Margin and is denoted by H_m. Thus

$$H_m = \left(\frac{-dC_M}{dC_L}\right)_{fixed} = h_m - h \tag{3.20}$$

The values of H_m and h_m are related to the corresponding static stability values by

$$\left.\begin{aligned} h_m &= h_n + \frac{\bar{V}a_1}{2\mu_1} \\ \text{and} \\ H_m &= H_n + \frac{\bar{V}a_1}{2\mu_1} \end{aligned}\right\} \tag{3.21}$$

The manoeuvre point can be considered as that point at which the total change in lift force acts in the manoeuvre with the controls fixed. On the basis of this simple theory the manoeuvre margin is always greater than the static margin (i.e. the manoeuvre point is behind the neutral point) by a small amount which is inversely proportional to μ_1, and which decreases with increase of height or wing loading.

2.3.2. ELEVATOR ANGLE AND STICK MOVEMENT PER APPLIED g

If at the bottom of the pull-out, the elevator is moved back to its initial trim position, the aircraft would recover from the pull-out because of the overall nose-down change in pitching moment produced as a result of its manoeuvre stability. To hold the aircraft in a steady pull-out at a constant rate of pitch, all the pitching moments must again be balanced out and the elevator must be deflected by an amount sufficient to give a nose-up moment exactly equal to the nose-down stabilizing one. Thus, the change in elevator angle (and hence the corresponding stick movement) required in a pull-out is a measure of the stick-fixed manoeuvre stability of the aircraft.

By comparison with eqn. (2.14) we have

$$\frac{-dC_M}{dC_L} = -\bar{V}a_2\frac{d\eta}{dC_L} \tag{3.22}$$

where, in this case, $-dC_M/dC_L = H_m$. Thus

$$\Delta\eta = -\frac{H_m}{\bar{V}a_2}\Delta C_L$$

Since $\Delta C_L = n_a C_L$,

$$\Delta\eta = -\frac{H_m}{\bar{V}a_2} n_a C_L$$

and hence

$$\text{Elevator angle per applied } g = \frac{\Delta\eta}{n_a} = \frac{-H_m}{\bar{V}a_2} C_L \tag{3.23}$$

Substituting $2W/\rho V^2S$ for C_L

$$\text{Elevator angle per applied } g = \frac{\Delta\eta}{n_a} = \frac{-2W}{\rho S}\frac{H_m}{\bar{V}a_2}\frac{1}{V^2}\text{rads} \quad (3.24)$$

The corresponding stick movement is given by

$$\text{Stick movement per applied } g = \frac{-2W}{m_e\rho S}\frac{H_m}{\bar{V}a_2}\frac{1}{V^2}\text{ ft} \quad (3.25)$$

where m_e is the stick gearing. The units in eqns. (3.24) and (3.25) depend on the units used for a_2; if this is expressed per radian, then the elevator angle will be in radians and m_e must be expressed in radians/foot to give the stick movement in feet. The sign convention employed is such that the minus signs indicate up elevator deflection and backward movement of the stick.

The following points should be noted about the results obtained from the simple theory

(i) the elevator angle (and stick movement) per g is independent of the amount of g applied (there are, of course, limitations to this; these limits are discussed in Chapter 4);
(ii) the elevator angle per g is directly proportional to the manoeuvre margin H_m and thus varies linearly with C.G. position;
(iii) the elevator angle per g varies inversely with ρV^2 (i.e. varies linearly with C_L). At a given V_i (i.e. C_L) the elevator angle decreases with height due to the reduction in H_m.

These results are illustrated in Figs. 3.4 and 3.5. By measuring the values of elevator angle per g at two different C.G. positions and extrapolating the line until it intersects the horizontal axis, the position of the stick-fixed manoeuvre point can be determined (Fig. 3.5). Forward movement of the C.G. increases the amount of elevator angle required for manoeuvring flight and this results in an alternative forward limit to the permissible C.G. range of movement that may be more critical than those produced from the considerations of level flight trim in Chapter 2.

2.4. Stick-free Manoeuvre Stability

If the analysis of the pull-out is again considered, this time with the stick and hence the elevator, free to move under the effect of the hinge moments acting on the control surface, then the control will normally float one way or the other and this will affect the value of the resulting pitching moment acting on the aircraft. Thus the stick-free manoeuvre stability will differ from the stick-fixed stability discussed above. If the control tends to float in the direction that it must be moved in order to manoeuvre the aircraft, then the floating helps the pilot and reduces the amount of stick force that he has to apply. On the other hand, if the control tends to float in the

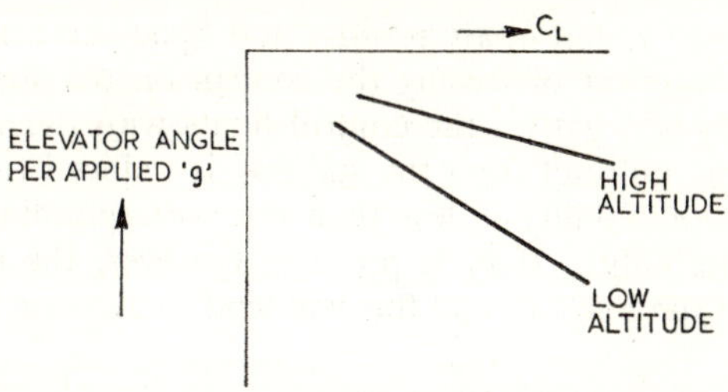

Fig. 3.4. Effect of altitude on the elevator angle per applied *g* curve

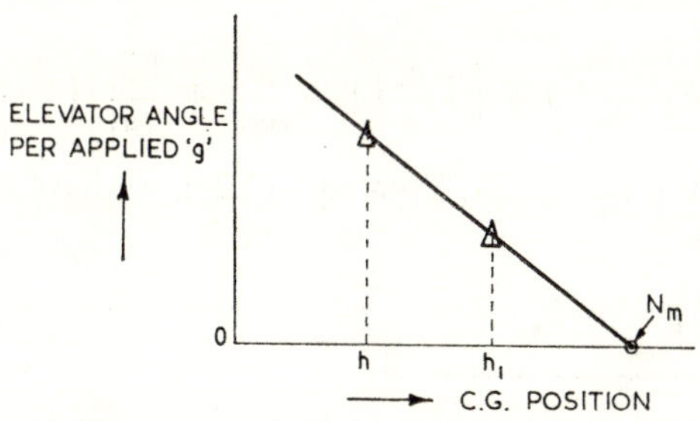

Fig. 3.5. Determination of the stick-fixed manoeuvre point

opposite direction to that required, the stick forces are made heavier.

If the control is free to float, the change in elevator angle is related to the change in the tailplane incidence by the equation

$$\Delta\eta = -\frac{b_1}{b_2}\,\Delta\alpha' \tag{3.26}$$

where b_1 and b_2 are the floating and heaviness parameters, respectively, of the elevator control. If this control movement is allowed for, then, as has been shown in Chapter 2, it has the effect of modifying the tailplane stabilizing contribution by a factor of

$$1 - (a_2 b_1 / a_1 b_2)$$

Equation (3.16) thus becomes modified for the case of stick-free manoeuvre stability to

$$\frac{-dC_M}{dC_L} = -(h - h_0) + \left[\bar{V}\frac{a_1}{a}\left(1 - \frac{d\varepsilon}{d\alpha}\right) + \frac{\bar{V}a_1}{2\mu_1}\right]\left(1 - \frac{a_2 b_1}{a_1 b_2}\right) \tag{3.27}$$

As before, since a_1 and a_2 are positive and b_2 negative, the sign of b_1 determines the effect of freeing the control on the stability of the aircraft. If b_1 is negative, the control floats with the wind (i.e. up elevator in the pull-out) and the stability is reduced (i.e. the stick-free manoeuvre stability is less than the corresponding stick-fixed manoeuvre stability). If b_1 is positive, however, the elevator will float down in the pull-out and this will tend to increase the stability.

2.4.1. STICK-FREE MANOEUVRE POINT

The particular position of the C.G. which gives neutral stability h_m' is called the stick-free manoeuvre point N_m'. Substituting in eqn. (3.27)

$$0 = -(h_m' - h_0) + \left[\bar{V}\frac{a_1}{a}\left(1 - \frac{d\varepsilon}{d\alpha}\right) + \frac{\bar{V}a_1}{2\mu_1}\right]\left(1 - \frac{a_2 b_1}{a_1 b_2}\right) \tag{3.28}$$

and subtracting eqn. (3.28) from eqn. (3.27), we have for any other C.G. position

$$\left(\frac{-dC_M}{dC_L}\right)_{free} = h_m' - h \tag{3.29}$$

Thus the stick-free manoeuvre stability is represented by the non-dimensional distance between the C.G. and the stick-free manoeuvre point. This distance is called the Stick-free Manoeuvre Margin and is denoted by the symbol H_m'. The values of H_m' and h_m' are related to the corresponding static stability values by

$$\left.\begin{aligned} h_m' &= h_n' + \frac{\bar{V}a_1}{2\mu_1}\left(1 - \frac{a_2 b_1}{a_1 b_2}\right) \\ H_m' &= H_n' + \frac{\bar{V}a_1}{2\mu_1}\left(1 - \frac{a_2 b_1}{a_1 b_2}\right) \end{aligned}\right\} \tag{3.30}$$

These equations should be compared with eqn. (3.21) for the corresponding stick-fixed relationships.

2.4.2. STICK FORCE PER APPLIED g

A measure of the stick-free manoeuvre stability is given by the difference between the elevator angle required for trim in the manoeuvre and the angle that the elevator would take up, under the same conditions, if the stick was left free (compare Chapter 2, para. 2.5.2). As the hypothetical stick-free position is not known, it is necessary to use some other quantity to measure the stability, and the stick force that the pilot has to apply is usually employed.

The difference between the floating and trim positions of the elevator is given mathematically by

$$\Delta\eta = -\frac{H_m'}{\bar{V}a_2}\Delta C_L \tag{3.31}$$

and the corresponding change in the hinge moment coefficient by

$$\Delta C_H = -\frac{H_m'}{\bar{V}a_2}\Delta C_L\, b_2 \tag{3.32}$$

The stick force per applied g is given by P/n_a where

$$\frac{P}{n_a} = \frac{m_e \Delta C_H \tfrac{1}{2}\rho V^2 S_\eta c_\eta}{n_a}$$

Substituting for ΔC_H from eqn. (3.32), $n_a C_L$ for ΔC_L and W/S for $C_L \tfrac{1}{2}\rho V^2$ we get

Stick force per applied g

$$= \frac{P}{n_a} = -m_e \frac{W}{S} S_\eta c_\eta \frac{b_2 H_m'}{\bar{V}a_2} \text{ lbs}/g \tag{3.33}$$

The following points should be noted about the result obtained from the simple theory

(i) the stick force per g is independent of the amount of g applied. (The limitations to this are also discussed in Chapter 4.)

(ii) the stick force per g is directly proportional to the stick-free manoeuvre margin H_m' and thus varies linearly with C.G. position and decreases with height.

(iii) the stick force per g is independent of forward speed. This result may be a little unexpected, but it can be explained in the following way. If a constant value of g is applied at different forward speeds, then, since C_L varies as $1/V^2$, so do the values of ΔC_L (eqn. 3.5) and ΔC_H (eqn. 3.32), but the corresponding stick force and thus the value of the stick force per g is unchanged.

These results are illustrated in Figs. 3.6 and 3.7. By measuring the values of stick force per g at two different C.G. positions and extrapolating the line until it intersects the horizontal axis, the position of the stick-free manoeuvre point can be determined (Fig. 3.6). Forward movement of the C.G. increases the stick force per g. Continued forward movement will eventually give rise to excessive values of stick force per g and this will set a limit to the permissible forward C.G. position. The variation of the stick force per g with forward speed at two different heights is shown in Fig. 3.7. The basic

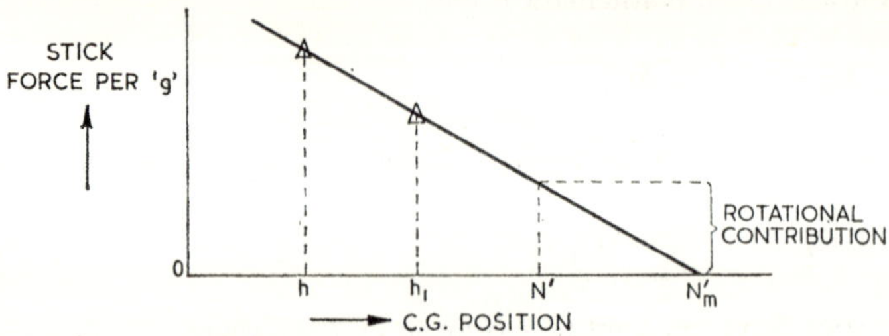

Fig. 3.6. Determination of the stick-free manoeuvre point

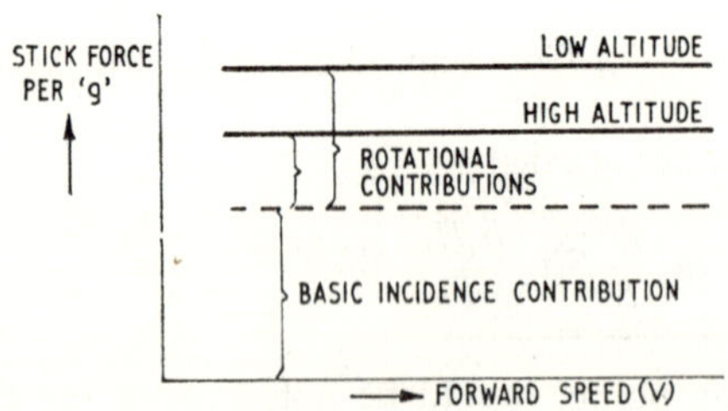

Fig. 3.7. Variation of stick force per g with forward speed at different heights

incidence contribution to the manoeuvre stability is independent of height, whereas the rotational contribution and hence the overall stick force per g decreases with increase of height.

If the C.G. is moved behind the stick-free manoeuvre point, the aircraft becomes unstable. Although a pull force would still be required to initiate a pull-out manoeuvre, the pilot would have to reverse the force in order to hold a steady pull-out and would, in fact, be pushing on the stick. These forces would be unconventional ones and would be considered unsatisfactory (see para. 4.1.).

3. General Treatment of Manoeuvre Stability

The preceding mathematical analysis of the pull-out manoeuvre is based on the same simplifying assumptions that were used in Chapter 2 (e.g. the values of a_1, b_1, etc., have been assumed to be constant, independent of forward speed). On this simple theory, the elevator angles per *g* and stick forces per *g* are independent of the amount of *g* applied and the positions of the manoeuvre points do not vary with forward speed. In practice, however, the values of the parameters vary with forward speed and, as in Chapter 2, they are then usually represented by the symbols A_1, B_1, etc. As a result of these variations, the elevator angle per *g* and the stick force per *g* values do not vary in exactly the way predicted by simple theory (see para. 4.3).

In the general theory of static stability, it was necessary to distinguish clearly between the static margin and the C.G. margin. When level flight trim is considered, the incidence (i.e. C_L) and forward speed are not independent variables; a change of incidence also involves a change of speed, and the overall effect is the sum of the two. At low speeds, only the incidence effects need to be considered and the static margin and the C.G. margin are identical. At high forward speeds, however, the pitching moment changes associated with the change in forward speed may be important and the two margins are no longer numerically equal to one another (Chapter 2, para. 3.2). In considerations of manoeuvre stability, this problem does not arise, because incidence and speed are, in this case, independent variables. This is because in the analysis of the pull-out the changes are assumed to take place at constant speed. Thus eqns. (3.20) and (3.29), deduced from the simple theory, are valid for all conditions.

4. Flight Test Measurements

4.1. Manoeuvrability Requirements

It is desirable that an aircraft should possess both stick-fixed and stick-free manoeuvre stability, since they determine the stick movements and stick forces, respectively, that are required to manoeuvre the aircraft in pitch. If an aircraft is a stable one, the stick movements and forces will be in the conventional sense (i.e. the stick has to be moved back and a pull force held in order to apply positive *g*).

The pilot is usually more interested in the stick forces involved, than in the corresponding stick movements and consequently stick-free stability is more important than stick-fixed. The requirements for military aircraft are laid down in Ref. 3. In order that the pilot should be able to manoeuvre the aircraft fairly easily the stick forces involved should not be too high, but to prevent him putting on excessive amounts of *g* the forces should not be too low. Thus the stick force per *g* values should lie between certain limiting values, the actual values depending on the type of aircraft being considered. In addition, the requirements lay down maximum and minimum values for the force needed to reach and maintain a specified high g value, which is a little below the design normal acceleration. These requirements are discussed in para. 3.4 of Chapter 10, and in Ref. 4, which also mentions possible requirements for the associated stick movements per g.

Ref. 5 also discusses stability requirements and mentions some investigations carried out on the divergent short period oscillations that are associated with negative manoeuvre stability. Tests using variable stability aircraft and ground simulators have shown that, although an aircraft is flyable with negative manoeuvre margins under certain conditions, it should not be permitted, even for emergency conditions.

4.2. Flight Test Techniques

There are three different flight test techniques that can be used for determining the longitudinal manoeuvring characteristics of an aircraft. They are

(i) *Steady pull-ups:* in this method the pilot pulls a constant specified g at the bottom of a dive at a given trimmed speed; the test is repeated for different g and speed values.
(ii) *Steady turns:* in this method the pilot pulls a constant specified g in a steady turn at a given trimmed speed; the test is repeated for different g and speed values.
(iii) *Wind-up turns:* in this method the pilot pulls a constant specified g while allowing the airspeed to fall off until the aircraft stalls or reaches some other limiting condition; the test is repeated for different g values.

These methods are discussed in Ref. 2.

The stick force per g values measured in a steady turn are higher than those measured in a pull-out. This is due to a difference between the basic mechanics of the two manoeuvres, the rate of pitch and hence the pitch damping contribution, corresponding to a given indicated g reading, being greater in a turn than in a pull-out. In the case of the pull-out, the aerodynamic lift and the weight of the aircraft act in the same straight line, and the resultant force $(L - W)$ acts upwards towards the centre of the pull-out, and the resulting rate of rotation (which is all pitch) is given by eqn. (3.7). In a steady turn, the lift and weight do not act in the same straight line and the force acting towards the centre of the turn is the component of the lift force $L \sin \phi$, where ϕ is the angle of bank. This produces a rate of rotation about the centre of the turn given by

$$\omega = n \sin \phi \frac{g}{V} \text{ rads/sec} \tag{3.34}$$

However, this is not all pitching motion and the rate of pitch in the turn is given by

$$q_t = \omega \sin \phi = n \sin^2 \phi \frac{g}{V} \tag{3.35}$$

In a steady turn, the vertical component of the lift force must balance the weight of the aircraft and hence

$$L \cos \phi = W$$

Since $L = nW$

$$\frac{L}{W} = n = \frac{1}{\cos \phi} \tag{3.36}$$

But $\sin^2 \phi = 1 - \cos^2 \phi$ and substituting $1/n$ for $\cos \phi$ gives

$$\sin^2 \phi = 1 - \frac{1}{n^2} \tag{3.37}$$

Substituting this in the equation for the rate of pitch (eqn. 3.35)

$$q_t = \left(n - \frac{1}{n}\right) \frac{g}{V} \tag{3.38}$$

This should be compared with the expression for the rate of pitch in a pull-out given by eqn. 3.7. Since the factor $n - (1/n)$ is always greater than $n - 1$, for all values of n greater than unity, it can be

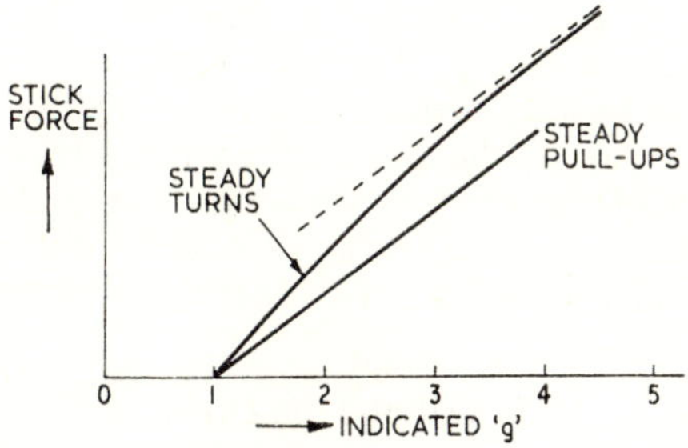

Fig. 3.8. Variation of stick force with g in pull-ups and steady turns

seen that the rate of pitch in a steady turn is always greater than that in a pull-out at the same values of n and V. The variation of stick force with applied g in a steady pull-out and a steady turn are shown in Fig. 3.8; the stick force per g in a steady turn is not independent of the applied g, but decreases slightly with increasing g.

4.3. Flight Test Results

The simple theory of the pull-out predicts that the stick movements and forces per applied *g* are independent of the amount of *g* applied, and that the stick movement per *g* is inversely proportional to V^2, whereas the stick force per *g* is the same at all speeds. In practice, these results are modified because of the changes that occur in the values of the basic aerodynamic parameters due to the various non-linearities that occur and also the effects of compressibility and aero-elasticity. The effects of different values of applied *g* and forward speed on the manoeuvrability characteristics of an aircraft can be determined by carrying out the tests using a constant applied *g* at various speeds and then by repeating the tests at various speeds

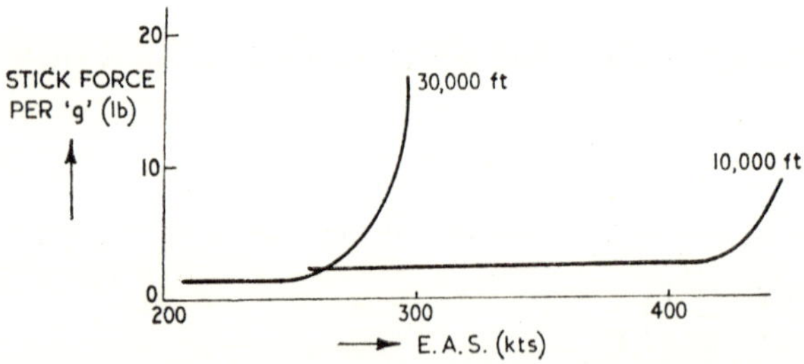

Fig. 3.9. Stick force per *g* measured in flight tests

using different values of applied *g*. The tests can then be repeated at different heights to show the effects of altitude. The curves of stick force per *g* against applied *g* show that the value usually decreases with increasing *g*. The stick force per *g* remains fairly constant at low forward speeds, increasing rapidly at the higher forward speeds due to the onset of compressibility (high Mach number) or aeroelastic (high E.A.S.) effects (Fig. 3.9). The stick force per *g* values are reduced by an increase of altitude, though the compressibility increase occurs at a lower value of E.A.S.

5. Out-of-Trim Dives

5.1. General

When a stable aircraft, trimmed in steady level flight, is put into a dive, a push force is required on the stick to hold the aircraft in the dive. When the stick is released, restoring moments will act on the aircraft to bring it out of the dive, and the normal acceleration will increase. The requirements lay down that, for a given applied out-of-trim stick force, the resulting pull-out on release shall not result in a dangerous flight condition, or a value of *g* outside the design limits. It is suggested in Ref. 4 that, in the case of a ground attack aircraft, almost neutral static stability is desirable, so that the stick forces to change speed will be small and consequently large accelerations will not result when the stick forces are relaxed.

The magnitude of the stick force to hold an out-of-trim dive depends upon the stick-free static stability of the aircraft and thus is proportional to the stick-free static margin K_n'. When this force is released, the g which builds up depends upon the ratio of the stick-free static margin and the stick-free manoeuvre margin H_m'

$$g \text{ on release} = \frac{\text{Stick force to hold the out-of-trim dive}}{\text{Stick force per } g} \propto \frac{K_n'}{H_m'} \tag{3.39}$$

Both the force to hold and the maximum g on release, increase with increase of speed from the trimmed value.

5.2. Effect of C.G. Movement

On the basis of simple theory, the stick-free manoeuvre margin is always slightly greater than the corresponding stick-free static margin.

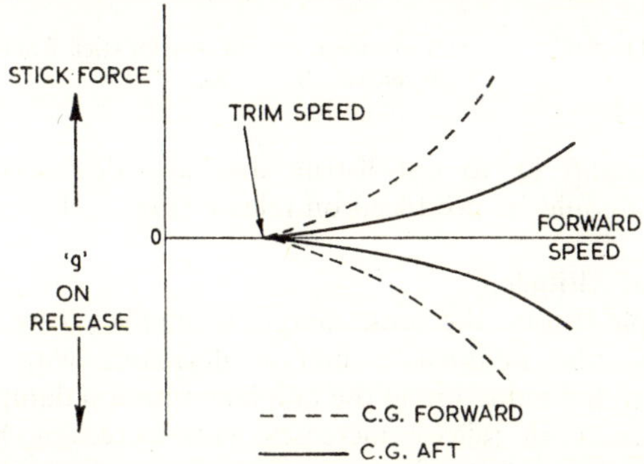

Fig. 3.10. Effect of C.G. movement on out-of-trim stick forces and g on release ($H_m' > K_n'$)

In these circumstances, a certain movement of the C.G. produces a larger percentage change in K_n' than in H_m' and thus the changes occurring in the stick force to trim and the g on release act in the same direction, i.e. forward movement of the C.G. results in increased out-of-trim forces and increased g on release, whereas rearward movement of the C.G. results in a decrease in the values of the forces and the g on release. This is the desirable variation, large values of g on release are then associated with large out-of-trim stick forces (Fig. 3.10).

In practice, however, it is possible for the stick-free manoeuvre margin to be less than the stick-free static margin (e.g. by the incorporation of a positive spring and a negative bob-weight, see para. 6, Chapter 10). In this case, a change in C.G. position produces a larger proportional change in H_m' than in K_n'. Thus a forward movement of the C.G. will increase the stick force required to trim, but will decrease the g on release. Alternatively, a rearward movement of the C.G. will decrease the stick forces, but will result in an increase in the g on release. This is a potentially dangerous situation because large normal accelerations will then be associated with small out-of-trim forces. Only small forces would be required to

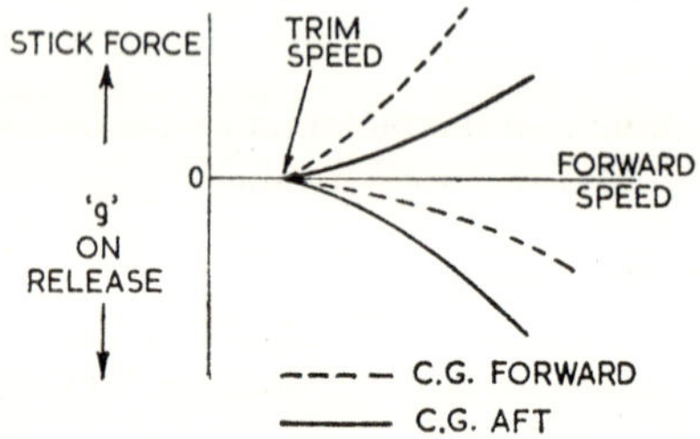

Fig. 3.11. Effect of C.G. movement on out-of-trim stick forces and g on release ($H_m' < K_n'$)

hold the aircraft in an out-of-trim dive and destructive normal accelerations could be produced on release (Fig. 3.11).

5.3. Effect of Altitude

On the simple theory, the static margin is unaffected by changes of altitude, but the manoeuvre margin decreases with increasing height, due to the reduction in the tailplane rotation damping. Thus the acceleration on release increases with increasing height. In practice, Mach number effects are superimposed on the altitude effect and the overall effect depends on the particular case being considered.

6. Conclusion

The work of this chapter has dealt with manoeuvre stability and the corresponding stick movements and forces required to manoeuvre an aircraft which is fitted with a simple manually operated longitudinal control system. The characteristics of this type of system can be modified by incorporating springs and bob-weights and by the use of spring-tab controls. The effects of these design features on the various control characteristics of the aircraft are discussed in Chapter 10. When the aircraft has power-operated controls, stick-free stability is not so important, the stick-forces involved being dependent on the stick-fixed stability and the particular type of feel system used. The control characteristics of aircraft with power controls are discussed in Chapter 12.

References

1. PERKINS, C. D. and HAGE, R. E.; *Airplane Performance, Stability and Control* (J. Wiley & Sons, New York, 1949).
2. *Agard Flight Test Manual* Vol. II, "Stability and Control."
3. *Design Requirements for Aircraft for the Royal Air Force and Royal Navy*, Av. P. 970, Ministry of Aviation.
4. ALLWRIGHT, H. J.; *Design Aims for Stability and Control of Piloted Aircraft*, Agard Report No. 337, 1961.
5. KOVEN, W. and WASICKO, R.; *Flying Quality Requirements for United States Navy and Air Force Aircraft*, Agard Report No. 336, 1961.

4 Manoeuvre Boundaries

1. Introduction

In the previous chapter, the stick movements and stick forces per applied g were determined. On the basis of the simple theory employed, these values are independent of the amount of g applied. In practice, the amount of g that can be developed on an aircraft in manoeuvring flight is limited for various reasons. One important limitation is on the grounds of structural considerations; the aircraft is designed to withstand a certain maximum value of g and the

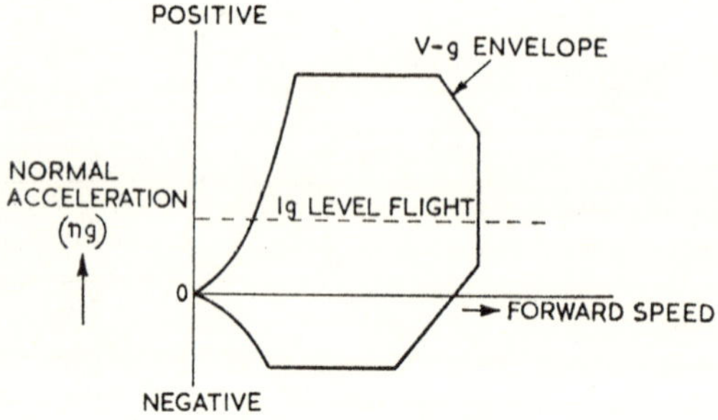

Fig. 4.1. *V–g* envelope

aircraft should be flown within the V–g envelope laid down (Fig. 4.1). At low altitudes, enough longitudinal control power and wing lift is usually available for the pilot to be able to exceed this limiting value of g. At high altitudes, however, it may not be possible for the limiting structural g to be reached, and the practical limit is set by other considerations. At the high wing incidences involved in manoeuvring flight, non-linearities are present in the lift and pitching moment curves and large increases of drag occur. The maximum g available for manoeuvring may, in these circumstances, be limited either by aerodynamic or engine thrust considerations. The graphs showing the way in which the maximum available g varies with forward speed and height are called the Manoeuvre Boundaries.

There are two types of manoeuvre boundary

(i) *Lift Boundary* In order to manoeuvre an aircraft the lift on the wing has to be increased and this additional lift is obtained by increasing the angle of attack of the wing. As the angle of attack is increased, the wing will eventually stall and this sets an absolute limit to the g that can be used for manoeuvring flight. Even before the stall is reached, however, the aircraft may experience severe buffeting or pitch-up and these characteristics may set an even lower limit to the amount of usable g. The lift boundary of an aircraft shows the maximum values of g that can be achieved at various speeds and heights, due to the aerodynamic limitations of buffet, pitch-up and stalling.

(ii) *Thrust Boundary* An aircraft in turning flight has a higher drag than in level flight and more thrust is required. Thus an increase in g requires an increase in thrust and eventually a condition is reached when all the thrust available is being used to sustain the aircraft in a level turn. Any further increase in g can only be achieved by allowing the aircraft to lose height during the turn. In level turning flight the thrust limitation may be reached before that of buffet, pitch-up, or stalling. The thrust boundary shows the maximum values of g that can be maintained in turning flight, at a given power setting, without losing either speed or height.

The importance of the manoeuvre boundaries depends on the type of aircraft being considered. In the case of a fighter aircraft they are of vital importance, since they will determine how successfully the aircraft is able to fulfil its particular role. In addition to having a large available g value, it is necessary for the handling characteristics of the aircraft to be satisfactory on the approach to, and at, the stall, in order that the pilot will be able to manoeuvre the aircraft to its limits, without fear of any serious handling problems occurring. In the case of a transport aircraft, the maximum values of g available are not so important, but the aircraft must still have satisfactory handling characteristics near the stall, since the aircraft may be flown in that condition during take-off or landing, or even, inadvertently, during the cruising phase of flight. This is a potentially dangerous situation and it is essential for all aircraft to have a good stall warning, satisfactory stalling characteristics and effective control at the stall.

On many modern aircraft designs, the natural stall warning and the natural stalling characteristics are not satisfactory and various design features have been incorporated to make them acceptable (e.g. wing boundary layer fences, vortex generators, stick shakers and stick pushers). The stall requirements laid down for civilian transport aircraft often involve the aircraft firm concerned in many hours of flight development work before it achieves the required standard.

In this chapter, only manoeuvrability at subsonic speeds will be considered, supersonic manoeuvrability is dealt with in Chapter 15.

2. Dynamics of Turning Flight

2.1. General Equations

In a steady horizontal turn (Fig. 4.2) the conditions of equilibrium

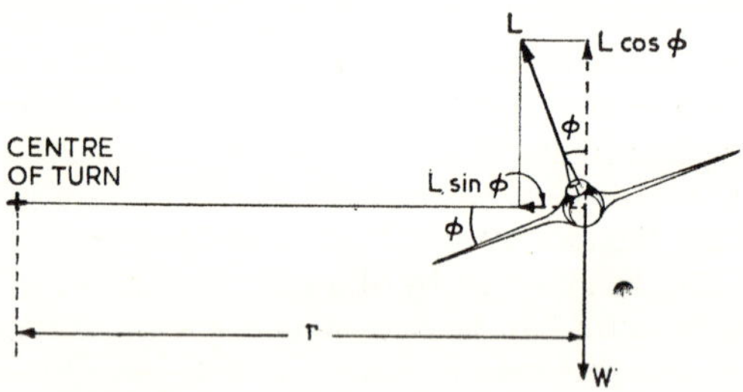

Fig. 4.2. Forces acting in a steady turn

can be expressed in the form

$$L \cos \phi = W \tag{4.1}$$

$$L \sin \phi = \frac{WV^2}{rg} \tag{4.2}$$

where L is the wing lift in lbs, W the all-up weight of the aircraft in lbs, V the true air speed in ft/sec, r the radius of turn in ft and ϕ the angle of bank. Dividing eqn. (4.2) by eqn. (4.1) we get

$$\tan \phi = \frac{V^2}{rg} \tag{4.3}$$

which is the basic turning equation relating V, r and ϕ. Once two of these variables are known, the other one can be determined. From eqn. (4.3), the radius of turn is given by

$$r = \frac{V^2}{g \tan \phi} \tag{4.4}$$

and the corresponding rate of rotation ω ($= V/r$) by

$$\omega = \frac{g \tan \phi}{V} \text{ rads/sec} \tag{4.5}$$

The load factor n in the turn is defined as the ratio L/W and hence from eqn. (4.1)

$$n = \frac{L}{W} = \frac{1}{\cos \phi} = \sec \phi \tag{4.6}$$

where n is the *indicated g*. The variation of n with angle of bank ϕ is shown in Fig. 4.3. Once the angle of bank is known, the value of

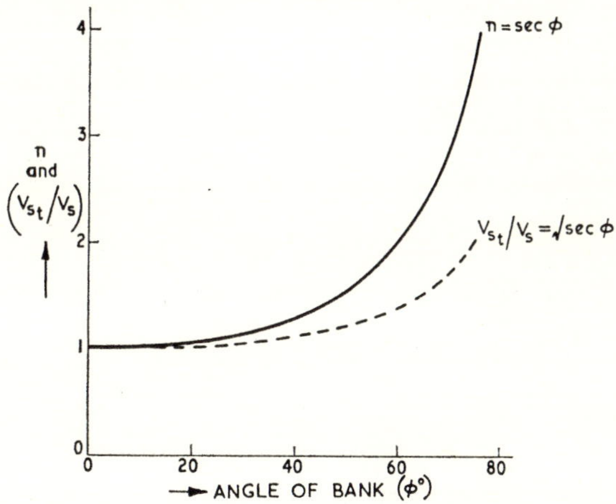

Fig. 4.3. Variation of n and V_{st}/V_s with angle of bank

n is fixed, irrespective of the type of aircraft being considered and independent of the speed and height at which the aircraft is flying.

Since $\tan^2 \phi = \sec^2 \phi - 1$, by substituting from eqn. (4.6) into eqn. (4.4), we get that

$$r = \frac{V^2}{g\sqrt{(n^2 - 1)}} \tag{4.7a}$$

Expressing this equation in terms of Mach number and substituting known values, it becomes

$$r = 1{\cdot}29 \times 10^4 \frac{M^2\theta}{\sqrt{(n^2 - 1)}} \text{ yds} \tag{4.7b}$$

where θ is the relative atmospheric temperature (i.e. Ambient temperature in °K/Standard sea-level temperature in °K). Nomograms relating V, ϕ (or g) and r are often used to determine the turning performance of an aircraft. Once two of these parameters are fixed, the other is known. Lines of constant rate of turn and Mach number can also be included in these nomograms if required.

2.2. Equations for Low Speed Turns

At very low speeds, before compressibility effects become appreciable, the value of C_{Lmax} can be assumed to remain constant. The maximum value of g that can be applied to an aircraft at a forward speed V is then given by

$$n_{max} = \frac{L}{W} = \frac{C_{Lmax}\frac{1}{2}\rho V^2 S}{C_{Lmax}\,\frac{1}{2}\rho V_s^{\,2} S}$$

$$n_{max} = \frac{V^2}{V_s^2} \tag{4.8}$$

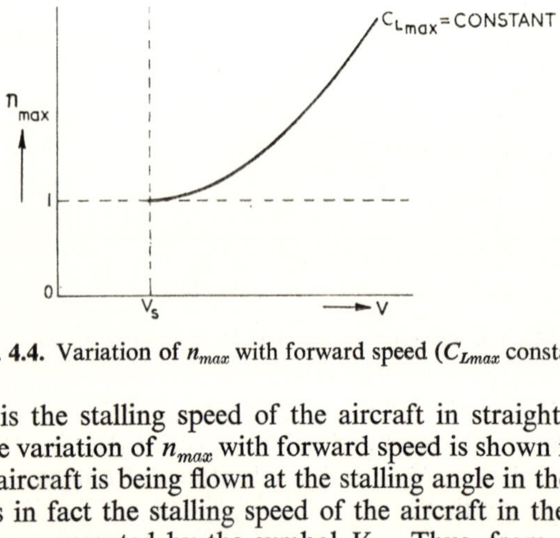

Fig. 4.4. Variation of n_{max} with forward speed (C_{Lmax} constant)

where V_s is the stalling speed of the aircraft in straight and level flight. The variation of n_{max} with forward speed is shown in Fig. 4.4. Since the aircraft is being flown at the stalling angle in the turn, the speed V is in fact the stalling speed of the aircraft in the turn and this can be represented by the symbol V_{st}. Thus, from eqns. (4.6) and (4.8)

$$V_{st} = V_s\sqrt{\sec\phi} \tag{4.9}$$

where ϕ is the angle of bank corresponding to the stalling condition. From eqn. (4.9) it can be seen that the stalling speed of an aircraft is always higher in a turn than it is in level flight (Fig. 4.3).

The radius of turn at any speed will be a minimum when the maximum g available is being used, i.e. when the aircraft is at the stalling angle of attack and $C_L = C_{Lmax}$. Substituting from eqn. (4.9) into eqn. (4.4) we get

$$r_{min} = \frac{V_s^2}{g \sin \phi} \tag{4.10}$$

and it should be noted that the only variable on the right hand side of this equation is ϕ, the angle of bank. Thus, if the aircraft is maintained at the stalling attitude, the value of the minimum possible radius of turn that can be obtained decreases with increase

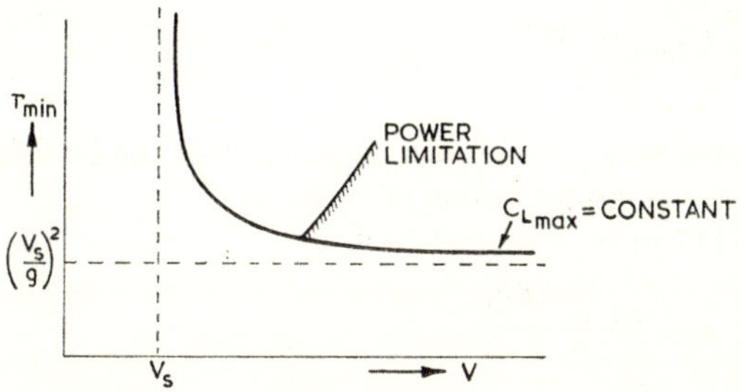

Fig. 4.5. Variation of the minimum radius of turn with forward speed (C_{Lmax} constant)

of forward speed (Fig. 4.5). When the aircraft is in level flight ($\phi = 0°$) at its stalling speed V_s, it is unable to execute a turn and $r_{min} = \infty$; as the forward speed is increased, the permissible value of ϕ increases and r_{min} decreases. In the limit, as V approaches to ∞, ϕ approaches to 90° and $\sin \phi$ approaches to 1. The corresponding radius of turn is given by V_s^2/g ft (Fig. 4.5) and this is, in theory, the smallest value that can ever be obtained for the radius of turn. It is interesting to note that this value of the radius of turn is only dependent on the level flight stalling speed. In practice, however, even this radius cannot be achieved, because, as the angle of bank increases, so does the amount of power required and eventually a limiting angle of bank is reached when all the available power is being used to maintain the aircraft in a level turn (Fig. 4.5). Larger values of angle of bank than this and correspondingly smaller values of the radius of turn, can only be achieved in descending flight.

2.3. Equations for High Speed Turns

Though the assumption of a constant value of C_{Lmax} simplifies the analysis, the assumption is true only over a very small range of forward speeds. The C_{Lmax} value of a wing begins to decrease at surprisingly low values of Mach number. The maximum value of g which can now be applied is given by

$$n_{max} = \frac{L}{W} = \frac{C_{Lmax}\,\frac{1}{2}\rho V^2 S}{W}$$

where C_{Lmax} is the particular value corresponding to the forward speed V. Expressing this equation in terms of Mach number we have, since $\gamma P/\rho = a^2$, where a is the speed of sound, γ the ratio of the specific heats and P is the atmospheric pressure at the height at which the aircraft is being flown

$$n_{max} = \frac{C_{Lmax}\gamma P M^2}{2W/S} \tag{4.11}$$

Once the value of n_{max} is known, eqn. (4.7 (*b*)) can be used to determine the corresponding radius of turn.

Eqn. (4.11) can be arranged to give

$$\frac{n_{max}W}{P} = \frac{C_{Lmax}\gamma M^2 S}{2}$$

and since γ and S are constants and C_{Lmax} can be assumed to be only a function of M, this equation can be expressed generally as

$$\frac{n_{max}W}{P} = f(M) \tag{4.12}$$

This equation can be written if required with the ambient pressure P replaced by the relative pressure p (i.e. Ambient pressure/Standard sea-level pressure). This functional relationship shows that, if the turning performance is represented graphically by plotting the quantity $n_{max}W/P$ against M, one unique curve will be obtained, irrespective of the height and all-up weight at which the aircraft is flying. This relationship is an extremely useful one and the results of flight test experiments are often plotted in this way (see para. 6).

Having determined the maximum available value of g at any one height, eqn. (4.12) can be used to predict the value at any other height. If the A.U.W. is constant, then at the same value of M (and hence the same C_{Lmax}) the maximum g values are related to the corresponding atmospheric pressures by

$$\frac{n_{max1}}{P_1} = \frac{n_{max2}}{P_2} \tag{4.13}$$

An increase of height results in a decrease of pressure and hence in a corresponding reduction of the maximum available g.

3. Stalling Characteristics

3.1. General

As we have seen in the previous paragraph, the manoeuvring performance of an aircraft may be limited by the phenomenon of stalling. Though the actual value of $C_{L_{max}}$ that can be achieved from the wing is important, the handling characteristics of the aircraft on the approach to, and at, the stall, are often of equal and sometimes more importance. When the angle of attack of the wing is increased, the flow eventually breaks down and the wing is said to "stall." Associated with this stall there is a loss of lift, an increase of drag and possible changes occurring in the pitching and rolling moment characteristics. The turbulent airflow moving rearwards around the fuselage and tail unit may also result in airframe buffeting. The resulting effects of the stall may be gentle or severe; in the latter case the aircraft may enter a spin or may exhibit violent random pitching, rolling and yawing motions (Ref. 4).

The stalling characteristics of an aircraft are very dependent on where the flow separation first occurs and the way it spreads over the wing. Various design techniques can be used to improve the stalling characteristics of an aircraft. The important factors involved in these considerations are

(*a*) the wing sectional shape (i.e. the t/c ratio and the position of the maximum thickness);
(*b*) wing planform and wing washout;
(*c*) position of the horizontal tailplane;
(*d*) high-lift devices (para. 3.3);
(*e*) power setting.

It is not possible to discuss all these aspects here, they are considered in more detail in various standard textbooks, e.g. Ref. 1, Chapter 2; Ref. 2, Chapter 11 and Ref. 3, Appendix C.

3.2. Effect of Wing Design and Tailplane Position

Three main types of stall exist on wing sections, the conventional trailing-edge stall (i.e. the "classical" stall), which occurs on the older thicker sectional shapes, the leading edge stall and the thin-aerofoil stall. The different types of stall are discussed in Ref. 13, Vol. 1, page 156. The type of stall which occurs depends on the shape of the section and the Reynolds Number of the flow (Ref. 5). In the trailing-edge stall, flow separation begins initially at the trailing edge of the wing, moving forward with further increase in angle of attack. On thinner sections with sharp leading edges, laminar separation may occur at the leading edge, followed by turbulent re-attachment further back along the surface. When the short "bubble" which forms underneath the separated layer eventually "bursts," the wing section stalls abruptly, with a sudden loss of lift (Fig. 4.6). This type of stall is unsatisfactory, since it is often associated with violent wing dropping.

On a rectangular wing, the stall usually starts at the root of the wing, and spreads outboard with further increase in angle of attack. These stalling characteristics are satisfactory, since they result in small rolling moment changes and the aileron control retains its effectiveness at the stall. Sweptback wings, on the other hand, suffer from the big disadvantage that they tend to stall first at the

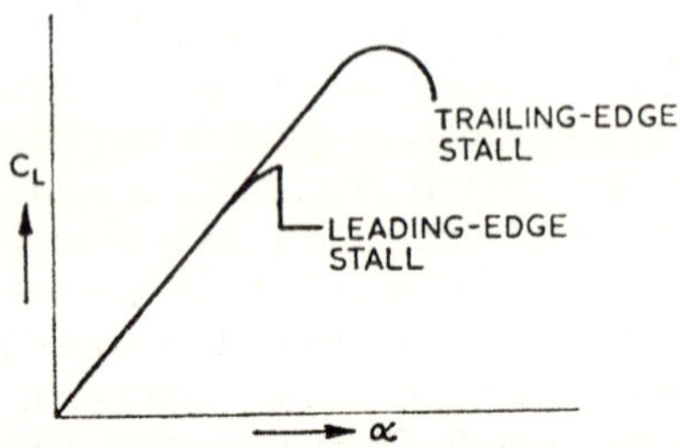

Fig. 4.6. Typical C_L/α curves for different aerofoil sectional shapes

wing tips, the stall spreading inwards with further increase in angle of attack. This is the result of the spanwise flow of air along the wing from the root towards the tip. The tip stall produces larger rolling moment changes, results in a reduction of lateral control effectiveness and tends to cause "pitch-up" problems.

When a wing is at a low angle of attack, the damping in roll is always positive (i.e. the damping in roll derivative l_p is negative). Under these conditions of flight, when a wing drops, the rolling moment produced opposes the disturbance. When the wing is near the stall, however, the damping in roll becomes negative and the wing is said to "autorotate." This phenomenon is an important factor in the balance of rolling moments when the aircraft is in a spin (see Chapter 13). When the loss of lift occurring at the stall is abrupt, violent wing dropping may occur.

At low angles of attack, the C_M/C_L curve is a straight line and the aircraft usually possesses positive static stability (i.e. dC_M/dC_L is negative). As the aircraft approaches the stall, the C_M/C_L curve becomes non-linear, and the aircraft pitches either nose-up or nose-down. These pitching moment changes result from the wing and the fuselage and from the effects of the downwash on the horizontal tailplane. A gentle nose-down pitching moment change is desirable, since the aircraft then has natural self-recovery stalling characteristics. Many modern high-speed aircraft designs, however, suffer from "pitch-up" problems and occasionally this effect may be serious enough to limit the manoeuvrability of the aircraft. The problem of pitch-up is discussed in more detail in Chapter 15.

As a result of pitch-up and the vertical sink of the aircraft arising out of the loss of lift on the wings, the angle of attack might be increased to a very large value, well above the stalling angle. The aircraft may become stabilized in this condition and recovery may be difficult, or even impossible, due to a loss in the elevator effectiveness (Fig. 4.7). This condition is known as the "Superstall" and

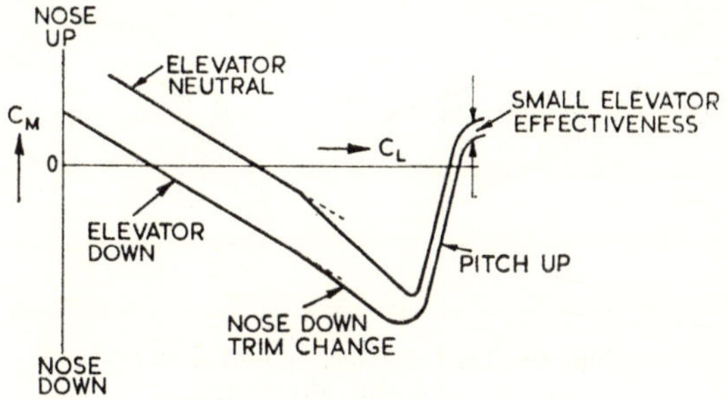

Fig. 4.7. Loss of elevator effectiveness at high C_L values

aircraft which have the horizontal tailplane set high up on the vertical fin are often prone to this particular problem. A detailed analysis of the stalling accident that occurred on one of the early BAC 111's is given in Ref. 15.

The stalling characteristics of an aircraft can be improved either by major design modifications (e.g. repositioning of the horizontal tailplane, incorporation of leading edge camber and washout, etc.) or by the use of simple "fixes" (e.g. boundary layer fences, vortex generators, leading edge spoilers, etc.). Many of these design features are discussed in Chapter 15. In the case of the BAC 111, the root section of the wing was modified to give a more nose-down pitching moment at the stall and power controls were fitted on the elevators to improve the control effectiveness.

3.3. High-lift Devices

Aircraft designed for high forward speeds have low aspect ratio swept wings with relatively thin symmetrical sectional shapes. The overall value of C_{Lmax} for these wings is fairly low and the stalling speed correspondingly high. In order to reduce the landing and take-off speeds to more suitable values, various devices are used to increase the usable value of C_{Lmax}. In addition to changing the stalling speed, these high-lift devices will also usually alter the stalling characteristics. The devices include

(*a*) leading-edge flaps and slots
(*b*) trailing edge flaps
(*c*) boundary layer control.

Leading-edge Flaps and Slots

The leading-edge slot (Fig. 4.8*a*) is a familiar means of increasing the value of C_{Lmax} of an aerofoil, though it has the disadvantage of

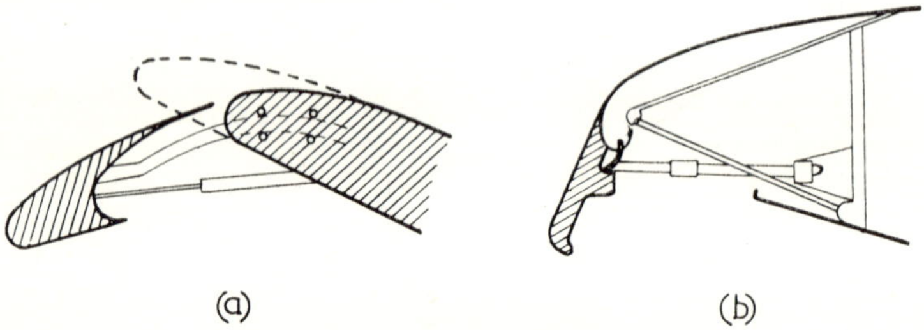

Fig. 4.8. Leading-edge high-lift devices
(*a*) Leading-edge slot. (*b*) Krueger-flap

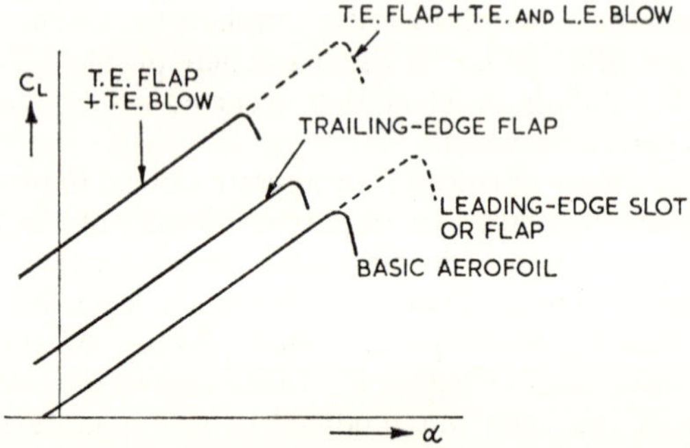

Fig. 4.9. Effect of flaps, slots and boundary layer control on the C_L/α curve

increasing the associated stalling angle (Fig. 4.9). Because the landing attitude of an aircraft is usually limited by considerations of undercarriage length, pilot view, or passenger comfort, it may not be possible to use this additional lift coefficient. The slot is a form of boundary layer control; the pressure difference occurring between the bottom and top surfaces of the wing produces a flow of air through the slot and out over the top surface. This flow re-energizes the boundary layer, and prevents separation from occurring until higher angles of attack are reached. On some swept-back wing aircraft, leading-edge slots have been used to smooth the airflow over the wing tips, so preventing the problems associated with wing-tip stalling.

On many modern aircraft, leading-edge slots have been replaced by leading-edge flaps (Fig. 4.8*b*), usually referred to as Krueger flaps. The effect of the leading-edge flap on the C_L/α curve is similar to that of the leading-edge slot (Fig. 4.9). In some cases, conventional slots are fitted on the outer parts of the leading-edge of the wing and Krueger flaps over the inboard portions (e.g. Boeing 727, D. H. Trident).

Trailing-edge Flaps

Nearly all aircraft incorporate some form of trailing-edge flap for use at take-off and landing. Many different types of flaps exist,

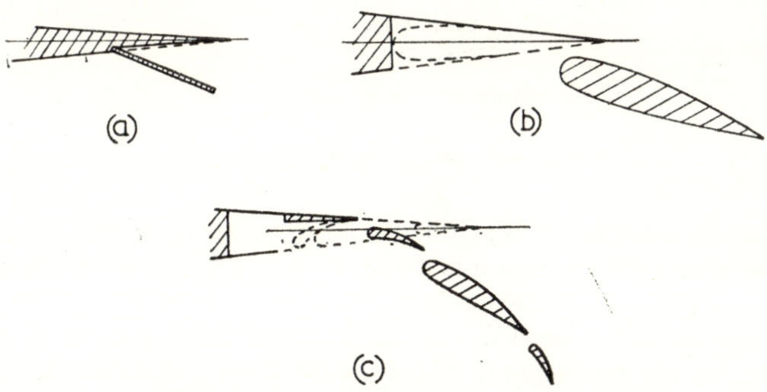

Fig. 4.10. Various types of trailing-edge flaps
(*a*) Split flap. (*b*) Fowler flap. (*c*) Triple-slotted flap

ranging from a simple split-flap, to the complicated triple slotted type of flap (Fig. 4.10). Flaps modify the lift characteristics of the wing as a result of the effective change of camber produced, though in some cases an additional effect is provided due to the increase in effective wing area obtained by the flaps moving backwards as they rotate. The flaps which move back and open up one or more slots as they rotate, are often referred to as Fowler-type flaps.

The effect of a flap on the C_L/α curve is shown in Fig. 4.9. The value of C_{Lmax} is increased without increasing the corresponding stalling angle and in some cases there may be a small reduction in the stalling angle. The increase in C_{Lmax} depends on the type of flap used; the characteristics of the different types of flaps are discussed in detail in Ref. 2, Chapter 14.

Boundary Layer Control (*B.L.C.*)
The characteristics of an aerofoil can be modified by applying various methods of boundary layer control. The ordinary leading-edge slot is a form of boundary layer control, though the term is usually reserved for methods involving blowing air out of slots, or sucking air away through slots, holes, or porous surfaces. There are two distinct forms of boundary layer control

(i) control to laminarize the flow and to reduce the drag,
(ii) control to prevent flow separation and to increase the lift. This is sometimes referred to as "Circulation" control.

Here we are only concerned with the second of these methods and for this purpose the boundary layer blowing technique is usually used (Fig. 4.11). When air is blown out of a narrow slot over the

Fig. 4.11. Boundary layer blowing

surface of the flap or aerofoil, the flow in the boundary layer is re-energized and separation is delayed. Fig. 4.9 shows the effect on the C_L/α curve of boundary layer blowing over the flap and at the leading-edge of the aerofoil. B.L.C. has been used on many aircraft during the last few years (e.g. F86 Sabre, Lockheed F104, Scimitar and Blackburn Buccaneer) and its use reduces the landing and take-off speeds by anything from about 10 to 30 kts.

When flap blowing is used a certain amount of blowing is required to make the flow remain attached to the surface of the flap. If the blowing is increased further than this, the lift produced by the wing continues to increase, but at a reduced rate. This increase of lift is a result of the "supercirculation" effect, which is the principle of the jet-flap scheme. In this case, the mechanical flap is removed and the changes in lift are produced by the deflection of a jet of air being blown out of a slot along the trailing edge of the wing. On the Hunting H126 jet-flap research aircraft, however, a very small mechanical flap has been retained. With a jet-flap, lift coefficients of the order of 12 are theoretically possible.

As well as producing changes in the lift characteristics of an aerofoil, these various high-lift devices produce corresponding changes in the drag and pitching moment characteristics. The pitching moment changes are particularly important because they modify the handling characteristics of the aircraft at the stall. Stalling tests thus have to be carried out for all possible aircraft configurations and power conditions.

High lift devices, boundary layer control and the jet flap scheme are discussed in more detail in Ref. 13.

4. Stalling Requirements

The stalling requirements for civil and military aircraft are laid down in Refs. 11 and 12 respectively. On modern designs it is becoming more difficult to satisfy these requirements by natural aerodynamic means and artificial methods are being used. If automatic devices are used, however, it is essential to ensure that they are safe and reliable.

4.1. Stall Warning

It is desirable that there should be a clear and distinct warning of the stall, consisting of buffeting or mild pitching or rocking. This warning can be produced naturally as a result of flow breakaway, or artificially by means of a "stick (or rudder pedal) shaker." This device is operated from an incidence measuring system and is usually arranged to give a warning of the stall about 10 to 12 per cent above the actual stalling speed. The system can also be designed to take account of the rate at which the stalling speed is being approached and to give the warning signal somewhat earlier when the speed is falling very quickly.

4.2. Stall Identification

The eventual stall should also be clear and unmistakable and it is preferable that it should consist of a clean nose drop with little buffeting. In practice, the limit may be set by other considerations (e.g. wing drop, loss of aileron control, etc.). However, a large amount of wing dropping is undesirable and it should be restricted to less than 20 or 30°. This guards against excessive angle of bank and loss of height and possible entry to a spin. If the desired characteristics cannot be obtained naturally, a "stick-pusher" can be incorporated to increase the required nose-down pitching moment. The stall identification then consists of a sharp, positive and unmistakable forward movement of the control column.

The layouts of stick-shaker and stick-pusher systems are shown in Ref. 14.

5. Lift Boundaries

If sufficient engine thrust and elevator control power are available, the manoeuvring performance of an aircraft will be limited by its aerodynamic characteristics. There are three types of lift boundary

(*a*) the stalling (or C_{Lmax}) boundary
(*b*) the buffet boundary
(*c*) the pitch-up boundary.

Any particular aircraft may possess some or all of these boundaries and may, in practice, be limited in the amount of usable g by any one of them. Older more conventional aircraft were normally limited by their stalling boundary and pitch-up did not occur. On modern designs, however, pitch-up may be serious enough to prohibit the aircraft from ever using the maximum lift which is available from the wing. This is very important from operational considerations, because it results in a loss of manoeuvrability.

5.1. Stalling (or C_{Lmax}) Boundary

From eqn. (4.11) it can be seen that the maximum g is obtained at any speed by flying at the stalling angle of attack. The shape of the

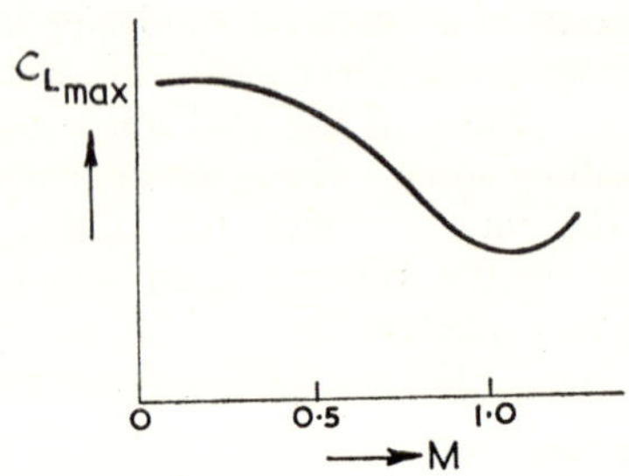

Fig. 4.12. Typical variation of C_{Lmax} with Mach number

curve of C_{Lmax} against Mach number depends on the particular aircraft being considered. It initially decreases with increase in forward speed, reaches a minimum value at transonic speeds and tends to increase again at supersonic speeds (Fig. 4.12). Using eqn. (4.11), the corresponding values of n_{max} can be determined for various heights. These are shown plotted against Mach number in Fig. 4.13. The variation of n_{max} should be compared with the curve obtained from eqn. (4.8). The available g decreases with height in accordance with eqn. (4.13), but the maximum available g occurs at the same Mach number at all heights. If these values are replotted in the form of $n_{max}W/P$ against M, then, as we have shown, one unique curve would be obtained.

The corresponding values of the radius of turn can be determined by substituting the values of n_{max} in eqn. (4.7*a*) or (4.7*b*); the

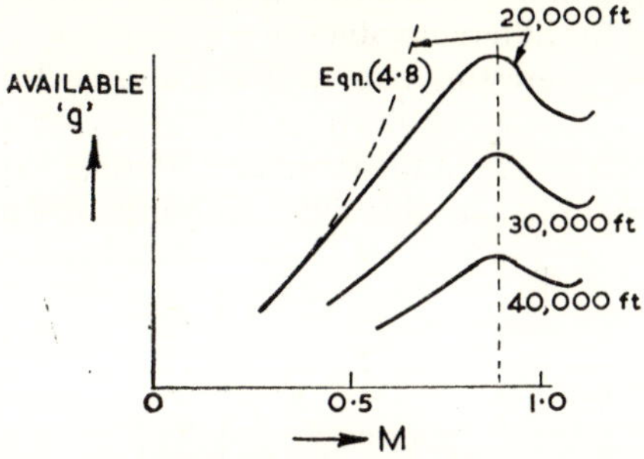

Fig. 4.13. Variation of available *g* with Mach number at various heights

variation with Mach number and height is shown in Fig. 4.14. The results for 20,000 ft should again be compared with those obtained from the simple theory. The radius of turn increases at the higher

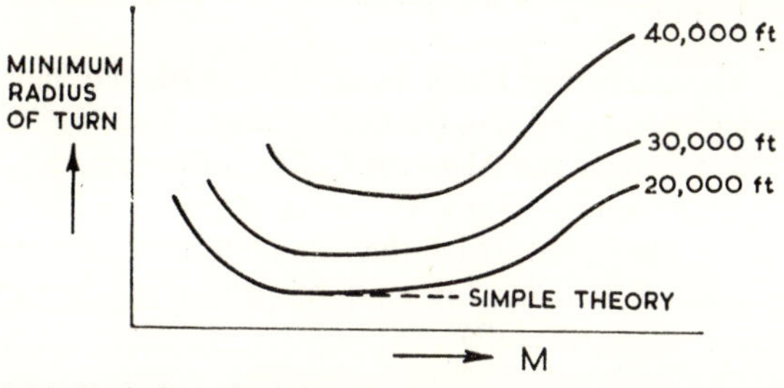

Fig. 4.14. Variation of minimum radius of turn with Mach number at various heights

Mach numbers due to the reduction in C_{Lmax}. As a result of this, the radius of turn has a minimum value, which occurs at a particular forward speed. The value of the minimum radius of turn and the corresponding optimum Mach number both increase with height (Fig. 4.15).

The condition for the maximum rate of turn is not the same as that for minimum radius of turn. The rate of turn ω is given by V/r and hence the maximum rate of turn corresponds to $(V/r)_{max}$ (i.e. $(r/V)_{min}$) and not to r_{min}. The optimum speed for maximum rate of turn can be determined by constructing the tangent from the origin to the curve of r against V (or M). From the curves of Fig.

4.14, it can be seen that this optimum speed will always be higher than the optimum speed for minimum radius of turn (Fig. 4.15).

In Fig. 4.13, the C_{Lmax} boundary is shown as the variation of n_{max} against M for different values of height. Alternative ways of expressing this information are available; a curve showing the variation of height with M for different values of g is shown in Fig. 4.18. This presentation is an interesting one, in that it shows clearly the reduction with height of the speed range available for any given value of g. The curve appropriate to the structural limiting value of g shows that, above a certain height, this limit cannot be reached. For each

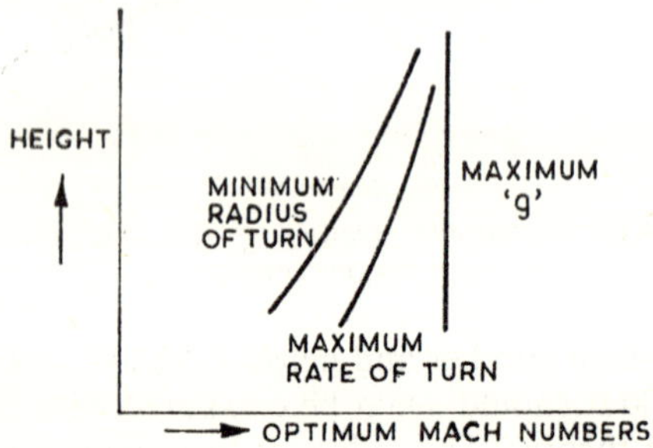

Fig. 4.15. Variation of the optimum Mach numbers with height

value of g, the curve reaches a peak at a particular height; this height is called the Lift Ceiling for that g value. The "$1g$" lift ceiling is usually referred to as the Absolute Ceiling of the aircraft. A more practical $1g$ flight ceiling limit is the so-called Service Ceiling, which is defined as that height at which the rate of climb of the aircraft has decreased to 100 ft/min (in the case of piston engine aircraft) or 500 ft/min (jet engine aircraft).

5.2. Buffet and Pitch-up Boundaries

As the aircraft approaches the stall, it may suffer from buffet and pitch-up. The corresponding boundaries may be drawn in the same way as the stalling boundaries discussed in para. 5.1. Depending on the characteristics of the particular aircraft concerned, it may be necessary to label these boundaries as "mild buffet," "severe buffet," etc., as appropriate.

6. Flight Tests

Wind tunnel tests can be carried out to determine the values of C_{Lmax}. The stalling characteristics are, however, very dependent on Reynolds Number effects and flight tests are a very important part of the overall programme. On military aircraft the main emphasis is on determining the maximum usable amount of C_L, whereas the main emphasis on civilian aircraft is on the qualitative handling characteristics at the stall.

6.1. Flight Test Techniques

The lift boundaries are determined by carrying out straight stalls and stalls in turning flight at various selected speeds and heights. For the straight stalls the aircraft is trimmed out with idling power at about 1·3 times the anticipated stalling speed and the speed slowly reduced at about $\frac{1}{2}$ knot per second. The stalls in turning flight are made at various forward speeds using power as required up to the limiting R.P.M. The aircraft is trimmed at the test height at a speed slightly above the required speed. It is then put into a steady turn and the g increased gently and smoothly until the aircraft stalls. At the higher speeds, it is necessary to lose height in the turn in order to maintain the selected speed and, in these cases, the turn has to be started at a height somewhat above the test height. The values of indicated g and the Mach number corresponding to the onset of buffet, pitch-up and the final stall are noted and qualitative assessments are made. The results obtained can be plotted in various ways, either by plotting g, h and speed directly, or by presenting the information non-dimensionally in the form of nW/p against Mach number. In order to indicate the areas of flow separation over the aircraft, the "wool tuft" method is often used. Small tufts are attached to the surface and photographs are taken of the flow patterns which are made visible. In the particular tests described in Ref. 7, nylon tufts 8 in. in length were used.

There are two methods available for determining the aircraft's thrust boundary. The first method is obvious and direct; it consists of carrying out a number of level turns at various speeds, at the selected height and power setting and measuring the maximum g that can be achieved. The second method is an indirect one and is based on the fact that the thrust boundary of an aircraft is determined by the amount of excess thrust available. This excess thrust can be used at a constant height either to maintain the aircraft in a turn, or to accelerate it forwards. The mathematical relationships for these two conditions of flight are known and if the acceleration of the aircraft in level flight is measured, the corresponding turning performance can be calculated. A variation of this method employs the "accelerated level—decelerated turn" technique. This test is made by performing an accelerated level run and a decelerated turn at a constant selected height. The speed is initially stabilized at a low value and then the throttle is opened quickly to give the required R.P.M. The aircraft is allowed to accelerate, at a constant height, until the speed has stabilized at its maximum value. When this condition has been reached, the aircraft is put slowly and smoothly into a turn, keeping the height and R.P.M. constant. The g applied has to be varied, so that the speed falls off slowly, at a rate not exceeding 1 knot per second, back to the original starting speed. To begin with the g has to be gradually increased, then, at the lower speeds, reduced again. By interpolating graphically at various

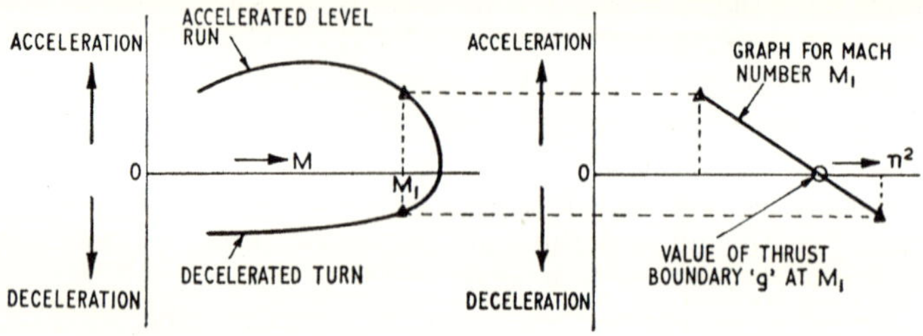

Fig. 4.16. Accelerated level-decelerated turn characteristics

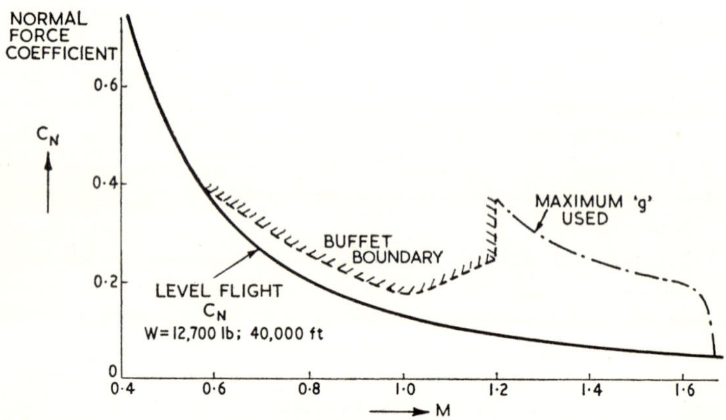

Fig. 4.17. Buffet boundary

speeds (Fig. 4.16) between the value of the acceleration under 1g conditions of flight and the value of the deceleration obtained with known amounts of applied g, the values of g which could just be sustained without the aircraft either accelerating or decelerating (i.e. the points on the thrust boundary) can be found.

6.2. Flight Test Results

The results of the buffet boundary tests of Ref. 7 are shown in Fig. 4.17. The buffet onset was very gradual and no pitch-up was

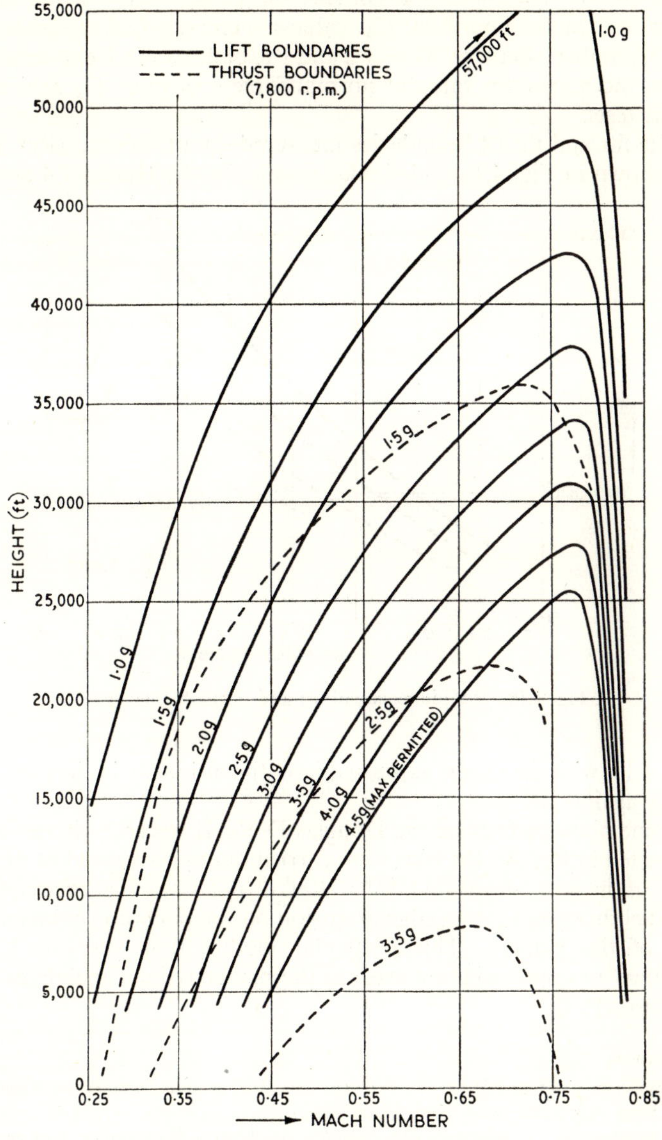

Fig. 4.18. Lift and thrust boundaries

experienced. The buffet margin (i.e. the margin between the level flight and buffet boundary C_N values) increased with increase of Mach number. Between $M = 1{\cdot}2$ and $1{\cdot}6$, buffet was not experienced up to the maximum value of normal force coefficient C_N achieved in the tests.

The lift and thrust boundaries measured on a Canberra (Ref. 10) are shown in Fig. 4.18 and similar results for the Hunter F4 are in

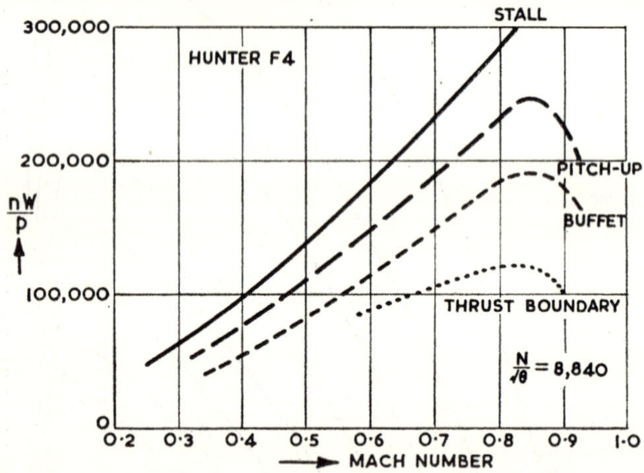

Fig. 4.19. Non-dimensional lift and thrust boundaries

Fig. 4.19, with the results expressed non-dimensionally in the form of nW/p against M.

Some pitch-up tests on the Douglas D-558-II research aircraft are discussed in Ref. 8; the tests were carried out up to a speed of $M = 1{\cdot}67$ at heights above 45,000 ft. Ref. 9 describes the effect of a parallel drooped leading edge extension on the pitch-up characteristics of the Hunter. This particular modification was found to produce an appreciable increase in the manoeuvring capabilities of the aircraft.

References

1. Perkins, C. D. and Hage, R. E.; *Airplane Performance Stability and Control* (John Wiley and Sons, New York, 1949).
2. Duncan, W. J.; *Control and Stability of Aircraft* (Cambridge University Press, 1952).
3. Kuethe, A. M. and Schetzer, J. D.; *Foundations of Aerodynamics* (John Wiley & Sons, New York, 2nd ed., 1959).
4. Scher, S. H.; *Post-stall Gyrations and their Study on a Digital Computer*, Agard Report No. 359, 1961.

5. WIMPENNY, J. C.; *Low-speed Stalling Characteristics*, Agard Report No. 356, 1961.
6. *Agard Flight Test Manual*, Vol. II, "Stability and Control," Chapter 7.
7. DEE, F. W. and NICHOLAS, O. P.; *Flight Determination of Wing Flow Patterns and Buffet Boundaries for the Fairey Delta 2 Aircraft at Mach Numbers between* 0·4 *and* 1·3 *and Comparison with Wind Tunnel Results*, R.A.E. Tech. Rep. No. 64012, 1964.
8. ANKENBRUCK, H. O.; *Determination of Longitudinal Stability in Supersonic Accelerated Manoeuvres for the Douglas D-558-II Research Airplane*, NACA RM L53 J20, 1954.
9. *The Effect of the Parallel Drooped Leading Edge Extensions and Centre of Gravity Movement on the Pitch-up Characteristics of Hunter Aircraft*, Unpublished M.O.A. Report.
10. *Lift and Thrust Boundaries without Wing-tip Tanks, Canberra B.2*, Unpublished M.O.A. Report.
11. *British Civil Airworthiness Requirements*, Air Registration Board.
12. *Design Requirements for Aircraft for Royal Air Force and Royal Navy.* Av.P.970, Chapter 603, Ministry of Aviation.
13. LACHMANN, G. V. (Ed.); *Boundary Layer Control and Flow Control*, Vols. 1 and 2 (Pergamon Press, 1961).
14. "Stalling—with Particular Reference to the Super-stall and to Stick-pushers," *Flight*, 18th March 1965, page 400.
15. *Aviation Week*, 26th April, 1965.

5 Subsonic Lateral and Directional Static Stability

1. Introduction

The concept of static stability can be applied to the lateral and directional motions of an aircraft, in a similar way to that in which it was used in considering the longitudinal stability characteristics. Directional static stability (or "Weathercock" stability) is the stability of the aircraft with respect to a disturbance in sideslip angle and is a measure of the tendency of the aircraft to align itself with the relative airflow. One of the major contributions to the overall stability, that of the vertical fin, can be analysed mathematically in the same way as the tailplane contribution to longitudinal static stability, with the angle of sideslip replacing the angle of incidence, the vertical fin replacing the horizontal tailplane and so on. Lateral static stability (or "Dihedral" effect) is, however, rather different, in that an initial disturbance in angle of bank does not produce any restoring moments and all aircraft are neutrally stable with respect to a disturbance of this type. When an aircraft rolls, however, a sideslip develops and rolling moments are produced which make the aircraft statically stable, the wings tending to recover their initial level position.

The lateral and directional static stability characteristics are extremely important in determining the overall lateral handling characteristics of an aircraft. The aileron and rudder deflections and forces required in steady sideslipping motion are dependent on the control-fixed and control-free static stabilities. They are also important factors in determining the general lateral dynamic characteristics of the aircraft, e.g. the Dutch roll oscillation and roll coupling motions.

2. Control-fixed Directional Static (or Weathercock) Stability

When an aircraft is sideslipping, as shown in Fig. 5.1, changes are produced in the aerodynamic forces acting on the wing, the fuselage, the vertical fin and on all the other parts of the aircraft. These forces produce yawing moments about the C.G. of the aircraft; the forces acting on the fuselage in front of the C.G. produce moments tending to increase the sideslip angle (i.e. destabilizing moments), while those acting on the fuselage and vertical fin behind the C.G.

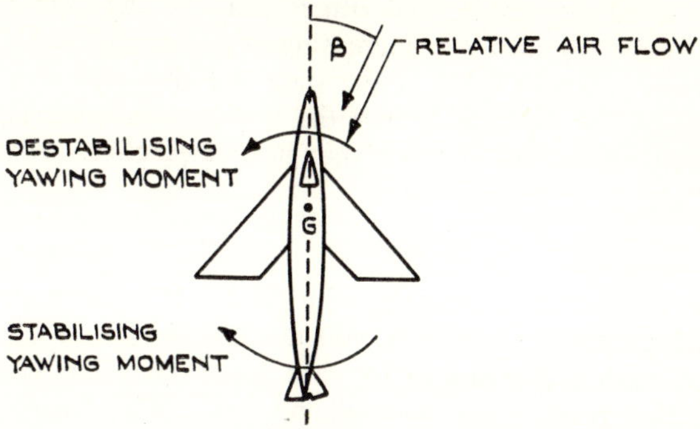

Fig. 5.1. Yawing moments acting on the aircraft in a sideslip

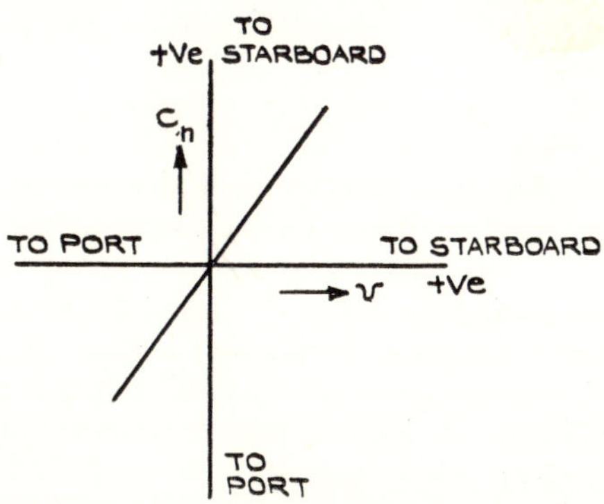

Fig. 5.2. Variation of yawing moment coefficient C_n with sideslip velocity v

produce moments tending to reduce the sideslip angle (i.e. stabilizing ones).

Longitudinal static stability is measured in terms of $dC_M/d\alpha$ (or dC_M/dC_L); directional static stability can be measured in a similar way in terms of $dC_n/d\beta$ (or dC_n/dv) [i.e. in terms of the derivative n_β or n_v (see Chapter 6)], where C_n is the yawing moment coefficient, β is the sideslip angle, and v is the sideslip velocity. The static stability can be represented graphically as the slope of the curve of C_n plotted against β (or v) [Fig. 5.2]. For positive stability, a positive sideslip (i.e. to starboard) should result in a positive yawing moment (i.e. to starboard) and hence the value of the slope of the C_n/v curve (i.e. dC_n/dv) should be positive.

Each different part of the aircraft contributes to the overall value of the static stability dC_n/dv. Some of these contributions may be small, or even negligible, and others may be dependent on the particular condition of flight, increasing or decreasing as the speed is changed. The overall value of the static stability will also be very dependent on the configuration of the aircraft; undercarriage, flaps and different conditions of power may have large effects on the value of dC_n/dv.

Wing Contribution

The contribution of a straight wing alone is usually negligible. Sweepback produces a stabilizing effect, which increases with increase in C_L (i.e. with a decrease in forward speed), whereas dihedral at the wing tips produces a destabilizing one. Wing/fuselage interference effects may be important and a high wing position is generally stabilizing. Engine nacelles on the wings produce a contribution that will depend on such factors as their size and position and the shape of the wing planform. On a straight wing, they usually produce a destabilizing effect.

Fuselage Contribution

The overall effect of the fuselage will depend on whether the destabilizing moments produced by the forces acting ahead of the C.G. are greater, or less, than the stabilizing moments produced by the forces acting behind the C.G. The resultant aerodynamic force will, however, act at a point which is about 25 per cent of the fuselage length from the nose and since the C.G. will normally be behind this point, the overall contribution is usually a destabilizing one. This is accentuated on recent aircraft designs, which have fuselages with high fineness ratios (i.e. length/diameter) and relatively far aft C.G. positions. The fuselage destabilizing effect may decrease at the higher angles of sideslip.

Fin Contribution

An aircraft should possess directional static stability and the fin is designed to balance out any resulting destabilizing effects from the other parts of the aircraft and to produce the required amount of stability. The fin contribution is obviously stabilizing, and its magnitude is dependent on such factors as the fin area and the fin arm (i.e. the fin volume coefficient), the fin lift curve slope and the sidewash factor. These are similar to the factors which determined the contribution of the horizontal tailplane to the longitudinal static stability and a mathematical analysis of this particular contribution can be made in a similar way.

The effect of the fin will be reduced at the higher angles of sideslip, due to fin stalling. This occurs in the same way as the stalling on a wing at high angles of attack. Stalling results in a reduction in the aerodynamic sideforce acting on the vertical fin and in a corresponding reduction of the stabilizing yawing moment. The aircraft should be designed so that the sideslip angle at which this occurs is outside the usable range of sideslip angles for the aircraft. If fin stalling is experienced on an aircraft, then the fin should be modified to increase the stalling angle. This can be done by the addition of a dorsal fin, or by replacing a single vertical fin by two or three fins of smaller aspect ratio.

Until fairly recently, it has not been difficult to design an aircraft with a suitable amount of directional static stability. Too much stability is not desirable, because, as we shall see later, this tends to make the aircraft spirally unstable. Too little stability is not desirable either, however, because under certain conditions of flight this will cause roll coupling problems. Unfortunately, the lift curve slope of any lifting surface decreases with increase of speed in the supersonic range and at these speeds the aircraft's directional static stability may become marginal. This is the reason for the large fins that can be seen on many present-day supersonic fighter aircraft and it also explains the numerous increases in fin size which often occur during the development period of these aircraft.

Undercarriage, Flap and Power Effects

Contributions to the overall stability will arise due to the various possible changes in the configuration of the aircraft, but it is difficult to generalize on these effects.

3. Control-fixed Lateral Static Stability

When an aircraft is disturbed in roll about its longitudinal axis, the angle of attack on the downgoing wing is increased and that on the upgoing wing is decreased. As long as the aircraft is not near the stall, this results in an increase of lift on the downgoing wing and a decrease on the upgoing wing. The rolling moment produced opposes the initial disturbance and gives a "damping-in-roll" effect. The rate of roll decreases from its initial value and eventually becomes zero. When this happens, however, the damping moment will also have reduced to zero, since it is always proportional to the rate of roll of the aircraft. The damping-in-roll moment cannot reverse the motion of the aircraft and cannot bring the wings back into the level position. In fact, if the roll reversed in direction, then the damping moment would also reverse, since it always opposes the rotation of the aircraft. Thus, if nothing else happened, an aircraft disturbed in roll would remain in this condition, with the wings out of the level position. An aircraft thus possesses neutral static stability with respect to an angle of bank disturbance.

Fortunately, however, something else does happen, and this gives an aircraft its lateral static stability. When an aircraft is disturbed laterally, it does not produce a pure rolling motion, because, as soon as the wings leave the level position, a component of the weight

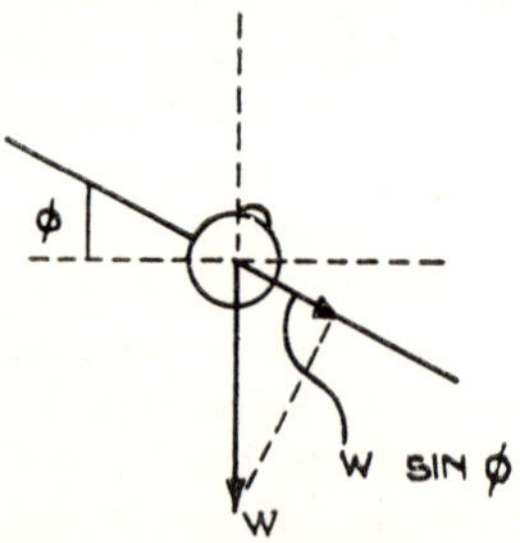

Fig. 5.3. Weight component acting in a banked sideslip

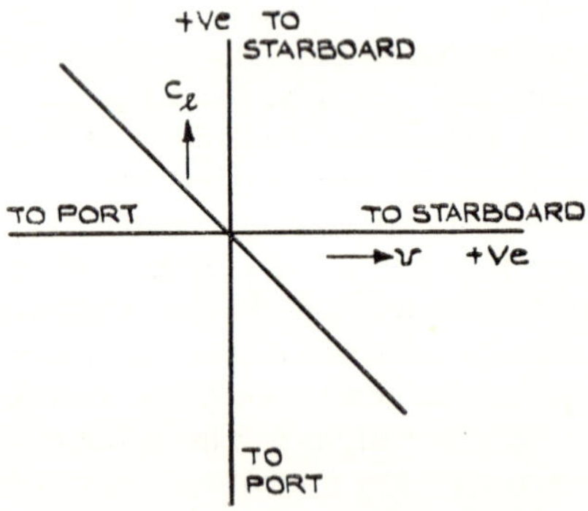

Fig. 5.4. Variation of rolling moment coefficient C_l with sideslip velocity v

acts sideways (Fig. 5.3), and a sideslipping motion develops. The forces arising on the different parts of the aircraft as a result of the sideslipping produce the rolling moment tending to restore the aircraft to its initial wings-level position.

Lateral static stability can be measured in terms of $dC_l/d\beta$ or dC_l/dv [i.e. in terms of the derivative l_β or l_v (see Chapter 6)], where C_l is the rolling moment coefficient and β and v are again the sideslip angle and the sideslip velocity, respectively. The stability can be represented graphically as the slope of the curve of C_l plotted against β (or v) (Fig. 5.4). For positive stability, a positive sideslip (i.e. to starboard) should result in a negative rolling moment (i.e. to port) and hence the value of the slope of the C_l/v curve (i.e. dC_l/dv) should be negative.

Each different part of the aircraft will contribute to the overall value of the static stability dC_l/dv. Once again, these contributions will be of different magnitudes and will depend on the actual condition of flight and the particular configuration of the aircraft.

Wing Contribution

When an unswept wing, with a positive dihedral angle of Γ, is sideslipping with a velocity v (Fig. 5.5), the wing pointing into wind has an increased effective incidence, whereas the incidence of the other wing is effectively reduced. This is because the sideslip velocity has a component $v \sin \Gamma$ normal to the wing surface, which is upwards on the into-wind wing and downwards on the other wing. As long as the aircraft is not near the stall, this produces an increase of lift on the into-wind wing, and a reduction of lift on the other wing and

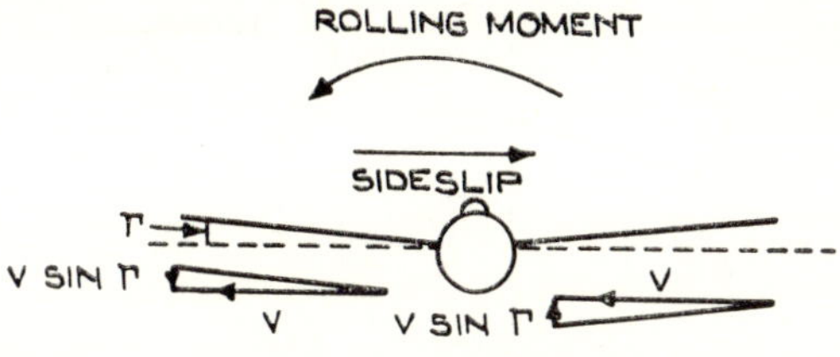

Fig. 5.5. Dihedral effect

the resulting rolling moment is a stabilizing one. This stabilizing contribution is proportional to the dihedral angle Γ and to the lift curve slope of the wing and is thus dependent on such factors as the wing aspect ratio, taper ratio, etc., being increased by an increase in aspect ratio and decreased by an increase in taper ratio. This is often one of the most important contributions to the overall stability, and for this reason the lateral static stability is often referred to as the "Dihedral Effect," although there are a number of other important contributions.

Wing sweepback has the effect of producing an additional stabilizing contribution, thus increasing the "effective" dihedral of the aircraft. If the wing is sweptback at an angle Λ, the sideslip increases the relative speed of the airflow normal to the leading edge of the wing pointing into wind and decreases that normal to the leading edge of the other wing (Fig. 5.6). Since the lift characteristics are chiefly determined by this normal flow component, the lift is increased on the leading wing and decreased on the trailing wing. An alternative way of looking at this, is to regard the effect of the sideslip as one of reducing the angle of sweep on the leading wing and increasing the sweep on the trailing wing. The changes in the lift curve slope associated with these changes in angle of sweep result in the variations in the lift forces on the two wings described above. Fig. 5.7 shows clearly that the contribution of sweep to the static stability becomes more important at the higher values of C_L (i.e. at the lower forward speeds). This is very important, because it means that the dihedral effect varies considerably over the speed range of the aircraft; at high speeds the stability is much less than at lower speeds. To reduce the stability to a more reasonable value at the higher

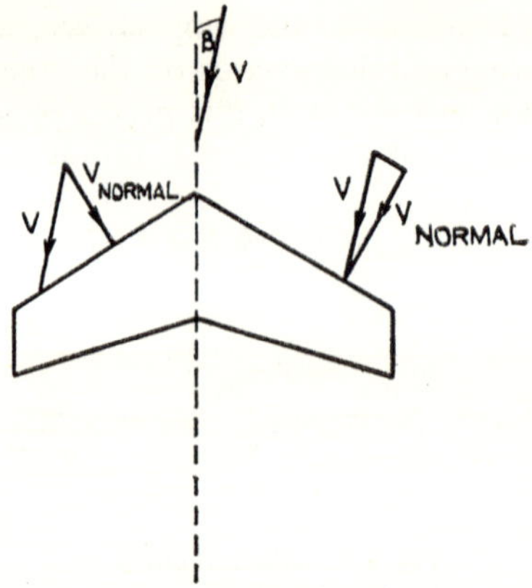

Fig. 5.6. Effect of sweepback on lateral static stability

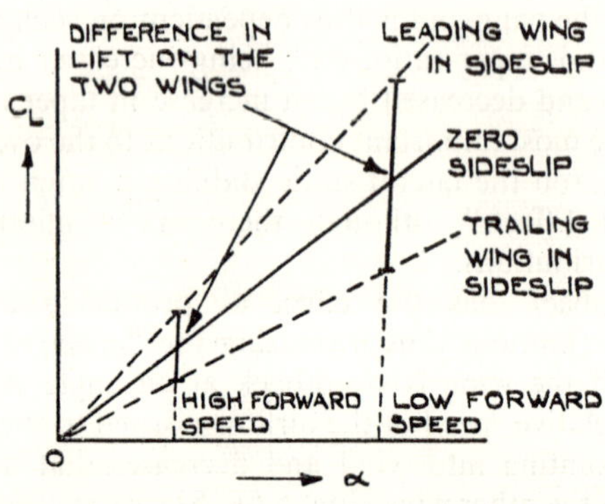

Fig. 5.7. Effect of sweepback and sideslip on the C_L/α curve

angles of attack, it may be necessary to incorporate some negative dihedral (i.e. anhedral) on the wing of the aircraft.

Wing/fuselage effects are important; the flow round the fuselage induces vertical velocity components that will alter the effective angle of attack of the wing. In the case of a high wing lay-out, the angle of attack of the leading wing will be increased and that of the trailing wing decreased (Fig. 5.8). This increases the effective dihedral of the wing. A low wing position, on the other hand, reduces the dihedral effect.

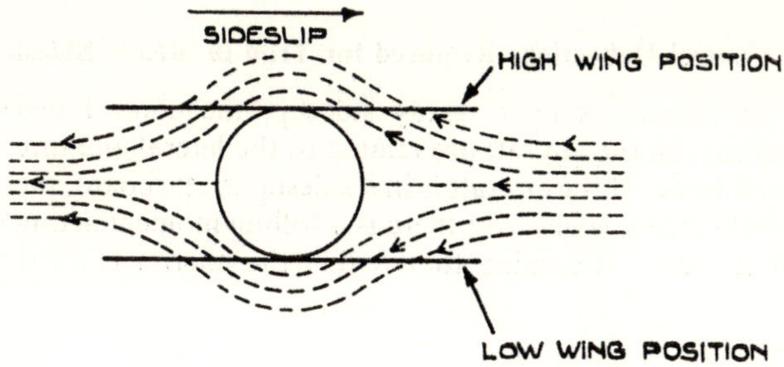

Fig. 5.8. Wing-fuselage interference effects

Fuselage and Vertical Fin Contribution
The fuselage and vertical fin produce an additional dihedral effect, the magnitude of which is dependent on such factors as the fuselage shape, the size and height of the vertical fin and the vertical position of the C.G. The contribution is usually small, but it may become appreciable on some of the present-day aircraft with high fins.

Undercarriage, Flap, and Power Effects
The overall stability will depend on the particular configuration of the aircraft. The effect of flap deflection will depend on the sweep of the flap hinge line; a sweptback hinge line will decrease the dihedral effect, whereas a sweptforward hinge line will increase it. The effect of power is generally destabilizing in the case of a propeller-driven aircraft, when the flaps are immersed in the slipstream. The effect is particularly important at low speeds.

It is desirable that an aircraft should possess lateral static stability. Rudder control can then be used to pick up a wing that has dropped as a result of a rolling disturbance; a starboard wing drop would require a starboard sideslip for recovery and hence a rudder deflection to port. This is particularly useful for controlling an aircraft which is near the stall, when application of aileron might aggravate the condition. However, the stability should not be too large, as this would lead to

(*a*) lateral oscillatory (i.e. Dutch roll) problems,
(*b*) large aileron control deflections and forces under asymmetric conditions,

(*c*) large rolling response to rudder deflection, requiring rapid aileron movements to counteract,

(*d*) the possibility of "Autorotation" under certain conditions of flight (see Chapter 16).

4. Control Deflections Required for Trim in Steady Sideslips

When an aircraft is in a steady sideslip, the control deflections required to trim the aircraft are related to the lateral and directional static stabilities. If the aircraft is in a sideslip to starboard, then if the stick-fixed value l_v is negative there is a rolling moment acting on the aircraft to port. Assuming that l_ξ [i.e. the aileron control power

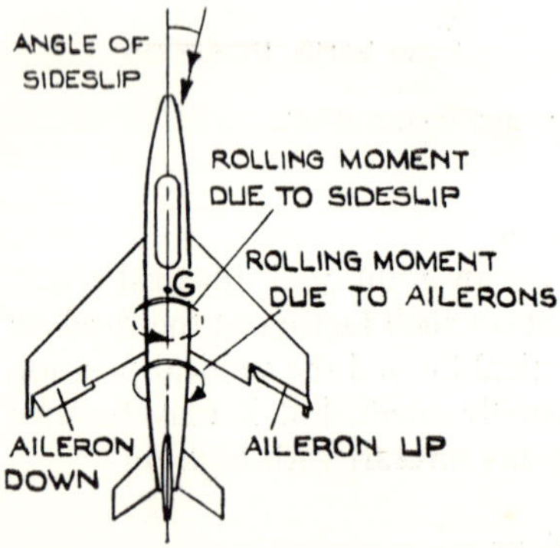

Fig. 5.9. Balance of rolling moments in a steady sideslip

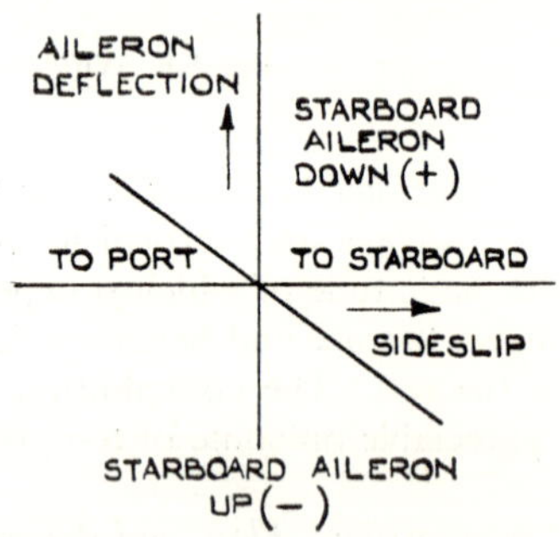

Fig. 5.10. Aileron angles required to trim in steady sideslips

derivative $dC_l/d\xi$ (see Chapter 6)] is negative, a negative aileron movement is required to balance this moment (i.e. the port aileron is down to keep the starboard wing down) (Fig. 5.9). Expressing this in symbols

$$l_\xi \xi + l_\beta \beta = 0 \tag{5.1}$$

hence

$$\xi = -\frac{l_\beta \beta}{l_\xi} \tag{5.2}$$

On simple theory (assuming that l_β and l_ξ remain constant throughout the range of sideslip angles) the aileron deflection required

(i) increases linearly with an increase in sideslip angle (Fig. 5.10),

(ii) increases with a numerical increase of l_β (or l_v),

(iii) decreases with a numerical increase in the aileron power derivative l_ξ.

Since both l_β and l_ξ are normally negative, the slope of the ξ/β curve is a negative one. Because the graph is a straight line, the slope of the line is a measure of the ratio of $-l_\beta/l_\xi$ (i.e. $d\xi/d\beta = -\ l_\beta/l_\xi$).

If the stick-fixed value of n_v is a positive one, then in a starboard sideslip there is a yawing moment acting on the aircraft to starboard. Assuming that n_ζ [i.e. the rudder control power derivative $dC_n/d\zeta$ (see Chapter 6)] is negative, a positive rudder is required to balance

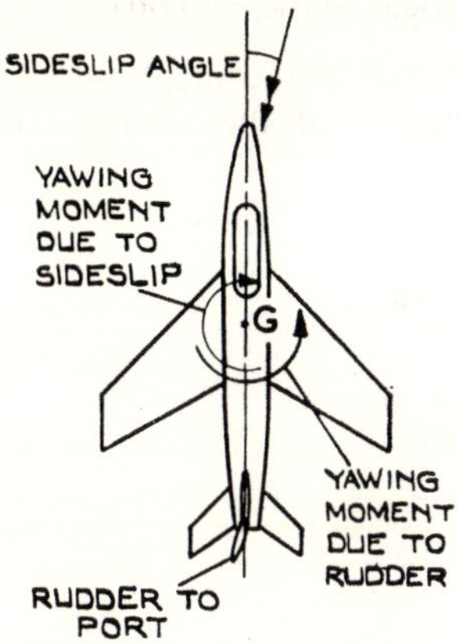

Fig. 5.11. Balance of yawing moments in a steady sideslip

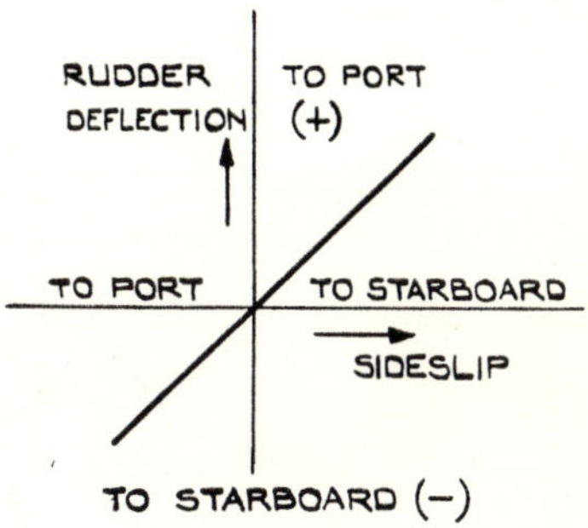

Fig. 5.12. Rudder angles required to trim in steady sideslips

this moment (i.e. the rudder is deflected to port to maintain a starboard sideslip) (Fig. 5.11). Expressing this in symbols

$$n_\zeta\zeta + n_\beta\beta = 0 \tag{5.3}$$

hence

$$\zeta = -\frac{n_\beta\beta}{n_\zeta} \tag{5.4}$$

Once again on simple theory (assuming that n_β and n_ζ are constants) the rudder angle required

(i) increases linearly with an increase in sideslip angle (Fig. 5.12),
(ii) increases with an increase in n_β,
(iii) decreases with a numerical increase in the rudder power derivative n_ζ.

However, because n_β is normally positive and n_ζ negative, the slope of the ζ/β curve is a positive one. The slope of the straight line is a measure of the ratio of $-n_\beta/n_\zeta$ (i.e. $d\zeta/d\beta = -\ n_\beta/n_\zeta$).

In these considerations we have assumed that the ailerons do not produce any yawing moments and that the rudders do not produce any rolling moments (i.e. $n_\xi = l_\zeta = 0$). On certain designs these derivatives may not be negligible ones and the extra moments may have to be allowed for (e.g. a large and high fin may produce an appreciable rolling moment from rudder deflection).

5. Control-free Lateral and Directional Static Stability

In paras. 2 and 3 it has been assumed that the control surfaces remain fixed during the sideslipping motion of the aircraft. If,

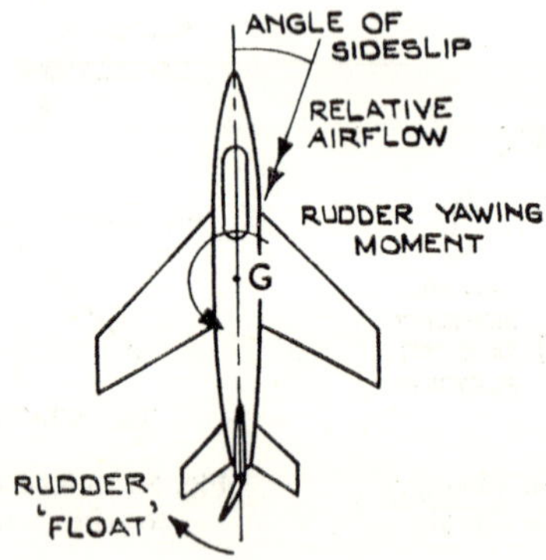

Fig. 5.13. Effect of rudder "float"

however, the controls are free to move and to take up their equilibrium position under the action of the various aerodynamic forces acting upon them (Fig. 5.13), then the stability characteristics of the aircraft will be modified by the control movements. This is similar to the effect of longitudinal stability of freeing the elevator, discussed in para. 2.5 of Chapter 2; the contribution of the tailplane to the longitudinal static stability of the aircraft was modified by a factor of $1 - (a_2b_1/a_1b_2)$ (eqn. 2.30).

In the lateral case, each of the control movements will affect both the rolling and the yawing moments, though, obviously, the greater effect of the aileron movement will be on the rolling moments (and hence on the value of l_v), and the greater effect of the rudder movement will be on the yawing moments (and hence on the value of n_v). The control-free values of l_v and n_v will be greater or less than the corresponding control-fixed values depending on the direction in which the controls float (i.e. on the b_1 value of the control surfaces). The effect of freeing the controls will also depend on such factors as the speed and height at which the aircraft is flying and also on the particular configuration being used (e.g. flaps, undercarriage, etc.). In the case of an aircraft fitted with non-powered controls, the rudder and aileron forces required in steady sideslips can be shown to be related to the corresponding control-free static stability [i.e. to the angle between the position of the control required for trim and the "floating" position (Fig. 5.14)]. With powered controls, on the

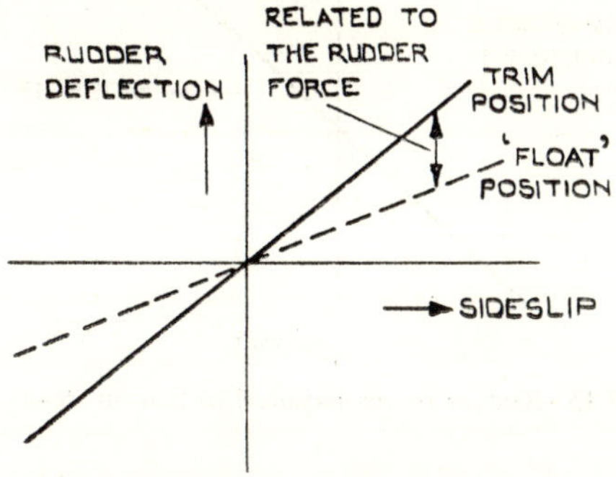

Fig. 5.14. Rudder angles to trim and rudder "float" in steady sideslips

other hand, no floating of the control surfaces occurs and the forces are related to the control-fixed stabilities.

Consider an aircraft sideslipping to starboard, with the rudder left free. If the b_1 of the rudder control is negative, then the rudder floats to port (Fig. 5.13). The rudder floats to the particular angle at which the floating hinge moment is just balanced by the hinge moment resulting from the control deflection (i.e. the b_2 effect)

$$\zeta_{float} = \frac{b_1}{b_2}\beta \tag{5.5}$$

Thus the floating angle can be reduced by making b_1 less negative or b_2 more negative. The effect of the rudder deflection is to produce an additional yawing moment to port and this is destabilizing. When b_1 is negative, the rudder-free direction stability will be less than the corresponding rudder-fixed value and the rudder forces will be reduced. The contribution of the fin to the rudder-free stability (n_{vf}') can be expressed in terms of the corresponding stick-fixed fin contribution (n_{vf}) by

$$n_{vf}' = n_{vf}\left(1 - \frac{a_2}{a_1}\frac{b_1}{b_2}\right) \tag{5.6}$$

where a_1 and a_2 are the fin lift curve slopes due to fin incidence and rudder deflection respectively, and b_1 and b_2 are the hinge moment characteristics of the rudder. If the numerical value of b_1 is large (or b_2 small) the control may tend to float to an angle greater than that required for trim. In this case the aircraft possesses directional instability with the rudder free. The rudder forces in steady sideslips

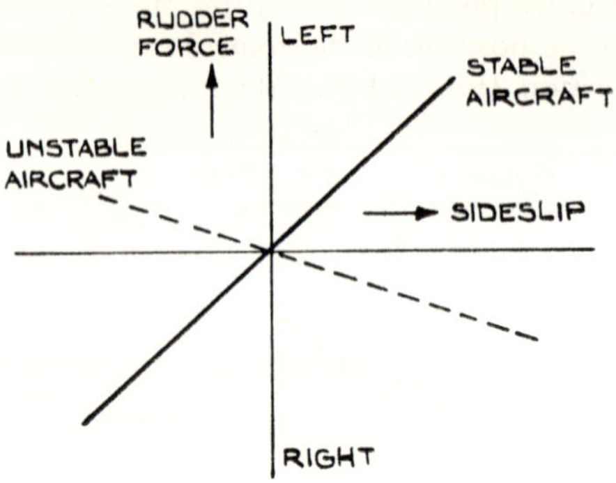

Fig. 5.15. Rudder forces required to trim in steady sideslips

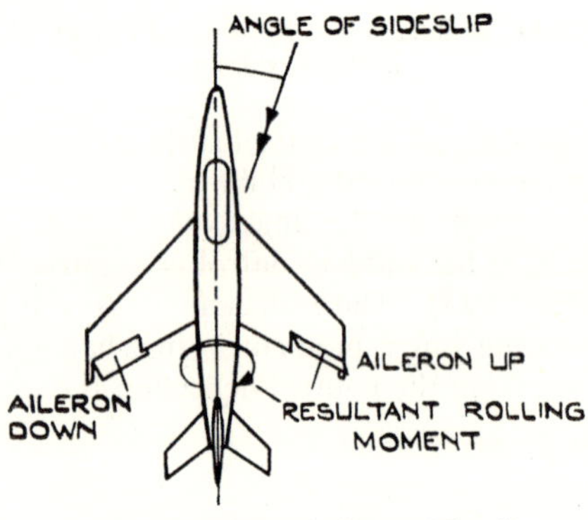

Fig. 5.16. Effect of aileron "float"

are then in the opposite direction to those normally obtained (e.g. a right rudder force would be required in a starboard sideslip) (Fig. 5.15).

If the b_1 of the rudder is made positive (e.g. by means of a large horn balance on the control), then the control floats in the opposite direction (i.e. against the wind). The control-free stability is then greater than that of the control-fixed stability and the rudder forces are correspondingly increased.

Similar considerations apply to the lateral static stability with the ailerons free. If, in a starboard sideslip, the starboard aileron floats up and the port aileron floats down (Fig. 5.16), then a rolling moment is produced acting to starboard. This is a destabilizing contribution, and the stick-free lateral stability is less than the corresponding stick-fixed value. As long as the floating angle is less than the aileron deflection required to trim, the stick forces for trim are in the usual direction and the aircraft possesses stick-free lateral static stability. If the control tends to float to a greater angle than that required for trim, the aircraft has lateral instability with the ailerons free and the stick forces required to trim are reversed in

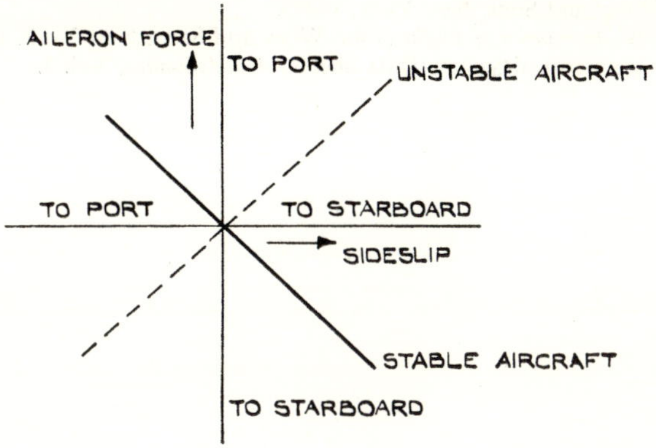

Fig. 5.17. Aileron forces required to trim in steady sideslips

direction, e.g. a left aileron force would be required in a sideslip to starboard (Fig. 5.17).

6. Conclusion

In this chapter we have dealt with the control-fixed and control-free lateral and directional static stabilities. As has been mentioned, these quantities are very important factors in any analysis of the lateral motion of an aircraft and we shall consider them at appropriate times in the following chapters. Since the steady sideslip tests form part of the overall tests on the lateral stability characteristics of an aircraft, a more detailed discussion of these will be left until Chapter 11.

From what has been said, however, it should be clear that the various stabilities are very dependent on the design of the particular aircraft being considered, and on the speed and angle of attack at which the aircraft is flying. Estimates of the numerical values for any particular design can be obtained from the appropriate charts in Ref. 3.

Compressibility effects are, of course, very important on the lateral and directional stabilities of transonic and supersonic aircraft. Some aircraft experience a reversal of the dihedral effect (i.e. starboard sideslip produces a roll to starboard) in the transonic speed range and all supersonic aircraft experience a gradual loss of directional stability in the supersonic speed range. The problems arising from these effects will also be considered in later chapters.

References

1. Perkins, C. D. and Hage, R. E., *Airplane Performance, Stability and Control* (John Wiley and Sons, New York, 1949).
2. Etkin, B., *Dynamics of Flight* (John Wiley and Sons, New York, 1959).
3. Royal Aeronautical Society Data Sheets, *Aerodynamics*, Vol. 3.

6 Stability and Control Derivatives

1. Introduction

The equations of motion relate the linear and angular accelerations of an aircraft to the resultant forces and moments acting on it. The various forces and moments which occur are either aerodynamic in nature, or arise as a result of the mass (i.e. inertia) characteristics of the aircraft. The aerodynamic forces and moments are dependent on the linear and angular motions of the aircraft and on the deflections of the control surfaces. The equations of motion can be simplified by assuming that the changes in the forces and moments are proportional to the small disturbances occurring in the motion of the aircraft. If an aircraft is disturbed, so that it is sideslipping with a velocity v, the change in the yawing moment ΔN produced as a result of the sideslipping motion can be expressed as

$$\Delta N = \frac{\partial N}{\partial v} v \tag{6.1}$$

where $\partial N/\partial v$ represents the rate of change of yawing moment with sideslip velocity. If, during the disturbance, the aircraft is also rolling at a rate p, yawing at a rate r and if the rudder deflection changes by an angle ζ, then the overall change in the yawing moment can be expressed as the sum of the effects of the separate disturbances

$$\Delta N = \frac{\partial N}{\partial v} v + \frac{\partial N}{\partial p} p + \frac{\partial N}{\partial r} r + \frac{\partial N}{\partial \zeta} \zeta \tag{6.2}$$

where $\partial N/\partial v$, $\partial N/\partial p$, $\partial N/\partial r$ and $\partial N/\partial \zeta$ represent the rate of change of yawing moment with sideslip, rate of roll, rate of yaw and rudder deflection, respectively. These terms are usually represented by the symbols N_v, N_p, N_r and N_ζ. Division by an appropriate quantity converts these derivatives into their non-dimensional form; they are then represented by n_v, n_p, n_r and n_ζ.

There are an extremely large number of Stability and Control Derivatives of various kinds and they can be classified in different ways. One method of classification is related to the nature of the quantity being considered, and its associated disturbance (e.g. "force-linear velocity" derivatives, "moment-control surface deflection" derivatives). An alternative method of classification is in terms of their effect on the aircraft (e.g. "damping" derivatives such as l_p, "control power" derivatives such as n_ζ, "cross" derivatives such as n_p, "direct" derivatives such as y_v and so on). The quantities b_1 and b_2, the elevator hinge moment characteristics, can also be described as control hinge moment derivatives, since $b_1 = \partial C_H/\partial\alpha'$ and $b_2 = \partial C_H/\partial\eta$.

The values of the derivatives depend on the actual case being considered, i.e. on the design of the aircraft, on the speed and height at which it is flying, on its particular configuration (e.g. flaps, undercarriage, etc.), on the type of power unit fitted to the aircraft and on the particular power setting being used. In some simple theoretical analyses many of the derivatives are assumed to be constants and the possibility of Reynolds Number, compressibility and aeroelastic effects is neglected. A number of the derivatives are, however, dependent on the speed (i.e. C_L) at which the aircraft is flying, e.g. the dihedral effect l_v of a swept-back wing aircraft. The values of the stability derivatives also depend on whether the control surfaces are "fixed" or "free"; in this chapter only stick-fixed derivatives will be considered.

The derivatives are also dependent on the type of motion being considered. The derivatives usually quoted are the "static" derivatives; these values are applicable to all steady motions (e.g. steady sideslips) and to relatively slow or exponential type oscillatory motions (Ref. 1, page 70–73). They are not applicable to high frequency oscillatory motions and in flutter analyses, for example, it is necessary to use the appropriate oscillatory values of the derivatives.

When reading any book or report on stability and control work, it should be remembered that different divisors can be used to define the non-dimensional derivatives and that different symbols are often used. The definitions that are being employed should be carefully noted.

2. Non-dimensional Derivatives

The derivatives are converted to their non-dimensional form by dividing by any suitable and convenient quantity, which has the same dimensions as the derivative itself. The main difference between the various systems is in the quantity used for the standard length. The divisors used in the British system (Ref. 1) for the most important force and moment derivatives are shown in Table 6.1. The quantity given in square brackets in this table indicates the dimensions of the divisor.

Table 6.1

	CONTROL DEFLECTION	LINEAR VELOCITY	ANGULAR VELOCITY
Force	$\rho V^2 S$ [M L T^{-2}]	ρVS [M T^{-1}]	ρVSs [M L T^{-1}]
Moment	$\rho V^2 Ss$ [M L^2 T^{-2}]	ρVSs [M L T^{-1}]	ρVSs^2 [M L^2 T^{-1}]

where s = a standard length, which is l' for the longitudinal derivatives and $\frac{1}{2}b$ for the lateral ones. *Note:* In the case of a tailless aircraft, the chord length of the wing may be used in place of l'.

In most of the work on stability and control, the longitudinal and lateral motions can be considered separately

(*a*) *Longitudinal Derivatives*

In longitudinal motion the forces and moments involved are those of X, Z and M and the major disturbances are in u, w, q and η. The corresponding derivatives are shown in Table 6.2.

Table 6.2

	u	w	q	η
X	x_u	$x_w(x_\alpha)$	x_q	x_η
Z	z_u	$z_w(z_\alpha)$	z_q	z_η
M	m_u	$m_w(m_\alpha)$	m_q	m_η

The vertical speed derivatives x_w, z_w and m_w, can also be referred to as angle of attack derivatives x_α, z_α and m_α, because a vertical motion w is equivalent to a change in angle of attack α

$$\alpha = \tan^{-1}\frac{w}{V} \tag{6.3}$$

where V is the forward speed.

Of the various derivatives in Table 6.2, however, some are much more important than others. A few of the derivatives are small enough to be neglected, others are only of relatively minor importance and the remainder are the important ones.

(b) Lateral Derivatives

In lateral motions the forces and moments involved are those of Y, L and N and the major disturbances are in v, p, r, ξ and ζ. The corresponding derivatives are shown in Table 6.3.

Table 6.3.

	v	p	r	ξ	ζ
Y	$y_v(y_\beta)$	y_p	y_r	y_ξ	y_ζ
L	$l_v(l_\beta)$	l_p	l_r	l_ξ	l_ζ
N	$n_v(n_\beta)$	n_p	n_r	n_ξ	n_ζ

The sideslip derivatives can be expressed either as sideslip *velocity* derivatives (i.e. y_v, l_v, etc.) or as sideslip *angle* derivatives (i.e. y_β, l_β, etc.) since β and v are related by

$$\beta = \tan^{-1}\frac{v}{V} \simeq \frac{v}{V}\text{ rads} \tag{6.4}$$

Once again, the relative importance of the derivatives in this table varies widely, some are negligible, and others are of major importance.

Note: The lateral derivatives are non-dimensional quantities and they are defined in such a way (see list of symbols) that the suffixes v, p and r can be regarded as representing the non-dimensional values of the sideslip velocity, rate of roll and rate of yaw, respectively (i.e. v/V, $pb/2V$ and $rb/2V$).

3. Stick-fixed Longitudinal Derivatives

The main longitudinal derivatives will now be described in more detail and it will be shown how each one arises as a result of the various forces acting on the aircraft. Their effects on the dynamic motion of the aircraft will be discussed in a later chapter.

3.1. Forward Velocity Derivatives

(a) x_u This derivative represents the rate of change in the force along the x-axis due to changes in the forward speed. It is usually negative and is a measure of the resistance to any changes of forward speed. It is sometimes referred to as the "speed-damping" derivative.

On simple theory it is related to the drag characteristics of the aircraft. By differentiating the drag equation $D = C_D \frac{1}{2}\rho V^2 S$ with respect to speed (assuming that C_D is independent of speed) and dividing through by ρVS, it can be shown that

$$x_u = -C_D \tag{6.5}$$

Thus, if the aircraft is disturbed so that the forward speed increases, the drag increases and this tends to reduce the speed again.

The forces and moments involved in the equations of motion have been said to be basically "aerodynamic" or "inertial" in origin. There are also, however, the thrust forces from the power unit. Any contributions from the thrust effects to the *changes* in the forces and the moments acting on the aircraft during a disturbance are usually included in the aerodynamic derivatives. Thus, if an increase in forward speed produces an increase in thrust for a constant engine setting, the speed-damping effect is reduced, whereas a reduction in thrust will increase the damping. In addition, the drag coefficient may itself vary with forward speed and, when these effects are both taken into account, the derivative x_u can be expressed more fully, in terms of Mach number, as

$$x_u = -C_D - \frac{M}{2}\left(\frac{\partial C_D}{\partial M} - \frac{\partial C_T}{\partial M}\right) \tag{6.6}$$

the value of $\partial C_T/\partial M$ depends on the type of power unit employed and the particular conditions of flight. This factor was, in fact, previously involved in our considerations of the more general theory of longitudinal static stability (Chapter 2).

(*b*) z_u This represents the rate of change in the force along the z-axis due to changes in the forward speed. These forces are the lift forces acting on the aircraft and by differentiating the lift equation $L = C_L \frac{1}{2}\rho V^2 S$ with respect to speed and making the result non-dimensional, it can be shown that

$$z_u = -C_L \tag{6.7}$$

As can be seen from eqn. (6.7), the value of z_u is usually negative. An increase in the forward speed produces an increase in the lift force, but due to the sign convention used, this is equivalent to an increase in the negative Z force. The above differentiation assumes, however, that the lift coefficient remains constant with change in forward speed (everything else, of course, including the angle of attack, remaining constant). This is only true in the absence of compressibility effects; when these effects are allowed for

$$z_u = -C_L - \frac{M}{2}\frac{\partial C_L}{\partial M} \tag{6.8}$$

(*c*) m_u This derivative represents the rate of change of the pitching moment with change in the forward speed. In the absence of aeroelastic and compressibility effects the value is zero; a change in the forward speed changes all the pitching moments acting on the aircraft in the same proportion and since they are all trimmed out in the initial condition, they remain trimmed out. As was pointed out in Chapter 2, when the simple theory of longitudinal static stability was being considered, although both angle of attack and speed changes were involved, it was possible to consider the stability as resulting only from the disturbance in angle of attack.

Aeroelastic and compressibility effects may, however, produce significant contributions to m_u (Ref. 2, page 151), and in the more general theory of static stability, it is necessary to take into account both angle of attack and speed effects (see Chapters 2 and 15).

3.2. Vertical Velocity (or Angle of Attack) Derivatives

(*a*) $x_w(x_\alpha)$ This derivative represents the rate of change in the X-force with respect to changes in vertical velocity or angle of attack. By considering the lift and drag forces acting on the aircraft, it can be shown that

$$x_w = \tfrac{1}{2}\left(C_L - \frac{\partial C_D}{\partial \alpha}\right) \tag{6.9}$$

A vertical upgust thus produces two components to the X-force, a positive one resulting from the forward inclination of the lift vector and a negative one resulting from the increased drag.

(*b*) $z_w(z_\alpha)$ This represents the rate of change of the Z-force with respect to vertical velocity or angle of attack changes. By similar considerations to those in (*a*) above, an increase in the vertical velocity produces two positive increments in the lift (i.e. negative Z) force, one resulting from the rotation of the drag vector and the other resulting from the increased lift coefficient

$$z_w = -\tfrac{1}{2}\left(C_D + \frac{\partial C_L}{\partial \alpha}\right) \tag{6.10}$$

Since C_D is small compared with $\partial C_L/\partial\alpha$, this is usually approximated to

$$z_w = -\tfrac{1}{2}\frac{\partial C_L}{\partial \alpha} \tag{6.11}$$

(*c*) $m_w(m_\alpha)$ This represents the rate of change of pitching moment with vertical velocity or angle of attack changes. This is a very important derivative, since it is a measure of the static stability of the aircraft with respect to incidence changes. It can be expressed in terms of the static margin H_n (see Chapter 2) as

$$m_w = \frac{1}{2}\frac{\bar{c}}{l'}\frac{\partial C_M}{\partial \alpha} = -\frac{1}{2}\frac{\bar{c}}{l'}aH_n \qquad (6.12)$$

3.3. Vertical Acceleration Derivatives

Though they have not so far been mentioned, it is, strictly speaking, necessary to allow for all the derivatives resulting from disturbances occurring in the linear accelerations, as well as those from the linear velocities. They are, however, only important is oscillatory motions; they do not affect the static stability, since this is only concerned with steady motions. As far as the longitudinal motion of an aircraft is concerned, only one of the acceleration derivatives is of importance. This derivative $m_{\dot{w}}$ is often called the "downwash-lag" derivative, because one of the main reasons for its existence is the fact that the downwash from the wing takes a finite time to reach the tailplane. There are, however, other possible contributions to the overall value of $m_{\dot{w}}$.

On a conventional aircraft, $m_{\dot{w}}$, is usually negative and it has the effect of increasing the tailplane damping coefficient. On a tailless configuration, the value of the derivative will be small, and at transonic and supersonic speeds it may even become positive (Ref. 5).

3.4. Rate of Pitch Derivatives

(*a*) x_q This represents the rate of change of the X-force with rate of pitch and is usually small enough to be neglected.

(*b*) z_q This represents the rate of change of the Z-force with respect to the rate of pitch. It is usually negative, the main contribution coming from the horizontal tailplane. In Chapter 3, when considering manoeuvre stability, it was shown that a positive rate of pitch produced an increase in the angle of attack of the tailplane, which increased the lift force acting on the tailplane. This tailplane contribution can be shown to be given by $z_q = -\frac{1}{2}a_1S'/S$, where a_1 is the tailplane lift curve slope and S' is the tailplane area.

The z_q derivative is not in itself very important and it is often neglected in the longitudinal equations of motion. The pitching moment produced by the change in the load on the tailplane, and hence its corresponding derivative m_q is, however, extremely important.

(*c*) m_q This is the "damping-in-pitch" derivative and it represents the rate of change in the pitching moment due to the rate of pitch

(Fig. 6.1). It is usually negative (i.e. a positive damping effect), the main contribution coming from the horizontal tailplane. This particular contribution to the overall value of m_q can be shown to be equal to $-\frac{1}{2}a_1S'/S$ (i.e. as a result of the definitions used, it is numerically equal to the tailplane z_q component above).

The wing produces a smaller contribution to m_q, and on a tailless configuration of course, this is all there is. The derivative is an important factor in considerations of longitudinal dynamic stability,

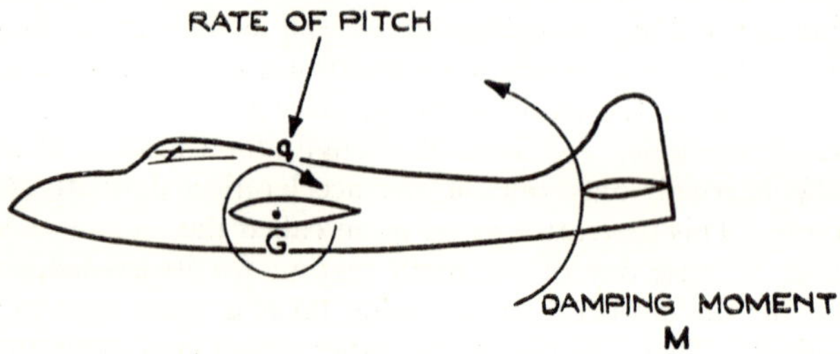

Fig. 6.1. Damping in pitch

and this explains why a tailless aircraft may have very poor damping characteristics under certain conditions of flight (see Chapter 15).

3.5. Elevator Deflection Derivatives

(*a*) x_η This represents the rate of change in the *X*-force with elevator deflection. It depends on the drag characteristics of the elevators, but is usually small enough to be neglected.

(*b*) z_η This derivative is a measure of the effectiveness of the elevator in producing a change in the lift force acting on the tailplane. A positive elevator deflection produces an increase of lift (i.e. a negative *Z*-force) and thus z_η is usually negative. It is given by $z_\eta = -\frac{1}{2}a_2S'/S$ where $a_2 = \partial C_L'/\partial\eta$. It is often possible to neglect the term containing z_η in the *Z*-force equation, but, in the case of a tailless aircraft, the change in wing lift produced by elevator deflection might become appreciable.

(*c*) m_η This derivative is numerically equal to that of z_η. The lift force on the tailplane resulting from the elevator deflection produces a pitching moment about the C.G. of the aircraft. m_η is a measure of the elevator control effectiveness or power. It is usually negative, a positive elevator deflection, for example, resulting in a nose-down (i.e. negative) pitching moment.

4. Control-fixed Lateral Derivatives

The most important lateral derivatives will now be discussed in a similar way.

4.1. Sideslip Velocity (or Sideslip Angle) Derivatives

(*a*) $y_v(y_\beta)$ This derivative represents the rate of change in the sideforce acting on the aircraft with change in sideslip velocity or sideslip angle. When an aircraft is sideslipping, sideforces are produced on all the various parts of the aircraft, the most important contributions being those from the fuselage and the vertical fin. y_v is negative, a positive (i.e. starboard) sideslip, for example, producing a negative sideforce (i.e. to port).

(*b*) $n_v(n_\beta)$ The above sideforce is also important in that it produces a yawing moment about the C.G. of the aircraft. The rate of change in the yawing moment with sideslip is called the "Weathercock effect"; it has already been discussed in Chapter 5. Sideforces acting ahead of the C.G. are destabilizing (i.e. n_v negative), because they tend to yaw the aircraft so as to increase the sideslip, whereas sideforces acting behind the C.G. are stabilizing (i.e. n_v positive).

The overall contribution of the fuselage is usually a destabilizing one, since the resultant sideforce acts at a point in front of the C.G. of the aircraft. The contribution of the vertical fin is obviously a positive one and the fin should be designed so that the complete aircraft possesses directional stability.

The effectiveness of the fin depends on the value of its lift-curve slope. This decreases at supersonic speeds and since the fuselage contribution remains almost unchanged, the directional stability of the aircraft will become low in this speed régime.

(*c*) $l_v(l_\beta)$ This represents the rate of change in rolling moment with sideslip velocity or angle. It is usually called the "Dihedral effect," and has already been discussed in Chapter 5. It is normally negative, the main contributions coming from the wing (due to its geometric dihedral and sweepback), the fuselage and possibly the vertical fin. On some aircraft, under certain conditions of flight, such as on the approach or at transonic speeds, the derivative reverses its sign and becomes positive. This means that the usual method of applying top rudder to pick up a dropped wing cannot be used. If the starboard wing drops, port rudder will produce a starboard sideslip which, in this case, will increase the starboard wing drop.

4.2. Rate of Roll Derivatives

(*a*) y_p This represents the rate of change in sideforce with rate of roll. It is usually small enough to be neglected, though a high vertical fin might produce a negative contribution.

(*b*) l_p This derivative represents the "damping-in-roll" effect, i.e. the rate of change of rolling moment due to a rate of roll. It is usually negative (i.e. damping), but, under certain conditions of

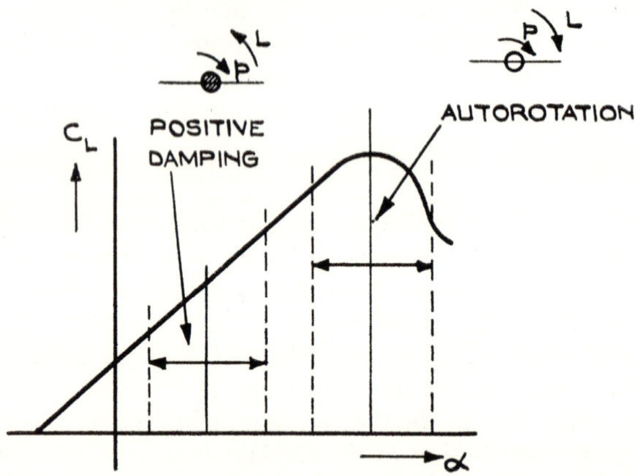

Fig. 6.2. Damping in roll

flight, it may become positive, and "autorotation" will commence, the change in the rolling moments produced assisting the rotation.

The main contribution is from the wing, though, once again, a high vertical fin might produce an additional effect. On the down-going wing the effective angle of attack is increased, and on the up-going wing decreased. As long as the aircraft is not flying near the stall, this changes the lift acting on the wings in such a way that a rolling moment is produced opposing the initial disturbance (Fig. 6.2). If the wings are near the stall, then the opposite effect is produced, and autorotation will occur (Fig. 6.2).

(*c*) n_p This represents the rate of change in the yawing moment due to a rate of roll. It is one of the cross-derivatives that cause the coupling of the rolling and yawing motions of an aircraft. The separate contributions to n_p consist of both positive (i.e. a roll to starboard producing a yaw to starboard) and negative (i.e. a roll to starboard producing a yaw to port) values; the overall value is usually negative, but it can be negligibly small, or even positive.

The main contribution comes from the wing, though the vertical fin may produce an appreciable positive contribution. The wing contribution can be considered to be made up of two parts, usually of opposite sign, the larger of the two being a negative one. When the aircraft is in a roll to starboard, the increased angle of attack on the downgoing wing increases both the lift and the drag, whereas on the upgoing wing the lift and drag are reduced. It should be noted, however, that the lift and drag forces are, by definition, perpendicular and parallel, respectively, to the local relative airflow on each wing. On the starboard wing the lift and drag forces are

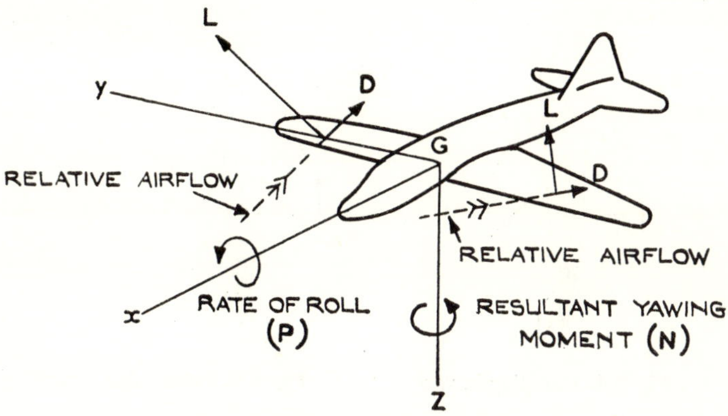

Fig. 6.3. Yawing moment due to roll

thus rotated in the nose-down direction and on the port wing in the nose-up direction (Fig. 6.3). The projections of the drag forces on the *xOy* plane usually produce a yawing moment to starboard (i.e. a positive n_p contribution), whereas the projections of the lift forces produce a yawing moment to port (i.e. a negative n_p contribution). The wing contribution is directly proportional to the value of C_L; it decreases with increasing speed, since, for a given rate of roll, the change in angle of attack will be less at the higher forward speed.

The overall value of n_p is thus usually negative, the numerical value decreasing with increasing speed and possibly, at the higher speeds, becoming positive.

4.3. Rate of Yaw Derivatives

(*a*) y_r This derivative represents the rate of change in the sideforce with respect to the rate of yaw. It is usually small enough to be neglected in the sideforce equation; areas behind the C.G. produce a positive contribution and areas in front of the C.G. a negative one.

(*b*) n_r Although the sideforces from y_r are neglected in the sideforce equation, they produce a very important moment about the C.G. of the aircraft. This is the "damping-in-yaw" moment, and n_r represents the rate of change in yawing moment with rate of yaw (Fig. 6.4). The direction of the sideforces in (*a*) above are such that the moment always opposes the direction of rotation, i.e. all contributions to n_r are negative ones. The main contributions come from the fuselage and the vertical fin of the aircraft. There is a further contribution to the overall value of n_r from the wing; the outer

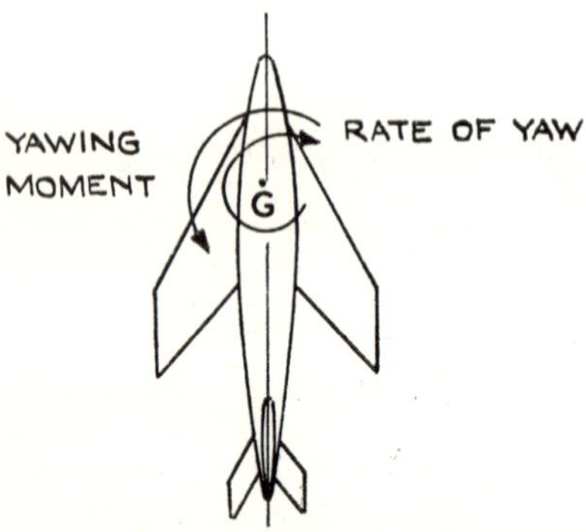

Fig. 6.4. Damping in yaw

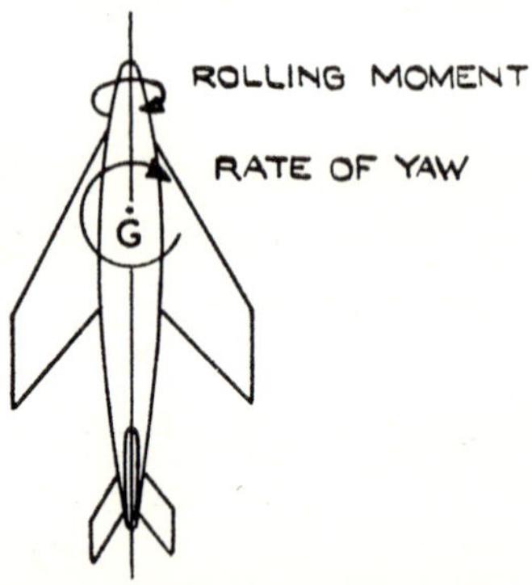

Fig. 6.5. Rolling due to yawing

wing moves faster than the inner wing and therefore has a greater drag acting on it. This produces an additional yawing moment opposing the yawing motion; this effect depends on the speed at which the aircraft is flying and increases with increase in C_L. Since, however, this is a relatively small contribution to n_r, the overall value of this derivative does not usually vary very rapidly with C_L.

(*c*) l_r This represents the rate of change in rolling moment with rate of yaw (Fig. 6.5). The main contribution comes from the wing, but the effect of a large vertical fin might be important. When an aircraft is yawing to starboard, the port wing moves faster than the starboard one. The change in the lift on the wings produces a rolling moment to starboard. Thus l_r has a positive value; the wing contribution is proportional to C_L and it increases with decrease of speed.

4.4. Aileron Deflection Derivatives

(*a*) y_ξ This derivative represents the rate of change in sideforce with change in aileron deflection; it is usually neglected.

(*b*) l_ξ This represents the change in the rolling moment with respect to aileron deflection; it is a measure of the aileron effectiveness or power. The sign convention is such that starboard aileron down is a positive deflection; l_ξ is thus usually negative (Fig. 6.6).

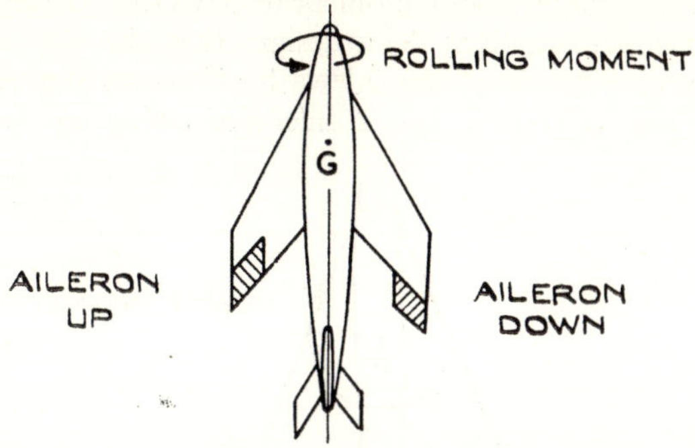

Fig. 6.6. Aileron power

Aileron reversal may occur due to aeroelastic effects, l_ξ is then positive.

(*c*) n_ξ This represents the change in the yawing moment due to aileron deflection. When an aileron is deflected downwards, both the profile and induced drag of the wing are increased. On the aileron-up wing, on the other hand, the induced drag will be decreased,

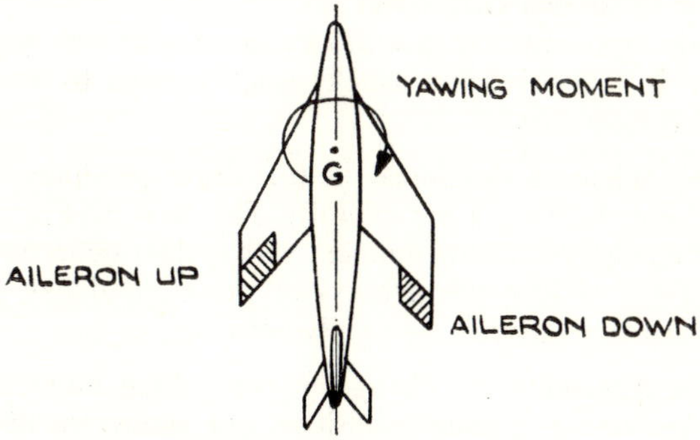

Fig. 6.7. Adverse aileron yaw

though the profile drag may still be increased. The changes in the drag forces acting are such as to produce a yawing moment. The derivative n_ξ is a positive one, since starboard aileron down produces a yaw to starboard (Fig. 6.7). This is called the "adverse aileron yaw" effect, because it causes the aircraft to yaw in the wrong direction when the ailerons are used to bank the aircraft into a turn. The adverse yaw is reduced by the use of differential and/or Frise-type ailerons (see Chapter 10). Often the rudder is geared to the ailerons and when the ailerons are deflected, the rudder moves so as to produce a suitable yawing moment (e.g. Victor, Crusader).

On some modern tailless delta designs (e.g. the B58 Hustler) n_ξ is negative, as a result of the sidewash effects on the vertical fin. When the port aileron is up, a sideforce acting to starboard is

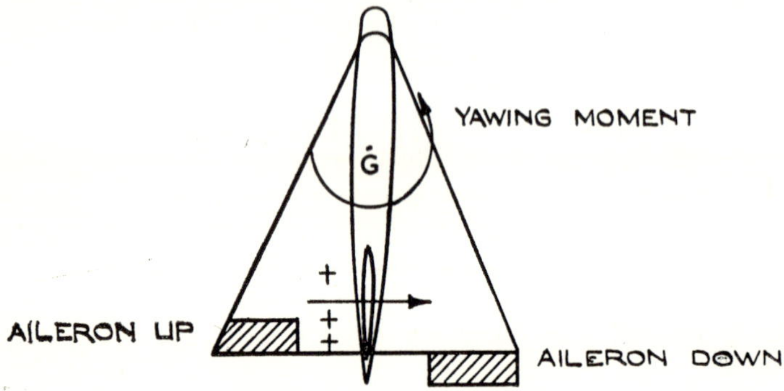

Fig. 6.8. Favourable aileron yaw

developed on the vertical fin and this produces a yaw to port (Fig. 6.8).

4.5. Rudder Deflection Derivatives

(*a*) y_ζ This represents the rate of change of sideforce with rudder deflection; it is positive and usually small enough to be neglected in the Y-force equation.

(*b*) n_ζ The sideforce developed in (*a*) above produces a yawing moment about the C.G. of the aircraft. The derivative n_ζ represents the rate of change of yawing moment with rudder deflection; it is a measure of the rudder effectiveness or power, and like the other control power derivatives it is usually negative.

(*c*) l_ζ This represents the change in the rolling moment due to rudder deflection. It is usually positive, but, unless the aircraft has a high vertical fin, it is often small enough to be neglected. The value of the derivative decreases with increase in C_L, since the increase in incidence will decrease the vertical distance between the resultant sideforce on the fin and the C.G. of the aircraft.

In cases where l_ζ is large (e.g. Belfast and Britannia) the aileron control surfaces are geared to the rudder to counteract the adverse rolling effect.

5. Various Factors Affecting the Derivatives

As was mentioned in the introduction, the derivatives are not constants. Even ignoring the Reynolds Number effects, the derivatives are still functions of

(i) *Aircraft design* Fairly obviously, from what has been said, the values of the various derivatives are very dependent on the design layout of the aircraft. Most of the discussion in this chapter has referred to the conventional rear tailplane layout, though the tailless delta configuration has been mentioned in one or two special cases. The derivatives are made up of a number of separate contributions, and, in the case of the longitudinal derivatives, the tailplane contribution is often the most important of these. On a tailless configuration the tailplane contributions are absent and the wing contributions become the predominant ones. In para. 4.4(*c*) above it was shown that a change of layout could, for instance, change the derivative n_ξ from its usual positive value to a negative one.

(ii) *Aircraft configuration and power* The values of many of the derivatives are very dependent on the configuration of the aircraft (e.g. flaps and undercarriage) and the power being used. These effects have been discussed for the l_v and n_v derivatives in Chapter 5.

(iii) *Speed and height* At low speeds, in the absence of aeroelastic and compressibility effects, many of the derivatives can be regarded as constants, independent of the speed and height at which the aircraft is flying (e.g. y_v, n_ζ). Other derivatives, however (or, to be precise, certain of the separate contributions of other derivatives) are, as we have seen, functions of C_L. The derivatives can then be expressed in equation form in terms of C_L or $C_L{}^2$, as appropriate. The longitudinal derivative z_u, for example, is given approximately by $-C_L$, and the expressions for the lateral derivatives l_v, l_r, n_p and n_r of the Jet Provost and Meteor 7 are given in Table 6.4.

Table 6.4

	JET PROVOST	METEOR 7
l_v	$-0{\cdot}05$	$-0{\cdot}052 - 0{\cdot}009C_L$
l_r	$0{\cdot}0322 + 0{\cdot}226C_L$	$0{\cdot}0295 + 0{\cdot}159C_L$
n_p	$-0{\cdot}03C_L$	$-0{\cdot}042C_L$
n_r	$-0{\cdot}077 - 0{\cdot}0121C_L{}^2$	$-0{\cdot}182 - 0{\cdot}025C_L{}^2$

The variation with C_L can be shown very conveniently in graphical form; graphs can be drawn showing the variation of the separate contributions, as well as that of the overall value, of the derivatives. The graphical presentation is more useful than the equational form of the derivatives, in that it can be used to show any tendency for the values of the derivatives (for instance, at the high angles of attack, near the stall) to depart from those predicted by the equation. Fig. 6.9 shows the variation of n_v with C_L for a swept-wing fighter-type

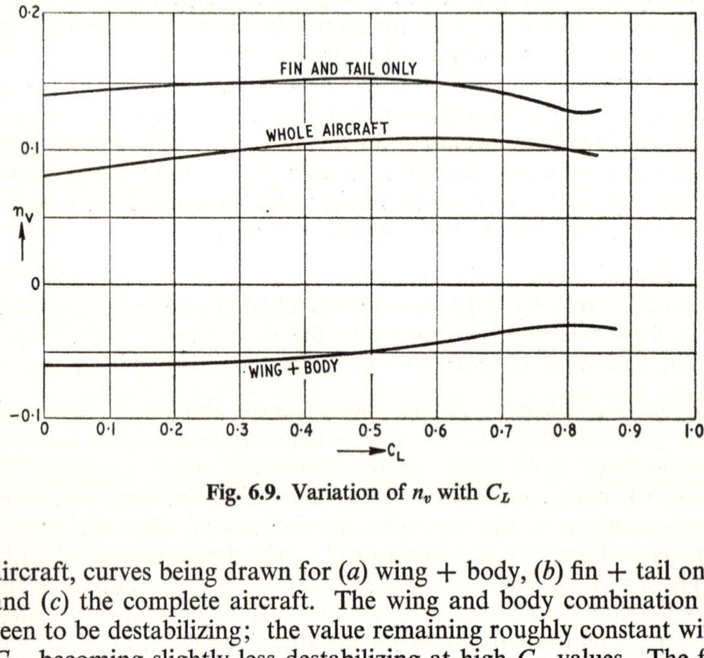

Fig. 6.9. Variation of n_v with C_L

aircraft, curves being drawn for (*a*) wing + body, (*b*) fin + tail only and (*c*) the complete aircraft. The wing and body combination is seen to be destabilizing; the value remaining roughly constant with C_L, becoming slightly less destabilizing at high C_L values. The fin + tail combination is stabilizing, the effect decreasing slightly at high values of C_L. The overall value of n_v for the complete aircraft varies little with C_L. The l_v curves for the same aircraft are shown in Fig. 6.10. The wing + body combination is slightly destabilizing (i.e. l_v positive) at very low values of C_L, but l_v soon becomes negative as C_L increases, becoming more and more negative until a high C_L value is reached. The fin + tail combination has a fairly small negative contribution to l_v at low values of C_L, becoming less negative (i.e. dihedral effect decreasing) with increase of C_L. The overall value for the complete aircraft varies in a similar way to that of the wing + body combination, but it remains negative for all values of C_L. The values of l_r, l_ζ, n_p and n_r for the complete aircraft are shown in Fig. 6.11. It is interesting to note that, in this case, n_p becomes positive as the speed is increased (see para. 4.2 (c)).

At the higher forward speeds, the problem becomes more complicated, because the derivatives are now functions of C_L, Mach number

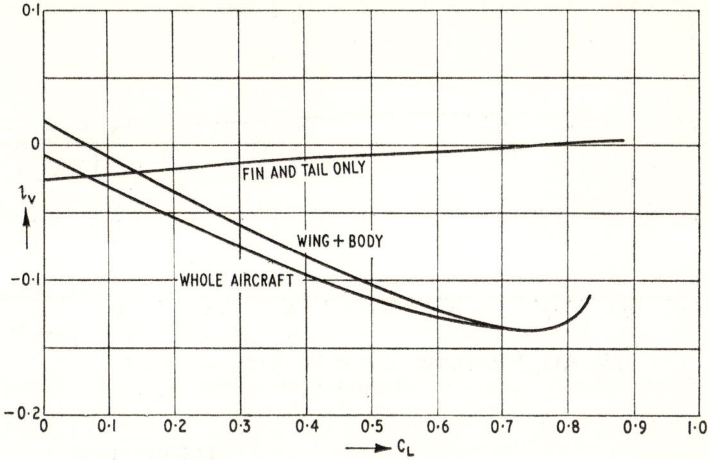

Fig. 6.10. Variation of l_v with C_L

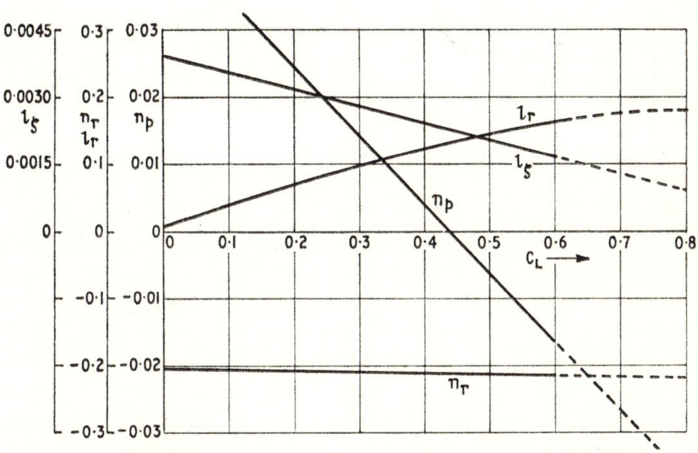

Fig. 6.11. Variation of l_r, n_r, l_ζ and n_p with C_L

and $\frac{1}{2}\rho_0 V_i^2$. It is very difficult to distinguish between compressibility and aeroelastic effects; an increase in speed at constant height will involve both effects. In an effort to separate the effects as far as possible, flight tests are carried out over the speed range at varying heights. Typical curves showing the variation of some of the longitudinal and lateral derivatives of the Fairey Delta 2 (taken from Refs. 6 and 7) are shown in Figs. 6.12 and 6.13.

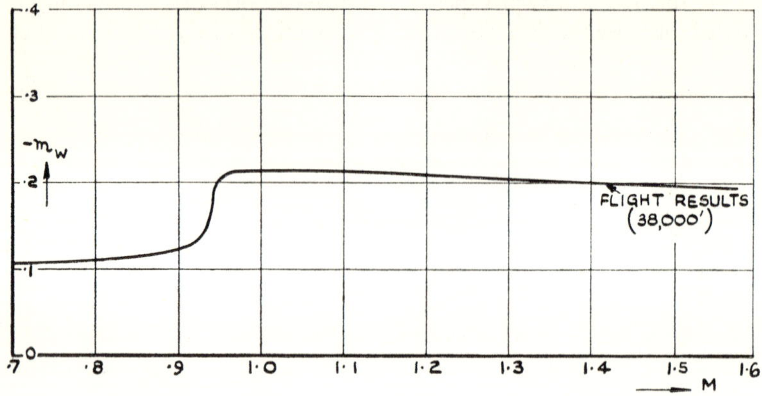

Fig. 6.12. Variation of m_w with Mach number for the F.D.2 (From Ref. 6)

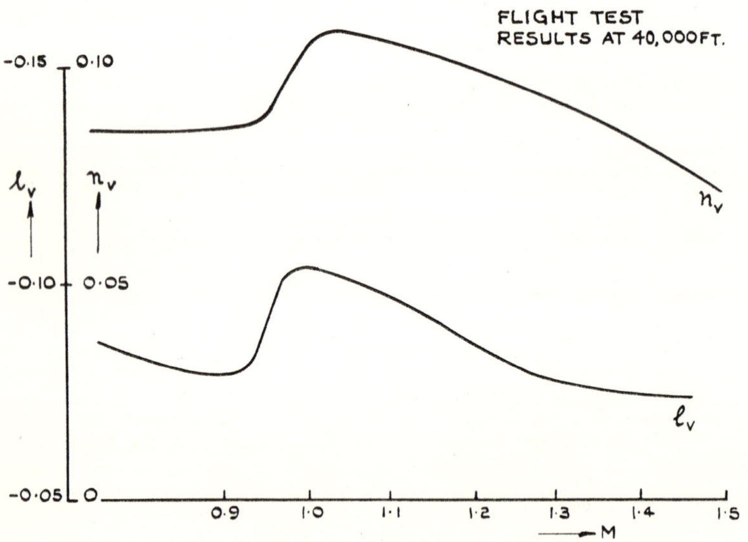

Fig. 6.13. Variation of l_v and n_v with Mach number for the F.D.2 (From Ref. 7)

As we shall see later (Chapters 15 and 16), the variation in the values of the derivatives at high Mach numbers may lead to serious stability and control problems.

6. Determination of the Values of the Derivatives

During the development stages of an aircraft it is important to know the numerical values of the derivatives of the aircraft under various conditions of flight. In the project design stage it will be necessary to depend on theoretical estimates, supported, as far as possible, by the results of wind tunnel and free-flight model experiments. Analogue computers (see Chapter 18) can then be used to calculate the likely response characteristics of the aircraft and to determine whether they will be satisfactory or not. When the aircraft has been built, flight tests can then be carried out to confirm, or otherwise, the theoretical and experimental values of the derivatives. The corrected values of the derivatives can then be used on an analogue computer to help to extend the flight test programme into the flight régimes where serious problems are anticipated. This method is particularly useful when the aircraft is being checked for occurrences which could turn out to be disastrous in flight, e.g. autopilot runaways, inertia or roll coupling effects, etc.

Flight test methods for determining the values of the derivatives are discussed in Chapter 17 (see also Refs. 4, 5, 6, 7, 8, and 9). Theoretical methods of estimation are discussed in Refs. 10 [this report contains a large bibliography of 158 relevant reports], 11, 12, 13 and 14, and wind tunnel methods are discussed in Ref. 15.

References

1. DUNCAN, W. J.; *Control and Stability of Aircraft* (Cambridge University Press, 1952).
2. ETKIN, B.; *Dynamics of Flight* (John Wiley and Sons, New York, 1959).
3. KOLK, W. R.; *Modern Flight Dynamics* (Prentice-Hall International, London, 1961).
4. PERRY, D. H., MORRALL, J. C. and PORT, W. G. A.; *Low Speed Flight Tests on a Tailless Delta Aircraft with Wings Swept Back* 44·5° (*Avro* 707*B*), Pt. 3, "Lateral Stability and Control," Unpublished M.O.A. Report.
5. DOETSCH, K. H.; *The Time-vector Method for Stability Investigations*, R.A.E. Rep. Aero 2495, R. & M. 2945, 1957.
6. ANDREWS, D. R.; *Measurements in Flight of the Longitudinal Stability Derivatives of a* 60° *Delta Wing Aircraft* (*Fairey ER* 103), R.A.E. Rep. Aero 2617, C.P. 639, 1959.
7. ROSE, R.; *Flight Measurements of the Dutch-Roll Characteristics of a* 60° *Delta Wing Aircraft* (*Fairey Delta* 2) *at Mach Numbers from* 0·4 *to* 1·5 *with Stability Derivatives Extracted by Vector Analysis*, R.A.E. Tech. Note Aero 2752, CP 653, 1961.
8. ZBROZEK, J. K.; *On the extraction of stability derivatives from full-scale flight data*, Agard Rep. No. 190, 1958.
9. WOLOWICZ, C. H. and HOLLEMAN, E. C.; *Stability-derivative determination from flight data*, Agard Rep. No. 224, 1958.
10. THOMAS, H. H. B. M.; *State of the art of estimation of derivatives*, Agard Rep. No. 339, 1961.

11. ACUM, W. E. A. and GARNER, H. C.; *The estimation of oscillatory wing and control derivatives*, Agard Rep. No. 340, 1961.
12. MALTHAN, L. V. and HOAK, D. E.; *Current progress in the estimation of stability derivatives*, Agard Rep. No. 341, 1961.
13. GERSTEN, K.; *Calculation of non-linear aerodynamic stability derivatives of aeroplanes*, Agard Rep. No. 342, 1961.
14. TOBAK, M. and LESSING, H. C.; *Estimation of rotary stability derivatives at subsonic and transonic speeds*, Agard Rep. 343, 1961.
15. WILEY, H. G. and BRASLOW, A. L.; *A method of accurately measuring dynamic stability derivatives in transonic and supersonic wind tunnels*, Agard Rep. No. 345, 1961.

7 Introduction to Dynamic Stability

1. Introduction

The static stability of an aircraft is concerned with the tendency of the aircraft to return to, or diverge from, its original equilibrium condition, following a small disturbance from that condition. This initial tendency is in fact only a small part of the overall dynamic motion of the aircraft. From a full analysis of the motion of the aircraft both the "static" and the "dynamic" stabilities of the aircraft can be determined. It is, however, convenient to consider the problem of static stability separately; this reduces the number of variables involved and enables a clearer understanding of the principles involved. The "static" derivatives (e.g. l_v, n_v) can then be obtained from steady flight tests (e.g. steady sideslips) and the values used to help to analyse the more complicated overall dynamic motion.

Before considering the dynamic stability characteristics of an aircraft in detail, we will discuss some of the basic principles of oscillatory motions and define some of the various ways in which an oscillation can be described in mathematical terms.

2. Simple Oscillatory Motions

2.1. General

The problem of the dynamic motion of an aircraft is a very involved one; the aircraft has complete freedom in space and its motion has to be referred to a chosen set of three mutually perpendicular axes. It has six degrees of freedom, three of linear motion along the axes and three of rotation about the axes. Various possible oscillatory modes occur, which involve these degrees of freedom in different proportions. Fortunately, when discussing any of the oscillatory modes of the aircraft, it is usually possible to neglect the majority of these degrees of freedom and thus to simplify the analysis. Very rarely, however, is it possible to reduce any of our problems to one involving only one degree of freedom, but some analyses using two or three degrees of freedom produce quite useful results.

These results are, of course, only approximations to the exact results that can be obtained by a full analysis, but they are extremely valuable in providing formulae which, though not exact, are often sufficiently accurate for the purposes required and also in providing a better understanding of the physical nature of the problem.

Before discussing motions involving two, three, or more degrees of freedom, it is very instructive to analyse a motion involving only one degree of freedom. The mathematical analysis involved is, in fact, applicable to various types of systems and it is possible to represent an aircraft by an analogous mechanical or electrical system (see Chapters 18 and 19). We shall see, for instance, that the moment of inertia of an aircraft oscillating about one of its axes is analogous to the mass in a simple spring-mass system or to the inductance in an electrical circuit.

2.2. Spring-Mass System with Zero Damping

The simplest example that can be considered is that of a mass free to oscillate vertically on the end of a spring (Fig. 7.1). If we assume that

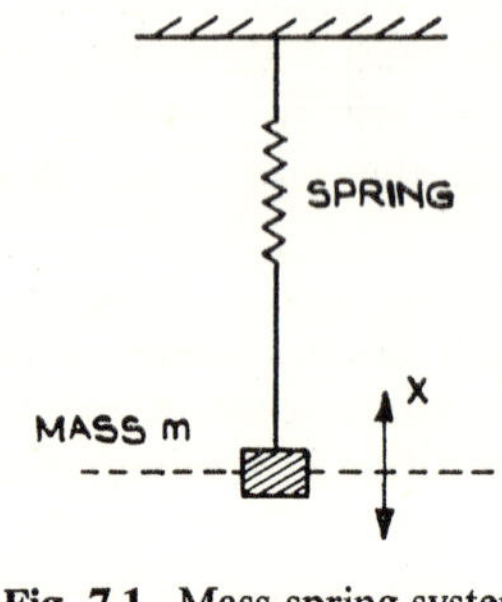

Fig. 7.1. Mass-spring system

the mass is restrained in some way so that it can only move up or down and that there is no damping whatsoever in the system, then, once the mass has been displaced and released, it will oscillate up and down, the amplitude of the oscillation remaining constant with time (Fig. 7.2). Obviously, this system is an imaginary one, since, in practice, the oscillations would gradually die away to zero, since damping cannot be eliminated altogether, but it is still a very useful system with which to begin our analysis.

Consider the spring in its unloaded condition (Fig. 7.3*a*). When a weight W is hung on the end, then, in the static condition, the spring will be extended (Fig. 7.3*b*) by an amount *e*. The stiffness coefficient *k* of the spring can be defined by

$$k = \frac{W}{e} \tag{7.1}$$

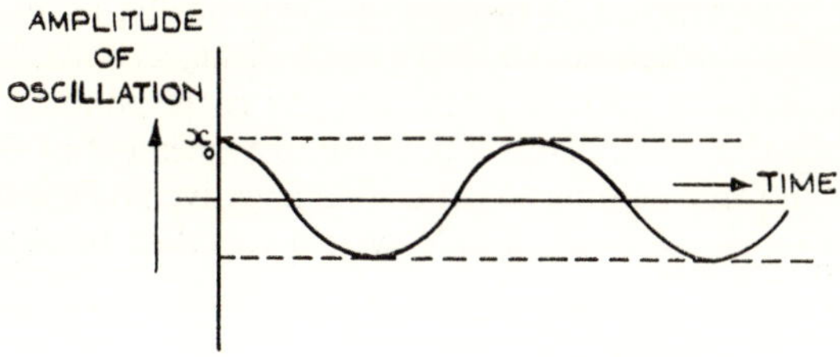

Fig. 7.2. An undamped oscillation

This equation relates the tension in the spring to the amount the spring is extended, since in the steady state shown in Fig. 7.3*b* the tension in the spring must be equal to the weight *W*.

If the weight is now displaced through a further distance *x* (Fig. 7.3*c*), the spring extends and the tension in the spring increases.

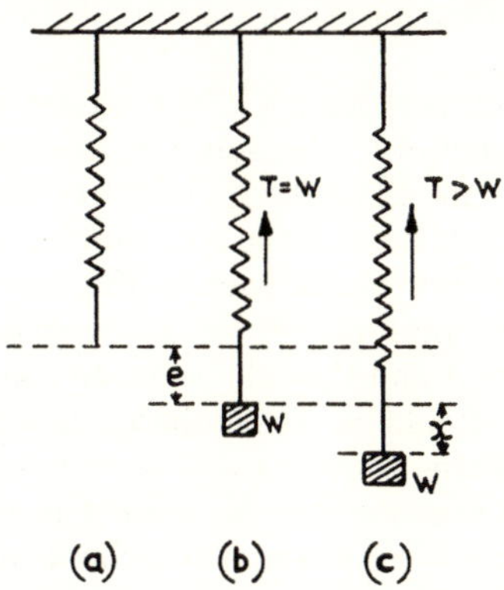

Fig. 7.3. Tension in the spring

This extra tension always acts as a restoring force, tending to return the weight towards the equilibrium position, so that, immediately the weight is released, it will move back towards the equilibrium position again. Thus it can be seen that in our analogy, the tension in the spring represents the static stability of an aircraft.

On reaching the equilibrium position the weight will continue moving upwards. The tension in the spring will now decrease, becoming less than the weight. A resultant force will now act vertically downwards, slowing the weight up, until eventually it comes to rest, and starts moving downwards again. With zero damping in the system, this will continue indefinitely, the amplitude of the motion being constant on either side of the equilibrium position, i.e. it is an undamped simple harmonic motion.

In order to represent this motion mathematically, let us, for convenience, select to denote all downward displacements, velocities, and accelerations as positive ones, and the corresponding upwards values as negative ones. Thus a positive displacement of the weight will result in a force tending to produce a negative acceleration. The equation of motion for this simple system can then be expressed in the form

$$kx = -\frac{W}{g}\ddot{x} = -m\ddot{x} \tag{7.2}$$

(i.e. restoring force = mass × acceleration).

Rearranging eqn. (7.2) we get

$$m\ddot{x} + kx = 0 \tag{7.3}$$

where m is the mass, k the spring stiffness coefficient (i.e. the restoring force per unit displacement), x the displacement, and $\ddot{x} = d^2x/dt^2$ the acceleration. To be consistent, m must be in slugs, $\ddot{x}$ in ft/sec², k in lb/ft and x in ft.

The equation of motion [eqn. (7.3)] is in the form of a second-order differential equation*. Its solution can be written down in various ways, which are all, in fact, closely related to one another. The function of x which satisfies the equation has got to be one that, when differentiated twice, comes back to the same function. Either a sine function, a cosine function, or an exponential function will satisfy this requirement and thus the solution can be written in any of these forms. It should be remembered however that a cosine function is simply a sine function with a phase change of 90° and that an exponential function can be written in terms of sines and cosines (e.g. $e^{i\theta} = \cos\theta + i\sin\theta$). For simplicity we will use the cosine form of the function

$$x = x_0 \cos\sqrt{\left(\frac{k}{m}\right)}\, t \tag{7.4}$$

* *Note.* Differential equations are categorized by "order" and "degree." The "order" is the largest number of times the dependent variable is differentiated and the "degree" is the highest power to which the variable is raised (or, when there is more than one variable, the largest number of variables in any product). Eqn. (7.3) is thus a second order differential equation of the first degree. First degree differential equations are often referred to as linear differential equations and analytical solutions can be found for them.

then x_0 represents the actual displacement at time $t = 0$. The frequency of the undamped motion is given by $\sqrt{(k/m)}$; representing this by ω_n we have

$$\omega_n = \sqrt{\left(\frac{k}{m}\right)} \text{ rad/sec} \tag{7.5}$$

and eqn. (7.4) can be written

$$x = x_0 \cos \omega_n t \tag{7.6}$$

Referring back to eqn. (7.3), we can see that, if the coefficient of $\ddot{x}$ is made equal to unity, then the coefficient of x is ω_n^2, i.e. $\ddot{x} + \omega_n^2 x = 0$.

The corresponding periodic time of the oscillation is given by

$$T = \frac{2\pi}{\omega_n} = \frac{2\pi}{\sqrt{(k/m)}} \text{ secs} \tag{7.7}$$

and it can be seen that the periodic time is directly proportional to $\sqrt{m}$ and inversely proportional to $\sqrt{k}$. It is thus dependent on the spring stiffness and is reduced by an increase in stiffness. This is an important result, since, in our analogy, spring stiffness is replacing the static stability. Hence, as we see later, the *static* stability of the aircraft is the major factor determining the periodic time of the oscillatory motion.

2.3. Spring-Mass System with Damping

Now let us consider the more practical case of a similar system containing a small amount of damping (Fig. 7.4); the corresponding

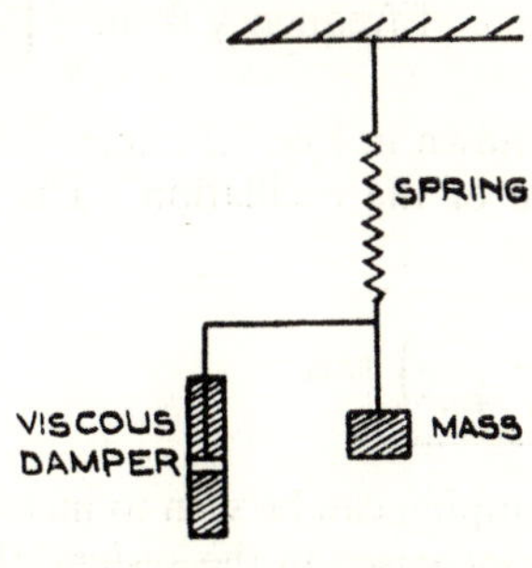

Fig. 7.4. Mass-spring-viscous damper system

motion of the mass will now consist of a damped oscillation (Fig. 7.5). If we assume that the damping force produced is proportional to the velocity of motion of the mass, then the equation of motion for this system can be written

$$m\ddot{x} + b\dot{x} + kx = 0 \qquad (7.8)$$

where b represents the damping coefficient (i.e. damping force per unit velocity). In this particular system, the force $b\dot{x}$ always opposes the motion; other cases exist in which the damping is negative.

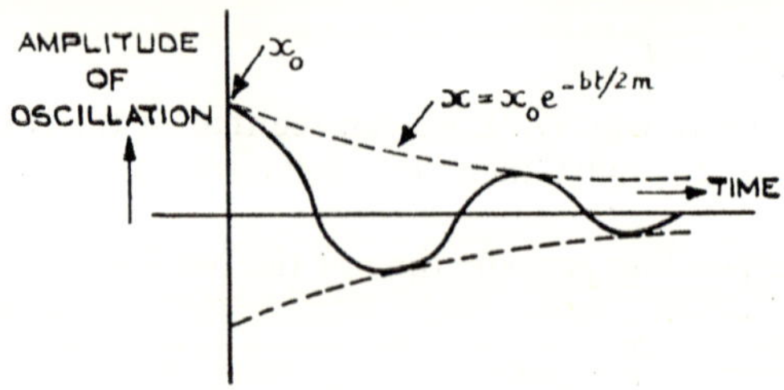

Fig. 7.5. A damped oscillation

The solution of eqn. (7.8) can be written in a similar form to eqn. (7.4)

$$x = x_0 e^{-(b/2m)t} \cos \sqrt{\left(\frac{k}{m} - \frac{b^2}{4m^2}\right)}\, t \qquad (7.9)$$

By comparing eqn. (7.9) and (7.4) we can see that the introduction of damping has

(i) resulted in an extra factor $e^{-(b/2m)t}$;

(ii) reduced the value of frequency from $\sqrt{\left(\frac{k}{m}\right)}$ to $\sqrt{\left(\frac{k}{m} - \frac{b^2}{4m^2}\right)}$.

The oscillation is shown in Fig. 7.5; $x_0 e^{-(b/2m)t}$ is the equation of the "envelope" curve of the oscillation. The periodic time of the oscillation is given by

$$T = 2\pi / \sqrt{\left(\frac{k}{m} - \frac{b^2}{4m^2}\right)} \text{ secs} \qquad (7.10)$$

and an increase of damping can be seen to increase the periodic time.

As the damping is increased in the system, the oscillatory motion will become more and more damped, until, eventually, a certain value of the damping will be reached at which the mass will no longer oscillate when released following a disturbance. Instead it will return very slowly towards the initial equilibrium position, the rate of approach decreasing the nearer it gets. This is a non-oscillatory or aperiodic motion (Fig. 7.6) and the smallest value of the damping

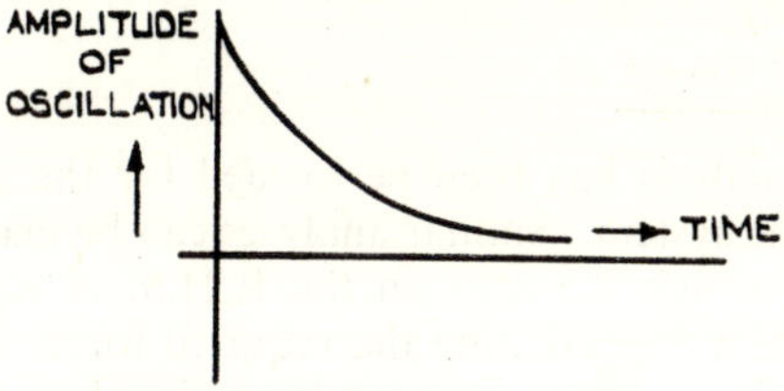

Fig. 7.6. "Dead-beat" motion

that will produce this result is called the critical damping b_{crit}; it is given by

$$\frac{b^2_{crit}}{4m^2} = \frac{k}{m} \tag{7.11}$$

$$b_{crit} = \sqrt{(4mk)} = 2\sqrt{(mk)} \tag{7.12}$$

Any further increase of damping results in a slower rate of decay of the aperiodic motion.

2.4. Alternative Forms of the Equations

One of the major difficulties with any work on stability and control is the fact that there are so many possible ways of writing the various equations involved. Two alternative methods, which occur frequently, are given below.

1st Alternative

Eqn. (7.8) can be rewritten in the form

$$\ddot{x} + 2\zeta\omega_n\dot{x} + \omega_n^2 x = 0 \tag{7.13}$$

or

$$(D^2 + 2\zeta\omega_n D + \omega_n^2)\, x = 0 \tag{7.14}$$

where $D^2 = d^2/dt^2$, $D = d/dt$, ω_n is the undamped natural frequency $= \sqrt{(k/m)}$ and ζ, the relative damping, is defined as $b/[2\sqrt{(mk)}] = b/b_{crit}$.

The corresponding solution then becomes

$$x = x_0 e^{-\zeta\omega_n t} \cos[\sqrt{(1 - \zeta^2)}]\omega_n t \tag{7.15}$$

2nd Alternative

Eqn. (7.8) can also be rewritten as

$$\ddot{x} + 2\mu\dot{x} + (\mu^2 + \omega^2)x = 0 \tag{7.16}$$

where μ is the damping index $= b/2m$ and ω is the damped frequency $= \sqrt{\left(\frac{k}{m} - \frac{b^2}{4m^2}\right)}$. The corresponding solution can then be expressed

$$x = x_0 e^{-\mu t} \cos \omega t \tag{7.17}$$

The above analysis has been performed for the so-called "free" oscillation of the system. Similar analyses can be made for "forced" oscillations, in which the zero on the R.H.S. of eqn. (7.8) can be replaced by a term representing the required force. However, most aircraft dynamic stability investigations involve the analysis of "free" or "transient" oscillations.

3. Oscillatory Motion of Aircraft

It is generally possible to break down the overall dynamic motion of an aircraft into the separate consideration of the longitudinal and lateral (including directional) motions. Though the problem is a more complicated one, it is basically similar. Both the longitudinal and lateral equations of motion can be reduced to a quartic equation in a quantity referred to as λ, by assuming that all the disturbances are exponential ones, of the form $x = x_0 e^{\lambda t}$. Thus, in each case, λ has four possible values and the values of λ determine the type of motion involved

(i) *λ real and negative*—this represents a damped (i.e. convergent) aperiodic motion.
(ii) *λ real and positive*—this represents a divergent aperiodic motion.
(iii) *λ complex, with a negative real part*—this represents a damped, or convergent, oscillation.
(iv) *λ complex, with a positive real part*—this represents an undamped, or divergent, oscillation.

When λ is complex it can be written in the form of $-(\mu \pm i\omega)$

$$x = x_0 e^{\lambda t} = x_0 e^{-(\mu \pm i\omega)t} \tag{7.18}$$

or, using the cosine function, as in eqn. (7.17). The quantity μ (the "real" part) represents the amount of damping and ω (the "imaginary" part) represents the frequency of the oscillation.

One further complication is involved, however, in the considerations of the dynamic stability of aircraft. The analyses are usually performed in terms of non-dimensional quantities. We have become familiar with non-dimensional lift and moment coefficients and with non-dimensional stability derivatives, but we must now also introduce what we can call a "non-dimensional weight" and a "non-dimensional time." The "non-dimensional weight," or the aircraft relative density parameter as it is called, can be defined for longitudinal motions by

$$\mu_1 = \frac{W}{g\rho S l'} = \frac{m}{\rho S l'} \tag{7.19}$$

where l' is the length of the tail arm. For lateral motions the definition is similar to this, but $s = \frac{1}{2}b$, the semi-span, replaces l' and the symbol μ_2 is used. The non-dimensional time in air-seconds is given by

$$\tau = \frac{t}{\hat{t}} \text{ air-seconds} \tag{7.20}$$

where t is the actual time in seconds and $\hat{t}$ is the "unit of aerodynamic time," also in seconds. This is defined by

$$\hat{t} = \frac{W}{g\rho S V} = \frac{m}{\rho S V} \text{ secs} \tag{7.21}$$

where V is the forward speed in ft/sec.

By comparing eqns. (7.19) and (7.21) we can see that

$$\hat{t} = \frac{\mu_1 l'}{V} \tag{7.22}$$

or for lateral motions

$$\hat{t} = \frac{\mu_2 b}{2V} \tag{7.23}$$

If eqn. (7.18) now represents the solution of the non-dimensional equations of motion, then the frequency and periodic time of the oscillation are given by

$$\left.\begin{aligned} \text{Frequency} &= \omega \text{ radians per air-second} \\ \text{Period } T &= \frac{2\pi}{\omega} \text{ air-seconds} \end{aligned}\right\} \tag{7.24}$$

or, in terms of ordinary time, by

$$\left.\begin{aligned} \text{Frequency} &= \frac{\omega}{\hat{t}} \text{ rads per sec} \\ \text{Period } T &= \frac{2\pi}{\omega}\hat{t} \text{ secs} \end{aligned}\right\} \tag{7.25}$$

The damping of the oscillatory motion can be expressed in a number of different ways; these are all related to one another and equations can be derived relating any one damping factor to any of the others. These relationships are given in Table 2 of Ref. 1.

(i) *Time to Half (or Double) Amplitude*
The equation of the envelope curve of the oscillation is given by

$$x = x_0 e^{-\mu\tau}$$

If μ is positive, the oscillation will be a damped one and the amplitude of the oscillation will have decreased to $\frac{1}{2}$ its initial value in a time $t_{1/2}$ given from

$$\tfrac{1}{2}x_0 = x_0 e^{-\mu\tau}$$

Hence

$$t_{1/2} = -\frac{\log_e \frac{1}{2}}{\mu}\hat{t} = \frac{\log_e 2}{\mu}\hat{t} \text{ secs} \tag{7.26}$$

If, on the other hand, μ is negative, the oscillation will be divergent. The time to double amplitude t_2 can be found in a similar way to be

$$t_2 = \frac{\log_e 2}{-\mu}\hat{t} \text{ secs} \tag{7.27}$$

Since $\log_e 2 = 0{\cdot}693$ we have

$$\text{Time to half (double) amplitude} = \frac{0{\cdot}693}{\pm\mu}\hat{t} \text{ secs} \tag{7.28}$$

(ii) *Time to 1/e of the Initial Amplitude*
This is often called the "natural" damping time or the "time constant" of the oscillation. Since $\log_e e = 1$, this definition eliminates the numerical constant 0·693 in (i) above, and we have

$$t_{1/e} = \frac{1}{\mu}\hat{t} \text{ secs} \tag{7.29}$$

(iii) *Logarithmic Decrement* δ
This quantity is often used to represent the characteristics of an oscillation. A pilot's assessment of the handling qualities is dependent

on both the damping and the periodic time and these two factors are combined in the definition of the log. dec.

$$\delta = \frac{T}{t_{1/e}} \quad \text{i.e.} \quad \frac{\text{the periodic time of the oscillation}}{\text{time for oscillation to decay to } 1/\text{e of initial amplitude}} \tag{7.30}$$

Thus we can express δ in a number of different ways

$$\delta = \frac{T}{t_{1/e}} = \frac{0{\cdot}693T}{t_{1/2}} = \frac{\mu T}{\hat{t}} = \frac{2\pi\mu}{\omega} \tag{7.31}$$

An alternative, but equivalent, definition of δ is given by

$$\delta = \log_e \left(\frac{\text{a peak value of the amplitude of the oscillation}}{\text{the succeeding peak value of the } \textit{same} \text{ sign}} \right) \tag{7.32}$$

(iv) *Attenuation Ratio per Half Cycle A*
This can be defined by

$$A = \frac{\text{a peak value of amplitude}}{\text{the preceding peak value of the } \textit{opposite} \text{ sign}} \tag{7.33}$$

and it is related to the log. dec. by

$$A = \mathrm{e}^{-\delta/2} \tag{7.34}$$

(v) *Relative Damping* ζ
This is defined by

$$\zeta = \frac{\text{the actual damping}}{\text{the critical damping}} \tag{7.35}$$

and is related to the log. dec. by

$$\zeta = \frac{\delta}{\sqrt{(\delta^2 + 4\pi^2)}} \tag{7.36}$$

(vi) *Number of Cycles to Half Amplitude* $N_{1/2}$
It can be shown that

$$N_{1/2} = \frac{\log_e 2}{\delta} \tag{7.37}$$

Table 1 shows some corresponding values of these various damping factors (taken from Ref. 1). The requirement laid down for oscillatory motion in Ref. 1 is that $N_{1/2}$ should not be more than 1 (i.e. δ should be greater than 0·693). In practice, this requirement is not always satisfied by modern aircraft, but they may still possess satisfactory handling characteristics.

Table 1

A	1·0	0·707	0·5	0
ζ	0	0·110	0·215	1·0
δ	0	0·693	1·39	∞
$N_{1/2}$	∞	1·0	0·5	0
	Neutrally damped oscillation	Satisfactorily damped oscillation	Well damped oscillation	Dead-beat oscillation

Reference

1. *Design Requirements for Aircraft for the Royal Air Force and Royal Navy*, Av. P. 970, Vol. 1, Ministry of Aviation, Leaflet 600/5.

8 Subsonic Longitudinal Dynamic Stability

1. Introduction

The dynamic stability characteristics of an aircraft with fixed controls can be represented by six equations of motion, three for the forces involved X, Y and Z and three for the moments L, M and N. Though these equations usually look rather complicated, they are basically fairly simple. The force equations relate the forces acting on the aircraft to the corresponding linear accelerations and the moment equations relate the moments to the corresponding angular accelerations. These equations can be expressed either in dimensional or non-dimensional form; the advantage of the non-dimensional method of representation is that the stability derivatives involved are independent of the linear scale of the aircraft and the system of units employed. This makes any comparison between different aircraft easier. Most standard textbooks perform the analysis of the motion of an aircraft in terms of non-dimensional parameters, but one exception to this is Ref. 6. The longitudinal equations of motion are given in non-dimensional form in Appendix B and in dimensional form in Appendix F.

It is usually possible to consider the longitudinal motion completely separately from the lateral-directional motion, by neglecting the various coupling terms. This reduces the number of equations that have to be considered to three (i.e. those for X, Z and M). The problem of coupling between the longitudinal and lateral motions of an aircraft is discussed in Chapter 16.

If the three variables involved (i.e. u, w and q) are assumed to be exponential functions of time (i.e. $q = q_0 e^{\lambda t}$, etc.), then the characteristic equation of the longitudinal stick-fixed motion is a quartic equation of the form

$$\lambda^4 + B_1\lambda^3 + C_1\lambda^2 + D_1\lambda + E_1 = 0 \tag{8.1}$$

where the coefficients B_1, C_1, D_1 and E_1 are functions of the longitudinal stability derivatives, the inertia of the aircraft about the y-axis, etc. The equations of motion expressed in non-dimensional form and the corresponding values of B_1, C_1, D_1 and E_1 are given in Appendices A and B.

Eqn. (8.1) has four roots; in the longitudinal case these generally form two pairs of complex roots, which represent two oscillatory

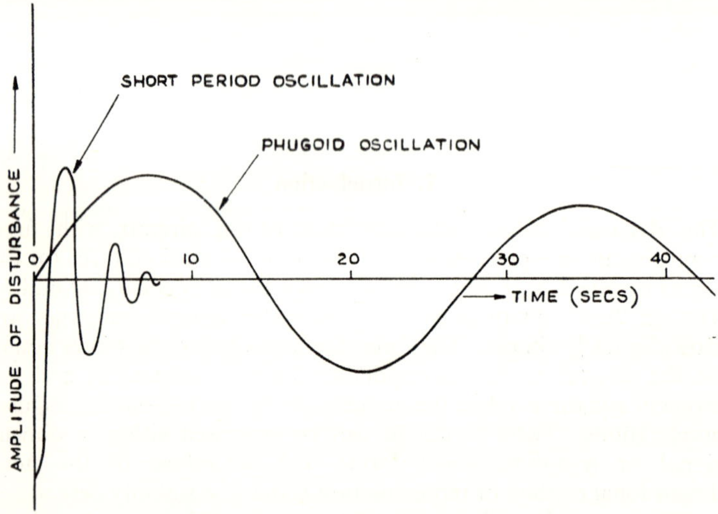

Fig. 8.1. The longitudinal oscillatory modes

motions. The overall motion of the aircraft is a combination of these two oscillatory modes, but, because the essential characteristics of these two oscillations are so widely different, it is normally possible to consider them separately. One of the modes usually consists of a very heavily damped short period oscillation (sometimes called the "rapid incidence adjustment") and the other is usually a fairly lightly damped oscillation with a much longer periodic time (the "Phugoid"), (Fig. 8.1).

The oscillatory characteristics of an aircraft are, however, dependent on its design and on the speed and height at which it is flying. Under certain conditions of flight, it is possible for the damping of the oscillations to become marginal and even for the classical short period and phugoid oscillations to disappear altogether and to be replaced by aperiodic (i.e. non-oscillatory) convergent and/or divergent motions.

Any deliberate or unintentional movement of the elevator control surfaces during the oscillation may modify the dynamic characteristics of an aircraft quite considerably. Thus it is necessary to consider both stick-fixed and stick-free dynamic stability. This is true not only for aircraft fitted with non-powered control systems, but also for those with powered controls, since, even with powered controls, it is unlikely that the control surfaces will remain absolutely fixed and even the smallest movements can have an appreciable effect on the dynamic stability characteristics. The oscillatory characteristics will also be modified by any automatic or pilot-induced control movements. Autostabilization systems can be used to improve the damping of the oscillatory motions. Control movements by the pilot, on the other hand, can result in a loss of damping due to the incorrect phasing of the movements and may cause serious short period "pilot-induced oscillations" (P.I.O.'s) to develop. Oscillations have been recorded in which the normal g on the aircraft has varied between $+8g$ and $-8g$, with a periodic time of less than 1 second.

Until a few years ago, the stick-fixed motions were of little importance, and very few flight tests were ever carried out to determine the longitudinal dynamic stability characteristics of an aircraft. If it possessed longitudinal static stability, then its dynamic characteristics were usually satisfactory also and a mathematical check was usually sufficient. Many of the references are, for this reason, theoretical ones (Refs. 7, 8 and 9). Stick-free stability was, however, more important, since, under certain critical conditions, the damping of the short-period oscillation could become small and even negative.

Within recent years, however, as a result of the increased speed and height range of modern aircraft, and the corresponding changes in their configurations, the longitudinal dynamic characteristics have become less satisfactory. The damping is often marginal in the transonic speed régime, and the phugoid stability on the approach to landing may constitute a problem. Pitch dampers are fitted into the control systems of many modern aircraft, to ensure satisfactory handling qualities under all conditions of flight.

The requirements that an aircraft should satisfy are laid down in the appropriate manuals. Av.P.970 states that for general flying the short period and phugoid oscillations should both be positively damped with an attenuation ratio per half cycle not greater than 0·7 (i.e. the relative damping should be greater than 0·11, and the oscillation should not take more than 1 cycle to damp to half amplitude). In the primary operating zone of the aircraft, the attenuation ratio for both the oscillations should not be greater than 0·30 or 0·45, depending on the role of the aircraft. These requirements are discussed in Ref. 11; similar requirements laid down for United States military aircraft are discussed in Ref. 12.

Flight tests can be carried out to determine the damping and the periodic time of the longitudinal oscillations, under different conditions of flight. In addition to showing that the aircraft possesses satisfactory dynamic stability characteristics, the results can also be analysed to give the numerical values of the various longitudinal derivatives (Ref. 9). The determination of the stability and control derivatives has become, within recent years, an important part of flight test work. The aircraft is trimmed at the selected speed and height and a longitudinal oscillation initiated by means of an abrupt control deflection. The control is returned to its initial position and then either held fixed or released as desired and the resulting oscillation allowed to damp out. By measuring the amplitude and phase relationships of the rate of pitch, the normal acceleration, and so on, the derivatives can be determined (see Chapter 17). These flight-determined values of the derivatives can then be used to check previous theoretical and wind-tunnel results, to assist in a cautious build-up programme investigating problem areas in the flight régime of the aircraft, in simulator studies and in the design of autostabilization systems.

2. Stick-fixed Longitudinal Dynamic Stability

2.1. General

The initial overall response of an aircraft following a longitudinal disturbance usually consists of an oscillatory motion involving changes in forward speed, vertical height, and rate of pitch (or normal acceleration), see note on p. 515. After a short time interval, it is generally found that the variation in normal acceleration has disappeared and the aircraft is left with a fairly long period oscillation involving only changes in speed and height. This long period oscillation is called the Phugoid oscillation; if it is filtered out of the initial response another oscillation remains. This is a very heavily damped short period oscillation mainly involving changes in normal acceleration. The analysis of longitudinal dynamic stability can be broken down into a consideration of these two oscillatory modes

(i) *The Phugoid Oscillation* This consists of a relatively long period oscillation of fairly low damping, involving slow variations of speed and height. Both the periodic time and the time to half amplitude may be of the order of minutes. Even though the damping of this oscillation may, under certain conditions, be low, it is not usually serious. Because of its long periodic time, the motion can normally be controlled fairly easily by the pilot.

(ii) *The Short Period Oscillation* This consists of a heavily damped short period motion involving changes in normal acceleration. No appreciable changes in speed and height occur and because of the heavy damping involved, the oscillation is not usually very important. Both the periodic time and the time to half amplitude may be only of the order of seconds. If the damping of this oscillation is reduced, then this is a more serious problem, because of the small periodic time of the oscillation.

The Time-vector method of representation (see Chapter 17) is extremely useful in showing the relative amplitudes of, and the phase differences between, the different terms in the equations of motion. It shows clearly the relative importance of the various quantities involved in the oscillatory motion (see also Ref. 1, page 205).

Because of the very different nature of the two oscillations, it is possible to split the stability quartic equation into two simpler quadratic equations; each of these quadratic equations then represents one of the oscillatory modes. Eqn. (8.1)

$$\lambda^4 + B_1\lambda^3 + C_1\lambda^2 + D_1\lambda + E_1 = 0$$

can be written as

$$(\lambda^2 + B_1\lambda + C_1)(\lambda^2 + \beta\lambda + \gamma) = 0 \tag{8.2}$$

where

$$\left.\begin{aligned} \beta &= \frac{C_1 D_1 - B_1 E_1}{C_1^2} \\ \gamma &= \frac{E_1}{C_1} \end{aligned}\right\} \tag{8.3}$$

The first quadratic on the left hand side of eqn. (8.2) (i.e. $\lambda^2 + B_1\lambda + C_1 = 0$) represents the short period oscillation and the second one (i.e. $\lambda^2 + \beta\lambda + \gamma = 0$) represents the phugoid oscillation. The expressions for the coefficients B_1, C_1, D_1 and E_1 given in Appendix A show that they are rather complicated functions, involving all the longitudinal derivatives. It can be shown that the coefficient C_1 is related to the manoeuvre stability of the aircraft and E_1 to the corresponding static stability. The periodic time and damping of the oscillatory motions can be expressed in terms of these coefficients by a similar method to that used for the simple spring-mass-viscous damper system discussed in Chapter 7. By substituting the expressions given in Appendix A for the coefficients, the periodic times and damping expressions can be written in terms of the longitudinal derivatives.

Though this approach is more exact, it is often extremely useful to analyse the motion of an aircraft, if possible, more simply. The simple analyses only give approximate solutions, within the limits of the assumptions made during the calculations. These solutions may, however, be sufficiently accurate for most purposes and the simpler analyses have the added advantage of giving a better physical appreciation of the nature of the problem.

It is not necessary to obtain a complete solution of the stability quartic equation in order to determine whether an aircraft is stable or not. If all the coefficients of the quartic are positive and if, in addition, a quantity called the Routhian Discriminant is positive, then the aircraft is stable. The Routhian Discriminant R is defined as follows

$$R = B_1C_1D_1 - B_1^2E_1 - D_1^2 \tag{8.4}$$

2.2. Short Period Oscillation

2.2.1. SIMPLE ANALYSIS

In any problem involving a number of variables it is always extremely useful to eliminate as many of the variables as possible, so reducing the analysis to one involving the minimum number of variables, as long as the resulting approximate answer is not too inaccurate and misleading. In this particular case, the oscillation is one involving large changes in pitch, with little variation in speed and height. Thus, instead of using the three equations of motion, an approximate solution can be obtained from the pitching moment equation of the aircraft, the oscillation being considered as a simple "one degree of freedom" type of motion.

If the aircraft is considered to possess only one degree of freedom that of θ, the angle of pitch, then the pitching moment equation can be written

$$I\ddot{\theta} + b\dot{\theta} + k\theta = 0 \tag{8.5}$$

This equation should be compared with eqn. (7.8) of Chapter 7. The problem is now basically identical to that of the spring-mass-viscous damper system. We are, however, in this case concerned with rotary motion and I represents the moment of inertia of the aircraft about the pitching (i.e. the 0y) axis. In eqn. (8.5) the terms $b\dot{\theta}$ and $k\theta$ represent the restoring moments acting as a result of the disturbance in pitch; k representing the aerodynamic stiffness (i.e. "spring") coefficient and b the aerodynamic damping coefficient.

The main contribution to both these restoring moments is obtained from the horizontal tailplane and in this simple analysis, the effects of the other parts of the aircraft are ignored.

As we have seen in para. 2.2 of Chapter 3, a pitching disturbance on an aircraft produces a change in the tailplane incidence of $\Delta\alpha'$, given by

$$\Delta\alpha' = \Delta\alpha\left(1 - \frac{d\varepsilon}{d\alpha}\right) + \frac{ql'}{V} \tag{8.6}$$

where $\Delta\alpha$ is the change in wing incidence and q is the rate of pitch. If only pitching motion is involved, then $\Delta\alpha$ and q can both be expressed in terms of θ, the change in the angle of pitch (assuming that θ is initially equal to zero), since $\Delta\alpha = \theta$ and $q = d\theta/dt$. Substituting in eqn. (8.6)

$$\Delta\alpha' = \theta\left(1 - \frac{d\varepsilon}{d\alpha}\right) + \frac{l'}{V}\frac{d\theta}{dt} \tag{8.7}$$

The corresponding change in the lift force acting on the tailplane is given by

$$\Delta L' = a_1 \Delta\alpha' \tfrac{1}{2}\rho V^2 S' \tag{8.8}$$

and the corresponding change in pitching moment by

$$\Delta M' = -l' a_1 \Delta\alpha' \tfrac{1}{2}\rho V^2 S' \tag{8.9}$$

Substituting for $\Delta\alpha'$ from eqn. (8.7)

$$\Delta M' = -l' a_1\left[\theta\left(1 - \frac{d\varepsilon}{d\alpha}\right) + \frac{l'}{V}\frac{d\theta}{dt}\right]\tfrac{1}{2}\rho V^2 S' \tag{8.10}$$

and separating out the terms in θ and $d\theta/dt$ (i.e. $\dot{\theta}$), we get

$$\Delta M' = -\left[\frac{(l')^2 a_1}{V}\tfrac{1}{2}\rho V^2 S'\right]\dot{\theta} - \left[l' a_1\left(1 - \frac{d\varepsilon}{d\alpha}\right)\tfrac{1}{2}\rho V^2 S'\right]\theta \tag{8.11}$$

By comparison with eqn. (8.5) we can see that

$$b = (l')^2 a_1 \tfrac{1}{2}\rho V S' \tag{8.12}$$

$$k = l' a_1\left(1 - \frac{d\varepsilon}{d\alpha}\right)\tfrac{1}{2}\rho V^2 S' \tag{8.13}$$

This result shows that the aerodynamic stiffness and damping coefficients are, as one would expect, dependent on such factors as the tailarm l', the tailplane area S', and the tailplane lift curve slope

a_1. In addition, they are dependent on the speed and height at which the aircraft is flying, the stiffness coefficient is proportional to ρV^2 (i.e. to $\rho_0 V_i^2$) and the damping coefficient to ρV [i.e. to $\sqrt{\rho} \times \sqrt{(\rho V^2)}$ or $\sqrt{\rho} \times \sqrt{(\rho_0 V_i^2)}$]. Thus, if a comparison is made at a constant value of V_i, the stiffness coefficient is independent of height and the damping coefficient decreases with increase in height.

By comparison with eqn. (7.18), the solution of eqn. (8.5) can be expressed in the form

$$\theta = \theta_0 e^{-(\mu \pm i\omega)t} \tag{8.14}$$

where the damping factor μ is given by $b/2I$ and the frequency ω by $\sqrt{\left(\frac{k}{I} - \frac{b^2}{4I^2}\right)}$.

2.2.2. EFFECT OF VARIATION OF SPEED AND HEIGHT

The characteristics of the short period oscillation are dependent on the speed and height at which the aircraft is flying. Even the simple theory of para. 2.2.1 above is sufficient to indicate the way in which the periodic time and damping of the oscillation will vary. From eqns. (8.12) and (8.13) it can be seen that b, the damping coefficient, is directly proportional to the forward speed V and that k, the spring coefficient, is directly proportional to V^2. Substituting these relationships in the expressions given above for μ and ω, it can be seen that both μ and ω are directly proportional to the forward speed of the aircraft. Thus, on the basis of this simple theory, the periodic time and the corresponding damping times of the oscillation are all inversely proportional to the forward speed; an increase in speed reduces the periodic time, and the time for the oscillation to decay to $\frac{1}{2}$ or $1/e$ of the initial amplitude.

The effect of an increase of height on the oscillatory characteristics depends on the particular conditions at which the comparison is made. If the comparison is made at the same E.A.S., then, as has already been mentioned, the value of the coefficient k is unchanged, whereas the coefficient b decreases with increasing height. Substituting in the expressions for μ and ω, it can be seen that μ also decreases with height, but that only the second term in the expression for ω contains any altitude effect. An increase in height thus reduces the value of μ and slightly increases the value of ω; the damping times are increased and the corresponding periodic times slightly reduced.

A typical flight test record of a longitudinal short period oscillation is shown in Fig. 8.2. The results shown in Fig. 8.3 plotted against Mach number show the tendencies predicted by the simple analysis.

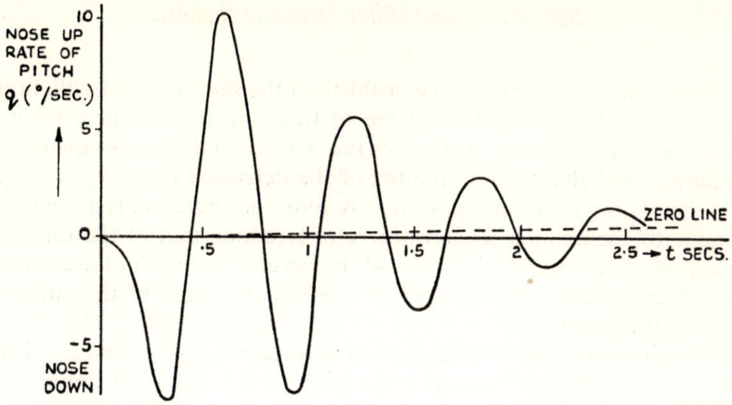

Fig. 8.2. Typical time history of a short period oscillation

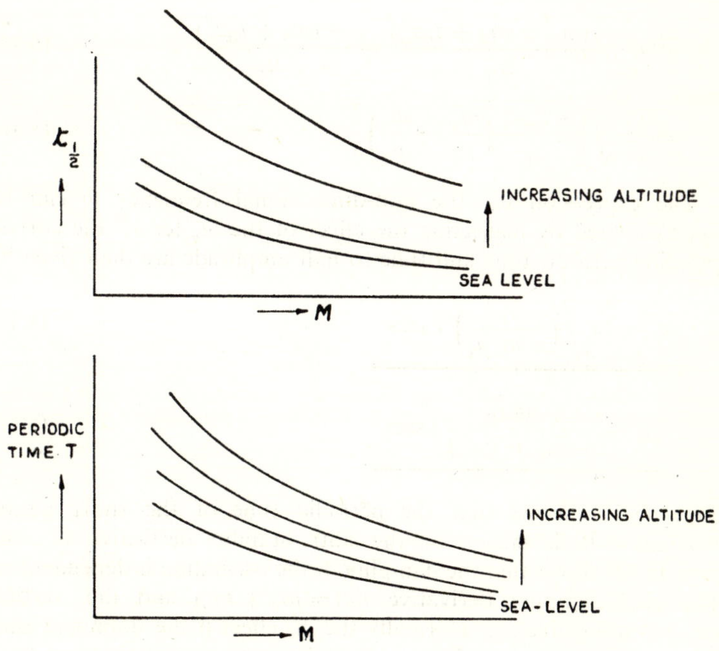

Fig. 8.3. Longitudinal short period oscillatory characteristics

2.2.3. MORE EXACT ANALYSIS

The simple theory reduces the problem of the short period oscillation to one involving only one degree of freedom, that of the pitching motion of the aircraft. It ignores the X-force and Z-force equations and assumes that a large number of the derivatives (e.g. x_u, x_w, x_q, z_u, z_w, z_q, m_u) are unimportant. A more accurate analysis can be made by including a second degree of freedom, that of the vertical motion of the aircraft. It is found, however, that even in these more exact calculations the changes in the forward speed of the aircraft can still be neglected.

The important derivatives are found to be m_w, m_q, $m_{\dot{w}}$ and z_w. The derivative $m_{\dot{w}}$ was discussed in Chapter 6; it is the most important of the acceleration derivatives. The sum of m_q and $m_{\dot{w}}$ is usually represented by $m_{\dot{\theta}}$. The results of this analysis are given in Appendix B, the non-dimensional values of μ and ω are given in terms of these derivatives by

$$\mu = \frac{-(m_{\dot{w}} + m_q + i_B z_w)}{2i_B} = \frac{-(m_{\dot{\theta}} + i_B z_w)}{2i_B} \tag{8.15}$$

$$\omega = \sqrt{\left[\frac{-z_w \mu_1}{i_B}\left(\frac{m_w}{z_w} - \frac{m_q}{\mu_1}\right)\right]} \tag{8.16}$$

The expression for the non-dimensional frequency ω can be approximated by neglecting the effect of the m_q term; the corresponding periodic time and time to half amplitude are then given by

$$T = 2\pi\sqrt{\left(\frac{i_B}{-m_w \mu_1}\right)}\ \hat{t} \text{ secs} \tag{8.17}$$

$$t_{1/2} = \frac{1{\cdot}386 i_B}{-(m_{\dot{\theta}} + i_B z_w)}\ \hat{t} \text{ secs} \tag{8.18}$$

Eqn. (8.17) shows that the periodic time of the short period oscillation is dependent on the static stability derivative m_w, and eqn. (8.18) shows that the damping of the oscillation is dependent on the pitch-damping derivative $m_{\dot{\theta}}(= m_q + m_{\dot{\omega}})$ and the vertical motion derivative z_w. Normally the $m_{\dot{\theta}}$ effect is the dominant one, but under certain conditions of flight (e.g. on tailless aircraft at transonic speeds) the value of $m_{\dot{\theta}}$ may be small or even negative and the z_w term may then be an extremely important one.

If it is assumed that the derivatives are constants, independent of speed and height, then since $\hat{t}$ $(= W/g\rho SV)$ is, by definition, proportional to $1/\rho V$ and $\mu_1(= W/g\rho Sl')$ proportional to $1/\rho$, it can be shown, by substituting in eqns. (8.17) and (8.18), that the effect of varying speed and height on the damping and periodic times is the same as that predicted by the simple theory.

It should be noted here that there are many different ways of expressing these results, depending on the notation and definitions employed and the degree of approximation that is used. These formulae are correct only as long as the assumptions which have been used to derive them are correct and before they are used it is necessary to check their validity for the particular case being considered (see Chapter 17).

2.2.4. EFFECT OF FACTORS

The oscillatory characteristics of an aircraft depend on various other factors, in addition to the speed and height at which it is flying. From eqns. (8.17) and (8.18) it is obvious that the periodic time and time to half amplitude are dependent on the pitching inertia of the aircraft i_B and on the derivatives m_w and m_q. An increase in pitching inertia increases both the periodic time and the time to half amplitude.

The static stability derivative m_w and the damping in pitch derivative m_q are both dependent on the position of the Centre of Gravity. The effects of C.G. movement on the longitudinal oscillatory characteristics are usually shown diagrammatically; these Stability Diagrams are discussed in more detail in para. 2.4.

2.3. Phugoid Oscillation

2.3.1. SIMPLE ANALYSIS

Once again, it is very instructive to analyse this motion in as simple a way as possible. In the phugoid oscillation the pitching motion of the aircraft is at all times very slow. This implies that the aerodynamic pitching moment is nearly zero and that the angle of attack is always very nearly that corresponding to zero pitching moment. In the simple analysis it is assumed that the angle of attack remains constant during the oscillation, that the drag effects are negligible and that the sum of the potential and kinetic energies remains constant. Any increase in height results in a reduction in speed and a loss of height in an increase in speed. In this way it is possible to reduce the problem to a consideration of one equation of motion, that involving speed and height.

If the height of the aircraft above the trimmed datum is denoted by z (ft) and the potential energy corresponding to the datum height is neglected (since it is common to both sides of the equation) we have that

$$\tfrac{1}{2}mV_0^2 = \tfrac{1}{2}mV^2 + mgz \qquad (8.19)$$

where V_0 is the initial trimmed speed (T.A.S.) in ft/sec. The equation of motion for the aircraft is given by

$$m\ddot{z} + (W - L) = 0 \tag{8.20}$$

where L is the instantaneous value of the lift on the aircraft. Since it is assumed that the angle of attack remains the same, $L \propto V^2$ and $W \propto V_0^2$, hence eqn. (8.20) can be written

$$m\ddot{z} + \text{constant}\,(V_0^2 - V^2) = 0 \tag{8.21}$$

From eqn. (8.19) it can be seen that

$$(V_0^2 - V^2) = 2gz \tag{8.22}$$

and substituting in eqn. (8.21) we get

$$m\ddot{z} + \text{constant} \times 2gz = 0 \tag{8.23}$$

The value of the constant can be written as W/V_0^2 (or mg/V_0^2), hence eqn. (8.23) becomes

$$m\ddot{z} + \left(\frac{2mg^2}{V_0^2}\right) z = 0 \tag{8.24}$$

Comparing this equation with eqn. (7.3) of Chapter 7 it can be seen that it represents an undamped simple harmonic motion, with a "spring" stiffness coefficient of $2mg^2/V_0^2$. By comparison with eqn. (7.4) of Chapter 7, the solution can be expressed in the form

$$z = z_0 \cos \sqrt{2}\,\frac{g}{V_0}\,t \tag{8.25}$$

representing an undamped oscillation with an amplitude of z_0 and a frequency equal to $\sqrt{2}g/V_0$. The corresponding periodic time is given by

$$T = \sqrt{2}\pi\,\frac{V_0}{g} = 0{\cdot}138\,V_0 \text{ secs} \tag{8.26}$$

where V_0 is the initial trimmed speed (T.A.S.) in ft/sec.

2.3.2. EFFECT OF VARIATION OF SPEED AND HEIGHT

On the basis of this simple theory the motion consists of an undamped (since all the damping terms have been ignored) oscillation, with a periodic time directly proportional to the true airspeed. This linear variation is usually verified in practice, but the value of the numerical coefficient in eqn. (8.26) is more nearly 0·178 than 0·138 (Ref. 4, page 402). An increase in altitude, for the same value of E.A.S., will not introduce any damping, but the periodic time will be increased by a factor of $1/\sqrt{\rho}$, because of the corresponding increase in the T.A.S.

2.3.3. MORE EXACT ANALYSIS

In a more exact analysis, it is necessary to allow for the changes in forward speed as well as for the pitching motion, though it is still possible to neglect the changes in angle of attack. The value of the periodic time is only slightly altered from the value given by the simple theory, but the oscillation now becomes a damped one. The amount of damping depends on the derivative x_u, i.e. on the drag characteristics of the aircraft. The time to half amplitude is often given in terms of C_D, the drag coefficient of the aircraft, as

$$t_{1/2} = \frac{1{\cdot}386}{C_D}\,\hat{t} \text{ secs} \tag{8.27}$$

Until fairly recently, the phugoid oscillation has not been of any great significance, the periodic time has been fairly long, and because of the relatively high drag characteristics of aircraft, the oscillation has been satisfactorily damped. Within recent years, however, the changes in the design of aircraft have been such as to reduce the drag coefficients and this has tended to make the phugoid oscillation more important.

A reduction in the forward speed of the aircraft increases both C_D and $\hat{t}$; this usually results in an increase in the time to half amplitude and a deterioration in the phugoid characteristics (Ref. 4, page 406). An increase in altitude at a constant E.A.S. increases $\hat{t}$ and since, in the absence of compressibility effects C_D remains unchanged, the corresponding time to half amplitude is also increased. The most serious flight régimes for the phugoid oscillation are thus when the aircraft is flying at low speeds and/or high altitudes.

2.3.4. EFFECT OF FACTORS

The expressions so far derived for the periodic time [eqn. (8.26)] and the time to half amplitude [eqn. (8.27)] do not indicate that there are any factors involved apart from the speed and height at which the aircraft is flying, its wing loading (since $\hat{t} = W/g\rho SV$) and C_D, the drag coefficient. However, there is inherent in the simple analysis an assumption which makes the expression for the periodic time applicable only for large values of the static margin and the effect of the damping in pitch has been neglected in the determination of the expression for the time to half amplitude. In practice, movement of the C.G. has an appreciable effect on both the periodic time and the damping of the oscillation.

The effect of C.G. movement depends on the relative positions of the stick-fixed manoeuvre and neutral points. If the manoeuvre point M.P. lies ahead of the neutral point N.P., then rearward movement of the C.G. may cause serious instabilities to occur. As the C.G. approaches the M.P., the phugoid oscillation may become unstable, since $K_n \gg H_m$ (Ref. 4, page 403; Ref. 3, Aircraft 00.00.03). Further rearward movement of the C.G. will produce a rapidly divergent motion. Usually the N.P. lies ahead of the M.P. and the phugoid oscillation does not become unstable before breaking down into aperiodic modes. The effects of C.G. movement on the oscillatory characteristics are often shown in the form of stability diagrams (see para. 2.4.) It should be noted that movement of the C.G. behind the N.P. does not necessarily coincide with any rapid deterioration in the aircraft's handling qualities and, with reasonable care, an aircraft can still be flown under these conditions. When the C.G. moves behind the M.P., however, the handling of the aircraft becomes extremely difficult.

An increase in wing loading decreases the damping (i.e. increases the time to half amplitude) and increases the periodic time, whereas an increase in pitching inertia decreases the damping, without having much effect on the periodic time.

2.4. Stability Diagrams and Iso-opinion Plots

The characteristics of the longitudinal stick-fixed oscillatory motions of an aircraft are dependent on a number of factors. The most convenient way of representing the effect of the variables involved is in diagrammatic form. Unfortunately, because of the large number of factors involved, it is not possible to show all the effects on one diagram. A selection has to be made and stability diagrams usually represent the effects of either one or two parameters. While these selected parameters are being varied, it is assumed that all the other factors involved are maintained constant.

2.4.1. ONE PARAMETER STABILITY DIAGRAMS

In these diagrams the effect of varying one parameter only is shown. The most important single factor determining the characteristics of the longitudinal oscillations is the position of the C.G. and its effect on the static margin. On the stability diagrams the factor is shown either as a variation in the C.G. position, in the derivative m_w, or in the static margin.

The stability diagrams on pages 132 and 133 of Ref. 5 illustrate the way in which the non-dimensional values of the frequency and damping of the short-period and phugoid oscillations vary with m_w, for different values of C_L. Rearward movement of the C.G. reduces the static margin, but the characteristics of the phugoid oscillation do not alter appreciably until a low value of the static margin is

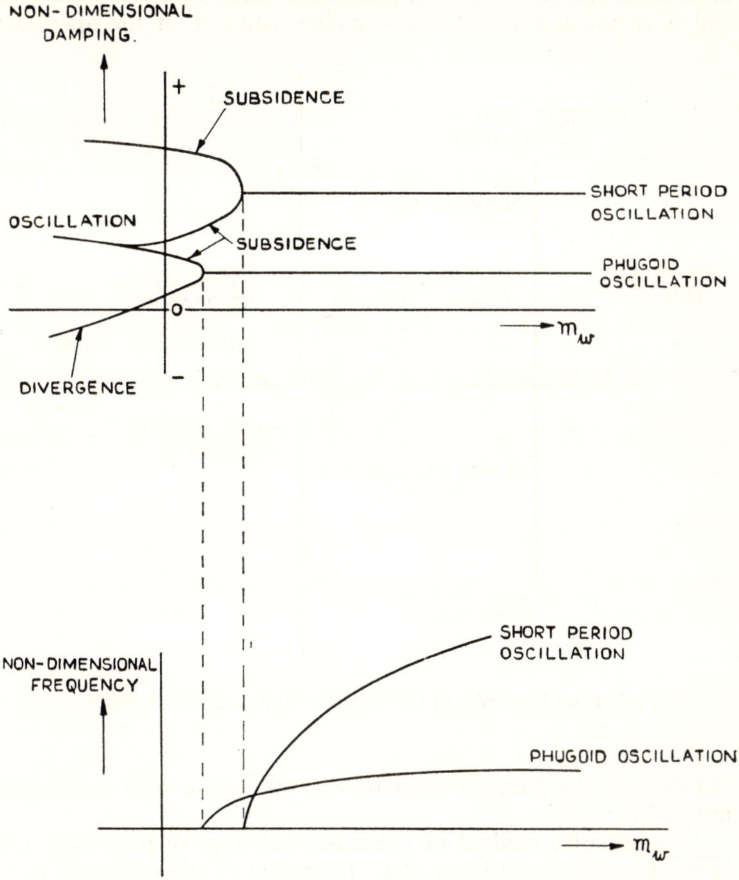

Fig. 8.4. Effect of variation of static margin on the oscillatory characteristics

reached (Fig. 8.4). The frequency then reduces rapidly with further reduction in the static margin and the phugoid oscillation breaks down into two subsidences. As the C.G. moves farther back the damping of one of the subsidences increases. The damping of the other subsidence, however, is reduced and, eventually, when the C.G. has moved behind the Neutral Point, this mode becomes divergent. The frequency of the short period oscillation decreases rapidly (i.e. the periodic time increases) as the static margin is reduced, but the damping does not vary much until the oscillation breaks down into two convergent modes. Further rearward movement of the C.G. causes the damping of one of these modes to decrease and it is possible that it will combine with one of the aperiodic

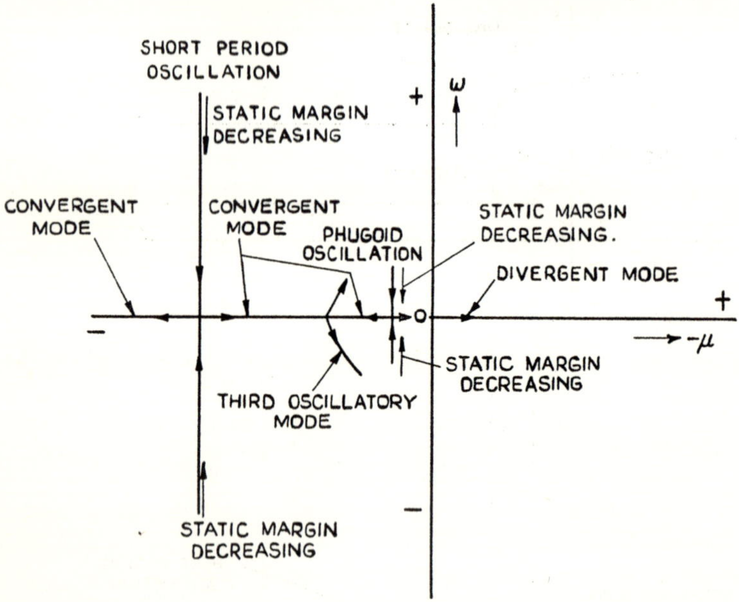

Fig. 8.5. Root locus plot of the longitudinal oscillatory characteristics

modes from the phugoid oscillation to produce another oscillatory mode (Fig. 8.4).

An alternative method of presenting this type of information is in the form of a root locus plot. Instead of plotting the frequency ω and the damping factor μ separately against the static margin, the axes are used to represent the values of ω and μ and each point is labelled with the appropriate value of the margin. The characteristics of the longitudinal oscillatory motions are shown in this form in Fig. 8.5. When the static margin is large, the values of μ and ω for the two oscillations are relatively large (though, of course, the numerical values of μ and ω are much higher for the short period oscillation [a high frequency, heavily damped oscillation] than for the phugoid oscillation [a low frequency, lightly damped oscillation]).

A reduction in the static margin decreases the frequency of both these oscillations, without appreciably changing the damping factor. The change in the frequency becomes more rapid at the lower values of the static margin. When the C.G. reaches a position close to the Neutral Point, the oscillations break down into convergent aperiodic modes and for very large negative values of the static margin the longitudinal motion may consist of one heavily damped convergent mode, one damped oscillation and one divergent mode (Fig. 8.5).

2.4.2. TWO PARAMETER STABILITY DIAGRAMS

In these diagrams the variation of the oscillatory characteristics with respect to two parameters is shown. Sometimes the diagrams indicate the variation in the frequency and the damping of the oscillation throughout the range of values of the parameters. Often, only the stability boundaries are shown, separating the stable regions from the unstable ones. All the two parameter diagrams in Ref. 1 (page 221), Ref. 4 (pages 405 and 406) and Ref. 5 (page 136) refer to the phugoid oscillation. The two parameters usually selected for these diagrams are the longitudinal derivatives m_w and m_q. On page 406 of Ref. 4, the phugoid stability boundaries are drawn for different values of C_L. On page 405 of this reference similar stability boundaries are shown using C_D and C_L as the two selected parameters, for various constant values of m_w and m_q.

Fig. 9.5 of Ref. 5 is a typical diagram showing the stability boundaries, and lines of constant frequency and damping, for the phugoid oscillation plotted against the parameters m_w and m_q.

2.4.3. ISO-OPINION PLOTS

The characteristics of an oscillation can be represented in terms of the damping and the frequency, and the pilot's assessment of the handling characteristics of an aircraft is dependent on both of these parameters. Iso-opinion plots are obtained from simulator or variable stability aircraft tests and show the regions of acceptability and unacceptability for the particular handling characteristics considered. Figs. 8.6 and 8.7 are taken from Ref. 12. Fig. 8.6 is for the short-period oscillation and shows that, in order to obtain good characteristics, the frequency should be greater than a certain minimum value (i.e. the periodic time should be less than a certain maximum value), irrespective of the amount of damping and that for optimum conditions a minimum level of damping is required. Fig. 8.7 is for the phugoid oscillation and it shows that the damping of the oscillation should not be allowed to become too negative.

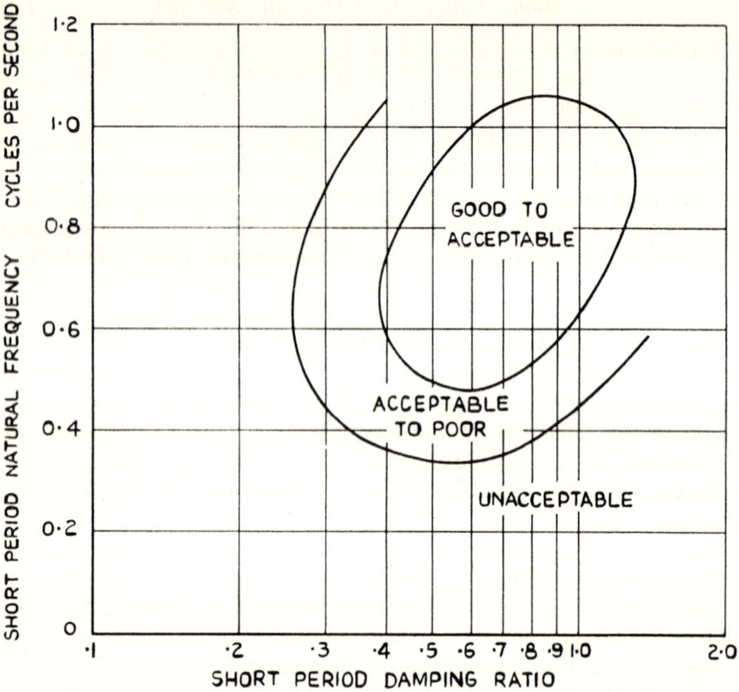

Fig. 8.6. Iso-opinion boundaries for the longitudinal short-period oscillation. (From Ref. 12)

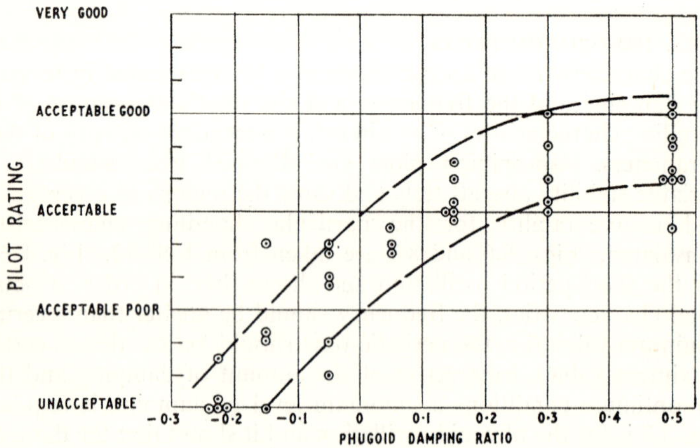

Fig. 8.7. Pilot opinion of phugoid damping. (From Ref. 12)

3. Stick-free Longitudinal Dynamic Stability

3.1. General

As has already been mentioned, freeing the elevator control may have an important effect on the longitudinal dynamic stability characteristics. The theory of the stick-free motion involves an additional degree of freedom (i.e. that of the moving elevator itself) and this complicates the problem. A number of new variables will be included in the equations of motion (e.g. the elevator hinge moment characteristics b_1 and b_2, the mass balance of the elevator and the friction in the control system) and this will make the problem a very difficult one to solve.

However, it is known that freeing the elevator has two main effects (Ref. 9). The first effect is the appearance of a completely new mode, consisting of the oscillation of the control surface about its hinge line. This is a very rapid oscillatory motion, with a much higher frequency and a much stronger damping than those of the short period mode. The second effect is that the basic short period characteristics are considerably modified. If this oscillation is poorly damped, it will make the behaviour of the aircraft unpleasant, particularly in rough air and when executing manoeuvres. The phugoid motion, on the other hand, is generally little affected, because the frequency of the forcing function is very much higher than the frequency of the oscillation (Ref. 1, page 223). Using this knowledge the stick-free dynamic stability characteristics can be obtained by means of a simplified analysis (Ref. 9).

If an aircraft is fitted with irreversible power operated controls only stick-fixed stability need usually be considered. However, it is sometimes found that elevator movements are occurring due to distortion in the airframe or the control system. These movements may have an appreciable effect on the damping of the motion and have to be allowed for in the analysis of the results.

3.2. Short Period Oscillation

Because the frequency of this mode is not so different from that of the elevator motion, important coupling effects may occur and under certain conditions the damping may become extremely weak and even negative. This divergent short period oscillation is sometimes called "porpoising" and the normal acceleration may reach dangerously high values.

The effects depend on the phase difference between the elevator motion η and the normal acceleration n, and the results are shown diagrammatically in Fig. 8.8. If the phase difference is between 0° and 180° (i.e. the elevator deflection "leads" the normal acceleration), then freeing the elevator increases the damping. This case will seldom occur in practice; the elevator usually lags behind the normal acceleration and reduces the damping. The frequency of the oscillation is always increased, as long as the elevator angle does not lead

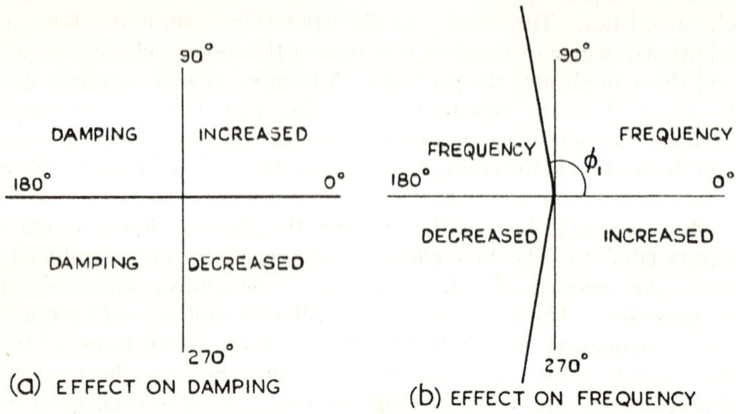

Fig. 8.8. Effect on the oscillatory characteristics of the control surface phase difference

or lag the normal acceleration by more than ϕ_1, where ϕ_1 is typically slightly greater than 90°.

3.2.1. EFFECT OF FACTORS

The effect of the motion of the elevator on the short period characteristics is dependent on a large number of factors. The effect of these various parameters can be represented diagrammatically by plotting the corresponding stability boundaries. The form of this presentation varies, depending on the particular factors which are selected as the variables.

The results of the analysis in Ref. 8 have been presented in the form of charts, with $-b_2$ plotted vertically against the weight moment in the elevator circuit horizontally, for different values of b_1; all the calculations having been made for one particular set of conditions of wing loading and height. These results show that an unstable oscillation may occur with close aerodynamic balance and either incomplete mass-balance or with an inertia weight in the circuit. If the elevator is not mass-balanced, then an unstable oscillation may occur with both negative and positive values of b_1, but if the elevator is mass-balanced this instability can only occur if b_1 is positive.

Friction in the control circuit is another important factor; it is likely to delay the onset of instability, but will increase the severity of the instability when it does occur.

The stability boundaries are often presented with b_1 values plotted vertically against b_2, for various values of the aircraft relative density parameter μ_1, the elevators being assumed to be mass balanced. These results (Fig. 8.9) also show that unstable oscillations are

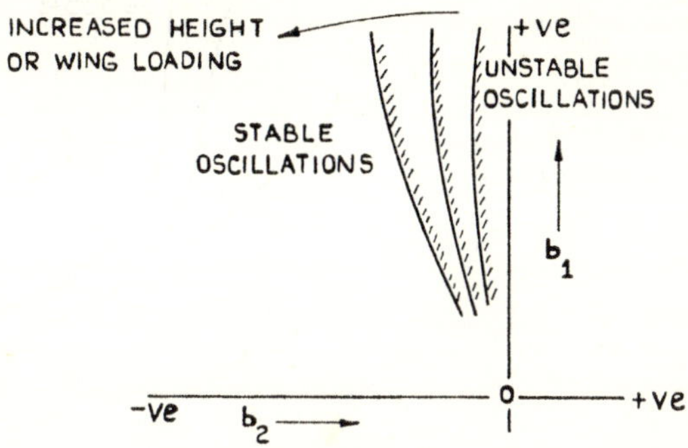

Fig. 8.9. Effect of the control surface parameters on the oscillatory characteristics

possible at low values of $-b_2$ when b_1 is positive. An increase of height or wing loading moves the stability boundary to the left, and increases the possibility of instability. Any mass unbalance will also move the boundaries in a similar way.

3.3. Phugoid Oscillation

To a first approximation, it is usually possible to neglect the influence of the free elevator on the phugoid motion, because the frequency of oscillation of the elevator is so high compared to that of the phugoid oscillation. This particular mode is then identical with the corresponding stick-fixed mode (Ref. 1, page 223).

Once again, however, a large number of factors are involved in the motion (Ref. 7). Poor phugoid damping may occur when a small static margin is combined with friction in the elevator circuit. The damping usually decreases with rearward movement of the C.G. and unstable oscillations or a divergent motion will occur if the C.G. goes behind the stick-free neutral or manoeuvre points.

The presence of springs and inertia weights in the control system may have an appreciable effect on the oscillatory characteristics. These effects are discussed in detail in Ref. 10.

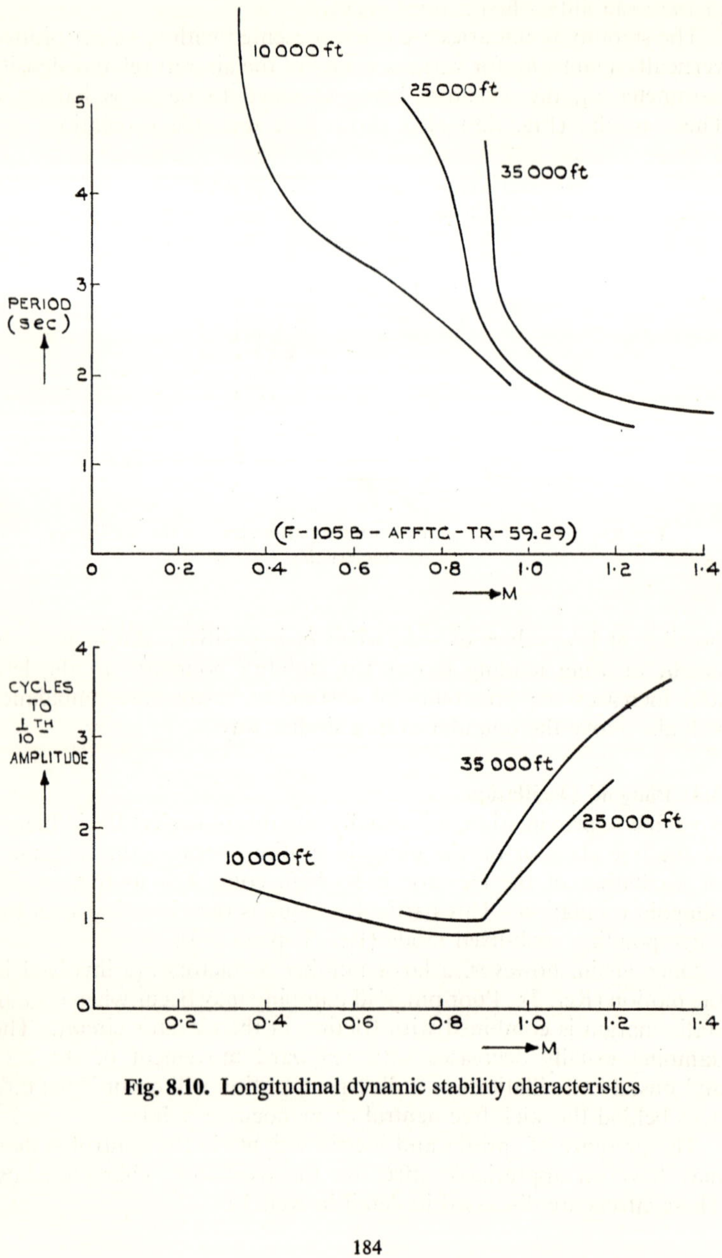

Fig. 8.10. Longitudinal dynamic stability characteristics

4. Flight Test Results

As mentioned in the introduction, the longitudinal dynamic stability characteristics have not posed any particular problem on low speed aircraft. For this reason, very few flight test results are available showing the variation of the measured values of periodic time, time to half amplitude, etc., throughout the speed and height range of a low speed aircraft. In the transonic and supersonic speed ranges, however, the longitudinal dynamic characteristics become more marginal, particularly at high altitudes, and more flight test results are available for these conditions of flight. Fig. 8.10 shows some actual flight test results; these measured results are in good agreement with the theoretical results discussed in para. 2.2.2, which predicted a periodic time decreasing with increase in forward speed and a number of cycles to $\frac{1}{2}$ (or $\frac{1}{10}$) amplitude remaining constant with speed (because according to the simple theory the values of T and $t_{1/2}$ vary at the same rate). These results are, of course, modified at high Mach numbers; the various problems that arise at high forward speed are discussed in more detail in Chapter 15.

References

1. Etkin, B.; *Dynamics of Flight* (John Wiley and Sons, New York, 1959).
2. Duncan, W. J.; *Control and Stability of Aircraft* (Cambridge University Press, 1952).
3. "Royal Aeronautical Society Data Sheets," *Aerodynamics*, Vol. 3.
4. Perkins, C. D. and Hage, R. E.; *Airplane Performance Stability and Control* (John Wiley and Sons, New York, 1949).
5. Babister, A. W.; *Aircraft Stability and Control* (Pergamon Press, Oxford, 1961).
6. Kolk, W. R.; *Modern Flight Dynamics* (Prentice-Hall International, London, 1961).
7. Lyon, H. M., Truscott, P. M., Auterson, E. I., and Whatham, J.; *A Theoretical Analysis of Longitudinal Dynamic Stability*, R.A.E. Report Aero, 1755, R. & M. 2075, 1942.
8. Whatham, J. and Lyon, H. M.; *A Theoretical Investigation of Dynamic Stability with Free Elevators*, Unpublished M.O.A. Report; R & M. 1980, 1943.
9. Neumark, S.; *Analysis of Short Period Longitudinal Oscillations of an Aircraft*, R.A.E. Report Aero. 2479, R. & M. 2940, 1956.
10. Goldberg, J.; *Effects of Springs and Inertia Devices on the Longitudinal Stability of Aircraft*, WADC-TR-53-350, 1954.
11. Allwright, H. J.; *Design Aims for Stability and Control of Piloted Aircraft*, Agard Report No. 337, 1961.
12. Koven, W. and Wasicko, R.; *Flying Quality Requirements for United States Navy and Air Force Aircraft*, Agard Report No. 336, 1961.

9 Subsonic Lateral Dynamic Stability

1. Introduction

If the coupling between the longitudinal and lateral motions of an aircraft is neglected, the stick-fixed lateral-directional motion can be analysed by considering only three equations of motion referred to the system of stability axes (i.e. those for Y, L and N). If the three variables involved (i.e. v, p and r) are assumed to be exponential functions of time (e.g. $p = p_0 e^{\lambda t}$), then the characteristic equation is a quartic of the form

$$A_2\lambda^4 + B_2\lambda^3 + C_2\lambda^2 + D_2\lambda + E_2 = 0 \qquad (9.1)$$

where the coefficients A_2, B_2, C_2, D_2 and E_2 are functions of the lateral stability derivatives, the rolling and yawing inertias of the aircraft, etc. The equations of motion are usually expressed in non-dimensional form and the corresponding expressions for A_2, B_2, C_2, D_2 and E_2 are given in Appendix A. If the product of inertia term in the expression for A_2 can be neglected (i.e. the $i_E{}^2/i_A i_C$ term) then $A_2 = 1$.

Eqn. (9.1) has four roots; in the lateral case these usually consist of two real roots and a pair of complex roots. The real roots represent two convergent (or divergent) motions and the complex pair of roots represents an oscillation. Because the characteristic times of these modes are widely different from one another, these three motions can be considered separately. The large real root corresponds to the roll subsidence mode, the small real root to the spiral stability mode and the oscillatory mode is usually referred to as the Dutch Roll oscillation.

The lateral characteristics of an aircraft are dependent on the speed and height at which it is flying and also on the particular configuration used (e.g. flaps, undercarriage, etc.). Under certain conditions of flight it is possible for the spiral mode to become divergent (i.e. the aircraft possesses spiral instability) and for the damping of the oscillatory motion to become small or even negative. In the case of an aircraft with a non-powered control system, it is necessary to consider both the control-fixed and control-free conditions, because the basic stability derivatives of the aircraft are modified when the control surfaces (i.e. aileron and/or rudder) are free to float.

The pilot is concerned with the way the aircraft responds to his control demands and to external disturbances, particularly in such conditions of flight as the approach to landing. The overall response of the aircraft is related to the three lateral modes; the roll subsidence mode determines the response to ailerons and lateral gusts, whereas the spiral mode determines the long-term stability and the lateral trim of the aircraft, and the Dutch Roll the long-term steadiness in flight through turbulent air.

Some of the recent design changes that have taken place on modern high-performance aircraft have made some of the assumptions of the "classical" approach invalid. The inertia coupling terms have become much more important and, under certain conditions of flight, it is no longer possible to consider the longitudinal and lateral motions separately. The problem of inertia coupling is considered in Chapter 16. The lateral characteristics of slender delta wing aircraft in particular may be very different from those of a conventional type of aircraft; in the analysis of the motion of these aircraft the principal inertia system of axes may be a more convenient one to use than the stability axes system (Ref. 11).

The stability requirements for military aircraft are laid down in Av.P.970. The Dutch Roll oscillation should damp to one half amplitude within one cycle (i.e. the log. dec. δ should be greater than 0·693). No specific requirement is laid down for the spiral stability mode, but, if it is divergent, the time to double amplitude should not be less than about 20 secs. Flight tests are carried out under various conditions of flight to determine the lateral characteristics of an aircraft. The Dutch Roll oscillation results can then be analysed to determine the numerical values of the lateral stability derivatives.

The effects of compressibility and aeroelasticity on the lateral dynamic stability characteristics will be considered in later chapters.

2. Control-fixed Lateral Dynamic Stability

2.1. General

The initial response of an aircraft following a lateral disturbance usually consists of an oscillatory motion involving changes in sideslip velocity, rate of roll and rate of yaw (Fig. 9.1), see note on p. 515. After a short time interval, it is generally found that these oscillations have damped out and the aircraft has either returned to its original condition of level flight (Fig. 9.2*a*), or is found to be diverging away in bank to one side or the other (Fig. 9.2*c*). The analysis of lateral dynamic stability can be broken down into a consideration of the three modes, which consist of one oscillatory mode and two aperiodic (non-oscillatory) ones. Because the characteristic times of these modes are usually widely different from one another, it is possible to deal with them separately. Each mode is then considered

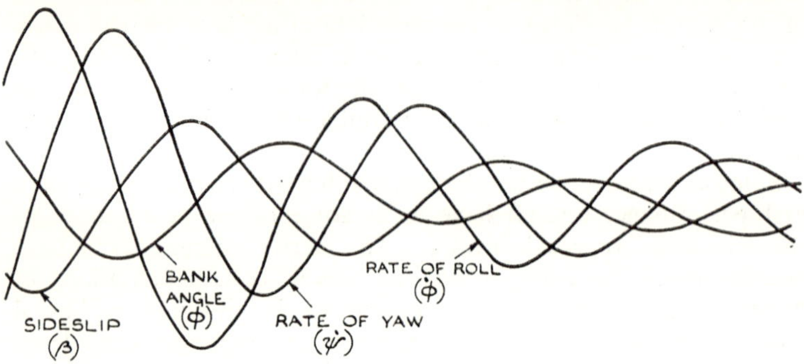

Fig. 9.1. Aircraft response to lateral disturbance

to be associated with a particular flight characteristic and it is assumed that the coupling between the modes is small. The three modes are

(i) *The Roll Subsidence Mode* This mode usually consists of a very heavily damped rolling motion, in which changes in angle of bank occur with practically no yaw or sideslip. It is very important in determining the roll response characteristics of conventional aircraft to atmospheric gusts and aileron deflections. The effect of this mode is discussed in more detail in the work dealing with lateral response (Chapter 11).

(ii) *The Spiral Mode* This is usually a weakly damped motion in bank and yaw, with negligible sideslip (i.e. approximately a correctly banked turn of increasing radius). Often this mode becomes unstable; the motion of the aircraft is then a tightening spiral one (i.e. approximately a correctly banked turn of decreasing radius) and this spiral motion gives this mode its name (Fig. 9.2*c*). It is important in determining the long term stability of the aircraft and affects the pilot's ability to achieve lateral trim in steady flight. Because a small amount of spiral instability does not detract very much from the handling qualities of an aircraft, and, since the satisfying of other more important stability requirements often tends to result in low spiral stability or even slight instability under certain conditions of flight, no quantitative standard of spiral stability is included in Av.P.970 (Vol. 1, Book 1, Pt. 6, Chapter 602), though the time to double amplitude should not be less than 20 secs.

(iii) *The Dutch Roll Oscillation* This consists of a relatively short period oscillation, which may be either damped (Fig. 9.2*a*) or divergent (Fig. 9.2*b*), involving rolling, yawing, and sideslipping motions.

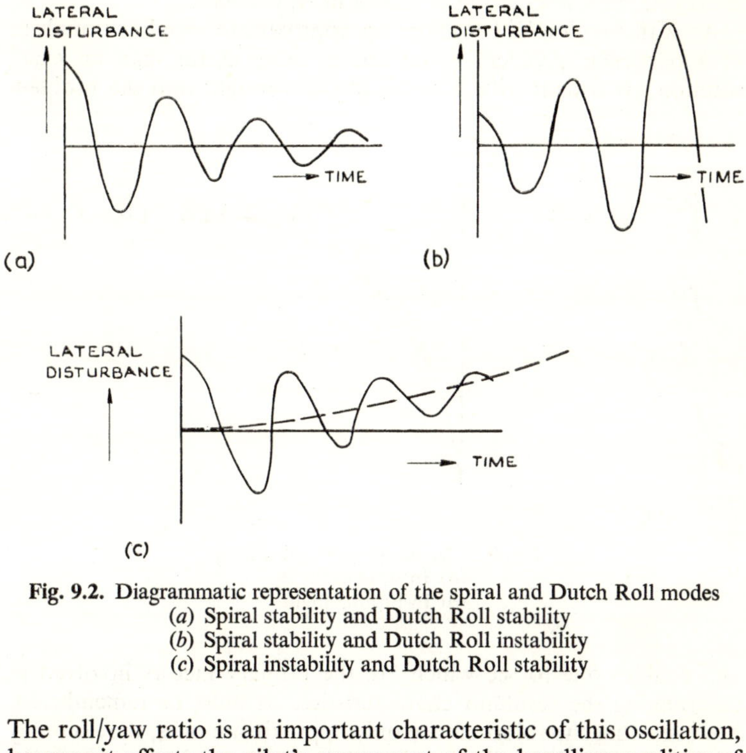

Fig. 9.2. Diagrammatic representation of the spiral and Dutch Roll modes
(*a*) Spiral stability and Dutch Roll stability
(*b*) Spiral stability and Dutch Roll instability
(*c*) Spiral instability and Dutch Roll stability

The roll/yaw ratio is an important characteristic of this oscillation, because it affects the pilot's assessment of the handling qualities of the aircraft. The ratio is usually large, and the motion is referred to as the Dutch Roll oscillation (presumably named after the typical motion of a Dutch ice skater); if the motion consists mainly of yawing, the oscillation is referred to as Snaking. The time-vector method of representation (see Chapter 17) is once again extremely useful in showing up the relative amplitudes of, and the phase differences between, the different terms in the equations of motion. It shows clearly the relative importance of the various quantities involved in the Dutch Roll oscillation.

The characteristics of the three lateral modes can be obtained by solving the lateral equations of motion, i.e. by determining the numerical values of the roots of eqn. (9.1). Having obtained the roots representing the roll and spiral modes, the corresponding factors can be divided into the stability quartic to obtain the quadratic representing the Dutch Roll oscillation. The roots of the quadratic can then be obtained and the characteristics of the oscillation (i.e. periodic time and damping) determined. A typical lateral stability calculation is shown in Appendix C.

As with longitudinal motion, an approximate simplified analysis of a particular problem is sometimes more useful than an exact solution. It usually give a better physical insight into the problem

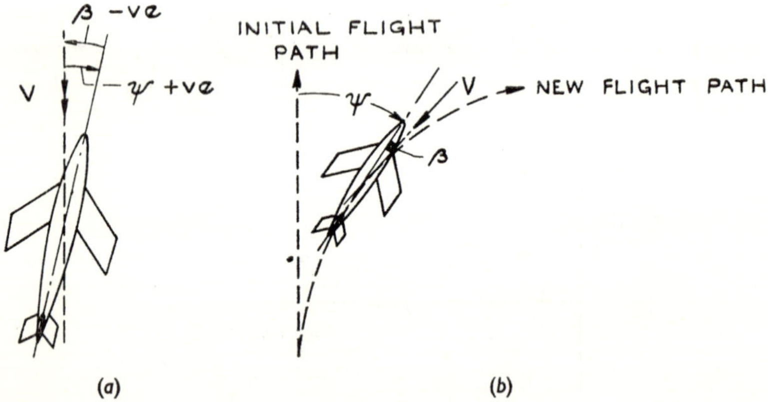

Fig. 9.3. Angles of yaw and sideslip
(*a*) In straight flight
(*b*) In turning flight

and enables one to see which are the primary factors involved in determining the resultant characteristics. It must be remembered, however, that the solutions are only approximate ones, the accuracy of the solutions depending on the validity of the various assumptions made. The basic assumption that the three lateral modes can be considered separately has been a valid one up to now, but on some modern and future designs (e.g. slender delta wing aircraft) this assumption may break down and there will be more coupling between the modes.

In any discussions on lateral stability it is important to distinguish carefully between the angle of sideslip β and the angle of yaw ψ. The angle of sideslip is defined as the angle between the relative airflow and the longitudinal axis of the aircraft (positive when the aircraft is sideslipping to starboard), and the angle of yaw is defined as the angle between the longitudinal axis of the aircraft and its initial undisturbed position (positive when the aircraft is yawed to starboard). When turning flight is not involved, these angles are numerically equal to one another (Fig. 9.3*a*) but opposite in sign, i.e. when the aircraft is yawed to starboard it is sideslipping to port. Thus $\beta = -\psi$ and there is a phase difference of 180° between them. When the aircraft is in a turn (Fig. 9.3*b*), the angles of sideslip and yaw are no longer numerically equal and they are completely unrelated to one another.

2.2. Roll Subsidence Mode

2.2.1. SIMPLE ANALYSIS

In its simplest form this mode is assumed to consist only of a rolling motion. The rolling moment equation can then be written as a one degree of freedom equation

$$A\ddot{\phi} - L_p\dot{\phi} = 0 \tag{9.2}$$

where A is the moment of inertia of the aircraft about the rolling (i.e. x) axis (slugs ft^2), L_p is the damping in roll coefficient (lb ft/radian per sec), ϕ is the angle of bank, $\dot{\phi}$ the rate of roll (rads per sec) and $\ddot{\phi}$ the angular acceleration in roll (rads per sec^2). This equation expresses the fact that when an aircraft is rolling at a rate p (i.e. $\dot{\phi}$), there is a damping-in-roll moment opposing the rolling motion (i.e. L_p is a negative quanity) and the aircraft is decelerating at a rate given by $\ddot{\phi} = L_p\dot{\phi}/A$. There is no displacement (i.e. ϕ) term in equation (9.2) and hence the mode is not an oscillatory one. This is because no restoring moment arises as a result of a displacement in angle of bank, i.e. an aircraft has neutral static stability with respect to angle of bank changes. In terms of p, the rate of roll, the equation can be written as

$$A\dot{p} - L_p p = 0 \tag{9.3}$$

and assuming a solution of the form $p = p_0 e^{\lambda t}$, the value of λ is given by

$$\lambda = \frac{L_p}{A} \tag{9.4}$$

and thus

$$p = p_0 e^{(L_p/A)t} \tag{9.5}$$

where t is the time in seconds. In this equation, the quantity L_p/A has the dimensions of 1/Time and it can be seen that $-A/L_p$ secs is the time that it takes for the initial rate of roll p_0 to decay to 1/e (i.e.

to approximately 37 per cent) of its initial amplitude (Fig. 9.4). This is known as the time constant of the motion. This value is also given by the intersection with the horizontal axis of the tangent drawn to the curve at time $t = 0$.

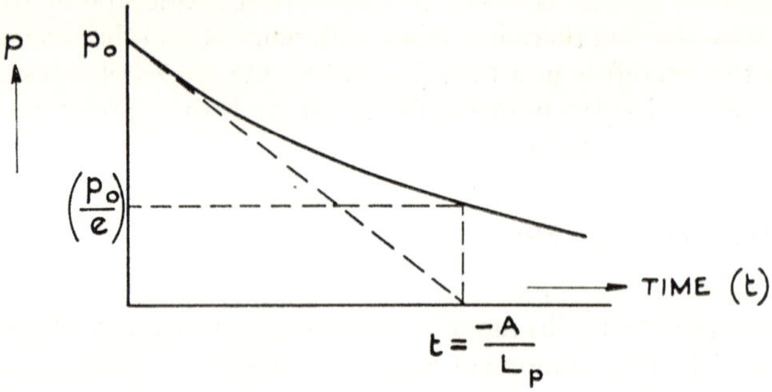

Fig. 9.4. Roll subsidence

The above result is often expressed in terms of the corresponding non-dimensional quantities l_p, i_A and τ where $l_p = \dfrac{L_p}{\rho VS(b/2)^2}$,

$$i_A = \frac{A}{m(b/2)^2} \text{ and } \tau = \frac{t}{\hat{t}} \text{ where } \hat{t} = \frac{m}{\rho SV}.$$

Since $\dfrac{L_p}{A}t = \dfrac{l_p}{i_A}\tau$ equation (9.5) becomes

$$p = p_0 e^{(l_p/i_A)\tau} \tag{9.6}$$

where τ is the time expressed non-dimensionally in air-seconds.

Because l_p is usually large and negative, this equation represents a rapidly convergent motion with a time to half amplitude given by

$$t_{1/2} = \frac{0{\cdot}693}{-l_p/i_A}\hat{t} \text{ secs} \tag{9.7}$$

The time to $1/e$ of the initial amplitude is given in a similar way by

$$t_{1/e} = \frac{\hat{t}}{-l_p/i_A} \text{ secs} \tag{9.8}$$

From equations (9.7) and (9.8) it can be seen that an increase in roll damping (i.e. an increase in the numerical value of l_p) decreases the time to half (or $1/e$) amplitude, whereas an increase in rolling inertia increases the time taken. The time constant is given by $-i_A/l_p$ airsecs, or $(-i_A/l_p)\hat{t}$ secs.

Since $t = m/(\rho SV) = W/(g\rho SV)$ and remembering that $V = V_i/\sqrt{\sigma}$, equation (9.8) can be written in the form

$$t_{1/e} = -13{\cdot}05 \frac{(W/S)}{V_i\sqrt{\sigma}} \frac{i_A}{l_p} \text{ secs} \tag{9.9}$$

where $13{\cdot}05 = 1/g\rho_0$, V_i is the E.A.S. in ft/sec, σ is the relative density and W/S the wing loading in lb/ft^2.

2.2.2. EFFECT OF VARIATION OF SPEED AND HEIGHT

The results of the simple analysis are sufficiently accurate to indicate the way in which the characteristics of this mode vary throughout

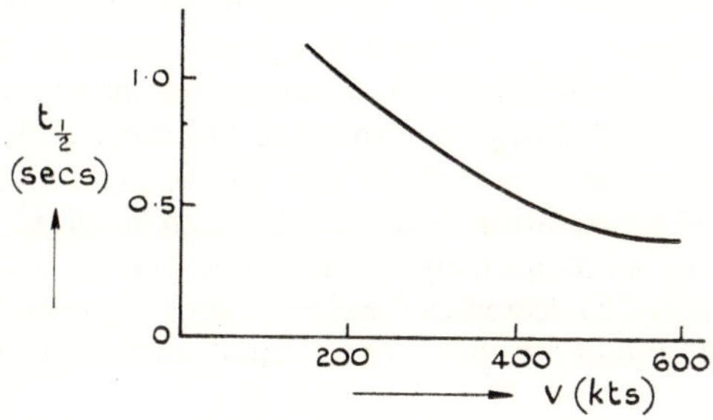

Fig. 9.5. Variation of time to half amplitude of the roll subsidence mode with forward speed

the speed and height range of the aircraft. Assuming that i_A and l_p are constants, independent of the speed and height, eqn. (9.9) shows that, at any height, the damping times decrease with increase of forward speed as $1/V$, or $1/V_i$ (Fig. 9.5), and that, at a given E.A.S., the times increase with height as $1/\sqrt{\sigma}$ or $1/\sqrt{\rho}$.

2.2.3. MORE EXACT THEORY

In practice, the motion will not be a simple one, involving only one degree of freedom. Some yawing and sideslipping will be involved in the motion, due to the presence of the coupling derivative n_p. A more exact solution for λ can be obtained from the lateral stability quartic of equation (9.1). If λ is large, then it is possible as a first approximation, to neglect all the terms in the quartic except the first two. The value of λ is then given approximately by the expression

$$\lambda = -\frac{B_2}{A_2} \tag{9.10}$$

Expressions for the coefficients A_2 and B_2 are given in Appendix A. If the terms containing the product of inertia i_E are neglected, eqn. (9.10) becomes

$$\lambda = \frac{l_p}{i_A} + \frac{n_r}{i_C} + y_v \tag{9.11}$$

This result should be compared with the one obtained from the simple theory (i.e. $\lambda = l_p/i_A$); it can be seen that the root is increased by the presence of the n_r (i.e. the damping in yaw) and the y_v (i.e. the sideforce due to sideslip) terms.

2.2.4. EFFECT OF FACTORS

Simple theory indicates that the two most important factors affecting the roll subsidence mode are the roll damping derivative l_p and the rolling inertia coefficient i_A. The result of the more exact analysis given in eqn. (9.11) shows that some of the other derivatives (e.g. n_r and y_v) may have important effects. Some derivatives not even contained in these equations may also have an effect on the time to half amplitude; an increase in dihedral effect (i.e. an increase in the numerical value of l_v) decreases the time slightly, whereas an increase in directional stability n_v produces a small increase in the characteristic time.

From eqn. (9.9) it can be seen that an increase in wing loading W/S increases the value of the time to 1/e amplitude.

2.3. Spiral Mode

2.3.1. SIMPLE ANALYSIS

If the numerical value of the root representing this mode is assumed to be small, then the λ^4, λ^3 and λ^2 terms in the stability quartic equation can be neglected. From the last two terms in the equation we can then obtain an approximate solution for λ

$$\lambda = -\frac{E_2}{D_2} \tag{9.12}$$

The expressions for E_2 and D_2 are extremely complicated (Appendix A); that for E_2 can be approximated to

$$E_2 = \frac{\mu_2 C_L}{2 i_A i_C}(l_v n_r - n_v l_r) \tag{9.13}$$

In order that the aircraft should possess spiral stability (i.e. for the motion to converge), λ must be negative and hence the ratio of E_2/D_2 must be positive. Since D_2 is normally a positive quantity, the spiral stability requirement can be stated by saying that the value of the E_2 term occurring in the stability quartic must have a positive value. From eqn. (9.13), the condition of spiral stability can be expressed as

$$l_v n_r - n_v l_r > 0 \tag{9.14}$$

or, alternatively, since l_v and n_r are usually negative, as

$$\left(\frac{-l_v}{n_v}\right) > \left(\frac{l_r}{-n_r}\right) \tag{9.15}$$

The numerical values of l_r and n_r are usually determined by other considerations and so eqn. (9.15) shows that $-l_v/n_v$ is the important

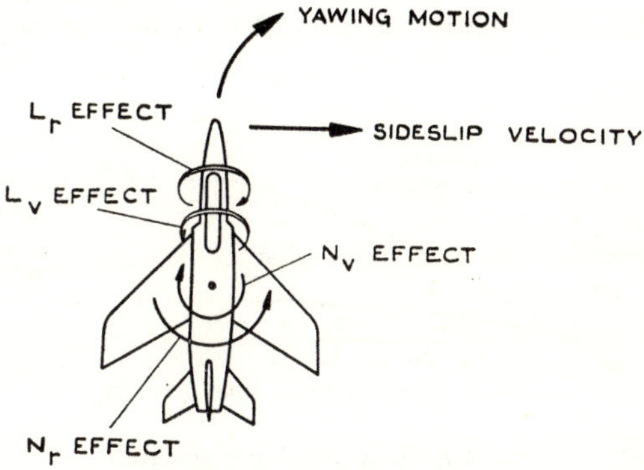

Fig. 9.6. Spiral stability

parameter determining the stability of the motion. To avoid spiral instability, the dihedral effect l_v must not be too small, nor the weathercock effect n_v too large.

The physical interpretation of this motion is straightforward. When an aircraft is in a banked turn, it is often yawing and sideslipping. In a turn to starboard, for example, it will be yawing and sideslipping to starboard. The l_v effect will tend to roll the aircraft out of the turn, whereas the l_r (i.e. the roll due to yaw) effect will tend to roll it in. Similarly, the n_r effect will tend to oppose the yawing motion of the aircraft, whereas the n_v effect will tend to increase the yaw. The condition for stability expressed in eqn. (9.14) states that the product of the derivatives which tend to bring the aircraft out of the spiral motion must be greater than the product of those tending to put it into the spiral (Fig. 9.6).

When the motion is a convergent one, the time to half amplitude is given by

$$t_{1/2} = \frac{0{\cdot}693}{E_2/D_2}\,\hat{t} \text{ secs} \tag{9.16}$$

When E_2 is negative, the motion is a divergent one, and the time to double amplitude is given in a similar way by

$$t_2 = \frac{0{\cdot}693}{(-E_2/D_2)}\,\hat{t} \text{ secs} \tag{9.17}$$

More exact solutions for the spiral mode can be obtained by increasing the accuracy of the root λ.

2.3.2. EFFECT OF FACTORS

From eqn. (9.14) it can be seen that the main factors influencing the spiral stability of an aircraft are the derivatives l_v, n_r, n_v and l_r. The speed and height at which the aircraft is flying does not affect the

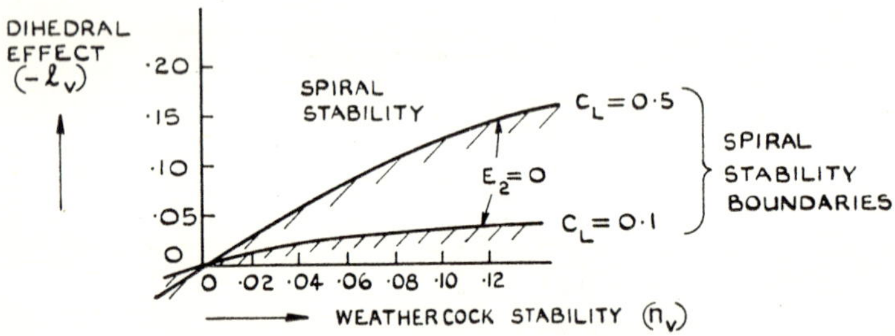

Fig. 9.7. Spiral stability boundaries

stability condition directly, except in as much as it affects the values of the four derivatives.

For given values of n_r and l_r, it is possible to draw the spiral stability boundary (i.e. $E_2 = 0$) on the $l_v - n_v$ diagram (Fig. 9.7). An aircraft is spirally stable for large values of $-l_v$ and for small values of n_v; a reduction in dihedral effect and an increase in weathercock stability both tend to make the aircraft more unstable.

However, because the values of the derivatives l_r and n_r vary with speed (i.e. with C_L), the position of the spiral stability boundary also varies. Because the main effect of increasing C_L is to increase the value of l_r, which is destabilizing in these considerations, the effect of reducing speed is to rotate the spiral stability boundary anti-clockwise. Thus an aircraft which possesses an n_v, l_v combination that is stable at high speeds, may become spirally unstable at low speeds (Fig. 9.8). For a given value of C_L (i.e. V_i), the spiral stability boundary is independent of height (i.e. of μ_2). It is also independent of the inertia characteristics (i.e. i_A and i_C) of the aircraft.

In practice, on some modern designs, the values of the l_v and n_v derivatives will vary quite appreciably throughout the speed range of the aircraft. Sweepback is equivalent to dihedral, the effect of sweepback on $-l_v$ increasing with C_L. Consequently, an aircraft with a moderate amount of l_v at high speeds will have too much at low speeds and alternatively an aircraft with a reasonable amount of l_v at low speeds will be deficient at high speeds. Too much dihedral leads to Dutch Roll oscillation problems (para. 2.4.5), but too little leads to spiral instability.

Lines of constant values of $t_{1/2}$ and t_2 can also be drawn on the l_v/n_v diagram. The characteristic times of the motion will, however,

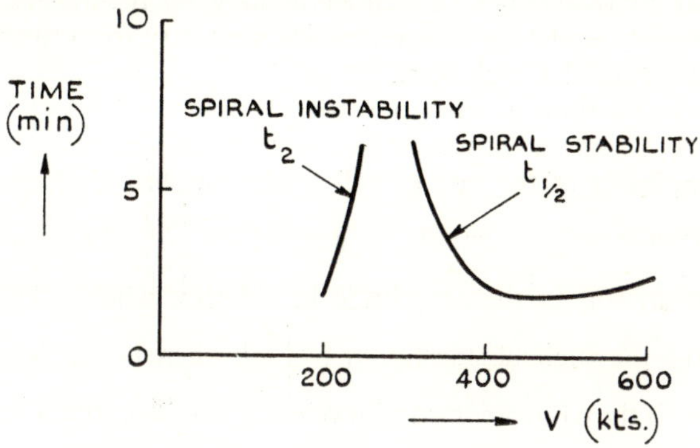

Fig. 9.8. Effect of forward speed on spiral stability

unlike the stability boundary itself, be dependent on the height of the aircraft, since, although the value of E_2/D_2 will vary little with height at a constant value of V_i, the unit of aerodynamic time will increase with height as $1/\sqrt{\rho}$.

The factors affecting spiral stability are discussed in more detail in Refs. 4 and 5.

2.4. Dutch Roll Oscillation

2.4.1. GENERAL

The Dutch Roll oscillation is a motion involving rolling, yawing and sideslipping motions. It is a very important factor in the pilot's assessment of the lateral handling characteristics of an aircraft. The period of oscillation is relatively short, usually too short to enable the pilot to control the motion satisfactorily, and, if the oscillation is poorly damped, then it may seriously detract from the basic handling qualities of the aircraft, quite apart from any interference with the operational role of the aircraft (e.g. gunnery, bombing or photography). Because it may be outside the control of the pilot, it is obvious that the oscillation can be a very dangerous one when it is undamped. For these reasons, the requirements of Av.P.970 (Vol. 1, Book 1, Part 6) state that any oscillations should be damped to half amplitude within one cycle (i.e. $\delta > 0{\cdot}693$).

A detailed analysis of the lateral oscillation is given in Ref. 11. This report discusses the "classical" approach to the lateral stability and control characteristics of conventional aircraft and shows how various approximate solutions for the oscillatory characteristics can be obtained by analysing motions with a limited number of degrees of freedom. The main purpose of the report, however, is to discuss the lateral characteristics of slender delta wing planforms, and to show how the oscillatory characteristics may vary from those of the more conventional Dutch Roll.

The roll/yaw ratio is an important parameter of the Dutch Roll oscillation. It can be expressed in the form of p/r, i.e. as the ratio of the amplitude of the rate of roll to the amplitude of the rate of yaw. Alternatively, it can be expressed as $\left|\frac{\phi}{\beta}\right|$, the ratio of the amplitude in roll to the amplitude in sideslip, when ϕ is the roll angle and β the angle of sideslip, or by $\left|\frac{\phi}{v_e}\right|$, where ϕ is in degrees and $v_e\,(=v\sqrt{\sigma})$ is the equivalent sideslip velocity in ft/sec. If β is measured in degrees, then, since $\frac{v}{V} \simeq \frac{\beta}{57{\cdot}3}$ rads

$$\left|\frac{\phi}{v_e}\right| = \frac{57{\cdot}3}{V\sqrt{\sigma}}\left|\frac{\phi}{\beta}\right| \text{ degrees/ft per sec} \tag{9.18}$$

where V is the forward speed (T.A.S.) in ft/sec and both ϕ and β are measured in degrees.

2.4.2. SIMPLE ANALYSIS

As mentioned above, approximate relationships for the periodic time and damping of the lateral oscillation can be obtained by analysing simple one degree or two degree freedom of motion equations. These simple theories are extremely useful in illustrating the physical principles involved and in showing how the various terms arise in the equations for the periodic time, the damping, and the roll/yaw ratio.

The simplest motion to analyse is one with only a single degree of freedom. If the aircraft is restrained so that the lateral oscillation consists of movement only in yaw, then the equation of motion for the aircraft can be written

$$C\ddot{\psi} + N_r\dot{\psi} + N_\beta\beta = 0 \tag{9.19}$$

where C is the yawing moment of inertia of the aircraft, N_r the damping in yaw coefficient, N_β (or N_v) the weathercock stability derivative, $\ddot{\psi}$ the angular acceleration in yaw, $\dot{\psi}$ (i.e. r) the rate of yaw, and β the angle of sideslip (which in this case is numerically equal to ψ, the angle of yaw).

This is, in effect, a second order differential equation with a single variable ψ; the N_r term represents the damping, and the N_β (or N_v) term the restoring (i.e. the spring) effect. By comparing this equation with that of the spring-mass-damper system considered in Chapter 7, the relationships for the periodic time and the damping of the oscillation can be derived. These results can then be expressed in terms of the non-dimensional derivatives. The periodic time of the oscillation is given by

$$T = 2\pi\sqrt{\left(\frac{i_C}{\mu_2 n_v}\right)}\,\hat{t} \tag{9.20}$$

the time to one half amplitude by

$$t_{1/2} = -1{\cdot}386\left(\frac{i_C}{n_r}\right)\hat{t} \tag{9.21}$$

and the corresponding log. dec. by

$$\delta = -\pi\sqrt{\left(\frac{i_C}{\mu_2 n_v}\right)}\,\frac{n_r}{i_C} \tag{9.22}$$

Since $V_i = V\sqrt{\sigma}$, $\mu_2 = \dfrac{W}{g\rho S(b/2)}$ and $\hat{t} = \dfrac{W}{g\rho SV}$

these equations can be rewritten in terms of V_i, $\sqrt{\sigma}$ and W/S

$$T = 16{\cdot}08\sqrt{\left(\frac{i_C b}{n_v}\right)}\sqrt{\frac{W}{S}}\,\frac{1}{V_i}\ \text{secs} \tag{9.23}$$

$$t_{1/2} = -18{\cdot}15\,\frac{1}{\sqrt{\sigma}}\,\frac{W}{S}\left(\frac{i_C}{n_r}\right)\frac{1}{V_i}\ \text{secs} \tag{9.24}$$

$$\delta = -0{\cdot}614\sqrt{\sigma}\sqrt{\left(\frac{i_C b}{n_v}\right)}\sqrt{\left(\frac{1}{W/S}\right)}\,\frac{n_r}{i_C} \tag{9.25}$$

In these equations V_i is the equivalent airspeed in ft/sec, σ is the relative density, b is the wing span in feet, and W/S the wing loading in lb/ft^2.

In this analysis, the roll/yaw ratio will, by definition, be zero.

Another simple analysis can be made by using the stability quartic equation (eqn. 9.1). The first two terms of this equation can be used to represent the roll subsidence mode and the last two terms to represent the spiral stability mode. By dividing the corresponding factors into the quartic a quadratic equation is left which represents the Dutch Roll oscillation. The two complex roots representing the oscillatory mode can be obtained from this quadratic equation in the usual way. If it is assumed that the real part of the roots is small, then some approximate relationships can be derived. If the complex pair of roots is written as $-\mu \pm i\omega$, then from this simplified analysis it can be shown (Ref. 3, page 173) that

$$\mu \propto (B_2C_2D_2 - A_2D_2^2 - E_2B_2^2) \tag{9.26}$$

and

$$\omega = \sqrt{\left(\frac{D_2}{B_2}\right)} \tag{9.27}$$

By substituting the expressions given in Appendix A for D_2 and B_2 and simplifying further (i.e. by neglecting all the terms other than the dominant n_v one) we can again obtain the relationship for periodic time given in eqn. (9.20).

Equation 9.26 shows that the damping of the lateral oscillation is dependent on the numerical value of a particular grouping of the coefficients of the stability quartic. This arrangement of coefficients is known as Routh's Discriminant, R

$$R = B_2C_2D_2 - A_2D_2^2 - E_2B_2^2 \tag{9.28}$$

and for the oscillation to be damped R must have a positive value (i.e. the real part of the root must have a negative value). A negative value of R represents a divergent oscillation and when $R = 0$ the oscillation is neutrally damped.

2.4.3. EFFECT OF VARIATION OF SPEED AND HEIGHT

If it is assumed that the lateral derivatives are constants, the results of the simple theory, expressed in eqns. (9.23), (9.24) and (9.25), can be used to derive the effect of the variation of speed and height on the oscillatory characteristics.

Equation (9.23) shows that the periodic time decreases with increase of forward speed, such that the product TV_i (or TV) remains constant. An increase of altitude is seen to have no effect on the periodic time at the same equivalent airspeed V_i. Some typical curves showing the variation in the periodic time are given in Fig. 9.9.

Equation (9.24) shows that the time to half amplitude varies with increase of forward speed in a similar way to the periodic time; $t_{1/2}V_i$ should remain constant. Altitude does have an important effect, however, the time to half amplitude increasing with increase of altitude as $1/\sqrt{\sigma}$ (i.e. $1/\sqrt{\rho}$) at a given value of V_i. Since the

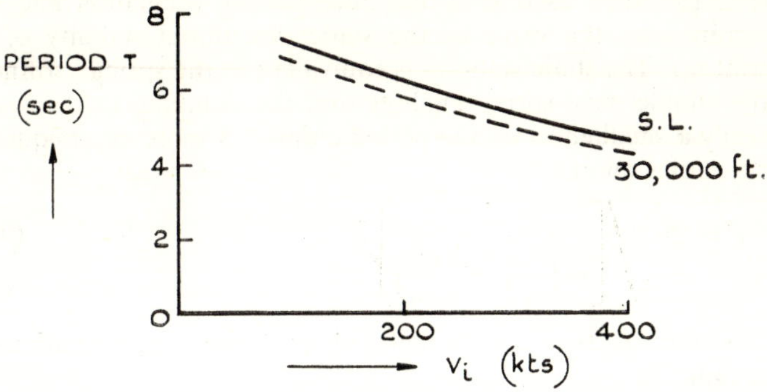

Fig. 9.9. Effect of speed and height on the periodic time of the Dutch Roll oscillation

periodic time and the damping time vary in the same way with change in forward speed, the log. dec. ($\delta = T/t_{1/e}$) is little affected and remains constant with increasing forward speed. Eqn. (9.25) shows, however, that $\delta \propto \sqrt{\sigma}$ and thus it decreases with increase

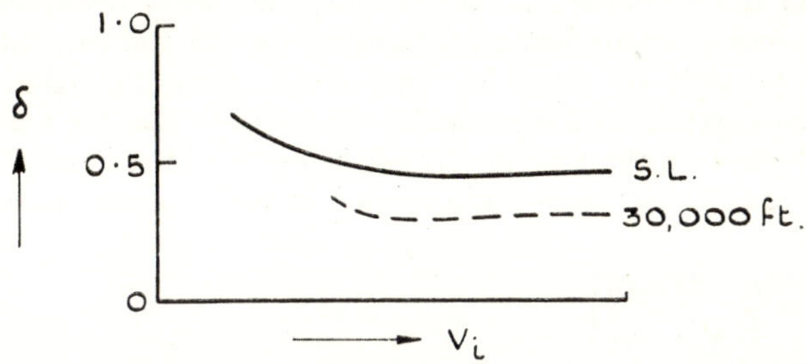

Fig. 9.10. Effect of speed and height on the damping of the Dutch Roll oscillation

of height. Fig. 9.10 shows some typical curves of δ values for various forward speeds and heights.

2.4.4. MORE EXACT THEORY

The results obtained from the simple theory of para. 2.4.2 can be improved by allowing for additional degrees of freedom in the oscillatory motion (or, alternatively, by leaving in more of the terms in the expressions for B_2, C_2 and D_2). The results obtained from these improved theories can be expressed in a number of different forms, depending on the actual assumptions made.

Equation (9.20) is usually found to be sufficiently accurate to give the periodic time of the Dutch Roll oscillation, i.e. the periodic time is the same as that of the pure yawing oscillation and it is determined by the value of the static directional stability of the aircraft n_v. The static stability is equivalent to the spring "stiffness" in the simple mass-spring analogy and the damping in the system has only a small effect on the periodic time. A more exact equation is sometimes quoted as

$$T = 2\pi \sqrt{\left[\frac{1}{\mu_2\left(\frac{n_v}{i_C}\right) + \left(\frac{i_E l_v}{i_C i_A}\right)}\right]} \hat{t} \tag{9.29}$$

Equation (9.22) is usually replaced by a more exact relationship (Appendix B)

$$\delta = -\pi\left[\sqrt{\left(\frac{i_C}{\mu_2 n_v}\right)}\right]\left(\frac{n_r}{i_C} + y_v\right) \tag{9.30}$$

This shows that the y_v (i.e. the sideforce due to sideslip) term is an important one in the damping of the lateral oscillation and that it is important to take the sideways motion of the aircraft into account in the analysis.

When rolling motion is allowed for, the theory determines the corresponding approximate relationship for the roll/yaw ratio. It is not possible to generalize this result, since the relationship obtained depends on the particular assumptions that are made and these depend on the type of aircraft being considered (see Chapter 17). The roll/yaw relationship quoted in Ref. 7 is the more usual one

$$\frac{p}{r} = \frac{l_v}{l_p}\sqrt{\left(\frac{\mu_2 i_C}{n_v}\right)} \tag{9.31}$$

but in Ref. 8 an alternative expression used for the Lightning aircraft is quoted as

$$\frac{p}{r} = \frac{i_C l_v / i_A n_v}{1 + i_E l_v / i_A n_v} \tag{9.32}$$

Eqn. (9.32) agrees with the relationship given in Ref. 15. The time-vector method is very useful in illustrating the need for different assumptions in different cases and in showing the validity, or otherwise, of the approximations used. The roll/yaw ratio can be expressed in terms of ϕ/β, a similar relationship to that in eqn. (9.31) usually being quoted

$$\frac{\phi}{\beta} = \frac{l_v}{l_p}\sqrt{\left(\frac{\mu_2 i_C}{n_v}\right)} \tag{9.33}$$

If the values of the derivatives are assumed to be constant, then these equations indicate that the roll/yaw ratio will remain constant independent of forward speed. Usually, however, on modern aircraft the value of $-l_v$ and hence the value of the roll/yaw ratio, decreases with increasing speed (Fig. 9.11). Eqn. (9.31) shows that on an aircraft obeying this relationship there is a direct height effect, the roll/yaw ratio increasing with height; this agrees with the results of Ref. 6. This variation is not shown by eqn. (9.32) and the results of Ref. 8; the only effect of height in this case would be due to the

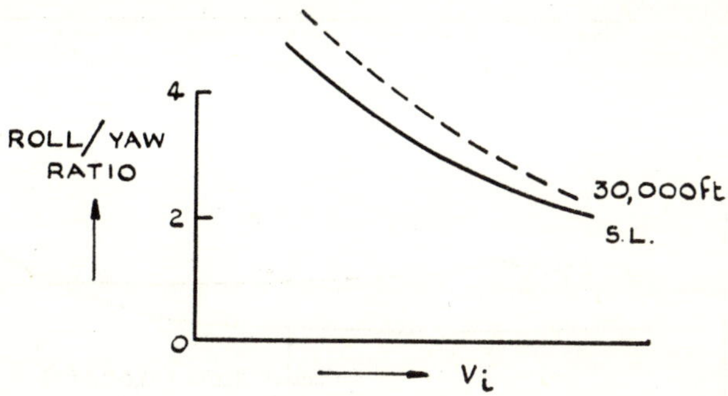

Fig. 9.11. Effect of speed and height on the roll/yaw ratio of the Dutch Roll oscillation

effect of variations of C_L and Mach number on the values of the derivatives.

Fig. 9.12, taken from Ref. 7, illustrates the amount of agreement between the results predicted by these approximate formulae and the results obtained from flight tests. More comparisons between theory and experiment are given in Ref. 6.

In order that an aircraft should possess satisfactory flying characteristics, δ should be greater than 0·693 (Av.P.970, Vol. 1, Book 1, Part 6, Chapter 602) and p/r less than about 3·0. Fig. 9.13 shows the boundaries between the satisfactory and unsatisfactory combinations of δ and p/r and the curves drawn show the calculated effects of speed and height on the Dutch Roll characteristics of a particular aircraft with rather poor basic oscillatory characteristics.

2.4.5. EFFECT OF FACTORS

The results of para. 2.4.3 were obtained on the basis of the simplifying assumption that the values of the derivatives are independent of speed and height. In practice, the values of the derivatives will vary and this variation will modify the simple results obtained (i.e. $TV =$ constant, $\delta \propto \sqrt{\sigma}$, etc.).

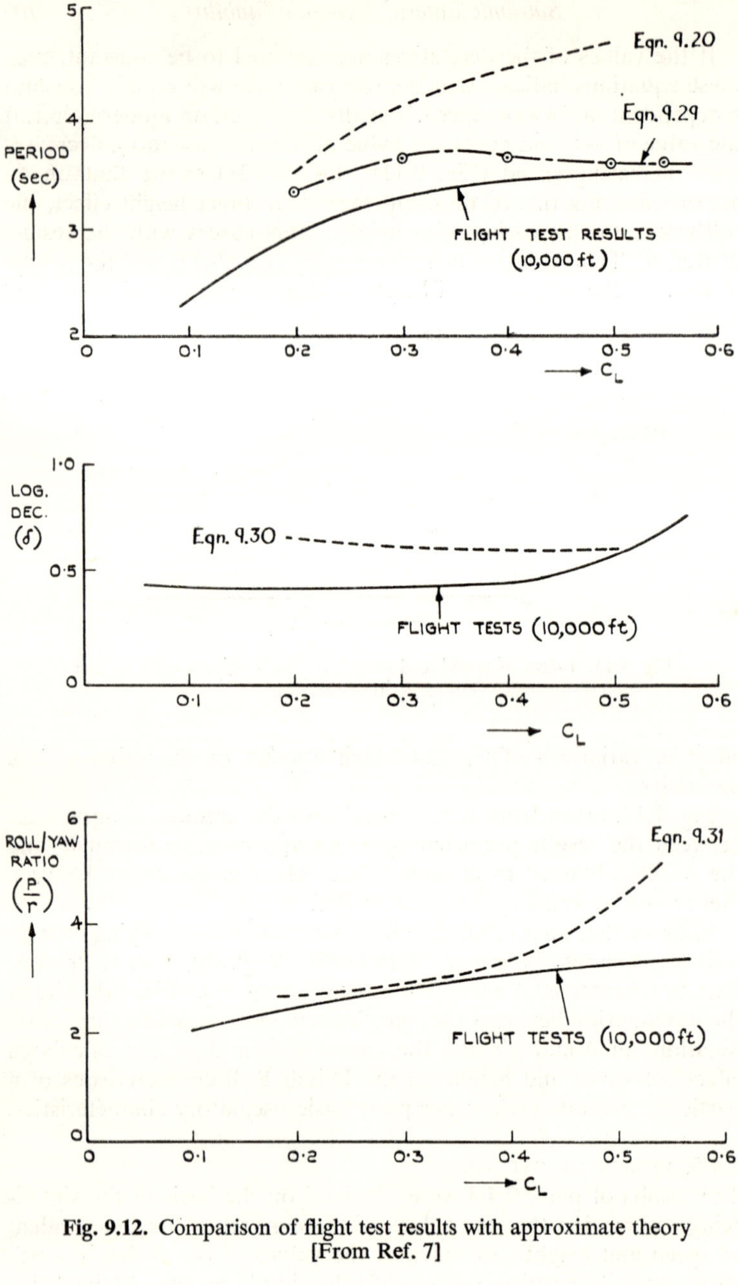

Fig. 9.12. Comparison of flight test results with approximate theory [From Ref. 7]

The effect of varying the values of the derivatives can be conveniently shown by the types of one and two parameter stability diagrams and iso-opinion plots discussed in para. 2.4 of Chapter 8. It is important to investigate the separate effects of the various derivatives on the oscillatory motion of the aircraft, in order to be able to determine the modifications that may be necessary to improve a particular undesirable characteristic. This may involve a change in the basic design of the aircraft (e.g. an increase or decrease in the

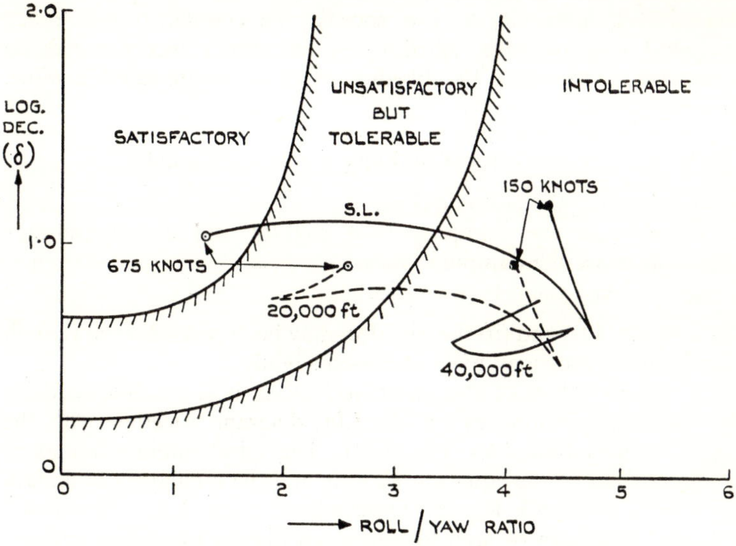

Fig. 9.13. Effect of speed and height on the oscillatory characteristics

dihedral effect, i.e. l_v), or the installation of an automatic damping system (e.g. a roll damper to increase the numerical value of l_p).

The most important derivatives affecting the lateral oscillation are l_v and n_v, but n_r, n_p and l_p may also be important factors. The effects of varying the values of these derivatives on the oscillatory characteristics of the Boulton Paul IIIA delta aircraft are discussed in Ref. 6. The actual effects will depend upon the particular aircraft being considered, but the results have been generalized as far as possible below.

(*a*) *Dihedral Stability*, l_v

An increase in dihedral effect (i.e. an increase in the value of $-l_v$) will usually

(i) decrease the periodic time slightly,

(ii) decrease the damping root and increase the damping time,
(iii) increase the roll/yaw ratio.

As a result of (i) and (ii) the log. dec. δ ($= T/t_{1/e}$) will be decreased, though this effect is not indicated by the approximate relationship of eqn. (9.30).

Modern sweptback wing aircraft have the disadvantage that l_v varies considerably through the speed range, the value of $-l_v$ increasing with increasing C_L. This often leads to problems of Dutch Roll instability at low speeds. To counteract this effect, anhedral (i.e. negative dihedral) is sometimes incorporated on swept wings, to reduce the dihedral effect to a more suitable value.

(*b*) *Weathercock Stability*, n_v

An increase in weathercock stability (i.e. n_v) will usually

(i) decrease the periodic time,
(ii) increase the damping root slightly and correspondingly decrease the damping times,
(iii) decrease the roll/yaw ratio.

As a result of (i) and (ii) the log. dec. may be decreased, even though the damping root may have increased slightly.

For given values of the other derivatives it is possible to draw the oscillatory boundary on the l_v/n_v diagram, together with the spiral stability boundary (Fig. 9.14). The spiral stability boundary is represented by the equation $E_2 = 0$, and the oscillatory boundary by $R = 0$, where R is Routh's Discriminant (para. 2.4.2). Too much dihedral effect can be seen to lead to oscillatory instability and too little to spiral instability. Similarly, too much weathercock stability leads to spiral instability and too little to oscillatory instability. The positions of the two boundaries depend on the value of C_L at which the aircraft is flying, an increase in C_L reducing the size of the stable region. In Ref. 13 (pages 265–6), the effects of variations of l_v and n_v on the oscillatory characteristics have been represented in the form of root locus plots.

(*c*) *Damping in Yaw*, n_r

An increase in the damping in yaw (i.e. in $-n_r$) will usually

(i) have little effect on the periodic time,
(ii) increase the damping root and decrease the damping time,
(iii) have little effect on the roll/yaw ratio.

As a result of (i) and (ii), the value of δ will be increased.

An increase in yaw damping will be beneficial for both spiral and oscillatory stability. It should be noted that any additional side area behind the C.G. increases both n_v and $-n_r$. In this case the period will be decreased because of the n_v effect ($-n_r$ having little effect on the period), but the value of δ will be increased due to the increase in $-n_r$ and decreased due to the increase in n_v. Any additional side area ahead of the C.G. increases $-n_r$ and decreases n_v, both the period and the value of δ will then be increased. The value of $-n_r$ can be increased, without changing the value of n_v, by incorporating an automatic yaw damper into the control system.

(*d*) *Damping in Roll,* l_p

An increase in the value of $-l_p$ will usually

(i) have little effect on the periodic time,
(ii) reduce the damping time slightly,
(iii) decrease the roll/yaw ratio.

As a result of (i) and (ii), the value of δ may be slightly increased.

(*e*) *Relative Density Parameter*

The parameter $\mu_2(= 2W/g\rho Sb)$ occurs in a number of the equations we have discussed. It represents the effect of both height and wing

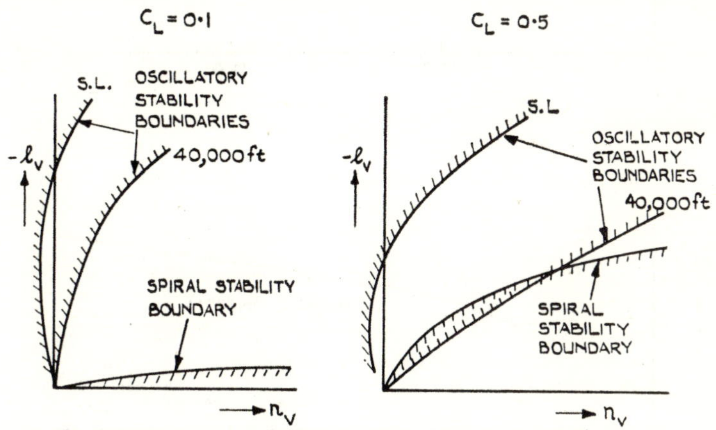

Fig. 9.14. Oscillatory and spiral stability boundaries

loading on the oscillatory characteristics. The effects of height have been discussed in para 2.4.3; Fig. 9.14 shows that an increase in height rotates the oscillatory boundary in a clockwise direction, further reducing the size of the stable region and making it more difficult for the designer to obtain values of l_v and n_v that will give satisfactory handling characteristics over the complete speed and

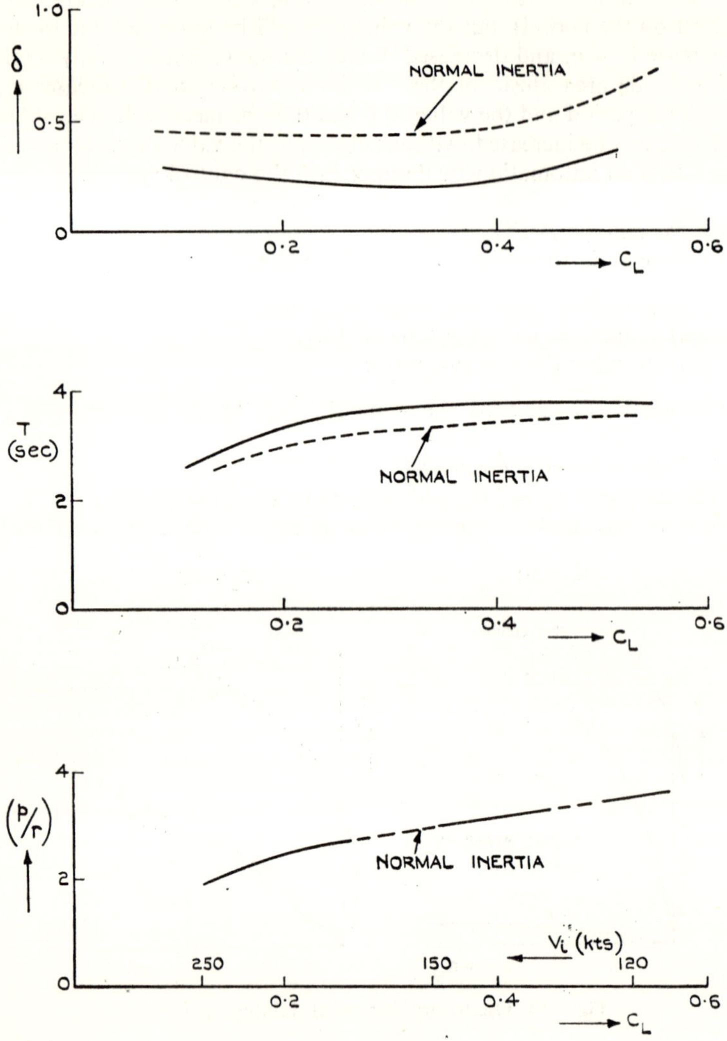

Fig. 9.15. Oscillatory characteristics of the Avro 707B at 10,000 ft [From Ref. 7]

height range of the aircraft. In this event some instability may have to be accepted under certain conditions of flight. Because spiral stability is not as critical as oscillatory stability, the aircraft is usually designed so that any instability that is present occurs in the spiral mode. In some circumstances (e.g. low speeds, flaps and undercarriage down) even the oscillatory stability may become weakly damped, or even slightly divergent.

An increase of wing loading has a similar effect to that of an increase of height in reducing the damping, and it also increases the periodic time slightly.

(*f*) *Rolling and Yawing Inertia Coefficients*

An increase in the yawing and rolling inertias of the aircraft may result in an increase or a decrease in the damping of the oscillation (Ref. 14, page 425). The period is little affected by a change in the rolling inertia but increases with increase in yawing inertia. Some flight test results showing the effect of increasing the rolling inertia by 60 per cent and the yawing inertia by 17 per cent are shown in Fig. 9.15, taken from Ref. 7. In this case, the damping is roughly halved, the period slightly increased and the roll/yaw ratio unaltered.

The terms containing the product of inertia i_E are usually neglected, though in certain cases they may have an important effect on the oscillatory characteristics.

The effect of the various factors on the oscillatory stability characteristics are discussed in more detail in Refs. 4 and 5.

3. Control-free Lateral Dynamic Stability

3.1. General

Freeing either the aileron or the rudder control surfaces may have an important effect on the lateral dynamic characteristics of an aircraft which is fitted with non-powered control systems. Freeing the ailerons does not usually affect the characteristics appreciably; the rolling, spiral and Dutch Roll motions are much as before, but, in addition, there are either two heavily damped convergent motions, or a heavily damped oscillatory one, involving roll and aileron deflection. Freeing the rudder, on the other hand, does not affect the rolling and spiral motions very much, but it may considerably reduce the damping of the lateral oscillation and in addition a heavily-damped short-period oscillation of the rudder will be present. In these circumstances, the lateral oscillation of the aircraft will consist largely of a yawing motion and the term "Snaking" is often used. Both cases are considered in Ref. 2 (pages 458–470); only the rudder-free problem will be considered here, however, since this usually is the more important of the two.

The analysis is similar to that of the control fixed case considered in para. 2 but the control-free values of the lateral derivatives have to be used instead of the control-fixed ones. Additional equations of motion are required, representing the balance of the inertial and aerodynamic hinge moments acting on the particular control surface and other factors become important (e.g. the control surface hinge moment characteristics, circuit friction, etc.).

The effect of the movement of the control surfaces on the oscillatory characteristics is determined by the magnitude of the control deflection, and the particular phasing of the control surface motion relative to that of the aircraft itself. Depending on the phase angle involved, the main effect of freeing the controls may be to increase or decrease either the corresponding static stability or the oscillatory damping. Any increase in static stability will tend to reduce the periodic time of the oscillation, whereas an increase in damping will reduce the time to half amplitude.

The effect of moving control surfaces on the oscillatory characteristics can be usefully employed in autostabilization systems (see Chapter 19). If the rudder surface is moved, with the appropriate phasing, in proportion to the measured rate of yaw, the yaw damping can be increased. Alternatively, if the rudder is moved in proportion to the sideslip angle, then the value of the weathercock stability can be altered as desired. In a similar way, by using the aileron control surfaces, the damping in roll and the dihedral effect can be increased or decreased as necessary.

The results of Ref. 7, however, show little overall effect resulting from freeing the aileron and rudder controls; freeing the rudder increased the damping slightly, but freeing the ailerons decreased the damping again to its initial control-fixed value and the period and roll/yaw ratio were unchanged.

3.2. Rudder-free Lateral Dynamic Stability

3.2.1. RUDDER-FREE LATERAL DERIVATIVES

When an aircraft is sideslipping and yawing, aerodynamic forces act on the rudder control, producing a moment about the hinge line of the surface. If the control is free to move, then it will rotate under the action of these hinge moments, additional hinge moments coming into effect (i.e. the b_2 effect) opposing the movement of the control. In steady conditions (e.g. steady sideslips), the control surface will reach an equilibrium position when the moments acting one way balance those acting in the opposite direction. When the aircraft is oscillating, the control surface will also oscillate at the same frequency, but the exact phasing of the two motions will depend on such factors as the inertia characteristics of the surface, and the friction in the circuit. If the control-free dynamic characteristics are unsatisfactory in any way, they can be modified by varying the magnitude and/or the phasing of the control surface movement.

When the aircraft is in a steady sideslip, the hinge moment acting on the rudder will be of magnitude $b_1\beta$, where b_1 is the "floating" parameter $\dfrac{dC_H}{d\beta}$ of the rudder control. If the rudder is free to move, it will move through an angle ζ, given by

$$\zeta = \frac{b_1}{b_2}\beta \tag{9.34}$$

Since control surfaces are designed so that b_2 always has a negative value, the direction that the rudder floats depends on the sign of b_1; if b_1 is negative, as is more usual, the control floats with the wind, if b_1 is positive, the control floats against the wind.

When the control floats with the wind, the aerodynamic sideforce produced by the fin and rudder combination is reduced. Since the fin produces a stabilizing contribution to the overall value of n_v, the effect of freeing the control is to reduce the overall directional static stability (i.e. the movement of the rudder produces an additional destabilizing contribution and the control-free directional static stability is less than the corresponding control-fixed value). In a similar way, when the control floats against the wind, the control-free stability is greater than the control-fixed value.

Considering only the fin and rudder contribution to the directional stability, the control-free value can be expressed as

$$n_{v\mathit{free}} = n_{v\mathit{fixed}}\left(1 - \frac{a_2 b_1}{a_1 b_2}\right) \tag{9.35}$$

where $n_{v\mathit{fixed}}$ is the corresponding contribution of the fin and rudder to the control-fixed stability. This equation should be compared with eqn. (2.30), which shows that the tailplane contribution to the longitudinal static stability is reduced by exactly the same factor (except, of course, that the a_1, a_2, b_1 and b_2 coefficients in eqn. (2.30) are the particular values for the elevator control).

If the aircraft is rotating at a steady rate of yaw, similar considerations show that aerodynamic hinge moments will again be produced. If the control floats with the relative wind the yaw damping is decreased, if it floats against the wind the yaw damping is increased.

3.2.2. EFFECT OF RUDDER MOTION ON THE OSCILLATORY CHARACTERISTICS

During a Dutch Roll oscillation the aircraft will be both yawing and sideslipping. If b_1 is negative, and there is no friction in the system, the control will move in phase with the aircraft (i.e. maximum

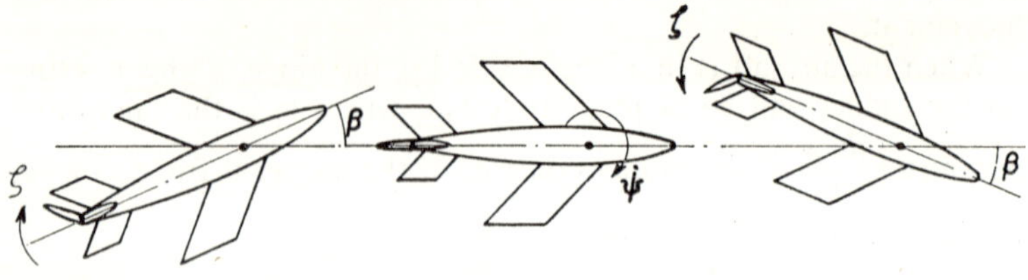

Fig. 9.16. Rudder and aircraft motion in phase

positive sideslip gives maximum positive rudder deflection, Fig. 9.16). In this case the rudder reduces the directional static stability and consequently the frequency of the snaking oscillation (i.e. the periodic time is increased). If, on the other hand, the rudder motion lags 90° behind the motion of the aircraft (i.e. the rudder deflection is a maximum in the positive direction when the aircraft is yawing to starboard at its maximum rate, Fig. 9.17), it is in phase with the

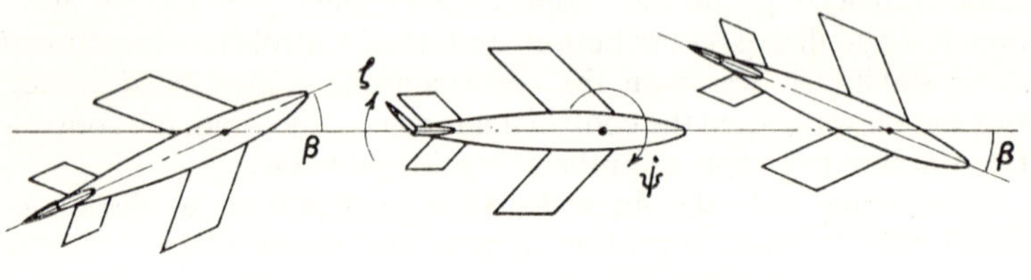

Fig. 9.17. Rudder motion lagging by 90°

yaw and since it opposes the motion the damping is increased. The results are reversed when b_1 is positive and an increase of friction reduces the damping of the oscillation.

The effect of freeing the control surface can be determined by solving the complete equations of motion, though, as in previous work, it is usually possible to simplify the analysis, and to consider only a limited number of degrees of freedom. Solutions for the rudder-free problem can be obtained by considering the motion to involve only two degrees of freedom, i.e. yawing oscillations of the aircraft along a straight flight path and rudder deflection (Ref. 2, page 459).

The important factors involved in this problem are the rudder hinge moment characteristics b_1 and b_2 and the friction, mass balance, and inertia characteristics of the rudder circuit. The results obtained are represented diagrammatically in Fig. 9.18. This shows that the control-free lateral oscillation tends to become unstable for small values of $-b_2$ (i.e. closely balanced controls) and positive values of b_1 (i.e. when the control tends to float against the wind). The instability is made worse (i.e. the oscillatory boundary

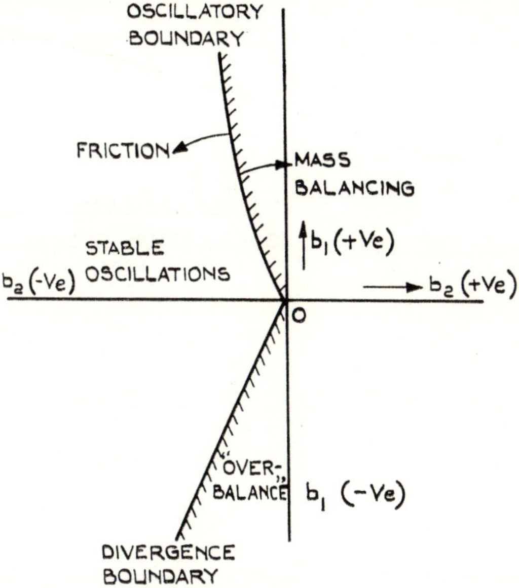

Fig. 9.18. Effect of hinge moment characteristics on the lateral oscillation

moves to the left) as a result of any friction or mass unbalance in the control system. When b_1 is large and negative and b_2 small and negative, rudder overbalance occurs; this problem will be discussed in Chapter 11.

3.2.3. EFFECTS OF POSSIBLE MODIFICATIONS

Various methods have been suggested for improving the characteristics of the snaking oscillation (Refs. 9 and 10). These obviously involve changing the values of b_1 and b_2, and modifying the friction and mass balance characteristics. If the oscillations are divergent the value of b_1/b_2 must be made more positive, either by making b_1

less positive or more negative, or by making b_2 more negative. These alterations can be made by

(*a*) Removing or reducing the amount of set-back hinge or horn balance.
(*b*) Reducing the trailing edge angle of the control surface.
(*c*) Fixing cords or metal strips to the trailing edge of the controls.

When b_1 is negative, the damping of the snaking oscillation can be improved by incorporating a viscous damper into the rudder

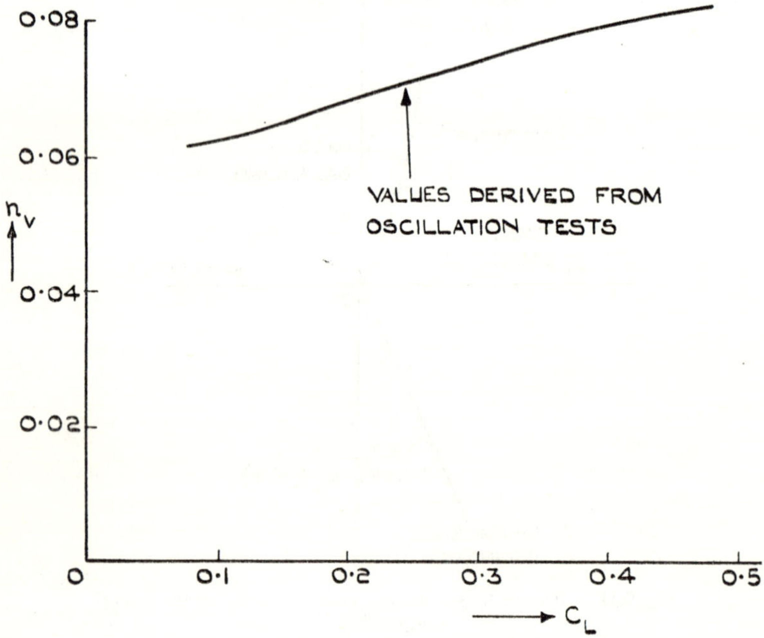

Fig. 9.19. Variation of n_v with C_L

control circuit (Ref. 9). This makes the control surface lag behind the motion of the aircraft and, as we have seen above, when b_1 is negative this produces an additional damping. Ref. 10 discusses the snaking problem on the Meteor and describes the installation of an auto-stabilization system to improve the damping.

4. Flight Test Results

It has generally been possible to design low speed aircraft so that they have satisfactory lateral handling characteristics. The Dutch Roll oscillation has, however, always been more of a problem than the longitudinal oscillatory motions and consequently more flight test results are available. Some typical variations of damping, periodic time and roll/yaw ratio are shown in Figs. 9.12 and 9.15; these results are in good agreement with the predictions of the simple theory discussed in para. 2.4.3. In Ref. 6 reasonably good agreement is obtained between the measured values of the oscillatory characteristics and the values estimated using the time-vector method.

In Ref. 7 the measured periodic times are used with eqn. (9.29) to determine the numerical values of the directional static stability. The results obtained are shown in Fig. 9.19; in the reference the n_v values are compared with those obtained by other methods (e.g. steady sideslip tests, wind tunnel tests, etc.).

As in the case of longitudinal dynamic stability, the recent design changes that have occurred on modern high-performance aircraft have resulted in a deterioration of the lateral oscillatory characteristics. Under certain conditions of flight (e.g. low speeds on the approach) the damping of the oscillation may be marginal or even negative. This has an important effect on the pilot's assessment of the handling characteristics of the aircraft. The problem cannot easily be eliminated by modification of the basic aircraft design and yaw dampers have to be incorporated into the control system.

References

1. ETKIN, B.; *Dynamics of Flight* (John Wiley and Sons, New York, 1959).
2. PERKINS, C. D. and HAGE, R. E.; *Airplane Performance Stability and Control* (John Wiley and Sons, New York, 1949).
3. DUNCAN, W. J.; *Control and Stability of Aircraft* (Cambridge University Press, 1952).
4. PRIESTLEY, E.; *A further investigation of lateral stability*, R.A.E. Report No. BA 1662, R. & M. 1989, 1941.
5. PRIESTLEY, E.; *An investigation of the periods and dampings of lateral motions*, R.A.E. Report No. Aero 1720, 1941.
6. NETHAWAY, J. E. and CLARK, J.; *Flight tests to investigate the dynamic lateral stability characteristics of a* 45° *delta, cropped to give three aspect ratios*, R.A.E. Tech. Note Aero 2671, R. & M. No. 3243, 1960.
7. PERRY, D. H., MORRALL, J. C. and PORT, W. G. A.; *Low speed flight tests on a tailless delta aircraft* (*Avro* 707*B*), *Part III Lateral Stability and Control*, Unpublished M.O.A. Report.
8. *Aerodynamics Technical Note Ae*76, English Electric Co., Ltd.
9. PINSKER, W. J. G.; *Flight tests on a Vampire Mk.*5 *with a hydraulically damped anti-snaking rudder*, R.A.E. Report Aero 2487, 1953.
10. DOETSCH, K. H.; *Meteor* 8 *snaking characteristics and possible means of improvement*, R.A.E. Tech. Note Aero 2069, 1950.
11. PINSKER, W. J. G.; *The lateral motion of aircraft and in particular of inertially slender configurations*, R.A.E. Report Aero 2656, 1961.
12. BISGOOD, P. L.; *A review of handling qualities research and its application to the handling problems of large aircraft*, R.A.E. Report Aero 2688, 1964.

13. Seckel, E.; *Stability and Control of Airplanes and Helicopters* (Academic Press, New York, 1964).
14. Babister, A. W.; *Aircraft Stability and Control* (Pergamon Press, Oxford, 1961).
15. Thomas, H. H. B. M. and Neumark, S.; *Interim note on stability and response characteristics of supersonic aircraft*, R.A.E. Tech. Note Aero 2412, 1955.

10 Aircraft Control Systems

1. Introduction

All aircraft have to be fitted with a control system that will enable the pilot to trim and manoeuvre the aircraft in flight about each of its three axes. On all conventional aircraft the controls are "aerodynamic" ones; the moments required about the C.G. of the aircraft being produced as a result of changes in the aerodynamic forces acting on the various parts of the aircraft. In the majority of cases, the changes in the aerodynamic forces are produced by means of flap-type control surfaces positioned at the extremities of the aircraft, the movement of the pilot's control causing the flap surfaces to move relative to the main structure of the aircraft. There are usually three separate control systems and three sets of surfaces, one set for controlling the aircraft about each of the axes. The pitching control is provided by the elevators, the rolling control by the ailerons, and the yawing control by the rudder. The elevators and rudder are usually part of the tail unit and the ailerons are usually at the wing tips, in order to provide the longest possible lever arm about the C.G. of the aircraft.

On some aircraft the effect of two of the controls is combined in a single set of control surfaces. On delta aircraft "elevons" are often used on the trailing edge of the wing; these surfaces deflect in the same direction to produce pitching control and differentially to produce rolling control. A few aircraft have employed V or "butterfly" tail units; on these aircraft the flap surfaces on the tail act together to provide pitching control and differentially to produce yawing control (and thus they can be referred to as "rudderators").

It is generally desirable that each set of control surfaces should produce a moment only about the corresponding axis. Unfortunately, in practice, moments are often produced about the other axes as well. The yawing moment produced by aileron deflection has always been an important effect, which has had to be allowed for, since the effect can generally be described as an "adverse" one.

Specially designed aileron surfaces have been used to reduce this adverse effect. On some modern aircraft, the rolling moment produced by rudder deflection may be important. These cross control problems can be eliminated by linking the separate control systems together. On the Belfast, for example, the rudder and aileron systems are linked together so that, when the rudder is moved by the pilot, the ailerons automatically deflect in such a way as to oppose the rolling moment produced by the rudder.

The aerodynamic forces acting on the control surface resulting from the application of the control oppose the motion of the surface. The forces produce a moment about the hinge line of the control surface which has to be overcome in order to position the control. The necessary power can be provided in various ways

(*a*) in a non-powered control system it is provided by the pilot exerting a force on his control column or rudder pedals,
(*b*) in a power control system it is obtained from either the hydraulic (or electrical) power supply of the aircraft,
(*c*) by a combination of these two methods, in which the pilot provides a proportion of the power required, the rest being obtained from the aircraft's hydraulic power supply.

The majority of high-performance civilian and military aircraft have fully powered control systems, a number have the power-boosted type of system, and many low performance aircraft still have manually operated systems. The characteristics of power controls are discussed in Chapter 12.

Irrespective of the way in which the control system is operated, the hinge moment characteristics of the control surfaces are very important. On a manual system, they determine the magnitude of the control forces that the pilot has to exert in order to trim and manoeuvre the aircraft over the speed and height range. Any control system must provide the pilot with satisfactory "feel" characteristics. Unfortunately, however, "feel" is not wholly quantitative, it is also subjective, and pilots do not agree with one another on what are "desirable characteristics." "Feel" is made up of a number of separate factors (e.g. the magnitude and direction of the control forces, the magnitude and direction of the control movements, the response of the aircraft to control application, control harmonization, etc.). Control forces are more important than control movements and in the "pressure-stick" system no control movements are involved. On conventional systems, the control forces and movements should lie between certain limits; the requirements for British military aircraft are laid down in Av.P. 970, and for civil aircraft in the B.C.A.R. publications. The forces must not be too light otherwise the pilot may overstress the aircraft, nor too heavy otherwise he will not be able to manoeuvre it. The stick movements must not be too small otherwise the controls are too sensitive, nor too large otherwise the controls will not fit into the restricted space of the cockpit. Manually operated control systems can be designed with satisfactory force and movement characteristics, as long as the range of possible flight conditions is not very large. On modern high-performance aircraft it becomes impossible to satisfy the requirements with this type of system and power operated controls are essential.

The size of the control movements involved can be modified by changing the mechanical gearing of the system and by incorporating non-linear gearings, etc. The control force characteristics can be improved by the addition of aerodynamic balancing to the control surfaces, by incorporating bob-weights and springs in the control system and by small modifications to the shape of the control surface. Even though the pilot does not have to provide the force required to move the control surface in a power control system, the size of the hinge moments involved are still important, because they determine the size of the hydraulic jack that is necessary. This must be optimized, so that the smallest and lightest possible jack can be used that will still operate the control surface over the required flight range of the aircraft. For this reason, aircraft fitted with power control systems still use some of the methods of aerodynamically balancing the control surfaces. This is, of course, essential if the system employs manual reversion in the event of a power system failure. The Vulcan has fully powered control systems, and has large internal sealed-nose balances on all the control surfaces. The Valiant had a fully powered system, with an internal sealed balance on the elevators and ailerons and a geared balance tab on the rudder.

One important problem associated with aircraft control systems is that of "flutter" (see Chapter 14). This can be prevented by "mass-balancing" the system. On a few low-performance aircraft the additional weight is added externally and can be seen attached to the surface of the control. Usually, however, the mass balance is stored internally, often inside the horn balance, in the leading edge of the control surface, or attached to the leading edge of the control surface on an arm which is inside the wing, the tailplane or the fin. In the case of power operated systems, the problem of flutter is eliminated if the control is a perfectly irreversible one, though, in practice, many fully duplicated power operated control systems still incorporate some mass balance.

The effectiveness of flap-type controls varies throughout the speed range of the aircraft. Since they are aerodynamic controls they depend on the dynamic pressure (i.e. $\frac{1}{2}\rho V^2$) and so increase in effectiveness with increase in forward speed, or, rather, with increase in the local airflow velocity. The elevator controls on a propeller-driven aircraft are, for example, made more effective by being in the high-velocity slipstream from the propeller. When the dynamic pressure becomes large, however, aeroelastic problems become important. There may be a loss of control effectiveness and even a reversal of control (Chapter 14). This may be particularly important for the lateral control system, and inboard ailerons have been used to reduce the effect (e.g. on the Gnat and on the Phantom). It is for this reason that the flap-type ailerons are sometimes replaced by spoiler controls, though, more usually, spoilers and ailerons are used in conjunction with one another. The application of the lateral control then deflects the aileron alone for some conditions of flight and the spoiler for others. On the F111 (TFX), for instance, the spoilers only operate when the wing is in the fully swept position.

The effectiveness of flap-type controls decreases at transonic speeds due to the formation of shock waves around the aircraft. Various methods are available for improving the control effectiveness under these conditions of flight (Chapters 15 and 16). The controls tend to regain some of their effectiveness again at supersonic speeds. On a number of high-performance military and civil aircraft, the conventional flap-type control surfaces have been replaced by "all-moving" surfaces. The most common occurrence of this has been on the pitch control system, where the tailplane/elevator combination has often been replaced by an all-moving or "slab" tailplane (with, possibly, a movable flap or elevator as well at low speeds). All-moving fins have been used on a small number of aircraft (e.g. on the X15 and the T.S.R.2). All-moving wing tip controls were tried on the Short Sherpa research aircraft. These controls are not very common on aircraft, though the large horn-balanced ailerons of the Bristol 188 research aircraft could almost be described as an all-moving wing tip control. On the T.S.R.2 the all-moving tailplane serves a dual purpose, when the surfaces act together they produce an "elevator" control, when they act differentially they produce "aileron" control. For this reason the surfaces are referred to as "tailerons."

The control surfaces are usually placed at the extremities of the aircraft to increase their effectiveness. The changes that have taken place in the design shapes of modern high-performance aircraft have tended to further reduce this effectiveness. The short tail arms have reduced the pitching effectiveness of the tailplane and elevator and the yawing effectiveness of the rudder, whereas the small span of the wings has reduced the aileron effectiveness. The loss of longitudinal control has been particularly serious on tailless delta aircraft and separate tailplanes have sometimes had to be added to the design in order to obtain the necessary manoeuvrability and spin recovery characteristics. Though the use of inboard ailerons reduces the effects of distortion, the lateral control effectiveness is also reduced. Spoilers, on the other hand, can produce a more effective lateral control without the problem of aeroelastic distortion.

At high forward speeds control "buzz" may occur; this consists of a very high frequency oscillation of the control surface, that is produced as a result of the movement of the shock waves around the control surface (see Chapter 14). This problem can be eliminated by the addition of metal strips on the control surfaces. Sometimes these strips are placed at right angles to the trailing edge of the control surface (e.g. the Meteor rudder), in other cases they have been positioned on the surface of the control a small distance ahead of the trailing edge (e.g. on the Gnat and the Hawker 1127 (Kestrel)).

The majority of control systems, both powered and non-powered ones, use mechanical signalling, the pilot's input to the control being transmitted by means of rods, cables, pulleys, levers, etc., to the control surfaces. Some aircraft have flown with electrical signalling systems (see Chapter 12) and this type of system is being designed for the BAC/Sud Concorde supersonic airliner (see para. 9, Chapter 12).

2. Basic Characteristics of Flap-type Control Surfaces

2.1. General

Fig. 10.1 shows an arrangement consisting of a movable control flap and tab attached to an aerofoil which can represent either a fixed

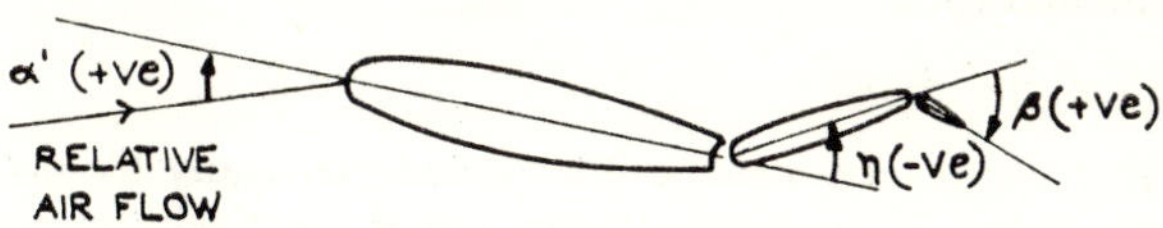

Fig. 10.1. Tailplane, elevator and trim tab system

tailplane, a vertical fin, or a wing, as desired. When the control surface is deflected relative to the aerofoil in a subsonic airflow, the pressure round both the aerofoil and the flap is affected. The resultant of all the aerodynamic forces acting on the aerofoil and flap produces a moment acting about the C.G. of the aircraft; this is the required pitching, rolling or yawing moment for which the controls have been designed. The resultant of those forces acting only on the control surface produces a moment about the hinge line of the control. This moment has to be balanced by an equal one provided either by the pilot himself, by the power control system of the aircraft, or by the deflection of the small tab (in this case a trim tab) at the trailing edge of the control. When the tab is deflected through an angle relative to the control surface, the pressure distribution round the tab is changed. The resultant force acting on the tab produces a moment about the hinge line of the control. By deflecting the trim tab the correct amount in the right direction, the moment of the tab can be made exactly equal and opposite to that produced by deflecting the control surface. This balances the hinge moments and the control surface is in equilibrium in the new position, with the control forces trimmed out.

2.2. Control Power (or Effectiveness) Characteristics

The object of the control surfaces is to produce a moment about the corresponding axis; the control powers being represented in derivative form by m_η, l_ξ, and n_ζ. If the system in Fig. 10.1 represents the tailplane, elevator and elevator tab combination, then the overall lift coefficient of the tailplane is given by

$$C_L' = a_0 + a_1\alpha' + a_2\eta + a_3\beta \tag{10.1}$$

where $a_1 = dC_L'/d\alpha'$, $a_2 = dC_L'/d\eta$ and $a_3 = dC_L'/d\beta$ are assumed to be constants, independent of the values of α', η, and β. Eqn. (10.1) shows that the overall lift is made up of three components, that due to the incidence of the tailplane, the additional component resulting from the deflection of the elevator and the third component resulting from tab deflection. If the main surface is symmetrical, or if the angle of attack of the aerofoil is measured from the no-lift condition, $a_0 = 0$, and since the $a_3\beta$ term is small eqn. (10.1) can be approximated to

$$C_L' = a_1\alpha' + a_2\eta \tag{10.2}$$

Neglecting the β term can be interpreted physically by saying that the small trim tab contributes little, either positively or negatively, to the overall lift of the complete tailplane system.

Eqn. (10.1) is represented diagrammatically in Fig. 10.2 and the effects of flap and tab deflection are also shown in Figs. 10.3 and 10.4. The slope of the basic C_L'/α' curve in Fig. 10.2 gives the value of the tailplane lift curve slope a_1. Flap deflection moves the curve vertically upwards by an amount $a_2\eta$, where η is the flap deflection and a_2, the elevator control lifting effectiveness, is represented by the slope of the straight line part of the curve in Fig. 10.3. Tab deflection moves the curve vertically upwards by a small amount $a_3\beta$, where β

is the tab deflection, and a_3 the tab lifting effectiveness is represented by the slope of the straight line part of the curve in Fig. 10.4. However, it should be noted that Fig. 10.2 is drawn for a positive deflection of both flap and tab. In practice, the tab is usually

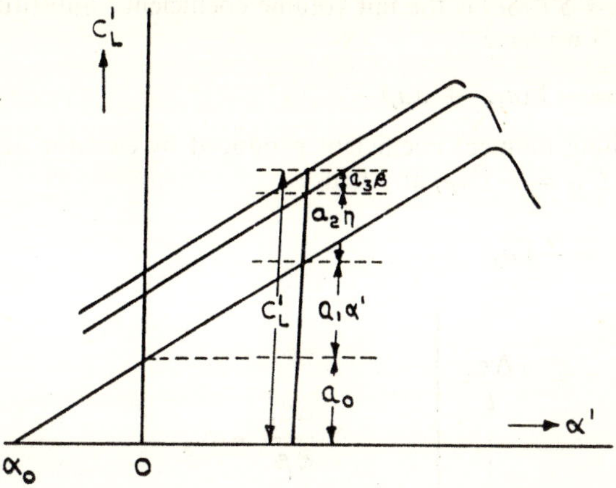

Fig. 10.2. Effect of elevator and tab deflection on the C_L'/α' curve

deflected in the opposite way to the flap in order to balance out the control hinge moments and, hence, when η is positive, β has a negative value (i.e. the tab deflection is such as to reduce the overall lift by a small amount).

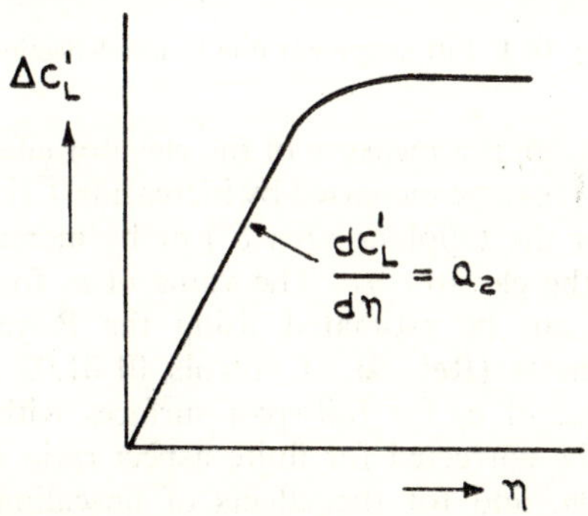

Fig. 10.3. Lift increment due to flap deflection

Eqn. (10.1) assumes that a_1, a_2 and a_3 are constant coefficients, independent of the angles of incidence and deflection. Figs. 10.2, 10.3 and 10.4 show that, in practice, this assumption is a valid one over the straight line parts of the curves, but that it breaks down at the higher values of incidence and flap and tab deflections.

In Chapter 2, it was shown that the corresponding pitching moment coefficient about the C.G. of the aircraft is given by

$$C_M = -\bar{V}C_L' \qquad (10.3)$$

where $\bar{V}(= S'l'/S\bar{c})$ is the tail volume coefficient. Substituting from eqn. (10.2) we have

$$C_M = -\bar{V}(a_1\alpha' + a_2\eta) \qquad (10.4)$$

The pitching moment coefficient produced by elevator deflection is given by $C_M = -\bar{V}a_2\eta$ and hence

$$\frac{dC_M}{d\eta} = -\bar{V}a_2 \qquad (10.5)$$

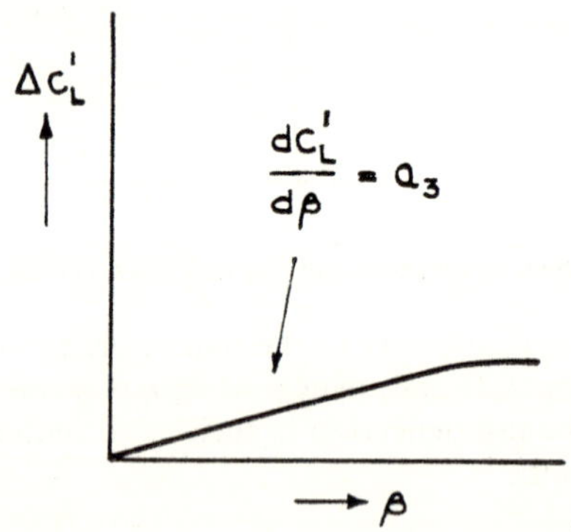

Fig. 10.4. Lift increment due to tab deflection

The value of $dC_M/d\eta$ is a measure of the elevator effectiveness or the elevator power; it can be increased by increasing $\bar{V}$ (i.e. by increasing the tail arm l' or the tailplane area S') or by increasing the lifting effectiveness of the elevator a_2. The value of a_2 for any particular control surface can be estimated using the Royal Aeronautical Society Data Sheets (Ref. 9). Controls 01.01.03 gives the two-dimensional value of a_2 for full-span surfaces with a sealed gap. The values can be corrected for finite aspect ratio effects by using Controls 01.01.06, and for the effects of unsealing the gap from Controls 01.01.04.

The elevator control effectiveness can also be represented in terms of the derivative m_η, where $m_\eta = M_\eta/(\rho V^2 S l')$. Since $M_\eta = (dC_M/d\eta)\frac{1}{2}\rho V^2 S\bar{c}$ we can see by substituting from eqn. (10.5) that

$$m_\eta = -\tfrac{1}{2}\bar{V}\frac{\bar{c}}{l'}a_2 = -\tfrac{1}{2}\frac{S'}{S}a_2 \qquad (10.6)$$

The rudder effectiveness is given in a similar way by

$$\frac{dC_n}{d\zeta} = -\bar{V}a_2 \tag{10.7}$$

where $\bar{V}$ and a_2 are the values appropriate to the fin (i.e. $\bar{V} = S_F l_F/S\bar{c}$, where S_F and l_F are the area and arm, respectively, of the

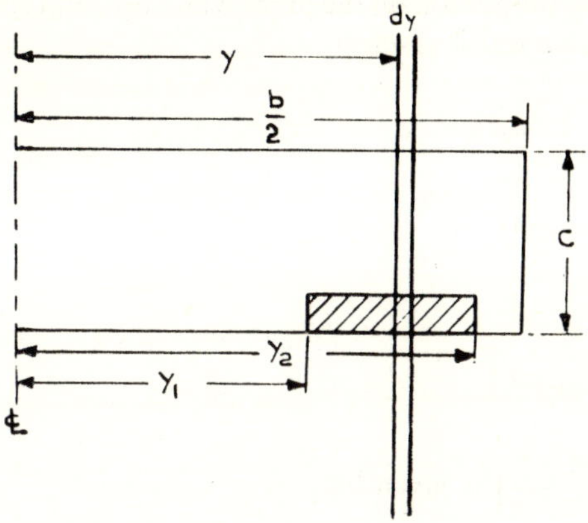

Fig. 10.5. Wing geometry notation

fin and $a_2 = dC_{LF}/d\zeta$). Since $N_\zeta = (dC_n/d\zeta)\frac{1}{2}\rho V^2 S\bar{c}$, the effectiveness of the rudder control can be expressed in terms of the derivative $n_\zeta\left(= \dfrac{N_\zeta}{\rho V^2 S(b/2)}\right)$ as

$$n_\zeta = -\frac{S_F l_F}{Sb}a_2 \tag{10.8}$$

and defining $\bar{V}_F$ as $\dfrac{S_F l_F}{Sb}$

$$n_\zeta = -\bar{V}_F a_2 \tag{10.9}$$

The aileron control produces the rolling moment required in a similar way, but in this case the effective lever arm is not a constant as it is on the elevator and the vertical fin. The part of the aileron near the wing tip produces a larger rolling moment than the inboard part. Fig. 10.5 shows the aileron on a wing; the additional lift produced as a result of deflecting the aileron on a small section of the wing (width dy) is given by

$$\text{Lift} = \Delta C_L \tfrac{1}{2}\rho V^2 c\, dy \tag{10.10}$$

where $\Delta C_L = a_2\xi$. The corresponding rolling moment is then given by

$$\text{Rolling moment} = -a_2\xi\tfrac{1}{2}\rho V^2 c\, y\, dy \tag{10.11}$$

(i.e. a positive aileron deflection produces a negative rolling moment). The rolling moment developed on one wing is obtained by integrating this expression across the span of the aileron surface. Allowing for the effect of the aileron on the other wing, the total rolling moment acting on the aircraft is given by

$$L = -2\int_{y_1}^{y_2} a_2\xi\tfrac{1}{2}\rho V^2 c\; y\, dy \tag{10.12}$$

$$L = -2a_2\xi\tfrac{1}{2}\rho V^2 \int_{y_1}^{y_2} c\, y\, dy \tag{10.13}$$

Thus

$$L_\xi = -a_2\rho V^2 \int_{y_1}^{y_2} c\, y\, dy \tag{10.14}$$

and $l_\xi \left(= \dfrac{L_\xi}{\rho V^2 S\tfrac{1}{2}b} \right)$ is given by

$$l_\xi = -\frac{2a_2}{Sb}\int_{y_1}^{y_2} c\, y\, dy \tag{10.15}$$

In the extremely simple case of full span ailerons on a wing of constant chord

$$l_\xi = -\tfrac{1}{4}a_2 \tag{10.16}$$

where a_2 is the lifting effectiveness of the aileron control surface.

2.3. Control Hinge Moment Characteristics

The hinge moment coefficient for the flap control shown in Fig. 10.1 can be represented by

$$C_H = b_0 + b_1\alpha' + b_2\eta + b_3\beta \tag{10.17}$$

where $b_1 = dC_H/d\alpha'$, $b_2 = dC_H/d\eta$ and $b_3 = dC_H/d\beta$ and b_0 is the value of C_H when $\alpha' = \eta = \beta = 0$. The value of b_0 depends on the camber of the aerofoil section, and is zero for a symmetrical aerofoil. The corresponding hinge moment is given by

$$H = C_H\tfrac{1}{2}\rho V^2 S_\eta c_\eta \tag{10.18}$$

where S_η and c_η are the area and chord length, respectively, of the control surface behind the hinge line of the control.

The coefficient b_1 is a measure of the "trailing" characteristics of the control surface. It can be negative or positive, floating with or against the wind, respectively (see Chapter 1). The coefficient b_2 is a measure of the control "heaviness" and controls should be designed so that this always has a negative value, the hinge moment produced as a result of control deflection opposing the motion of the control and tending to return it to the neutral position. If b_2 becomes positive, the hinge moment assists the movement and the control tends to go over to the stop. The control is then said to be "overbalanced" (see Chapter 1). The coefficient b_3 is a measure of the effectiveness of the trim tab in balancing out the hinge moments acting on the control. Though the value of lift produced by the tab $a_3\beta$ is small, it acts a relatively long way behind the hinge line

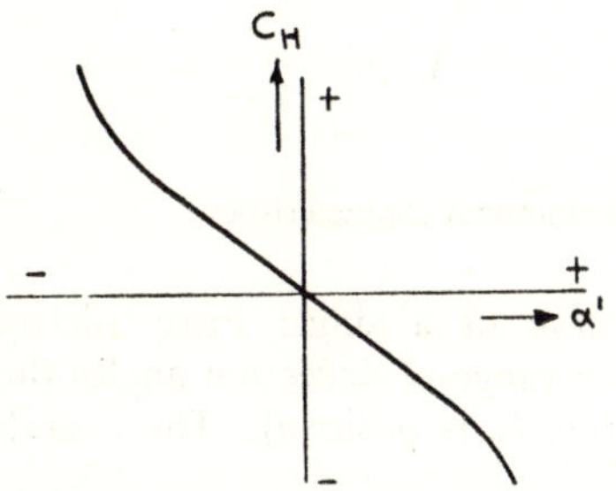

Fig. 10.6. Variation of elevator hinge moment coefficient C_H with tailplane incidence

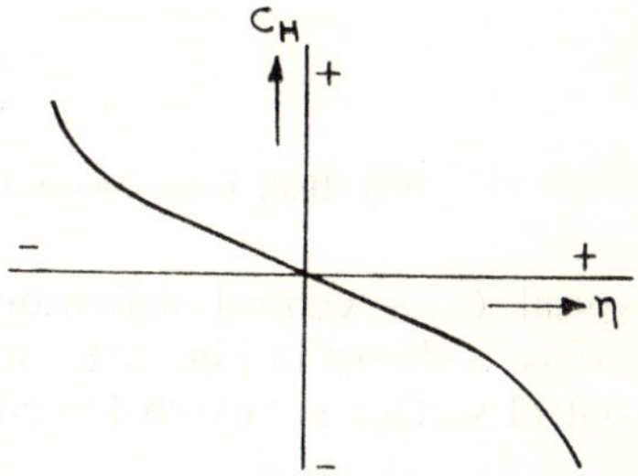

Fig. 10.7. Variation of elevator hinge moment coefficient C_H with elevator deflection

of the control and hence produces an important $b_3\beta$ term in eqn. (10.17).

The values of the coefficients are very dependent on the design of the particular control system being considered. They can fairly easily be changed, in order to produce more satisfactory aircraft handling characteristics, by modifying the sectional shapes used and the size of the gap between the aerofoil and the flap and by incorporating methods of aerodynamic balancing on the control surfaces (para. 4). The basic hinge moment characteristics of control surfaces can be estimated from the Royal Aeronautical Society Data Sheets 04.01.01, 04.01.02 and 04.01.08.

In eqn. (10.17) the values of the coefficients b_1, b_2 and b_3 are assumed to be constants, independent of the angles of attack and flap deflection. The values of b_1 and b_2 are represented by the slopes of the C_H/α' and the C_H/η curves, respectively and Figs. 10.6 and 10.7 show that the linear assumption is usually only valid over a limited range of incidence and deflection angles.

Some controls are deliberately designed with non-linear characteristics. In order to prevent the adverse yaw associated with aileron deflection, Frise ailerons are sometimes used (see para. 4*e*). A

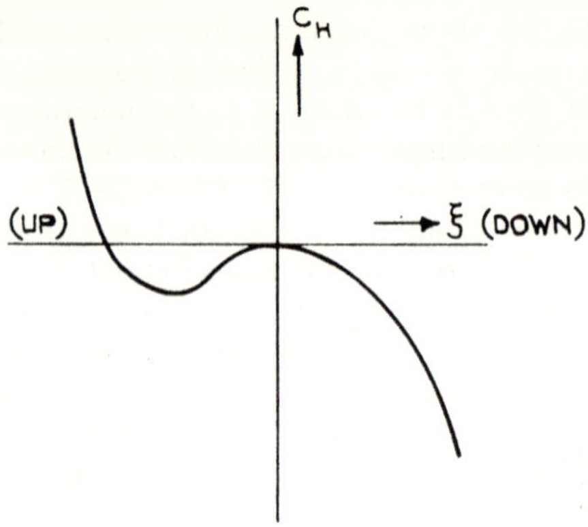

Fig. 10.8. Frise aileron hinge moment characteristics

typical C_H – control deflection curve of a single Frise aileron surface is shown in Fig. 10.8; over a range of deflection angles the control surface is "overbalanced" (i.e. b_2 is positive). The overall

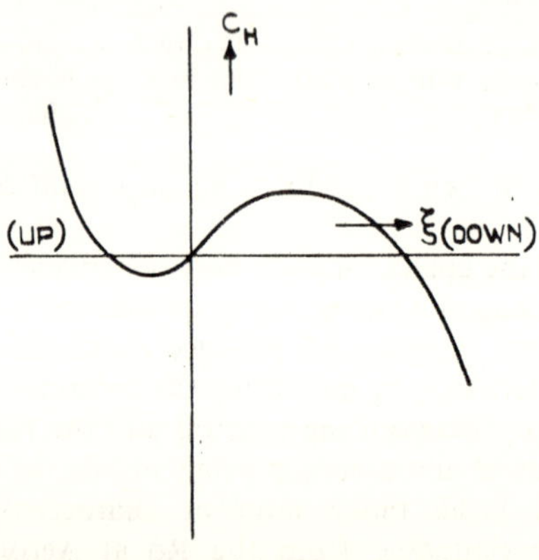

Fig. 10.9. Frise aileron hinge moment characteristics

hinge moment characteristics of the aileron surfaces on both wings depend on how the surfaces are rigged. If the neutral position is the one shown in Fig. 10.8, then the up-aileron is overbalanced and the down-aileron is balanced; the total hinge moment is the sum of the two effects. The control surfaces can be rigged in such a way (Fig.

10.9) that both the up-aileron and the down-aileron are overbalanced over a certain range of deflection angles.

There are also moments acting about the hinge line of the tab; the overall tab hinge moment C_{Ht} can be represented by

$$C_{Ht} = c_0 + c_1\alpha' + c_2\eta + c_3\beta \tag{10.19}$$

where $c_1 = dC_{Ht}/d\alpha'$, $c_2 = dC_{Ht}/d\eta$, etc. The tab hinge moment characteristics are particularly important when a servo-tab is being used (para. 4*h*), because in this case these are the moments that the pilot has got to overcome by applying his control forces.

2.4. Cross-control Effects

As has been mentioned, it is desirable that control application should only produce a moment about the appropriate axis. When

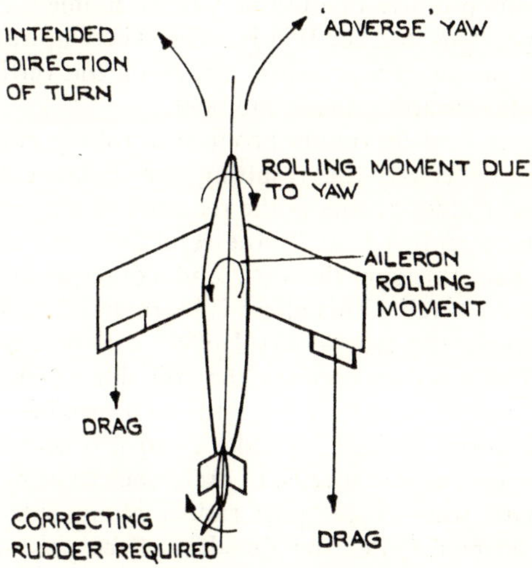

Fig. 10.10. Adverse aileron yaw

the ailerons are deflected, however, drag variations are produced on the two wings and a yawing moment results. Consider a positive aileron deflection (Fig. 10.10); on the starboard wing, where the aileron is down, the lift of the wing is increased, and so are the profile and induced drags. On the aileron-up wing, however, the lift and hence the induced drag component is reduced. The difference in the drag acting on the two wings produces a yawing moment. If the ailerons are being used to turn the aircraft to port, then this moment acts to starboard and for this reason it is described as an "adverse aileron yaw." The aileron yaw effect is represented by the derivative n_ξ and, if the effect is adverse, n_ξ has a positive value.

This adverse aileron yaw also tends to reduce the rolling performance of the aircraft, because the yawing produces a sideslip and the dihedral effect produces a rolling moment opposing the aileron rolling moment. The aileron system is usually designed to eliminate or reduce this effect by the use of differential or Frise ailerons (para. 4*e*). In these designs it is arranged for the down-aileron to produce less drag as a result either of the decreased deflection or the smoother airflow pattern, whereas the up-aileron produces additional drag because of the increased deflection or the turbulent flow created. The problem can also be eliminated by linking the aileron and rudder control circuits, so that when the ailerons are deflected, the rudder moves in such a way as to oppose the yawing moment produced by the ailerons (Fig. 10.10).

On some modern designs the aileron-yaw problem is not present and the ailerons produce favourable yawing moments. This occurs on delta wing designs, e.g. the B58 Hustler, where upwards deflection of the aileron produces a pressure change on the vertical fin which results in the favourable yawing moment.

The rudder control inevitably produces a rolling moment. When the rudder is deflected, the resulting aerodynamic force on the vertical fin and rudder produces the required yawing moment about the C.G. of the aircraft. Usually the resultant force acts above the longitudinal axis through the C.G. and consequently produces a rolling moment as well. This effect is represented by the derivative l_ζ and once again the cross-control effect can be described as an "adverse" effect since it opposes the roll due to sideslip or yaw effect that the rudder may have been put on to produce (e.g. in order to pick up a dropped wing). Rudder control is often used to pick up a dropped wing at low speeds, in preference to using the ailerons. If the starboard wing drops, port rudder is applied. The aircraft sideslips to starboard and the dihedral effect produces a rolling moment which tends to raise the lower wing. The direct rolling effect of the rudder discussed above would, however, tend to roll the aircraft farther to starboard. Usually the l_ζ effect is small and it has not been a problem on many aircraft. On some present-day aircraft the effect is important (e.g. Belfast) and it is eliminated by linking up the rudder and aileron control circuits so that, when the rudder is deflected, the ailerons deflect in the direction to oppose the l_ζ effect.

The Northrop T38A (Talon) has ailerons at the mid-span position on the trailing edge of the wing and a rudder-aileron interconnect system which is designed to produce adverse yaw when the ailerons are deflected. When an aircraft rolls it tends to yaw, due to yaw with roll effect n_p, and sideslip. To prevent this happening at high speeds and to reduce the corresponding vertical tail loads, the rudder-aileron interconnect comes in at about 370 knots (E.A.S.), the rudder deflecting by an amount equal to $0{\cdot}22\xi$ in the direction opposite to that required for a co-ordinated turn.

3. Control Requirements

3.1. Control Power Requirements

Control systems are required to be sufficiently effective over the speed and height range of the aircraft under all the various possible conditions of flight (e.g. take-off, landing, asymmetric conditions, etc.) and the control forces and movements involved must be within the limits laid down in the appropriate aircraft requirement regulations (Refs. 10 and 11).

The elevator control is required so that the pilot can trim the aircraft in level flight and so that he can manoeuvre the aircraft about its lateral axis. The control must be effective enough to do this and for military aircraft the forces must be within the limits laid down in Chapter 601 of Av.P.970 (Ref. 10). The lateral control is required to trim the aircraft laterally and to enable rolling manoeuvres to be performed. The control must be effective enough to produce not less than a certain rate of roll at high forward speeds (Ref. 10, Chapter 602, para. 4.1), and not less than a certain value of $pb/2V$ at low forward speeds (Ref. 10, Chapter 602, para. 4.2). The value of $pb/2V$ quoted depends on the type of aircraft being considered, but the minimum desirable value of $pb/2V$ is usually given as 0·07. The rolling performance of aircraft is discussed in more detail in Chapter 11; on some modern aircraft designs alternative ways of expressing the rolling performance are found to be more preferable. The rudder control must be powerful enough to trim the aircraft directionally and to balance out any asymmetric yawing moments produced by engine failure on multi-engined aircraft. Aileron and rudder controls are used together in most cases of lateral manoeuvring flight (e.g. steady turns, steady sideslips, etc.) and also, with the elevator, to enable the aircraft to be recovered from a spin.

3.2. Maximum Pilot Effort

Table 1 (from Ref. 4) lays down the various maximum forces and movements that are to be demanded from the pilot, the values quoted depending on the type of control column used and the period of time involved. The table shows that the maximum comfortable forces for a short time with one hand are 20 lb on the aileron, 30 lb on the elevator and 60 lb on the rudder, though much bigger forces can be exerted for short time intervals.

Table 1

CASE		AILERON		ELEVATOR (PUSH/PULL)		RUDDER (PUSH)
		STICK	WHEEL	STICK	WHEEL	
Maximum all out effort for very short time (2 hands)		90	120	180	220	400
Maximum permissible force for short time	2 hands	—	80	100	110	200
	1 hand	50	50	70	70	
Maximum comfortable force for short time	2 hands	—	30	—	40	60
	1 hand	20	20	30	30	
Largest hand or foot movement for full travel		±10 in.	±20 in.	±9 in.	±9 in.	±5 in.

(All forces are in pounds, and blank spaces indicate that 2 hands should not be demanded.)

It is essential that the control forces should be of the correct magnitude. If the control forces are "light" for normal flying, then the pilot may possibly overstress the aircraft by misuse of the controls. On the other hand, a "safe" aircraft may be unacceptably heavy for normal flying. Ref. 5 gives some experimental results of the maximum effort of which a pilot is capable, the magnitude of the control forces that he can maintain during various periods of time and the maximum rate at which he can build up control forces.

3.3. Control Harmonization

Before discussing some of the particular control requirements in more detail, mention must be made of the problem of "harmony" of the controls. It is generally agreed that this is an important factor in the pilot's assessment of the overall handling characteristics of the aircraft, though, since it is so subjective, it is not possible to lay down precise quantitative requirements. The values sometimes quoted as the desirable ones for the ratios of the aileron to elevator to rudder control forces are 1:2:4. The pilot's "feel" of the aircraft, however, is related not only to the stick forces and movements that he has to apply, but also to the way in which the aircraft responds to application of the controls.

3.4. Stick Force per g Characteristics

The stick force per g requirements are discussed in Av.P.970, Chapter 601, para. 8.1. The forces should always be in the conventional direction and should conform to certain limitations. In the

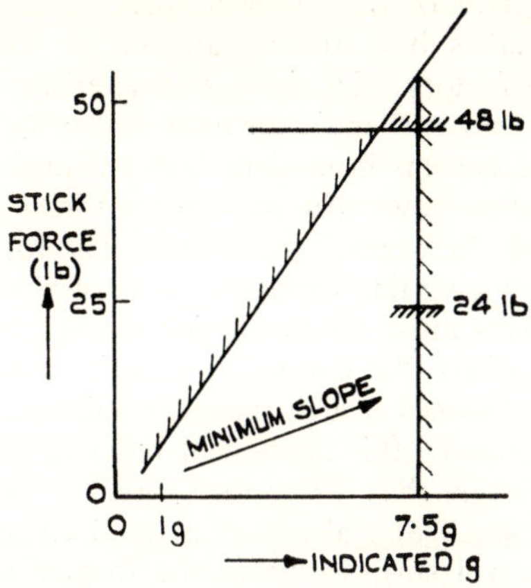

Fig. 10.11. Stick force per g requirements

case of a particular fighter aircraft the limitations are given below and illustrated in Fig. 10.11.

(*a*) the stick force required to achieve the maximum permissible g should not exceed 48 lb and should not be less than 24 lb.

(*b*) the stick force per g should not increase at a slower rate than 1·85 lb/g at any time.

(*c*) the stick force should never exceed a value which increases linearly with g in a manner specified in detail in Ref. 10.

4. Aerodynamic Balancing

The control forces that have to be exerted by the pilot are dependent on the hinge moment characteristics of the control surfaces, on the mechanical gearing of the control system, on the dimensions of the control surface and on the speed and height at which the aircraft is flying. If a control surface, say an elevator, is deflected through an angle η, then, in the absence of any response effect (see para. 5), the control hinge moment is given by

$$H = C_H \tfrac{1}{2}\rho V^2 S_\eta c_\eta = b_2 \eta \tfrac{1}{2}\rho V^2 S_\eta c_\eta \text{ lb ft} \qquad (10.20)$$

where S_η and c_η are the area and chord of the control surface respectively. The corresponding stick force is given by

$$P = m_e b_2 \eta \tfrac{1}{2}\rho V^2 S_\eta c_\eta \text{ lb} \qquad (10.21)$$

where m_e is the elevator-stick gearing in radians per foot. The dynamic pressure term $\frac{1}{2}\rho V^2$ in these equations can be replaced by $\frac{1}{2}\rho_0 V_i^2$ and eqn. (10.21) shows that, in the absence of compressibility and aeroelastic effects, the stick force for a given control deflection varies as the square of the forward speed (E.A.S.). This "speed-squared" law makes it extremely difficult to produce a control system with satisfactory stick force characteristics over the whole speed range of the aircraft and is the main reason for the introduction of power control systems in modern high-performance aircraft.

Eqn. (10.21) also shows that the control forces are proportional to b_2, the control "heaviness" parameter. The forces involved can be reduced by reducing the numerical value of b_2 (i.e. by making b_2 less negative). This must not be carried too far, however, because if b_2 becomes positive the control surface is "overbalanced." To reduce the control forces at high speeds to suitable values, $-b_2$ may have to be very small; this method of reducing the control forces has the disadvantages that the control forces at low speeds will be correspondingly small and also that a small change in the design shape of the control surface due to the formation of ice, or any damage or production imperfections, might result in an over-balanced control.

The numerical value of b_2 can be altered by the various methods of aerodynamic balancing available; some of these methods also modify the value of b_1, the "floating" parameter of the control. The methods of aerodynamically balancing a control reduce the hinge moments involved, either by moving the resultant force acting on the control surface nearer to the hinge line, or by producing a moment about the hinge line of the control acting in the opposite direction. When the control surface hinge moment characteristics are modified for any reason, it is important to investigate the effects of the changes on the corresponding stick-free static and dynamic stability characteristics of the aircraft.

Some common forms of aerodynamic balance are now described.

(*a*) *Set-back Hinge Balance*

When the hinge line of the control surface is at the leading edge of the control, b_1 and b_2 are both negative. If the hinge line is moved farther back (Fig. 10.12*a*) the values of b_1 and b_2 are made more positive (Fig. 10.12*c*) because the pressure changes arising on the control surface ahead of the hinge line produce a hinge moment opposing the one produced by the pressure changes acting behind the hinge line (Fig. 10.12*b*), i.e. the distance between the hinge line and the resultant force acting on the control surface is reduced.

It should be noted that this type of balance does not reduce the value of the control effectiveness (i.e. a_2); all the pressure changes assist in producing the required moment about the C.G. of the aircraft.

The shape of the leading edge of the control surface, the width of the gap between the control and the aerofoil and whether the gap

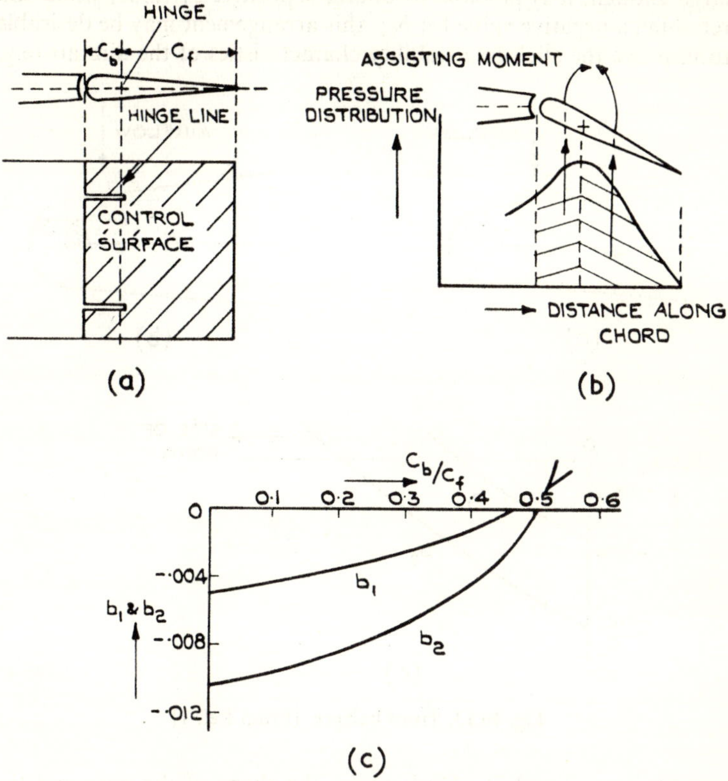

Fig. 10.12. Set-back hinge balance [From Ref. 1]

is sealed or not, are important factors in determining the values of b_1 and b_2. Unsealing the gap and blunting the nose of the control tend to increase the balance, making b_1 and b_2 less negative, but a blunt nose may produce non-linear characteristics (Ref. 9; Controls 04.01.00 and 04.01.03).

(b) Horn Balance

This type of aerodynamic balance is similar to the set-back hinge above, except that all the area ahead of the hinge line is concentrated on one part of the surface. If the horn extends to the leading edge of the aerofoil it is referred to as an "unshielded" horn (Fig. 10.13*a*); if not, it is called a "shielded" horn (Fig. 10.13*b*). The horn balance makes b_1 and b_2 less negative, though the effect on b_1 is larger than in the case of the set-back hinge (Fig. 10.13*c*). By making the horn large enough, it is possible to obtain a positive b_1 value, while still retaining a negative value for b_2; this arrangement may be desirable to improve the stick-free stability characteristics of the aircraft (e.g.

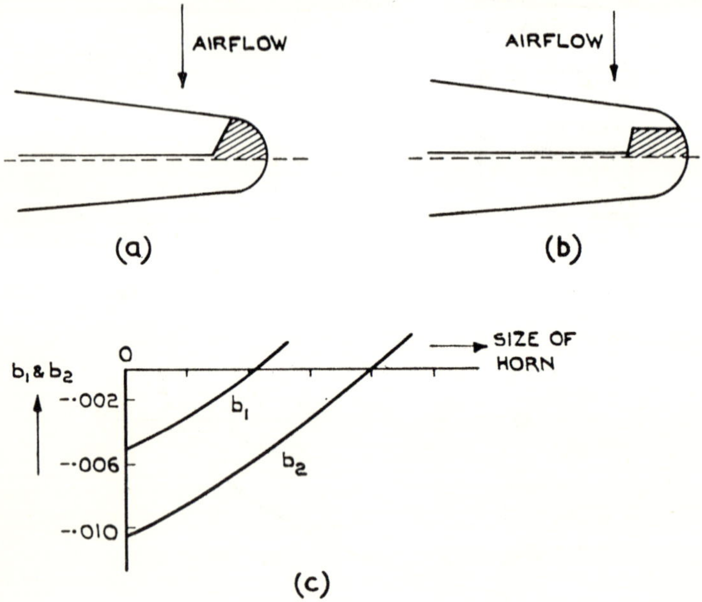

Fig. 10.13. Horn balance [From Ref. 1]

see Chapters 2 and 3). Once again, the shape of the nose and the size of the gap are important factors, and the effects of these are referred to in Ref. 9; Control 04.01.07.

(*c*) *Westland–Irving Internal Balance*

This is quite a common form of aerodynamic balance, though it is not an obvious one, since the balance area is contained inside the contour of the aerofoil (Fig. 10.14*a*). It is really only a modification of the set-back hinge type of balance; the inside of the aerofoil has to be vented to the external pressures, and these, acting on the balancing area or "beak" of the control, provide the necessary balance.

The effectiveness of this balance can be increased by sealing the gap between the leading edge of the control surface and the structure of the aerofoil by means of a flexible sealing strip and the amount of balance can be adjusted by venting the seal. This particular type of balance has a greater effect on b_2 than on b_1 (Fig. 10.14*b*). Some data on this type of balance is given in Ref. 9; Controls 04.01.04.

This type of balance is used on all the control surfaces of the Avro Vulcan, on the ailerons and aileron tabs of the Canberra T4

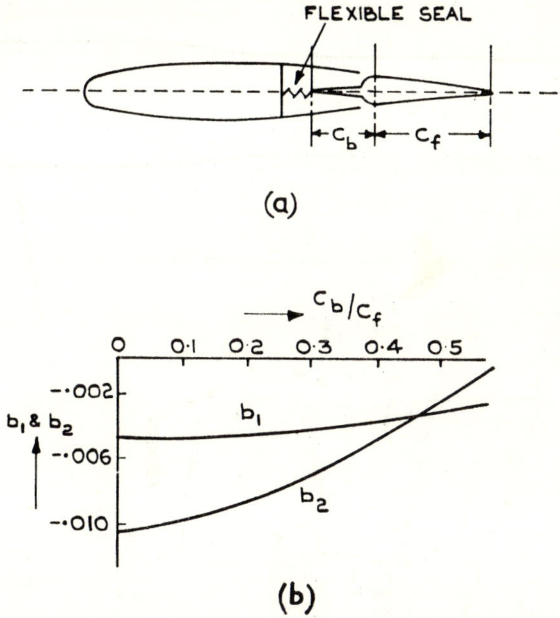

Fig. 10.14. Internal (Westland-Irving) balance [From Ref. 1]

and on the ailerons of the Varsity T1. In the case of the Canberra ailerons (Fig. 10.15) the gaps are not sealed completely, but the clearances between the aileron beak and the mainplane rear wall and between the aileron tab beak and the rear web of the aileron are extremely small. The Varsity T1 is worth mentioning because it has flexible fabric sealing strips, with pre-set variable-area leaks provided so that the degree of balance can be adjusted.

(d) Trailing-edge and Control Contour Modifications

The shape of the control surface is a very important factor in determining the hinge moment characteristics of the control. A

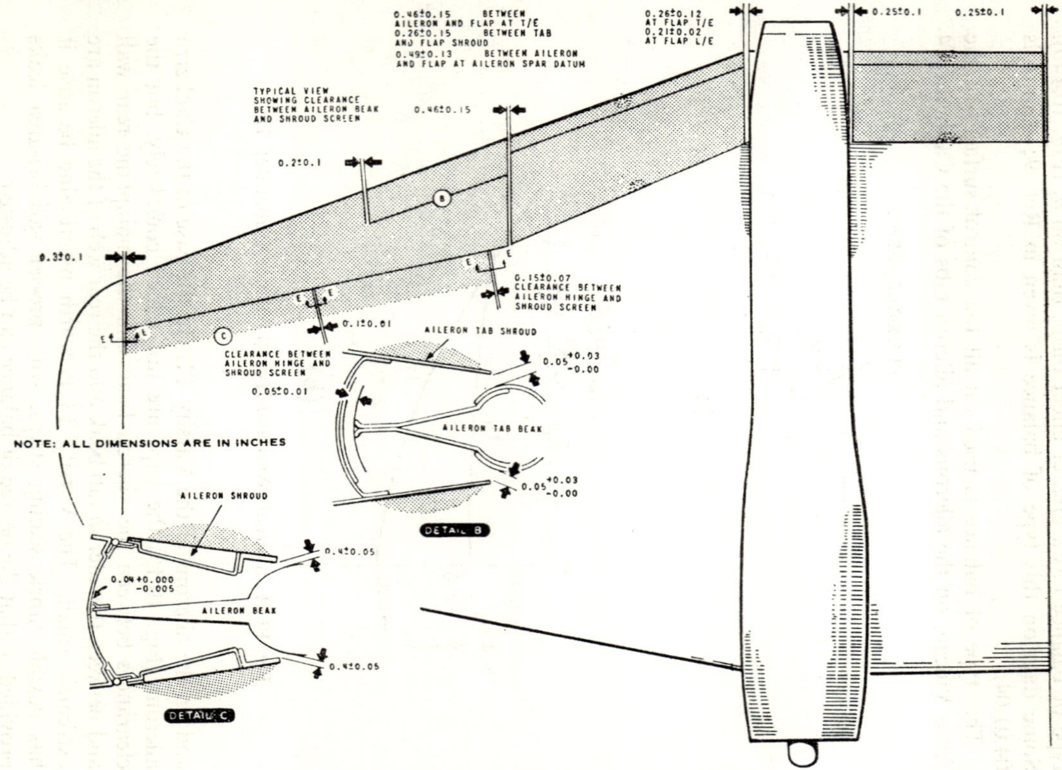

Fig. 10.15. Canberra T4 Mainplane [Crown copyright]

control surface which bulges outward (i.e. a convex surface) due either to its design or to aerodynamic effects has some undesirable characteristics (e.g. non-linear b_2 characteristics and a dangerous short period oscillation tendency) and straight-sided and concave control surfaces are to be preferred.

The values of b_1 and b_2 can be very easily modified by making small changes to the trailing edge of the control surface (e.g. by the addition of strip or cord, or by bevelling the trailing edge).

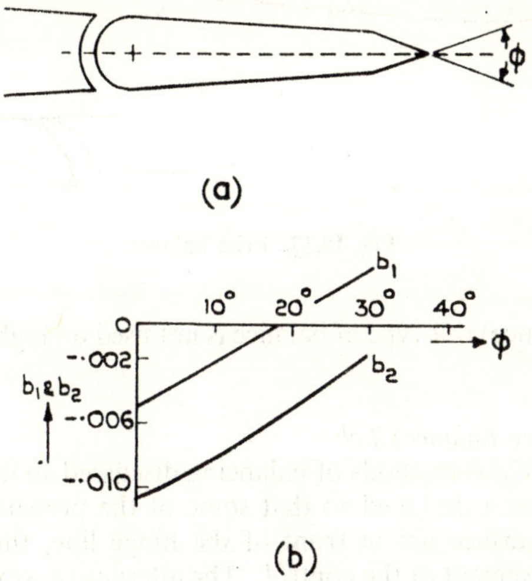

Fig. 10.16. Bevelled trailing edge control [From Ref. 1]

These changes affect the airflow pattern round the trailing edge of the surface and are effective because of the relatively long lever arm about the hinge line of the control.

The effect of bevelling the trailing edge is shown in Fig. 10.16; b_1 and b_2 are made less negative. Large trailing edge angles may, however, result in non-linearities in the hinge moment characteristics (Ref. 9; Controls 04.01.00).

The addition of strip or cord to the trailing edge of the control surface is a very simple and convenient way of modifying the hinge moment characteristics; it makes the control "heavier," i.e. b_1 and b_2 become more negative.

(*e*) *Frise Balance*

This type of aerodynamic balance (Fig. 10.17) is occasionally used on the ailerons of some low performance aircraft, in order to eliminate the adverse yaw effects of the ailerons. The design was used on the ailerons of the Lancaster, Spitfire and Hurricane aircraft (Ref. 12). The non-linear hinge moment characteristics of this type of control have already been discussed in para. 2.3. Because it suffers from a number of possible disadvantages (e.g. aileron buffet

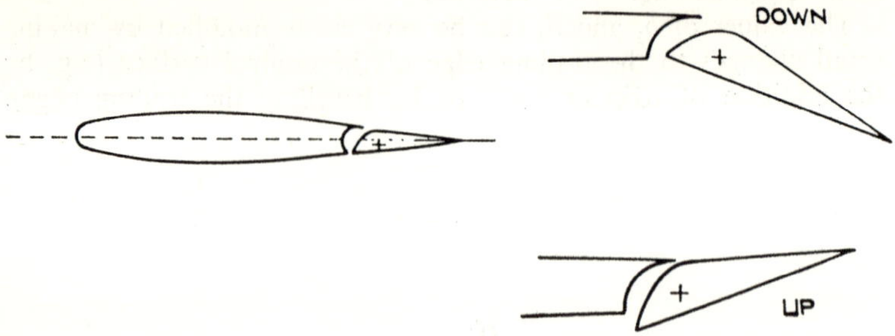

Fig. 10.17. Frise balance

and overbalance) this type of balance is not used on high performance aircraft.

(*f*) *Geared* (*or Balance*) *Tab*

In all the previous methods of balancing discussed so far, the control surface has been designed so that some of the pressure changes on the control surface act in front of the hinge line, thus tending to assist the movement of the control. The alternative way of balancing the control surface is by deflecting a tab on the trailing-edge of the control surface, the deflection of the tab being in the appropriate direction to assist the control movement.

A balance tab is one in which the tab is geared to move, at a fixed ratio, in the opposite direction to the movement of the control surface (Fig. 10.18*a*). Though the tab is small, the pressure changes caused by its deflection produce an appreciable moment about the hinge line of the control and hence reduce the value of $-b_2$. Because the tab only deflects when the main control surface moves, it has no effect on the value of b_1 (Fig. 10.18*b*).

The tab moves in the opposite direction to the control surface and, as a result of this, produces a small loss of lift (i.e. a small reduction in the control effectiveness). Since, however, a_3 is usually very small compared with a_2, this effect is only a small one.

The control forces can be increased by using an anti-balance tab; in this case the tab is geared to move in the same direction as the control surface. An anti-balance tab is fitted on the Devon elevator and on the port elevator of the Vickers Viscount.

Often it is necessary to combine two types of aerodynamic balance to achieve the overall desired result. In order to produce the required value of b_1, for example, it might be necessary to fit a horn balance. This also makes b_2 more positive, however, and it may be necessary to heavy up the control again by incorporating

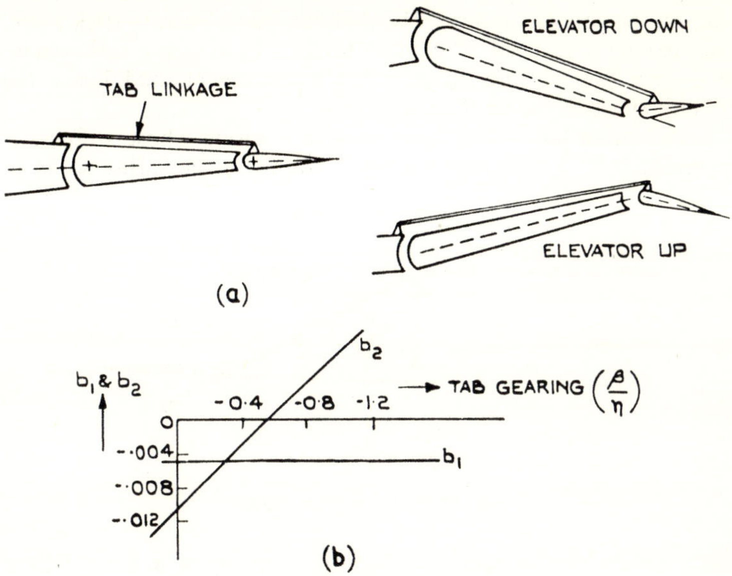

Fig. 10.18. Geared balance tab [From Ref. 1]

an anti-balance tab. This, as we have seen, has the effect of making b_2 more negative, whilst leaving b_1 unchanged.

(*g*) *Spring Tab*

This is a modification of the gear tab system above, in which a "spring" is incorporated in the circuit so that the effective gearing of the tab increases with an increase in the applied control forces. A simple spring tab system is shown diagrammatically in Fig. 10.19. The pilot's input rod is connected to the control crank, which is free to pivot about the control hinge line, being restrained only by the presence of the spring. If the input rod moves forward, the control surface moves up, the spring extends and the tab moves downwards, thus assisting the movement of the control surface. The input force is also transmitted through the spring to the control surface and the control surface moves as the result of the combined effect of the input force and effect of the tab. The pilot's control force is less than it would have been with a rigid crank and no tab.

This particular system has the advantage that it cannot produce overbalance; if the control is underbalanced before the spring tab is fitted, addition of the tab cannot cause overbalance. This is because there cannot be any tab deflection without there being a load on the control surface. Overbalance implies a zero stick force, followed by a reversal of control forces. In a spring tab system, zero stick force results in zero tab displacement, and hence the balance effect of the tab becomes zero.

If the spring tab system is fitted with a spring whose stiffness is represented by λ in/lb and the tab is linked to the spring so as to

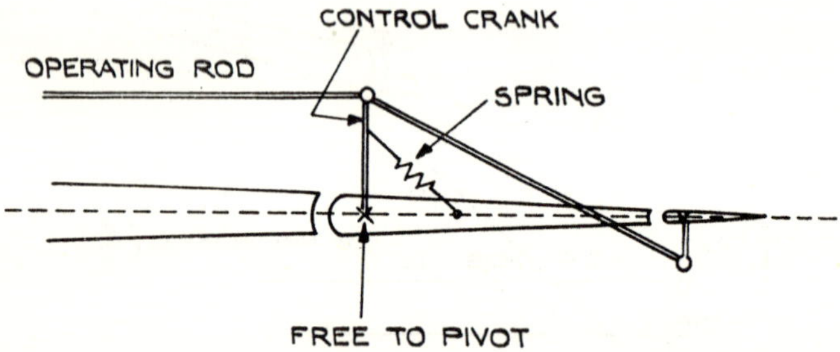

Fig. 10.19. Spring tab system

move θ rads/inch of spring deflection, then, when a force P lb is applied, the tab deflects through an angle $P\lambda\theta$ radians. The stick force is then given by

$$P = m_e(b_2\eta + b_3P\lambda\theta)qS_\eta c_\eta \tag{10.22}$$

and solving for P

$$P = \frac{m_e b_2 \eta q S_\eta c_\eta}{1 + kq} \tag{10.23}$$

where $k(= -m_e b_3 \lambda\theta S_\eta c_\eta)$ is a positive quantity, since b_3 is usually negative.

Comparing eqns. (10.21) and (10.23) the effect of the spring tab can be regarded either as producing

(*a*) A change in the basic speed law, the force being proportional to $q/(1 + kq)$ instead of to q. Because k is a positive quantity this results in a reduced rate of increase of force with forward speed. As k increases (i.e. as the spring is made "weaker") the force per unit control displacement tends to become less dependent on speed (Fig. 10.20*a*); or (*b*) A change in the effective value of b_2 from b_2 to $b_2/(1 + kq)$, where b_2 is the basic tab-locked value. Obviously, if b_2 is negative, the spring tab cannot produce overbalance (i.e. the factor $1/(1 + kq)$ cannot change the sign of b_2).

Spring tabs may be preloaded to prevent them from coming into operation until the stick force exceeds a pre-determined value. This

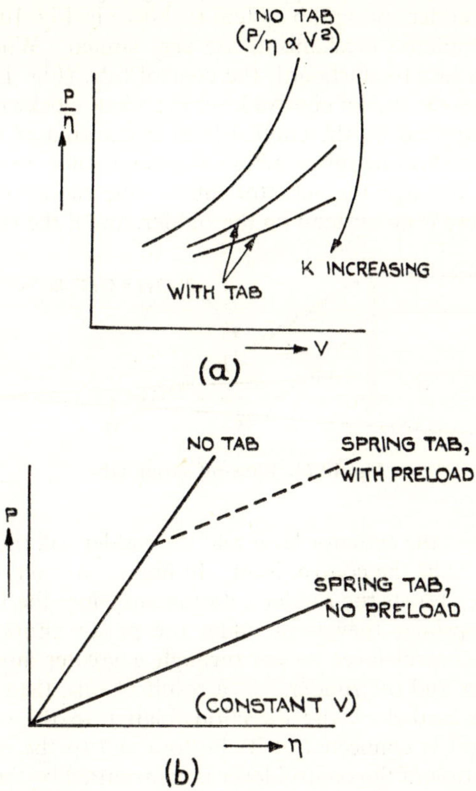

Fig. 10.20. Effect of spring tab on the stick force characteristics

is done to keep the spring tab out of action at low speeds, thus avoiding excessive lightening and lack of feel. The corresponding variation of stick force with control deflection is shown in Fig. 10.20*b*; a disadvantage of this arrangement is the sudden change in stick force gradient that occurs when the tab comes into action.

Another modification of the simple spring tab system is the "sprung-spring tab" or "blow-off spring tab." In this system ,an auxiliary spring is placed in the tab operating rod (Fig. 10.21). This has the effect of giving the tab a gearing which varies with the speed of the aircraft. As the speed increases, the aerodynamic loads on the tab will increase and, therefore, for a given displacement of the control crank, the tab deflection will decrease with increasing speed.

The Canberra T4 is fitted with blow-back spring tabs on all the control surfaces; in these systems the "springs" consist of torque tubes. The rudder spring tab system is shown in Fig. 10.22, and the elevator and aileron mechanisms are very similar. When the pilot moves the rudder to starboard, the control tube (Fig. 10.22) moves aft and rotates the rudder control lever in an anti-clockwise direction. The torque applied by the control level is transmitted through the inner and outer torque tubes, acting as a single tube, to the actuator lever which, through the actuator, moves the rudder over to starboard. If there is no air load on the rudder, and if the rudder hinges

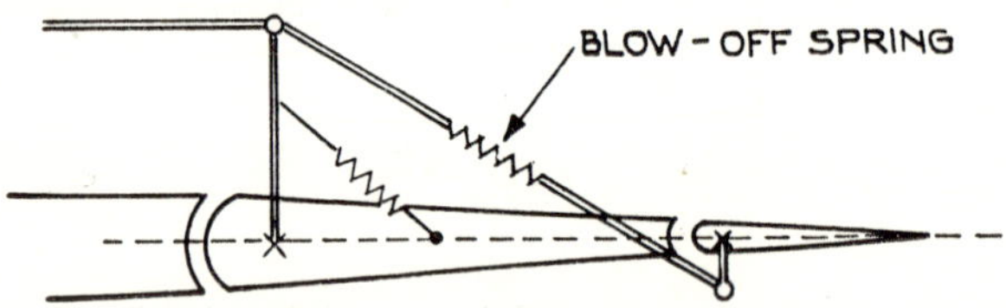

Fig. 10.21. Blow-off spring tab

are frictionless, the actuator lever and the rudder will rotate through the same angle as the control lever. In flight, however, the air load on the rudder resists the rudder rotation and since the torque tubes are torsion springs, they twist under the pilot's effort and consequently the control lever moves through a greater angle than the actuator lever and the rudder. As a result of this, the control lever rotates in an anti-clockwise direction relative to the rudder. The blow-back rod is connected at its bottom end to the control lever and any rotation of the control lever is transmitted by the blow-back rod to the tab actuating lever, which will also rotate in the anti-clockwise direction. The rotation of the tab actuating lever is transmitted to the tab by the tab actuating rod and this moves the tab to port. Thus the air load on the tab helps to move the rudder to starboard. In addition to its function as a driving shaft between the control lever and the tab actuating lever, the blow-back rod has a safety role. Under the action of the high air loads resulting from large tab deflections at high speeds, the blow-back rod twists in the direction opposite to that applied by the pilot to the control lever, with consequent reduction of the tab and rudder deflection angles.

The Varsity T1 has a spring tab on the rudder surface. The mechanism contains a pre-loaded spring which prevents the tab acting as a balancing tab until the load applied to the rudder bar is

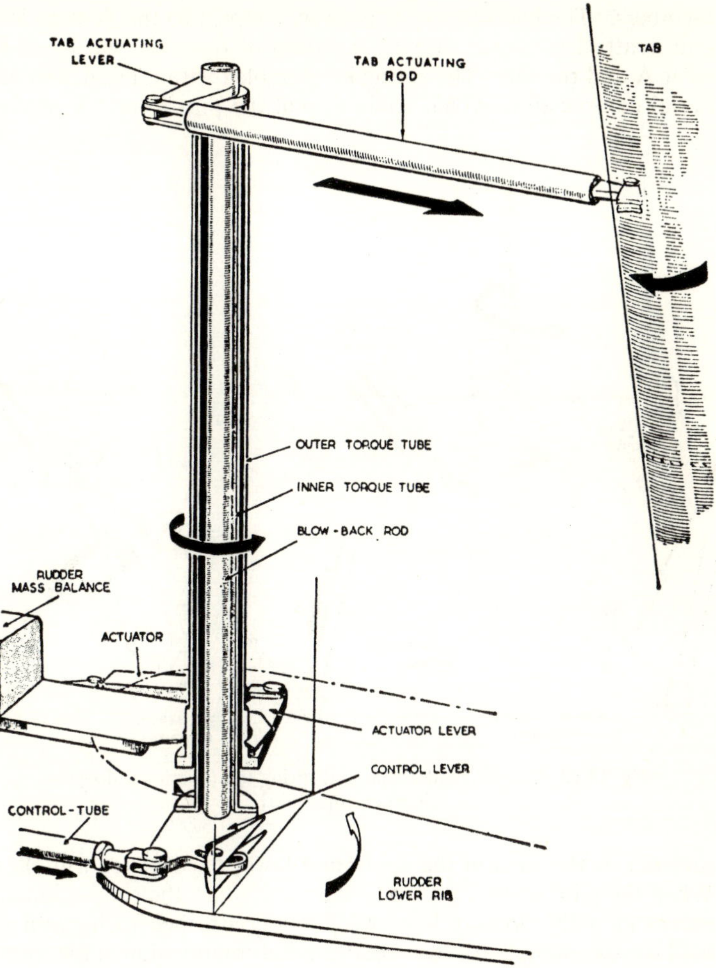

Fig. 10.22. Canberra T4 rudder spring tab mechanism [Crown copyright]

great enough to overcome the initial compression of the spring (Fig. 10.23). The main operating lever for the rudder is located horizontally between two fixed plates attached to the rudder post. The operating lever itself is a floating one, and it is arranged to hinge about the centre of the post. The operating lever consists of an upper and lower plate, trunnions, a rod, and a compression spring. Each trunnion is provided with a retaining pin (D and E), which extends through slots in the floating lever plates and the fixed members. The end of lever C is freely coupled to the floating lever and is attached to the arm B; rotation of this arm will move the lever A and the tab. The spring is assembled with a pre-determined initial compression, which tends to pull the pins D and E as close

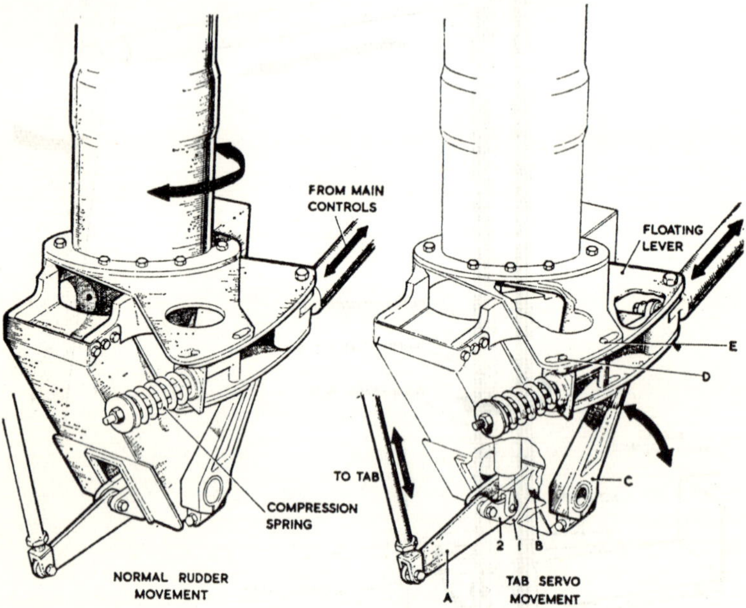

Fig. 10.23. Varsity rudder trim and spring tab functional diagram [Crown copyright]

together as the slots in the upper and lower fixed plates will allow. When the pilot applies a load to the rudder bar, the rudder initially moves as if the floating lever were rigidly attached to it, until the load is great enough to overcome the initial compression of the spring. As the load is further increased, the floating lever moves, relative to the upper and lower plates. When the floating lever is moved in either direction from its mid-position in relation to the remainder of the mechanism, D and E are moved farther apart by the ends of the slots of one or other pair of plates through which they pass. Thus any relative movement of the floating lever increases the compression of the spring and at the same time moves the tab from its neutral position in the appropriate direction to assist the movement of the rudder.

(h) Servo-tab

A servo-tab system is similar to the simple spring tab system with the spring removed; the pilot's input rod is connected directly to the tab and the main control surface is free to rotate about its hinge line (Fig. 10.24). There is no direct rigid connection between

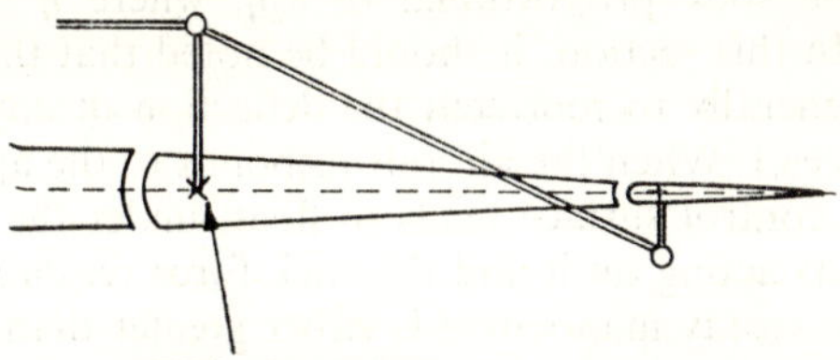

Fig. 10.24. Servo-tab system

the pilot and the main control surface. The pilot's control input deflects the tab and the moment produced about the hinge line of the main control surface causes this surface to "float" to its position of equilibrium, given by

$$\eta = \frac{-b_3}{b_2}\beta \tag{10.24}$$

where β is the deflection of the servo tab. The floating control will then produce the required moment about the C.G. of the aircraft. The control forces that the pilot has to apply can be considerably reduced in this way, because the pilot only has to balance the hinge moments acting on the tab, which are much less than those on the main control surface.

The servo-tab system is not used on many modern aircraft. The original BAC 111 aircraft were fitted with servo-tab elevators, but these were replaced during the development of the aircraft by a power control system. On the Bristol Britannia and the Short Belfast, however, all the control surfaces are servo-tab operated. The Belfast control systems are described briefly in para. 8.

5. Response Effect

The control hinge moments and the pilot's stick forces are determined by the amount of control deflection, the forward speed, and the particular control surface characteristics. The most important control surface parameter determining the hinge moments is b_2 (i.e. the control "heaviness" parameter); the moments may, however, be increased or decreased due to the effect of b_1 (i.e. the control "floating" parameter). This effect is usually termed the "Response Effect"; it is the modification in the hinge moment and stick force characteristics of a control, resulting from the response of the aircraft to the control movement.

When the pilot initiates a particular manoeuvre, he has to apply a force to the stick proportional to $b_2\eta$, where η is the control deflection. (In this section, it should be noted that the symbol η is being used generally to represent the deflection of any of the three control surfaces.) When the aircraft responds to the applied control moment, the control surface tends to float, under the action of the hinge moments acting on it and the stick force required to hold the aircraft in the steady manoeuvre is either greater than, or less than, the force required to initiate the manoeuvre, depending on the sign of b_1. If b_1 is negative the control floats with the wind (i.e. in the direction required) and the stick force is reduced; if b_1 is positive, on the other hand, the stick forces are increased. A favourable response effect is usually assumed to be one in which the initiation force is slightly greater than the steady force required. In Ref. 6 it is stated that the force to initiate a manoeuvre should not be greater than the subsequent force required to maintain the manoeuvre by more than 20 per cent.

If a change in control deflection $\Delta\eta$ produces a change in the aerofoil angle of attack of $\Delta\alpha'$, then the corresponding change in the hinge moment coefficient is given by

$$\Delta C_H = b_1\Delta\alpha' + b_2\Delta\eta \tag{10.25}$$

In the absence of any response effect, the hinge moment coefficient would be given by

$$\Delta C_H = b_2\Delta\eta \tag{10.26}$$

If the Response Factor K is defined by

$$K = \frac{\text{Effort with response}}{\text{Effort without response}} \tag{10.27}$$

then

$$K = \frac{b_1\Delta\alpha' + b_2\Delta\eta}{b_2\Delta\eta}$$

$$K = 1 + \frac{b_1}{b_2}\frac{\Delta\alpha'}{\Delta\eta} \tag{10.28}$$

The control surface tends to float through an angle $\Delta\eta_f$ given by

$$\Delta\eta_f = -\frac{b_1 \Delta\alpha'}{b_2}$$

and thus eqn. (10.28) can be rewritten in the form

$$K = 1 - \frac{\Delta\eta_f}{\Delta\eta} \tag{10.29}$$

From either eqn. (10.28) or eqn. (10.29) it can be seen that if the control tends to float in the same direction as it is being deflected (i.e. $\Delta\eta_f$ and $\Delta\eta$ have the same sign and b_1 is negative), then the control force is reduced (i.e. $K < 1$). Alternatively, if the control tends to float against the wind (i.e. $\Delta\eta_f$ and $\Delta\eta$ have opposite signs and b_1 is positive), then the control forces are made heavier (i.e. $K > 1$).

Response effect has already, in effect, been discussed in the work on longitudinal stability in Chapter 2. The stick forces are related to the stick-free static stability, and the effect of freeing the stick depends on the sign of b_1. The stick forces can be increased by making the value of b_1 more positive (e.g. by incorporating a horn balance into the design of the control surface). In many cases K is approximately unity, but on those controls which have a large value of b_1/b_2, the response may have an appreciable effect on the control feel. The value of b_1/b_2 usually lies between $+1{\cdot}0$ and $-1{\cdot}0$, though, in the case of geared tabs, the value may increase to about $2{\cdot}0$.

6. Methods of Modifying the Stability and Control Characteristics of an Aircraft

6.1. Effect of Varying the Control Surface Characteristics

The handling characteristics of an aircraft are very dependent on the values of the parameters of the control surfaces. By modifying these parameters it is possible to alter the aircraft's handling characteristics. However, it is very important to remember that modifying any one parameter may result in widespread changes in the various stability and control characteristics of the aircraft.

6.1.1. Stick-free Static Stability

As we have seen, the ratio b_1/b_2 is a very important factor in the stick-free stability, the stability being increased or decreased by a factor $1 - (a_2 b_1/a_1 b_2)$ (Chapter 2). If b_1 is negative the stick-free stability is less than the corresponding stick-fixed stability, but if b_1 is positive the stick-free stability is increased.

Many of the devices used for aerodynamic balancing (e.g. set-back hinge, horn balance, etc.) modify the values of both b_1 and b_2. The stick forces involved depend on both the amount of stick-free stability possessed by the aircraft and on the particular control characteristics. A reduction in the numerical value of b_2 decreases the stick forces involved, whereas an increase in the stick-free stability increases the forces.

The addition of a spring tab to an elevator circuit has no effect on the value of b_1, nor on the stick-free stability, since, when the control is free to float, no forces are involved and the elevator takes up its equilibrium position without the tab deflecting. The spring tab has the effect of reducing the numerical value of b_2 and hence reducing the stick forces required to trim.

6.1.2. STICK-FREE MANOEUVRING STABILITY

The manoeuvring stability is also modified by the factor $1 - (a_2b_1/a_1b_2)$ when the controls are left free (Chapter 3), and the

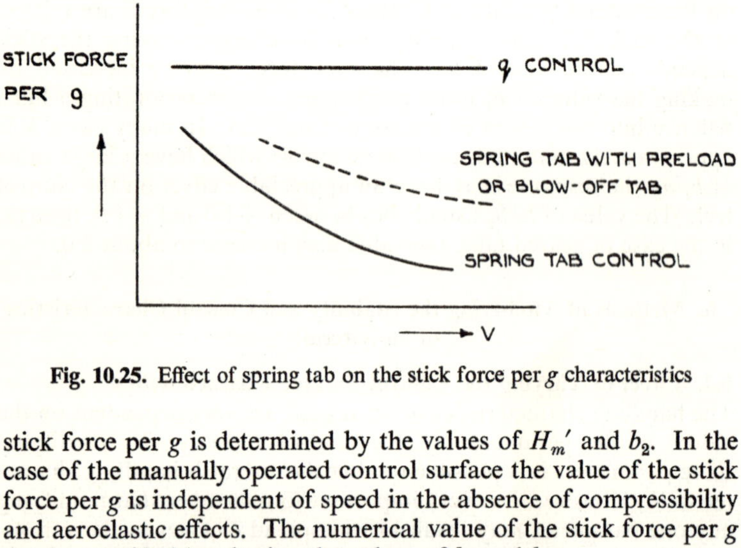

Fig. 10.25. Effect of spring tab on the stick force per g characteristics

stick force per g is determined by the values of H_m' and b_2. In the case of the manually operated control surface the value of the stick force per g is independent of speed in the absence of compressibility and aeroelastic effects. The numerical value of the stick force per g can be modified by altering the values of b_1 and b_2.

As in para. 6.1.1, aerodynamic balancing devices affect the values of both H_m' and b_2. In the case of a spring tab system, the stick force for a given control displacement increases with speed as $q/(1 + kq)$, i.e. at a slower rate than q (eqn. 10.23). The variation of stick force per g with speed is shown in Fig. 10.25. The low values of stick force per g which occur at high forward speeds may be undesirable and they may be increased either by preloading the spring or by using a blow-off spring tab.

6.2. Effect of Incorporating Springs and Inertia Weights in the Control System

The addition of springs and inertia weights into the control system of an aircraft will modify the longitudinal stick-free static and manoeuvre stabilities; they will have no effect, however, on the corresponding stick-fixed characteristics. The effects are discussed in more detail in Ref. 13. In these systems, the "spring" is a very soft spring, often in the form of a long bungee cord. The tension of the spring remains approximately constant as the stick is moved and there is thus no direct spring feel effect on the controls.

6.2.1. STICK-FREE STATIC STABILITY

The addition of a positive weight or spring (i.e. by definition, one that pulls the elevator down) increases the stick-free static stability and hence the stick forces required to trim. On the simple theory of Chapter 2, which only allowed for incidence disturbances, the presence of a weight or a spring would have no effect at all. However, in the more general theory, it was shown that disturbances in forward speed could also have important effects and it is this contribution to the overall stability that is affected by the presence of springs and weights.

Consider an aircraft in steady level trimmed flight. To balance the elevator-down moment of the weight or spring, an additional downward trim tab movement is required. This additional trim tab angle will, however, only balance the effect of the weight (or spring) at the particular selected trimmed speed. If the aircraft is disturbed, and the forward speed increases, the moment from the trim tab will increase, since it is proportional to V^2, whereas the moment of the weight (or spring) will be unaffected. This will make the elevator move up, so tending to oppose the increase in speed (i.e. an additional stabilizing effect). If the speed is decreased in the disturbance, the elevator tends to move down, once again tending to oppose the speed disturbance.

6.2.2. STICK-FREE MANOEUVRING STABILITY

The addition of a positive weight will also increase the stick-free manoeuvre stability and hence the stick force per g, but a spring will have no effect, Consider an aircraft in a pull-out at a constant speed. There will, in this case, be no effect due to the additional trim tab angle since the forward speed is unchanged, but the effect of the weight will depend upon the applied normal acceleration. The result will be to pull the elevator down with a force and hence a hinge moment, that increases with increase of g. This tends to oppose the pull-out and consequently increases the stick force per g.

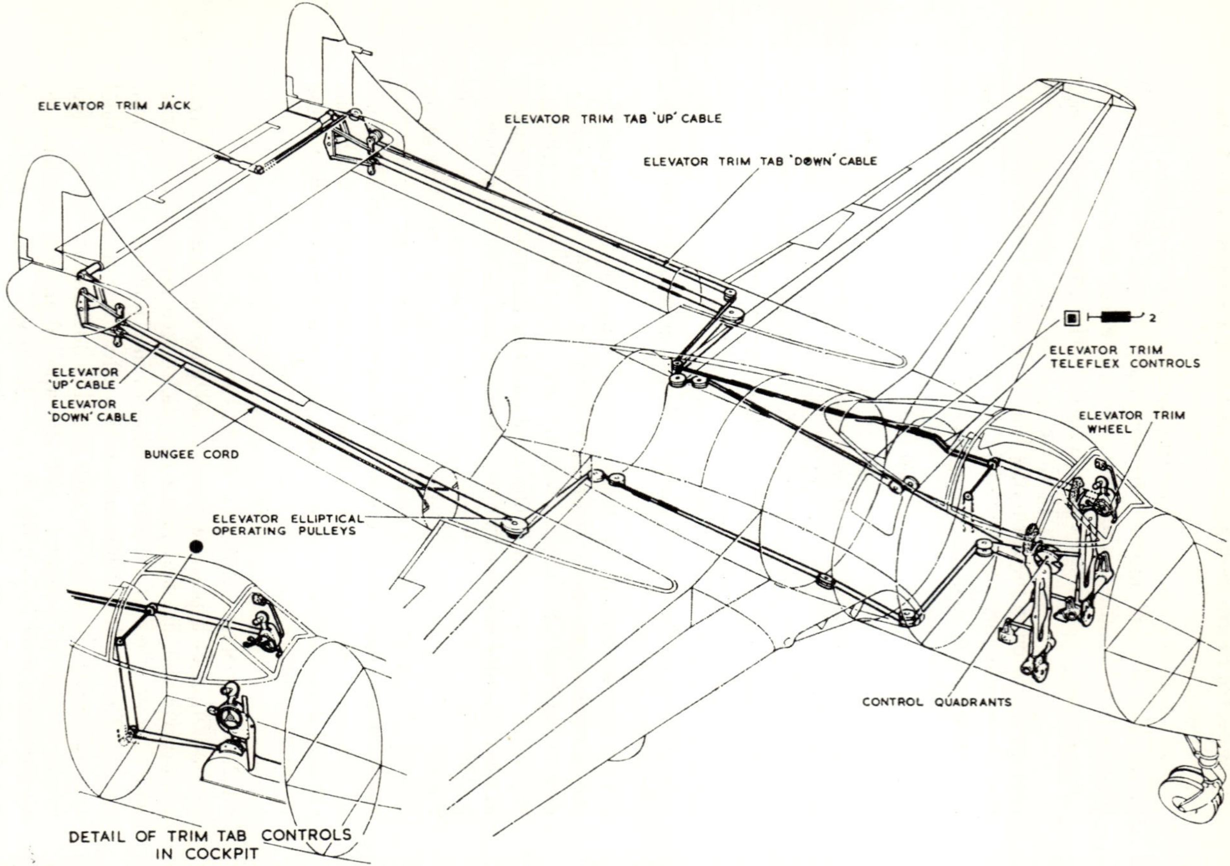
ELEVATOR TRIM JACK
ELEVATOR TRIM TAB 'UP' CABLE
ELEVATOR TRIM TAB 'DOWN' CABLE
2
ELEVATOR TRIM TELEFLEX CONTROLS
ELEVATOR TRIM WHEEL
ELEVATOR 'UP' CABLE
ELEVATOR 'DOWN' CABLE
BUNGEE CORD
ELEVATOR ELLIPTICAL OPERATING PULLEYS
CONTROL QUADRANTS
DETAIL OF TRIM TAB CONTROLS IN COCKPIT

The overall effects of weights and springs on the aircraft characteristics are summarized in the table below (Ref. 14, Table 81.3(4)).

EFFECT ON	VARIES AS	POSITIVE WEIGHT	POSITIVE SPRING	POSITIVE WEIGHT AND EQUAL NEGATIVE SPRING	POSITIVE WEIGHT AND EQUAL POSITIVE SPRING
Stick force to change speed	K_n'	Increase	Increase	No change	No change
Stick force per g	H_m'	Increase	No change	Increase	Decrease
Acceleration on release of stick in out-of-trim condition	$\dfrac{K_n'}{H_m'}$	Usually small increase	Large increase	Decrease	Large increase

Springs have been used in the elevator circuits of the Vampire T Mk. 11 (Fig. 10.26), the Hastings and the Twin Pioneer to increase the longitudinal stability. The addition of an inertia weight in the circuit to increase the value of stick force per g is a very common method on both powered and non-powered control systems. It has been used on the Vampire Mk. 3, on the Vampire T11 and on the Provost Mk. 1.

6.2.3. STICK-FREE DYNAMIC STABILITY

The incorporation of springs and bob-weights affects the positions of the stick-free neutral and manoeuvre points and hence the longitudinal stick-free dynamic stability characteristics. The effect of the relative positions of the C.G., the neutral point and the manoeuvre point have been discussed for the short period and phugoid oscillations in Chapter 8, para. 2.3.4 (though the discussion was for the stick-fixed condition, it is equally applicable for the stick-free condition, if the stick-fixed points are replaced by the corresponding stick-free ones). The presence of a weight or a spring in the elevator circuit usually reduces the damping of the phugoid oscillation, particularly for an aircraft with a small tailplane or at high altitudes (Ref. 14, Table 81.7). The effects of weights and springs are discussed in more detail in Ref. 8. Instability of the short period oscillation in the case of an aircraft fitted with power controls is discussed in Ref. 7.

7. Alternative Types of Control Surfaces

7.1. Spoilers

7.1.1. GENERAL

In order to improve the effectiveness of the controls on high performance aircraft, flap-type controls are often replaced by spoiler-type or all-moving control surfaces. On conventional aircraft lateral control is usually achieved by flap-type aileron surfaces positioned at the trailing edge of the wings. The ailerons must be designed to give satisfactory rolling response and, in the case of manual controls, to have satisfactory control force characteristics. These control requirements are discussed in Chapter 6 of Ref. 15. Trailing-edge ailerons suffer from two main disadvantages

(i) They limit the amount of trailing-edge span that is available for high-lift flaps,
(ii) They usually produce an "adverse" yawing moment; this causes the aircraft to sideslip and the roll due to sideslip opposes the rolling moment of the ailerons, thus producing a loss in the overall effectiveness of the control.

It is essential that the ailerons should produce satisfactory characteristics throughout the speed and height range of the aircraft. At low speeds, large aileron surfaces and large aileron deflections may be necessary to produce the rolling power required. At high speeds much smaller aileron movements may be sufficient. As the forward speed increases, however, flap-type ailerons begin to lose their effectiveness due to both compressibility and aeroelastic effects. In the latter case, the aerodynamic forces produced by the deflection of the aileron twists the wing in such a way that the rolling power is reduced (Chapter 14). Other forms of lateral control are possible and it is desirable to have a control that will improve the rolling performance of the aircraft at both high and low speeds and one that will leave the trailing-edge of the wing free for the high-lift flaps which are essential on modern high-performance aircraft. Various types of lateral control are discussed in Chapter 6 of Ref. 15.

One possible alternative form of lateral control is the spoiler-type control; some of the various basic types available are shown in Fig. 10.27. Spoiler-type controls have been experimented with for some time, both in wind-tunnel and in flight tests, but it is only fairly recently that they have been used to any great extent on aircraft. Within recent years, however, they have been employed on many civil and military aircraft (e.g. McDonnell Phantom, Hawker Siddeley Trident). The reason for the slow development of this form

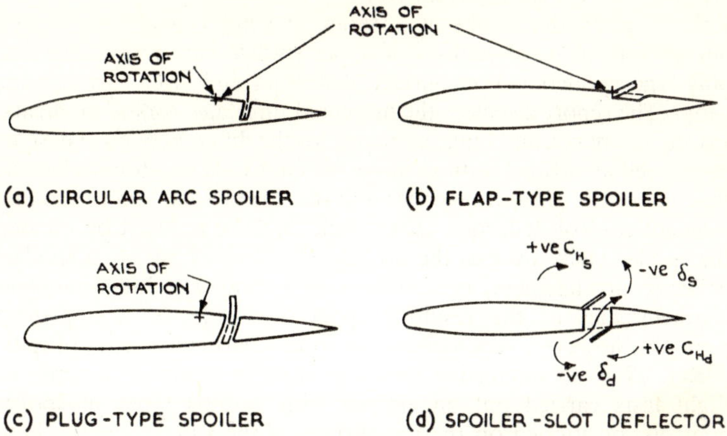

Fig. 10.27. Various types of spoiler controls

of control was that initially spoilers suffered from a number of disadvantages which included

(*a*) A non-linear rolling moment effectiveness, with a "dead-zone" for small control deflections.
(*b*) A loss of control effectiveness at high angles of attack.
(*c*) An appreciable time lag between the operation of the control and the response of the aircraft. (In para. 6.2 of Ref. 15 it is stated that the peak value of the rolling acceleration should occur not later than 3·75 $\bar{c}/V$ seconds or 0·1 seconds, whichever is the greater, after the aileron reaches full deflection.)

In addition, spoiler type controls have the obvious disadvantage that they work by destroying lift, rather than by creating extra lift. Thus lift is destroyed on the wing which has the spoiler deflected, and a wings level condition is regained by dropping the high wing, rather than by picking up the low wing. This is an important difference when the aircraft is operating close to the ground.

Many of the disadvantages associated with spoilers can be eliminated by suitable design, and the spoiler provides a very useful form of lateral control. It provides a very effective control throughout the subsonic, transonic, and supersonic speed ranges, the aeroelastic effects are small, and the associated yaw is usually favourable. The hinge moments can also be made small, though in practice this is less important on modern aircraft, since the spoiler controls are usually power operated.

7.1.2. SPOILER CHARACTERISTICS

Ref. 16 gives the results of a series of wind tunnel tests carried out on various combinations of upper surface flap, retractable and plug-type spoilers, lower surface deflectors and slots through the wing. The report discusses the measured time lags, rolling moments, yawing moments, and hinge moments for the different control designs positioned at various stations along the chord. It was found that the time lag of the aircraft response and the "dead-zone" on the rolling moment-control deflection curve could both be reduced by moving the spoiler back towards the trailing edge and by having a spoiler/deflector arrangement, preferably with a slot. The hinge moment characteristics of the resulting spoiler/slot/deflector arrangement were not, however, very satisfactory in this particular series of tests.

Ref. 17 is an interesting report giving the results of a series of flight tests carried out on aircraft with various types of spoiler controls, positioned on the top surface of the wing between about 60 and 75 per cent of the chord, just ahead of the trailing-edge flaps. These results included those from tests on the Northrop P-61A aircraft fitted with

(*a*) a vented circular arc spoiler,
(*b*) a vented circular arc spoiler with an aileron,

and on the Vought-Sikorsky OS2U-2 aircraft fitted with

(*a*) a vented plug-type spoiler,
(*b*) a circular arc spoiler,
(*c*) a flap-type spoiler.

Measurements were again made of the time lag, the rolling and yawing moments and the hinge moment characteristics of the control. The aircraft response was found to be good, apart from a small "dead-zone" in the rolling moment-control deflection curves and the time lags were equal to or less than the required value, which in this report is quoted as $3{\cdot}75\bar{c}/V$ sec or 0·2 sec. The rolling effectiveness was good at the stall and increased with increase in forward speed. The yawing moments were slightly adverse at low speeds, becoming favourable ones at the higher forward speeds. The control forces were smooth, but increased at an increasing rate with control deflection. The feel was poor at small control deflections and some negative aerodynamic moments (i.e. moments tending to open the control) were measured. Feel could be improved either by using a spring in the system, or by using a combined aileron-spoiler system. This particular arrangement is also discussed in Ref. 18 and it is commonly used on many present-day aircraft.

Refs. 19, 20, 21 and 22 give the results of more recent series of wind tunnel tests carried out on various spoiler/slot/deflector arrangements on swept wings over the speed range of about $M = 0{\cdot}6$ to $M = 2$. In these experiments, the spoiler surface occupies about 15 per cent of the wing chord, between 55 per cent and 70 per cent or 60 per cent and 75 per cent of the chord. When the spoiler and deflector are operated, the pressure distribution round the wing is changed. The effect for a wing operating at a subsonic Mach number and at a moderate angle of attack is shown in Fig. 10.28;

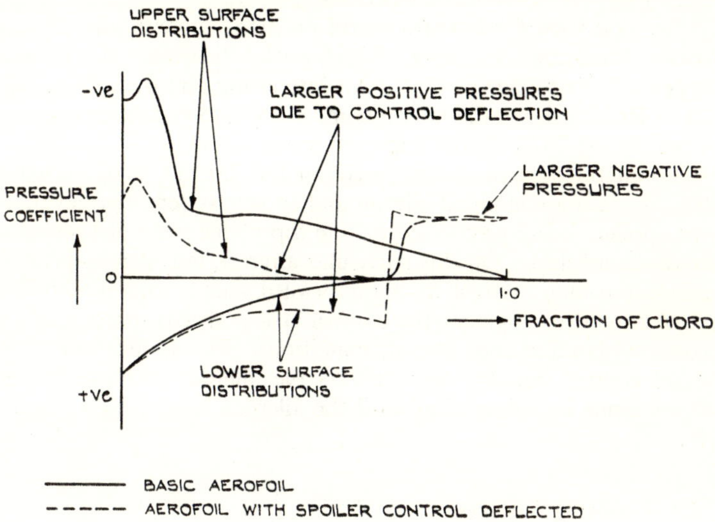

Fig. 10.28. Effect of a spoiler on the pressure distribution round an aerofoil

the pressure is increased in front of both the spoiler and the deflector and reduced behind both of the surfaces. The increase in pressure ahead of the spoiler produces the desired loss of lift, whereas the reduction in pressure behind the spoiler has an adverse effect. The effect of the slot is to increase the pressure recovery behind the spoiler, thus increasing its effectiveness. The increase of pressure ahead of the deflector increases the lift on the wing and is, for this reason, an adverse effect. This is more than balanced, however, by the effect of the reduced pressure region behind the deflector which has a beneficial effect on the rolling effectiveness particularly at the higher angles of attack. This type of lateral control arrangement is found to be very effective throughout the speed and angle of attack range. The maximum rolling moment is obtained by having a deflector deflection δ_d to spoiler deflection δ_S ratio between 0·5 and

1·0, the optimum ratio increasing with increasing angle of attack. The positive (i.e. closing) hinge moments increase with spoiler deflection and usually decrease with deflector deflection, though at high angles of attack the deflector hinge moment may be in the closing direction (i.e. positive) for small deflector deflections. The minimum total hinge moment is generally obtained with a δ_d/δ_S ratio of about 0·5.

7.1.3. PRACTICAL SPOILER SYSTEMS

Spoilers are used for lateral control on many modern aircraft (e.g. North American Vigilante, McDonnell Phantom II, Hawker Siddeley Trident, Boeing 707). An interesting photograph on page 221 of Ref. 23 shows the lateral control and flap system on the wing of the Boeing 707 with the flap-type spoilers deflected.

The lateral control system used on the Trident is described in Ref. 24. It is a combined aileron and spoiler system, the flat plate-type spoilers being positioned on the top of the wing just ahead of the outboard flaps. On this particular aircraft, the ailerons produce sufficient rolling control at high forward speeds and the spoilers are only used to augment the control at low speeds, particularly in crosswind landing and take-off conditions. The application of the lateral control initially only deflects the ailerons and the spoilers do not come into operation until the ailerons have moved through 15°.

7.2. All-moving Surfaces

7.2.1. GENERAL

The effectiveness of the flap-type control decreases at high speeds and many high performance aircraft are fitted with all-moving or slab control surfaces which have a greater effectiveness under these conditions of flight. The all-moving tailplane is very commonly used for longitudinal control on many modern aircraft (e.g. Lightning, Northrop T38 Talon, Republic F105, Lockheed F104, Hawker Siddeley Trident). All-moving fins have been used on the North American X-15, the North American Vigilante A3J(A5), and the BAC TSR2. These three aircraft have all-moving tailplanes that also produce lateral control by means of differential operation (i.e. "tailerons"). In the case of the X-15 and the TSR2 this is the only form of lateral control; the Vigilante also has a spoiler system on the wing. All-moving wing tips have not been used very much for the lateral control of aircraft. They were fitted to the Short Sherpa research aircraft and the large horn-balanced ailerons of the Bristol 188 research aircraft could be regarded as all-moving tip controls.

7.2.2. ALL-MOVING TAILPLANES

The stability and control characteristics of an aircraft fitted with an all-moving tailplane are represented in the same way as those of an aircraft fitted with a conventional elevator control. The sign convention is retained (i.e. a positive control movement produces a negative pitching moment) and hence tailplane leading-edge up is defined as a positive control movement, and tailplane leading-edge down is negative. A typical trim curve showing tailplane η_T against lift coefficient C_L, is given in Fig. 10.29. High performance aircraft

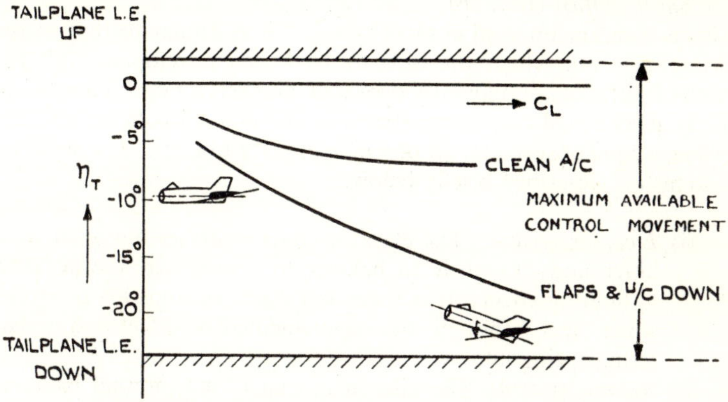

Fig. 10.29. Tailplane angles required to trim against C_L

are designed so that at high speeds the tailplane angle required to trim is very small. As the speed is reduced, more leading-edge down or negative tailplane movement is required, until at low speeds the tailplane is deflected at a large negative angle.

The longitudinal control system of the Trident is described in Ref. 24. The tailplane is an all-moving surface with a geared slotted trailing-edge flap which deflects upwards for negative (i.e. leading-edge down) tailplane angles, to improve the effectiveness of the control at low speeds.

7.2.3. ALL-MOVING WING TIPS

Little information is available on the characteristics of all-moving wing tip controls. Ref. 25 discusses the various theoretical methods available for determining the characteristics of this type of control on a delta wing at subsonic, transonic and supersonic speeds. These theoretical results are in good agreement with the rolling moments and hinge moments measured from free-flight rocket model tests on a 60° sweptback delta wing over the speed range $M = 0{\cdot}69$ to $1{\cdot}50$.

8. Some Typical Aircraft Control Systems

As we have seen, there are many different types of control systems in use on modern aircraft, from the manually operated flap-type systems to the power operated all-moving surface systems. A few of the more interesting types of control systems are described very briefly below. Power-operated systems are also discussed in Chapter 12.

(*a*) *Short Belfast* (Ref. 26)
This is a rather unusual control system; it is similar to that of the Britannia from which the aircraft has been developed. All the control surfaces are manually operated with servo tabs. There is no direct mechanical connection between the control column and the primary control surfaces, all of which are free to float. The separate systems are described briefly below.

(i) *Elevator system* The elevator control surface has a set-back hinge and enough horn balance to compensate the negative b_1 floating effect. It is servo tab operated and this produces some "feel," though it is supplemented by q-feel and a centering spring unit.

(ii) *Aileron system* The aileron has an Irving internal balance, and is servo tab operated. It does not float entirely freely because a spring is provided to resist upfloat of the aileron at high incidences. The aileron feel is almost entirely aerodynamic in origin.

Simple plate spoilers are incorporated to supplement the rolling power of the ailerons. They are positioned on the top surface of the wing just ahead of the outer flaps and inboard of the ailerons. They have a small dead zone on either side of the neutral position of the stick, to avoid unnecessary movement of the spoilers in cruising flight when the ailerons provide adequate control.

(iii) *Rudder system* The rudder is servo tab operated and a spring unit is incorporated to improve the feel characteristics. A torsion spring is introduced into the servo tab circuit to allow the tab to "blow back" at high speeds, in order to prevent excessive sideslip angles.

(*b*) *Vickers VC* 10 (Ref. 27)
This system is a duplicated power control system without manual reversion and the control surfaces are split into a number of independent sections.

(i) *Elevator system* The elevator is split into four sections, each with its own power control unit.

(ii) *Aileron system* There are two aileron surfaces on each wing, operated by separate power control units. The rolling power is supplemented by the incorporation of three vented spoiler sections on each wing which hinge upwards to a maximum angle of 50°. These are positioned on the top surface of the wing ahead of the outer flap sections. The spoilers have a dual function, operating either in conjunction with the ailerons to provide roll control, or functioning as speed brakes.

(iii) *Rudder system* The rudder is split into three sections, each with its own power control unit.

(*c*) *B.A.C. TSR* 2 (Ref. 28)

The control system of the TSR 2 is rather unusual in that it has two all-moving control surfaces, i.e. the "tailerons" to produce pitch and roll control and an all-moving fin for directional control. The controls are power operated.

(i) *Elevator system* For pitch control the taileron surfaces operate together, as in the case of an all-moving tailplane. Though it does not possess conventional elevators, the taileron does have power-operated trailing-edge flaps.

(ii) *Aileron system* There are no ailerons on the wing and for lateral control the taileron surfaces operate differentially.

(iii) *Rudder system* The yawing control is provided by the power-operated all-moving vertical fin.

References

1. DOMMASCH, D. O., SHERBY, S. S., and CONNOLLY, T. F.; *Airplane Aerodynamics* (Pitman, New York & London, 1951).
2. DUNCAN, W. J.; *Control and Stability of Aircraft* (Cambridge University Press, 1952).
3. ETKIN, B.; *Dynamics of Flight* (John Wiley and Sons, New York, 1959).
4. MORGAN, M. B. and THOMAS, H. H. B. M.; "Control System Design," *Journal of the Royal Aeronautical Society*, 49, 431, 1945.
5. T. VAN OOSTEROM; *Measurements on the relation between magnitude and duration and on the rate of application of the control forces achieved by pilots in simulated manoeuvres*, Agard Report 241, 1959.
6. DICKINSON, R. P.; *An Introduction to the Subject of Feel in Power Operated Control Systems*, Unpublished M.O.A. Report.
7. NEUMARK, S., COLLINGBOURNE, J., THOMAS, H. H. B. M.; *A Case of Longitudinal Stick-free Dynamic Instability of an Aircraft fitted with Power-operated Control, g-Restrictor and Spring Feel*, R.A.E. Rep. Aero, 2517, R. & M. 3094, 1954.
8. GOLDBERG, J.; *Effects of Springs and Inertia Devices on the Longitudinal Stability of Aircraft*, WADC-TR-53-350, 1954.

9. Royal Aeronautical Society Data Sheets, *Aerodynamics.*
10. *Design Requirements for Aircraft for the Royal Air Force and Royal Navy*, Av.P.970. Vol. I. Book II.
11. *British Civilian Airworthiness Requirements*, Air Registration Board.
12. THOMAS, H. H. B. M. and CRABBE, E. R.; *Note on the Frise Aileron*, R.A.E. Report Aero 2163, R. & M. 2502, 1952.
13. AUTERSON, E. I. and LYON, H. M.; *The Effect of Combined Springs and Weights on Longitudinal Stability and Manoeuvrability*, R.A.E. Report Aero 1804.
14. Boscombe Down A. & A.E.E. *Technical Guide.*
15. *Agard Flight Test Manual*, Vol. II, "Stability and Control."
16. WENZINGER, C. J. and ROGALLO, F. M.; *Wind Tunnel Investigation of Spoiler Deflector and Slot Lateral-Control Devices on Wings with Full-span Split and Slotted Flaps*, NACA Report No. 706, 1941.
17. SPAHR, J. R.; *Lateral-Control Characteristics of Various Spoiler Arrangements as Measured in Flight*, NACA Tech. Note No. 1123, 1947.
18. ASHKENAS, I. L.; *The Development of a Lateral-Control System for use with Large-span Flaps*, NACA Tech. Note No. 1015, 1946.
19. HAMMOND, A. D. and BROWN, A. E.; *Results of an Investigation at High Subsonic Speeds to Determine Lateral-Control and Hinge-Moment Characteristics on a Spoiler-Slot-Deflector Configuration on a* 35° *Sweptback Wing*, NACA RM L57 C20, 1957.
20 LORD, D. R. and MORING, R.; *Aerodynamic Characteristics of a Spoiler-Slot-Deflector Control on a* 45° *Sweptback Wing at Mach Numbers of* 1·61 *and* 2·01, NACA RM L57 E16a, 1957.
21 WEST, F. E., WHITCOMB, C. F. and SCHMEER, J. W.; *Aerodynamic Load* . *Distribution over a* 45° *Swept Wing having a Spoiler-Slot-Deflector Aileron and other Spoiler Ailerons for Mach Numbers from* 0·60 *to* 1·03, NACA RM L57 J11, 1957.
22. FOSTER, G. V.; *Effects of Spoiler-Slot-Deflector Control on the Aerodynamic Characteristics at a Mach Number of* 2·01 *on a Variable-Wing-Sweep Configuration with the Outer Wing Panels Swept Back* 75°, NASA TM X-273, 1960.
23. GUNSTON, W. T. (Ed.); *Flight Handbook* 6th Edition (Iliffe Books Ltd., London, 1962).
24. *Aircraft Engineering*, June, 1964.
25. THOMAS, H. H. B. M. and MANGLER, K. W.; *All-Moving Wing-Tip Controls at Subsonic, Sonic and Supersonic Speeds. A Discussion of Available Theoretical Methods and their Application to a Delta Wing with Half-Delta Controls*, R.A.E. Rep. Aero 2499, R. & M. No. 3086, 1953.
26. *Aircraft Engineering*, September, 1963.
27. *Aircraft Engineering*, June, 1962.
28. *Aircraft Engineering*, November, 1964.

11 Subsonic Longitudinal and Lateral Control

1. Introduction

In the previous chapter we considered the general characteristics of flap-type controls; the tailplane and elevator were used as an example, but most of what was said was applicable to all the controls. Control is required about the three axes of the aircraft, in order to trim and manoeuvre the aircraft about these axes. This control is usually achieved by having three separate sets of control surfaces (i.e. elevators, ailerons and rudder) though sometimes the action of two of the controls is combined in a single set of surfaces (e.g. elevons, ailerators, etc.). In these cases, the separate surfaces act together to produce one form of control and differentially to produce the other.

Ideally, each control is meant to produce a response about one axis only. Ailerons, for example, are provided to produce rolling moments and rolling motion and not to produce yawing or pitching effects. Control cross-coupling effects do occur, however, particularly on the lateral controls. An aileron deflection generally produces a yawing moment as well as a rolling one. Depending on the design of the ailerons, this yawing moment can have adverse or favourable effects on the manoeuvring capabilities of the aircraft. Similarly the rudder will produce a rolling moment as well as a yawing moment, and the effect may be important on an aircraft with a large vertical fin and rudder. When an aircraft responds to a control deflection, the resulting motion may be a very complicated one, since aerodynamic effects arise which may produce rolling, pitching and yawing motions. If an aircraft is rolling, for example, there is a yawing moment due to roll effect n_p as well as a damping in roll effect l_p. Similarly, if an aircraft is yawing there is a rolling moment due to yaw effect l_r as well as a damping in yaw effect n_r. These close lateral/directional interactions make it impossible for these motions to be separated from one another and this is why a discussion of the lateral dynamic characteristics is more involved than that of the corresponding longitudinal ones.

In addition to the aerodynamic coupling effects there are also inertia coupling effects. These are always present, but under normal conditions of flight, the inertia moments are small compared to the aerodynamic ones and can be neglected. The inertia coupling effects may, under certain conditions, cause large yawing or pitching moments to develop on the aircraft. There is thus a coupling existing between the longitudinal and lateral motions of an aircraft. Strictly, it is not permissible to separate out the longitudinal and lateral motions. It is, however, very convenient to do this and, as long as the inertia coupling effects are small, the results obtained are sufficiently accurate. Inertia coupling effects will be considered in detail in a later chapter.

The response of an aircraft to the movement of the controls is very closely related to the basic static and dynamic characteristics of the aircraft, see note on p. 515. The overall response can be divided into two parts and it is convenient to consider these separately.

(i) *The transient or dynamic response* This is the response of the aircraft in the very short time interval immediately following the application of the control. The transient response may be an oscillatory one. After a time the oscillations die away and the aircraft is once again in a steady state. This part of the response is closely related to the aircraft's dynamic stability characteristics.

(ii) *The static or steady-state response* This is the final steady response of the aircraft, after the transient response has died away. It is related to the static stability characteristics of the aircraft. The steady state response to longitudinal control has already been discussed in previous chapters; in Chapter 2, the elevator control movements and forces required to trim the aircraft at different forward speeds and incidences (i.e. at different steady state conditions) were determined and in Chapter 3 expressions for the control movements and forces required to apply steady g values to the aircraft were derived by a similar analysis. These control movements and forces were found to be related to the stick-fixed and stick-free static and manoeuvre stability characteristics of the aircraft. In Chapters 2 and 3 only the steady-state response characteristics were considered; the transient characteristics were ignored at that time.

The general response characteristics of various systems to different control inputs are discussed in Chapter 19. The standard types of input that are usually considered in dealing with aircraft response to control inputs are the step and sinusoidal inputs. The response of the aircraft to random inputs from atmospheric gusts is also important, but this is outside the scope of the work of this chapter.

Consider the response of an aircraft to a step input of a control (Fig. 11.1*a*). Depending on the particular control considered, the

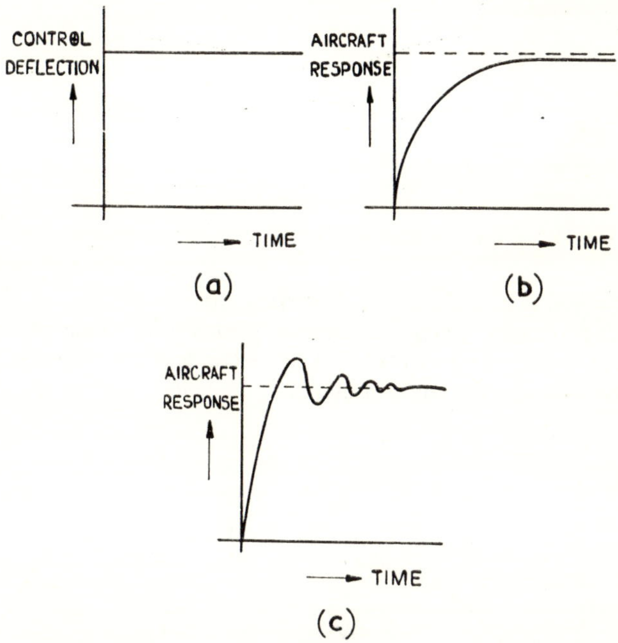

Fig. 11.1. Diagrammatic representation of aircraft response to a step control input

response may be in displacement, rate of change of displacement, or in acceleration. If the aircraft behaves as a simple order system, then the response will be as shown in Fig. 11.1*b*. If on the other hand it behaves as a second order system, then the response will be as shown in Fig. 11.1*c*, the magnitude of the oscillations and the rate at which they decay depending upon the damping in the system. Usually, however, the response of an aircraft is more complicated than either of these simple systems, though, in some cases, the aircraft response can be approximated to them.

In this chapter the response characteristics of flap-type controls at subsonic speeds will be considered; the effects of aeroelasticity and compressibility will be dealt with in later chapters.

2. Longitudinal Control Response Characteristics

The longitudinal control is required on an aircraft to trim the aircraft out at different speeds in straight and level flight and to produce

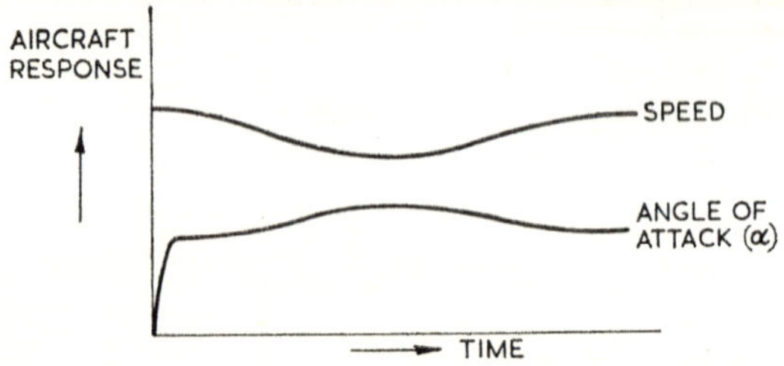

Fig. 11.2. Typical aircraft response to an up elevator displacement

longitudinal (i.e. pitching) manoeuvres. If the elevator is given a step displacement, then the aircraft responds in angle of attack, forward speed, and *g*. In the first few seconds of the response, large changes occur in the angle of attack and *g* values, but the associated

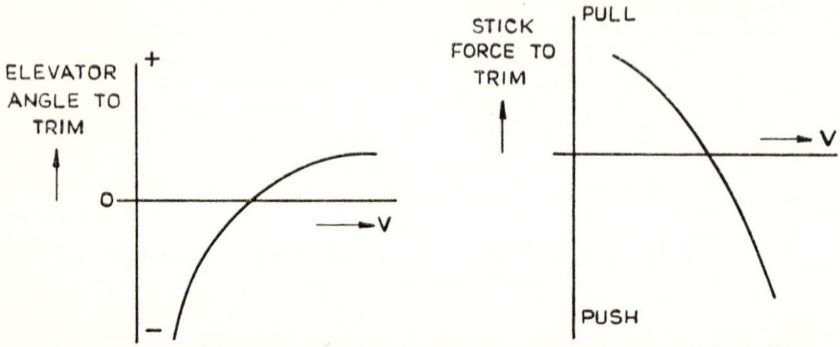

Fig. 11.3. Elevator angle and stick force required to trim against forward speed

change in speed is small. In the later part of the transient motion the velocity oscillates slowly about the final steady value, but, over this range, the variation in the angle of attack and *g* values is small. Eventually, if the aircraft is a stable one, it settles down at the trim speed and angle of attack corresponding to the new elevator position. A typical response for an up elevator displacement is shown diagrammatically in Fig. 11.2. The first part of the oscillatory response is related to the characteristics of the longitudinal short period mode of the aircraft and the latter part to the characteristics of the phugoid mode.

The magnitude of the steady state response in angle of attack and speed depends upon the amount of static stability that the aircraft possesses. As the C.G. of the aircraft moves rearwards, the static stability decreases and the response increases. Eventually a position is reached at which the aircraft tends to diverge in angle of attack

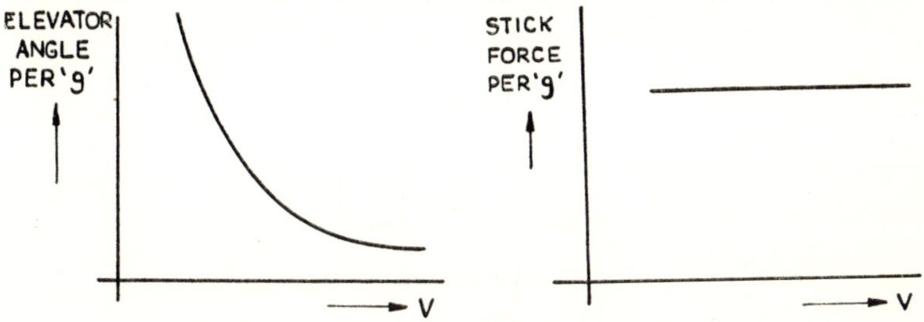

Fig. 11.4. Elevator angle per g and stick force per g against forward speed

and speed following a step displacement of the elevator. The elevator angles and forces required to trim an aircraft in level flight at different forward speeds have already been determined in Chapter 2; the results are shown in Fig. 11.3.

The longitudinal manoeuvrability characteristics of an aircraft were discussed in Chapter 3. Typical curves showing the variation with forward speed of the control movement per g and the stick forces per g derived from simple theory are shown in Fig. 11.4.

The transient longitudinal response characteristics are dealt with in more detail in References 4, 5 and 6.

3. Lateral Control Response Characteristics

3.1. General

The lateral control of an aircraft is achieved by means of aileron and rudder controls. The ailerons are required to trim the aircraft in the wings level position and to manouevre the aircraft about the rolling axis. The rudder is required to trim out any asymmetric yawing moments arising from slipstream effects or asymmetric power conditions and to manoeuvre the aircraft about the yawing axis. In certain manoeuvres only one of these controls is applied, but often the manoeuvre is made by a co-ordinated movement of both the aileron and rudder controls.

3.2. Damping in Roll Derivative l_p

In para. 3.3 it will be found that the roll response characteristics of an aircraft are very dependent on the value of the damping in roll derivative. This derivative has already been discussed in Chapter 6, where it was shown that the damping moment was produced as a result of the changes in angle of attack on the wings of the aircraft during the roll. The change in angle of attack at a section of the

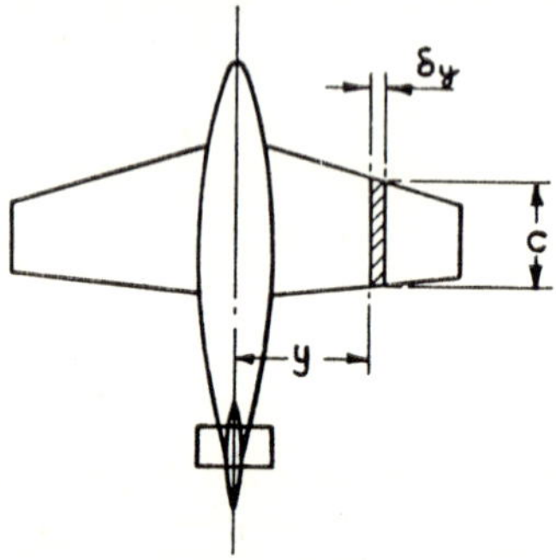

Fig. 11.5. Wing geometry notation

wing at a distance y ft from the axis of rotation is given approximately by

$$\Delta\alpha = \frac{p \times y}{V} \text{ degrees} \tag{11.1}$$

where p is expressed in degrees/sec and V is the forward speed (T.A.S.) in ft/sec.

The rolling moment opposing the rolling motion produced by a small element of the wing with a chord length c and width δy (Fig. 11.5) is obtained by multiplying $\Delta\alpha$ by $a\frac{1}{2}\rho V^2 c y \delta y$, where a is the lift curve slope of the wing section. However, since $\Delta\alpha \propto 1/V$, the resulting rolling moment is only proportional to ρV and not to ρV^2. Since $\rho V = \sqrt{\rho}\sqrt{\rho}V = \sqrt{\rho}\sqrt{(\rho_0 V_i^2)}$, we can see that, at a given E.A.S., the damping moment is proportional to $\sqrt{\rho}$.

As a result of this simple theory, the damping moment corresponding to a given rate of roll

(i) increases with an increase of forward speed, but only as the first power of the speed;
(ii) decreases with an increase of height at the same E.A.S.

3.3. Response of Aircraft to Aileron Control

When the aileron controls are deflected, asymmetric lift forces are produced on the wings of the aircraft. As long as the wings are not close to the stall, the lift is increased on the aileron-down wing and decreased on the aileron-up wing. This produces a rolling moment L and the aircraft begins to accelerate at a rate $\dot{p}_0$ given by

$$\dot{p}_0 = \frac{L}{A} \tag{11.2}$$

where $\dot{p}_0$ is the rolling acceleration of the aircraft at time $t = 0$, and A is the moment of inertia of the aircraft about the rolling, $0x$, axis.

As the rate of roll increases, the "damping-in-roll" moment (the l_p effect) increases and the corresponding rolling acceleration

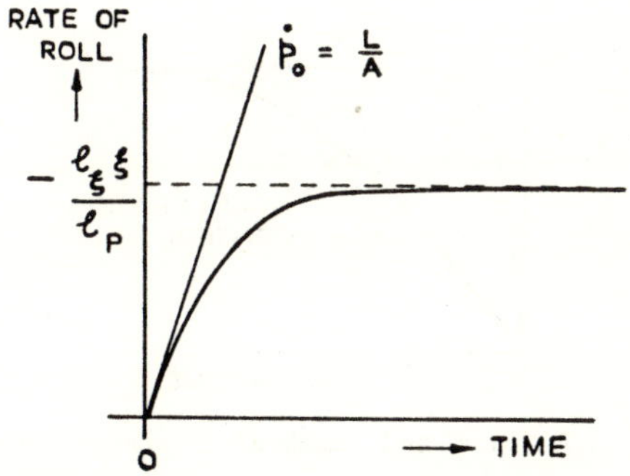

Fig. 11.6. Roll response to aileron deflection

decreases. The transient response of the aircraft depends upon the lateral characteristics of the aircraft. The time involved is so short, however, that the effects of the Dutch Roll and spiral modes are small, and the motion can be considered as a one degree of freedom motion in angle of bank, related to the roll convergence mode of the aircraft. Eventually, after a fairly short time interval, the rolling moment produced by the ailerons will be exactly balanced by the damping moment and the aircraft will attain a steady rate of roll (Fig. 11.6) given by

$$p = \frac{-l_\xi \xi}{l_p} \tag{11.3}$$

where l_ξ and l_p are the aileron power and roll damping derivatives, respectively, and ξ is the aileron deflection. The minus sign in eqn. (11.3) has to be there, because the sign convention is such that a positive aileron deflection (i.e. starboard aileron down) produces a negative rolling moment and a negative rate of roll (i.e. a roll to port).

Note. The symbol p on the left-hand side of eqn. (11.3) represents the non-dimensional rate of roll, i.e. $pb/2V$ (see note in Chapter 6, para. 2*b*) thus the actual rate of roll is given by

$$p = -\frac{l_\xi \xi}{l_p}\frac{2V}{b}$$

The shape of the response curve shown in Fig. 11.6 depends upon the time constant of the rolling convergence mode of the aircraft. Usually this is very short (possibly of the order of $\frac{1}{2}$ sec or so) and the rate of roll builds up to its steady value very quickly. Thus,

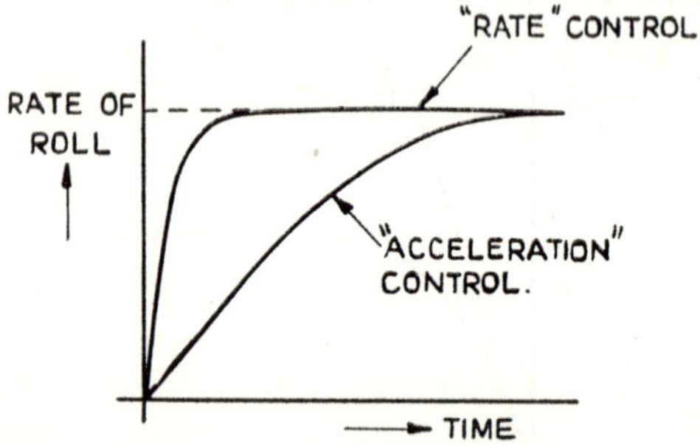

Fig. 11.7. "Rate" and "acceleration" aileron controls

over most of the time that the aileron control is being used, it is giving a steady rate of roll response, and, for this reason, it is described as a "rate control." On some modern designs (e.g. slender delta aircraft) the response time is increased (Fig. 11.7), and by the time the steady rate of roll is obtained (or even *before* this), the aileron is being taken off again. In these circumstances the aileron control is producing a rate of roll response which is always increasing, and it is then described as an "acceleration control." This type of response is different from that of the elevator control. The elevator can be described as a "displacement control," because an elevator deflection produces a steady-state change in angle of attack, i.e. a steady change of position. Thus a control is classified as a "displacement," "rate," or "acceleration" control, depending on the type of response it produces during the time interval it is being applied.

There are two important design criteria for the aileron controls. The first is for control at low speeds; this usually determines the aileron size, since the limitation is one of maximum aileron deflection.

The second is for control at high speeds; this determines the amount of aerodynamic balancing required on non-powered controls, since the limitation is one of maximum stick force. Other design limitations may possibly arise associated with inertia coupling problems.

3.4. Effect of Forward Speed and Altitude on Aircraft Response

The value of the steady rate of roll produced by a given aileron deflection will depend upon the speed and height at which the aircraft is flying. Additional effects (e.g. aeroelastic and compressibility effects) may also be present which will modify the values of the basic stability and control derivatives and hence the roll response of the aircraft.

(i) *Effect of Forward Speed*

If a constant aileron deflection is maintained (Fig. 11.8*a*) and the l_ξ and l_p derivatives assumed constant, then the applied rolling moment increases as V^2. The damping in roll moment, however, only increases as the first power of speed and hence, on this simple theory, the rate of roll increases linearly with forward speed (Fig. 11.8*b*). The Av.P.970 requirements are expressed in terms of the value of $pb/2V$, though these requirements are not suitable for some modern designs. Expressed in this way, the rate of roll response becomes a constant, independent of forward speed (Fig. 11.8*c*). The quantity $pb/2V$ can be regarded as the wing tip helix angle, since it is a measure of the change in angle of attack of the wing at the tips.

The aileron force required for a given control deflection with non-powered controls will also increase as V^2 (Fig. 11.8*d*) and eventually the pilot's control force limitation will be reached. If this force is then maintained at higher forward speeds (Fig. 11.8*d*), the corresponding aileron deflection will decrease in proportion to $1/V^2$ (Fig. 11.8*a*). Thus, in this speed range, the applied aileron rolling moment will be constant, independent of the forward speed, and since the damping in roll is still proportional to V, the rate of roll will decrease as $1/V$ (Fig. 11.8*b*). The value of $pb/2V$ will decrease more rapidly as $1/V^2$ (Fig. 11.8*c*), due to the limitation of the aileron deflection. If the results are expressed in the form of $(pb/2V)/\xi$, then, on the above simple theory, the value would remain constant throughout the complete speed range. In practice, aeroelastic distortion effects are present; these tend to reduce the aileron control effect and they are shown up by a reduction in the value of $pb/2V\xi$ (Fig. 11.8*e*).

(ii) *Effect of Altitude*

The applied rolling moment produced by a constant aileron deflection depends upon the dynamic pressure ($\frac{1}{2}\rho_0 V_i^2$) and hence for a

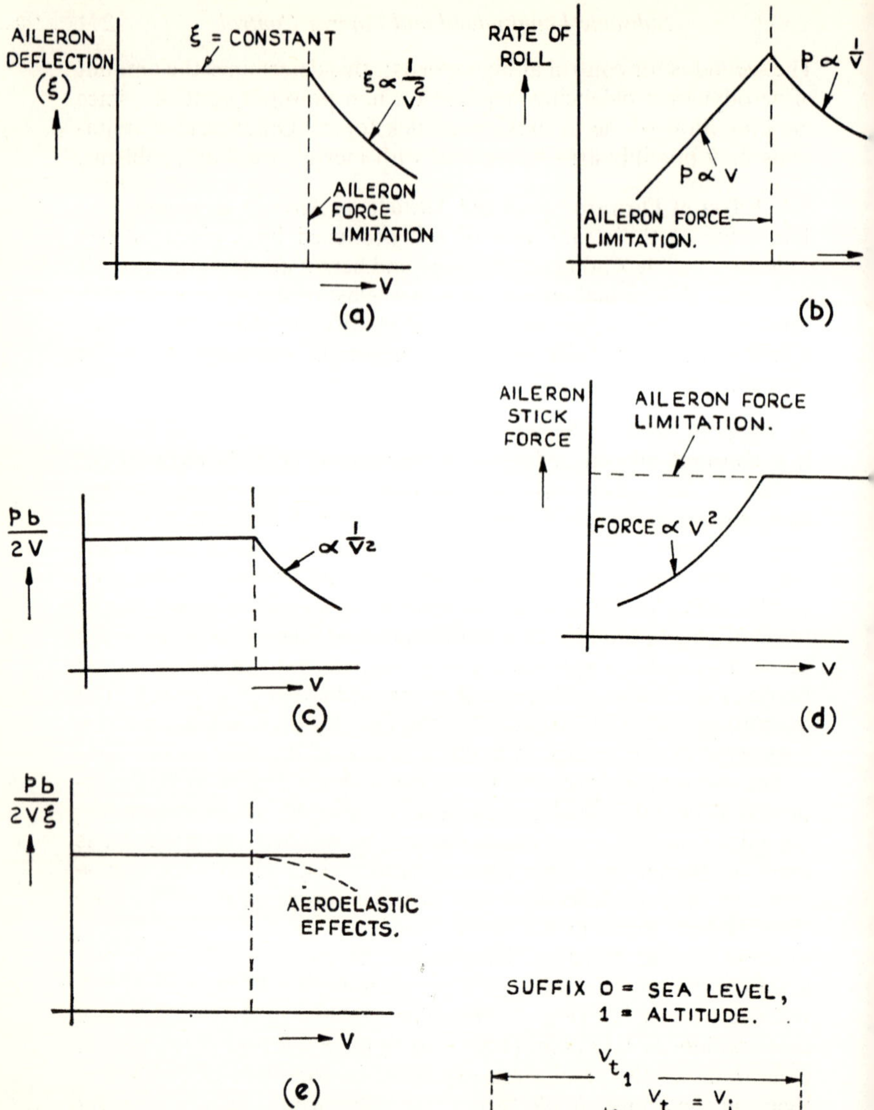

Fig. 11.8. Variation of aileron deflection and aircraft rolling characteristics with forward speed

(*a*) Constant aileron deflection
(*b*) Rate of roll increases linearly with forward speed
(*c*) Rate of roll response constant in terms of $pb/2V$
(*d*) Aileron force for a given control deflection
(*e*) Illustrating aeroelastic distortion effects

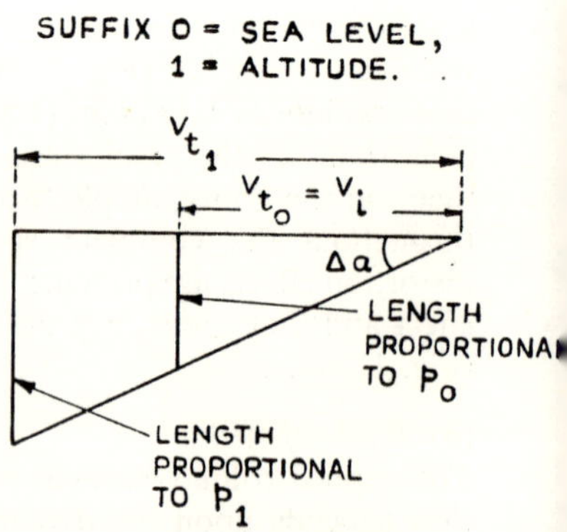

Fig. 11.9. Effect of altitude on the steady rate of roll

given E.A.S. is independent of height. The damping in roll moment on the other hand decreases with increase of height at the same E.A.S. On the basis of the simple theory used so far in this analysis, the steady value of the rate of roll will increase with increase of height in accordance with

$$p = \frac{\text{constant}}{\sqrt{\rho}} \tag{11.4}$$

An alternative explanation is shown in Fig. 11.9. Irrespective of the height at which the aircraft is flying, the change in the angle of attack of the wing corresponding to the condition of the steady rate of roll is a constant, depending on the value of the applied rolling moment. Since the T.A.S. is greater at height, for a given E.A.S., the value of the steady rate of roll will also increase.

4. Lateral Manoeuvres

4.1. General

The aileron and rudder controls are used to manoeuvre the aircraft about its longitudinal and directional axes. In any steady manoeuvre all the forces and moments acting upon the aircraft must be balanced and the aircraft must stabilize itself at the appropriate values of sideslip, angle of bank, etc., corresponding to the particular control deflections. In the following paragraphs the balance of the forces and moments occurring in various steady conditions of flight will be discussed

(i) steady sideslips,
(ii) co-ordinated turns,
(iii) steady turns using only one lateral control.

The control movements and forces required in these steady manoeuvres will be shown to be related to the corresponding values of the control-fixed and control-free stabilities. These manoeuvres can be used in flight test work to determine the lateral stability characteristics of the aircraft. It should be emphasized that we are only considering the steady manoeuvring conditions of flight; the initial response of the aircraft to control application on entry to the turn or sideslip, and the response on recovery, are also important, and they can also be analysed to determine some information on the stability and control characteristics of the aircraft (e.g. aileron adverse yaw characteristics, etc.).

4.2. Steady Sideslips

4.2.1. SIMPLE THEORY

The control movements and forces required to maintain steady sideslipping motion on an aircraft with a non-powered control system have already been discussed in Chapter 5. On simple theory (assuming that the derivatives remain constant and that the control cross-derivatives are zero) the deflection and force curves are linear ones (Fig. 11.10). If the aircraft possesses stick-fixed and stick-free lateral static stability the aileron curves have a negative slope and if it possesses control-fixed and control-free directional static stability

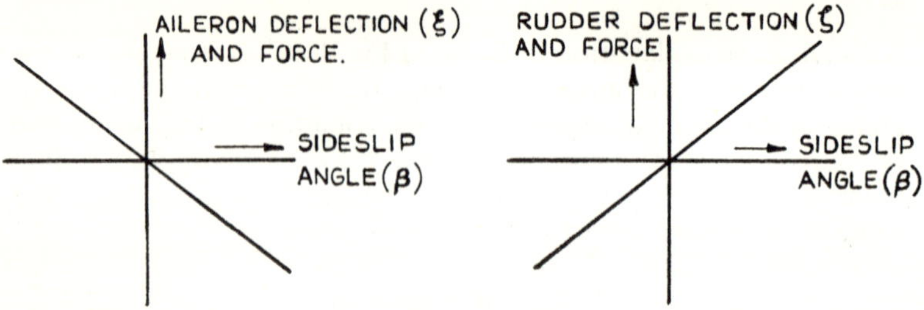

Fig. 11.10. Aileron and rudder angles and forces to trim in steady sideslips

the rudder curves have a positive slope. The slope of the aileron deflection curve determines the ratio of l_β/l_ξ since

$$\frac{d\xi}{d\beta} = -\frac{l_\beta}{l_\xi} \tag{11.5}$$

and the slope of the rudder deflection curve determines the ratio of n_β/n_ζ

$$\frac{d\zeta}{d\beta} = -\frac{n_\beta}{n_\zeta} \tag{11.6}$$

These formulae are, however, only applicable if the graphs are straight line ones.

If the values of the control powers have been determined from flight tests using asymmetric weights on the wings and wing tip parachutes (see Chapter 17), or if they can be estimated by any other method (Refs. 1 and 2), these equations can be used to calculate the values of the control-fixed lateral and directional static stabilities (l_β and n_β). The force curves are related to the corresponding control-free stabilities.

4.2.2. CONTROL DEFLECTIONS REQUIRED FOR TRIM

In practice, the curves of control deflections against sideslip angle may not be straight lines over the whole range of sideslip angles. They will usually be linear over a limited range of sideslip angle, but will often become non-linear at the larger angles of sideslip. This non-linearity may be the result of a change in any of the stability derivatives involved. The aileron deflection curve, for example, will become non-linear if changes occur in either l_β or l_ξ; an increase in dihedral will increase the deflection required, whereas an increase in aileron control power will decrease the deflection. On the non-linear part of the curve (e.g. point A on Fig. 11.11) the ratio ξ/β is still equal to $(-l_\beta/l_\xi)$. Thus the ratio of the derivatives is given by the slope of the line drawn to the point on the curve *from the origin.* It should be noted that, although we have previously talked about the slope of the curve being equal to $-l_\beta/l_\xi$, it is not the local slope

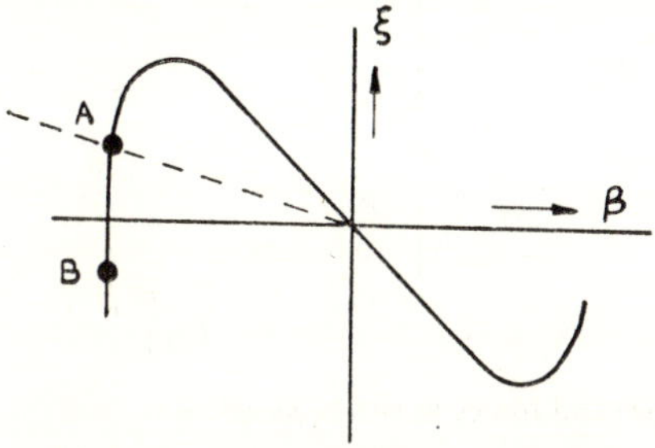

Fig. 11.11. Non-linear aileron deflection characteristics

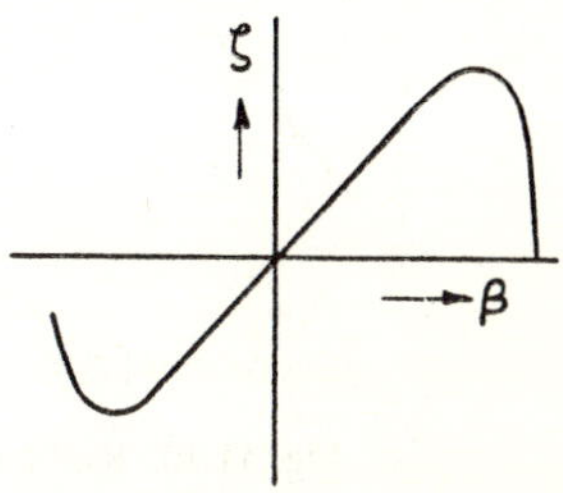

Fig. 11.12. Non-linear rudder deflection characteristics

of the curve. This is important when there is a reduction in ξ at the higher sideslip angles. The local slope of the curve is now positive, but this does not mean that l_β has reversed in sign (assuming l_ξ is still negative). The aileron still has to be held on to starboard in a starboard sideslip and the aircraft still possesses stick-fixed lateral static stability, though the stability is less than it was at smaller angles of sideslip. It is not until the control deflection required actually reverses (e.g. point B on Fig. 11.11) that the aircraft has become unstable. (By "reverse" in this context it is meant that the control position required to hold a new trim condition is in the opposite direction to that required to initiate the change.) The slope of the line drawn from point B to the origin of the graph is now a positive one and the aircraft possesses stick-fixed lateral instability.

Similar results often occur with the corresponding rudder deflection curves. Initially the rudder deflection increases linearly and the control-fixed directional stability remains constant. The rudder deflection may then fall off at the higher sideslip angles (Fig. 11.12) and the corresponding directional stability will decrease (though still remaining positive). Finally, if the rudder curve crosses the β-axis, the control has reversed and under these conditions the aircraft has negative control-fixed directional stability.

4.2.3. CONTROL FORCES REQUIRED FOR TRIM

When an aircraft is sideslipping, the controls tend to float from their neutral position, the direction of float depending upon the b_1 characteristics of the control. If the rudder has a negative value of b_1, then, in a starboard sideslip, the rudder would tend to float to port, i.e. in the same direction as that required to trim the aircraft (Fig. 11.13*a*). The force that the pilot has to apply to the rudder

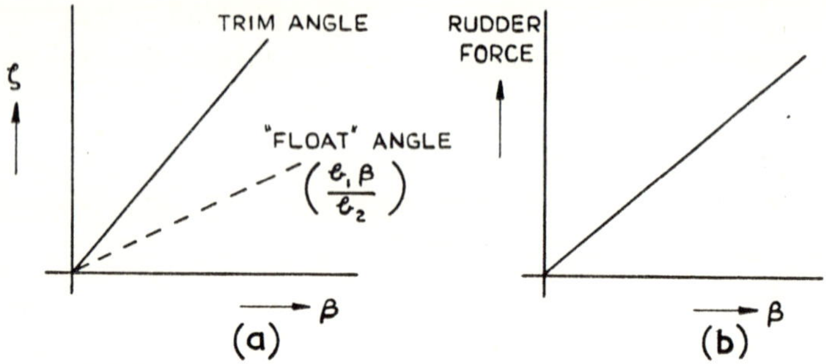

Fig. 11.13. Rudder angles and forces in steady sideslips

pedals is dependent on the value of the rudder hinge moment coefficient given by

$$C_H = b_2\zeta - b_1\beta \tag{11.7}$$

and in the case discussed above the floating characteristics of the control are such as to reduce the hinge moment. The rudder floats to an angle ζ_f are obtained from eqn. (11.7) by putting $C_H = 0$

$$\zeta_f = \frac{b_1\beta}{b_2} \tag{11.8}$$

and if b_1 and b_2 are both negative, it can be seen that ζ_f and β have the same sign (i.e. a positive sideslip to starboard produces a positive rudder float to port). From eqn. (11.8) the slope of the curve of ζ_f against β is given by b_1/b_2. By substituting from eqn. (11.8) into eqn. (11.7) the hinge moment coefficient is given by

$$C_H = b_2(\zeta - \zeta_f) \tag{11.9}$$

and thus the rudder force is directly proportional to the angular difference between the rudder angle required for trim and the floating angle. This angular difference and hence the rudder force is also a measure of the control-free static stability of the aircraft.

As long as the two rudder deflection curves remain linear, the rudder force increases linearly with sideslip angle (Fig. 11.13*b*). Any non-linearity in either of the curves results in a non-linearity in the rudder force curve. The important derivatives determining the slopes of these curves are n_ζ, n_β (n_v), b_1 and b_2. If the value of the rudder b_1 (the floating parameter) increased, or if the value of the rudder b_2 (the heaviness parameter) decreased, then the floating angle corresponding to a given sideslip angle would increase (Fig. 11.14*a*). The rudder force would correspondingly decrease, and if this effect continued until the rudder floating angle became equal to the angle required for trim, then the rudder force would decrease to zero. At even higher angles of sideslip, the control would tend to

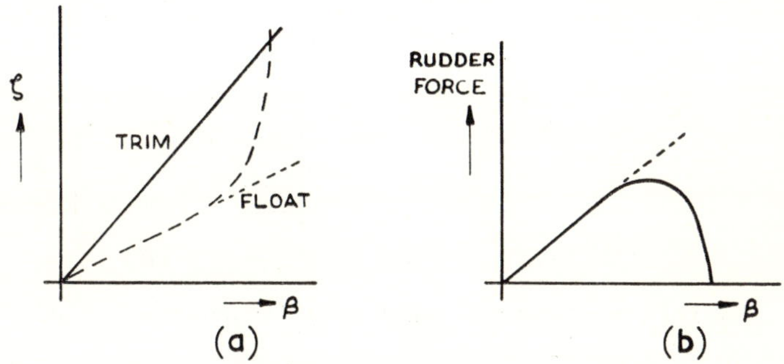

Fig. 11.14. Non-linear rudder force characteristics

float to a larger angle than that required for trim. In this case, the control would have to be brought back to the trim position and a force in the opposite direction to the normal one would have to be applied to hold it there. The rudder is then said to have "over-balanced" (Note that on the part of the curve where the force is reducing, but still in the usual direction, there is said to be a "tendency to overbalance".) The word "overbalance" is used in this context more widely than to mean simply that the b_2 of the control has gone positive; it implies here that the value of $(b_2\zeta - b_1\beta)$ has reversed in sign.

The same effect on the rudder force can be produced by a non-linearity in the curve of rudder angle to trim (or, of course, by non-linearities in both the trim and floating rudder deflection curves). Since the trim curve is determined by the values of n_β and n_ζ, a reduction in rudder angle would result from either a reduction in n_β or an increase in n_ζ (Fig. 11.15*a*). The rudder force is again reduced to zero (Fig. 11.15*b*) when the rudder "float" angle becomes equal to the angle required to trim.

Thus, assuming that the rudder power is constant, a reduction and overbalance of the rudder forces may be due to either

(i) a reduction in the directional static stability (n_β or n_v), or
(ii) changes in the rudder control surface characteristics (b_1 and b_2).

A reduction in n_v at the larger sideslip angles may result from "fin stalling"; this is often accompanied by some airframe buffeting. A serious overbalance of the rudder may result in the control going right over to the control stop. If the force involved is too large for

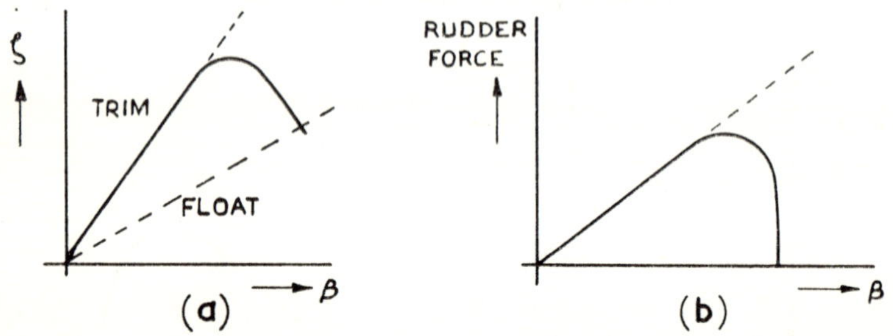

Fig. 11.15. Non-linear rudder angle and rudder force characteristics

the pilot to move the rudder back again, "rudder locking" is said to have occurred.

Similar overbalance characteristics may occur on the aileron controls.

4.2.4. METHODS AVAILABLE FOR MODIFYING THE CONTROL CHARACTERISTICS

It is important to distinguish between the two basic causes of control overbalance. If the control deflection curve is a straight line and the control force curve is a non-linear one, then the basic control characteristics are at fault. If, on the other hand, neither curve is straight, then the fault probably lies with the directional stability of the aircraft. To eliminate any adverse control force characteristics it is obviously essential to choose the appropriate method of correction.

(a) Modification of the Control Characteristics

As we have already seen, a reduction in rudder force may result due to an increase in b_1 or a reduction in b_2. The derivative b_2 is the heaviness or balance parameter of the control surface. As long as the curve of hinge moment against control deflection is a straight line,

b_2 is constant. The slope of this line can be modified in various ways; it can be reduced by any of the familiar methods of aerodynamic balancing (e.g. set-back hinge, horn balance, etc.). At the higher

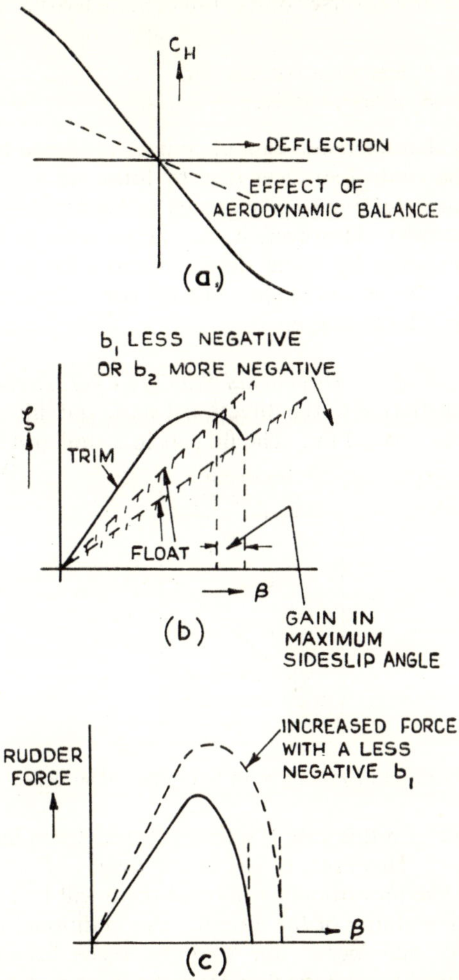

Fig. 11.16. Effect of hinge moment characteristics on the rudder angle and rudder force to trim curves

control deflections, the hinge moments may reduce, and b_2 will become less negative (Fig. 11.16*a*). It should be noted that the value of b_2 is given by the slope of the line drawn from the point considered to the origin. When the hinge moment decreases to zero and reverses, the control surface is said to be overbalanced and on this occasion we are using this word in its usual limited sense, to mean that b_2 has gone positive.

To eliminate any adverse control force characteristics it is necessary either

(i) to make b_1 less negative, and/or
(ii) to make b_2 more negative.

These control characteristics can, of course, be altered by a complete redesign of the control surfaces (e.g. by fitting an anti-balance tab, with a non-linear gearing to give greater tab deflections at the larger control deflections). However, it may be possible to make a large enough modification by some simple means such as trailing edge cord or strips. The effects of changing the control characteristics are shown in Figs. 11.16*b* and *c*.

(*b*) *Modification of the Directional Stability Characteristics*

The main contribution to the directional static stability of an aircraft comes from the vertical fin. The fin acts in a similar way to a wing,

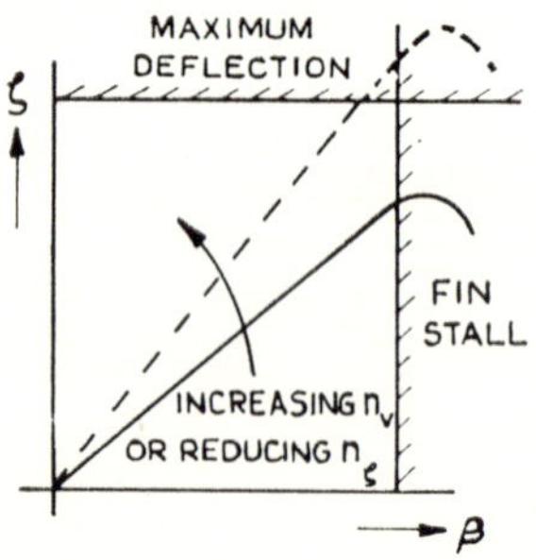

Fig. 11.17. Effect of varying n_v or n_ζ on the rudder deflection curve

producing "lift" (in this case a sideways force) when inclined to the relative airflow. However, at the larger angles of sideslip, the fin will stall just like an ordinary wing and there will be a reduction in the directional stability of the aircraft. An additional loss of directional stability may occur due to interference between the flow round the vertical fin and the flow from the front part of the aircraft. Vortices from the wing or from the nose of a high-fineness ratio fuselage may act in such a way as to produce a reduction in n_v. The loss in n_v can be prevented

(i) by preventing the aircraft from reaching its fin stalling angle (Fig. 11.17) by

(*a*) increasing the fin area (and hence n_v), or

(*b*) reducing the rudder power (e.g. by reducing the rudder area).

(ii) by increasing the fin stalling angle (Fig. 11.18) by

(*a*) adding a dorsal fin. This has the effect of increasing n_v at the larger sideslip angles.

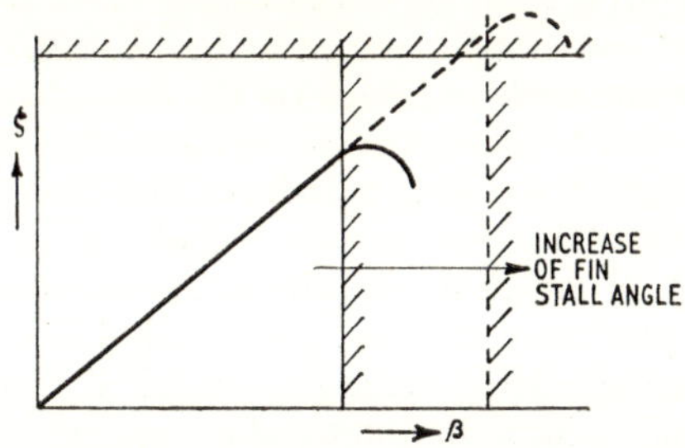

Fig. 11.18. Effect of increasing the fin stalling angle on the rudder deflection curve

(*b*) using a number of smaller fins of smaller aspect ratio rather than one large fin. This modification produces the desired result, because the stalling angle of a lifting surface is usually increased by a decrease in the aspect ratio of the surface.

In making any modification, it is, of course, necessary to consider the effect on the overall stability and control characteristics of the aircraft. If an increase of n_β at all angles of sideslip is being considered, then the effects on the spiral stability and Dutch Roll characteristics must also be investigated, to see whether any other adverse effects would arise as a result of this modification.

4.3. Steady Coordinated Turns

The steady coordinated turn is discussed only briefly here, because it does not form one of the flight tests used for determining the lateral stability and control characteristics of an aircraft. It is discussed in more detail on pages 298–304 of Ref. 5.

In a coordinated level turn the value of the sideslip and the rate of roll are zero; the aircraft is left with only pitching and yawing motions. Since the motion is a steady one all the forces and moments must be balanced out. The rudder has to be held into the turn to balance the damping in yaw n_r effect, though this balance is modified by any aileron yaw n_ξ effect. The aileron has to be held *out* of the turn to balance the rolling moment due to yaw l_r effect, which acts into the turn. Thus in this manoeuvre an aircraft with normal stability characteristics exhibits aileron position instability, i.e. the stick has to be moved over in the normal direction to initiate the manoeuvre and then brought back beyond the original position to hold the aircraft in the steady turn. This is sometimes referred to as "holding off the bank."

When the aircraft is climbing or descending in the turn it has a rate of roll, as well as one of pitch and yaw. In this manoeuvre the wing will have a different angle of attack on the two sides, the change in angle of attack always being greatest on the inner wing. In a climb the angle of attack of the wing is reduced and thus the outer wing develops more lift and it is necessary to hold off more bank. In a descent the angle of attack of the wing is increased and in this case the inner wing produces more lift and less holding off of bank is required. Thus the aileron deflection required is increased in a climbing turn and reduced in a descending one.

4.4. Turns on One Control

4.4.1. GENERAL

The method of carrying out steady turns using only one of the lateral controls, with the other either held fixed or left free, is quite a useful technique for determining some of the lateral stability and control characteristics of an aircraft. It is not, however, a coordinated turn; the aircraft will be sideslipping during the manoeuvre, and the "ball" will not be in the centre. Because the turns are level turns, no rates of roll will be involved. Since the turn is a steady manoeuvre, all the forces and moments acting on the aircraft must be balanced. The tests consist of

(i) turns on rudder alone, with the aileron fixed or free,
(ii) turns on aileron alone, with the rudder fixed or free.

These tests are discussed in more detail in paras. 5.12 and 5.13 of Ref. 3. If the aircraft is fitted with powered irreversible controls then there is no difference between the control fixed and free cases. From measurements of the aileron (or rudder) deflections and forces required to hold the aircraft in a steady turn, it is possible to make certain deductions concerning the control-fixed and control-free stability characteristics of the aircraft.

4.4.2. TURNS ON ONE CONTROL WITH THE OTHER CONTROL FIXED

These are the easiest cases to analyse, because the contribution to the moments from the fixed control can be assumed to be zero.

(i) *Turns on Rudder Alone*

Consider a turn to starboard being initiated on an aircraft using only the rudder control. The rudder is deflected to starboard and the aircraft yaws to starboard and sideslips to port. These aircraft motions produce a rolling moment to starboard and the aircraft banks to starboard. As a result of this, a component of the weight acts to starboard, the port sideslip is reduced and a sideslip to starboard develops. Eventually the aircraft settles down in a steady turn, the deflection of the rudder being adjusted as necessary to trim the aircraft out. In a turn to starboard, the aircraft will usually finish up yawing to starboard, sideslipping to starboard, banked to starboard, and with the rudder deflected to starboard.

The forces and moments acting on the aircraft will be balanced as follows—

(*a*) *Rolling Moments* If the rudder is assumed not to produce any rolling moment, then there is a simple balance between the roll due

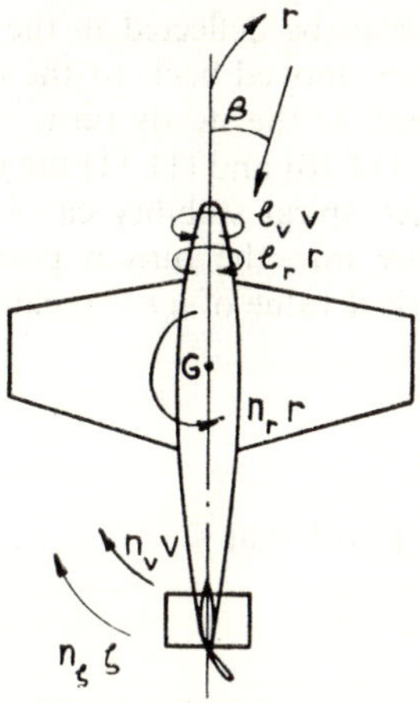

Fig. 11.19. Balance of rolling and yawing moments in a rudder-only turn

to yaw effect acting into the turn and the roll due to sideslip effect out of the turn (Fig. 11.19). It is for this reason that the aircraft must be sideslipping *into* the turn. In symbols this balance is expressed by saying that

$$l_r r + l_v v = 0 \tag{11.10}$$

Any rolling moment effect from the rudder will reduce the necessary sideslip.

Note. In many of the equations in this chapter it should be remembered that the symbols p, r and v can be regarded as representing the non-dimensional values of the rate of roll, the rate of yaw and the sideslip velocity respectively. (See note in Chapter 6, para. 2*b*.)

(*b*) *Yawing Moments* There are three main yawing moments acting on the aircraft, one from the yawing motion (the n_r effect), one from the sideslipping motion (the n_v effect) and one from the rudder deflection (the n_ζ effect) (Fig. 11.19). These all add up to zero,

$$n_r r + n_v v + n_\zeta \zeta = 0 \qquad (11.11)$$

In a starboard turn the n_r moment acts to port, opposing the turn, and the n_v moment acts to starboard, assisting the turn (assuming a starboard sideslip). The rudder deflection required to trim depends on which of these two effects is the greater. If the n_r effect is the greater of the two, the rudder deflection is *into* the turn; if the n_v effect is the greater, the rudder deflection is *out* of the turn. If it so happened that the n_r and n_v effects exactly balanced one another, the rudder would have to be deflected in the usual way to initiate the manoeuvre and then moved back to the original trim position again to hold the aircraft in the steady turn.

By combining eqns. (11.10) and (11.11) the necessary requirement for the aircraft to have spiral stability can be determined. If the aircraft requires rudder into the turn it possesses spiral stability. This is so if the numerical value of $n_r r$ is greater than that of $n_v v$, i.e.

$$|n_r r| > |n_v v|$$

or

$$|n_r r| - |n_v v| > 0 \qquad (11.12)$$

From eqn. (11.10) we have* that

$$v = -\frac{l_r r}{l_v} \qquad (11.13)$$

and substituting this in eqn. (11.12) gives the stability requirement as

$$|n_r r| - \left|\frac{n_v l_r r}{l_v}\right| > 0$$

Dividing through by r, removing the "numerical value" symbols and rearranging, gives

$$\underline{\underline{(n_r l_v - n_v l_r) > 0}} \qquad (11.14)$$

* See note in section 2b of Chapter 6; in dimensional form eqn. 11.13 becomes

$$v = -\frac{l_r r}{l_v}\frac{b}{2}$$

This is the requirement for an aircraft to possess spiral stability. The equation states the condition irrespective of the direction of the turn; it states that the product of the two "anti-turn" derivatives must be greater than the product of the two "pro-turn" derivatives. The derivative n_r is a measure of the opposition of the aircraft to the rate of yaw and the effect of l_v is to roll the aircraft out of the turn, whereas the effect of l_r is to roll the aircraft into the turn and that of n_v to yaw the aircraft into the turn.

The values of the derivatives depend upon the speed and height at which the aircraft is flying and the particular configuration (e.g. flaps, undercarriage, etc.) being used. The derivative l_r is dependent on the value of C_L and increases as the forward speed is reduced; this tends to decrease the spiral stability. On swept wing aircraft the value of l_v increases at the low speeds, this counteracts the l_r effect and tends to make the aircraft more spirally stable (and increases the possibility of Dutch Roll instability).

(*c*) *Sideforces* The various sideforces acting on the aircraft must also balance, but this is not such a useful balance as the ones involving the moments. The sideforces acting are the component of the weight acting into the turn, the aerodynamic sideforce (the y_v effect) acting out of the turn and the "inertia" sideforce acting out of the turn. This can be looked at another way, by saying that the weight component and the aerodynamic sideforce must produce a resultant acting into the turn sufficient to produce the necessary rate of yaw (i.e. the necessary acceleration for the aircraft in a curved flight path).

(ii) *Turns on Aileron Alone*

The turns on aileron alone with the rudder fixed can be analysed in a similar way. To initiate a turn to starboard, the port aileron is moved down and the starboard aileron up. The aircraft rolls to starboard and sideslips to starboard. As a result of this, the aircraft will also begin to yaw to starboard. Finally, the aircraft settles into a steady turn, the aileron being deflected as required to balance out the moments on the aircraft.

(*a*) *Yawing Moments* If the aileron yaw is assumed zero, there will be a simple balance between the damping in yaw moment (the n_r effect) and the yawing due to sideslip (the n_v effect) (Fig. 11.20)

$$n_r r + n_v v = 0 \tag{11.15}$$

In a starboard turn, once again, the aircraft must be sideslipping to starboard to obtain the required starboard yawing moment from n_v. Any aileron yawing moment will modify this simple result.

(*b*) *Rolling Moments* There are three main rolling moments acting on the aircraft, one from the yawing motion (the l_r effect), one from the sideslipping motion (the l_v effect) and one from the aileron deflection (Fig. 11.20). They must all add up to zero,

$$l_r r + l_v v + l_\xi \xi = 0 \tag{11.16}$$

In a starboard turn the l_r moment acts to starboard and the l_v effect acts to port (assuming a starboard sideslip). The aileron deflection required depends on which of these two effects is the greater. If the

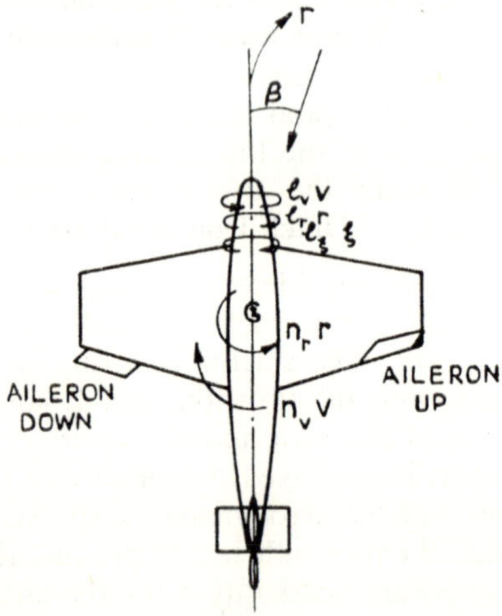

Fig. 11.20. Balance of rolling and yawing moments in an aileron-only turn

l_v effect is greater than the l_r effect, aileron deflection is required *into* the turn; if the l_r effect is greater than the l_v effect, aileron deflection is required *out* of the turn. If aileron deflection is required into the turn, the aircraft possesses spiral stability, and by an analysis similar to that in para. 4.4.2 (i), eqn. (11.14) can again be derived from eqns. (11.15) and (11.16).

4.4.3. TURNS ON ONE CONTROL WITH THE OTHER CONTROL FREE The analysis of these cases is somewhat more difficult than the control fixed ones, because the additional moment produced by the movement of the free control has to be allowed for. Thus, even assuming that the cross control effects (i.e. n_ξ and l_ζ) are zero, there are still going to be three separate moments in each of the rolling moment and yawing moment equations; there is no longer a simple balance between two moments in either of the equations.

(i) *Turns on Rudder with Aileron Free*

The balance of moments is given by the following equations

$$n_r r + n_v v + n_\zeta \zeta = 0 \tag{11.17}$$

$$l_r r + l_v v + l_\xi \xi_f = 0 \tag{11.18}$$

where ξ_f is the floating angle of the aileron. The floating aileron can be regarded as modifying the dihedral stability of the aircraft; if in a starboard sideslip the starboard aileron floats up the dihedral is effectively reduced, if on the other hand the starboard aileron floats down the dihedral is effectively increased. The $(l_v v + l_\xi \xi_f)$ term can be replaced by $l_v' v$, where l_v' represents the stick-free lateral static stability. The condition for spiral stability can still be expressed in the form of eqn. (11.14), except that l_v must be replaced by l_v'.

(ii) *Turns on Aileron with Rudder Free*

Neglecting the n_ξ and l_ζ effects, the balance of moments is given by

$$n_r r + n_v v + n_\zeta \zeta_f = 0 \tag{11.19}$$

$$l_r r + l_v v + l_\xi \xi = 0 \tag{11.20}$$

where ζ_f is the floating angle of the rudder. This floating of the rudder can be regarded as modifying the directional stability of the aircraft and the term $n_v v + n_\zeta \zeta_f$ can be replaced by $n_v' v$, where n_v' is the control free directional stability. Eqn. (11.14) again gives the spiral stability requirement, with n_v replaced by n_v'. If the rudder floats to port in a starboard sideslip, the directional stability is effectively reduced; the control-free value of n_v is then less than the corresponding control-fixed value and the spiral stability is increased.

5. Flight Tests

The lateral manoeuvres discussed in para. 4 can be used in flight test work to assist in the determination of the lateral stability and control characteristics of the aircraft. The tests usually carried out are steady sideslips, turns on one control and oscillatory tests. The analysis of the oscillatory tests is discussed in detail in Chapter 17. The assessment of the lateral derivatives is made easier if the values of the control powers (i.e. l_ξ and n_ζ) are known. These values can be determined from the *Data Sheets* of the *Royal Aeronautical Society*, or obtained from flight test experiments using wing tip parachutes and asymmetric weights. These experiments are described in para. 6.1 of Chapter 17 and on pages 5.19 and 5.25 of Ref. 3.

From the steady sideslip results of aileron angle and rudder angle required to trim, the values of n_v and l_v can be calculated using eqns. (11.5) and (11.6). If aileron is required to hold the leading wing down, the aircraft possesses positive lateral static stability l_v and if port rudder is required to hold a starboard sideslip then the aircraft also possesses positive directional static stability n_v.

In steady turns on one control with the other control fixed, if control is required into the turn then (as long as the l_v and n_v characteristics are normal) the aircraft possesses control-fixed spiral stability.

Some useful tables are given in Ref. 2. Table 82.4 indicates the various lateral stability and control features and the particular tests that provide information on them. Table 82.5 lists various possible lateral deficiencies and indicates the common causes of these deficiencies, the means of identifying the particular cause and typical methods of remedial action. Finally, Table 82.6 shows how deductions about the relative sizes of l_v and n_v can be made from the amount of sideslip, rate of turn, magnitude of control deflections, etc., measured during the various flight tests.

6. Flight on Asymmetric Power

6.1. General Considerations

When discussing the lateral and directional control characteristics on asymmetric power, two aspects have to be considered

(*a*) the regaining of control immediately following the failure of an engine on a multi-engined aircraft (i.e. the "dynamics" of the motion);

(*b*) the adequacy of the controls in steady straight flight with an engine out (i.e. the "statics" of the motion).

When an engine fails on a multi-engined aircraft yawing and rolling moments are produced on the aircraft

(*a*) A yawing moment is applied due to the loss in thrust. This yawing moment can be very large in comparison with the aerodynamic moments, particularly at full throttle at low speeds. This is the primary cause of trouble and hence the pilot's first concern should be to eliminate it, either by throttling back the opposite engine, or by application of opposite rudder, or both.

(*b*) In the case of propeller aircraft, a rolling moment is applied owing to the reduction in slipstream velocity on one side, leading to a loss of lift. This may be important at extremely low speeds, but otherwise it is well within the power of the ailerons to cope. However, aileron deflection can add to the adverse yawing moment and its contribution at low airspeeds is such that the use of aileron should not be made in the initial stages.

If corrective action is not taken immediately, the aeroplane will yaw and roll in the direction of the dead engine, the resulting sideslip tending to increase the roll. Should the yaw attain large proportions, then fin stalling and rudder locking may follow. When flying at low airspeeds (high C_L) during take-off and on the approach, there is also a danger of wing tip stalling due to the rate of roll and to any aileron deflection applied. These both increase the effective angle of attack of the dead-engined wing.

6.2. Control in Steady Straight Flight with Asymmetric Power

In steady flight in a straight line with asymmetric power, the forces and moments must, as usual in steady flight, all be balanced out. This balance can be achieved by various combinations of angles of bank and sideslip (i.e. of aileron and rudder deflections).

The main forces acting on the aircraft in the plane of the wings are

(i) the sideforce on the body and fin due to sideslip,
(ii) the sideforce on the rudder when it is applied,
(iii) the component of the weight in the plane of the wings,
(iv) the thrust of the live engine,
(v) the drag (including windmilling drag).

By rearranging that the forces add up to zero and also produce a zero moment, we can immediately sketch three principal conditions of flight at constant heading on asymmetric power, using the following symbols

T = live thrust
D = drag in line of flight
$R.C.F.$ = rudder control force
$S.F.$ = side force due to sideslip velocity
$Wt.C.$ = weight component

(*a*) *Wings Held Level* (no angle of bank)

If the wings are held level, then the rudder must be deflected to produce a yawing moment about the C.G. of the aircraft. Since this would produce an unbalanced sideforce, it is necessary to sideslip the aircraft towards the dead engine to produce a sideforce which, together with the sideways component of the thrust, will balance the sideforce on the rudder. The balance of forces is shown in Fig. 11.21; they must form a closed vector polygon. The yawing moments from the thrust, the rudder, and the sideforces must also be balanced.

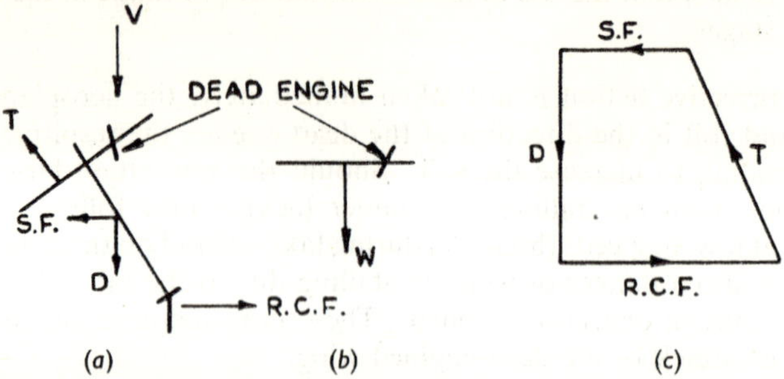

Fig. 11.21. Balance of forces under asymmetric conditions—wings held level

In this case the turn and slip indicator gives zero indication of sideslip and the compass does not give the true heading of the aircraft. The sideslip angle required is usually small ($< 5°$).

(b) No Sideslip and a Small Angle of Bank

If there is no sideslip, then the yawing moment of the rudder must exactly balance the yawing moment from the asymmetric thrust.

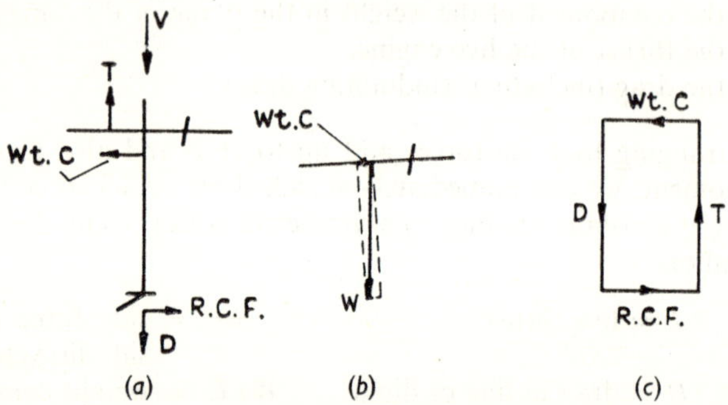

Fig. 11.22. Balance of forces under asymmetric conditions—zero sideslip

The only way of balancing out the resulting sideforce is to use bank. The aircraft has to be banked towards the live engine, so that the weight component exactly balances the rudder sideforce. The balance of forces is shown in Fig. 11.22. The angle of bank required is usually small ($< 3°$).

(*c*) *Rudder Free and an Increased Angle of Bank*
This final method avoids the necessity of holding on a rudder force, but the angle of bank involved may be large (of the order of 15°). The balance of forces is shown in Fig. 11.23; in practice it will depend on the floating characteristics of the rudder. This method can be potentially dangerous because, at the large sideslip angles

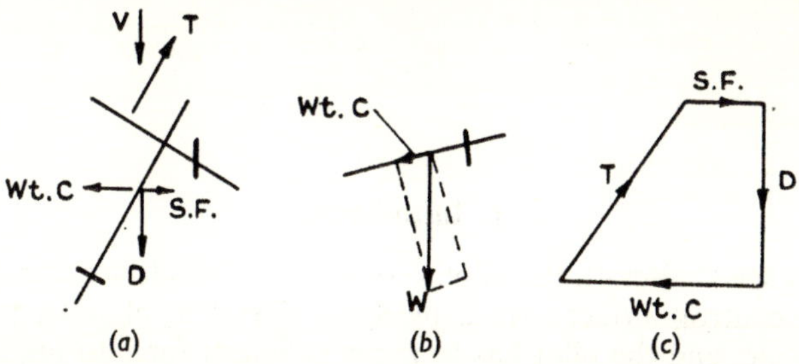

Fig. 11.23. Balance of forces under asymmetric conditions—rudder free

involved, the rudder may be near to locking and/or the fin to stalling, particularly as the speed decreases.

The usual method employed is somewhere between cases (*a*) and (*b*) above, with the live-engine-wing down about three degrees.

7. Conclusion

In this chapter we have discussed the balance of forces and moments involved in various steady lateral manoeuvres. These manoeuvres, together with the oscillation tests discussed in other chapters, can be used in flight test work to determine the lateral stability and control characteristics of an aircraft. The work of this chapter is chiefly related to an aircraft fitted with conventional non-powered control surfaces.

References

1. *Royal Aeronautical Society Data Sheets*, "Aerodynamics."
2. Boscombe Down A. & A.E.E. *Technical Guide.*
3. *Agard Flight Test Manual*, Vol. II, "Stability and Control."
4. PERKINS, C. D. and HAGE, R. E.; *Airplane Performance, Stability and Control* (John Wiley and Sons, New York, 1949).
5. ETKIN, B.; *Dynamics of Flight* (John Wiley and Sons, New York, 1959).
6. BABISTER, A. W.; *Aircraft Stability and Control* (Pergamon Press, Oxford, 1961).

12 Power Control Systems

1. Introduction

On low speed aircraft, the pilot's control movements are transmitted to the control surfaces via a mechanical system of rods, pulleys, cables, etc. and the pilot has to apply sufficient force to his control stick or rudder pedals to overcome the aerodynamic forces acting on the control surfaces. At low speeds these forces are within the pilot's limitations. As the forward speeds have increased over the years, the corresponding aerodynamic forces have also increased and various methods of aerodynamic balancing (e.g. set-back hinge balance, geared tabs, etc.) had to be used to keep the pilot's loads to reasonable proportions. Eventually, these methods became inadequate for some aircraft and hydraulically-operated power controls began to be installed. One of the first applications of an hydraulically operated control was on the aileron system of the Lockheed Lightning (P-38J) aircraft (Ref. 10).

The earliest power controls were of the power boosted type, in which only a certain proportion of the aerodynamic loads were fed back to the pilot, the hydraulic system overcoming most of the load. In these systems, although the pilot did not have to provide all the force required, he retained the natural feel of the controls and the stick forces experienced increased as the square of the forward speed. These power-boosted systems were eventually replaced by fully-powered control systems, in which none of the aerodynamic load on the control surfaces was fed back to the pilot. The only force that the pilot need exert is the force required to overcome the friction, etc., in the mechanical system, all the necessary power is supplied by the aircraft's hydraulic or electrical system. However, a pilot's assessment of the handling characteristics of an aircraft is very dependent on the control forces that he has to apply and thus it was necessary to incorporate "artificial" feel. This can be done in different ways, depending on the particular control force characteristics that are required. The control system of an aircraft must be safe and reliable and therefore a power control system must be duplicated or capable of reversion to manual operation.

On modern large high-performance aircraft, the conventional mechanically operated type of power control system is becoming extremely complicated and it suffers from a number of disadvantages. Electric signalling control systems avoid many of the difficulties associated with mechanical signalling, and allow easier integration of autostabilization systems and the incorporation of manoeuvre demand and adaptive control systems.

2. Power Control Systems

2.1. General

An aircraft's power-operated flying control system is a remote position servomechanism, a movement of the stick by the pilot

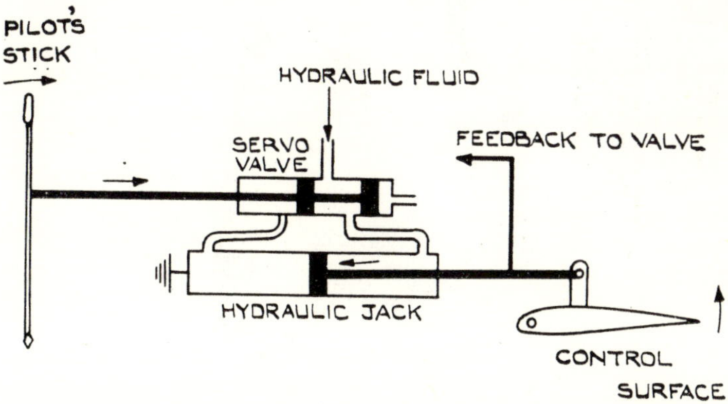

Fig. 12.1. Diagrammatic representation of a simple power control system

demanding a corresponding deflection of the control surface. A diagrammatic representation of a simple system is shown in Fig. 12.1. The pilot's input signal is transmitted, usually by a mechanical system of rods and cables, to a servo-valve and the opening of the valve allows high pressure hydraulic fluid to enter the cylinder of the jack. Relative movement between the piston and the body of the jack, which is anchored to the structure of the aircraft, causes the control surface to move in the required direction. The movement of the control surface is fed back through a differential lever system (the "follow-up" linkage) to the servo-valve and this closes the valve off again, leaving the control surface in a new equilibrium position which is dependent on the magnitude of the pilot's input. This use

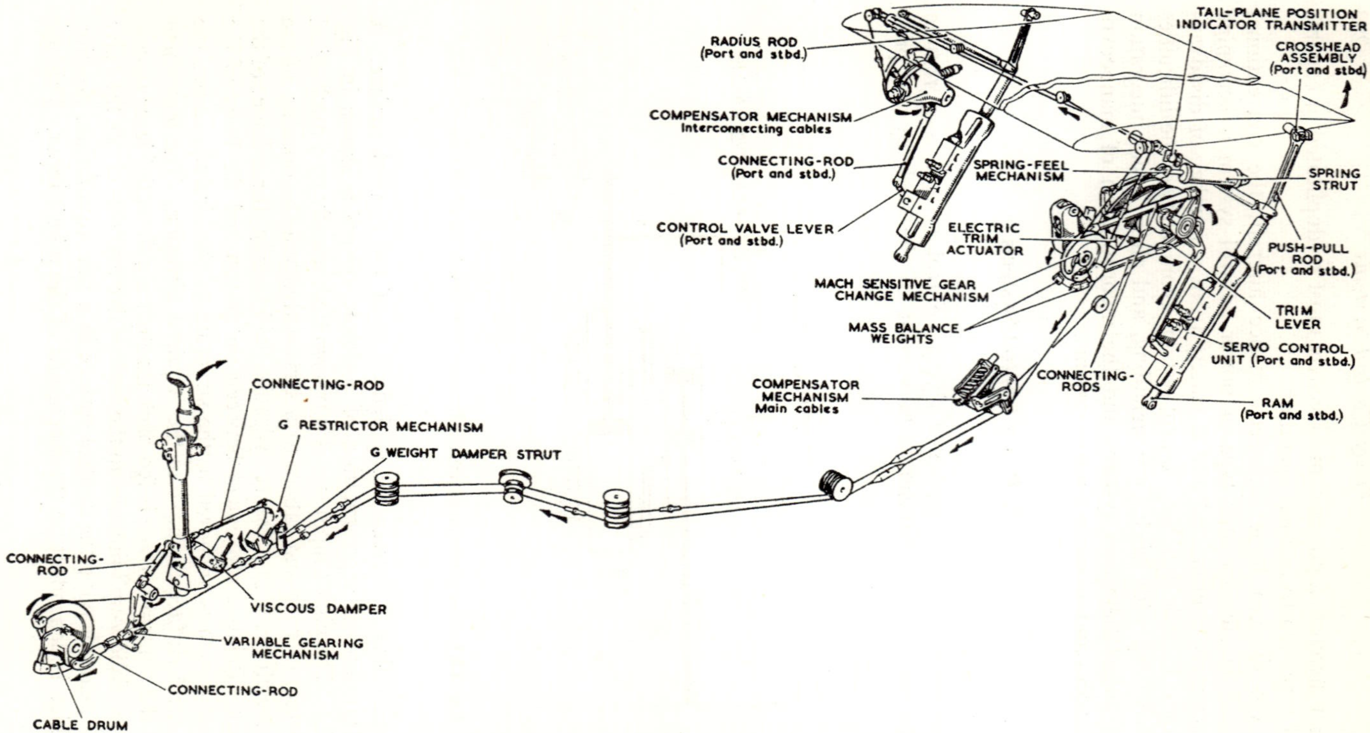

Fig. 12.2. Tailplane control circuit [Crown copyright]

of "feedback" is an important basic principle of all servomechanisms (see Chapter 19). The force that has to be exerted by the pilot to open the servo-valve is very small and the hydraulic system of the aircraft provides the power required to move the control surfaces against the large aerodynamic loads exerted on them. Thus the system produces a large power amplification.

The basic control system described above is complicated, in practice, by the need to provide the pilot with satisfactory handling characteristics throughout the flight envelope of the aircraft and by the need to make the system safe and reliable. The requirements result in the incorporation into the overall system of a number of the following devices—

(i) an artificial feel system,
(ii) a variable gearing and a gear change system,
(iii) a trimming system,
(iv) a viscous damper,
(v) a bob-weight (or *g*-restrictor),
(vi) an autopilot and an autostabilizer system,
(vii) a duplication of the power control system, or the provision of manual reversion for emergency conditions.

The control systems on some modern aircraft are extremely complicated. Fig. 12.2 shows the longitudinal control system of a particular aircraft (the Sea Vixen); it incorporates spring feel with variable gearing and an automatic gear change, together with a "*g*-restrictor," mass balance weights and a viscous damper. The aileron circuit of the same aircraft has spring feel, with variable gearing and a manual gear change, and the rudder circuit contains *q*-feel, a mass-balance system and a yaw damper.

Though most present-day power control systems have mechanical transmission and hydraulic power operation, systems can be made with electrical transmission and electrical power operation. Even on conventional mechanical systems, the signals initially produced by the autostabilization systems are electrical ones and some power control units are capable of accepting both mechanical and electrical signals.

2.2. Types of Power Control Systems

The two main types are

(i) Hydromechanical systems,
(ii) Electro-hydraulic-mechanical systems.

In hydromechanical systems, the inputs and outputs are mechanical ones and the power supply is the hydraulic system of the aircraft.

In electro-hydraulic-mechanical systems the inputs and outputs are again mechanical ones, but the power is taken from the aircraft's electrical supply in order to run a hydraulic generator to produce the necessary hydraulic power.

Electro-mechanical systems are often used for trimming purposes and in these systems the control surface is driven by an electric motor. The main power control system, however, usually requires the final drive to be a hydraulic one.

The power control system is usually taken off the main hydraulic supply of the aircraft, though on some aircraft a separate hydraulic system may be provided. The pumps of these hydraulic systems are driven off the engines of the aircraft, and the systems usually operate at pressures in the range 2,000 to 4,000 lb/in^2.

2.3. Basic Components of Power Control Systems

The basic components of any power control system are the servo-valve and the hydraulic jack. Fig. 12.1 shows the layout of a simple power control system. The pilot's input signals are transmitted by mechanical linkages to the servo-valve and movement of the control column (or rudder pedals) opens the valve and allows high pressure hydraulic fluid to flow into the cylinder of the jack.

There are two types of hydraulic jack available, the piston-type and the screw type. Fig. 12.1 represents a piston-type jack; when the high pressure fluid enters the cylinder of the jack, the piston and body of the jack move relative to one another, the actual motion depending on which part of the jack is earthed to the structure of the aircraft. In the case of the screw-type jack, the opening of the servo-valve allows high pressure hydraulic fluid to flow to a hydraulic motor which operates the screw jacks.

When the servo-valve is opened, the control surface will continue to move until the control stop is reached, unless the movement of the control is fed back in such a way that, as it moves, it closes the servo-valve. When feedback is used in this way, the control surface will move to a new position of equilibrium and the control deflection will be proportional to the pilot's input signal. When the jack body is earthed and the piston moves, this feedback must be provided by means of mechanical linkages. When the piston is earthed and the jack body moves, then movement of the jack body automatically closes the servo-valve and no additional feedback linkages are necessary.

2.4. Basic Requirements of Power Control Systems

A power control system must meet all the various basic requirements of any closed loop servomechanism. It must have a satisfactory performance, with good response and stability characteristics, and it must be a safe and reliable system.

(*a*) *Performance* The performance should be such that the system is capable of producing the necessary thrust required to move the control surfaces throughout the flight envelope of the aircraft. The required performance can be obtained by suitably designing the jack, since the thrust that can be obtained is related to the area of the piston and the pressure of the hydraulic supply. If the aerodynamic loads acting on the control surface become too large for the jack to produce the necessary thrust, "jack stalling" is said to occur. This will limit the amount of control that can be used during certain conditions of flight and will correspondingly reduce the manoeuvrability of the aircraft.

(*b*) *Response* The response characteristics of the system should be satisfactory to give accurate and precise control under all conditions of operation. The control surfaces must be positioned accurately and quickly in accordance with the input signal and consequently the system should be a sensitive one and such features as "backlash," time lags, "break-out" forces and "dead-zones," should be reduced to a minimum. Any deficiency in these respects would lead to the possibility of over-controlling, pilot-induced oscillations, (P.I.O.s) etc.

(*c*) *Stability* Since power control systems are basically high-gain servomechanisms, every care is necessary to ensure that the system is a stable one (see Chapter 19). The stability requirement is usually satisfied by designing the system with a high natural frequency, the inertia of the moving parts being made small in relation to the thrust produced by the jack, and by connecting the power control unit directly to the control surface.

(*d*) *Stiffness* (*Irreversibility*) In a perfectly irreversible system the aerodynamic loads acting on the control surfaces are not fed back into the system. In this case, the problem of flutter does not arise, and mass balance weights can be eliminated from the system. In practice, however, no system is perfectly irreversible and the problem of flutter is still an important one, even on aircraft fitted with fully powered control surfaces (see para. 7.4, Chapter 14). On some aircraft mass balance weights are still incorporated in the power control systems, but in others they are omitted, with a consequent saving in weight. In the case of systems which incorporate manual reversion, it is important to retain the mass balance weights. To prevent flutter, the mass balance weights must be positioned on the output side (i.e. "downstream") of the power control unit.

Mass balance weights are also used in a control system (e.g. in the Lightning Mk. 5 rudder system) to overcome the effects of inertia in the system during longitudinal acceleration or deceleration of the aircraft. This problem may be a particularly important one in the case of naval aircraft while they are being launched from the catapult of an aircraft carrier.

(*e*) *Safety and Reliability* Power control systems should be very reliable and for greater safety must be either fully duplicated or provided with facilities for manual reversion.

3. Artificial Feel Systems

The pilot's assessment of the handling characteristics of an aircraft is very dependent on the magnitude of the control movements and the control forces that he experiences during flight. In the case of an aircraft fitted with non-powered controls, the stick forces increase (on simple theory) with the square of the forward speed. This results in a variation of stick force characteristics (e.g. stick force per g, etc.) that pilots have become accustomed to. In a perfectly irreversible fully-powered control system, the aerodynamic loads are not fed back to the stick. The only force that the pilot has to exert is that required to open the servo-valves on the power control jacks. Thus the pilot loses one of the most important of the various cues that go to make up the "feel" of the aircraft. "Feel," however, is essentially a subjective rather than an objective quantity. It depends on the response characteristics of the aircraft and on the mechanics of the control system. It is very difficult to get a number of pilots to agree on the desirable characteristics for a power control system and the optimum design will depend on the particular system being considered. Though control forces and movements are both important factors in the overall "feel" of the aircraft, the forces are more important than the movements and "pressure (force) controls" have been experimented with, in which no movements are involved. This type of system can be used with a miniature side-stick controller and the conventional large centre stick can then be eliminated.

With the advent of fully powered controls it was necessary to incorporate artificial (force) feel into the system. The great advantage of power control systems is that it is possible to incorporate different types of feel variation and to give the pilot more or less what he wants in the way of control force characteristics. The type of feel system used in any particular case depends upon the handling characteristics that are considered to be desirable and the usual factors of complexity, cost, safety and reliability. The systems used are mechanical, inertial, hydraulic or air pressure operated devices and the main types are

(*a*) *Spring-feel System* The simplest type of feel system is one incorporating a spring. Movement of the controls compresses the spring and the pilot experiences a force which is dependent on the

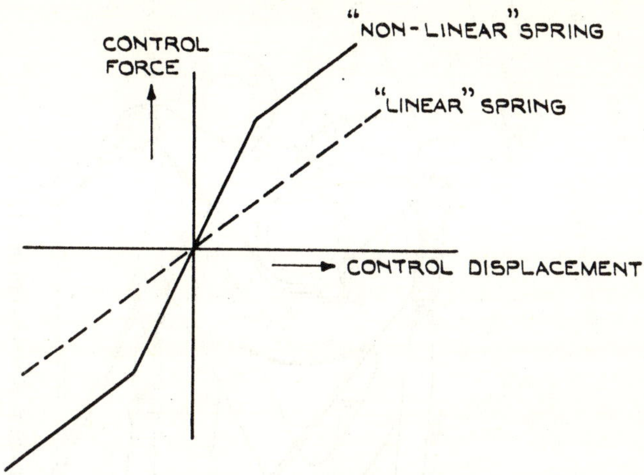

Fig. 12.3. Linear and non-linear spring-feel characteristics

strength of the spring and the amount of control displacement, but independent of the forward speed of the aircraft.

The spring may be a linear one, in which case the force increases proportionally to the control displacement. Often, however, non-linear springs are used which have a steeper gradient around the zero stick position in order to give the required self-centering characteristic (para. 4). Linear and non-linear spring characteristics are shown in Fig. 12.3. In some cases (e.g. Lightning aileron system) the conventional "spring" may be replaced by torsion bars.

The spring feel type of system is commonly used on the ailerons (e.g. Hunter, Sea Vixen, Buccaneer), though it can also be used in the other control circuits. The spring feel system of Fig. 12.2 is shown in more detail in Fig. 12.4. When the control column is moved either side of the neutral position, a cam system causes a plunger to be forced into the spring feel strut and the stick forces correspondingly increase.

The spring feel system tends to give high stick forces at low speeds and very sensitive control at high speeds. The force characteristics of the simple system can be improved by further complicating the system with variable feel and a gear change (para. 5).

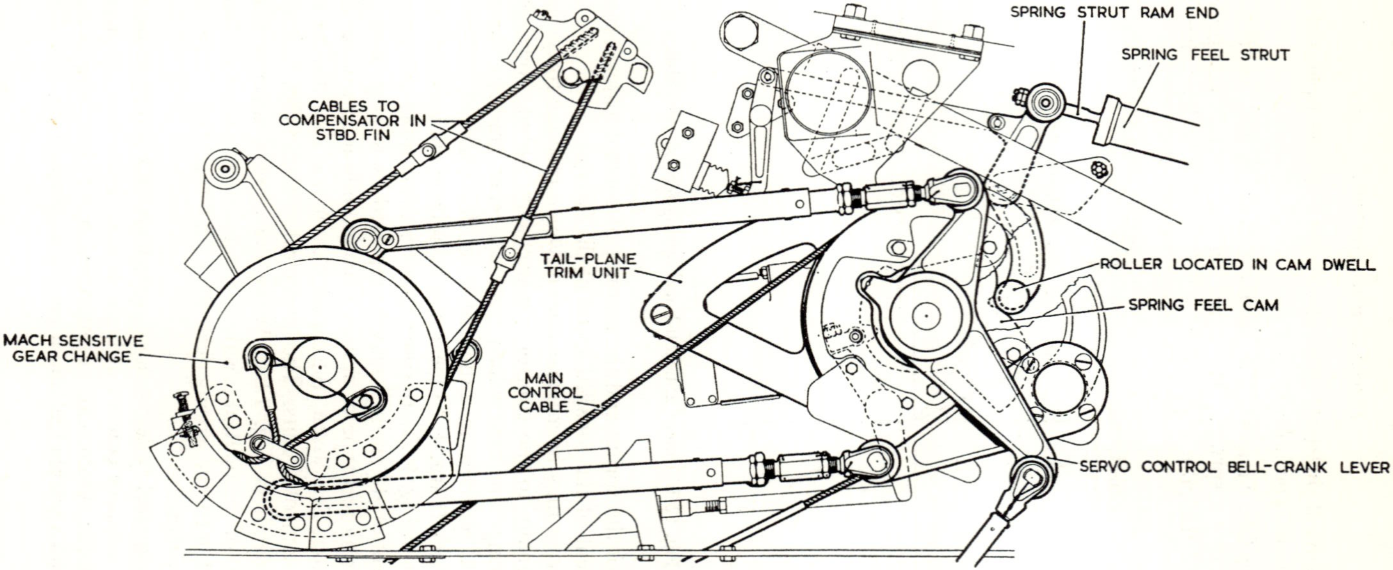

Fig. 12.4. Spring-feel mechanism [Crown copyright]

(*b*) *q-feel* In this system the feel force varies with the dynamic pressure of the air, $\frac{1}{2}\rho V^2$ or q (i.e. with V_i^2), as in an ordinary non-powered system. This is a very commonly used system, particularly in elevator and tailplane circuits and it has the advantage over the spring system of producing forces that are dependent on both the amount of control movement and the forward speed of the aircraft.

The q-feel system used in the rudder circuit of a particular aircraft (i.e. the Sea-Vixen) is shown in Fig. 12.5. The mechanism consists of a "q-pot" assembly which has a large piston operating within a sealed cylindrical chamber and linked by a chain and sprocket arrangement to the rudder circuit.

One problem associated with a simple q-feel system is that if there is a failure and a loss of dynamic pressure, the pilot loses his feel. Many systems (e.g. Buccaneer tailplane and rudder) have a spring incorporated in the system (in some cases this combination is almost equivalent to a V_i system), so that, in the event of a pressure failure, the spring will still provide sufficient stick forces. These systems may also be fitted with Mach number cut-outs (e.g. Lightning) so that the stick forces do not continue to increase with speed above a selected value of Mach number (Fig. 12.6).

The feel may be produced directly by feeding the pitot and static pressures into the q-pot. An alternative method, used on the Lightning Mk. 5 rudder and tailplane circuits, is to use the measured pitot and static pressures to meter hydraulic fluid under pressure to a feel unit which consists of a piston in a cylinder. When the controls are zero, the forces are zero, deflection of the controls in either direction causes the piston to move against the pressure exerted by the fluid.

(*c*) *V and V^3 feel* If the q-feel system is modified by the incorporation of a mechanical system consisting of various arrangements of non-linear springs, cams, etc., the control forces can be made to increase with forward speed at a slower or a faster rate than V^2. The system can be designed to have approximately V or V^3 variations.

V^3-feel is not often employed, but it has been used occasionally on the rudder circuits of some aircraft (e.g. Avro Vulcan and B.A.C. 111), in order to prevent large control deflections and corresponding large sideslip angles at high speeds.

(*d*) *g-feel* Longitudinal control forces can be modified by incorporating a bob-weight into the control system. This produces an additional stick force which is independent of the forward speed of the aircraft and proportional to the amount of extra g being pulled.

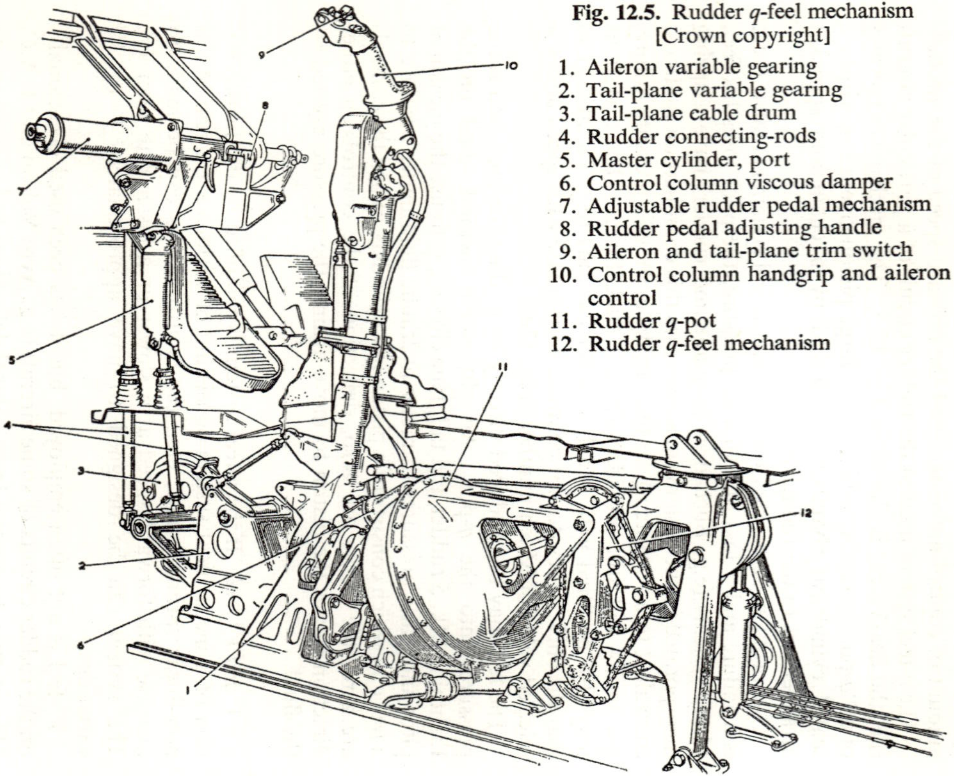

Fig. 12.5. Rudder *q*-feel mechanism
[Crown copyright]

1. Aileron variable gearing
2. Tail-plane variable gearing
3. Tail-plane cable drum
4. Rudder connecting-rods
5. Master cylinder, port
6. Control column viscous damper
7. Adjustable rudder pedal mechanism
8. Rudder pedal adjusting handle
9. Aileron and tail-plane trim switch
10. Control column handgrip and aileron control
11. Rudder *q*-pot
12. Rudder *q*-feel mechanism

This increases the values of stick force per *g* and prevents the application of excessive amounts of *g* during manoeuvring flight. Bob-weight systems are often referred to as "*g*-restrictors" or "response-feel" systems. They can of course also be used to improve the stick force per *g* characteristics of non-powered control systems.

The longitudinal control system shown in Fig. 12.2 has a bob-weight or *g*-restrictor mechanism. The McDonnell Phantom also has one with a gearing of about 5 lb/*g*.

One disadvantage of the bob-weight system is the time lag in the response of the aircraft to control movements and in the build-up

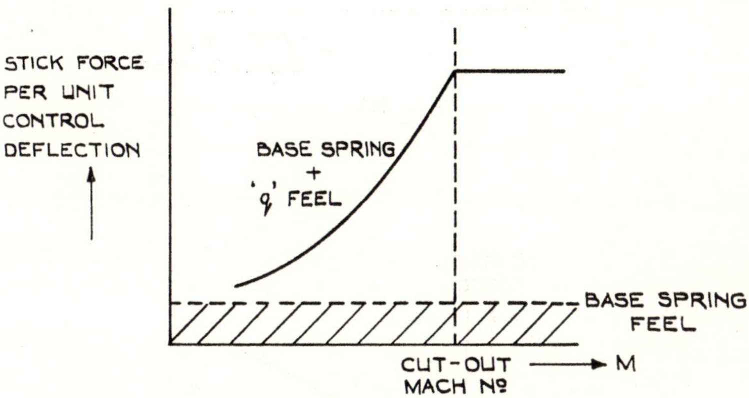

Fig. 12.6. Stick force characteristics of a system with base spring, *q*-feel and Mach cut-out

of the stick forces. By applying very rapid control movements the pilot can "beat" the bob-weight, and large *g* values may inadvertently be developed. Bob-weights have the further disadvantage that they respond to atmospheric turbulence.

In practice, control systems often contain both bob-weights and springs, and whether a system is described as a *g*-feel or a spring feel system depends on the relative importance of these two contributions. These particular systems may suffer from problems of dynamic instability (see para. 8.3.)

4. Friction, Backlash, and Preloading

If the power control system is an irreversible one, none of the aerodynamic forces are fed back to the system and there is no natural aerodynamic self-centering tendency. A spring can be incorporated to produce the necessary self-centering. In the case

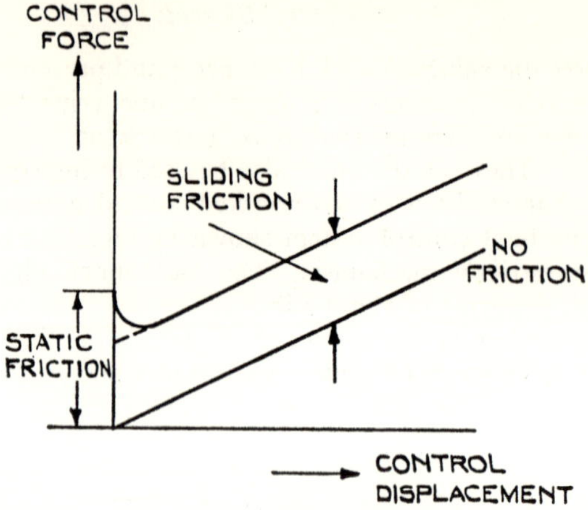

(a)

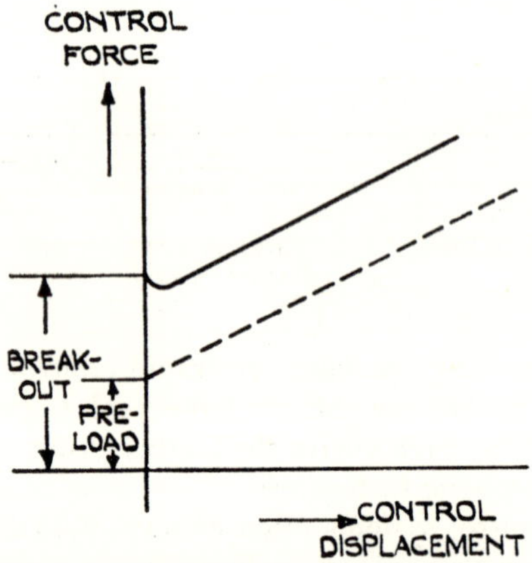

(b)

Fig. 12.7. Control force characteristics of systems with friction, stiction and preloading
(*a*) Without preload; (*b*) with preload

of a system using a spring feel system, the force gradient of the spring can be increased around the zero stick position (Fig. 12.3). The self-centering spring has to be strong enough to overcome the friction and stiction (i.e. static friction) that exists in the operating valve and in the remainder of the control circuit.

The self-centering of the controls can be improved by preloading the spring, but if the friction is high, then the preloading required to produce the necessary self-centering will also be high. This will cause heavy break-out forces, which are awkward when the pilot is attempting to apply small precise control movements (Ref. 7). The control force-control displacement characteristics of various types of control systems are shown in Fig. 12.7.

In practice, there will always be some backlash in any mechanical control system, i.e. a small "dead region" exists in which movement of the control column does not result in any movement of the control surface.

Friction, stiction, and backlash are undesirable features in a control system. They result in undesirable stick force characteristics, over-controlling, a lag in response, and pilot induced oscillations. The characteristics of a control system can be improved by reducing them to a minimum. This will allow for much more accurate and precise control by the pilot.

5. Variable (Non-linear) Gearing and Gear Change Systems

5.1. Non-linear Gearing

The simplest type of control system is one with a linear relationship between the stick movement and the corresponding control surface deflection. In this case, the control surface movement is proportional to the stick displacement, the proportionality remaining constant throughout the range of stick displacement. This type of system does not produce satisfactory characteristics, however, because if the control surface/stick gearing is suitable for small displacements, the stick movements tend to be excessive for large control movements. In practice, a control system will contain some non-linearity, but the characteristics of the system can be improved by deliberately designing non-linearities into the circuit (Ref. 7). Fig. 12.8 shows the stick movement-control surface displacement relationship for a non-linear system. Around the neutral control position, a particular stick movement will produce a relatively small displacement of the control surface (i.e. a low control surface to stick gearing). Near the extremities of the stick movement, however, the same stick movement will produce a much larger displacement of the control surface (i.e. a high control surface to stick gearing). This type of system has been used on a number of naval aircraft (e.g. Buccaneer and Sea Vixen) to reduce the excessive stick movements on round-out and to improve the handling characteristics of the aircraft in turbulence at low airspeeds and on carrier approaches.

Variable gearing can be obtained by a suitable mechanical system of rods and levers in the control circuit, such that the movement of the cable drum, and hence, the control surface, increases its rate of

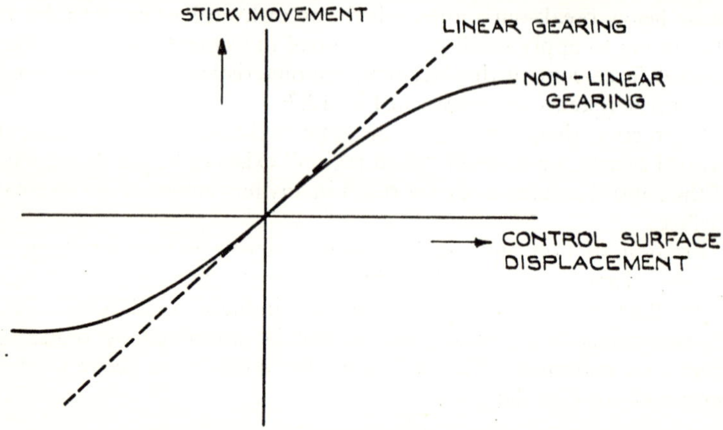

Fig. 12.8. Linear and non-linear stick gearings

rotation as the stick is moved from the neutral position to either of the full travel positions.

A linear gearing also has the disadvantage of producing a control system which is very sensitive at high speeds and rather sloppy at low speeds. When the aircraft is fitted with a feel-trim system (see para. 6) in which the level flight trim position of the stick varies with forward speed, a non-linear system can be designed by incorporating a mechanical system (e.g. a suitably shaped cam) which changes the gearing of the system dependent on the trim position of the stick. In this way the control to stick gearing can be varied from a low value when the stick is forward, to a higher value when the stick is aft (Fig. 12.9).

The control gearing relationship can also be varied at different forward speeds by means of a q-sensitive or Mach number-sensitive automatic gear change unit. In this case, the speed signals are fed to a hydraulic jack, which operates to change the mechanical linkage ratios in the system, so reducing the control surface movements at high forward speeds. In order to obtain the required characteristics, it may be necessary to incorporate both a mechanical variable gearing unit and an automatic speed-sensitive unit.

During the development programme of an aircraft, a number of changes may have to be made to the variable gearing mechanism (e.g. to the shape of the cam, etc.), in order to obtain the best compromise over the whole speed range. In some cases, a satisfactory

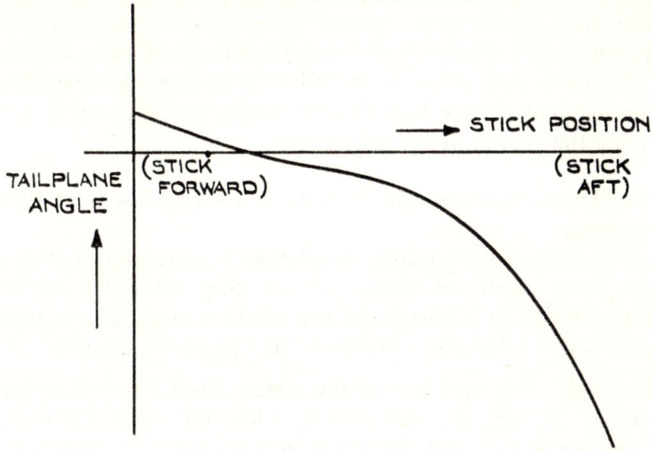

Fig. 12.9. Non-linear variation of control to stick gearing

compromise may not be possible, and a gear change mechanism may be inevitable.

5.2. Gear Change Systems

To obtain satisfactory stick to control surface gearings in all conditions of flight, a gear change mechanism may be necessary. This type of system has generally been confined to elevator (or tailplane) and aileron channels only, and usually takes the form of a simple two-position gearing (i.e. a high speed one and a low speed one).

A gear change unit is useful on elevator circuits fitted with spring feel because of the relatively small stick force per *g* values obtained at high forward speeds. By changing the stick to control surface gearing at these speeds, the stick force per *g* values can be increased to a more satisfactory level (Fig. 12.12).

To obtain good lateral control at low speeds and, for example, on the approach under turbulent conditions, full aileron control deflection is required from full stick deflection. At high speeds, on the other hand, when the roll response of the aircraft is high, full aileron deflection is not required. A reduction in the control surface/stick gearing at these speeds will reduce the value of the rate of roll that can be achieved, and decrease the possibility of roll coupling problems.

The gear change can be operated either manually by the pilot (e.g. Sea Vixen and Buccaneer ailerons, Comet C Mk. 2 elevators), or automatically. In the latter case the gear change mechanism is coupled to a speed (or Mach number) sensitive device (Fig. 12.2) or, in some cases, to the undercarriage system, so that the gear change occurs when the undercarriage is raised or lowered. The automatic gear change system relieves the pilot of having to remember to operate the gear change himself and eliminates the possible danger of leaving the system in the wrong gear.

5.3. Problems Associated with Non-linear Gearing and Gear Change Systems

Though systems incorporating non-linear gearing and gear-change systems have improved characteristics, they naturally introduce some problems, in addition to the obvious ones of mechanical complexity and reliability. Some of these problems are (Ref. 7)

(*a*) *Backlash* Any backlash in the system should be reduced to a minimum. At high forward speeds, when the control surface to stick gearing is low, any backlash in the control system "downstream" of the gearing unit, will be magnified back at the pilot's stick. To reduce this stick movement, the gearing unit should be positioned as far "downstream" in the system as possible (though in the particular system shown in Fig. 12.2 it is up near the stick).

(*b*) *Loss of Feel by the Pilot* One possible objection to a gear change system is that the pilot tends to lose his "feel" around the gear change condition. However, the gear change does not normally occur within the cruising speed range of the aircraft, since it is often tied in with the operation of the undercarriage and this objection is not normally a very valid one.

(*c*) *Reduction of Available Control Surface Movement* In some systems there may be a restriction in the amount of control surface movement available at any one time. On aileron controls this may be desirable, since it will limit the rate of roll response at high speeds. On elevator control systems, however, a reduction in the control surface movement may be a serious disadvantage.

6. Feel Trim and Datum Trim Systems

In any control system it is essential to give the pilot a method of trimming out his stick and pedal forces for steady conditions of flight. On non-powered control systems this is usually done by deflecting small trimming tabs on the control surfaces, the tabs being operated by moving the trimming wheels provided in the cockpit.

Trim systems are also provided on power control systems, and these are usually operated electrically from switches on the control column. There are two types of trim systems

(i) Feel trim systems,
(ii) Datum trim systems.

(i) *Feel Trim Systems* These are similar to the conventional systems used on non-powered controls. The control forces are trimmed out in such a way that there is a unique relationship between the position of the stick and the position of the control surface. At high speeds the stick is forward and as the trim speed is reduced the stick is moved back, the forces being continually trimmed out. In the case of a system incorporating spring feel, for example, the stick forces are trimmed out by moving the spring-feel box to adjust the tension in the spring.

(ii) *Datum Trim System* This system is less frequently used than the feel trim system, but it has been incorporated in the longitudinal control systems of some aircraft (e.g. Gnat Fighter and Buccaneer). The essential difference in this system is that the stick is always in the same trim position, irrespective of the speed at which the aircraft is flying. If, for example, the aircraft is initially in trim at a high forward speed and the speed is then reduced, the stick has to be pulled back, but as the stick forces are trimmed out, the stick moves back to its original trim position. Thus the trim position of the stick gives no indication of the speed of the aircraft.

In the feel trim system above, the stick has to be moved to move the control surface and the resulting forces are then trimmed out. In the datum trim system, operation of the trimmer causes the control rod in the circuit to vary in length, thus producing the necessary movement of the control surface, the stick force becoming zero again when the stick is in exactly the same position in the cockpit.

The datum trim system is simple from an engineering point of view, but it does suffer from a number of disadvantages. To make the design of the gearing mechanism easier, the system may be designed so that the stick is only capable of moving the control surface over a limited range of angles. In this case, full movement of the stick only moves the tailplane over a limited range (e.g. about 2/3rds) of the full available tailplane movement. The amount of control surface movement available in each direction at any time depends on the position of the datum trim unit. If the aircraft is trimmed nose up, then a small amount of back stick (from the neutral trimmed position) will produce the remaining nose up control movement, but the full forward movement of the stick will not produce the full nose-down control surface movement. The system has been satisfactory on those aircraft where only small ranges of elevator movement have been required (e.g. Hunter 6, Valiant), but it is not so satisfactory on aircraft that are designed to land and take off at high lift coefficients (e.g. Buccaneer), since it may be desirable on this type of aircraft to have the full nose-down control immediately available at all times.

7. Viscous (or Hydraulic) Dampers

A viscous damper is often incorporated in a power control system to eliminate the possibility of flutter or pilot induced oscillations. At high forward speeds, the longitudinal control system of an aircraft is usually very sensitive, very small movements of the stick producing very large values of g. Values of stick movement per g of 0·2 in. may be typical and under these circumstances 1 in. of stick movement would produce an additional normal acceleration of $5g$. This high sensitivity leads to overcontrolling and the possibility of pilot induced oscillations (P.I.O.S). A viscous damper is often fitted to oppose the motion of the stick and to prevent abrupt control surface movements by increasing the stick forces associated with rapid stick movements. The presence of the damper may, however, have an adverse effect on the general handling characteristics of the aircraft, particularly at low speeds. It opposes the movement of the stick, and consequently introduces a lumpiness and a time lag into the system which will reduce the aircraft's response. These adverse effects can be reduced by correctly positioning the viscous damper in the system. In some cases, the magnitude of the damping is made to vary with forward speed. When the hydraulic damping is obtained from the movement of oil through ports of a particular size, the system can be designed so that the size of the ports is controlled by a q-pot. In this way the damping can be made to increase with increase of forward speed.

8. Stability and Control Characteristics of Aircraft with Power Operated Controls

8.1. Control Surface Hinge Moment Characteristics

On a conventional non-powered control system, the stick-free stability characteristics are different from the stick-fixed ones (see

Chapter 2), the difference depending on the control surface floating parameter b_1 and the heaviness parameter b_2. The stick-fixed static and manoeuvre stabilities determine the stick movements required for level flight trim and manoeuvring flight, respectively, whereas the stick-free stabilities determine the corresponding stick forces.

In the case of a power operated control system, the control surfaces will not move (i.e. "float") under the action of the forces acting upon them, as long as it assumed that

(i) the control system is perfectly irreversible,
(ii) there is no aeroelastic distortion of the airframe,
(iii) there is no inertia effect (i.e. mass unbalance) in the control system.

On these simple considerations, there is no difference between the stick-fixed and stick-free stabilities. The stick-fixed stability will then determine both the stick movements and the stick forces required and the stick-fixed and stick-free neutral points will be identical. The stick-fixed and stick-free manoeuvre points will also coincide with one another. The b_1 and b_2 values are still important quantities, however, because they determine the size of the hydraulic jack that it is necessary to install in the control system. The jack must be large enough to produce the loads which are necessary to balance the aerodynamic forces on the control surfaces and yet it must not be too large, otherwise there will be a weight penalty involved. If the jack is too small, however, then, under certain conditions of flight, jack "stalling" may occur. Tests must be performed to determine the values of the hinge moment characteristics, in order that the jack design can be optimized for the particular aircraft.

In practice, as a result of distortion occurring in the control circuit or in the airframe, some movement of the control may occur, and it may be necessary to take this movement into account (para. 8.3). If bob-weights are added to the system, the stick-force per g values are increased, and the stick-free manoeuvre point will then lie behind the corresponding stick-fixed point.

8.2. Control Force Characteristics

Typical control force characteristics obtained with the different basic feel systems (spring, V and q) are shown in Fig. 12.10. These results can be modified by the addition of variable gearing, bob-weights, etc., and the values of the forces tend to increase at the higher values of V_i due to the effects of aeroelasticity and compressibility.

8.2.1. LONGITUDINAL CONTROL

Fig. 12.10*a* shows the variation in the values of stick force per *g*. On the simple theory of Chapter 3, the *SF*/*g* of an ordinary non-powered or *q*-feel control system remains constant, independent of forward speed, in the absence of aeroelastic and compressibility

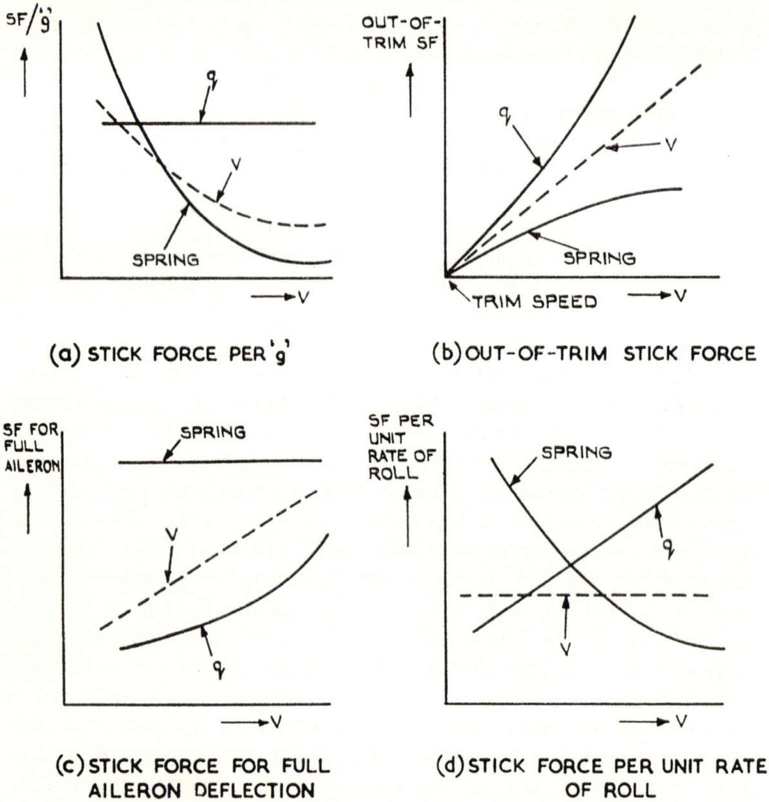

Fig. 12.10. Stick force characteristics of spring, *V* and *q*-feel systems

effects. The stick movement per *g*, however, decreases with increase of forward speed as $1/V^2$ (Chapter 3) and since the forces in a linear spring feel system are directly proportional to the stick displacement, the stick force per *g* values also decrease as $1/V^2$. The *V*-feel system lies between these two systems, the stick force per *g* value decreasing as $1/V$. As we have already seen, a *V*-feel system can be obtained by combining together a spring feel and a *q*-feel system.

On simple theory, the value of *SF*/*g* is independent of the amount of applied g. If a non-linear spring is fitted, however, with a steeper force gradient around the trim stick position, then the stick force per g value tends to decrease with increasing g (Fig. 12.11). Once again, these results may be modified by compressibility effects.

A constant value of SF/g, independent of forward speed, is often taken to be the desirable characteristic of a longitudinal control system. For this reason, the q-feel system is often used in the elevator or tailplane system, perhaps with a Mach number cut-off to prevent excessively large forces at high speeds. The disadvantage of the simple spring feel system is the very small SF/g values that might be

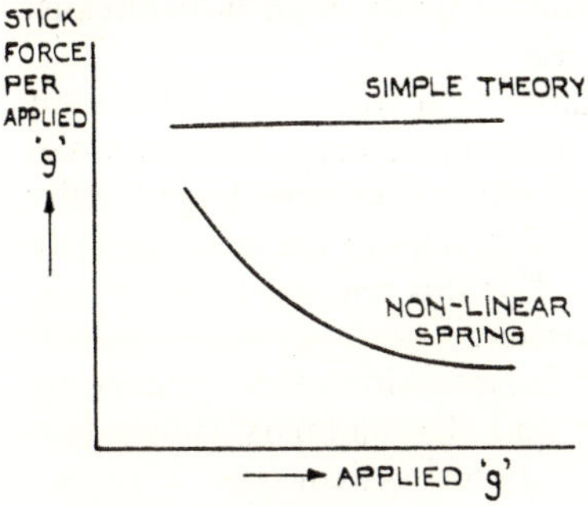

Fig. 12.11. Variation of stick force per g with applied g

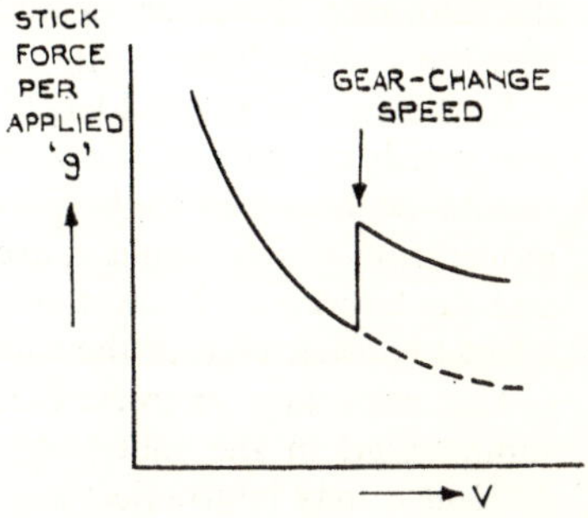

Fig. 12.12. Effect of gear change on stick force per g characteristics

obtained at high forward speeds, though aeroelastic and compressibility effects tend to counteract this to some extent. This disadvantage can be overcome by some of the methods already discussed, e.g.

(*a*) variable gearing,
(*b*) gear change (Fig. 12.12),
(*c*) bob-weights.

Some authorities have, however, suggested that it might be preferable to have a longitudinal control system that produces a constant rate of pitch per unit force at all speeds.

The variation in the longitudinal out-of-trim forces is shown in Fig. 12.10*b*. The amount of stick movement required to trim was determined in Chapter 2 and the curve corresponding to a spring feel system has a similar shape. The V and q-feel curves are then obtained from the spring feel curve by multiplying the force values by V and V^2, respectively.

8.2.2. LATERAL CONTROL

The stick forces required for full aileron deflection are shown in Fig. 12.10*c*. The value for the spring-feel system is obviously a constant, independent of forward speed and the variation of the forces for the other two systems are in proportion to V and V^2.

The stick forces per unit rate of roll are shown in Fig. 12.10*d*. Since, on simple theory, the aileron angle required for unit rate of roll varies inversely with forward speed, the curve for the spring feel system must vary in a similar way ($\propto 1/V$). By multiplying these forces by V and V^2, it can be seen that the V-feel system has constant values of stick force per unit rate of roll and the q system has values that increase linearly with speed ($\propto V$).

If it is decided that a constant stick force per unit rate of roll is the desirable characteristic of aileron controls, then it is obvious that the V-feel system is the one to use.

In the case of the rudder, reasonably light forces are required at low speeds in order to allow large rudder movements, but the forces should be heavy at high speeds in order to prevent large rudder movements which would cause excessive sideslip angles to develop and the aircraft to be over-stressed. For this reason, q-feel or even V^3-feel is often used in the rudder circuit.

The important characteristics of these various feel systems are summarized in the following table and the summary shows how difficult it may be to select a satisfactory feel system—

	q-FEEL	V-FEEL	SPRING-FEEL
Stick force per g	Constant with speed	Large at low speeds, and small at high speeds. At high speeds compressibility and distortion effects may alleviate the low forces	Similar to V-feel, but rather more exaggerated
Out-of-trim elevator forces	Excessive forces for large change of speed	Reasonable	Gradient decreases with speed. At high speeds only small forces are required for large speed changes
Stick force for full aileron	If designed for reasonable forces at high speeds, the forces may be too light at low speeds	Reasonable	Constant with speed. If a reasonable value is selected at high speeds, this may feel unnaturally heavy at low speeds
Stick force per unit rate of roll	May be too heavy at high speeds	Constant with speed	Large at low speeds and small at high speeds

8.3. Dynamic Stability

When an aircraft is fitted with irreversible power-operated controls, it is usually assumed that the control surfaces do not move when the aircraft is oscillating and that only the condition of stick-fixed dynamic stability need be considered. In practice, control surface deflections often occur due to signals being fed to the power control units as a result of

(i) aeroelastic distortion of the aircraft structure (see Chapter 14);
(ii) the effect of a bob-weight in the control system.

Bob-weights are used to increase the magnitude of the stick force per *g* values (see para. 8.2.1), but during a longitudinal oscillation, variation of normal acceleration will displace the bob-weight and produce control surface deflections. The oscillations of the control surface may couple together with the oscillations of the aircraft in such a way as to cause a serious reduction in the damping of the short-period oscillatory motion. Any attempt by the pilot to prevent the oscillatory motion may in certain cases make the condition worse and pilot induced oscillations (P.I.O.S) may develop, during which very large positive and negative values of *g* may be experienced.

Some severe cases of stick-free dynamic instability have been experienced in flight on aircraft fitted with power operated controls. Ref. 9 describes the analysis of a particular control system incorporating spring-feel and a bob-weight, which suffered from this particular problem. It is found that the two important factors which influence the stability of the system are the longitudinal position of the bob-weight and the elevator-to-bobweight gear ratio.

9. Electric Signalling Control Systems

Nearly all present-day aircraft depend upon a mechanical system of rods and cables to transmit the pilot's control inputs to the control surfaces. Initially the pilot also supplied all the necessary force, but, as we have seen, power-operated control systems are now widely used. On modern high-performance aircraft it is essential that the control surfaces should be capable of being positioned very accurately. Very precise control is very difficult on large aircraft fitted with mechanical signalling, because of the problems associated with aeroelastic distortion and differential thermal expansion of the structure. Compensating devices can be incorporated to correct for these effects, but the control systems of large aircraft have become extremely complicated.

Many of the problems associated with mechanical signalling can be overcome by the use of electric signalling controls (Ref. 10).

These systems have the additional advantage that the electric signals from the autostabilizer can be mixed more easily with the pilot's input signal and the problem of feedback of the autostabilization signals to the pilot is eliminated. They also lend themselves very readily to such features as

(i) the use of a miniature side-stick controller,
(ii) the incorporation of manoeuvre-demand and adaptive-type control systems (see Chapter 19).

The development of electric signalling control systems is described in Ref. 10. They have been used on a number of research vehicles, including the Rolls Royce Flying "Bedstead," the Tay Viscount, the Short S.C.1 vertical take-off aircraft and the Avro 707C. The electric signalling control systems of the Avro 707 C delta wing research aircraft are shown in Fig. 14 of Ref. 10. This aircraft has been used at the R.A.E. for flight research on electric signalling and manoeuvre demand control systems.

The BAC/Sud Concorde supersonic airliner has been designed with electric signalling controls. The proposed system consists of elevons and rudder, the surfaces being power-operated and controlled from the cockpit through dual conventional control columns and rudder pedals. The movements of the stick and rudder pedals will be converted to electric signals for transmission to the servo valves of the hydraulic jacks which are adjacent to the control surfaces. The electric control system is fully duplicated and there is a stand-by conventional mechanical system. A layout of the proposed electro-hydraulic-mechanical control system is shown in Ref. 11.

References

1. Lyons, D. J.; "Present Thoughts on the Use of Powered Flying Controls in Aircraft," *Journal of the Royal Aeronautical Society*, March 1949, page 253.
2. Joy, C. F.; "Power Controls for Aircraft," *Journal of the Royal Aeronautical Society*, January 1952, page 7.
3. Dickinson, R. P.; *An Introduction to the Subject of Feel in Power Operated Control Systems*, Unpublished M.O.A. Report.
4. Dickinson, R. P.; *A Collection of Pilots' Experiences with Various Forms of Feel and Power Control Systems.* Unpublished M.O.A. Report.
5. Lang, D. A.; *Control System Requirements for Military Aircraft in Relation to the Human Operator*, Unpublished M.O.A. Report.
6. Kolk, W. R.; *Modern Flight Dynamics* (Prentice-Hall International, 1961).
7. Lang, J.; *Feel in Power Control Systems*, Lecture to the Empire Test Pilots' School, Farnborough, unpublished.
8. Turner, H. W.; *Flight Testing of Automatic Stabilization and Control Systems in Manned Aircraft.* Lecture to the Empire Test Pilots' School, Farnborough, unpublished.

9. NEUMARK, S., COLLINGBOURNE, J. and THOMAS, H. H. B. M.; *A Case of Longitudinal Stick-free Dynamic Instability of an Aircraft Fitted with Power-operated Control, g—Restrictor and Spring Feel*, R. & M. No. 3094, 1958.
10. HARLE, T. F., HOWELL, G. C. and TWINEY, F. J.; "The Use of Electric Signal Transmission in Aircraft Flight Control Systems," *Proceedings of the Joint Conference on the Importance of Electricity in the Control of Aircraft*, held by the R.Ae.S. and I.E.E., 26–28th February, 1962.
11. *Flight International*, 4th March 1965, page 325.

13 Spinning

1. Introduction

When an aircraft is accidentally or deliberately stalled, the motion of the aircraft may in some cases develop into a spin. The spin is a more or less steady manoeuvre, involving pitching, rolling, yawing and sideslipping motions. The important characteristics of a spin are that

(i) the aircraft is descending along a steep helical path about the vertical spin axis,
(ii) the incidence of the wing is well above the stalling angle,
(iii) the aircraft has a high rate of rotation about the vertical axis,
(iv) looked at from above, the aircraft executes a circular path about the spin axis, and the radius of the helix is usually less than the semi-span of the wings,
(v) the aircraft may be in the "erect" or "inverted" position in the spin.

The characteristics of the particular spin that an aircraft executes depend on such factors as the aircraft's geometry and mass distribution and on the control surface deflections. In practice, an aircraft may rotate for two or three turns before finally settling down to a reasonably steady spin, but in some cases the spin may continue to be oscillatory.

The steady spin is a complicated manoeuvre involving pitching, rolling and yawing motions. In the previous work on stability and control it has been possible to neglect the inertia coupling terms and to consider the longitudinal and lateral motions separately. In the analysis of the spinning motion, however, it is essential to take these terms into account.

In a steady spin all the forces and moments acting on the aircraft are in equilibrium. This must be true, irrespective of the particular set of axes chosen to formulate the equations of motion. In the spin it is usual to consider the conditions of equilibrium with respect to two different sets of axes,

(i) the spin axes,
(ii) a set of body axes $0xyz$, fixed relative to the chord line of the wing,

and the balance of forces and moments determines the equilibrium values of the radius of turn, the sideslip, the rate of descent, the angle of attack, etc., of the aircraft. In order to obtain a better physical appreciation of the motion involved, each condition of equilibrium will be considered separately. Unfortunately, these conditions are all inter-related with one another, and it is sometimes rather difficult to see how the various factors involved influence the overall motion. In practice, the aircraft will have to "solve its own equations of motion" simultaneously, and the resulting motion in the spin will be that motion for which all the conditions are balanced. If an equilibrium condition does not exist, then the aircraft will not attain a steady state, and the spin will be an oscillatory one.

An aircraft can be recovered from a spin by the correct use of the controls, though the optimum control movements depend upon the particular aircraft configuration being considered.

The spinning characteristics and the optimum spin recovery techniques of an aircraft can be determined from wind tunnel tests, free-flight model tests and full-scale flight tests. Wind tunnel tests are subject to Scale Effect, but it is possible to obtain reasonable correlation between the various experimental methods.

2. Attitude of an Aircraft in a Spin

Before analysing the balance of forces and moments in an erect spin, it is essential to consider the various attitudes that the aircraft can take up. Consider an aircraft in a positive spin (i.e. one in which the aircraft is rotating in the clockwise direction looking downwards along the spin axis), and assume for simplicity that the wings are in the horizontal position. From Fig. 13.1, it can be seen that in this case the aircraft is sideslipping towards the leading or outer wing (which in the case being considered is the port wing). The angle of sideslip β is numerically equal to the helix angle of the spin which is given by

$$\gamma = \tan^{-1}\left(\frac{\Omega R}{V}\right) \tag{13.1}$$

where Ω (rads/sec) is the rate of spin about the spin axis, R (ft) the radius of spin and V (ft/sec) the vertical rate of descent. However, since γ is by definition positive and a sideslip towards the port wing is conventionally negative, these two quantities are of opposite sign, i.e.

$$\beta = -\gamma \tag{13.2}$$

The helix angle is relatively small (possibly only of the order of 3 to 5°), and it decreases with increase of angle of attack in the spin,

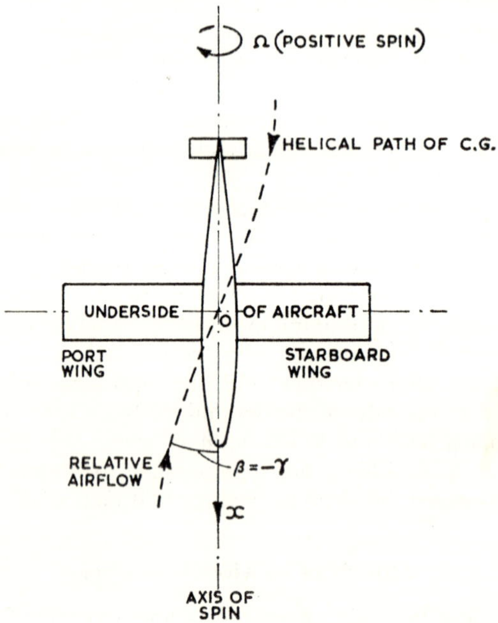

Fig. 13.1. Aircraft spinning with wings horizontal

due to the corresponding reduction in the value of the radius (see para. 3). The resultant relative airflow is thus approximately vertical and the angle of attack of the wing is the angle between the vertical and the chord line of the wing (Fig. 13.2).

The rotary motion of the aircraft about the spin axis can be resolved into its components relative to the set of axes fixed in the body. From Fig. 13.3, it can be seen that these components are given by

$$\left.\begin{aligned} p &= \Omega \cos \alpha \\ q &= 0 \\ r &= \Omega \sin \alpha \end{aligned}\right\} \tag{13.3}$$

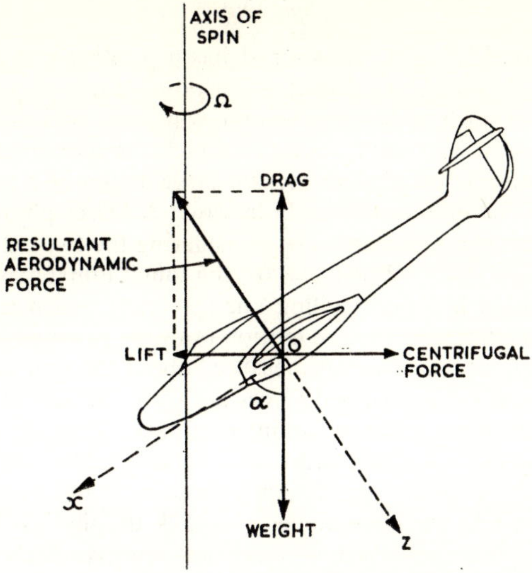

Fig. 13.2. Balance of forces in a spin

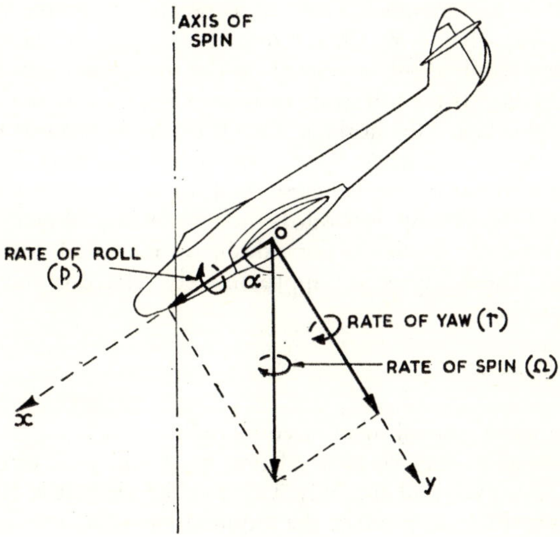

Fig. 13.3. Rates of rotation in a spin

Thus, in a positive spin, the aircraft has a positive rate of roll (i.e. the starboard wing going down) and a positive rate of yaw (i.e. the nose going to starboard), but when the wings are horizontal the rate of pitch is zero. For small angles of attack the rate of yaw is small and the motion is mainly in roll; for large angles of attack, on the other hand, the motion is mainly in yaw. A better physical appreciation of this can be obtained by considering the two extreme cases (i.e. $\alpha = 0°$ and $90°$). When $\alpha = 0°$, the longitudinal axis is vertical and the motion is a pure rolling one ($p = \Omega$); when $\alpha = 90°$ the aircraft is in a flat spin and the motion is a pure yawing one ($r = \Omega$).

The sum of all the moments measured about the spin axis must be zero. Since all the centrifugal forces act from the axis of spin, they cannot produce any moment about that axis. Thus the sum of all the aerodynamic moments must also be zero about the spin axis and hence the resultant aerodynamic force itself must pass through the spin axis. At the large angles of attack usually involved in the spin, the resultant aerodynamic force acts approximately normal to the wing chord, i.e. the relatively small sideforces acting on the aircraft can be neglected and the normal to the chord line of the wing (i.e. the $0z$ axis) must pass through the axis of spin.

In the more general case, however, the wings will be tilted out of the horizontal by an amount θ_y (defined as positive when the starboard wing is down). This wing tilt is required to adjust the sideslip angle to the value required for balancing up the rolling moment equation (see para. 4.2*b*). If the following or inner wing (in this case the starboard wing) is below the horizon, then the sideslip towards the outer wing is reduced and when $\theta_y = \gamma$ the sideslip is zero (Fig. 13.4). When the starboard wing is below the horizon by an angle greater than γ, then the sideslip is towards the inner wing. When the inner wing is above the horizon, the sideslip is increased towards the port wing. These angles are approximately related to one another by the equation

$$\beta = \theta_y - \gamma \tag{13.4}$$

which is a more general form of eqn. (13.2).

The normal to the wing chord line must still pass through the spin axis, however, and the only way that the aircraft is allowed to change its attitude to produce the required wing tilt is by a rotation about the normal axis. It is not permitted to change its attitude by rolling about the $0x$ axis. (It should be emphasized here that we are considering the way in which the aircraft establishes its own equilibrium set of conditions; the response to pilot control inputs will be considered later (para. 5).) To produce a wing tilt of θ_y, the aircraft has to rotate about the normal axis through an angle χ, where θ_y, χ and α are related by the equation

$$\sin \theta_y = -\cos \alpha \sin \chi \tag{13.5}$$

The angle χ is defined as positive when the inner wing is above the horizon (i.e. the aircraft has been yawed in the positive direction).

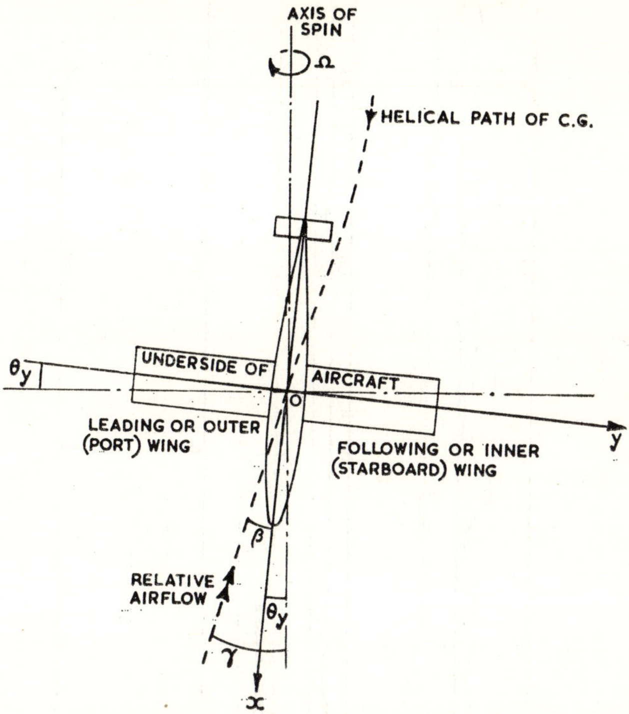

Fig. 13.4. Normal spinning attitude of modern aircraft

Ω positive
θ_y positive
χ negative
β negative ($\beta \simeq \theta_y - \gamma$)
γ positive

Once again a better physical appreciation of the formula can be obtained by considering the two extreme cases $\alpha = 0$ and $\alpha = 90°$. When $\alpha = 0°$ ($\cos \alpha = 1$) the aircraft is rolling about the vertical axis and θ_y and χ are numerically equal to one another (i.e. to tilt the wings out of the horizontal, the aircraft has to rotate about the normal axis through exactly the same angle). When $\alpha = 90°$ ($\cos \alpha = 0$) there is no relationship between θ_y and χ (i.e. in a flat spin yawing the aircraft about the vertical axis has no effect on θ_y).

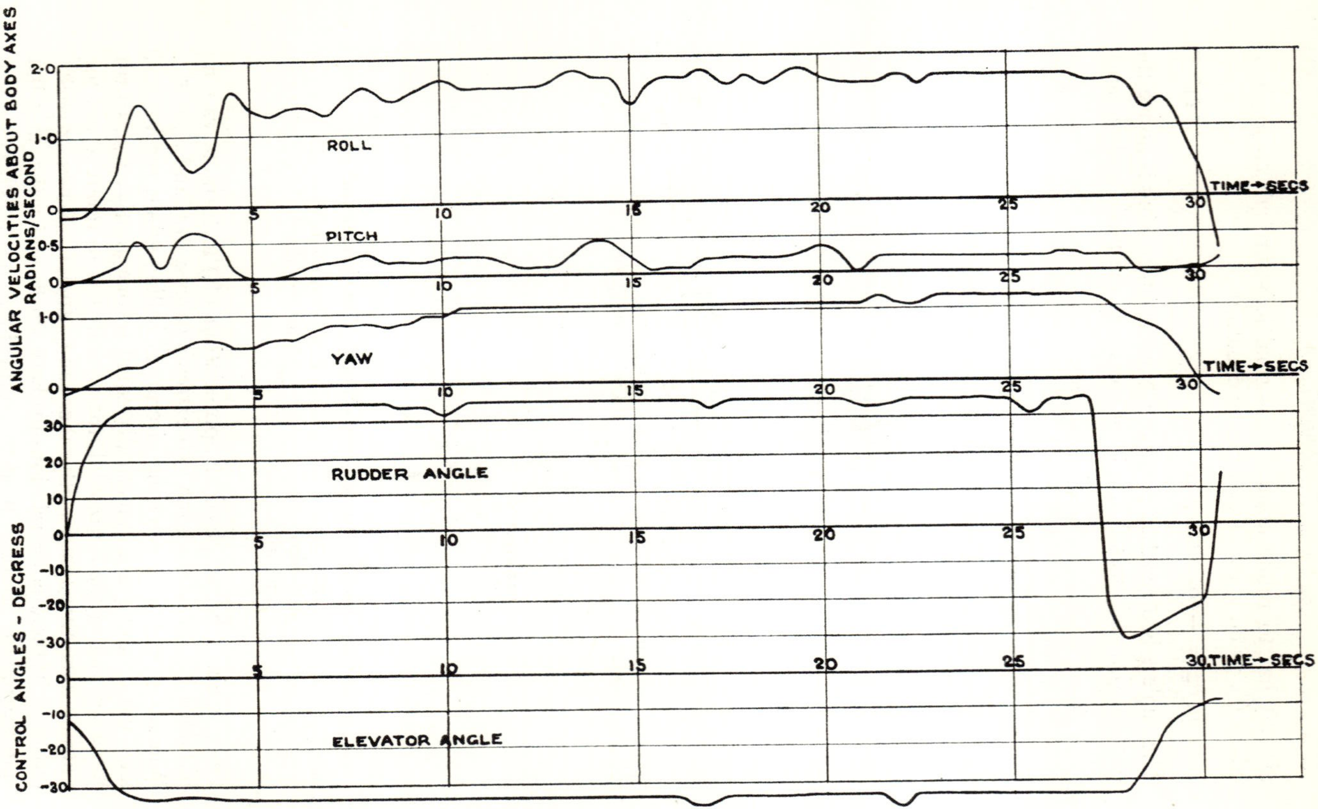

Fig. 13.5. Time history of a spin to starboard

In between these two extreme cases, the wings can be tilted out of the horizontal as a result of the aircraft yawing about the normal axis through a greater angle (i.e. $|\chi| > |\theta_y|$, except at $\alpha = 0°$ when $|\chi| = |\theta_y|$).

The total rate of spin can once again be resolved into components about the body axes and a more general form of eqn. (13.3) results

$$\left.\begin{aligned} p &= \Omega \cos \alpha \cos \chi \\ q &= -\Omega \cos \alpha \sin \chi \\ r &= \Omega \sin \alpha \end{aligned}\right\} \qquad (13.6)$$

By comparing eqn. (13.3) and eqn. (13.6) it can be seen that the rate of yaw remains the same (because the $0z$ axis remains in the vertical plane), but that, as a result of the rotation of the aircraft about the normal axis, the original "rolling" angular velocity is now split up into a rate of roll and a rate of pitch component. As long as the angle χ is small, however, little pitching motion is involved in the spin. From eqn. (13.6) it can also be seen that when the inner wing is down (i.e. χ negative) a positive spin can be resolved into a positive rolling motion, a positive yawing motion, and a positive (i.e. nose up) pitching motion about the body axes (Fig. 13.5). The majority of modern aircraft designs spin with the inner wing down and this particular attitude is assumed for much of the following analysis (Fig. 13.4).

3. Balance of Forces

The three main forces acting on an aircraft in a spin are the resultant aerodynamic force, the centrifugal force and the weight. In para. 2 it was shown that the resultant aerodynamic force acts approximately normal to the wing chord line. This normal force can be resolved into a vertical and a horizontal component. Since the resultant relative airflow is approximately vertical, the vertical component of the normal force can be referred to as the "drag" and the horizontal component as the "lift" (Fig. 13.2). The conditions of equilibrium are that the drag equals the weight and the lift equals the centrifugal force. Expressing the normal force in terms of the corresponding normal force coefficient C_N, these force relationships can be re-arranged to give

$$V^2 = \frac{2W}{C_N \rho S \sin \alpha} \qquad (13.7)$$

and

$$R = \frac{g}{W\Omega^2} C_N \tfrac{1}{2} \rho V^2 S \cos \alpha \qquad (13.8)$$

Substituting from eqn. (13.7) into eqn. (13.8)

$$R = \frac{g \cot \alpha}{\Omega^2} \tag{13.9}$$

Eqn. (13.9) shows that, for a given rate of spin, the radius of spin decreases with increase of incidence (i.e. as the spin flattens, the radius of spin decreases). From a consideration of the balance of pitching moments, it can be shown that the rate of spin also tends to increase with an increase in angle of attack and this produces an additional reduction in the radius of spin. If it is assumed that C_N does not vary appreciably with incidence, then eqn. (13.7) shows that an increase in incidence also decreases the rate of descent.

The balance of the forces acting in the spin is a relatively unimportant factor and it is the balance of the various moments that is of vital importance in determining the final equilibrium conditions of sideslip, angle of attack, etc. However, if it is assumed that the angle of attack and rate of spin have been determined by other considerations, then the balance of forces tends to determine the radius of the spin and the vertical rate of descent.

4. Balance of Moments

4.1. Balance of Moments about the Spin Axis

In a steady spin, all the moments acting on the aircraft must be balanced out, irrespective of the system of axes selected. Consider the motion first of all with respect to the spin axis. Since the centrifugal forces on all parts of the aircraft act radially outwards from the axis of the spin, they cannot produce any moment about this axis. Consequently the various aerodynamic forces acting must also produce zero resultant moment about the spin axis (i.e. if the aerodynamic pitching, rolling, and yawing moments acting on the aircraft are measured with respect to the body axis, then the sum of the components of the moments taken about the spin axis must be zero). As we shall see later, the rolling moments provide the main pro-spin moments and the yawing moments the main anti-spin moments. The components of the pitching moment about the spin axis will usually be small.

4.2. Balance of Moments about the Body Axes

The overall balance of moments can be considered by analysing the conditions of equilibrium about each axis in turn

(a) Balance of Pitching Moments

The balance of the pitching moments acting on the aircraft is shown diagrammatically in Fig. 13.6. The centrifugal forces acting on the imaginary concentrated masses of the aircraft due to the rotation about the spin axis act so as to flatten the spin and increase the angle of attack. The aerodynamic forces acting on the wing and the horizontal tailplane must therefore produce a nose-down moment exactly equal and opposite to that produced by the centrifugal forces.

A more exact analysis of the inertia coupling terms (see Chapter 16, para. 7) shows that the inertia pitching moment is given by

$$M_i = (C - A)rp \qquad (13.10)$$

(It should be noted that, since they have to balance one another, the *aerodynamic* pitching moment will be equal in magnitude but

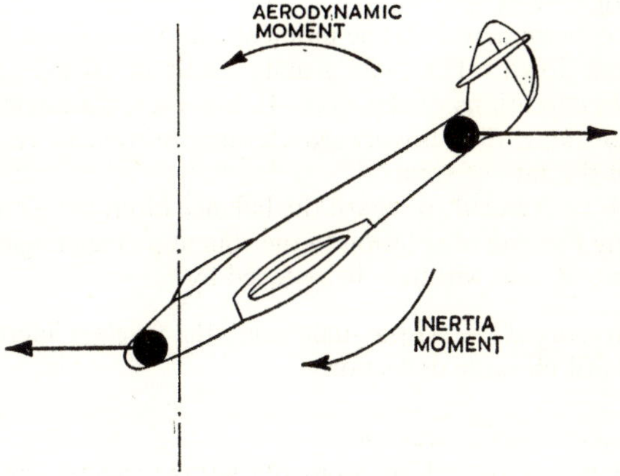

Fig. 13.6. Balance of pitching moments

opposite in sign. The suffix i has been used to emphasize that the equations express the relationships for the inertia moments and not the aerodynamic ones.) An alternative physical explanation of the inertia coupling terms can be given based on the principles of gyroscopic procession (see Chapter 16, para. 7). Consider a gyroscope of moment of inertia C positioned in the aircraft with the axis of rotation in the $0z$ direction and rotating at a rate r in the clockwise direction from above. When the aircraft rolls to starboard, it precesses in the nose-up direction and the nose-up pitching moment produced is equal to Cpr. Similarly, if a gyroscope of moment of inertia A is positioned with the axis in the direction $0x$ and rotating at a rate p, clockwise from the rear, then when the aircraft yaws to starboard, it precesses in the nose-down direction. The pitching moment produced is equal to $-Apr$. Thus the inertia coupling effects always produce two moments which act in opposite directions.

In the case of the pitching moments, since $C > A$, the resulting moment always acts in the nose-up direction and is given by $(C - A)pr$ (eqn. 13.10).

Substituting from eqn. (13.6) into eqn. (13.10) we get

$$M_i = \tfrac{1}{2}(C - A)\Omega^2 \sin 2\alpha \cos \chi \tag{13.11}$$

and hence assuming that $\cos \chi \simeq 1$, the inertia moment is mainly determined by the angle of attack α and the rate of spin Ω. The aerodynamic moment, on the other hand, is mainly dependent on α and the elevator control deflection. If the angle of attack is determined by other considerations, then, for a given elevator setting, the balance of the pitching moments will determine the equilibrium value of the rate of spin Ω.

The nose-down aerodynamic moment can be increased by moving the elevator down. This will usually result in changes in all the equilibrium conditions of the spin. If, however, the angle of attack remains constant, then downward elevator movement results in an increase in the rate of spin.

Thus it is convenient to regard the balance of pitching moments as determining the rate of rotation in the spin at a given angle of attack and the rate of spin will then be reduced by

(i) increasing $(C - A)$ [i.e. increasing the fuselage inertia],
(ii) upward elevator deflection.

(*b*) *Balance of Rolling Moments*

The balance of the rolling moments acting in a spin cannot easily be represented in the form of a diagram with concentrated masses, as in Fig. 13.6 for the case of pitching moments. The mathematical equation representing the inertia rolling moments is still of the same form, however

$$L_i = -(C - B)rq \tag{13.12}$$

and the gyroscopic explanation can still be used. With the gyroscopic axis along $0z$, a positive rate of yaw and a positive rate of pitch results in a rolling precession to port (i.e. a rolling moment of $-Crq$). Similarly, with the gyroscope axis along the $0y$ axis, a positive rate of yaw and a positive rate of pitch results in a precessional roll to starboard (i.e. a rolling moment of Brq). The resulting inertia rolling moment is thus given by a moment $(B - C)rq$ [or since $C > B$, $-(C - B)rq$] acting to port.

Thus when the aircraft is in the normal spinning attitude (i.e. inner wing down and pitching nose up), the inertia moment acts in the anti-spin direction, tending to roll the aircraft out of the spin until the wings are again horizontal. The anti-spin inertia moment has to be balanced by an exactly equal pro-spin aerodynamic moment. When the ailerons are in the neutral position, this aerodynamic moment is made up of two main parts, that arising from the rolling motion of the aircraft (i.e. the l_p contribution) and that arising from the dihedral effect (i.e. the l_v contribution).

Under normal flight conditions, l_p is negative, i.e. the rolling moment produced opposes the rate of roll and the effect is referred to as "damping-in-roll." At and beyond the stall, however, it is possible for the sign of the derivative to become positive and the rolling moment produced acts to assist the rolling motion of the aircraft. This effect is described as "autorotation" (see Chapter 6, para. 4.2). The sign of l_p in any particular spin will depend upon the equilibrium value of the angle of attack and the rate of rotation of the aircraft. It is most unlikely, however, that the l_p rolling moment will exactly balance the inertia rolling moment and sideslip one way or the other will normally be required to produce zero overall rolling moment. Thus the balance of the rolling moments will determine the equilibrium value of the sideslip angle.

As we have seen in para. 2, when the wings of the aircraft are horizontal, it will be sideslipping towards the leading wing by an amount equal to the helix angle. To reduce the amount of outward sideslip, the inner wing has to be below the horizon. Modern aircraft designs usually need a small amount of outward sideslip to balance up the rolling moments, but the sideslip angle required is usually less than the helix angle. Thus they take up an equilibrium attitude with the inner wing tilted below the horizon (Fig. 13.4).

When the ailerons are deflected, an additional rolling moment is produced and this will modify the amount of sideslip required to balance the rolling moments acting and hence will change the wing tilt of the aircraft. These secondary effects are extremely important when discussing the effect of aileron controls on spin recovery (see para. 5).

(*c*) *Balance of Yawing Moments*

The equilibrium value of wing tilt is also very important in determining the balance of the yawing moments acting and it is this balance which plays the major part in determining the final equilibrium conditions of an aircraft in a spin.

The inertia moments involved can be conveniently shown by the concentrated mass representation and Fig. 13.7 shows the yawing moments acting when the inner wing is below the horizon. The fuselage inertia moment is an anti-spin one (since it tends to yaw the aircraft to port, i.e. a negative rate of yaw) and the wing inertia moment is a pro-spin one (since it tends to yaw the aircraft to starboard). The resultant inertia yawing moment can be expressed as $(A - B)pq$, or, since B is usually greater than A, as

$$N_i = -(B - A)pq \tag{13.13}$$

Thus, for negative wing tilt, when $B > A$, the resulting inertia moment is an anti-spin one, and when $A > B$ a pro-spin one. It should be noted that the yawing inertia moment is proportional to

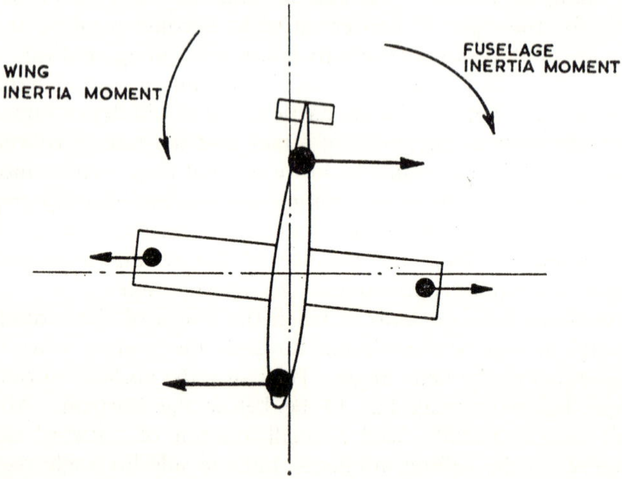

Fig. 13.7. Balance of yawing moments

q, the rate of pitch, and consequently is zero if the wings are not tilted out of the horizontal.

The gyroscopic explanation is similar to that already given for the other moments and is the same as that given for "yaw divergence" in para. 7 of Chapter 16; for positive rates of roll and pitch the two inertia yawing moments are given by $-Bpq$ to port and Apq to starboard.

The resulting inertia moment has to be balanced by an equal and opposite aerodynamic moment and since the inertia moment is an anti-spin one (if $B > A$), the aerodynamic moment must be a pro-spin one. The overall aerodynamic moment is made up of a large number of separate parts, some arising out of the yawing motion of the aircraft (i.e. the n_r contribution) and some arising out of the sideslipping motion (i.e. the n_v contribution). The n_v contribution is usually a small one (Ref. 1, page 283) and when the sideslip is towards the leading wing it normally acts in the anti-spin direction. The overall n_r effect can also be regarded as made up of two contributions, one from the wing, and one from the fuselage and vertical fin. The latter contribution acts in the usual way, i.e. as a damping-in-yaw or anti-spin effect, opposing the yawing motion of the aircraft in the spin. The wing contribution to n_r is more difficult to determine; it may act in either direction, but at large angles of attack usually acts in the pro-spin direction.

The total aerodynamic yawing moment depends on the angle of attack of the aircraft and the final balance of the yawing moments is achieved by the aircraft taking up the appropriate angle of attack at which the aerodynamic moments just balance the inertia moments. This particular angle of attack also has to be associated with the appropriate rate of spin required to balance the pitching moments and the appropriate angle of sideslip required to balance the rolling moments.

The damping-in-yaw effects of the fuselage and vertical fin depend on the design configuration of the aircraft. The fuselage contribution is very dependent on the shape of the fuselage cross-section; a cross-section with a round bottom and a flat top has, for example, over four times the effect of a circular fuselage (Table 8.1, Ref. 3). The damping effect of the fuselage can also be appreciably increased by the addition of fuselage strakes.

The damping-in-yaw effect of the vertical fin depends on the design of the tailplane-fin combination. When the aircraft is in a spin at a very high angle of attack, part of the vertical fin is immersed in the turbulent wake of the tailplane, and thus has a reduced effectiveness. In extreme cases, the presence of the tailplane may even change the anti-spin effect of the fin into a pro-spin one. This effect may be particularly serious with low set tailplanes and the damping contribution can be increased by correctly positioning the horizontal tailplane.

Deflection of the rudder control will produce an additional yawing moment, opposite rudder resulting in an anti-spin yawing moment.

5. Methods of Spin Recovery

The standard method of spin recovery is given as—"Apply full opposite rudder and then move the stick forward until the spin stops, maintaining the ailerons in the neutral position." On swept and delta wing aircraft, however, the ailerons are a very effective spin recovery control and the recovery is often made easier by the application of pro-spin or in-spin aileron. Application of out-spin aileron may prohibit spin recovery altogether, even if the elevator and rudder controls are used correctly. The effectiveness of each control is discussed separately below.

(*a*) *Rudder Control*

The rudder is always effective as a recovery control, opposite rudder (i.e. rudder deflected in the direction in which the aircraft is yawing) being used to produce an anti-spin aerodynamic yawing moment. This tends to reduce the rate of rotation and the angle of attack of the aircraft in the spin. It also tends to increase the amount of wing tilt by tilting the inner wing more below the horizontal. When $B > A$ this effect tends to increase the anti-spin inertia yawing

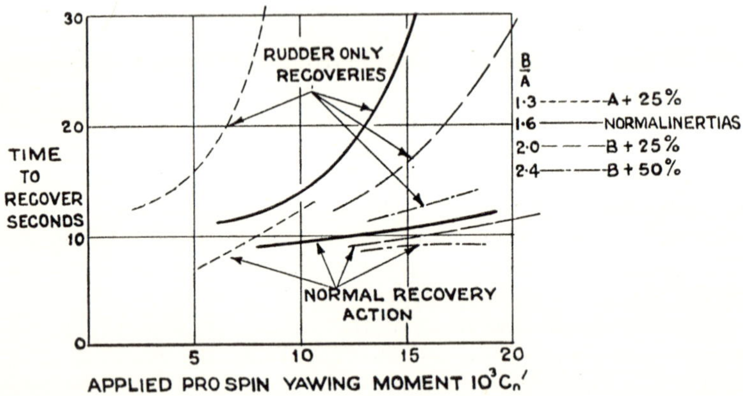

Fig. 13.8. Results of spinning tests on a Meteor 8

moment and in this case the secondary effect of the rudder also produces an anti-spin contribution. Fig. 13.8 shows the results of some spinning tests carried out on a Meteor with various B/A ratios. The curves of rudder-only recoveries show that the rudder effectiveness increases with increase of B/A. This change in control effectiveness is also represented in Fig. 13.9.

(*b*) *Elevator Control*

The elevator is also effective as a recovery control, down elevator being used to produce a nose-down anti-spin aerodynamic pitching moment. This tends to reduce the incidence of the aircraft in the spin. Once again, however, there are also some important secondary effects. Down elevator movement causes the rate of rotation of the aircraft to speed up (see para. 4.2*a*) and this increased rate of rotation can only be achieved by an increase in the in-spin rolling moment. This requires an increase in the amount of outward sideslip and thus, in this case, the tilt of the wings is reduced. If $B > A$, the consequent reduction in the anti-spin inertia yawing moment reduces the overall effectiveness of the elevator control. When $B < A$, on the other hand, the elevator will be a much more effective

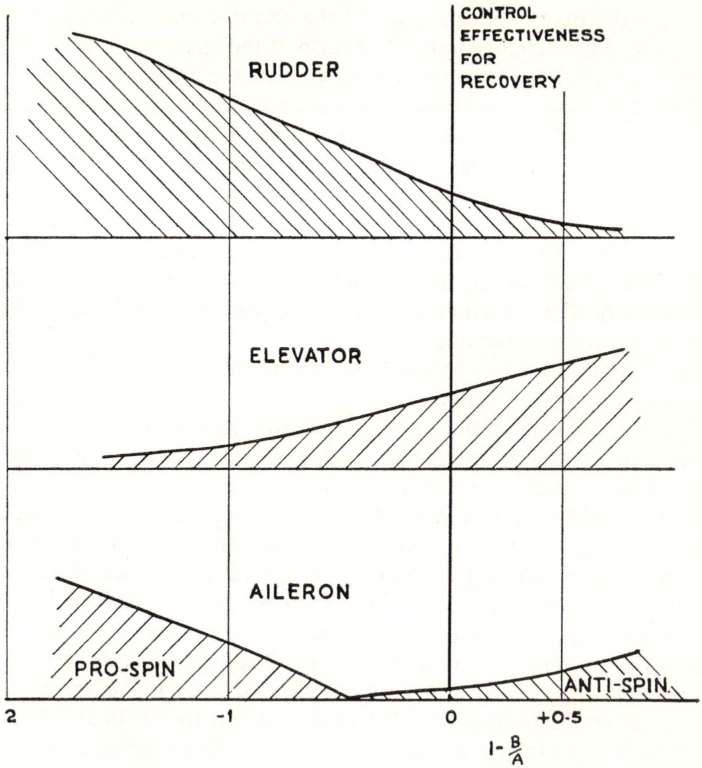

Fig. 13.9. Pictorial representation of control effectiveness for recovery from the spin on a conventional aircraft

control. On Fig. 13.8 the vertical separation between a "rudder-only" recovery curve and the corresponding "normal recovery" (i.e. rudder and elevator) curve shows the effectiveness of the elevator at the various B/A ratio values. The results are shown pictorially in Fig. 13.9; the elevator is more effective at positive values of $(1 - B/A)$, i.e. for $B/A < 1$.

It should be noted that down elevator tends to increase the shielded area of the fin and rudder, thus reducing the important anti-spin aerodynamic yawing moments obtained from these parts of the aircraft. It is for this reason that there is normally a small time delay in the recovery sequence between the pilot applying the opposite rudder control and moving the stick forward.

(*c*) *Aileron Control*

Though the overall effectiveness of the elevator and rudder controls is modified by a change in the B/A ratio of the aircraft, these controls always act in the same sense, i.e. down elevator and opposite rudder are always required to recover from a spin. The overall sense of the aileron control will, however, change with a change in B/A, i.e. in some cases the application of "in-spin" aileron (i.e. stick to the right in a spin to the right) will prevent recovery from a spin, in other cases it will be a very valuable recovery control. This change in the overall effectiveness of the control is due to the various secondary effects of the aileron control, which have a more powerful influence on the overall conditions in the spin than the direct rolling moment effects of the aileron.

When ailerons are applied in the in-spin direction, they produce a pro-spin aerodynamic rolling moment. They also produce important changes in the aerodynamic and inertia yawing moments. The change in the aerodynamic yawing moment is due to the aileron yaw effect, which is generally in the sense described as "adverse" (i.e. port aileron down yaws the aircraft to port). In this case, it produces an anti-spin aerodynamic moment. The change in the inertia yawing moment is due to a change in the attitude of the aircraft in the spin. Since the rolling moments must remain balanced, an increase in the amount of pro-spin rolling moment reduces the necessary amount of outward sideslip. Thus application of in-spin aileron control tends to increase the tilt of the wings (i.e. inner wing more below the horizontal). When $B > A$ this results in an increase in the anti-spin inertia yawing moment, and this is a more important effect than the direct effect of the ailerons in producing a pro-spin rolling moment. Wind tunnel tests (Ref. 4) have shown that in-spin ailerons are effective recovery controls when $1 - (B/A) < -0{\cdot}37$ (i.e. $B/A > 1{\cdot}37$) and that anti-spin aileron should be used when $B/A < 1{\cdot}37$ (Fig. 13.9). This reference also shows that ailerons are very effective spin recovery controls on delta wing aircraft.

6. Oscillatory Spin

When an aircraft initially enters a spin, oscillations occur in the values of the rates of pitch, roll and yaw. In many cases these oscillations damp out by about the second or third turn and a steady spin results (Fig. 13.5). In some cases, however, the oscillations do not damp out and in a fully developed oscillatory spin the oscillations may become quite violent (Fig. 13.10). The oscillatory type of spin is possibly caused by the aircraft not being able to reach a condition

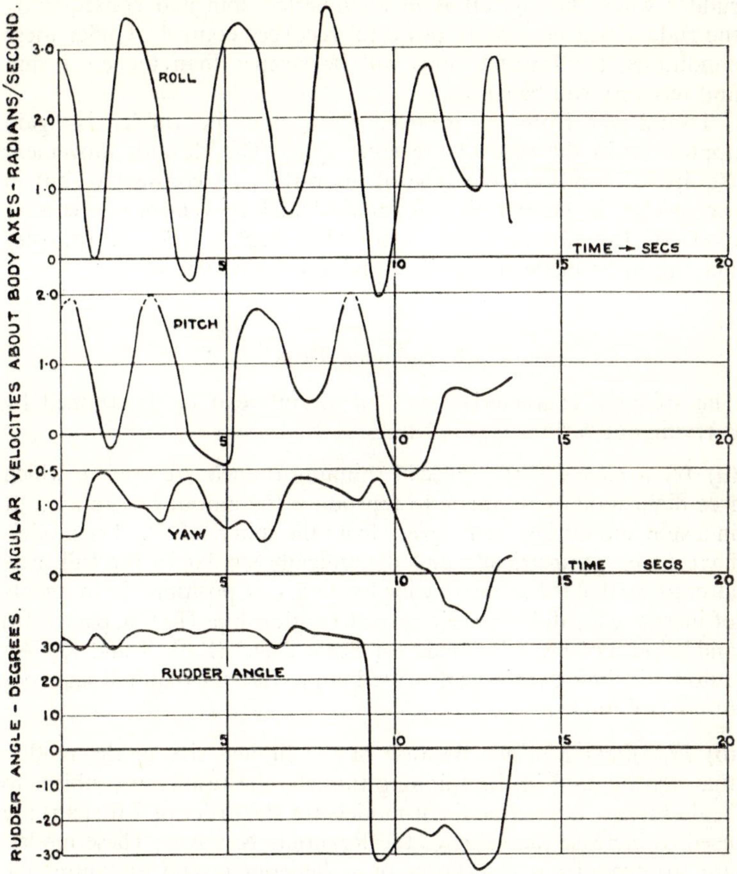

Fig. 13.10. Time history of an oscillatory spin to starboard

of equilibrium about its yawing axis and this type of spin is associated with high W/S and high B/A ratios. Recovery from an oscillatory spin may be relatively easy, however, particularly if the controls are applied as the nose of the aircraft is falling relative to the horizon.

7. Inverted Spin

An inverted spin is basically the same as an erect one and all the forces and moments acting must again be balanced. The differences are caused by the change in the relative positions of the fin, the fuselage and the horizontal tailplane. In the case of an aircraft with a low set tailplane, there will be less shielding of the fin and rudder when the aircraft is in an inverted spin and consequently the rudder will be a more powerful recovery control. Under these conditions, the inverted spin will be steeper than the erect spin and recovery will be easier.

To recover from an inverted spin, opposite rudder is again applied as in the case of the erect spin. The elevator movement required depends on the type of aircraft being considered, but in the case of the Hunter aircraft forward stick movement is necessary (Ref. 9). Initially, the ailerons should be kept neutral, but recovery may again be made easier by applying in-spin aileron.

8. Spinning Tests

The spinning characteristics of an aircraft can be determined by carrying out various types of tests.

(*a*) *Wind Tunnel Tests* Special spinning tunnels are used in which free-flight models are made to reproduce the motions of an aircraft in a spin and during the recovery from the spin (Ref. 5). The models have to be geometrically and dynamically similar to the full-scale aircraft, so that the values of wing loading, C.G. position, the moments of inertia, etc., all have their correct relationships (Ref. 6, page 8.16) and allowance must be made for Scale Effect (Refs. 4 and 7). An automatic timing device can be used to produce the required sequence of control movements.

(*b*) *Free-flight Models* Because of the limited size of the models that can be used in the spinning tunnels, the results are subject to Scale Effect. Large free-flight models (of the order of 7 ft span) are used to increase the value of the Reynolds Number. These models are dropped from a balloon or a helicopter, and an automatic timing device can again be used to produce the necessary control movements.

(*c*) *Full-scale Flight Testing* Though not fully reliable (Ref. 5), the results of the wind tunnel and free-flight model tests are an extremely useful guide to the spinning and control recovery characteristics of the full-scale aircraft. However, the model tests do have to be backed up by spinning tests carried out on the aircraft for all the various possible configurations (e.g. flaps down, tanks and stores fitted, etc.) These full scale results can then be compared with the values predicted from the model tests and this comparison can be used to improve the methods of prediction. The flight test techniques and the instrumentation required are discussed in Part II of Chapter 8 of Ref. 6. A comparison is made in Ref. 7 of the full-scale and model results obtained on a number of aircraft. Ref. 9 describes the results of erect and inverted spinning tests on the Hunter aircraft.

References

1. Duncan, W. J.; *Control and Stability of Aircraft* (Cambridge University Press, 1952).
2. Babister, A. W.; *Aircraft Stability and Control* (Pergamon Press, Oxford, 1961).
3. Kerr, T. H.; "General Principles of Spinning," *Agard Flight Test Manual*, Vol. II, Stability and Control, Chapter 8, Pt. I.
4. Harper, D. J.; *The Influence of Rolling Moments on Spin Recovery as Observed in Model-spinning Tests*, R. & M. No. 2831, 1954.
5. Gough, M. N.; *Notes on Correlation of Model and Full-scale Spin and Recovery Characteristics*, Agard Report No. 27, 1956.
6. Shorr, M.; "Spin Flight Tests," *Agard Flight Test Manual*, Vol. II, Stability and Control, Chapter 8, Pt. II.
7. Kerr, T. H.; *A New Standard for the Prediction of Full-scale Spin and Recovery Characteristics for Model Tests*, Agard Report No. 25, 1956.
8. Marx, A. J.; *Some Results of Comparison of Model and Full-scale Spinning Tests*, Agard Report No. 26, 1956.
9. Merewether, H. C. H.; *Erect and Inverted Spinning with Particular Reference to the Hunter, Journal of the Royal Aeronautical Society*, December 1965, page 835.

14 Effects of Aeroelasticity on the Stability and Control Characteristics of Aircraft

1. Introduction

In the previous work on stability and control the aircraft has been considered to possess a rigid structure, though it has been mentioned that any aeroelastic effects will in fact modify the results of the various simple theories employed. No structure is perfectly rigid and an aircraft is no exception, since the structure of an aircraft has to be designed to be as light as possible, consistent with the various structural requirements. This results in the aircraft being a fairly flexible structure, the amount of flexibility depending on the design configuration of the aircraft. A delta wing layout, for example, is usually less flexible than an aircraft with a fairly large aspect ratio sweptback wing. The wing of the American B52 bomber is an extreme example of aircraft flexibility, the wing can be loaded to give a total deflection at the wing tip of about 30 ft (Fig. 1.3, Ref. 2).

Aeroelastic effects become important at the higher forward speeds, the magnitude of the effects depending on the dynamic pressure (i.e. $q = \frac{1}{2}\rho_0 V_i^2$) and hence on the value of the equivalent air speed (E.A.S.). As the forward speed increases, however, the corresponding Mach number also increases and compressibility effects occur together with the aeroelastic effects. The overall result on any particular characteristic is thus the sum of these two separate effects and it is not directly obvious how much is due to aeroelasticity and how much to compressibility. The effects can be separated out to some extent by carrying out flight tests over a range of speeds and heights. If tests are carried out at the same Mach number at two different heights, then the E.A.S. is greater at the lower altitude and the difference is most likely due to aeroelastic effects. On the other hand, if the tests are carried out at the same E.A.S. at two different heights, the Mach number is greater at the higher altitude and any difference is probably due to the effects of compressibility.

Though aeroelastic effects are nowadays more associated with high performance aircraft, aeroelasticity has always been an important factor in aircraft design from the earliest days of flight. Sometimes flexibility is deliberately designed into the aircraft in order to provide a means of control, or to produce smoother flight in turbulent conditions, but more often it has detrimental effects on the stability and control characteristics of an aircraft. The Wright brothers initially obtained their lateral control by means of "wing-warping," and this method of control has been used more recently on a German glider design. On the other hand, the failure of Langley's monoplane in 1903 may have been due to wing torsional divergence, while the H.P. 0/400 bomber suffered from tailplane flutter in 1916 and the Fokker D.8 of World War I also had wing torsional divergence problems (Refs. 1 and 6).

The effects of aeroelasticity must be considered during the design of any high performance aircraft. In an exact analysis it is necessary to consider the overall effect of the aerodynamic, inertia, and elastic forces acting on the aircraft. In the work dealing with static stability and control in Chapters 2 to 5, only the effects of the aerodynamic forces were considered. The effects of the inertia forces are important when dynamic motions are being considered (e.g. in the analysis of dynamic stability in Chapters 7 to 9, spinning in Chapter 13, and inertia coupling in Chapter 16). In order to include the effects of flexibility, it is also necessary to allow for the elastic forces in the analysis. It is convenient to discuss the various possible stability and control problems with reference to Collar's Aeroelastic Triangle (or more aptly, perhaps, the "Aero-inertia-elastic" Triangle) shown in Ref. 6. As had been mentioned above, only the aerodynamic forces need to be considered in the analysis of the static stability characteristics of a rigid aircraft. Inclusion of the associated aeroelastic forces in the analysis allows the problems of control effectiveness, structural divergence and the static stability characteristics of a flexible aircraft to be investigated. The rigid aircraft dynamic stability characteristics and its response to control application and gusts are analysed by including the aerodynamic effects and the inertia effects, and the inertia and elastic loads are used to analyse mechanical vibrations and the effect of impacts. The problems in the middle of Collar's triangle require all the aerodynamic, inertial, and elastic forces to be included in the analysis; these problems include the dynamic stability and response characteristics of a flexible aircraft, flutter, and control surface buzz.

More recently, another factor which has become important in

these considerations is the effect of aerodynamic (or kinetic) heating on the elastic properties of the structural materials. Garrick has suggested (Ref. 10) that Collar's triangle should be modified to produce an Aero-inertia-thermo-elastic tetrahedron. Thermal effects are discussed in Ref. 9.

Aeroelastic problems can be divided into two groups

(i) Static (including Quasi-static) problems.
(ii) Dynamic or Unsteady problems.

(i) *Static problems* These problems are the ones which are concerned with steady state (i.e. non-oscillatory) aerodynamic loads and the associated steady state distortions. The aerodynamic loads acting on the aircraft produce a distortion of the structure, and this distortion produces corresponding elastic forces. The distortion produces an additional aerodynamic loading and this process is continued until either an equilibrium condition is reached, or until a structural failure occurs. In the analysis of these problems only the aerodynamic and elastic forces have to be included and the numerical values of the stability and control derivatives occurring in the equations of motion have to be modified to allow for the effects of aeroelasticity. When the oscillatory motions of the aircraft are slow (i.e. of low frequency) aircraft dynamic stability problems can also be considered under this general heading and the analysis is done in a similar way by modifying the values of the "rigid" aircraft derivatives. The longitudinal and lateral dynamic stability characteristics of an aircraft can usually be determined by this "quasi-static" method, since the frequencies involved are relatively low, and the distortions of the structure can be assumed to be in phase with the motion of the aircraft. According to Chapter 10 of Ref. 11, however, this result may not be a valid one in the case of tailless aircraft, when unsteady flow effects may be important.

(ii) *Dynamic problems* These problems are the ones involving high frequency oscillatory motions, in which the inertia effects become very important. In these cases, the concept of derivatives can still be used (Ref. 12); the values are, however, different from the corresponding static values and are dependent on the frequency of oscillation. One of the most important of these problems is that of flutter; this is concerned with the oscillatory motion of part or parts of the aircraft (e.g. wing, aileron, etc.), relative to the remainder of the structure. The phasing of the various motions involved is the important factor and it may be such that coupling will occur between the motions, producing serious, and even disastrous, flutter characteristics.

As we have seen above, dynamic stability can normally be dealt with as a quasi-static problem. However, as the structural frequencies of aircraft approach nearer to the normal rigid body characteristic frequencies, the structural deformations may be large enough to have a significant effect on the dynamic stability characteristics of the aircraft (see para. 6).

Most of the standard textbooks on aeroelasticity are chiefly concerned with the second class of problems. This is a very specialized field and will only be considered briefly here. The work of this chapter deals mainly with the static effects of aeroelasticity on the basic stability and control characteristics of aircraft.

The adverse effects of aeroelasticity can be eliminated by the use of various design techniques. One obvious preventative measure is to increase the bending and torsional stiffnesses of the wing, fuselage, tailplane, etc., though this method has an associated structural weight penalty. Control surface flutter can be prevented by increasing the stiffness of the control circuits (see para. 7.4) and by incorporating viscous dampers and mass balancing weights on the control surface. The problems associated with control systems on large flexible aircraft has been one of the reasons behind the proposal for the introduction of electric signalling controls (see Chapter 12). The aeroelastic design requirements for military aircraft are given in Part V of Ref. 13, and methods of predicting flutter characteristics are discussed in Ref. 24.

2. Effects of Aeroelasticity on the Values of the Stability and Control Derivatives

The stability and control characteristics of an aircraft can be expressed in terms of the stability and control derivatives. In Chapter 6 it was stated that the numerical values of all these derivatives are modified as a result of both compressibility and aeroelastic effects. These variations in the values of the derivatives affect all the stability and control characteristics of the aircraft at high Mach numbers and high E.A.S. values. The aeroelastic effects result from the structural deformation of the aircraft and depend on the relative importance of the various structural modes of distortion. The following discussion describes how the various derivatives are modified under "static" conditions (i.e. steady or low frequency oscillatory motions); these modified values of the derivatives are the values that have to be used in the analysis of any static or quasi-static problem. The measurement of the corresponding values of the derivatives under dynamic conditions is not dealt with here; it is discussed in Refs. 14 and 15.

The aeroelastic effects experienced are dependent on the particular design configuration of the aircraft being considered. A conventional aircraft lay-out, consisting of a wing-fuselage-tailplane combination, suffers from various possible modes of structural distortion, including

(*a*) vertical and lateral fuselage bending,
(*b*) wing twisting and bending,
(*c*) tailplane twisting,
(*d*) control surface distortion.

The effects of the changes in the numerical values of the derivatives on the overall stability and control characteristics are discussed in later sections of this chapter and also in Chapters 15 and 16.

2.1. Lift and Hinge Moment Characteristics

The values of the basic lift curve slopes and hinge moment characteristics of the various aerofoil and control surfaces are themselves derivatives and are important factors in the expressions for many of the other derivatives. In the simple mathematical analysis of longitudinal static and manoeuvre stability considered in Chapters 2 and 3, these quantities were represented by the symbols a, a_1, a_2, b_1, b_2, etc., and the values were assumed to be constant, independent of forward speed. In the more general theory, the values were assumed to vary with speed, due to the effects of compressibility and aeroelasticity and the symbols A, A_1, A_2, B_1, B_2, etc., were used. In this section the effects of aeroelasticity on the values of these parameters will be discussed.

The lift curve slope (i.e. $A = dC_L/d\alpha$) of any aerofoil is modified due to the effects of aeroelasticity, as a result of the twisting of the wing produced by the aerodynamic load acting on it. The value of the lift curve slope is either increased or decreased, depending on whether the aerodynamic centre of the wing lies in front of or behind the flexural axis. When the aerodynamic centre lies ahead of the flexural axis, the aerofoil twists nose up and the lift is increased, producing further twist. This leads in some cases to the problem of torsional divergence, a problem which is considered in more detail in para. 3. On a finite wing the effect of aeroelasticity on the value of the lift curve slope depends on the aspect ratio and sweep of the wing. On an unswept wing of moderate aspect ratio the lift curve slope tends to increase with increase in forward speed. When the wing is sweptback the value of A tends to be reduced, due to the washout (i.e. nose-down twist) associated with the wing bending (see para. 3). A reduction in the value of the wing lift curve slope due to aeroelastic effects is shown by the results of Refs. 16 and 17, both of which are concerned with high-subsonic speed sweptback wing aircraft. On low aspect ratio delta wings the effect is usually small. The twisting of a wing is reduced to some extent at supersonic speeds, as a result of the rearward movement of the aerodynamic centre of the wing which occurs in going from subsonic to supersonic speeds.

The value of the tailplane lift curve slope A_1 may also be reduced because of the twisting of the tailplane, but it is more likely to be effectively reduced as a result of fuselage bending. When the attitude of the aircraft is increased, an additional load acts on the horizontal tailplane. If the fuselage is a flexible one, and assuming for the purposes of this discussion that the wing and tailplane are rigid structures, then it will bend vertically upwards, thus reducing the effective change in the angle of attack of the tailplane. If the determination of the tailplane lift curve slope is based on the actual change in the tailplane angle of attack, then the value of A_1 is in fact unchanged, but, because of the distortion of the fuselage, the tailplane is less effective as a stabilizing device and if A_1 is based on the attitude change of the aircraft or on the change in angle of attack of the wing, then the value is decreased.

The quantity A_2 is a measure of the lift produced by the particular control surfaces. These values are modified by the aeroelastic effects, due to the distortion of the airframe structure and also that of the control surfaces. When an elevator is deflected downwards, for example, additional lift is produced on the tailplane and elevator. This aerodynamic loading produces, amongst other things, a vertical bending of the fuselage and a twisting of the elevator control surface. Both these distortional modes tend to reduce the up load acting on the tailplane and elevator. If the determination of the value of A_2 is based on the corresponding amount of elevator deflection in the case of a rigid aircraft, then both these effects produce a reduction in the lifting effectiveness of the control and hence in the corresponding control power (para. 2.2). The effects of the various possible distortional modes on the value of A_2 are shown separately in Fig. 22 of Ref. 18.

Similar effects are produced by the aileron controls. When an aileron is deflected downwards, the increase of lift on the wing causes it to bend and twist, thus producing a reduction in the lifting effectiveness of the aileron. This is a particularly important problem which is considered in more detail in para. 4.

The values of the control hinge moment characteristics are also affected by aeroelastic distortion. In Ref. 18 the value of B_2 is made large and negative as a result of both compressibility and aeroelastic effects; in this particular case the latter effects are the dominant ones. In Fig. 27 of this reference, it is shown that fuselage distortion makes B_2 slightly more negative, but that the major effect is due to the flexibility of the tailplane and elevator.

It has been mentioned above, that a wing twists and bends under the action of an aerodynamic load, the amount of distortion depending on the particular configuration, and varying across the span of the wing. Thus when a flexible wing is producing a certain amount of lift, it will not have the same lift distribution as the corresponding rigid wing producing the same amount of lift. In the case of a sweptback wing, the amount of nose-down twist will increase towards the tips, thus producing a loss of lift, and the angle of attack of the centre part of the wing will have to be increased by a certain amount to counteract this and to produce the same overall lift as the rigid wing. Thus, on the flexible wing, the lift is more concentrated in the centre part of the wing and hence farther forward.

The change in the overall distribution of the wing lift affects two factors which are particularly important in stability and control work, the position of the wing aerodynamic centre and the downwash characteristics. From what has been said above, the twisting of a sweptback wing tends to move the aerodynamic centre farther forward.

The basic formula used in Chapter 2 for the analysis of the longitudinal trim and static stability characteristics of an aircraft, i.e.

$$C_M = C_{M0} + (h - h_0)C_L - \bar{V}[a_1(\alpha - i - \varepsilon) + a_2\eta + a_3\beta] \quad (14.1)$$

can be seen to include many of the factors which have been discussed in this section, and it should be obvious that the effects of aeroelasticity on the separate factors involved will have an important overall effect on the trim and static stability characteristics of an aircraft (see para. 5).

2.2. Control Power Derivatives

The effects of aeroelasticity on the control power derivatives (i.e. l_ξ, m_η and n_ζ) follow directly from the effects discussed in para. 2.1, since the moments produced by the control surfaces about the C.G. of the aircraft are directly related to the lifting effectiveness of the corresponding surface. If A_2 is the tailplane lift curve slope of the elevator control surface, then the pitching moment effectiveness of that control can be represented by $\bar{V}A_2$. As we have seen, movement of the elevator will produce an aerodynamic load on the tailplane and the resulting aeroelastic distortion of the structure will result in a reduction in the effectiveness of the control.

Similar reductions in effectiveness will occur on the aileron and rudder controls. When the ailerons are deflected, the consequent bending and twisting of the wing will result in a loss and even, possibly, a reversal, of aileron effectiveness (para. 4). Deflection of the rudder increases the sideload on the vertical fin and rudder, and the resulting bending of the fuselage decreases the effectiveness of this control. The variation with speed at high subsonic Mach numbers of the values of these control derivatives on a particular aircraft is shown in Ref. 16. Alternative forms of control which are used to increase the effectiveness at high forward speeds are discussed in Chapter 10.

2.3. Longitudinal and Lateral Stability Derivatives

An aircraft obtains its static stability as a result of the stabilizing moments produced by the changes in the aerodynamic forces acting on the wing, the tailplane, and the vertical fin, associated with a change in angle of attack or sideslip. The changes in aerodynamic force are dependent on the corresponding values of the lift curve slopes and these, as we have seen, are modified by the effects of aeroelasticity.

The longitudinal static stability of an aircraft is reduced as a result of vertical fuselage bending, for example, the increase of tailplane load resulting from a nose-up disturbance producing a bending of the fuselage tending to reduce the effective angle of attack of the tailplane and hence the tailplane stabilizing contribution. This problem is discussed in more detail in para. 5.

The value of the directional static stability derivative n_v will be reduced in a similar way, as a result of lateral fuselage bending. Loss of n_v at high forward speeds is particularly important and may result in a number of stability and control problems. Fig. 30 of Ref. 16 shows the effect of flexibility on the n_v value of a swept-wing bomber at high subsonic speeds. The results of the tests carried out on the Fairey Delta 2 aircraft discussed in Ref. 19 also indicate a reduction in n_v at transonic speeds due to aeroelasticity.

The main contribution to the lateral static stability of an aircraft comes from the dihedral effect. When an aircraft is sideslipping the increase in lift on the leading wing results in wing bending and twisting. The bending produces an effective increase in dihedral, whereas the twisting may be in the nose-down direction. In the case considered in Ref. 16, the overall effect of flexibility is to increase the dihedral effect (i.e. to increase the value of $-l_v$).

The damping derivatives m_q, l_p and n_r obtain their main contributions from the changes in angle of attack of the tailplane, wing, and vertical fin, respectively, produced by the rotary motion of the aircraft. From what has already been said, the values of these derivatives will generally be reduced as a result of the effects of aeroelasticity. The results of Ref. 19 indicate a small reduction in l_p and a large reduction in n_r at transonic speeds. The results of Ref. 16, on the other hand, show a large reduction in l_p, the damping in roll. The effects of aeroelasticity on the dynamic stability characteristics of an aircraft are discussed in para. 6.

Graphs showing the variation with speed and height of the values of some stability and control derivatives are given in Chapters 15 and 16.

3. Wing Torsional Divergence

Wing torsional divergence occurs when the aerodynamic loads acting on the wing of the aircraft produce structural distortions tending to increase the aerodynamic forces, thus further increasing the amount of aeroelastic distortion. Consider the effect of an aerodynamic load on a particular wing section. If an upload is applied to the wing at its trailing edge the wing will twist in the nose-down direction, whereas if the load is applied at the leading edge, the section will twist nose-up. There is consequently a point, somewhere between the leading and trailing edges of the wing, at which an application of load will not produce any twisting. This point is called the Flexural Centre of the wing section. On a finite wing the locus of the sectional flexural centres is called the Flexural (or Elastic) Axis. A load applied to a wing on its flexural axis will not produce any twist at the particular section at which the load is applied, though it may cause twisting at other inboard or outboard sections.

On a straight wing the aerodynamic centre is usually in front of the flexural axis. When the angle of attack of the wing is increased, the additional aerodynamic load (i.e. lift) acting at the aerodynamic centre twists the wing in the nose-up direction. The distortion of the wing produces elastic torsional moments which oppose the aerodynamic moments, thus tending to limit the amount of distortion. The twisting of the wing, however, produces an additional lift force and this tends to distort the wing still farther. The extra lift produced as a result of wing distortion increases the effective lift curve slope of the wing, based on the incidence at the wing root. At low forward speeds the aerodynamic and the elastic twisting moments eventually balance one another and a condition of equilibrium is reached. As the forward speed increases, however, the aerodynamic moments associated with a given change of incidence increase as the square of the speed, while the elastic torsional stiffness of the wing remains the same. Eventually, a speed is reached at which equilibrium is no longer attained and the process described above continues until the wing eventually fails. The wing is said to diverge and the speed at which this occurs is known as the Critical Wing-torsional Divergence Speed. The value of this speed can be increased

(*a*) by increasing the torsional stiffness of the wing (but only at the expense of weight),

or

(*b*) by designing the internal structure of the wing so that the flexural axis is well forward, thus reducing the aerodynamic twisting moments. This can be achieved by designing the wing with a strong D-type leading edge structure. An "iso-clinic" wing is one which is specially designed so that it will not twist under load.

The critical divergence speed is increased by sweepback, and decreased by sweepforward (Ref. 20); the effect of sweep is shown

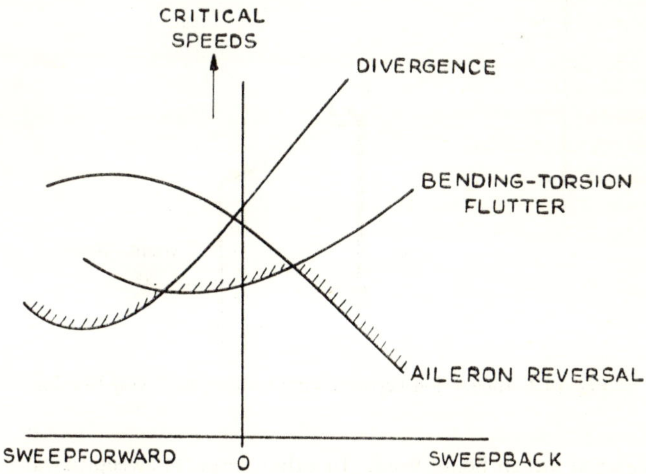

Fig. 14.1. Effect of sweep on the critical speeds

in Fig. 14.1. The aerodynamic loads acting on the wing cause it to bend and this bending results in a nose-down twist. The reason for the interaction between wing bending and wing twisting can be seen by considering a sweptback wing of moderate or large aspect ratio (Fig. 14.2). The structure of a wing is such that it tends to bend so that all the points along a line AB normal to the main spar move vertically through equal amounts. When the movement of a wing section parallel to the free stream flow CD is considered, it can be seen that the point D on the trailing edge, since it is outboard of the point B, must bend up through a greater amount than the point C at the leading edge of wing, which is inboard of point A. Thus the effect of bending on a swept-back wing is to twist the wing in the nose-down direction. This opposes the direct twisting effect of the aerodynamic forces which is in the nose-up direction and thus the critical divergence speed is increased.

Sweepforward on the wing produces the opposite effect and the critical divergence speed is reduced. This is the great disadvantage of sweepforward and is the main factor which prevents large angles

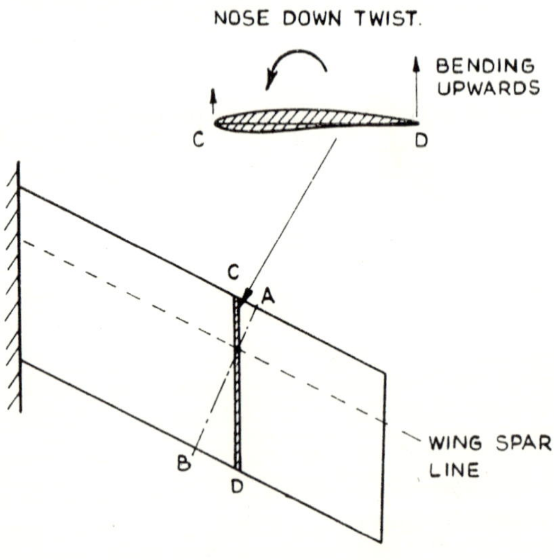

Fig. 14.2. Interaction between wing bending and wing twisting

of forward sweep being used. In other respects sweepforward may have some advantages; it reduces the effects of compressibility in the same way as sweepback, without some of the problems associated with wing-tip stalling. It has some structural and aerodynamic advantages and for this reason has been used on the twin-jet Hansa HFB 320 aircraft which is described in Ref. 21. This aircraft made its maiden flight on 21st April, 1964. It was designed for a speed of about $M = 0{\cdot}76$ and incorporated about 15° of sweepforward. The wing divergence problem was eliminated by designing the wing structure so as to reduce the twisting due to bending, and by installing wing tip tanks with the centre of gravity well forward so that the effects of inertia counteract the divergent tendency.

This sweepforward design is limited to fairly low values of forward speed, though M and W shaped wings, which incorporate both sweepforward and sweepback, were proposed some years ago for aircraft designed to cruise at low supersonic speeds.

4. Control Effectiveness and Control Reversal

Control is usually achieved by means of a flap-type control surface at the trailing edge of an aerofoil, which may be the wing, the

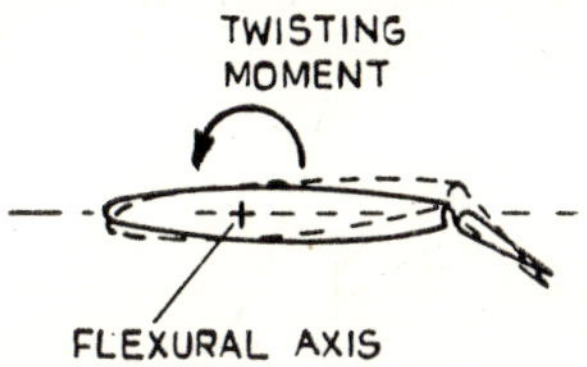

Fig. 14.3. Twisting due to control surface deflection

horizontal tailplane, or the vertical fin. When a control surface is deflected downwards at the trailing edge of an aerofoil an additional aerodynamic loading is produced. The resulting force will usually act behind the flexural axis of the main surface and thus it will tend to produce a nose-down twisting moment. If the surface is a

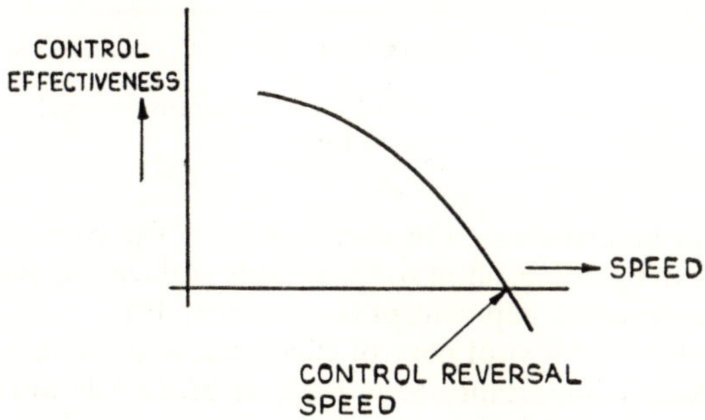

Fig. 14.4. Control reversal speed

flexible one then it will twist and the resulting change in the aerodynamic load acts to oppose the effect of the control surface deflection (Fig. 14.3). This results in a reduction in the effectiveness of the control surface. The loss increases with increase of forward speed, until eventually a speed is reached at which the control effectiveness is zero (Fig. 14.4). At higher speeds than this, the distortion of the structure produces an effect which is greater than that of the initial control surface deflection, and control reversal is said to occur.

Sweepback has an adverse effect in this case, because of the additional nose-down twisting effect on the aerofoil, resulting from the bending produced by the aerodynamic loading. Thus sweepback decreases the critical aileron control reversal speed, and sweep-forward increases it (Fig. 14.1).

Once again, the critical speed can be increased by increasing the torsional and bending stiffnesses of the wing, though a weight

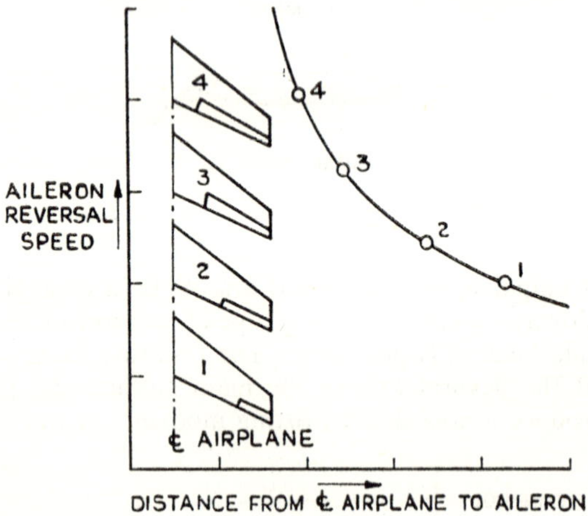

Fig. 14.5. Effect of aileron design on the control reversal speed [From Ref. 9]

penalty will be involved. The effectiveness of the controls can also be increased by using all-moving or slab surfaces in place of the fixed aerofoil with a flap control (see Chapter 10).

This problem of loss of control effectiveness and control reversal is a particularly important one in the case of the lateral (or rolling) control of the aircraft, because of the thin-wing, moderately high aspect ratio, swept-back configuration used on many modern aircraft. The aileron effectiveness is often represented in terms of the value of $pb/2V$ (see Chapter 11) and this decreases with speed due to the aeroelastic effects. Fig. 14.5, taken from Ref. 9, shows that the aileron reversal speed can be increased by changing the configuration of the aileron controls and the same reference shows that by suitably designing the wing and by moving the ailerons inboard the control effectiveness can be made to increase slightly with increase in forward speed. Inboard ailerons are used on the Folland Gnat, and on the McDonnell Phantom II, while the T38 Talon has ailerons at the semi-span position. Often aircraft are fitted with two sets of aileron surfaces, the outboard ones being locked at high forward speeds and only the inboard ones being used for rolling control under these conditions of flight. The aileron control surfaces of the Boeing 707 are shown in the photograph on page 221 of Ref. 22.

The problem of loss of aileron effectiveness at high speeds can be eliminated by using spoiler-type control surfaces in place of the flap-type controls. The characteristics of these lateral spoiler controls are discussed in more detail in Chapter 10.

5. Effects of Aeroelasticity on the Longitudinal Static and Manoeuvre Stability

5.1. General

The simple theory of static and manoeuvre stability, discussed in Chapters 2 and 3, neglected the effects of compressibility and aeroelasticity. On the basis of this theory, the static margin and the C.G. margin were shown to be identical to one another and it was also shown that the stick-fixed and stick-free static stabilities are related to the stick movements and forces required to trim, whereas the stick-fixed and stick-free manoeuvre stabilities are related to the stick movement per g and stick force per g, respectively. In the general theory, the compressibility and aeroelastic effects are allowed for, and this modifies the simple results. The static margins and C.G. margins are no longer identical, but are only proportional to one another, the factor of proportionality varying with forward speed (Chapter 2, para. 3).

When aeroelastic distortion of the airframe is allowed for, the resulting effects on the stability characteristics of the aircraft depend on which of the various possible modes is the predominant one. The problem is a very involved one, and only the following modes are discussed here

(*a*) wing bending and twisting,
(*b*) fuselage bending,
(*c*) tailplane and elevator twisting.

The effects of the various modes are discussed in detail in Ref. 26. These effects are rather difficult to generalize, since they are dependent on a large number of variable factors which include the C.G. position, the forward speed, the value of $B_1(= dC_H/d\alpha')$ and the values of the tailplane and/or the elevator angles required for trim. In the following paragraphs an attempt has been made to describe simply some of the possible effects. It will be seen that a change in the value of the elevator angle to trim can change the effect of a particular mode of distortion from a stabilizing contribution to a destabilizing one. It is also important to note that, to be strictly correct, any distortion in the elevator control circuit makes it necessary to distinguish carefully between "stick-fixed" and "elevator-fixed" conditions of flight.

5.2. Effect of Wing Bending and Twisting

Aerodynamic forces acting on the wing of an aircraft cause it to bend and twist, the wing bending being particularly noticeable on aircraft with high aspect ratio flexible wings. The twisting of the wing may be either a direct effect, resulting from the action of the aerodynamic forces, or an indirect effect, resulting from the bending of the wing. The direct twisting effect depends on the distance between the torsional axis of the wing and the point of application of the resultant aerodynamic force. Bending does not produce twisting on a straight wing, but on a swept wing it does. When a sweptback wing bends upwards, it twists nose down and the angle of attack of the section of the wing is reduced (para. 3). The "elastic washout" increases towards the tips of the wing and will be particularly important on sweptback wings of large aspect ratio. The resulting changes in angle of attack will modify the lift distribution on the wing and consequently the pitching moment characteristics.

The longitudinal static stick-fixed stability is related to the elevator angle required to trim the aircraft in level flight. Consider the effect of a change in the trimmed speed on a rigid aircraft with a flexible wing. When the forward speed is increased, the angle of attack of the wing has to be reduced (i.e. the value of the lift coefficient C_L has to be reduced, so that the lift is still equal to the weight), the centre of pressure on the wing moves rearwards and a nose-down twisting moment acts on the wing. This moment tends to twist the wing in the nose-down direction, the twist increasing towards the tip. This reduces the overall lift produced by the wing, and thus in order to trim the aircraft in level flight (i.e. to make the lift again equal to the weight) the attitude of the whole aircraft will have to be increased to a value greater than the one that would be required at the same speed by the completely rigid aircraft. Though the total lift is once again the same, it will not be distributed along the span in the same way. The lift will be more concentrated at the root of the wing and on a sweptback wing this will result in the resultant lift force moving forward, thus producing a nose-up change in pitching moment. This additional effect is a stabilizing one, since it tends to increase the angle of attack, and hence, to reduce the forward speed back to its initial value.

However, because the whole attitude of the aircraft has been increased, the incidence of the tailplane has been correspondingly increased and, assuming that the effectiveness of the tailplane is unchanged, this produces a nose-down (i.e. destabilizing) change in pitching moment. In practice, these separate effects are added together; since they act in opposite directions they tend to cancel one another out and the overall effect on the stability of the aircraft depends on which of the two effects is the predominant one. The

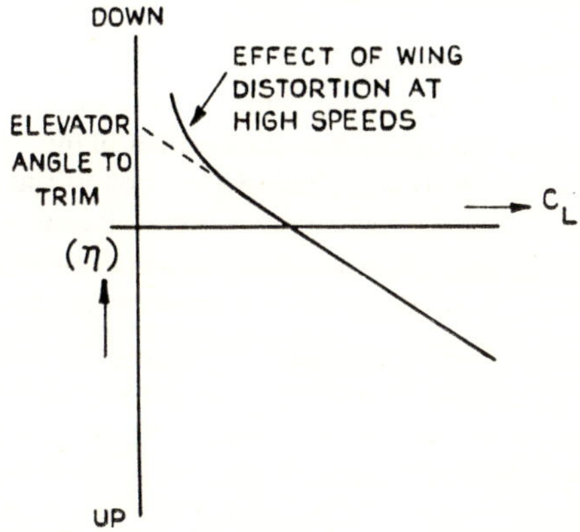

Fig. 14.6. Effect of wing flexibility on the elevator angle to trim curve

resulting effect is generally small; there is usually a small increase in the stability and more down elevator is required to trim (Fig. 14.6).

It is worthwhile emphasizing here that this explanation is the technical one used to explain the effect of wing flexibility on the shape of the η/C_L curve obtained from flight tests; the pilot does not observe any of these effects in flight, apart from recording elevator angles to trim which are different from those predicted by rigid aircraft analysis.

The longitudinal manoeuvring stability is related to the stick movements and forces per applied g. In a pull-out at constant speed, the angle of attack is increased and the centre of pressure moves forwards on the wing. The direct torsional effect of the aerodynamic load produced will be to twist the wing in the nose-up direction. This increases the angle of attack of the wing, and hence, further increases the aerodynamic load (i.e. a destabilizing contribution). If the wing is sweptback, however, bending of the wing will cause it to twist nose-down. This reduces the angle of attack and the corresponding lift and moves the aerodynamic centre forward. To develop a certain amount of g the overall lift produced by the wing must be the same and thus the aircraft with the flexible sweptback wing would have to be at a larger attitude than the rigid aircraft. The tailplane, which in this case is assumed to be rigid, would also be at an increased angle of attack and, as a consequence, would produce an additional nose-down (i.e. stabilizing) pitching moment. The overall effect is the resultant of all the separate contributions, but, in practice, there is usually a reduction in the amount of manoeuvre stability due to wing flexibility.

5.3. Effect of Fuselage Bending

If the fuselage is a flexible one, then a load on the tailplane will produce bending of the fuselage in the vertical plane. If the tailplane is considered to be rigidly attached at the root to the fuselage, then there will be a change in the angle of the tailplane setting relative to the wing chord line. It is convenient to consider the total load on the tailplane to be made up of two separate parts

(i) The upload which is required to balance the nose-up pitching moment produced by the lift force on the wing considered to be acting at the aerodynamic centre. Since the lift must always equal the weight in a level flight, then, assuming the aerodynamic centre to be a fixed point, this part of the tailload is a constant, for a given C.G. position, independent of forward speed.

(ii) The download which is required to balance the nose-down pitching moment due to C_{M0}, the wing zero-lift pitching moment coefficient. This part of the tailload increases as the square of the forward speed, but is independent of the position of the C.G.

The effect of the upload in (i) above is to bend the fuselage vertically upwards, thus making the tailplane setting more negative. To maintain the value of the tailplane lift required for trim, the elevator has to be moved down, by an amount which is independent of forward speed (Fig. 14.7). Rearward movement of the C.G. increases the upload on the tailplane; this increases the fuselage bending and the additional amount of down elevator movement required for trim. The stick fixed static margin is given by $K_n = -\bar{V}A_2 d\eta/dC_R$ and though in this case $d\eta/dC_R$ remains unchanged for a given C.G. position, the value of $\bar{V}A_2$ is reduced by fuselage bending and this loss in elevator effectiveness reduces the static margin.

The effect of the download in (ii) will be to bend the fuselage downwards. More up elevator angle will be required to trim the aircraft, the amount of additional up elevator movement required increasing as the square of the forward speed (Fig. 14.7). This reduces the slope of the η/C_R curve and is also destabilizing. Thus the static margin of an aircraft with a flexible fuselage is less than that of the rigid aircraft; the reduction in stability is made up of two components, one which is constant throughout the speed range, the other increasing as the square of the speed. The corresponding elevator tab angle to trim curves will show a similar reduction in the stick-free static stability.

In a steady pull-out the additional load on the tailplane has to be in the upward direction, in order to balance the increase in the lift

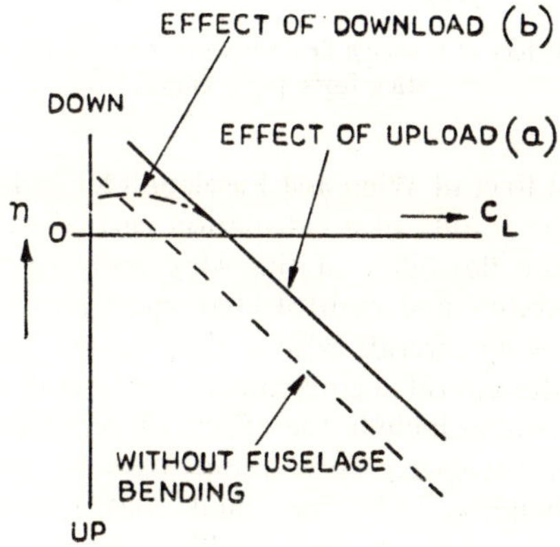

Fig. 14.7. Effect of fuselage flexibility on the elevator angle to trim curve

force acting on the wing of the aircraft, although it should be noted that elevator control has to be deflected in the upwards direction. This additional upload will depend on the g being pulled on the aircraft and it will cause the fuselage to bend upwards. This has the effect of reducing the tailplane incidence and assisting the pull-out (i.e. destabilizing). Thus the manoeuvre margins are decreased, both the stick movement per g and the stick force per g being reduced. The reduction in stick movement per g is constant throughout the speed range (Fig. 14.8*a*), whereas the reduction in stick force per g increases as the square of the forward speed (Fig. 14.8*b*). This reduction in SF/g due to fuselage flexibility may become serious at high speeds, particularly for aft C.G. positions.

In the above analysis only the effects of the aerodynamic loads have been considered. In Ref. 28 it is suggested that the effect of the inertia loads acting on the fuselage during the pull-out may more than counteract the effect of the aerodynamic loads and the overall manoeuvre stability may be increased.

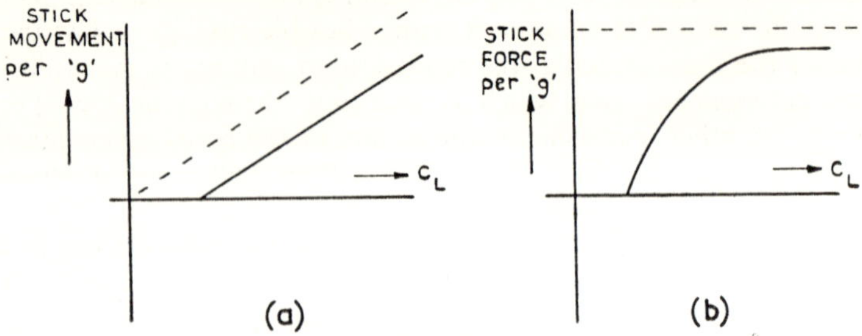

Fig. 14.8. Effect of fuselage flexibility on the stick movement and stick force per g curves

5.4. Combined Effect of Wing and Fuselage Flexibility

Ref. 17 gives the results of a calculation carried out to investigate the effects of the flexibility of the wing and the fuselage on the longitudinal stability and control characteristics of a hypothetical 35° sweptback wing aircraft with a rigid all-moving tailplane (the dimensions of the aircraft being similar to those of a Boeing B.47). The calculations also include the effects of compressibility and the results are given for speeds from about $M = 0{\cdot}4$ up to $M = 0{\cdot}8$ at two different heights, 1,000 mets and 10,000 mets. Graphs are given showing the separate and combined effects of distortion on the tailplane angle to trim, the tailplane angle per g and the manoeuvre margins, the results for the flexible aircraft being compared with those of the corresponding rigid aircraft.

The effect on the tailplane angle to trim curve is similar to that shown in Fig. 14.7, more nose-down control movement (i.e. more positive η_T) being required, the additional movement being constant throughout the speed range. The greatest effect is, however, due to the compressibility effects, the effects of the flexible wing and the flexible fuselage are relatively small and roughly equal to one another. These results are reasonably consistent with those already discussed in paras. 5.2 and 5.3; in the case of the flexible wing, the effect of the forward movement of the aerodynamic lift on the wing and the effect of the increased angle of attack on the tailplane tend to cancel one another out. The corresponding static margins have not been determined, since they tend to lose practical significance at high speeds.

The effect on the corresponding tailplane angle per g curves for the hypothetical aircraft is similar to that shown in Fig. 14.8*a*. The curves are moved parallel to one another, the effects of compressibility and flexibility requiring more nose-down tailplane movement, i.e. the nose-up control required is decreased and thus there is a loss of stability. Once again the main effect is that of compressibility, the effect of the flexible wing and the flexible tailplane are small and roughly equal to one another. However, these changes do mean that the tailplane angle required per *g* is roughly halved at the high speed end due to the combined effects of compressibility

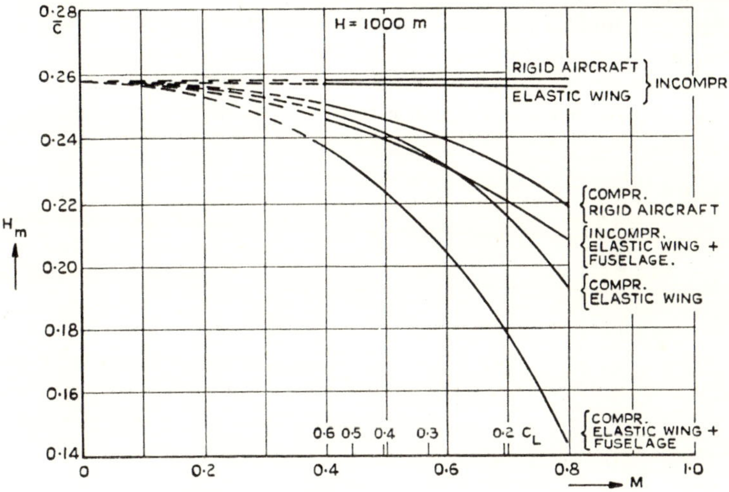

Fig. 14.9. Effects of aeroelasticity and compressibility on the manoeuvre stability [From Ref. 17]

and flexibility, the effect decreasing with decrease of speed. The corresponding manoeuvre margins are shown in Fig. 14.9, taken from Reference 17; the values decrease appreciably with increase in forward speed and the separate effects of compressibility and the distortional modes are indicated.

5.5. Effect of Tail Unit Distortion

The overall effect of tailplane and elevator twist is rather complicated; it depends on the relative stiffness of the tailplane and elevator, and the various elevator control surface characteristics (e.g. hinge moment characteristics, mass balance, etc.). It is convenient to consider the two effects separately

(i) the effect of a flexible tailplane with a rigid elevator,
(ii) the effect of a flexible elevator with a rigid tailplane.

(i) *Tailplane Twist*

The effect of the tailplane twist on the static stability depends on the elevator angle required to trim at a given speed; this can be varied in the case of an aircraft fitted with an adjustable tailplane. The change in stability is due to the forward speed changes involved. If the elevator is down for trim, then an increase in speed will increase the nose-down twist of the tailplane. This produces an additional download on the tailplane and this has the effect of producing a nose-up (i.e. a stabilizing) pitching moment, since it tends to return the speed of the aircraft back to its original value. If the elevator angle to trim is a negative one, a nose-up twist is produced on the tailplane and this results in a destabilizing contribution.

In addition to the effect on the speed stability discussed above, there is a further effect which is always a stabilizing one. To trim

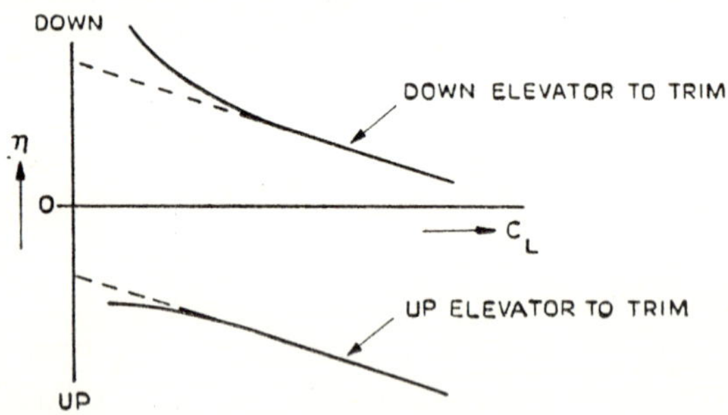

Fig. 14.10. Effect of tailplane twist on the elevator angle to trim curve

the aircraft out at a higher speed, it is necessary to move the elevator in the downwards direction. This produces an additional moment on the tailplane tending to twist it in the nose-down direction, thus reducing the effective tailplane incidence. If the elevator is assumed to be rigid, the angle of the elevator relative to the tailplane will be increased. Because of the reduction in tailplane incidence, there will be a loss of tailplane lift and effectiveness (the A_1 contribution), though some of this loss will be recovered due to the increase in elevator angle relative to the tailplane (the A_2 contribution). As a result of this apparent loss of control effectiveness, more down elevator will be required to trim the aircraft at the higher speeds and the effective stability is increased.

The overall trim curves are shown in Fig. 14.10. When down elevator is required to trim, the two effects discussed above act in the same way, both requiring additional down elevator to trim, the effect increasing with increase in forward speed. The overall effect is an increase in static stability. When up elevator is required to trim the two effects tend to counteract one another; if the effect of tailplane twist is greater than that of loss of control effectiveness, there will be a small reduction in the overall static stability of the aircraft. Tailplane twist is found to have similar effects on the stick-free static stability.

Rearward movement of the C.G. requires a larger upload on the tailplane to trim at a given speed and more down elevator. For a given change of speed the restoring moment will be greater than for the more forward C.G. position and the stabilizing contribution will be greater. This is opposite to the usual effect of C.G. movement, discussed in Chapter 2, which is destabilizing. If the tailplane is sufficiently flexible, there may be a critical speed above which the stabilizing contribution from the tailplane twist outweighs the normal destabilizing contribution from rearward C.G. movement. This result would be exactly the opposite of what one would normally expect.

In a pull-out at constant speed, the up elevator movement required causes an additional nose-up tailplane twist. This counteracts the nose-up pitching moment from the elevator and tends to bring the aircraft out of the manoeuvre. Thus tailplane twist has the effect of increasing the manoeuvre stability and hence increasing the corresponding stick movements and stick forces per g. Once again, the effects of the inertia forces might be important, causing either an increase or decrease in the manoeuvre stability.

(ii) *Elevator Twist*

The effects of elevator twist are rather complicated, depending on the mass balance and hinge moment characteristics of the control. In Ref. 26 the effect is discussed for a mass balanced control with $B_1 = 0$. If the elevator angle required to trim is a positive one, an increase in forward speed will tend to rotate the elevator back towards its neutral position. This produces an additional nose-up pitching moment on the aircraft and this is a stabilizing contribution. If the elevator angle to trim is negative, then elevator twist produces a destabilizing contribution. As in (i) above, in addition to this effect which may be stabilizing or destabilizing depending on the position of the elevator, there is an additional effect which is always stabilizing. This occurs because the elevator must be moved in the downwards direction to trim the aircraft out at a higher speed. The actual movement of the elevator will, however, be reduced due to its twisting and there will be an apparent loss in control effectiveness. This will necessitate increased stick movements to achieve the desired trim and an apparent increase in the stick fixed stability.

The final trim curves are similar to those shown in Fig. 14.10, except that, in this case, the elevator angle plotted vertically must be the angle equivalent to the measured stick movement in the absence of distortion. When the elevator is down for trim the two effects add on to one another, more down elevator is required to trim and the effect increases with speed. When the elevator is up to trim the two effects counteract one another. If the effect of speed stability is greater than that of the loss of control effectiveness, more up (or less down) elevator will be required at the higher speeds.

5.6. Overall Effect of Distortion

Other distortion effects (e.g. control surface distortion, control circuit stretch, etc.) are dealt with in detail in Ref. 26. The overall effect of distortion on the stability and control characteristics of an aircraft depends on which of the various modes is the dominant one. It is difficult to generalize the results, but the results are summarized as far as possible in the table below. It can be seen that the effects of wing and fuselage distortion are usually destabilizing ones and that the tailplane and elevator effects are usually stabilizing (particularly when the elevator is down for trim), but there are some exceptions to this generalization.

Table 1. Effects of Aeroelastic Distortion on Stability

TYPE OF DISTORTION	TRIM	STATIC MARGIN		MANOEUVRE MARGIN	
		STICK FIXED	STICK FREE	STICK FIXED	STICK FREE
Wing distortion	—	Increase (small)	Increase	Decrease	Decrease
Fuselage bending	—	Decrease	Decrease	Usually decrease	Usually decrease
Tailplane twist (elevator rigid)	η positive	Increase	Increase	Increase	Increase
	η negative	Decrease	Decrease		
Elevator twist (tailplane rigid)	η positive	Increase	Negligible effect	Increase	Negligible effect
	η negative	Decrease			

Ref. 18, first published in 1946, is an interesting report in that it gives the results of some flight tests carried out on an aircraft with an appreciable amount of distortion. The tests were performed at 8,000 ft over a range of speeds from 140 to 440 mph (E.A.S.) and measurements were made of the actual distortion of the structure of the aircraft and of the stick forces and elevator angles to trim and the stick forces and elevator angles per *g*, for various C.G. positions and power settings, and with two different tailplane settings.

The elevator angle and hinge moment to trim curves were analysed in this report to determine the variation of A_2 and B_2 and the corresponding static margins. The value of A_2 decreases rapidly with

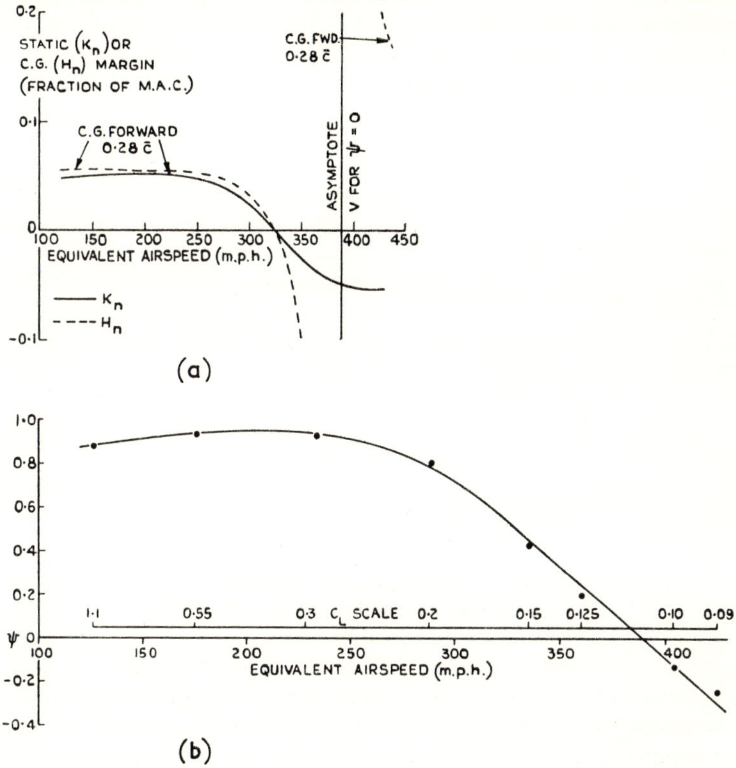

Fig. 14.11. Flight test measurements of the effects of aeroelasticity on static stability [From Ref. 18]

increasing forward speed; the effect of the various distortion modes are shown and one of the biggest effects is seen to be that of the flexible tailplane. The value of B_2, the control heaviness parameter, becomes rapidly more negative at the higher forward speeds, mainly due to the effects of the flexibility of the tailplane and elevator. The variation of the stick-fixed static margin is shown in Fig. 14.11*a*; for a C.G. position of $0{\cdot}28\bar{c}$, it remains fairly constant up to about

250 mph, but above this speed decreases rapidly, tending to zero at about 325 mph. The results were also analysed to obtain the numerical values of ψ and χ (see Chapter 2, para. 3.2). At low speeds the stick-fixed static margin and the C.G. margin are roughly equal to one another in accordance with simple theory (i.e. $\psi \simeq 1$). At higher speeds, however, ψ tends to zero at about 390 mph (Fig. 14.11*b*), and the C.G. margin tends to infinity, though the static stability remains finite. At the speed where $\psi = 0$, forward or rearward movement of the C.G. has no effect whatsoever on the stability of the aircraft. Under these conditions of flight, the concept of neutral points and C.G. margins lose their validity. At higher

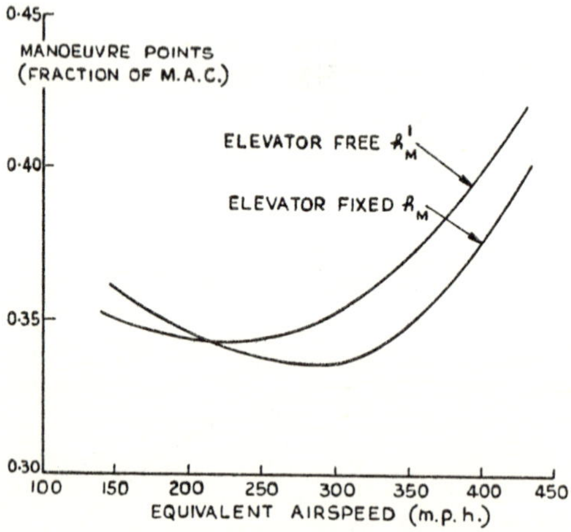

Fig. 14.12. Flight test measurements of the effects of aeroelasticity on manoeuvre stability [From Ref. 18]

speeds ψ becomes negative, and under these conditions of flight the effect of C.G. is reversed, rearward movement producing an increase of static stability. The stick-free static margins were also determined; the values increase initially with increasing forward speed, reaching a maximum value at about 300 mph, and decreasing again at the higher forward speeds. At the very lowest speeds the stick-free static and C.G. margins are approximately the same, again in accordance with simple theory (i.e. $\chi \simeq 1$). The values quickly diverge from one another however and χ decreases rapidly, becoming zero at about 290 mph. At this speed the C.G. margin becomes infinite, but the stick-free static stability remains finite, C.G. movement having no effect on stick forces to trim. At higher speeds, rearward C.G. movement increases the stick force values.

The elevator angle per *g* decreases initially with increasing forward speed, but begins to increase again at the top speed end. The stick force per g values remain constant initially but increase rapidly at high speeds. These results were also analysed to give the corresponding stick-fixed and stick-free manoeuvre points. The results are shown in Fig. 14.12, the manoeuvre points initially move forward slightly with increasing speed, up to about 300 mph, and then move rapidly back at the higher speeds.

Ref. 25 describes some flight test measurements made on the swept-wing tailless De Havilland 108 aircraft in order to investigate the effects of aeroelasticity on the stability and control characteristics at high forward speeds. Measurements were made of the magnitude of the wing bending and twisting involved and of the elevon and tab angles and the stick forces required to trim the aircraft over a range of speeds up to $M = 0{\cdot}98$ at 38,000 ft. A wing bending deflection of more than 0·4 in. was measured at the tip of the wing and a corresponding change of wing incidence of about 0·4 degrees.

6. Effects of Aeroelasticity on the Longitudinal and Lateral Dynamic Stability

When an aircraft is experiencing longitudinal or lateral oscillatory motion, changes occur in the aerodynamic forces acting on the various parts of the aircraft. These forces produce the static stability and the damping that are required to give satisfactory oscillatory characteristics. If the aircraft is a flexible one, the values of the derivatives will be modified by the structural distortions, which will generally be in phase with the motion of the aircraft. Under these conditions, the quasi-static values of the derivatives can be used. The natural frequencies of the aircraft structure depend on the design configuration, however, and if the design is such that the structural frequencies become comparable with the frequencies of the oscillatory motions (i.e. the longitudinal short period oscillation and Dutch Roll oscillation) of the rigid aircraft, then coupling can occur between the motions and this coupling will have significant effects on the stability characteristics of the aircraft. This problem will become more important as aircraft become larger, with consequent reduction in the natural frequencies of the structural modes (Ref. 11, para. 10.10). A review of the literature dealing with the effects of aeroelasticity on aircraft dynamic stability is given in Ref. 27.

Ref. 16 gives a comparison between the predicted and experimental longitudinal and lateral dynamic stability characteristics of a large flexible swept-wing aircraft (the Boeing B-47A) at an altitude of

35,000 ft, at speeds between $M = 0{\cdot}6$ and $M = 0{\cdot}85$. The aircraft was excited by pulse-type control inputs and the dynamic response of the aircraft was measured. The structural response of the aircraft was also measured at various stations on the wing and fuselage. Typical time histories are given in the report and the various structural modes discussed (e.g. wing bending modes, fuselage modes, etc.). The analysis of the longitudinal responses shows that the principal effect of flexibility is to decrease the natural frequency of the short period mode, the damping remaining almost unchanged. The predicted results are compared with those obtained from flight tests in Fig. 17 of the reference, and reasonable agreement is obtained. The effects of the flexibility of the aircraft on the Dutch Roll characteristics are shown in Fig. 29 of the reference. The natural frequency of the oscillation is slightly reduced, and the damping is appreciably decreased.

The effects of aeroelasticity on dynamic stability are also discussed in Ref. 9. One of the cases analysed is that of the dynamic stability characteristics of a delta wing aircraft with a canard control, for different amounts of fuselage flexibility, the calculations being performed for a cruise condition of $M = 3{\cdot}0$ at 60,000 ft. As the fuselage stiffness is reduced, the structural frequencies decrease, until, eventually, a value of stiffness is reached at which the structural frequencies become comparable with the frequency of the longitudinal short period mode. At very low values of fuselage stiffness, the damping of both the short period mode and one of the structural modes goes negative, indicating a dynamic instability.

The dynamic stability characteristics of an aircraft may be affected by the feedback of the aeroelastic distortion effects into the control system of the aircraft. This may produce unintentional control movements which will modify the oscillatory characteristics. In the Dutch Roll oscillation tests discussed in Ref. 19, the pilot kept his feet off the rudder pedals during the oscillation. Appreciable rudder movement was observed, however, which had to be allowed for in the analysis. The rudder motion was not due to a failure of the rudder jack to operate irreversibly, but was thought to be an aeroelastic effect resulting from fin distortion under sideslip loads. The rudder jack was at the base of the fin and fin distortion caused the valves to open and the jack to operate. The rudder operated in counter-phase with the sideslip angle and thus was in the sense to increase the effective value of n_v, while leaving the damping unchanged.

This problem of structural feedback into the control system may be particularly important when the aircraft is operating under automatic control. The sensors which are feeding signals into the control system are positioned in various places on the aircraft and structural distortion of the airframe will result in unintentional signals being fed back into the control system. The coupling between the distortion and the control surface movements could be such as to produce dynamic instability and, possibly, structural failure. This problem is discussed in Refs. 9, 23 and 39 (Chapter 11); it is particularly important in missile design and the sensor locations have to be carefully selected (Ref. 23). Other methods of eliminating the effects of structural feedback are mentioned in Refs. 9 and 39.

The elevator and rudder circuits of the Handley Page Victor incorporate "compensating levers" which compensate for the effects of temperature changes and fuselage distortion; they prevent spurious input signals reaching the power control units.

7. Flutter and Control Surface "Buzz"

7.1. General

The field of dynamic elasticity or unsteady aerodynamics is a very specialized one and it will not be discussed in any detail here. It is dealt with in the various standard textbooks (Refs. 1, 2 and 3), and in the Agard Manuals on Aeroelasticity (Ref. 12). A number of general reviews of the subject are given in Refs. 4, 5, 6, 7, 8 and 29, and these references contain large bibliographies of additional references. However, because the problems of flutter and buzz are important factors in determining the overall design of the aircraft, the control surfaces and the control system, a few of the basic principles will be described briefly.

In this work we are concerned with the oscillatory motions of various parts of aircraft and we are particularly interested in the conditions under which these oscillatory modes tend to become divergent (i.e. increase in amplitude), because this may result eventually in a structural failure of the aircraft. Some actual instances in which aircraft have been damaged and even destroyed because of flutter are described in Refs. 6 and 24. Ref. 5 also mentions a case of flutter that occurred on a jet fighter aircraft with tip tanks. Ref. 30 discusses a number of non-aeronautical problems of dynamic instability, including the well-known example of the destruction of the Tacoma Narrows Bridge.

The problem of flutter has been made worse on high-performance aircraft because of the design shapes that are used (e.g. thin wings, slender fuselages, etc.). It can be prevented by suitable design techniques; these include increasing the stiffnesses of the various component parts of the aircraft, arranging a suitable mass distribution, mass balancing the control surfaces, etc. A compromise is, of course, necessary in the design, since most of these methods result in an increase in the weight of the aircraft, which must always be kept a minimum.

7.2. Basic Principles of Flutter

Flutter is a particular type of resonance phenomenon, in which energy is extracted from the airstream in order to build up the amplitude of the oscillatory motion. Every physical system has its own natural frequency (or frequencies). Consider the simple spring-mass system dealt with in Chapter 7. If the spring is fixed to a stationary support and the mass disturbed and then allowed to oscillate freely, the resulting undamped oscillation can be represented by the equation

$$x = x_0 \cos \sqrt{\left(\frac{k}{m}\right)}\, t \tag{14.2}$$

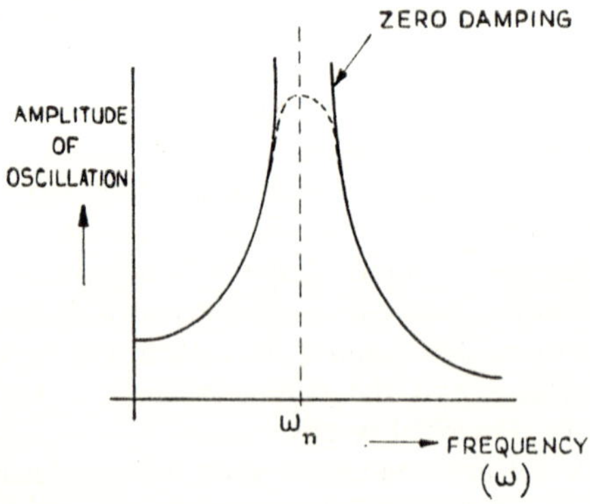

Fig. 14.13. Frequency response of a simple system

This system will always oscillate at the same frequency and this particular value is called the natural frequency ω_n of the system, where

$$\omega_n = \sqrt{\left(\frac{k}{m}\right)} \text{ rads/sec} \tag{14.3}$$

If the top support of the spring is made to oscillate with a constant amplitude at a frequency of ω, then the mass will oscillate at the same frequency. The amplitude of oscillation of the mass in this "forced oscillation" will vary with the frequency of the excitation. As the applied frequency is increased, the amplitude of oscillation of the mass increases, until it reaches a maximum when the forcing frequency is equal to the natural frequency of the system. At higher frequencies the amplitude of the mass again decreases. Fig. 14.13 shows the frequency response of this particular system, with the ratio of the output amplitude to the input amplitude plotted vertically and the applied frequency horizontally. We have assumed that there is no damping in the system and, in this case, the amplitude of the oscillation at the resonant frequency would be theoretically infinite. In practice, there must be some damping in the system and the resonant amplitude will be limited to a large but finite value.

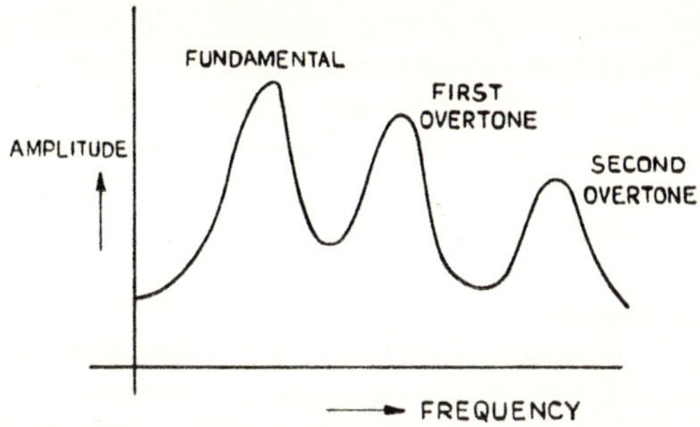

Fig. 14.14. Frequency response of a more complicated system

The frequency response of a second order system with damping is shown in Fig. 19.17. It should be noted that an increase in damping reduces both the resonant amplitude and the resonant frequency.

If we now consider the rather more complicated case of a uniform elastic beam, then, when it is subjected to forcing oscillations of varying frequency, it will exhibit a number of resonant frequencies

Fig. 14.15. Possible oscillatory modes of a uniform beam

corresponding to various possible oscillatory modes (Fig. 14.14). As the forcing frequency is increased, the first resonant frequency corresponds to the so-called "fundamental" frequency and a completely free beam will oscillate in the way shown in Fig. 14.15*a*. At a higher frequency, the second resonant condition (i.e. the "first overtone") will be produced and the beam will oscillate in a rather more complicated way (Fig. 14.15*b*). Further increase of frequency will produce second, third, etc., overtones. The "frequency spectrum" of any structural component is very important and tests must be carried out on the system to make sure that there are no forcing frequencies that would result in either a rapid or a slow structural failure of that particular component.

Finally, we must consider the aircraft itself. This is a very complicated elastic structure and the frequency response of this system has a large number of resonant peaks corresponding to the various oscillatory modes. In flight it is possible for these structural modes to be produced as a result of mechanical (e.g. propeller or compressor rotation), or aerodynamical (e.g. aerodynamic buffeting), forcing excitations. In certain circumstances, however, even in the absence of forcing excitations, it is possible for two of the structural modes to couple together and for the oscillations to become divergent (i.e. flutter). Occasionally, one degree of freedom oscillations may also become divergent (i.e. control surface "buzz").

7.3. Types of Flutter

Flutter is the oscillatory instability of the structure of an elastic body in an airstream involving coupling between two oscillatory modes, in which energy is taken from the airstream in order to build up the amplitude of the oscillation. Under certain conditions, this instability will be produced by the smallest disturbance. The amplitude of the oscillation will build up until it is limited by a non-linear effect, or until the structure breaks. Various types of flutter may exist; only two of the more familiar ones will be discussed here.

7.3.1. Wing Flexural-Torsional Flutter

The wing of an aircraft will oscillate when disturbed in bending or torsion, each of these oscillatory modes having a natural frequency determined by the bending stiffness and torsional stiffness, respectively, of the wing. At zero forward speed these two natural frequencies are generally appreciably different from one another. As the forward speed is increased, however, the aerodynamic effects modify both the stiffness and the damping of the oscillations. The bending stiffness is increased by the aerodynamic "stiffness" effects because the aerodynamic forces oppose the motion and the frequency of the bending oscillation is correspondingly increased. The torsional stiffness, on the other hand, is usually reduced by the aerodynamic effects, because the aerodynamic load acting on the wing tends to twist it in the direction to further increase the load (para. 3) and thus the torsional oscillation frequency is reduced. Because the aerodynamic centre, the flexural axis, and the centre of mass of the wing do not coincide with one another, aerodynamic-structural-inertia coupling effects are present. As a result of this, a speed is eventually reached at which the bending and torsion modes couple together and the amplitude of the oscillations becomes divergent. This is called the Critical Flutter Speed.

The two oscillations can only help one another to increase in amplitude if they are coupled together with an appropriate phase difference between the motions. One particular phase difference that enables the twisting oscillation to sustain the bending, and the bending oscillation to sustain the twisting, is shown in Fig. 14.16. Under these conditions the maximum nose-up twist occurs when the wing is moving upwards through the equilibrium position, and the maximum nose-down twist when the wing is moving downwards through the equilibrium position. It can be seen that the change in

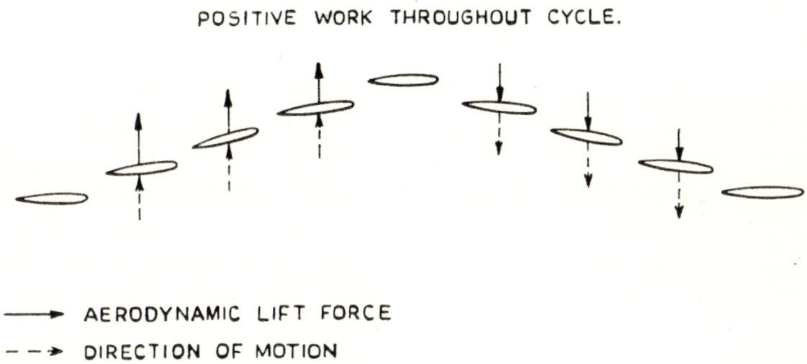

Fig. 14.16. Wing flexural-torsional flutter

the aerodynamic lift force always acts in the direction in which the wing is moving. Thus work is done on the wing and energy is removed from the airflow to enable the oscillations to increase in amplitude.

The phase difference can be shown to be a function of the forward speed. It is not necessary for the phase difference to be exactly that depicted in Fig. 14.16, for flutter to occur. In this case work is being done on the wing during part of the cycle and work is being done by the wing during the rest of the cycle. As long as the net work done on the wing is positive, the amplitude of the wing will increase.

7.3.2. CONTROL SURFACE FLUTTER

This type of flutter arises as the result of coupling between the rotation of a control surface and the bending or torsional motion of the wing, tailplane or vertical fin. As an example of this we will consider the case of wing flexural-aileron flutter, in which the motions involved are wing bending and aileron rotation. Once again, flutter is produced when these two motions couple together with an appropriate phase difference.

When a wing is oscillating in bending, then, if the C.G. of the control surface is behind the hinge line of the control, the inertia effects are such that as the wing moves upwards the control tends to

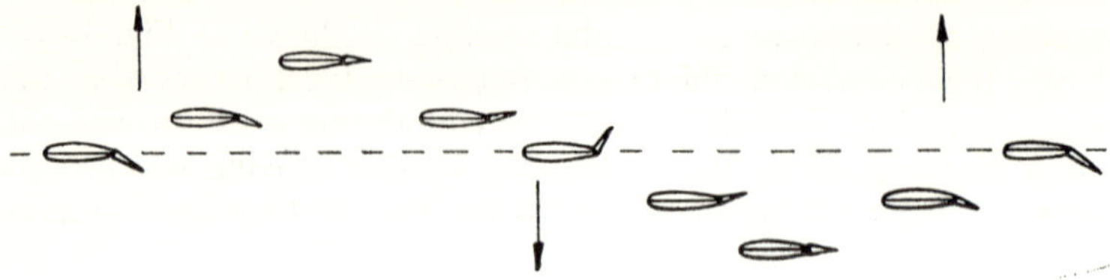

Fig. 14.17. Control surface flutter

rotate downwards, and as the wing moves downwards the control tends to rotate upwards. If the resulting phase difference (which is, in practice, itself a function of forward speed) is as shown in Fig. 14.17, then the motion of the aileron assists the bending motion of

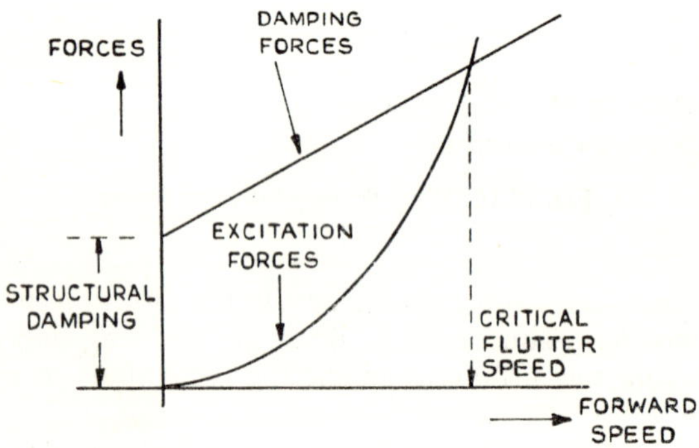

Fig. 14.18. Effect of forward speed on the excitation and damping forces

the wing. This motion is opposed by the elastic forces of the wing and the aerodynamic damping effects and at low forward speeds flutter does not occur. As the forward speed is increased, however, the aileron excitation forces increase as the square of the speed, whereas the aerodynamic damping effects only increase linearly with speed. Thus, above a certain critical speed, flutter will occur (Fig. 14.18).

7.4. Effect of Factors on the Flutter Characteristics

The flutter characteristics are dependent on the speed and height at which the aircraft is flying, on its configuration and structural design and on the characteristics of the particular type of control system used on the aircraft. Various methods are available for modifying the flutter characteristics of an aircraft.

The effect of Mach number on wing flutter is discussed in Ref. 5. As the Mach number is increased, the critical flutter speed initially decreases. On a straight wing aircraft, it reaches a minimum value at about $M = 1{\cdot}0$ and then increases again at supersonic speeds. Sweepback has the effect of increasing the value of the Mach number at which the minimum critical flutter speed is reached.

The mass distribution of the wing structure is an important factor and the critical flutter speed of a wing is sensitive to the position of its inertia axis (Ref. 4). Heavy concentrated masses (e.g. external stores, fuel tanks, etc.) also have an important effect on the wing flutter speed (Refs. 4 and 5). Ref. 31 shows how forward movement of the C.G. of the tip tanks can be used to increase the flutter speed.

Control surface flutter can be prevented by the addition of mass balance to the control surface. On some older aircraft designs the mass balance consists of a concentrated mass externally mounted on the control surface. Usually however, the mass is housed internally inside the structure of the aircraft and it consists either of a concentrated or a distributed mass. Often the mass is housed inside the horn balance or leading edge balance, which are used to aerodynamically balance the surface. The purpose of the mass balance is to move the C.G. of the control surface up nearer to the hinge line of the control. The effect on the flutter speed is shown in Fig. 14.19. In general there are two critical flutter speeds, an upper one and a lower one, of which, in practice, the lower value is the important one. As the amount of mass balance is increased, these two critical speeds come closer together and, eventually, a value of mass balance is reached at which the two speeds become equal to one another. A further amount of mass balance will eliminate the possibility of flutter. It can be seen from Fig. 14.19 that the amount of mass balance required to balance the control dynamically is usually less than that required for static balance. In practice, the problem is, of course, more complicated than this. It is possible to attach the mass balance in a way that makes the flutter more serious (Ref. 4, page 52) and, in some cases, too much mass balance may produce flutter. The problems of mass balancing are discussed in more detail in Ref. 32.

Another way of eliminating control surface flutter is by increasing the control circuit stiffness. This can be achieved with power operated control systems which, under ideal conditions, are irreversible (i.e. the force acting on the control surface is not fed back to the pilot). In practice there will always be some flexibility in the system and flutter is still possible. In some cases mass balance and viscous dampers are still used, though in other cases they are eliminated altogether. According to Ref. 4, the Saunders-Roe Princess flying boat was the first British aircraft to rely on the high circuit stiffness of the irreversible control system, without any mass balance for the prevention of flutter. The Vickers VC 10 has a power operated control system with no mass balance or dampers in the system (Ref. 8, page 362). In many cases, however, mass balance is still retained in the control system (see Chapter 12). In the case of

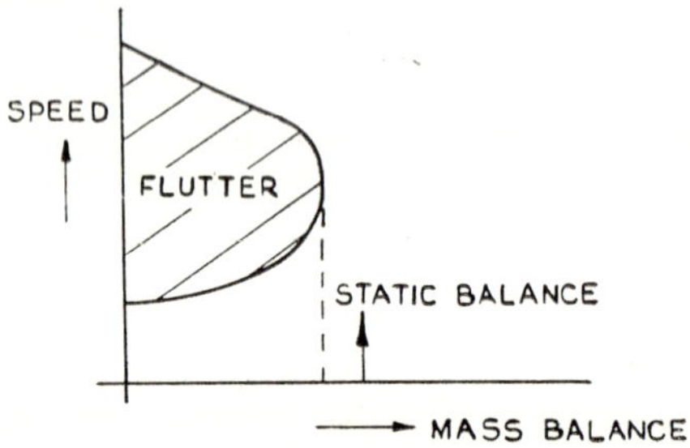

Fig. 14.19. Effect of mass balance on the critical flutter speed

the Handley Page Victor, the mass balances in the elevator and rudder control circuits are to overcome the inertia effects in the control runs which occur when the aircraft is accelerating or decelerating, and not for the prevention of flutter.

7.5. Determination of Flutter Characteristics

The flutter characteristics of an aircraft can be determined by various methods

(*a*) by theoretical methods, which may involve the use of analogue or digital computers,
(*b*) by carrying out wind tunnel and free-flight model tests,
(*c*) by carrying out ground resonance and flight flutter tests on the aircraft.

All these various methods are discussed in Ref. 4, 6, 8 and 33, and some experimental results are given in Refs. 34, 35 and 36. Ref. 37 describes the results of a comprehensive investigation of the determination of the fin flutter speed of the Handley Page Victor, which included flutter calculations, model tests and flight tests.

The flutter clearance programme is one of the earliest to be covered in the flight test programme of a new aircraft. A method commonly used to excite the oscillations in flight is by using small rockets called "bonkers." The techniques used are described in more detail in Ref. 38.

7.6. Control Surface Buzz

Buzz is a one degree of freedom flutter involving only rotation of the control surface. It is experienced at high subsonic and at supersonic speeds and was first observed in about 1945. Experimental work has shown that there are various types of instability and these are described in Refs. 5 (page 468), 7 (page 141) and 8 (page 364).

The first two types of instability are produced at subsonic speeds just less than $M = 1{\cdot}0$ and they are associated with the shock waves produced around the aerofoil and the control surface. The third type of instability occurs at low supersonic speeds, when the shock waves have reached the trailing edge, and this is due to the aerodynamic damping of the control becoming negative. Negative pitch damping can also occur on all-moving surfaces at supersonic speeds (Refs. 5 (page 470) and 7).

Mass balancing is not usually very effective against buzz and the methods most commonly used to eliminate the instability are

(i) by the addition of dampers to the control system (Ref. 8, page 365),
(ii) by an increase in the circuit stiffness.

Anti-buzz strips have also been fitted on the surface and at the trailing edge of the rudders of a number of aircraft. These strips modify the shock wave-surface rotation interaction patterns and when suitably positioned can eliminate the "buzz."

References

1. Bisplinghoff, R. L., Ashley, H. and Halfman, R. L.; *Aeroelasticity* (Addison-Wesley, Reading, Massachusetts, 1955).
2. Bisplinghoff, R. L. and Ashley, H.; *Principles of Aeroelasticity* (John Wiley & Sons, New York, 1962).
3. Fung, Y. C.; *An Introduction to the Theory of Aeroelasticity* (John Wiley and Sons, New York, 1955).
4. Broadbent, E. G.; *The Elementary Theory of Aeroelasticity*, Aircraft Engineering Monograph, 1954.
5. Broadbent, E. G.; "Aeroelastic Problems and High Speed Flight," *Journal of the Royal Aeronautical Society*, July, 1956.
6. Collar, A. R.; "Aeroelasticity—Retrospect and Prospect," *Journal of the Royal Aeronautical Society*, January, 1959.
7. Jones, W. P.; "Trends in Unsteady Aerodynamics," *Journal of the Royal Aeronautical Society*, March, 1963.
8. Hitch, H. P. Y.; "Modern Methods of Investigating Flutter and Vibration," *Journal of the Royal Aeronautical Society*, June, 1964.

9. Runyan, H. L., Pratt, K. G. and Bennett, F. V.; *Effects of Aeroelasticity on the Stability and Control Characteristics of Airplanes*, Agard Report No. 348, 1961.
10. Silverstein, A.; "Researches in Space Flight Technology," *Journal of the Royal Aeronautical Society*, December 1961.
11. *Agard Flight Test Manual*, Vol. II, "Stability and Control."
12. *Agard Manual on Aeroelasticity*, Vol. I; Introductory Survey.
13. *Design Requirements for Aircraft for the Royal Air Force and Royal Navy, Ministry of Aviation Av.P.*970, Volume I, Book I.
14. Acum, W. E. A. and Garner, H. C.; *The Estimation of Oscillatory Wing and Control Derivatives*, Agard Report No. 340, 1961.
15. *Agard Manual on Aeroelasticity*, Vol. II, "Aerodynamic Aspects."
16. Cole, H. A., Brown, S. C. and Holleman, E. C.; *Experimental and Predicted Longitudinal and Lateral-Directional Response Characteristics of a Large Flexible* 35° *Swept-Wing Airplane at an Altitude of* 35,000 *ft*, N.A.C.A. Rep. No. 1330, 1957.
17. Kalkman, C. M.; *The Influence of Aeroelasticity on the Longitudinal Stability of a Swept Wing Subsonic Transport Aircraft*, Agard Report No. 351, 1961.
18. Bisgood, P. L. and Lyons, D. J.; *Preliminary Report on the Flight Measurement of Aeroelastic Distortion in relation to its Effects on the Control and Longitudinal Stability of a Mosquito Aircraft*, R. & M. 2371, 1949.
19. Rose, R.; *Flight Measurements of the Dutch Roll Characteristics of a* 60 *Degree Delta Wing Aircraft* (*Fairey Delta* 2) *at Mach Numbers from* 0·4 *to* 1·5 *with Stability Derivatives Extracted by Vector Analysis*, R.A.E. T.N. Aero. 2752, A.R.C., C.P. No. 653, 1963.
20. Kussner, H. G.; "Aeroelastic Problems on Aircraft Construction," *Advances in Aeronautical Sciences*, Vol. II (Pergamon Press, 1959).
21. *Aircraft Engineering*, August 1965.
22. Gunston, W. T. (Ed.); *Flight Handbook 6th Edition* (Iliffe Books Ltd. 1962).
23. Prince, L. T.; *The Influence of Structural Elasticity on the Stability of Airplanes and Multi-Stage Missiles*, Agard Report No. 349, 1961.
24. Broadbent, E. G.; *Flutter Prediction in Practice*, Agard Rep. No. 44, 1956.
25. Zbrozek, J. and Mettam, A. R.; *Some Measurements of Stability and Aeroelastic Distortion on the* DH 108, R.A.E. Report Aero 2397, 1950.
26. Lyon, H. M. and Ripley, J.; *A General Survey of the Effects of Flexibility of the Fuselage, Tailunit, and Control Systems on Longitudinal Stability and Control*, R.A.E. Report Aero. No. 2065, R. & M. 2415, 1945.
27. Taylor, A. S.; "Present Status of Aircraft Stability Problems," *Journal of the Royal Aeronautical Society*, April 1959.
28. Taylor, A. S.; *The Present Status of Aircraft Stability Problems in the Aeroelastic Domain*, R.A.E. Tech. Note Aero 2538, 1957.
29. Templeton, H.; *A Review of the Present Position on Flutter*, Agard Report No. 57, 1956.
30. Rocard, Y.; *Dynamic Instability* (Crosby Lockwood, London, 1957).
31. Garrick, I. E.; *Aerodynamic Theory and its Application to Flutter*, Agard Report No. 34, 1956.
32. Templeton, H.; *Massbalancing of Aircraft Control Systems* (Chapman and Hall, London 1954).
33. Templeton, H.; *Flutter Research at the Royal Aircraft Establishment, Farnborough*, Agard Report No. 4, 1955.
34. Rainey, A. G.; *Interpretation and Applicability of Results of Wind Tunnel Flutter and Control Surface Buzz Investigations*, Agard Report No. 219, 1958.

35. MAZET, R.; *Some Aspects of Ground and Flight Vibration Tests*, Agard Report No. 40-T, 1956.
36. SCRUTON, C. and WINDSOR, E. P. L.; *Flutter Investigations in High Speed Tunnels*, Agard Report No. 222, 1958.
37. BALDOCK, J. C. A.; *The Determination of the Flutter Speed of a T-tail unit by Calculations, Model Tests and Flight Flutter Tests*, Agard Report No. 221, 1958.
38. *Several Techniques for Flight Flutter Testing*, Agardograph No. 56, 1960.
39. KOLK, W. R.; *Modern Flight Dynamics* (Prentice-Hall Space Technology Series, 1961).

15 Longitudinal Stability and Control Problems of High Speed Aircraft

1. Introduction

In previous chapters we have considered the longitudinal stability characteristics of aircraft at subsonic speeds. In this chapter we will consider the effects of increasing the forward speed of the aircraft through the high subsonic and transonic ranges and into the supersonic flight régime. Most of the work will be concerned with the compressibility effects associated with the high Mach numbers involved, though aeroelastic effects are also important when the corresponding E.A.S. values are high.

In order to fly at these high speeds, fundamental changes have to be made in the basic design shape of the aeroplane. An aircraft which has been designed to fly efficiently at a Mach number of 2, looks very different from one designed for a top speed of 300 knots. As well as considering the aerodynamic changes resulting from the increase in Mach number, it is essential to consider the effects arising out of these design changes. The configuration of an aircraft designed for low speeds usually consists of a high aspect ratio straight wing and a fuselage, with a horizontal tailplane and a vertical fin at the rear of the aircraft for the purposes of trim and stability. As speeds have increased, the wings have become smaller in area, with a swept or delta planform and a low value of aspect ratio. The weight of the aircraft is now concentrated in the fuselage, which has tended to become long and slender. These changes in configuration have altered the inertia characteristics of aircraft; the inertia in pitch and yaw becoming large compared to the rolling inertia. This has resulted in some important changes in the stability and control characteristics of aircraft. These problems will be discussed in more detail in later chapters. Inertia Coupling will be discussed in Chapter 16 and the problems associated with the very slender delta planforms in Chapter 21. In this chapter we will mainly restrict our considerations to those applicable to more-or-less conventional swept or delta planforms. The final configuration of a supersonic aircraft designed for a conventional take-off and landing must always be a compromise between the requirements of high-speed and low-speed flight. Variable wing sweep can be used in an attempt to obtain satisfactory characteristics at both ends of the speed range. If VTOL techniques are employed, however, the aircraft can be optimized for high speed flight, though the low speed aerodynamics of the aircraft are still of vital importance.

Aircraft designed for low speeds are usually fitted with non-powered control systems. With these systems, the control forces increase as the square of the forward speed. At the higher forward speeds the forces become excessive and powered control systems are essential. This considerably modifies the control force characteristics experienced by the pilot, since the forces are dependent on the control displacements and on the particular type of feel system incorporated. Stick-free static stability is not of any importance; both the control displacements and forces are now related to the stick-fixed stability.

There are many problems associated with high speed aircraft which we cannot discuss in this chapter. Some of these are performance problems. These can be solved by designing the aircraft to have a low drag at high speeds, and by installing sufficient engine thrust. The "sonic boom" is a problem which cannot be avoided with an aircraft flying at supersonic speeds. It is associated with the shock wave pattern produced on an aircraft flying at these high speeds. The effect on the ground is dependent on the size of the aircraft and the speed and height at which it is flying. The magnitude of the boom can only be significantly reduced by flying at a high altitude, since the size of the aircraft and the speed are usually fixed by other considerations. Any changes in the design shape of the aircraft have relatively little effect.

Many of the problems associated with aircraft designed for high speeds actually occur at the low speed end of the flight range, or at high angles of attack. This is because the design requirements are different for different conditions of flight. However, it is essential that an aircraft should have satisfactory handling characteristics in low speed flight and during take-off and landing. To improve the characteristics of the aircraft in these phases of flight, it may be desirable to incorporate some automatic stabilization, but the system must be safe and reliable, and the aircraft should still be controllable even when the automatics have failed.

The basic approach to the stability and control of an aircraft at high speeds is similar to that employed at low speeds. The equations of motion are derived, the appropriate values of the derivatives and the inertias substituted and the solutions (approximate or exact ones as required) are obtained. The derivatives are, however, functions of Mach number and their values will generally be different at supersonic speeds. It should also be remembered that approximations which are valid for low speed flight are not necessarily correct at supersonic speeds. In this speed range different assumptions may be necessary.

2. Design Characteristics of High Speed Aircraft

The final design of any aircraft is always a compromise between many conflicting requirements. It will depend on the type of aircraft being considered and the speed and height ranges over which it will be required to operate. In order to obtain satisfactory characteristics over the whole flight régime, it may be necessary to incorporate a number of special devices such as variable sweep, high-lift flaps, boundary layer control, and so on.

During the last twenty years there has been a tremendous increase in the speeds and heights over which aircraft have been operating. Before this, most aircraft had the conventional type of layout consisting of a moderately high aspect ratio unswept wing planform, with a relatively large thickness-chord ratio and cambered sectional shape and a load-carrying fuselage with a tail unit at the rear of the aircraft. This design was an optimum for the speeds then used because the large aspect ratio kept the vortex (or induced) drag down to a reasonably low value. It was also suitable for flying at fairly low speeds and the stalling characteristics were generally satisfactory. The stalling speed could easily be reduced by the use of high-lift devices (e.g. flaps or slots). The airflow pattern round this configuration was of the "classical" type, with vortices occurring at the wing tips of the aircraft and flow separation commencing at the trailing edge of the wing and moving forward as the angle of attack increased.

As aircraft speeds increased towards the speed of sound, large increases of drag were experienced (Fig. 15.1). This drag rise is associated with the formation of shock waves round the aircraft and it was found that the onset of these compressibility problems could be delayed to a higher forward speed by sweeping the wings either forwards or backwards, so reducing the component of the forward speed perpendicular to the leading edge of the wing. Because of the aeroelastic problems associated with sweptforward wings, sweptback wings are generally used, though these particular planforms also have a number of problems associated with them because of their wing-tip stalling characteristics. In addition to their sweep, wings designed for high subsonic speeds have much smaller aspect ratios and thickness-chord ratios and more symmetrical sectional shapes. The low speed performance of these aircraft is less satisfactory, but the stalling speed can still be kept down to reasonable values by the use of special high-lift devices.

When the flight speeds were increased into the low supersonic speed régime, it was necessary to increase the angle of sweep still further, in order to keep the component of the forward speed perpendicular to the leading edge less than the speed of sound. The wing t/c ratios also had to be made smaller and most of the load

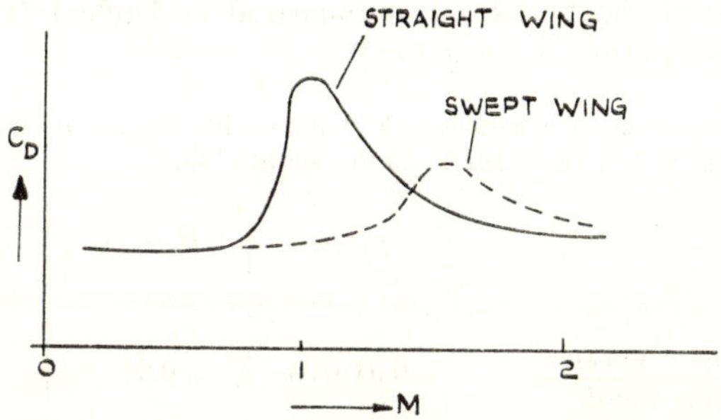

Fig. 15.1. Variation of drag coefficient with Mach number

(fuel, etc.) had to be carried in the fuselage. To keep the drag down, the fuselages were made long and slender, i.e. of a high fineness ratio. Associated with these changes in the aircraft's external configuration, there were large changes in the inertia characteristics; the inertia in roll becoming small compared to the inertia in pitch and yaw. Because of the reduction in the aerodynamic effects associated with flight at the high speeds and high altitudes involved, inertia coupling problems became much more important.

Some designers at this time were using the delta wing configuration for high subsonic speeds. This layout incorporates the sweepback on the leading edge of the wing that is necessary and it enables physically thick wings to be designed with quite low t/c ratio values (because of the large root chord associated with a delta wing). In this case, the fuselage can often be merged, at least partially, into the wing, with a corresponding reduction in drag. In addition, it is often possible to eliminate the horizontal tailplane, thus reducing the drag even farther. Whether the tailplane can be eliminated or not on any particular design is dependent on the requirements of the aircraft with respect to the trimming of any high-lift devices used, and the amount of stability and manoeuvrability required to fulfil its particular role.

At speeds above $M = 1\cdot5$ some of the advantage of sweepback is lost, because the zero lift wave drag of the straight wing decreases to a lower value of drag coefficient than that of the swept wing of the same t/c ratio (Fig. 15.1). The lift-dependent drag of this configuration will be large, however, and only a few aircraft (e.g. F.104, Bristol 188) have been designed with thin low aspect ratio unswept wing planforms. Very little space is available inside the wings, and most of the fuel and equipment in these aircraft has to be stored in the high fineness ratio fuselage.

The various design changes occurring from the time of the Spitfire to the time of the F-104 are summarized in Table 1 (taken from *Aircraft Engineering*, August 1957).

Table 1. Comparison of Characteristic Values with Regard to Geometry and Loading for the F-104, the F-102A and the Spitfire Mk. V

	F-104	F-102A	SPITFIRE
$\dfrac{\text{Wing max. thickness}}{\text{Wing chord}}$	0·03–0·04	0·04	0·12
Aspect ratio $\left(\dfrac{\text{Wing span}^2}{\text{Wing area}}\right)$	2·5	2·2	5·6
$\dfrac{\text{Fuselage length}}{\text{Wing span}}$	2·5	1·6	0·8
$\dfrac{\text{Fuselage length in front of C.G.}}{\text{Wing span}}$	1·6	1·0	0·23
$\dfrac{\text{Vertical tail arm}}{\text{Fuselage length in front of C.G.}}$	0·44	0·35	2·1
$\dfrac{\text{Vertical tail surface area}}{\text{Wing area}}$	0·21	0·11	0·06
Max. wing loading, lb/ft^2	75	45	25
$\dfrac{\text{Moment of inertia in roll}}{\text{Moment of inertia in pitch}}$	$\frac{1}{10}$	$\frac{1}{6}$	$\frac{3}{4}$

The very thin unswept wing planform does, however, have a number of other disadvantages associated with it. To be reasonably efficient at the high speed end of the range, it must have a very small t/c ratio and have a sharp leading edge. This leads to problems at low speeds; the flow separates from the leading edge of the wing at a fairly low incidence and reattaches again near the trailing edge forming what is known as a "long bubble." On the Bristol 188, strakes were added to the leading edge of the wing adjacent to the fuselage and engine nacelles to increase the incidence at which this separation occurs. The stalling speed and landing speed associated with this configuration are high. At the high speed end of the range, the drag of the aircraft will be increased due to interference effects between the fuselage, the wing, and the engine nacelles.

If the design requirement is for an efficient long range aircraft, cruising at a supersonic speed of $M = 2$ or 3, then the optimum configurations are of the highly swept or slender delta planforms. That no one shape is the best, even for this particular requirement, can be seen from the various existing design studies put forward for the supersonic transport. The B.A.C./Sud Concorde aircraft will be a slender delta aircraft, as are two of the early American proposals (i.e. the North American and Lockheed designs). The Boeing design, on the other hand, is for a highly swept wing planform incorporating variable sweep.

In order to maintain a subsonic leading edge at $M = 2$, a leading-edge sweep of 60° to 70° is required. This large sweep results in important changes in the flow pattern round the wing and the so-called "classical" flow pattern no longer exists. The flow separates from the wing at very low angles of attack and separated flow has to be accepted even for the cruise condition of flight. However, to ensure that the characteristics of the flow and the corresponding aerodynamic forces vary smoothly with changes in incidence, it is essential that the separation should be controlled. It is undesirable for the separation line to lie partly on the top surface and partly on the bottom surface—this would imply vortices crossing the leading edge from the upper to the lower surface. The wing has to be designed so that the separation takes place only on one surface. Thus there must be one angle of attack (called the attachment angle) at which no separation occurs. At any lower incidence than this, the flow separates from the bottom surface, at any higher incidence the flow separates from the top surface. This is achieved by using a wing with a very sharp leading edge, separation then occurs above a degree or two of incidence. The aerodynamics of slender delta wings will be discussed in more detail in Chapter 21.

For speeds much above $M = 3$ it will not be possible to increase the leading edge sweep in order to maintain a subsonic leading edge, since angles of sweep approaching 80° would be required. This would lead to too slender a wing planform, with serious problems at low speeds. Research on the optimum configurations for hypersonic speeds ($M > 5$) is being carried on at the present time.

3. Longitudinal Stability and Control Derivatives

Stability and control problems of aircraft are analysed using the concept of derivatives. In the introductory chapter on derivatives (Chapter 6), it was pointed out that they could be expressed in terms of the ordinary aerodynamic forces (e.g. lift and drag) acting on the aircraft, and that, since the aerodynamic effects vary with forward speed the derivatives are also functions of Mach Number. As a result of this the numerical values of the derivatives will vary considerably throughout the speed range. Basically the analysis is no different from that at subsonic speeds, since the same derivatives are involved. The only difference is that some of the derivatives which were important at subsonic speeds are not now so important, and others which were zero or negligible at subsonic speeds have now become large and significant.

Contributions to the overall values of the derivatives come from all the various parts of the aircraft—the wing, the fuselage, the tail unit, etc. On a conventional layout, the most important contribution to the longitudinal derivatives comes from the horizontal tailplane. On a tailless layout, however, obviously the wing and fuselage contributions are the dominant ones.

Numerical values of the derivatives can be obtained by theoretical, experimental (wind tunnel and free-flight model techniques) and flight test methods. Initial estimates are made in the early design stages of an aircraft and these values are compared later with those determined from flight tests. The corrected values are then used in calculations and on analogue computers and simulators to extend the "flight" envelope of the aircraft into possible danger areas.

Theoretical methods of determining the values of the derivatives are discussed in various references (e.g. Ref. 1). Many factors influence the values of the derivatives (e.g. wing sweepback, shock wave-boundary layer interaction, etc.), and some of these effects are not very amenable to theoretical treatment. Many modern designs have very low aspect ratio wings and large fuselages, and the interference effects between them are very important. Some of the methods give reliable estimates for the values of the derivatives, though the methods tend to break down in the transonic flight régime.

3.1. Stability Derivatives

The values of the derivatives at supersonic speeds can be derived from similar formulae to those used at subsonic speeds. Only the more important derivatives are discussed below.

(a) $dC_L/d\alpha = a$, *the lift curve slope*

The lift curve slope is a very important basic parameter. The values for the wing and the tailplane are factors in many of the formulae for the longitudinal derivatives and that for the vertical fin comes into the expressions for a number of the lateral ones. The value varies throughout the speed range; it increases gradually at subsonic speeds (in accordance with the Glauert Law) and decreases gradually at supersonic speeds (in accordance with the Ackeret Law). In the transonic zone it may reach a peak value at about $M = 1$, or it may decrease abruptly, and then increase again, before finally decreasing again at supersonic speeds (Fig. 15.2). The exact

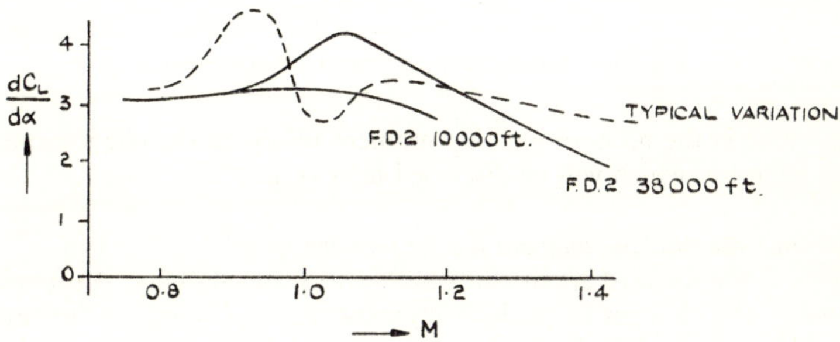

Fig. 15.2. Variation of lift curve slope with Mach number
[From Ref. 7]

shape of the $dC_L/d\alpha$ curve depends on the way that the shock waves form round the wing and the way in which the shock wave pattern changes as the forward speed is increased. Some actual results for the F.D.2, taken from Ref. 7, are also shown in Fig. 15.2.

(b) $m_\alpha(m_w)$, *the pitching moment due to incidence*

This is an important derivative in the consideration of static stability which is discussed more fully in para. 4. On the simple theory of Chapter 2, $-dC_M/d\alpha$ (or $-dC_M/dC_L$) was equal to the static margin, but in the more general theory the static stability must include not only an incidence term but also a speed term.

The value of this derivative is dependent on the position of the wing aerodynamic centre. Rearward movement of the A.C. increases the value of the static stability with respect to incidence and hence the value of m_α. On a wing section, an increase in speed from a subsonic to a supersonic value, moves the A.C. rearwards from the $\frac{1}{4}$ chord to the $\frac{1}{2}$ chord point. The corresponding A.C. movement on a finite wing depends upon the planform of the wing, but on a slender delta wing the movement may be as low as 7 or 8 per cent of the root chord.

At low angles of attack the C_M/C_L curve is a straight line, but at higher incidences the curve often becomes non-linear. An abrupt

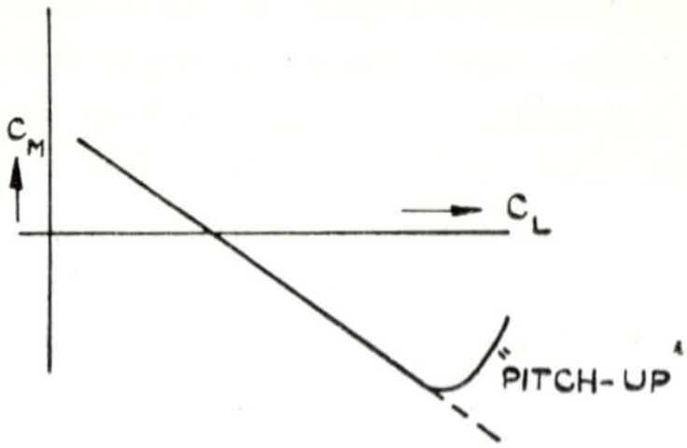

Fig. 15.3. Typical pitching moment characteristics

increase in the nose-up pitching moment results in the phenomenon of pitch-up, which will be discussed later (Fig. 15.3).

(c) m_u, the pitching moment due to forward speed

This is the forward speed contribution to static stability. In simple theory the changes in pitching moment due to changes in forward speed were neglected, i.e. m_u was assumed to be zero. At higher

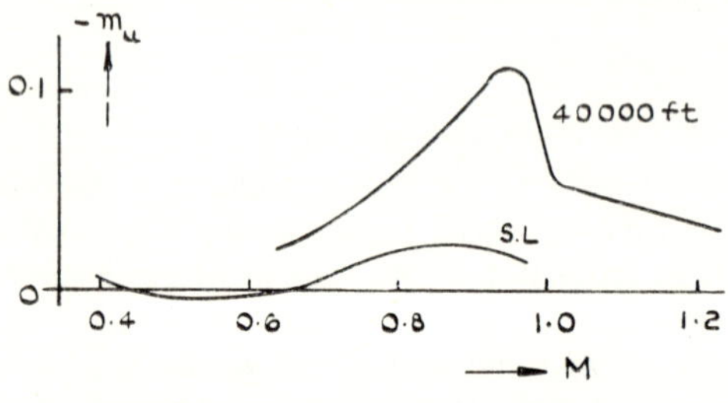

Fig. 15.4. Variation of $-m_u$ with Mach number

Mach Numbers this assumption is no longer valid and the effects of speed changes have to be taken into account. Positive values of m_u are stabilizing, negative values indicate "tuck-under" effects. The variation of m_u for a particular fighter-type aircraft is shown in Fig. 15.4.

(d) m_q, the damping in pitch

This derivative represents the pitching moment produced due to the pitching motion of the aircraft. In the case of a conventional aircraft, fitted with a horizontal tailplane, this contribution is usually the predominant one in the overall damping of the longitudinal pitching oscillations of an aircraft. When the tailarm is small, or when the aircraft has no separate tailplane, the value of m_q may be relatively small and other damping contributions become important. Typical variations of m_q with Mach Number are shown

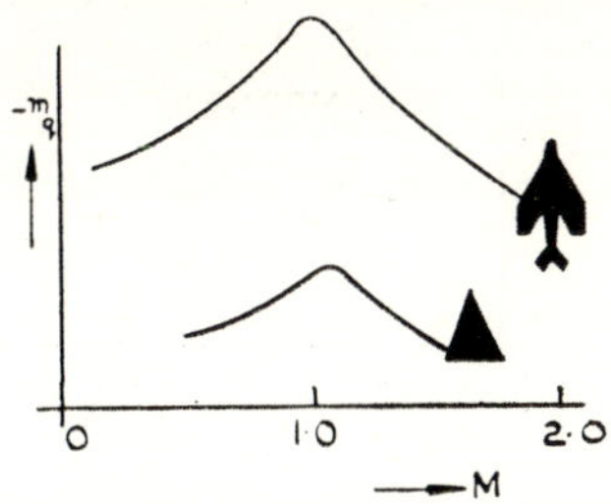

Fig. 15.5. Variation of $-m_q$ with Mach number

in Fig. 15.5. On some designs the damping in pitch may be small in the transonic speed régime.

(*e*) $(m_{\dot\theta} + i_B z_{\dot w})$, *the total pitch damping derivative*
The overall pitch damping derivative can be expressed as the sum of the total moment derivative $m_{\dot\theta}$ and the so-called "plunging" component derivative $i_B z_{\dot w}$, where the total moment derivative is given by

$$m_{\dot\theta} = m_q + m_{\dot w} \tag{15.1}$$

This overall derivative relates the damping of the pitching oscillation of an aircraft to its rotation in pitch q and its vertical motion w and acceleration $\dot w$. The $\dot w$ term exists because of interference effects between the different parts of the aircraft. One important contribution to this term is the "downwash lag" effect; this arises because the downwash at the tail lags behind the lift on the wing. At transonic speeds the value of $m_{\dot\theta}$ may go positive (i.e. unstable) on some designs. In these cases only the effect of a negative z_w keeps the pitching oscillations of the aircraft from becoming divergent (Fig. 15.6).

3.2. Control Derivatives

The main longitudinal control derivatives are x_η, z_η and m_η. The derivative x_η represents the drag produced by control surface deflection, z_η the corresponding lift and m_η the resulting pitching moment about the C.G. of the aircraft. The main requirement is for m_η, but all-moving surfaces and large elevator surfaces may produce appreciable values of z_η. These "lift" forces produced by the elevator control may lead to problems of "trim drag" on supersonic aircraft.

The derivative m_η represents the control effectiveness. At low speeds longitudinal control is obtained by flap-type elevators. The

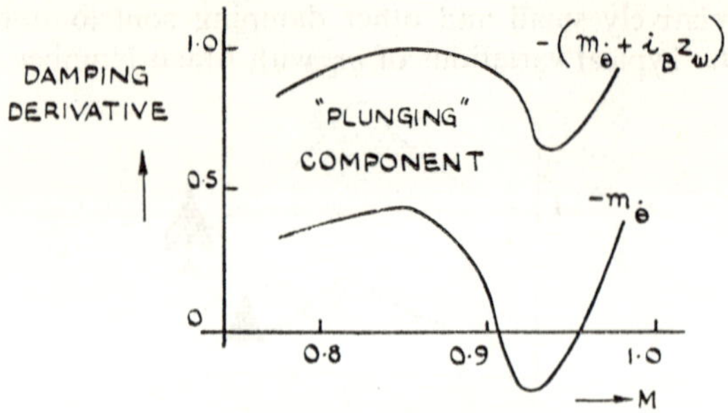

Fig. 15.6. Typical pitch damping characteristics

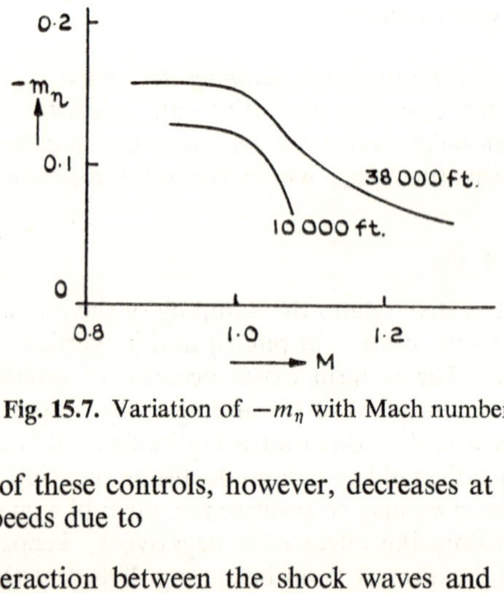

Fig. 15.7. Variation of $-m_\eta$ with Mach number

effectiveness of these controls, however, decreases at transonic and supersonic speeds due to

(i) the interaction between the shock waves and the boundary layer, which produces a turbulent wake of air surrounding the control surface,

(ii) the effect of the control deflection on the pressure distribution being limited to that round the control surface itself. The variation of the elevator power of the F.D.2 is shown in Fig. 15.7.

The loss in elevator power is greater on elevators with large trailing-edge angles, and the effectiveness can be increased by using small trailing-edge angles (Fig. 15.8). The same effect can be achieved with blunt trailing-edge control surfaces, since these are effectively

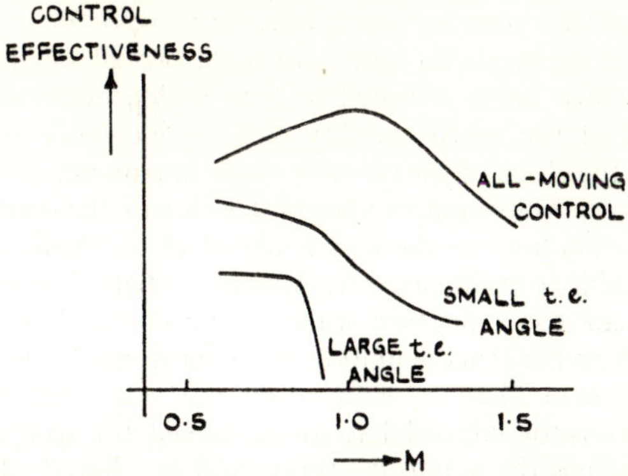

Fig. 15.8. Effect of trailing edge angle on control effectiveness

surfaces with small trailing-edge angles, but with the rear part of the surface removed. Longitudinal control at high Mach Numbers can be further improved by using an all-moving tailplane. The

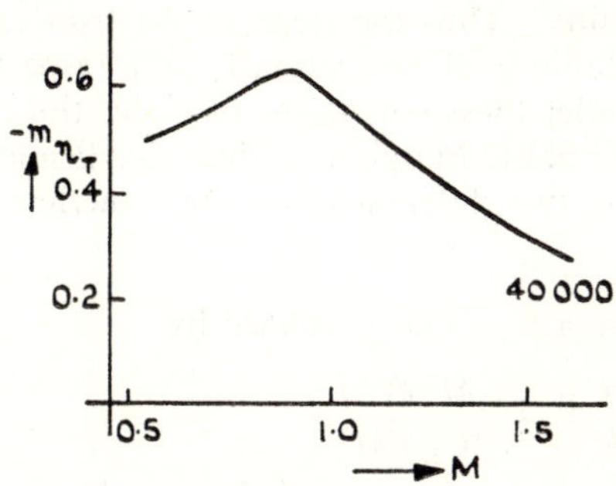

Fig. 15.9. Variation of $-m_{\eta T}$ with Mach number

variation of the tailplane effectiveness $m_{\eta T}$ with Mach Number for a fighter-type aircraft is shown in Fig. 15.9.

4. Longitudinal Trim and Static Stability

4.1. Static Margin and Restoring Margin

In the earlier work of Chapter 2 it was shown that the elevator angles required to trim an aircraft at different forward speeds in level flight were related to the amount of longitudinal static stability possessed by the aircraft. At low speeds the change in pitching moment with forward speed (i.e. m_u) could be neglected and the static stability was analysed as a pure incidence effect (i.e. m_α or m_w). In general, however, it is necessary to take both these two effects into account, and the elevator angle required to trim is dependent on the combined effect of the speed stability and the incidence stability. At low speeds, the slope of the elevator angle to trim curve is a measure of the static stability and the Neutral Point of the aircraft can be determined. As long as the C.G. is ahead of the Neutral Point the aircraft is stable to incidence disturbances. Whether or not the slope of the trim curve is a measure of the "static stability" at high speeds is dependent on the exact definition of the terms used. As long as the pitching moment changes with speed are small the direction in which the elevator control has to be moved for trim will not be changed, though the actual deflection will be altered. If the m_u is negative it has a destabilizing effect and if it increases numerically with speed, then a speed may be reached at which the pitching moment changes associated with the speed changes are greater than those associated with the incidence changes and the elevator angle to trim is reversed in direction. Though the aircraft may still be stable to incidence disturbances, the elevator angle curve indicates a form of instability. Thus the slope of the trim curve no longer represents the stability of the aircraft with respect to incidence disturbances. Under these conditions of flight, the concepts of the Neutral Point and Static Margin lose their significance.

For this reason, two different terms are sometimes used, defined as follows

(i) *Static Margin K_n* This is defined by

$$K_n = -\frac{dC_M}{dC_L} + \frac{M}{2C_L}\frac{dC_M}{dM} \tag{15.2}$$

The first term on the right-hand side of this equation represents the incidence stability and the second term the speed stability. On the basis of this definition, the static stability represents the overall stability of the aircraft and it is related to the slope of the elevator angle trim curve.

(ii) *Restoring Margin K_m* This is defined by

$$K_m = -\frac{dC_M}{dC_L} \tag{15.3}$$

and thus only represents the incidence stability contribution.

Unfortunately, in practice, this latter term is not used very much, and the expression "static stability" is used very loosely to cover both these forms of stability. Thus, whenever static stability is referred to, it is essential to make sure whether the term is being used to include both the speed and incidence effects, or only the incidence effects. As we have mentioned, it is quite possible for an aircraft to be "stable" with respect to incidence changes, but "unstable" with respect to combined speed and incidence effects. Those people who use the term static stability to refer to the incidence effects only, describe the additional effect as a change in trim with increasing speed or Mach Number.

4.2. Trim Changes at High Mach Numbers

The basic longitudinal trim equation for a conventional aircraft is given in eqn. (2.6) of Chapter 2

$$C_M = 0 = C_{M0} + (h - h_0)C_L - \bar{V}(A_1\alpha' + A_2\eta + A_3\beta) \quad (15.4)$$

where $\alpha'(= \alpha - i - \varepsilon)$ is the local tailplane incidence. For a given C.G. position h and control deflections η and β, the factors affecting the trim of the aircraft can be seen to be

(i) C_{M0}, the wing zero-lift pitching moment,
(ii) C_L, the wing lift coefficient,
(iii) h_0, the position of the wing aerodynamic centre,
(iv) ε, the downwash angle at the tailplane,
(v) A_1, A_2, A_3, the tailplane lifting effectiveness due to tailplane incidence, elevator deflection and tab deflection, respectively.

These factors are all functions of Mach Number, though the way in which they vary is very dependent on the particular speed range being considered and the design of the aircraft. The changes in trim may be nose-up or nose-down ones, the overall effect at any speed being the sum of the separate effects. It is difficult to generalize the overall effects completely, and we will first of all discuss each effect separately.

(i) C_{M0}, *the wing zero-lift pitching moment*

At low speeds, the value of C_{M0} is a constant, depending on the wing sectional shape, and the twist and sweep of the planform. A cambered wing section has a negative value of C_{M0}, though this can be made positive by the use of a reflex camber at the trailing edge. The overall value of C_{M0} for the aircraft must be positive, and this is usually achieved by fitting a horizontal tailplane with longitudinal dihedral.

As the forward speed is increased at low subsonic speeds little change occurs in the shape of the pressure distribution round the wing. The angle of attack for zero-lift remains constant and the value of C_{M0} changes only gradually in accordance with the Glauert Law. When the speed is reached at which shock waves form on the wing, large and abrupt changes can occur in the pressure distribution round the wing and in the corresponding values of the C_{M0} (Fig. 15.10) and the zero-lift angle (Fig. 15.11). The value of C_{M0} becomes

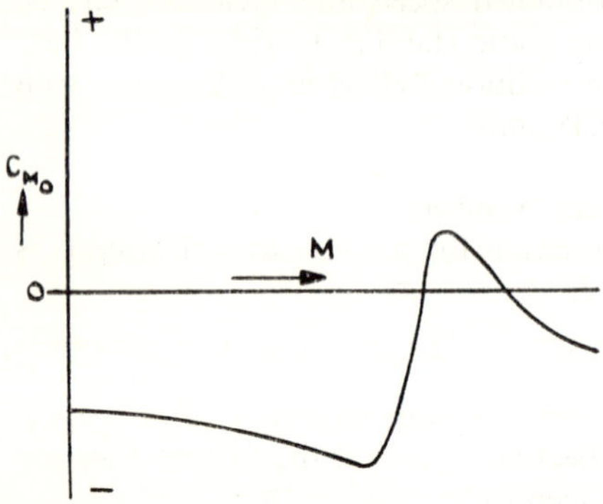

Fig. 15.10. Zero-lift pitching moment characteristics

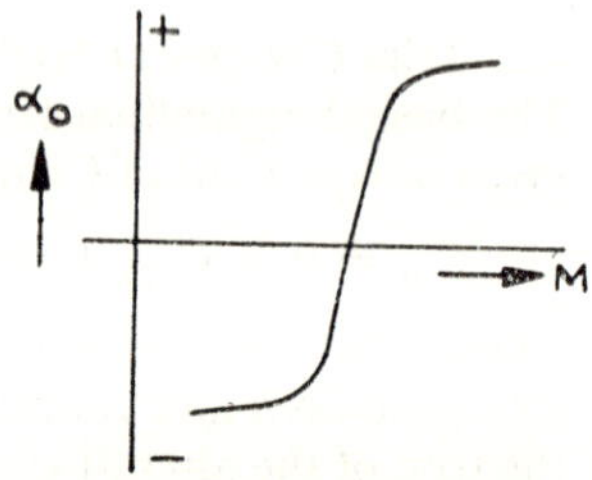

Fig. 15.11. Effect of Mach number on the no-lift angle

less negative, and may even become positive. From the C_{M0} effect alone a nose-up change of trim is produced. Following the abrupt change at transonic speeds, the value of C_{M0} usually decreases slowly at supersonic speeds, back to a negative value.

(ii) C_L, *the wing lift coefficient*
At the higher Mach Numbers, the slope of the wing lift curve decreases (Fig. 15.2) and the no-lift angle becomes positive (Fig. 15.11). Thus a particular value of C_L will require a greater wing incidence. This will result in a corresponding increase in the incidence of the tailplane and (assuming for these considerations that there are no changes in the downwash angle or the tailplane effectiveness) a *nose-down* trim change results.

(iii) h_0, *the position of the wing aerodynamic centre*
The aerodynamic centre is the point at which the resultant lift force due to incidence effects can be considered to act. At subsonic speeds it is around the $\frac{1}{4}$ chord point on a wing section or on a rectangular wing planform. As the forward speed is increased, the A.C. moves gradually forward from the $\frac{1}{4}$ chord point. This causes a small *nose-up* change in the pitching moment.

At supersonic speeds, however, the pressure distribution round an aerofoil section is much more symmetrical, and the A.C. is round about the $\frac{1}{2}$ chord point. Theoretically on biconvex and double-wedge sectional shapes it would be at the $\frac{1}{2}$ chord point, but because of boundary layer effects it is usually round about the 45 per cent chord position. Thus in going from subsonic to supersonic speeds the overall movement of the A.C. must be a rearward one (Fig. 15.12).

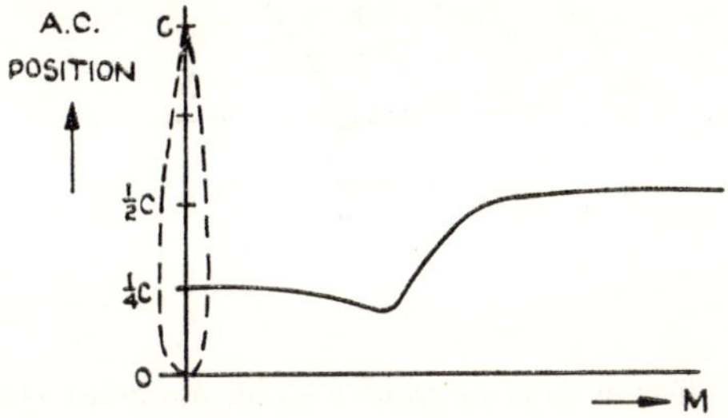

Fig. 15.12. Effect of Mach number on the position of the aerodynamic centre

This rearward movement takes place in the transonic speed range, and causes a large *nose-down* change in pitching moment.

The movement of the aerodynamic centre is one of the most important factors producing the trim changes. It can be reduced by suitably designing the aircraft. As we have seen, on a slender-delta configuration the movement between subsonic and supersonic speeds, expressed as a percentage of the chord, is small, though the actual physical distance may be quite large. The large nose-down moment produced requires a large up-elevator movement for trim, with the associated disadvantages of trim drag. In the design of supersonic transport aircraft, it is necessary to reduce the A.C. movement and the corresponding trim drag to as low a value as possible. To counteract the A.C. movement, a method of fuel transfer may be used to move the C.G. back as the A.C. moves back, in order to keep the distance between them small. This system is used on the B58 Hustler and is also proposed for the Concorde.

(iv) ε, *the downwash angle at the tailplane*

At low speeds the downwash from the wing is directly proportional to C_L, the wing lift coefficient. As the forward speed is increased into the transonic régime, the formation of shock waves produces a change in the spanwise lift distribution, and, hence, in the wing downwash pattern. Depending on the shock wave pattern, the downwash may be either increased or decreased. If the downwash angle is reduced, the effective angle of attack of the tailplane will be increased, and this will produce a *nose-down* change in pitching moment.

At supersonic speeds the deflection of the flow around an aerofoil section is confined to the region lying between the shock wave and

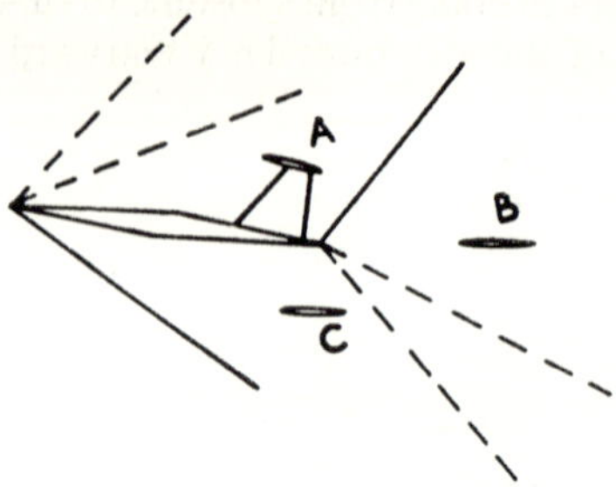

Fig. 15.13. Effect of tailplane position on the downwash characteristics

expansion region régimes at the leading and trailing edges of the wing. On simple theory, the downwash behind a supersonic wing section is zero. In practice, there may be a small upwash. On a finite wing, the downwash is confined within the Mach cones at the wing tips. As long as the tailplane lies outside the Mach cones, the

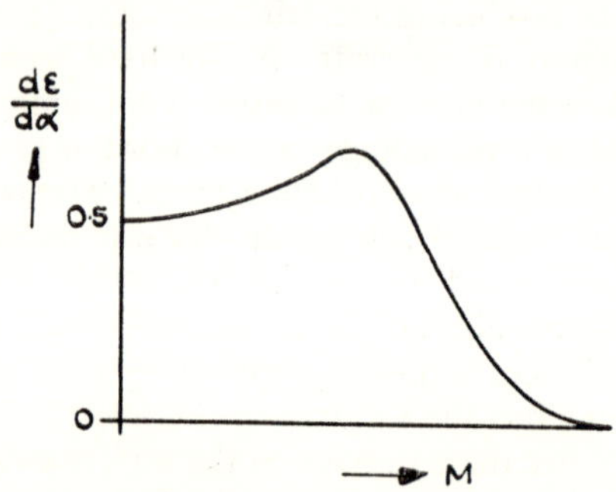

Fig. 15.14. Variation of $d\varepsilon/d\alpha$ with Mach number

downwash will decrease to zero at supersonic speeds. Thus from subsonic to supersonic speeds there will be a reduction in downwash and a *nose-down* change in pitching moment.

The value of $d\varepsilon/d\alpha$ is an important factor in considerations of stability. At subsonic speeds the value is about $\frac{1}{2}$. The value under supersonic conditions depends on where the tailplane is positioned with respect to the flow pattern round the wing. In positions *A* and C on Fig. 15.13, $d\varepsilon/d\alpha = 1{\cdot}0$ and the tailplane produces no stabilizing contribution. At B, $d\varepsilon/d\alpha = 0$ and the tailplane has a stabilizing effect. For a layout with a tailplane behind a wing of reasonable aspect ratio, the value of $d\varepsilon/d\alpha$ will decrease from about 0·5 at subsonic speeds to zero at a high supersonic speed (Fig. 15.14).

(v) A_1, A_2, A_3, *the tailplane lifting effectiveness derivatives*
The values of these derivatives will vary with Mach Number. The value of A_1, the tailplane lift curve slope, will vary in a similar way to the wing lift curve slope discussed in para. 3.1*a*. Any increase in A_1 will produce a larger moment from the tailplane, and the stabilizing contribution will be increased. The derivative A_2 represents the control effectiveness which has also been discussed in para. 3.2. The control effectiveness determines the amount of control deflection required to trim out the various pitching moments involved.

In passing through the complete speed range from subsonic to supersonic values, there must be an overall *nose-down* trim change, due mainly to the rearward movement of the aerodynamic centre and the reduction in downwash. Initially, at subsonic speeds, only small, gradual, changes of trim will occur with increasing forward speed. These trim changes will be associated with the changes in the wing and tailplane lift curve slopes, in the A.C. movement, and in the changes in the downwash characteristics. In the transonic régime the trim changes may be large and abrupt and may involve both nose-up and nose-down changes of trim. The aircraft may be difficult to fly under these conditions, because

(*a*) small speed changes may result in large trim changes,
(*b*) large and abrupt changes in the control movements and forces may be required. Over a certain speed range, the forces and movements may have to be reversed.

4.3. Elevator (or Tailplane) Angle and Force Required to Trim
Most of the aircraft designed for high speed flight are fitted with power-operated controls. In this case the longitudinal control deflections and forces are both related to the stick-fixed static (speed and incidence) stability of the aircraft, though the forces are also dependent on the particular type of feel system incorporated in the aircraft. The variation of the basic pitching moment characteristics are shown in Fig. 15.15. The curves of C_M against C_L at constant Mach Numbers represent the incidence stability contribution, and the slopes of these lines increase at the higher Mach Number values to show the increased incidence stability resulting from the rearward

movement of the A.C. The lines are also spaced vertically, moving downwards at the higher Mach Numbers to represent the nose-down change in trim which occurs at these speeds.

A curve showing the variation of the pitching moment with forward speed in level flight is superimposed on this figure. At low forward

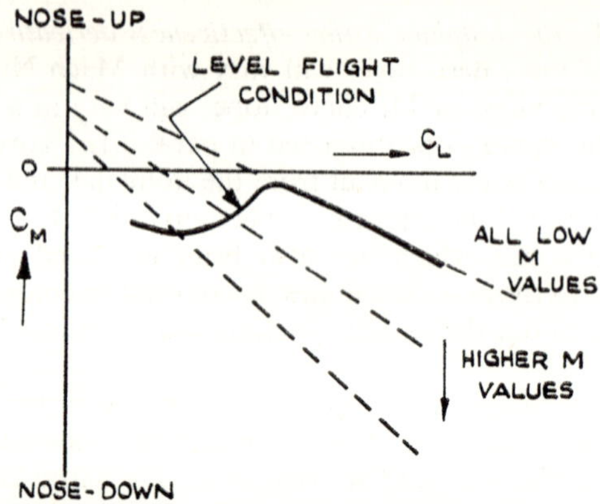

Fig. 15.15. Pitching moment characteristics in level flight

speeds it coincides with the C_M/C_L curve applicable to low Mach Numbers. In this speed régime increasing down-elevator is required to trim the aircraft at increasing values of forward speed. Above a certain value of M, the level flight trim curve turns over, and more

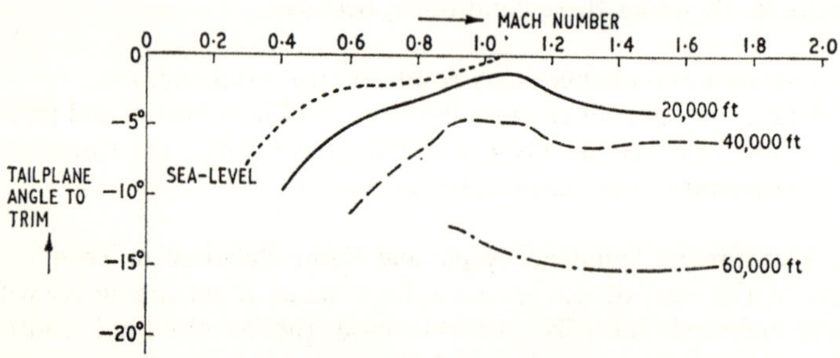

Fig. 15.16. Typical tailplane angle required to trim curves

up-elevator is now required. At still higher speeds, the curve may reverse once again, more down-elevator being required to trim out at increasing speeds. Fig. 15.16 shows typical curves of tailplane angle required to trim at various altitudes throughout the complete speed range of a supersonic aircraft.

Fig. 15.17 shows the variation of the static margin with Mach Number for these conditions of flight, determined from the tailplane trim curves and measured values of the tailplane effectiveness.

Large reversals of stick movements and forces within the transonic speed range may produce unfavourable handling characteristics. These may be eliminated by modifying

(i) the basic pitching moment characteristics of the aircraft,
(ii) the control system characteristics.

However, since most of the trim changes are unavoidable, it is not possible to modify the pitching moment values to produce the

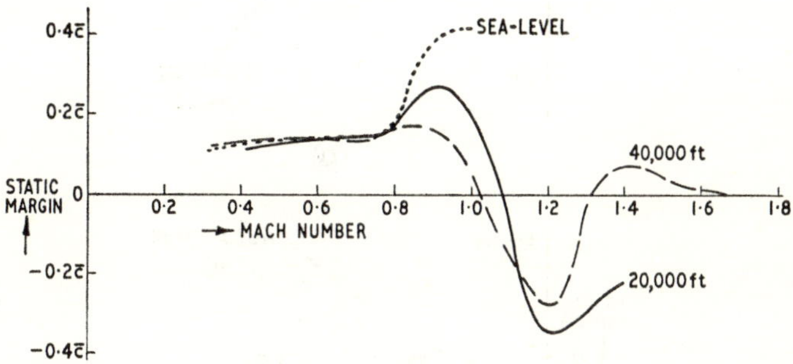

Fig. 15.17. Variation of static margin with Mach number at various heights

required handling characteristics. Both the Vulcan and the Victor experience the reversal of control surface movement in the transonic range, but the unusual stick movements and forces (but *not*, of course, the control surface movements) are eliminated by the Auto-Mach trimmer. Thus, although the control surface is moving in the necessary direction for trim at increasing forward speeds, the pilot is experiencing a continually increasing push force with the stick moving forward. An expanding link in the control system enables the stick to continue moving forwards, while the control surface is moving upwards. The results in Fig. 15.18 are taken from Ref. 6; they show the effects with and without the auto-trimmer operating.

4.4. Pitch-up

Though aircraft are generally stable with respect to incidence disturbances at low angles of attack, some designs do exhibit serious instabilities at high angles of attack. This will correspond to flight at low speeds or to manoeuvring flight at high speeds. If the instability consists of an abrupt nose-up trim change, it is usually referred to as "pitch-up." Pitch-up is shown as a non-linearity in the curve of C_M against C_L at constant speed or Mach Number (Fig. 15.3).

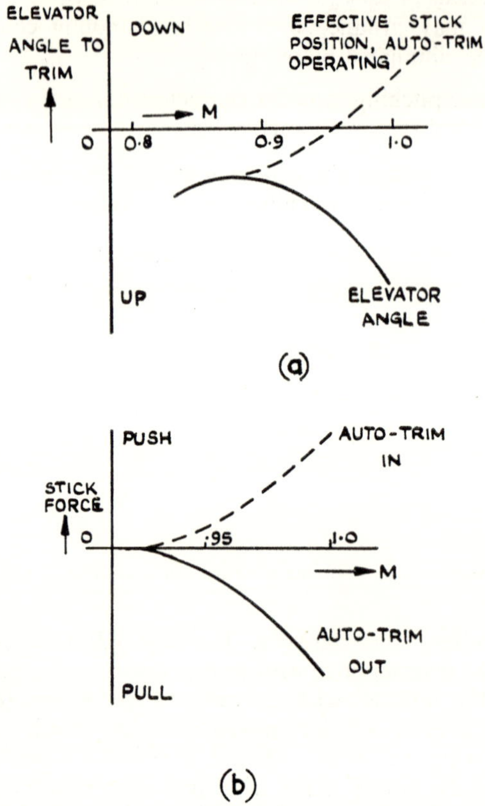

Fig. 15.18. Effect of the auto-Mach trimmer on the elevator angle to trim and stick force curves

Pitch-up is generally associated with sweptback wings, though it may also be a problem on low aspect ratio straight wing aircraft. The main causes of pitch-up on sweptback wings are

(i) The stalling characteristics of this wing planform are such that it tends to stall first at the wing tips. The aerodynamic centre (i.e. the centre of lift) moves forward on the wing and this causes a nose-up pitching moment to be produced.

(ii) Because of the redistribution of the lift spanwise along the wing and the greater concentration of lift inboard, there is a corresponding increase in the downwash behind the centre part of the wing. This increased downwash striking the tail-plane produces a further nose-up pitching moment.

Other nose-up pitching moment changes may result from the changes in the fuselage pitching moments occurring at high angles of attack, and due to the effect of the fuselage vortices influencing the down-wash striking the tailplane. On short-span straight wing aircraft,

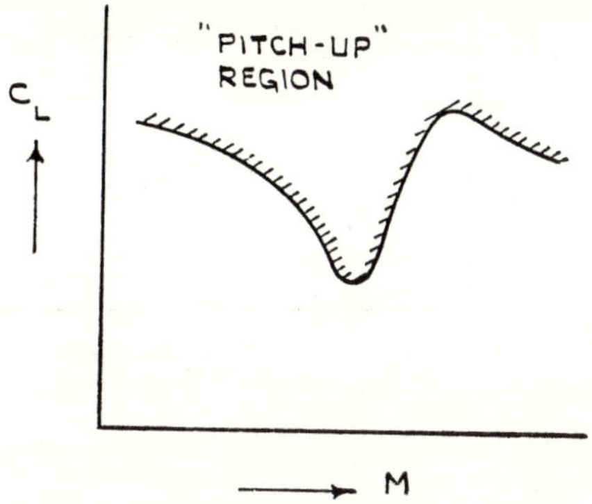

Fig. 15.19. Variation of C_L for pitch-up with Mach number

pitch-up may be produced due to the effects of the wing-tip vortices on the tailplane downwash.

The value of C_L at which pitch-up occurs varies with Mach Number. It usually decreases to a minimum in the transonic speed range, and then increases again at supersonic speeds (Fig. 15.19). Pitch-up will probably not be a limitation on an aircraft at supersonic speeds, since the g (i.e. C_L) available is usually limited by other considerations. It is very dependent on the planform of the aircraft, high values of both sweep and aspect ratio have an adverse effect. Fig. 15.20 shows that the value of the aspect ratio required to avoid pitch-up decreases with increase of sweep. The crescent wing plan-form (e.g. the Handley-Page Victor) can also be used to reduce the pitch-up.

Pitch-up can also be reduced or eliminated by either modifying the design of the wing, or by positioning the tailplane correctly with respect to the wing downwash.

(*a*) *Wing Modifications* Various modifications can be made to the wing

(i) conical camber,
(ii) wash-out,

(iii) boundary layer fences, saw-tooth leading edges, or saw-cut leading edges,
(iv) reduced thickness-chord ratio at wing tips,
(v) vortex generators.

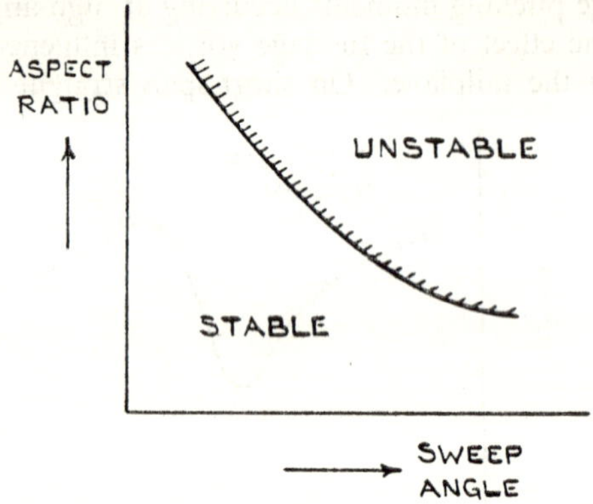

Fig. 15.20. Effect of aspect ratio and sweep on pitch-up

(*b*) *Tailplane Position* The tailplane must be positioned out of the wing downwash, by putting it high up on the vertical fin or low down on the fuselage (Fig. 15.21). The low position is the better of the

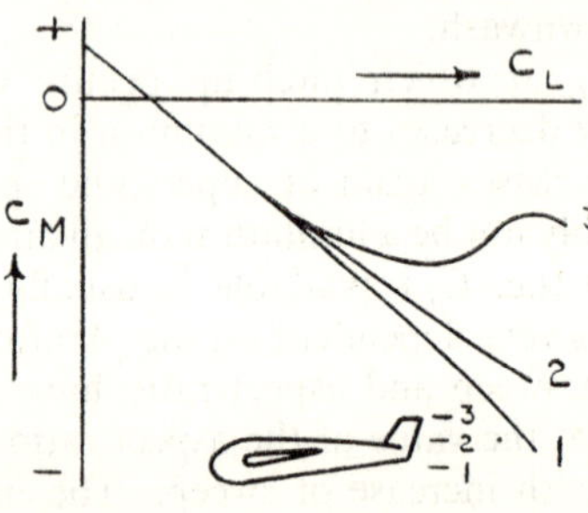

Fig. 15.21. Effect of tailplane position on pitch-up

two, but it may, in practice, be ruled out because of the engine position, jet efflux, or other considerations, and it is then put higher up on the fin. This T-tail configuration is potentially dangerous, pitch-up may still occur and the aircraft may become stabilized in what is called the "super-stall" or "deep-stall" condition. The elevator effectiveness may also be reduced, and the pilot may not be able to recover the aircraft from this condition.

5. Longitudinal Manoeuvring Stability

5.1. Manoeuvre Margin

In Chapter 3 the manoeuvre stability of an aircraft was shown to be the stability with respect to incidence and *g* changes at a *constant forward speed.* Thus the problem is not complicated by the changes in pitching moment resulting from changes in forward speed and no confusion exists as it does with the definition of static margin.

The manoeuvre stability depends on many of the same factors that affect the static stability (e.g. aerodynamic centre movement,

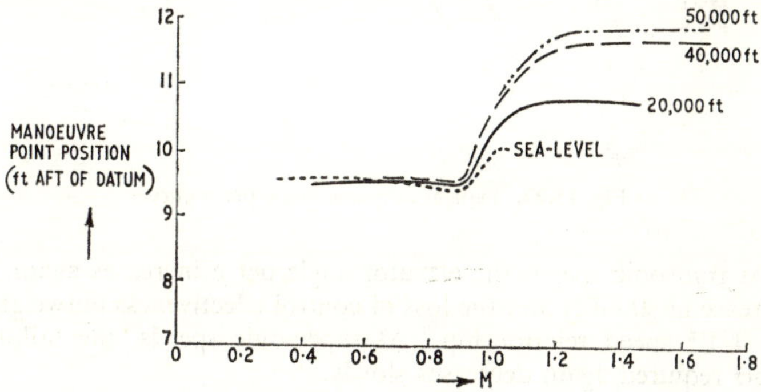

Fig. 15.22. Variation of manoeuvre point position with Mach number at various heights

etc.), though the overall variation is determined by the design of the aircraft being considered. The manoeuvre stability usually decreases slightly with increasing speed in the subsonic range, increases rapidly at transonic speeds and finally, at supersonic speeds, decreases slowly again. Some typical results, determined from the tailplane angle per *g* curves are shown in Fig. 15.22.

5.2. Elevator (or Tailplane) Angle per *g*

The elevator or tailplane angle (and hence the stick movement) required per *g* is dependent on the stick-fixed manoeuvre stability of the aircraft and the effectiveness of the elevator control. The elevator angle required is increased by an increase in stability, and decreased by an increase in control effectiveness. A typical variation of the tailplane angle per *g* at various heights throughout the speed range is shown in Fig. 15.23. The angle required decreases initially at subsonic speeds; on the simple theory of Chapter 3, the elevator angle required to trim decreases with increasing speed as $1/V^2$. This assumes that the manoeuvring stability and the control effectiveness remain constant. If, in practice, there is a small loss in stability (due to the forward movement of the A.C. at subsonic speeds) then the amount of elevator angle required decreases more rapidly than the simple law predicts.

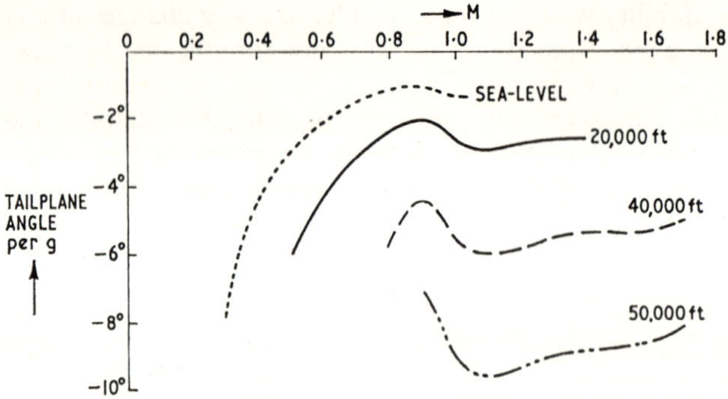

Fig. 15.23. Typical tailplane angle per *g* curves

At transonic speeds the elevator angle per *g* increases again, the increase in stability and the loss of control effectiveness outweighing the $1/V^2$ speed relationship. At supersonic speeds, the tailplane angle required again decreases slowly.

In addition to knowing the stick movement per *g* required, it is also important to know the maximum *g* that can be obtained on

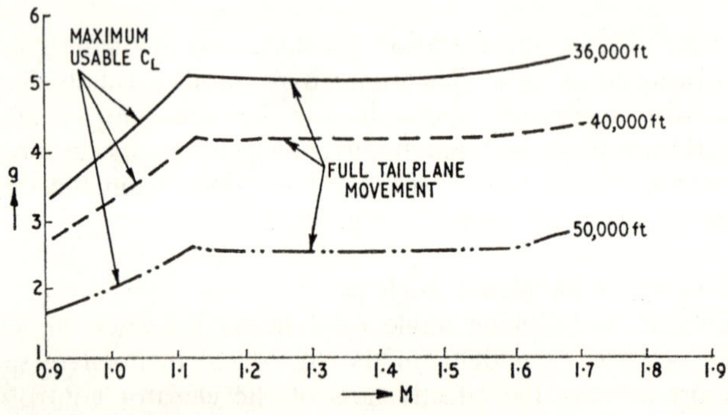

Fig. 15.24. Effect of speed and height on the maximum available *g*

the aircraft under different conditions of flight. The maximum *g* may be limited in various ways; it may be limited by structural considerations or by pilot force limitations. In some cases it is limited by the amount of stick travel (i.e. elevator or tailplane movement) available to the pilot. This is particularly so at high Mach Numbers, where there is a considerable reduction in the amount of up-elevator movement which is available for manoeuvring. This loss of available elevator is the result of the increased amount of up-elevator required in this speed range to trim out the various nose-down pitching moments that have already been discussed. At high altitudes and speeds this will limit the amount of g that can be pulled on an aircraft, to a value less than that of the structural limit of the aircraft (Fig. 15.24). This also explains why pitch-up may not be a serious limitation on an aircraft's manoeuvring performance at supersonic speeds, because it may not be possible to pull the amount of g at which pitch-up would occur.

5.3. Stick Force per *g*

In the case of an aircraft fitted with a non-powered control system, the stick force per g characteristics are dependent on the stick-free

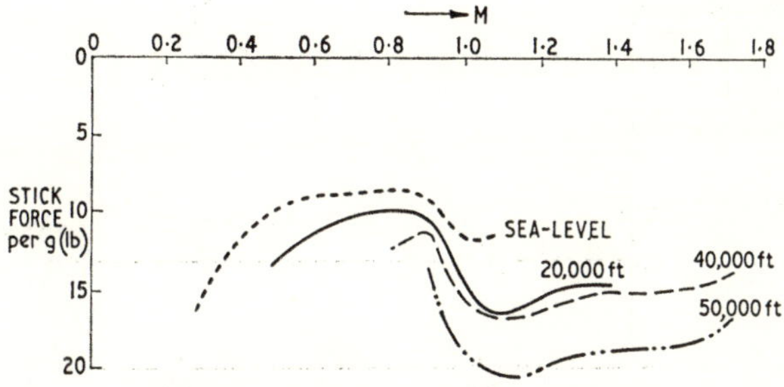

Fig. 15.25. Typical stick force per g curves

manoeuvre stability of the aircraft and the control characteristics. On the simple theory of Chapter 3, the stick force per g values are independent of forward speed. In practice, the stick force per g may vary with speed due to the changes in the stick-free manoeuvre stability. At transonic speeds the values will increase abruptly, due to the increase in manoeuvre stability and the loss of elevator effectiveness.

Most high speed aircraft are fitted with power-operated controls. On these aircraft the stick forces are related to the stick-fixed manoeuvre stability and are dependent on the type of feel system incorporated into the controls. If a simple spring-feel system is used, then the stick force per g curves are similar to the corresponding stick movement (or elevator angle) per g curves. Many aircraft are fitted with q (i.e. $\frac{1}{2}\rho_0 V_i^2$) feel on the elevators and in some cases this is combined with a spring feel. A Mach number cut-off may also be incorporated, which cuts out the q-feel at the higher Mach Numbers.

Typical SF/g curves are shown in Fig. 15.25. The values decrease at subsonic speeds due to the reduction in elevator (or tailplane) angle required, and the possible loss of manoeuvre stability. In the transonic régime, the values of SF/g increase abruptly because of the large increase of manoeuvre stability. At supersonic speeds the values again decrease gradually due to the gradual reduction in the tailplane angle (or stick movement) per g. It should be noted that the curves of Figs. 15.23 and 15.25 are very similar to one another.

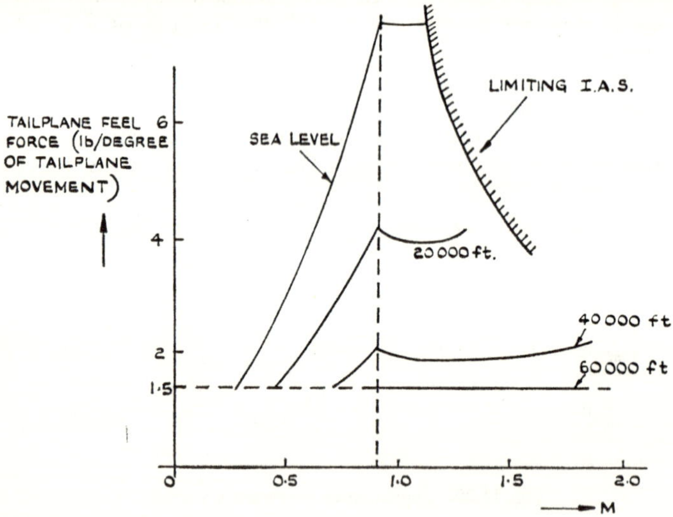

Fig. 15.26. Tailplane feel characteristics

In the early Lightning aircraft a feel system with the following characteristics was incorporated—

(*a*) a light self-centering spring with a rate of $1\frac{1}{2}$ lb/degree of tailplane movement,
(*b*) q-feel up to $0{\cdot}9M$, then held constant at higher speeds.

The resulting tailplane feel characteristics for various heights are shown in Fig. 15.26. The stick-force–normal acceleration characteristics can then be determined by combining information from these curves with that of the amount of tailplane movement required to produce g. The resulting stick force curves are given in Fig. 15.27; the curves showing the stick forces required to produce the limiting acceleration of 7·5 or the stick forces corresponding to full tailplane movement, as appropriate. The curves corresponding to the $7{\cdot}5g$ value are similar to those of Fig. 15.25, but at 40,000 ft and 60,000 ft full tailplane movement will not produce this value of g at supersonic speeds. At 40,000 ft full tailplane movement only produces about $6g$ and at 60,000 ft only about $2\frac{1}{2}g$. From these curves it can

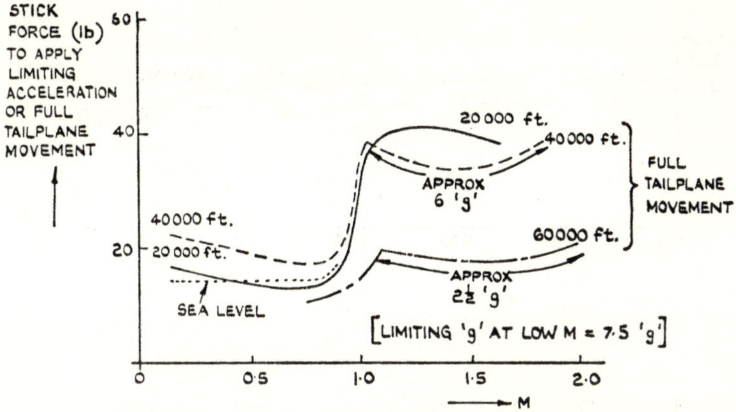

Fig. 15.27. Stick force-normal acceleration characteristics

be seen that, with this particular system, the SF/g values are maintained fairly constant within the range of 5 to 8 lb/g at all speeds above $M = 1$, up to a height of 60,000 ft.

6. Longitudinal Dynamic Stability Characteristics

6.1. General Considerations

On conventional subsonic aircraft the longitudinal dynamic stability characteristics do not usually present any problem. The motion can be separated into two oscillatory modes called the Short Period Oscillation and the Phugoid Oscillation (Chapter 8). Both these oscillations are usually well damped. Under certain conditions of flight, the phugoid damping may become low, but, because the periodic time is so large, this does not cause any handling difficulties.

At supersonic speeds the dynamic motion can still be analysed in the same way. The two oscillations still exist, though the characteristics of periodic time and damping are somewhat changed. The same derivatives are involved in the mathematical analysis of the problem as at subsonic speeds, though the relative importance of these derivatives is different because of the effects of compressibility and the changes in the basic design of the aircraft.

6.2. Short Period Oscillation

The variation of the characteristics of the short period oscillation depends on the way that the derivatives vary through the speed range.

These derivatives have been discussed in para. 3. Because of the rearward movement of the aerodynamic centre, an aircraft usually possesses a large amount of static stability. The damping in pitch,

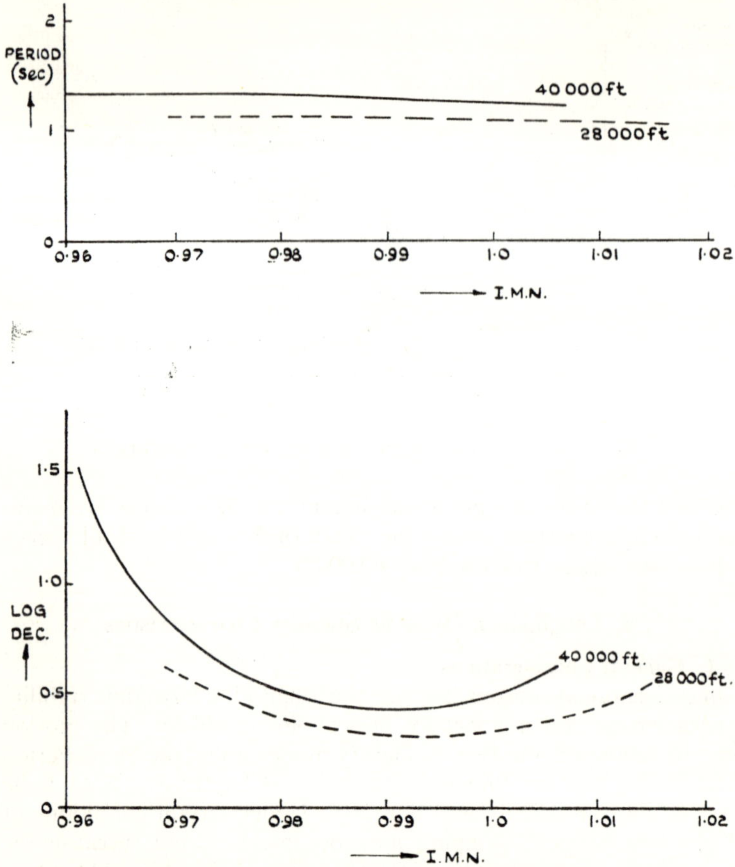

Fig. 15.28. Flight test measurements of the longitudinal short period oscillatory characteristics of an aircraft at transonic speeds

on the other hand, may become very small at transonic speeds, particularly in the case of tailless aircraft.

Fig. 15.28 shows some flight test results on the Vulcan, measured at two different heights in the transonic speed range. In these tests the pitch damper was inoperative.

The variations of the periodic time and the damping of the short period oscillation of the F.D.2 are shown in Fig. 15.29, taken from Ref. 7. The periodic time decreases with increasing speed and

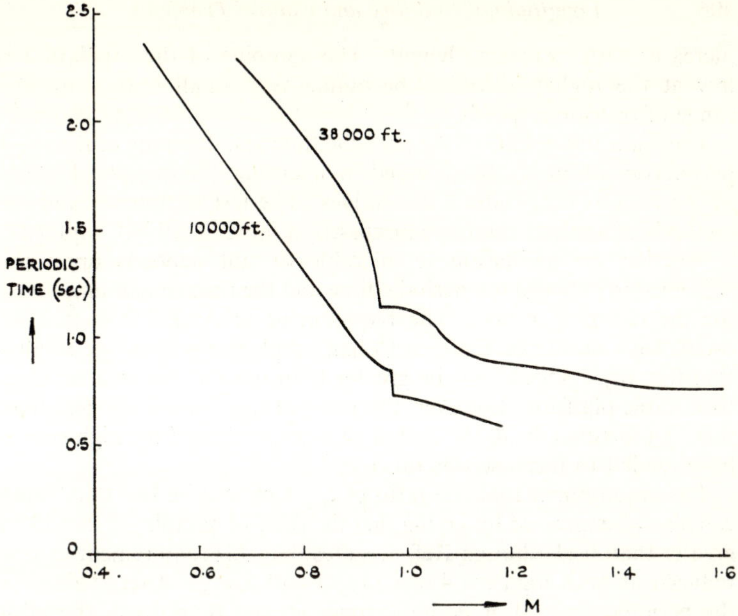

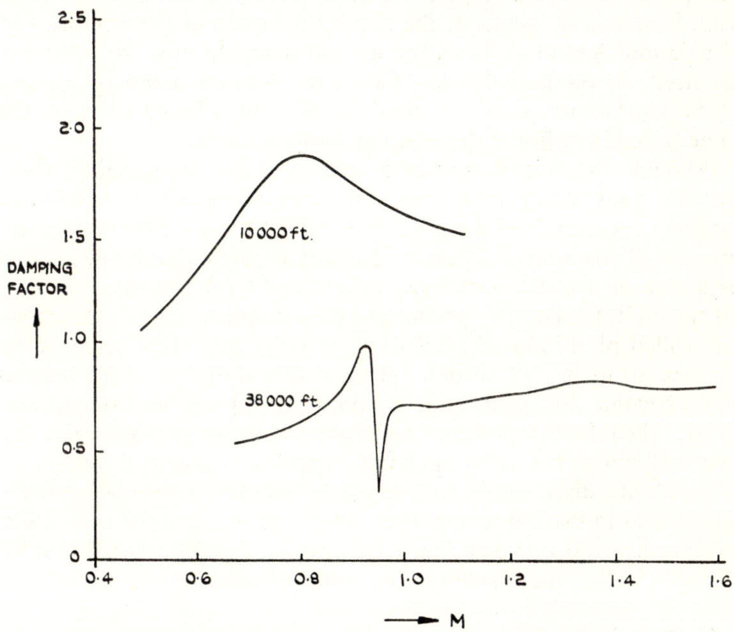

Fig. 15.29. Longitudinal short period oscillatory characteristics of the F.D.2 [From Ref. 7]

increases with increasing height. The damping of the oscillation is low at the higher altitudes, becoming very small over a narrow range of transonic speeds.

The flight test results of the periodic time and damping of the short period oscillation can be analysed to determine the numerical values of the various longitudinal derivatives. The test techniques and the methods of analysis used, are discussed in more detail in Chapter 17.

Whether an oscillation is tolerable or not depends upon the relationship between the periodic time and the time to half amplitude (or the damping ratio). The requirement in Av.P.970 is that all oscillations should be damped to half amplitude within one oscillation (i.e. $t_{1/2}/T$ should not be greater than unity). To be a satisfactory stable platform, however, the value of $t_{1/2}/T$ should be less than this. Unfortunately all the design changes in aircraft for high speeds have tended to increase this ratio.

The requirement that the ratio of $t_{1/2}/T$ should be less than unity can also be expressed by saying that the relative damping ζ should be greater than 0·11, though Ref. 4, states that this requirement can be reduced at both high ($>$ 4 secs (app)) and low ($>$ 1 sec) values of the periodic time. For periodic times greater than 4 secs the pilot can, if necessary, augment the natural damping of the aircraft, but for periods less than 4 secs the natural damping must be adequate by itself. If the period is less than 1 sec, a lower damping appears to be satisfactory, since the pilot's assessment is based more on the time to half amplitude than on the damping ratio.

Though the aircraft may be inherently stable, longitudinal short period dynamic instability may still occur, under certain conditions of flight, as a result of the pilot attempting to control the oscillatory motion of the aircraft, thereby inducing negative damping (i.e. this is a case of dynamic instability of a closed loop system, consisting of the pilot, the control system, and the airframe). These oscillations are called pilot-induced oscillations (P.I.O's), and they often occur on aircraft flying at transonic speeds at low altitudes. If the oscillations become divergent, large g values can be produced on the aircraft. The pilot can prevent P.I.O's from occurring by either releasing the stick altogether or by applying a steady constant pull force.

P.I.O's are discussed in Ref. 8, which describes the possible effects of changes in the feel spring, the control gearing, and the bobweight effectiveness. It is shown that P.I.O's can be significantly reduced by changes in the components of the control system.

6.3. Phugoid Oscillation

The characteristics of the phugoid oscillation also vary throughout the speed range. Under some conditions of flight the damping may become small or even negative, though, as long as the periodic time remains large, this is not a dangerous situation. On some designs there is a possiblity of a major change in the longitudinal oscillatory characteristics of the aircraft resulting from the phugoid breaking down into two aperiodic modes.

Since the phugoid motion involves speed and height changes, with little change in g, one of the most important derivatives in the damping of the motion is x_u. The value of this derivative is determined by the way in which the airframe drag and engine thrust characteristics vary with forward speed. If an increase in forward speed produces an increase in drag and no change in thrust, then x_u is negative and this helps to damp the phugoid motion. Any increase of thrust, however, will reduce the numerical value of x_u, and, hence, the damping of the motion. At supersonic speeds the thrust and drag tend to vary in a similar way with variation of forward speed and the damping is correspondingly reduced.

The results of a computer study on the dynamic stability characteristics of tailless delta aircraft at supersonic speeds are given in Ref. 9. This shows the effect of varying the Mach number, height, wing sweep, C.G. position and wing loading. The phugoid motion tends to instability at high Mach numbers and high altitudes. The short period oscillation, on the other hand, is almost independent of Mach number changes, but tends to instability at high altitudes.

6.4. Response to Longitudinal Gusts

This is a convenient place to mention briefly the problem of gust response, which is becoming more important on modern aircraft designs. The problem is dealt with in more detail in Refs. 12, 13 and 14. Though the air is in random motion, the gust response analysis is usually performed for hypothetical sharp-edged or ramp-type gusts. Typical longitudinal responses for conventional and slender delta aircraft configurations are shown in Ref. 14.

On simple theory, when an aircraft meets a sharp-edged gust of vertical velocity w_g ft/sec, then the corresponding change in angle of attack is given approximately by w_g/V, where V is the forward speed (T.A.S.) in ft/sec. Assuming that the increase in lift occurs instantaneously on the wing of the aircraft, the increment in normal acceleration ($\Delta n = \Delta L/W$) is given by

$$\Delta n = \frac{w_g \, a \rho V S}{2W} \tag{15.5}$$

where a is the lift curve slope (per radian) and W the all-up weight (lb). In practice, the increment in g is reduced by a gust alleviation factor and can be further reduced by installing a gust alleviator system. These features are both discussed in Ref. 13 and additional references are given to flight test experiments carried out with gust alleviators.

7. Stability Augmentation Systems

As we have seen, the design of a high-performance aircraft is such that, under certain conditions of flight, the aircraft may become statically or dynamically unstable. Though an aircraft can be flown with a limited amount of instability, this is not very satisfactory, and the stability of the aircraft is often improved by means of automatic stability augmentation systems. Various methods of stability augmentation exist, some of which (e.g. the incorporation of springs or bob-weights into the control system) have already been discussed in previous chapters.

In order to reduce the trim drag occurring on supersonic transport aircraft, it may be necessary to design the aircraft so that it is inherently statically unstable at subsonic speeds. This type of instability will also occur on some missiles which are designed without stabilizing surfaces in order to save weight. This problem is discussed in Ref. 10, and the reference describes three methods which are available for compensating the longitudinal static instability. These methods are

(*a*) a mechanical stability augmentation system,
(*b*) a fuel transfer system,
(*c*) a conventional autopilot system.

A fuel transfer system has already been used on the American B58 Hustler and it is proposed for the BAC/Sud Concorde. Since the problem is caused by the rearward movement of the wing aerodynamic centre which occurs in going from subsonic to supersonic speeds, the fuel transfer system is used to move the C.G. of the aircraft rearwards as the aircraft accelerates and forwards again as it slows down to subsonic speeds. In this way the static stability is maintained at a satisfactory level throughout the complete speed range of the aircraft. Ref. 10 discusses, for a hypothetical aircraft, the quantity of fuel to be transferred, the required fuel flow rates and the mechanism of the fuel transfer system. The reference also describes a conventional autopilot system, using pitch attitude and pitch rate signals to move the longitudinal control surface, that can also be employed to stabilize a statically unstable aircraft.

The damping of the longitudinal short period oscillation may become very low at high speeds and high altitudes (see para. 6.2). The characteristics of the longitudinal oscillation can be improved by the incorporation of an autostabilization system (i.e. a "pitch damper"). Autostabilizers operate by measuring the various response characteristics of the aircraft, and feeding them back in such a way that the resulting automatic control surface movements oppose the motion of the aircraft, and increase the damping of the oscillatory modes. In the case of a pitch damper, a rate gyro is used to measure the rate of pitch q of the aircraft, and the system is designed to move the elevator control surface in accordance with a control law of the form

$$\eta = Kq \tag{15.6}$$

where K is called the gearing of the pitch damper system. Some automatic stabilization systems are designed so that the value of K varies automatically with the speed and height at which the aircraft is flying.

Ref. 11 describes an investigation of the effect of a pitch damper on the longitudinal stability characteristics of the Fiat G 91 aircraft. An analogue computer study was initially made of the handling characteristics of the aircraft for various conditions of flight and pitch damper characteristics. The results showed that the pitch damper substantially improved the longitudinal short period stability characteristics of the aircraft, and these results were confirmed by flight test experiments.

References

1. THOMAS, H. H. B. M.; *Notes on Stability and Control Derivatives at Subsonic and Transonic Speeds.* Unpublished M.O.A. Report.
2. *Flight Test Note* AFN/P1/57 Pt. 1, English Electric Co. Ltd.
3. HAMILTON, J. A., HUNT, G. K. and HARMER, R. G.; *Free Flight Measurements at Transonic Speeds of the Dynamic Longitudinal Stability Characteristics of a Delta Winged Bomber* (*Vulcan*). Unpublished M.O.A. Report.
4. ZBROZEK, J. K.; *Handling Characteristics of Supersonic Aircraft*, R.A.E. Tech.Note. Aero 2422, 1955.
5. JUKES, J., SMITH, K.; *Flight Measurements of the Short Period Longitudinal Oscillation of a* 50° *Delta Aircraft at high Mach numbers* (*Avro* 707A), R.A.E. Tech.Note. Aero 2319, 1954.
6. *Trials for Initial C.A. Release—Vulcan B Mk.* 1. Unpublished M.O.A. Report.
7. ANDREWS, D. R.; *Measurements in Flight of the Longitudinal Derivatives of a* 60° *Delta Wing Aircraft* (*F.D.* 2), R.A.E. Rep. Aero 2617 (C.P. 639), 1959.
8. LEVI, O. A. and NELSON, W. E.; "An Analytical and Flight Test Approach to the Reduction of Pilot Induced Oscillation Susceptibility," *Journal of Aircraft*, Vol. 1, No. 4. July–August 1964, page 178.
9. KALKMAN, C. M.; *Investigation of the Dynamic Longitudinal Stability of Tailless Delta Aircraft at Supersonic Speeds*, Holland NLR-TN V1913 (Amsterdam), 1963.

10. DUBLIN, M.; *An Approach to the Control of Statically Unstable Manned Flight Vehicles*, Agard Report No. 364, 1961.
11. MAUTINO, R.; *Investigation on the Improvement of Longitudinal Stability of a Jet Aircraft by the Use of a Pitch Damper*, Agard Report No. 370, 1961.
12. ETKIN, B.; *Dynamics of Flight* (J. Wiley and Sons, New York, 1959).
13. ZBROZEK, J. K.; "Atmospheric Gusts," *Journal of the Royal Aeronautical Society*, January, 1965.
14. PINSKER, W. J. G.; *Features of Large Transport Aircraft Affecting Control During Approach and Landing*, Agard Report No. 421 ,1963.

16 Lateral Stability and Control Problems of High Speed Aircraft

1. Introduction

The low speed lateral stability and control characteristics of aircraft have been dealt with in previous chapters (i.e. Chapters 5, 9 and 11). We will now consider the changes that occur in the lateral characteristics due to the effects of compressibility and the modifications in design associated with high-performance aircraft. The effects of aeroelasticity at high speeds are dealt with separately in Chapter 14. The changes in the basic design features of modern high-performance aircraft have already been discussed in the previous chapter; the wings are thin, with a highly swept or delta shaped planform, the tailarm is small, and, occasionally, the horizontal tailplane is eliminated altogether. In addition, the mass distribution is considerably different from that of older aircraft; the inertia in roll A is small, and the inertias in pitch B and yaw C large.

The basic approach to the problems of lateral stability and control is similar to that employed at low speeds. The lateral derivatives are, however, functions of Mach number and their values vary considerably throughout the subsonic, transonic and supersonic speed ranges. Though the characteristics of the lateral modes vary through the speed and height range of the aircraft, they usually retain their classical forms (i.e. the overall motion can be broken down into the roll subsidence mode, the spiral mode and the Dutch Roll oscillatory mode, in the manner discussed in Chapter 9).

High-performance aircraft tend to experience a number of lateral stability and control problems at the top end of their speed range (i.e. a loss of directional stability and a reduction in the oscillatory damping, etc.). These problems can be reduced or eliminated by suitable modifications in the design of the aircraft, or by the installation of artificial autostabilization systems (e.g. a roll or yaw damper). Many of the problems associated with aircraft designed for high speed actually occur, however, at the low speed end of the flight range (e.g. Dutch Roll problems on the approach).

As a result of the changes occurring in the inertia characteristics of modern designs, these aircraft tend to roll about their principal inertia axis rather than around the direction of their flight path. This leads to problems in the roll response characteristics of the aircraft (para. 4.3), and also to Inertia (or Roll) Coupling (para. 7). When an aircraft is rolling rapidly it tends to diverge either in pitch or in yaw, consequently developing large angles of attack or side-slip, which in some cases may result in a structural failure.

2. Lateral Stability and Control Derivatives

Separate contributions to the overall values of the lateral derivatives arise from all the various parts of the aircraft, including the wing, the fuselage and the vertical fin. In the case of an aircraft designed for high speeds, with very low aspect ratio wings and high fineness ratio fuselages, the fuselage-fin and wing-fin interference effects are, however, relatively more important than on a conventional low-speed design.

The values of the derivatives of any project design can be determined by theoretical methods, from wind-tunnel tests and from free-flight model tests. Eventually, when the aircraft has been built and flown, flight tests can be carried out and the values of the derivatives again determined. These values can then be compared with those obtained from the methods employed earlier. The theoretical methods that are available for estimating the values of derivatives are discussed in Refs. 1 and 2. Ref. 1 deals with both longitudinal and lateral derivatives and covers the subsonic, transonic and supersonic speed ranges. The report also includes an extensive bibliography with 158 additional references. Ref. 2 presents the results of the theoretical calculations of the stability derivatives of a high-speed interceptor-type aircraft over the speed range of $M = 1{\cdot}4$ to $M = 3{\cdot}0$. Further discussion on the estimation of derivatives can be found in Ref. 3 (Part II, Chapter 10) and Ref. 4 (Chapter XV).

A number of reports are available showing the measured variation of the lateral derivatives of an aircraft with Mach number obtained from wind tunnel tests, free-flight model tests and flight tests. Ref. 5 gives the results of tests carried out with free-flight models of the Bristol 188 supersonic research aircraft over the speed range of $M = 1{\cdot}2$ to $2{\cdot}6$. A comparison is also made with the results obtained from tests carried out in the R.A.E. Bedford 8 ft × 8 ft supersonic wind tunnel. The results of the flight tests carried out on the Fairey Delta 2 over the Mach number range of $M = 0{\cdot}4$ to $1{\cdot}5$ are given in Ref. 6. The lateral derivatives were determined using the Time-vector method of analysis (see Chapter 17), and the way that some of these derivatives vary with Mach number is shown in Figs. 6.13 and 16.3.

A few of the more important lateral derivatives are discussed separately below.

(a) $dC_L/d\alpha = a$, *the Lift Curve Slope*

The lift curve slopes of the wing and the vertical fin are important basic parameters in the equations for many of the lateral derivatives. The way in which the lift curve slope varies with Mach number has already been discussed in Chapter 15, and Fig. 15.2 shows how the value may vary through the speed range of a supersonic aircraft. At subsonic speeds, the value increases in accordance with the Glauert law, but at supersonic speeds there is a gradual decrease in the value of $dC_L/d\alpha$ for all "lifting" surfaces. This decrease is in accordance with the Ackeret law, and it has an important effect on the values of such lateral parameters as l_p, the damping-in-roll derivative, and n_v, the weathercock stability derivative.

(b) l_v, *the Dihedral Effect*

This derivative represents the lateral static stability of the aircraft. The most important contribution to l_v comes from the wing and the magnitude of this contribution is mainly dependent on the amount of geometric dihedral and sweepback on the wing. Sweepback increases the dihedral effectiveness (i.e. increases the value of $-l_v$), the effect of the sweep increasing with increase of incidence (i.e. with increase of C_L). This effect is particularly important, because it means that the l_v value of a swept-wing aircraft must vary widely over the speed range of the aircraft. At low forward speeds, when the aircraft is at a high angle of attack, the value of $-l_v$ will be relatively large; at high speeds, when the angle of attack is small, the value of $-l_v$ will be correspondingly low. In order to reduce the value of $-l_v$ to a suitable magnitude at low forward speeds, the wings of some high-performance aircraft incorporate negative dihedral (i.e. anhedral). At high Mach numbers, compressibility effects will further modify the value of this derivative. On some designs, the vertical fin and the horizontal tailplane may produce important contributions to the overall value of l_v, and the effects of wing-body interference may also be important (Ref. 1).

The variation of l_v with Mach number obtained from the flight tests on the Fairey Delta 2 aircraft is shown in Fig. 6.13 of Chapter 6. The curve, taken from Ref. 6, shows a gradual reduction in the numerical value of l_v at high subsonic speeds, due to the corresponding reduction in angle of attack, followed by an abrupt increase between $M = 0{\cdot}9$ and $1{\cdot}0$ due to the effects of compressibility. Above $M = 1{\cdot}0$ the dihedral effect decreases gradually with increasing speed.

The loss of dihedral effect at high forward speeds is typical of many high-performance aircraft; in some cases l_v may reverse in sign (i.e. become positive). The theoretical aspects of this are considered in Ref. 7. When l_v reverses in sign on an aircraft, the pilot can no longer employ the technique of using top rudder instead of aileron control to pick up a dropped wing. Using this technique, if the starboard wing drops, port rudder is applied. This causes the aircraft to yaw and sideslip, and if the dihedral effect acts in the usual direction, the starboard wing would come up. If l_v is positive, however, sideslipping to starboard would cause the starboard wing to drop even farther. This phenomenon is known to occur on the Handley Page Victor bomber at speeds around $M = 0{\cdot}94$ (Ref. 8), and on the X-15 at speeds above $M = 2{\cdot}0$ (Ref. 9).

(c) n_v, the Weathercock Stability

This is one of the most important of the lateral derivatives, because the way in which it varies with Mach number and incidence may lead to serious effects on the stability and control characteristics of the aircraft. The two main contributions to the overall value of n_v are those of the wing-body combination and the vertical fin. The wing-body contribution is usually a destabilizing one, and it remains fairly constant throughout the speed range of the aircraft, variation of Mach number having little effect, particularly for high fineness ratio fuselages. The vertical fin contribution, on the other hand, is a stabilizing one, but since it is dependent on the value of the lift curve slope of the fin, its value increases at subsonic speeds, reaches a maximum at about $M = 1{\cdot}0$ and then decreases again at supersonic speeds (Fig. 16.1). The vertical fin should be large enough for the stabilizing effect of the fin to counteract the destabilizing effect of the wing-body combination and to produce the required amount of directional stability. However, since the n_v contribution of the vertical fin continues to decrease with increase of forward speed above $M = 1{\cdot}0$, a speed is eventually reached at which the overall value of n_v falls below the minimum permissible value allowed from considerations of roll coupling. If the aircraft is required to fly at still higher speeds, the area of the vertical fin must be increased. The size of the vertical fin should be kept as small as possible, however, because an increase in area also increases the overall drag of the aircraft. The continual increase in the area of the vertical fin has become quite a familiar feature in the development of many modern high-performance aircraft. On some aircraft, the natural stability is insufficient at the higher Mach numbers, and automatic stabilization systems are used to increase the effective value of n_v.

The variation in the value of n_v with Mach number for the F.D.2 is shown in Fig. 6.13 of Chapter 6. The value remains fairly constant at high subsonic speeds, increases to a maximum at about $M = 1{\cdot}0$ and then decreases continuously up to $M = 1{\cdot}5$.

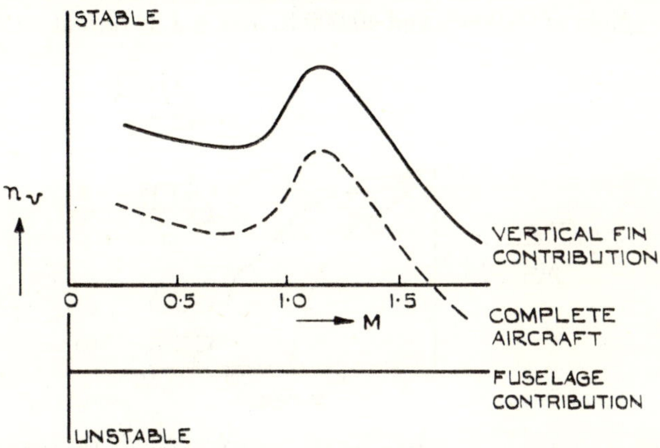

Fig. 16.1. Typical variation of the n_v derivative with Mach number

Incidence may have an appreciable effect on the weathercock stability of an aircraft, an increase of incidence resulting in a reduction of n_v. This reduction may be the result of

(i) a loss in the dynamic pressure of the air around the vertical fin,
(ii) wing-fin interference effects,
(iii) the effect of vortices shed from the nose of high fineness-ratio fuselages or from canard control surfaces.

The effect of vortices shed from the wing or the fuselage are discussed in Ref. 10. The magnitudes of the sidewash velocities and the corresponding load distributions on the fin are calculated and Fig. 16.2, taken from the reference, shows the overall effect of an increase of incidence on the normal force acting on the vertical fin in a sideslip at a high subsonic Mach number ($M = 0{\cdot}80$), for various vertical fin shapes. The results of some low speed tunnel tests on sharp-nosed fuselages at high incidences described in Ref. 11, also show the effect of the fuselage vortices on the yawing moment characteristics.

The effects of incidence can be estimated from flight tests by carrying out the particular tests for a certain Mach number either at a different height or at the same height in turning flight. If the tests are carried out at different heights, however, it must be remembered that the value of the E.A.S. will also be different for a given Mach number, so that some of the difference observed may be due to aeroelastic effects. The tests on the F.D.2 were carried out in level (i.e. 1g) flight at 10,000 ft and 40,000 ft, and in a 2g turn at 40,000 ft.

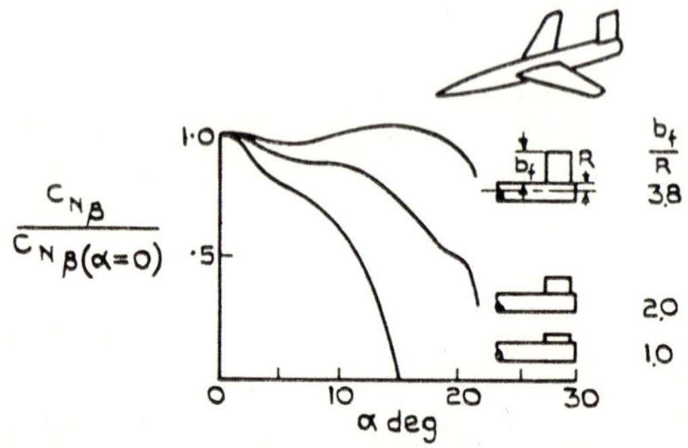

Fig. 16.2. Effect of incidence on $C_{N\beta}$ (or n_v) for various fin shapes [From Ref. 10]

The few results obtained above $M = 1{\cdot}0$ at 40,000 ft indicated that n_v was reduced by an increase of incidence.

The unavoidable loss of n_v experienced at supersonic speeds has resulted in a number of innovations on some modern aircraft designs. Additional ventral fins have been added; in some cases these are small and are permanent fixtures, but in other cases they are so large that they have to be folded or jettisoned for landing. These ventral fins may be particularly effective in increasing the value of n_v, because they are in a region of high dynamic pressure. On the XB-70 Valkyrie, folding wing tips are used; these are folded down at supersonic speeds to increase the value of n_v and also to reduce the corresponding longitudinal static stability (see Chapter 20). Directional stability problems on the X-15 aircraft were investigated in wind tunnel model tests (Ref. 12). It was found that the stability could be improved by the use of a ventral fin and strakes along the side of the forward fuselage. On some canard designs twin vertical fins have to be used to give an adequate amount of directional stability (e.g. XB-70 Valkyrie). The preliminary designs of the BAC/Sud Concorde also incorporated a canard surface, but this was eliminated because of the directional stability problems associated with it.

(d) l_p, the Damping-in-roll Derivative

The most important contribution to the overall value of l_p comes from the wing of the aircraft. It is dependent on the value of the lift curve slope of the wing and is consequently small on modern designs, due to the small aspect ratio and large sweepback angles that are used. The variation of l_p with Mach number follows the same trend as that of the lift curve slope; it increases gradually at

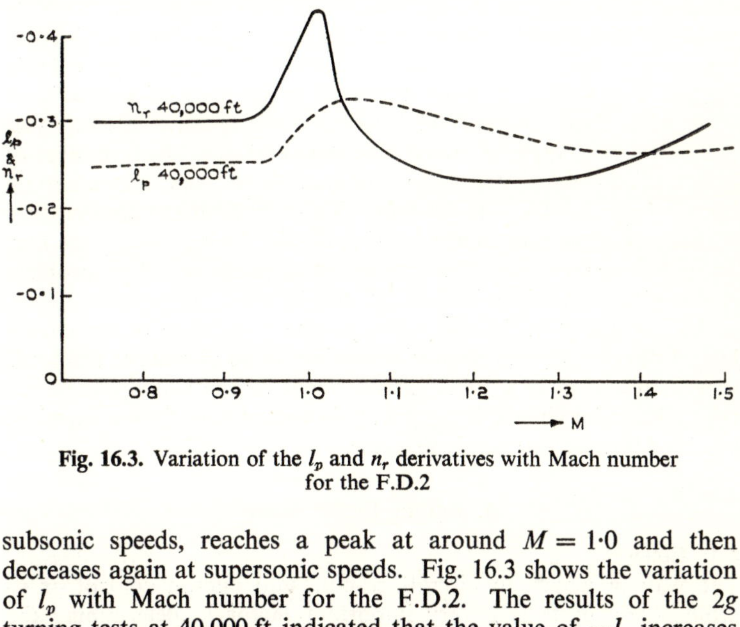

Fig. 16.3. Variation of the l_p and n_r derivatives with Mach number for the F.D.2

subsonic speeds, reaches a peak at around $M = 1{\cdot}0$ and then decreases again at supersonic speeds. Fig. 16.3 shows the variation of l_p with Mach number for the F.D.2. The results of the 2*g* turning tests at 40,000 ft indicated that the value of $-l_p$ increases appreciably with incidence at supersonic speeds.

(e) n_r, the Damping-in-yaw Derivative

The wing contribution to this derivative is usually small and the main effects are produced by the fuselage and the vertical fin. This is an important derivative because it usually determines the damping of the lateral oscillation. It tends to be low on modern designs, due to the short arm of the vertical fin and many aircraft are fitted with automatic yaw dampers in order to improve the oscillatory characteristics of the aircraft. The variation of n_r with Mach number for the F.D.2 is shown in Fig. 16.3.

(f) l_ξ and n_ζ, the Aileron and Rudder Control Power Derivatives
The effectiveness of flap-type control surfaces tends to decrease at high speeds and these controls are often replaced by all-moving or spoiler-type control surfaces. These alternative forms of control are discussed in more detail in Chapter 10.

3. Static Lateral-directional Stability

The variations in the lateral stability derivatives discussed in para. 2 produce a number of stability and control problems. In para. 2(*c*) it was shown that the directional static stability n_v may become very small or even negative at high speeds and high incidences. Negative values of directional stability will lead to a rapid directional divergence occurring with consequent development of large sideslip angles. Fig. 9.16 of Ref. 14 shows an actual trace of a directional divergence that occurred on a swept-wing research aircraft (see Ref. 13) as the angle of attack was increased at a Mach number of 0·76 and Fig. 9.6 of the same reference shows that over a certain range of speeds this particular problem may limit the amount of g that can be pulled on an aircraft (i.e. the directional divergence may occur before the onset of severe buffet or pitch-up). This problem is also discussed briefly in Refs. 15 and 16.

Ref. 14 (page 9.27) also refers to the possibility of an abrupt lateral-directional trim change occurring on an aircraft at transonic speeds, resulting in a small displacement of the sideslip angle for trim. The trim change is probably associated with flow separation, but is of little structural significance.

4. Rolling Performance

4.1. General

The way in which an aircraft responds to aileron control or to atmospheric gusts is a very important factor in determining the handling characteristics of an aircraft. The basic lateral response characteristics have already been discussed in Chapter 11. An aircraft usually responds very quickly to aileron deflection and soon reaches a steady rate of roll given by $-\dfrac{2V}{b}\dfrac{l_\xi \xi}{l_p}$ (eqn. 11.3). For this reason, the aileron controls are usually described as "rate controls."

The rate of roll produced by the ailerons should be sufficient to enable the pilot to have satisfactory lateral control over the aircraft throughout the speed range. The requirements can be expressed in terms of

(i) the maximum or average rate of roll produced,

(ii) the value of the parameter $pb/2V$,
(iii) the time for the aircraft to roll through a specified angle of bank,
(iv) the angle that can be rolled through in a given time.

The requirement is often expressed in terms of the parameter $pb/2V$ (see Chapter 11), though this criterion does lead to rather excessive values of p at high speeds. On more modern designs, the requirement is often expressed in the form of (iii) or (iv). In these cases, the conditions at the beginning and end of the manoeuvre are usually defined and the values are often quoted for the condition of "roll and stop." The lateral handling characteristics are discussed in Refs. 14 and 17 and in the reports mentioned in the bibliographies of these references. The requirements laid down for aircraft are continually being revised in the light of further knowledge obtained from simulator experiments and from flight tests on variable stability aircraft.

The pilot's assessment of the lateral handling characteristics of an aircraft is also very dependent on the cross-control effects. When the ailerons are deflected to roll the aircraft, a yawing moment is also usually produced. On conventional aircraft this is often described as "adverse," because it tends to reduce the rate of roll and oppose the direction of turn. On some modern designs, however, the effect may be a favourable one.

4.2. Lateral Control Requirements

The roll requirements for British military aircraft are laid down in Ref. 18. These requirements state the minimum rate of roll performance that is permissible at both high and low forward speeds and also lays down the minimum value for the rate of growth of rolling velocity. The low speed requirement is stated in terms of the parameter $pb/2V$ (i.e. the tangent of the wing tip helix angle); under various specified conditions the value of $pb/2V$ should not be less than 0·07 radians at sea-level. In the case of naval aircraft this lower limit is increased to 0·09 radians, because a greater degree of lateral control is required on these aircraft. The corresponding high speed rate of roll requirement is stated in the form of an equation expressed in terms of V_D, the design diving speed and M, the Mach number.

The lateral control requirements are also discussed in para. 6.2 of Ref. 14. The requirements for a bomber-type aircraft are shown in Fig. 6.1*b* of the reference; at low speeds the minimum value of $pb/2V$ is again quoted as 0·07 and at higher speeds the limiting value decreases with increasing speeds. This reference also states that the peak value of the rolling acceleration shall not occur later than $3{\cdot}75\bar{c}/V$ seconds or 0·1 second, whichever is the greater, after reaching full aileron deflection and that the initial response to lateral control should always be in the correct direction (i.e. there should be no initial reversal of response to control movement due to adverse aileron yaw).

A more recent discussion of the lateral control requirements of aircraft is given in Ref. 17. The response of an aircraft to lateral control is shown in Fig. 16.4. The rolling characteristics of the aircraft can be defined in terms of the time constant of the motion T_R (i.e. the time taken for the rate of roll to reach 63 per cent of its

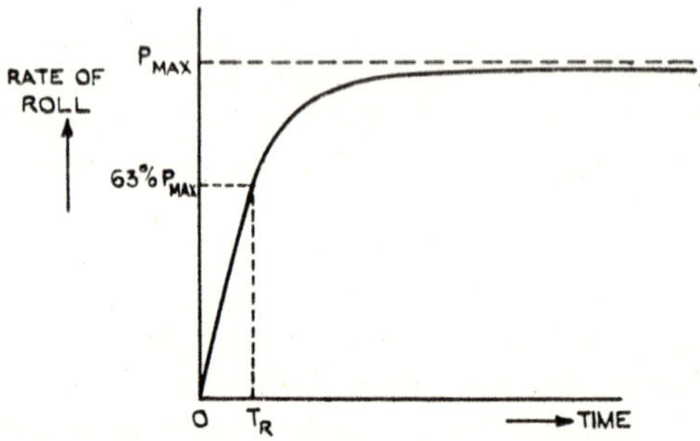

Fig. 16.4. Rate of roll response characteristics

final steady value) and the final steady rate of roll p_{max}. As we have seen in Chapter 11, if the time constant of the motion is small compared with the time interval over which the control is applied, then the ailerons are said to act as "rate" controls, whereas, if the two times are of the same order, they act as "acceleration" controls. In the case of a fighter aircraft, the response can usually be described as satisfactory if the time constant is less than about 1·0 or 1·5 sec. The corresponding time for a bomber or transport aircraft is of the order of 2 or 3 sec. The control characteristics are likely to be unacceptable if T_R exceeds about 6 sec. An acceleration-type control is not very satisfactory, because the increase in effort and concentration required to make rapid and precise adjustments to attitude causes pilot opinion to deteriorate. An acceleration control can be compared to trying to steer a marble around on a horizontal sheet of cardboard.

If the time constant of the motion is less than the limiting value quoted above, then the pilot's assessment of the lateral handling characteristics of an aircraft depends upon the control power available. Fig. 16.5, taken from Ref. 17, shows the various possible limitations for a fighter-type aircraft (Fig. 4 of the reference shows the corresponding limitations for a bomber or transport aircraft). The lower limit of the "satisfactory" area of the graph can be seen to correspond to a steady rate of roll of about 60°/sec. Ref. 17 shows that this limitation also corresponds approximately to a

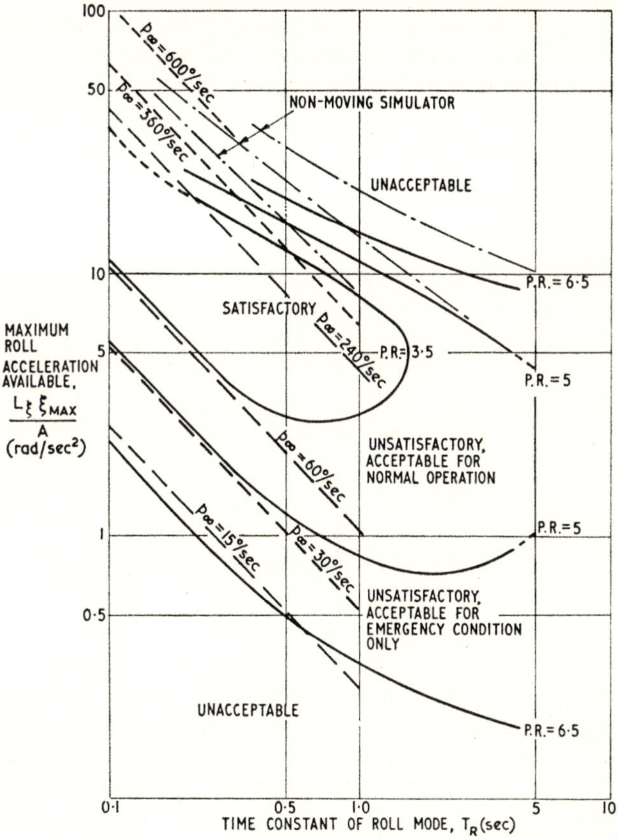

Fig. 16.5. Roll response. Pilot opinion boundaries for fighter-type aircraft [From Ref. 17]

definite bank-and-stop manoeuvre. This is a more practical type of manoeuvre, particularly in the case of a bomber/transport aircraft, where the requirement for this type of manoeuvre on the approach will probably determine the necessary aileron power. In the case of a fighter-type aircraft, Ref. 17 shows that the lower limit of 60°/sec corresponds approximately to a bank-and-stop manoeuvre of

1·5 radians in 2 sec. The report also discusses possible requirements for lateral control forces and movements.

4.3. Steady Rate of Roll Characteristics

The steady rate of roll performance of an aircraft can be expressed in terms of either the maximum rate of roll that can be achieved or the corresponding value of $pb/2V$ (see Chapter 11). The variation of the steady rate of roll performance of a fighter-type supersonic aircraft at two different heights is shown in Fig. 16.6. The rate of

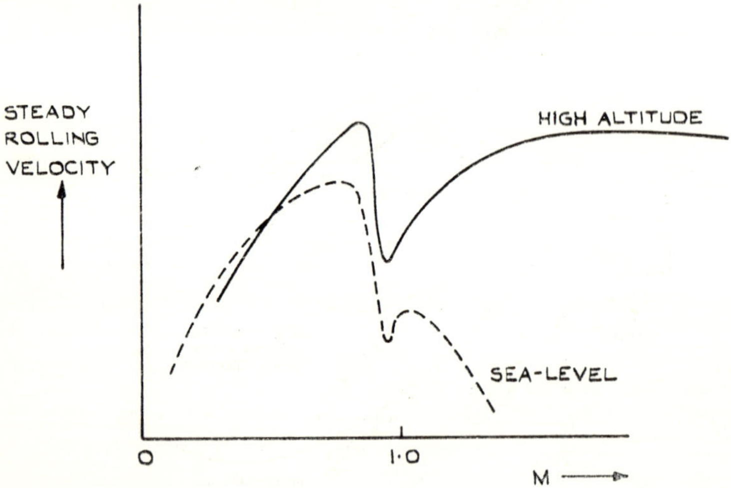

Fig. 16.6. Typical steady rate of roll performance for a supersonic aircraft

roll increases with speed in the subsonic range and then decreases abruptly at transonic speeds. At high altitudes, the rate of roll increases at supersonic speeds and tends to level off at a high value of rate of roll. At sea-level, the rate of roll is reduced by the effects of aeroelasticity. In this case, the steady rate of roll achieved at supersonic speeds at a high altitude is large compared to the corresponding sea-level values.

The roll response characteristics of an aircraft may be appreciably reduced by the effects of adverse aileron yaw and inertia coupling. In each case, when the aircraft rolls a sideslip develops, and the resulting rolling moment arising from the dihedral effect opposes the motion of the aircraft. Thus the rolling effectiveness of the lateral control is reduced. The problem of roll coupling is described in detail in para. 7; the adverse sideslip develops as a result of the tendency of some modern high-performance aircraft designs to roll around the principal inertia axis. On slender delta aircraft the adverse rolling effect arising from the sideslipping motion may be so large that the aircraft may be reluctant to roll beyond a certain angle of bank, an equilibrium condition being reached at which the applied aileron rolling moment and the dihedral rolling moment just balance one another (see Chapter 21). In this rather unusual case, the aileron control can be said to act as a "position-" or "displacement-" type control. In practice, oscillation may be superimposed on the rate of roll response and, after a time delay, the angle of bank may continue to increase.

The results of some low-speed carrier approach tests carried out on the Douglas F4D-1 are described in Ref. 19. When the ailerons were moved slowly, the adverse aileron yaw caused sideslip to develop, with little or no rolling motion. Abrupt aileron displacements resulted in a rolling motion about the highly inclined longitudinal axis of the aircraft. Large sideslip angles were developed and the rate of roll actually reversed in direction.

Flight tests on the Convair 106A (Ref. 20) showed that the sideslip developed during rolling motions was adverse under some conditions of flight, and favourable under others. A turn coordinator was incorporated in the control system of the aircraft in order to maintain the sideslip at zero. The system senses the rate of yaw, the rate of roll, and the position of the elevons and automatically applies rudder to coordinate the manoeuvre. When adverse sideslip is present it is reduced by the turn coordinator and the rate of roll is correspondingly increased. When favourable sideslip exists, the turn coordinator again applies corrective rudder, but this time the rate of roll is reduced.

4.4. Bank-and-Stop Manoeuvre

As we have seen, a more useful criterion of rolling performance on modern aircraft is the bank-and-stop (or the roll-and-stop) type of manoeuvre. This particular manoeuvre is analysed theoretically in Ref. 21 (pages 342–6), and an example calculation is performed in Ref. 22 (Data Sheet EG8/2). The calculation is made for a 90° banked turn from straight and level flight, and a comparison is made between the results obtained for a Spitfire flying at 188 kts at sea-level, and a hypothetical supersonic aircraft flying at $M = 2{\cdot}5$ at 53,000 ft. In the case of the Spitfire, the initiating aileron input can be held for a longer time and the overall time to 90° is less than that for the supersonic aircraft. In order that the aircraft should finish up at the required angle of bank with zero rate of roll, the ailerons must be reversed at some point during the manoeuvre, i.e. the ailerons must initially be moved in the conventional direction to begin the roll and then reversed in order to check the manoeuvre and to reduce the rate of roll to zero. The relative duration of the checking input depends upon the design of the aircraft and on modern aircraft relatively more checking is required than on older designs.

An alternative way of expressing the rolling performance of an aircraft is in terms of the angle that the aircraft would roll through in a given time in a roll-and-stop manoeuvre. Fig. 16.7 shows the

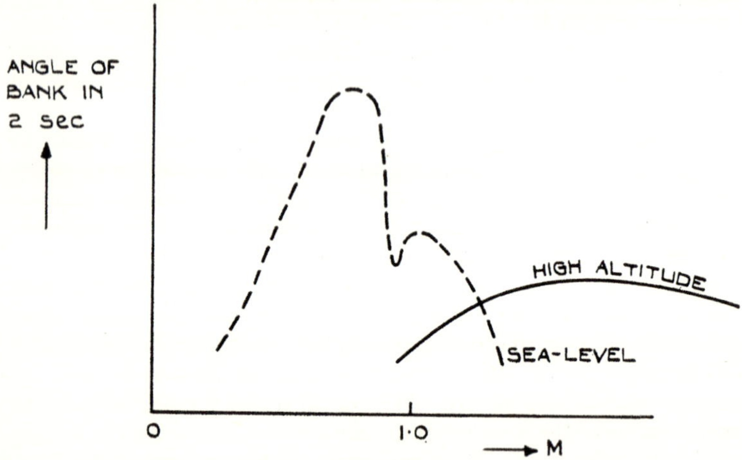

Fig. 16.7. Typical roll-and-stop manoeuvre performance for a supersonic aircraft

angle rolled through in 2 sec by a hypothetical supersonic aircraft. These values should be compared with those of the corresponding steady rates of roll shown in Fig. 16.6. Although the aircraft can achieve high values of steady rates of roll at high altitudes, its bank-and-stop manoeuvring capability is poor compared to that at sea-level. The reason for this is that at high altitude the ailerons have to be reversed much earlier than at sea-level.

5. Lateral Dynamic Stability

5.1. General

In Chapter 9 it was shown that a conventional aircraft design possessed three lateral modes

(i) a rapid rolling subsidence mode,
(ii) a spiral mode,
(iii) a Dutch Roll oscillatory mode.

The characteristics of these modes are determined by calculating the roots of the lateral stability quartic. The values of the coefficients of the quartic equation are dependent on the values of the lateral stability derivatives and the rolling and yawing inertias of the aircraft. These parameters are themselves dependent on such factors as the forward speed of the aircraft, the height at which it is flying and the particular configuration being used. The characteristics of the lateral modes will thus vary considerably over the speed and height range of a high-performance aircraft. Generally, however, the lateral modes retain their "classical" forms, and the changes that occur are only in the numerical values and, possibly, the signs, of the time constants, the damping, etc.

The rolling subsidence mode determines the nature of the response of the aircraft to lateral control; this has already been discussed in para. 4.2. The spiral stability of an aircraft is dependent on the value of the parameter $(l_v n_r - n_v l_r)$; for stability this parameter should be positive. The effect of Mach number on the separate derivatives, and hence on the overall value of the spiral stability parameter, will depend on the design configuration of the aircraft concerned. A reduction in the value of $-l_v$ (due, possibly, to a reduction in the sweepback contribution at high speeds) and an increase in l_r would result in a loss of spiral stability. Spiral instability is not a serious problem, however, unless the time to double amplitude is less than about 20 sec and the main problems usually arise with the Dutch Roll oscillatory mode.

5.2. Dutch Roll Oscillation

The characteristics of the Dutch Roll oscillation at high speeds can be analysed by the same methods that are used at low speeds (see Chapters 9 and 17). The factors on which the Dutch Roll characteristics depend and the way in which they can be varied in order to modify the oscillatory characteristics are discussed in Chapter 9 (para. 2.4.5). The changes in the design configurations of modern aircraft, and the increased speeds and heights at which they fly, result in an increase in μ_2, the relative aircraft density parameter, and in the value of the ratio of the inertias i_C/i_A, as well as varying the values of the basic stability derivatives.

In Ref. 23, a theoretical analysis is made of the lateral stability characteristics of three possible supersonic aircraft designs over a range of speeds and heights. A comparison is made of the results obtained from exact and approximate theories. The results are presented graphically in the form of the periodic time, the number of cycles to half amplitude and the roll/yaw ratio plotted against Mach number. A number of graphs are given showing the effect of varying the individual parameters (e.g. l_v, n_r, i_E, i_A, i_C and l_p) on the oscillatory characteristics of the aircraft at a Mach number of 2·5 at

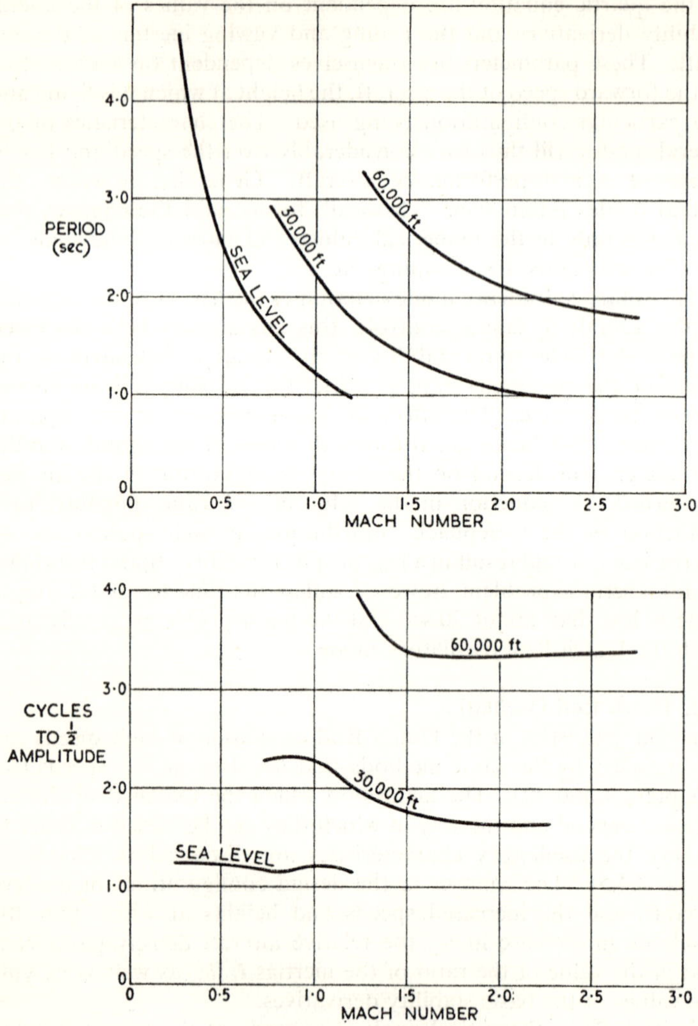

Fig. 16.8. Lateral oscillation periods and dampings of a supersonic fighter showing effects of height and speed [From Ref. 23]

53,000 ft. Figs. 16.8 and 16.9, taken from this reference, show the general effects of speed and height on the oscillatory characteristics and the effects on the damping of varying the three derivatives

n_r, l_v and l_p. From Fig. 16.8 it can be seen that in the case considered the periodic time decreases with increasing Mach number and increases with increasing height. The number of cycles to half amplitude generally decreases slightly with increasing Mach number and increases with height. Fig. 16.9 shows that an increase in $-n_r$, the damping-in-yaw derivative, has an appreciable effect on the damping of the oscillation, whereas variations in l_p and l_v do not have much effect.

The results of a similar theoretical analysis are discussed in Ref. 24. Methods are determined for estimating the effects of changes in the

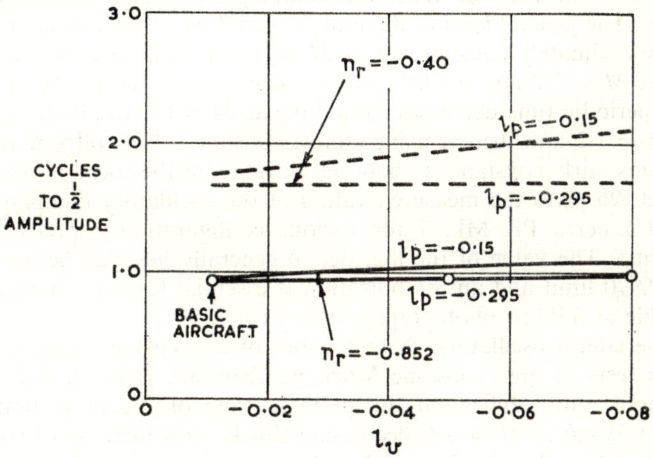

Fig. 16.9. Effect of varying l_v and n_r on the period and damping of the lateral oscillation ($M = 2{\cdot}5$, height 53,000 ft) [From Ref. 23]

lateral derivatives and the aircraft inertias on the characteristics of the roll subsidence mode, the spiral stability, and the Dutch Roll oscillation of an aircraft. The methods are used to calculate the actual effects on four different aircraft and the various results are all shown graphically. The results show the relative importance of the different parameters in determining the characteristics of the various lateral modes, and indicate the best way of modifying these characteristics. They also indicate that the effect of a change in a particular parameter on one aircraft may be different from the effect of the same change on another aircraft.

In Ref. 25 the Time-Vector Method has been used to determine the characteristics of the Dutch Roll oscillation throughout the speed and height range of the aircraft. The periodic time decreases with increase of speed at subsonic speeds and remains fairly constant or increases slightly up to $M = 1{\cdot}7$. It increases appreciably with height. The value of the log. dec. decreases with height and increases gradually at supersonic speeds. The roll/yaw ratio decreases with speed at subsonic speeds and increases gradually at supersonic speeds, the values increasing with increase of height.

The results obtained from some actual flight tests on various aircraft are given in Refs. 6, 26 and 27. The results of the Dutch Roll oscillation tests carried out on the Fairey Delta 2 at 10,000 ft and 40,000 ft over the speed range of $M = 0{\cdot}4$ to $1{\cdot}5$ are given in Ref. 6. Fig. 16.10, taken from this reference, shows the measured values of the periodic time, the damping ratio, and the roll/yaw ratio. The general level of damping is very low; the damping ratio is approximately constant up to $M = 0{\cdot}9$, it then decreases up to about $M = 1{\cdot}2$ and finally increases slowly again up to $M = 1{\cdot}5$. The periodic time decreases from 3 sec at $M = 0{\cdot}7$ to about $1\frac{1}{2}$ sec at $M = 1{\cdot}2$ and then remains almost constant. The roll/yaw ratio remains fairly constant at about 4·0 throughout the speed range.

Ref. 26 gives the measured values of the oscillatory damping of the Canberra PR Mk. 7 for various configurations, speeds and heights. The value of the log. dec. δ generally lies well below the Av.P.970 limit and autostabilization is essential for this aircraft to be able to fulfil its photographic survey role.

The lateral oscillatory characteristics of the Vulcan measured in flight tests at high subsonic Mach numbers are given in Ref. 27. Without autostabilization the periodic time of the oscillation is about $5\frac{1}{2}$ sec at $M = 0{\cdot}8$, decreasing slowly with increase of speed and the log. dec. is about 0·2 at $M = 0{\cdot}8$, increasing slowly with speed. The aircraft does not meet the Av.P.970 damping requirement but, because of the long periodic time involved, the aircraft is not difficult or unpleasant to fly. Under approach conditions (i.e. 150 kts, undercarriage down and airbrake open) the damping is very low ($\delta \simeq 0{\cdot}3$), and the roll/yaw ratio high ($p/r > 3{\cdot}0$). The comments of the pilot on the handling characteristics of the aircraft were again not unduly adverse, however, presumably because of the long periodic time ($T > 8{\cdot}0$ sec). The yaw damper was very effective, particularly at high altitudes, and provided a value of damping very close to the Av.P.970 requirement.

5.3. Fuel Surge

The problem of fuel surge (or fuel slosh) is discussed in para. 9.5*g* of Ref. 14. Under certain conditions of flight, coupling may occur between the motion of the fuel in the tanks and the rigid-body aircraft modes. This may lead to a reduction in the damping of the Dutch Roll oscillation, and it may be a particularly important

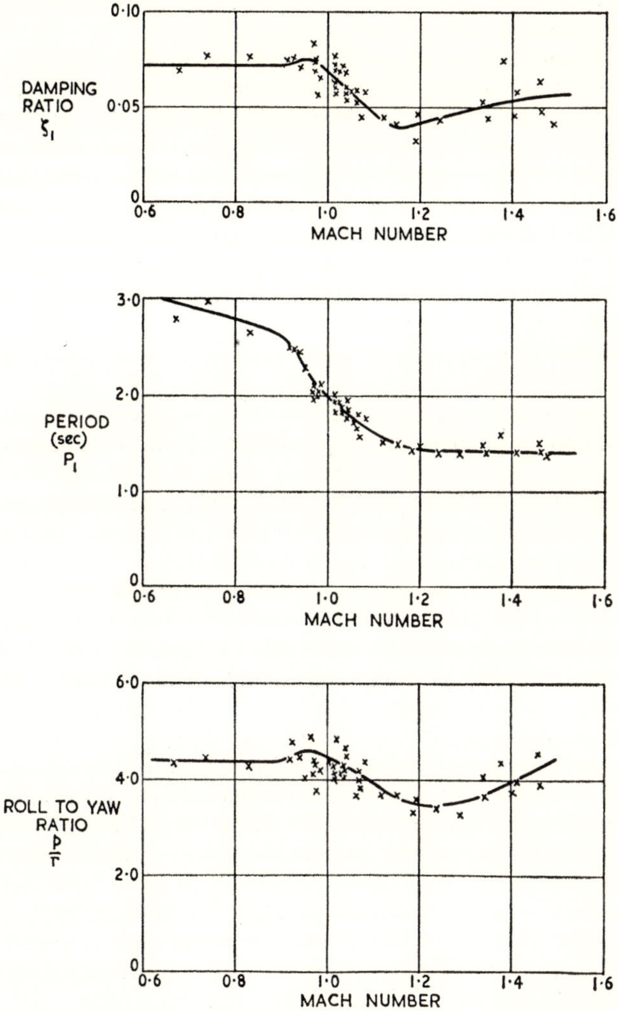

Fig. 16.10. Characteristics of the Dutch Roll oscillation in level flight at 40,000 ft. Fairy Delta 2. [From Ref. 6]

problem on modern jet aircraft because of the large amounts of fuel carried and the unfavourable position of the fuel tanks.

5.4. Response to Lateral Gusts

The dynamic stability characteristics of an aircraft also determine its response to a lateral gust. This problem is discussed in Refs. 17 and 28; the analysis is simplified by assuming that the aircraft's response to gusts is mainly determined by the roll subsidence mode. An important parameter in the response is the value of the acceleration in roll produced by a given sidegust. Since $\dot{p} = L/A$, where L is the rolling moment produced by the lateral gust velocity v_g and A is the rolling inertia, then, substituting $i_A ms^2$ for A and $l_v v_g \rho VSs$ for L, the rolling acceleration per unit gust velocity is given by

$$\frac{\dot{p}}{v_g} = \rho g \frac{l_v}{i_A} \frac{V}{(W/S)\,(b/2)} \tag{16.1}$$

This formula shows that the magnitude of the aircraft's response is increased by an increase in dihedral effect l_v and true airspeed V, and also by a decrease in rolling inertia i_A, wing loading W/S and wing span b. Fig. 13 of Ref. 28 illustrates the results for various aircraft designs; the calculated values of $\dot{p}/v_g$ range from about 0·0025 rads/sec^2 per ft/sec in the case of the Britannia, to 0·0065 for the Boeing 707, to over 0·02 for some project designs.

When the pilot senses the response of the aircraft to a sidegust, he opposes it by application of his aileron control. Ref. 28 shows the way in which the maximum bank angle of the aircraft produced by a sharp-edged gust varies with the pilot's delay time and the aileron control power.

Conventional straight wing aircraft are less sensitive to gusts and thus the control of atmospheric disturbances is not an important factor in determining the amount of lateral control power required. This is usually determined by considerations of manoeuvrability. Modern designs are much more responsive to gusts, however, and the reasons are shown clearly in eqn. (16.1). Consequently, on these aircraft, the amount of lateral control power that is required is dictated more by the problems of crosswind landings and atmospheric turbulence.

Ref. 17 discusses these results, and relates them to the pilot's handling assessment of the aircraft. The report suggests, as a rough guide, that, under approach conditions, a large aircraft can be classified as at least "acceptable for normal operation" provided that the bank following a 10 kt velocity sidegust can be limited to 5° by the use of not more than one-half aileron. In some cases very powerful ailerons may be required to satisfy this criterion and the control power available may be very close to the upper limit required for manoeuvring flight. This may lead to a conflict between the control requirements for manoeuvring flight and those for atmospheric gusts.

6. Autostabilization Systems

6.1. General

The basic stability of an aircraft varies throughout the speed and height range and, under certain conditions of flight, it may become small or even negative. The characteristics of the basic lateral modes were discussed in Chapter 9 and simple analyses were made to show the effects of the major aerodynamic derivatives. The characteristics of the modes can be altered by changing the values of the derivatives. If this cannot be done aerodynamically, by modifying the design of the aircraft, then it has to be done artificially by installing an automatic control system (see Chapter 19). In order to modify a particular lateral derivative, the motion of the aircraft is measured and the resulting signal used to deflect the appropriate rudder or aileron control surface to produce the desired effect (e.g. to modify the value of n_r, the rate of yaw is measured, and the signal used to deflect the rudder in the appropriate direction).

The effect of an automatic control system on the lateral characteristics can be determined by substituting the modified values of the derivatives in the complete equations of motion, or in the approximate equations derived from them.

In Chapter 9 it was shown that the roll subsidence mode was determined by the derivative l_p and the spiral mode by the derivatives l_v, n_r, n_v and l_r. In the Dutch Roll oscillation, the periodic time is mainly determined by n_v, the damping by n_r and the roll-yaw ratio by either l_v, l_p and n_v (eqn. 9.31) or l_v and n_v (eqn. 9.32). The effects of varying the different parameters were discussed in para. 2.4.5 of Chapter 9. An increase in dihedral effect (i.e. an increase in $-l_v$) was found to increase the roll-yaw ratio of the oscillation and at the same time improved the spiral stability. An increase in yaw damping (i.e. an increase in $-n_r$) is obviously beneficial in increasing the damping of the oscillation and it also improves the spiral stability. An increase in roll damping (i.e. $-l_p$), on the other hand, increases the damping of the roll subsidence mode and decreases the roll-yaw ratio of the oscillatory mode (eqn. 9.31). Ref. 29 shows that an increase in the damping derivatives n_r and l_p produces an increase in the total damping of the aircraft, whereas a change in the values of the cross-derivatives l_r and n_p only produces a redistribution of the overall damping.

The effects of an automatic control system on the lateral characteristics of an aircraft can be represented by any of the various graphical methods discussed in Chapter 9. One-parameter diagrams may be used which show the variation of the oscillatory characteristics T and $t_{1/2}$ (or μ and ω) plotted against the particular derivative being varied. Alternatively, root locus plots showing how the roots of the stability equation vary, or two-parameter diagrams showing how the point representing the aircraft moves with respect to the stability boundaries, may be used.

In addition to the effects of the automatic system on the oscillatory, spiral and roll subsidence modes, it is important to assess the effects on the overall handling characteristics of the aircraft. Many systems produce an improvement in oscillatory damping, only at the expense of some other unsatisfactory handling characteristic. Ref. 29 shows that a large value of $-l_p$, for example, results in poor rolling manoeuvrability, and that large values of n_p and l_r result in severe apparent spiral instability.

Unsatisfactory handling characteristics may arise as a result of cross-control effects. In practice, when the ailerons are deflected, the resulting motion does not consist of a pure roll. There are generally yawing moments associated with the ailerons as well as rolling moments. If the yawing moments are adverse, then when the aircraft is rolled using the ailerons, a sideslip develops and the resulting rolling moment produced by the dihedral effect opposes the applied rolling moment from the ailerons. To eliminate this effect, an automatic aileron-rudder interconnect system is often used, so that any aileron movement automatically produces some rudder movement in the appropriate direction to cancel out the aileron yawing moment.

Within recent years, a number of research aircraft have been fitted with automatic control systems which enable many of the stability and control derivatives to be varied over a wide range. These variable stability aircraft (see Chapter 19) have enabled investigations to be carried out of the effect of all the various derivatives.

6.2. Effect of Varying Derivatives

A number of reports (Refs. 29 and 30) discuss the effects of varying the numerical values of a number of derivatives. Ref. 29 describes the results of an analytical and free-flight model series of investigations on the effect of variations in the values of the lateral rotary derivatives (i.e. n_r, l_p, n_p and l_r) on the lateral stability and control characteristics of a sweptback wing fighter-type aircraft. Most of the calculations were based on an idealized automatic system with zero time lag, though some calculations allowed for the effect of a small constant time lag (see para. 6.6). The different derivatives were varied separately over a wide range of values on either side of the basic condition. The resulting effects on the lateral oscillatory and general handling characteristics of the aircraft are summarized diagrammatically in Fig. 16.11, taken from Ref. 29. The only

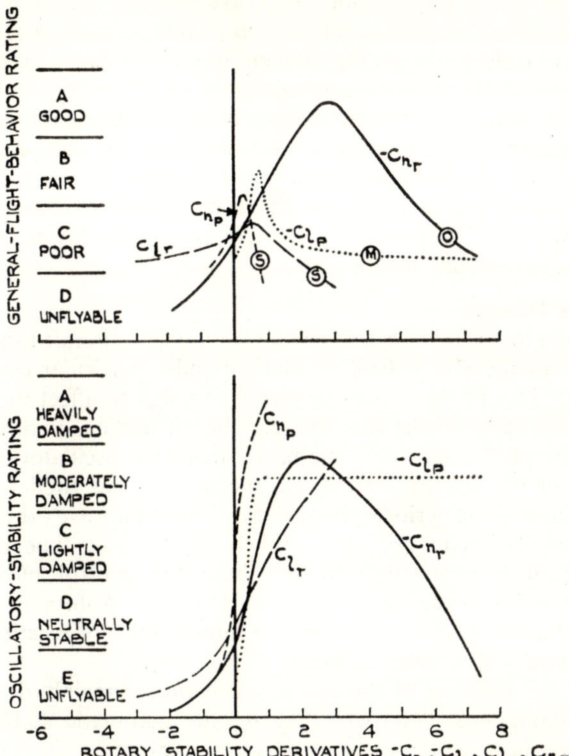

Fig. 16.11. Effect of varying the rotary derivatives on the lateral characteristics [From Ref. 29]

derivative that provides an increase in the damping of the oscillatory mode, without adversely affecting other handling characteristics, is n_r, the damping-in-yaw derivative. The results show that an increase in yaw damping (i.e. in $-n_r$) initially improves the damping of the oscillatory and spiral modes, the roll convergence mode becoming slightly less damped. Too much yaw damping, however, causes the lateral oscillatory characteristics to deteriorate again, while the two convergent modes coalesce to produce a long period very heavily damped oscillatory mode. The general handling characteristics of the aircraft also improve initially with increase of yaw damping, but deteriorate at very high values of $-n_r$.

Ref. 30 discusses the results of an idealized theoretical investigation of the effect of auxiliary damping systems on the lateral stability and control characteristics of a high-speed aircraft. The derivatives varied are n_r, n_p, l_p, etc., and the corresponding response characteristics of the aircraft (i.e. rate of roll p, rate of yaw r and sideslip angle β) to rolling and yawing moment inputs are shown graphically. The effects of the various derivatives on the characteristics of the oscillatory, spiral and roll subsidence modes are compared in tables in the reference. An increase in $-n_r$ increases the damping of the lateral oscillatory and spiral modes, without having much effect on the roll subsidence mode or on the general handling characteristics of the aircraft. The possibility of oscillatory instability at very high values of $-n_r$ is also mentioned.

6.3. Yaw Dampers

One of the main stability deficiencies on modern aircraft is a lack of damping of the Dutch Roll oscillation under particular conditions of flight. This problem may be serious enough to affect the operational capability of the aircraft (e.g. bomb aiming, photography, etc.). The pilot can modify the basic stick-fixed oscillatory characteristics of the aircraft by moving the aileron and rudder control surfaces and, if the periodic time is large, he can improve the damping of the oscillation in this way. When the periodic time is short, however, it becomes difficult to phase the control movements properly and the instability may be made worse. Automatic control systems have the advantage over the human pilot of reacting more quickly and with greater accuracy.

Since the damping of the lateral oscillation is mainly dependent on the damping-in-yaw derivative n_r, the automatic stabilization system needs to be designed to make n_r more negative. This type of system is referred to as a "yaw-damper." A block diagram of a simple yaw-damper system is shown in Fig. 16.12. A rate gyro is used to measure the rate of yaw response of the aircraft and the signal is fed back in such a way that the rudder moves with the appropriate phase to increase the damping. The control law may be expressed in its simplest form by

$$\zeta = K_r r \tag{16.2}$$

where K_r is the gearing of the control surface. The behaviour of the complete autostabilizer/airframe system can be analysed in the various ways discussed in Chapter 19, each "block" in the system being represented by the appropriate transfer function. Due to the large change in the stability characteristics and the control effectiveness over the speed and height range of the aircraft, it is usual to incorporate automatic switches that will vary the magnitude of the yaw damper gain as required.

The effects of a yaw-damper on the low speed stability of a large sweptback wing aircraft are discussed in Ref. 31. The basic aircraft

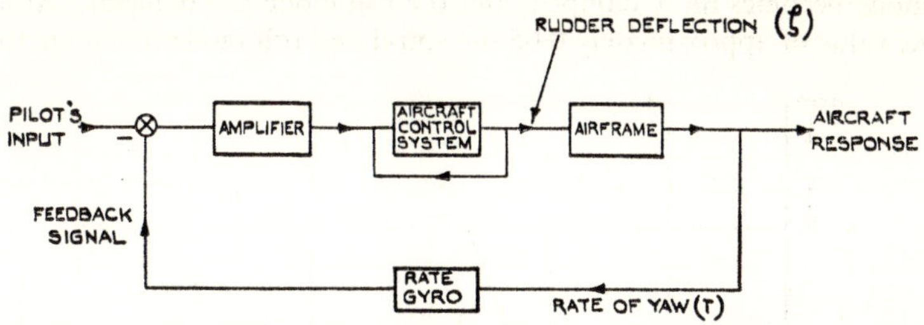

Fig. 16.12. Block diagram of a yaw-damper system

is spirally unstable and has a lightly damped Dutch Roll oscillation, but with the damper operating, the aircraft is spirally stable and the damping of the oscillation is considerably improved. The effect of

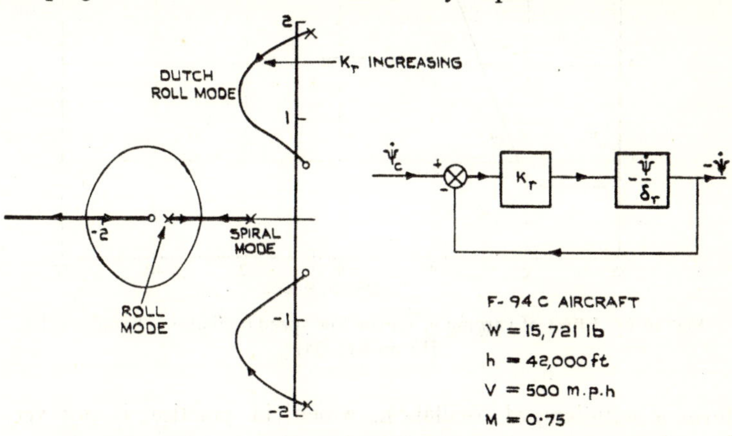

Fig. 16.13. Effect of a yaw damper on the lateral characteristics of an aircraft [From Ref. 32]

the yaw damper is to increase the value of n_r by a factor of about 3, while leaving the value of n_v unchanged.

Ref. 32 discusses the theoretical effects of a yaw damper system on the lateral characteristics of the F94C aircraft for a particular condition of flight (h = 42,000 ft, T.A.S. = 500 m.p.h., M = 0·75, W = 15,721 lb). The results are presented in the form of a root locus plot. Fig. 16.13, taken from Ref. 32, shows that, at the selected condition of flight, the spiral mode of the aircraft is convergent and the Dutch Roll divergent. An increase of yaw damper gain K_r to a value of 0·18 degrees of rudder movement per degree per second of yaw rate makes the Dutch Roll stable and an increase of K_r to 1·0 makes the oscillation well damped. At the same time, the spiral mode becomes more damped, and the roll mode less damped. At a K_r value of approximately 0·65 the spiral and roll modes combine to

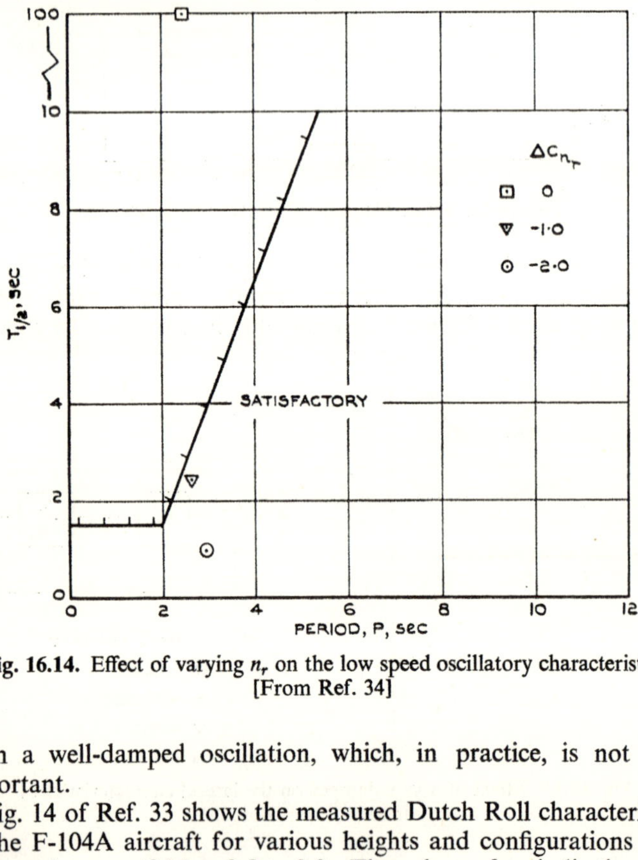

Fig. 16.14. Effect of varying n_r on the low speed oscillatory characteristics [From Ref. 34]

form a well-damped oscillation, which, in practice, is not very important.

Fig. 14 of Ref. 33 shows the measured Dutch Roll characteristics of the F-104A aircraft for various heights and configurations over the speed range of $M = 0{\cdot}5$ to 2·0. The values of periodic time and damping are given for both yaw damper "on" and yaw damper "off" conditions of flight.

In an idealized yaw-damper, the rudder motion related to the rate of yaw will only affect the n_r derivative and there will be no time lag between the motion of the aircraft and the corresponding rudder deflection. In practice there will be a small time lag in the system (see para. 6.6), and, because the gyro spin axis is not aligned with the wind axis, some variations in the derivatives n_p, l_r and l_p will also occur. The results of a theoretical investigation on the lateral characteristics of an aircraft are discussed in Ref. 34. The analysis is carried out for a low speed and for speeds of $M = 0{\cdot}8$ and $1{\cdot}2$. The effect of varying n_r on the low speed oscillatory characteristics is shown in Fig. 16.14, taken from Ref. 34; the periodic time is not altered appreciably, but the time to half amplitude is considerably reduced. The effect of a constant time lag in the system, assumed to be 0·1 sec, was investigated and was shown to have a negligible effect.

6.4. Roll Stabilizers

A roll-damper is an automatic system which modifies the value of the damping-in-roll derivative l_p. The effects of a roll damper on the oscillatory characteristics are shown in Fig. 16.15 taken from Ref. 35. The damping of the oscillation is expressed in the form of the reciprocal of the number of cycles to half amplitude $1/C_{1/2}$, and the roll/yaw ratio by ϕ/v_e. The installation of the roll-damper has little effect on the damping of the oscillation, but the roll/yaw ratio is considerably reduced. On the slender delta designs of the future, however, the Dutch Roll oscillation may involve more rolling motion and a roll-damper may be more effective than a yaw damper in improving the damping of the lateral oscillation.

Ref. 32 discusses the theoretical effect on the rolling motion of a B-47 aircraft of a system in which the aileron deflection is proportional to the measured error in the roll angle, i.e. $\delta_a = K_a(\phi_c - \phi)$ where δ_a is the aileron deflection, ϕ_c is the required angle of bank, ϕ the actual angle of bank and K_a the gain of the system. The effect of varying the value of the gain is shown in Fig. 16.16. The basic aircraft possesses oscillatory stability and spiral divergence. An increase in K_a improves the spiral mode, but has little effect on the Dutch Roll oscillation, and increases the time constant of the rolling convergence mode.

6.5. Effect of Cross-control Derivatives

Ref. 36 presents the results of a theoretical investigation of the lateral dynamic stability characteristics of a variable sweep aircraft design with the wings swept at 75° flying at $M = 3{\cdot}0$ at two different heights (60,000 ft and 70,000 ft). The motion is analysed using the classical linearized equations of motion based on the principal inertia system of axes. The values of the periodic time, the damping and the roll/yaw ratio of the oscillation are determined, and the effects of varying the roll damper and yaw damper gains investigated. The effects of the cross-control derivatives (i.e. l_ζ and n_ξ)

are found to be particularly important. The effect of the aileron yaw is to transfer some of the damping from the Dutch Roll and spiral modes to the roll mode, and the oscillatory damping may decrease with an increase in the gain of the roll damper. The effect

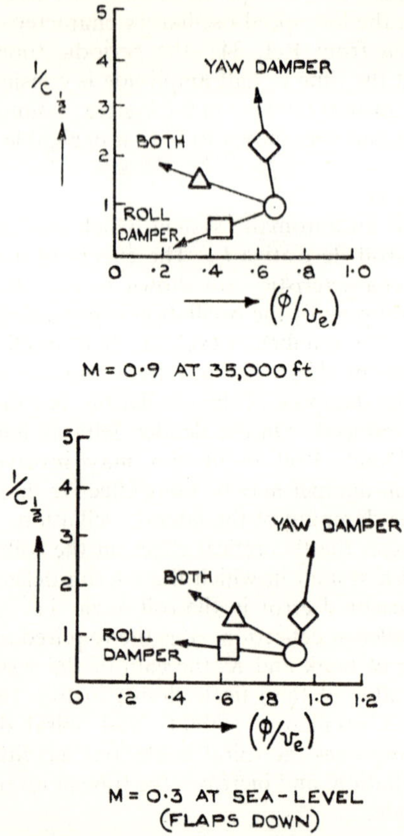

Fig. 16.15. Typical effects of a roll and yaw damper on the lateral oscillatory characteristics [From Ref. 35]

of the roll with rudder is also to redistribute the damping between the various modes, but in this case the Dutch Roll damping increases with increase of yaw damper gain. The low value of n_v of the particular aircraft considered may also result in some roll-coupling problems.

Ref. 37 discusses the results of a similar theoretical investigation of the effects of the cross-control derivatives on the characteristics of an aircraft flying at 125,000 ft at $M = 6{\cdot}0$. When roll and yaw dampers are incorporated, the cross-control derivatives may have destabilizing effects and, under certain conditions, the lateral motions become unsatisfactory.

Ref. 38 discusses the results of a theoretical investigation carried out on possible aircraft configurations designed for $M = 3{\cdot}0$. The

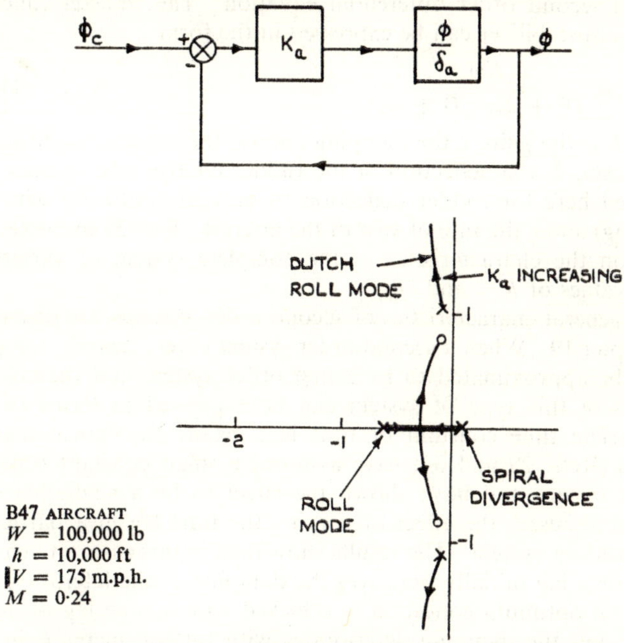

Fig. 16.16. Effect of a roll stabilizer on the lateral characteristics [From Ref. 32]

analysis employs the linearized lateral equations of motion referred to body axes to determine the periodic time, the time to half amplitude, and the roll/yaw ratio of the lateral oscillation over a range of angles of attack at a height of 75,000 ft. The analysis is extended to include a study of various roll and yaw damper gains and the cross-control effects. The graphical presentations show the effect on the roll mode, the spiral mode and the oscillatory mode of varying the major derivatives (e.g. l_p and n_r) and also indicate the acceptability or otherwise of the aircraft's handling qualities.

6.6. Effect of Characteristics of the Autostabilizer System

In many of the theoretical analyses of autostabilizer systems, it is assumed that the control surface moves instantaneously with respect to the measured response of the aircraft (i.e. it is assumed to be a perfect system of constant gain with a zero time lag, etc.). In a more accurate analysis of a complete autostabilizer/airframe system, however, it is necessary to allow for the dynamic characteristics of the autostabilizer itself. Ref. 39 shows that the actual frequency response characteristics of most yaw dampers can be approximated to by a second order differential equation. The transfer function of the autostabilizer can be expressed in the form

$$\frac{\delta_r}{r} = \frac{K\omega_n^{\,2}}{D^2 + 2\zeta\omega_n D + \omega_n^{\,2}} \tag{16.3}$$

where K is the gain, ζ the damping and ω_n the natural frequency of the system, δ_r the deflection of the rudder control (the symbol ζ is not used here for rudder deflection to prevent confusion with the damping) and r the rate of yaw of the aircraft. Ref. 39 discusses the effect on the characteristics of the complete system of variations in the values of K, ζ and ω_n.

The general characteristics of second order systems are discussed in Chapter 19. When a second order system is very heavily damped, it can be approximated to by a first order system and the characteristics of this type of system can be expressed in terms of the appropriate time constant or time lag. Many theoretical investigations (Refs. 29 and 34) have assumed a small constant time lag for the system and have shown the effect to be a negligible one. Ref. 40 discusses the effect of varying the time lag in a particular autostabilizer system. The results show that, in this case, an increase in the time lag initially improves the damping of the lateral oscillation. An optimum condition is achieved with a time lag of about 0·2 sec, but the damping deteriorates with further increase in the lag.

7. Inertia Coupling

7.1. General

In the majority of our work on stability and control, we have assumed that the longitudinal and lateral motions can be considered separately (i.e. all the cross-coupling terms can be neglected). This has been permissible until fairly recently (except in the analysis of the spin), because the inertia effects involved have been small, in comparison with the associated aerodynamic ones. However, Phillips (Ref. 43) has pointed out that the equations of motion of an aircraft contain terms which couple the longitudinal and lateral motions together and that when an aircraft with large pitching and yawing inertias and a small rolling inertia is executing rapid rolls, the inertia coupling effects can result in pitch or yaw divergence. Large angles of incidence and sideslip may develop on the aircraft in the rolling manoeuvre, or in the recovery from it, and the aerodynamic loads produced may overstress the structure and even, possibly, cause a structural failure.

The main reasons for the increased relative importance of the inertia effects are

(*a*) The change in the basic design shape of modern aircraft. Typical supersonic aircraft have very small, thin wings, and the weight of the aircraft tends to be concentrated in the fuselage. This leads to a very small value of the rolling inertia (I_x or A) and large values of the pitching (I_y or B) and yawing (I_z or C) inertias.

(*b*) A reduction in the aerodynamic restoring moments (i.e. the longitudinal static stability [M_α or M_w] and the lateral static stability [N_v]) under the conditions of flight involved in the manoeuvres (e.g. high Mach number, high altitude, large incidences and sideslip angles). The value of N_v is reduced at high Mach numbers due to the decrease in the lift-curve slope of the vertical fin and at high angles of attack due to the possibility of the fin operating in a turbulent airflow, or in the vortices produced by the high fineness ratio fuselages used on supersonic aircraft. Similarly at high incidences, the value of M_α may decrease and even, under certain conditions, reverse (i.e. "pitch-up").

The magnitude of the inertia coupling effects depend on the speed and height at which the aircraft is flying, and the amount of g which is being pulled. To clear a particular aircraft for service use, it is necessary to co-ordinate computer studies with flight tests to determine any particular critical flight areas. If any such areas exist, then it may be necessary to lay down certain limitations for the aircraft (in the form of maximum permissible aileron angles, maximum angles of roll, etc.), since it is quite impossible for the pilot to co-ordinate his controls in such a way as to eliminate the coupling effects. In addition to yaw and pitch divergence, another type of instability may arise due to inertia coupling effects. This instability is called autorotation; it is not as yet a serious problem because it is only liable to happen under rather adverse conditions (e.g. a rapid roll executed during a push-over manoeuvre).

7.2. Simple Considerations

7.2.1. SPECIAL CASES

Before going on to consider the more complicated aspects of inertia coupling, it is extremely useful and instructive to consider first of all some simple cases, in order to get the basic principles firmly established. As has been mentioned in para. 7.1, the vital factor involved in inertia coupling considerations is the ratio of the inertia effects to the aerodynamic effects. Let us consider the rolling characteristics of an aircraft in two very special idealized cases, firstly the case of large inertia and zero aerodynamic effects and, secondly, that of zero inertia and large aerodynamic effects.

(*a*) *Predominant Inertia Effects*

When the moments of inertia of an aircraft are large, and the aerodynamic stabilities zero, then, when the aircraft rolls, it will roll about its principal axis of inertia. If this axis is inclined to the flight path, there will be a cyclic interchange of incidence. (*Note.* This incidence is the incidence of the principal inertia axis) and sideslip (Fig. 16.17.)

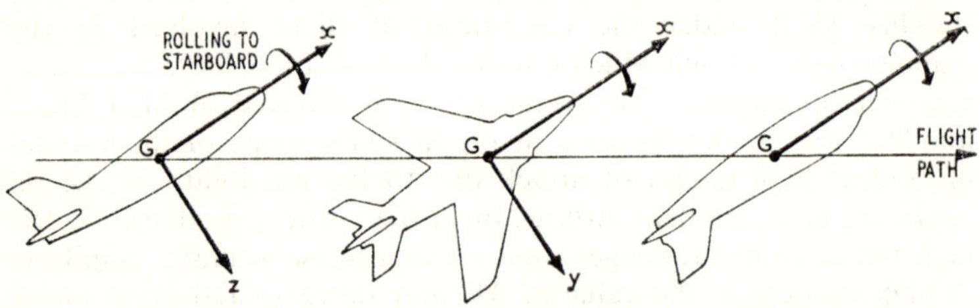

Fig. 16.17. Roll response—inertia effects predominant

In this motion, large sideslips may be developed, and the maximum value of the sideslip will be numerically equal to the initial incidence. The variation of the incidence α and slideslip β angles can be represented by

$$\alpha = \alpha_0 \cos pt \tag{16.4}$$

$$\beta = \alpha_0 \sin pt \tag{16.5}$$

where α_0 is the initial incidence, p is the rate of roll, and t is the time.

Though the values of α and β can show wide variations, the motion is still a stable one—no divergence or instabilities of any sort are present.

(*b*) *Predominant Aerodynamic Effects*

We are, however, rather more familiar with the opposite state of affairs in which the aerodynamic stabilities are large, and the inertia effects are zero. In this case (Fig. 16.18), a rolling aircraft will roll around the flight path, maintaining throughout a constant angle of attack (because of the strong longitudinal static stability) and a constant (i.e. zero) sideslip angle (because of the strong directional stability). Once again, the motion involved is a stable one.

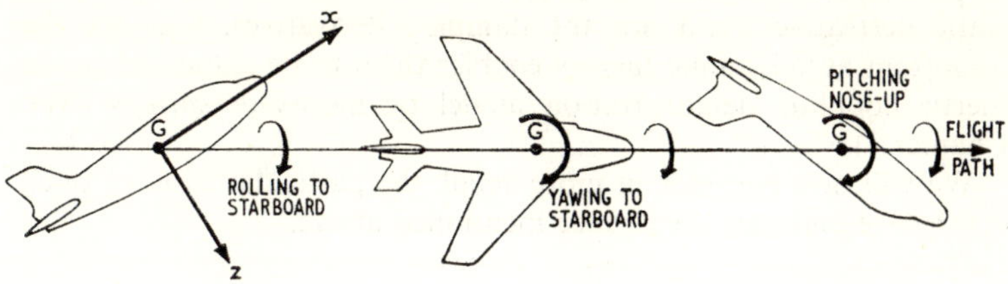

Fig. 16.18. Roll response—aerodynamic effects predominant

7.2.2. GENERAL CASES

In both the above special cases, the motion involved is a stable one. In practice, however, an aircraft lies somewhere between these two extreme cases, because the relative importance of the inertia and aerodynamic effects varies, depending on the design of the aircraft and the speed and height at which it is flying. In these more general cases, involving both aerodynamic and inertia effects, large variations may occur in the angles of incidence and sideslip and, under certain conditions, divergence may occur in either pitch or yaw. The primary requirements for the early development of inertia coupling effects are

(*a*) a large value of pitching inertia B and hence of yawing inertia C, compared to the roll inertia A. [For simplicity, in the mathematical analysis, it is often assumed that $C \simeq A + B$],

(*b*) a marked difference between the longitudinal static stability M_α or M_w and the directional static (or weathercock) stability N_v. When M_α is large and N_v is small, the aircraft will tend to be divergent in yaw and when M_α is small and N_v is large, the aircraft will tend to be divergent in pitch.

Before going on to consider these cases in more detail, it should be noted that the aircraft possesses certain natural frequencies of oscillation. The natural frequency of the uncoupled pitching oscillation is dependent on the longitudinal *static* stability $[\propto\sqrt{(-M_\alpha)}]$ and the natural frequency of the uncoupled yawing or directional oscillation is dependent on the directional *static* stability $[\propto\sqrt{(N_v)}]$. It will be shown later that divergence only occurs when the value of the rolling frequency lies *between* these two natural frequencies. When the rolling frequency is very low (below both the natural frequencies), or very high (above both the natural frequencies), then divergence does not occur and the motion is stable. The results obtained would be the same as those described in para. 7.2.1*b* and *a* respectively. Finally, the fact that it is the static derivatives, and not the damping derivatives, that are the important aerodynamic factors enables them to be simulated on an inertia coupling demonstration model by means of springs (Ref. Chapter 7).

We will now consider in more detail, the particular cases of pitch divergence and yaw divergence mentioned above.

(*a*) *Pitch Divergence*

When the longitudinal stability is small and the weathercock stability is large, inertia coupling effects result in a divergent motion in pitch.

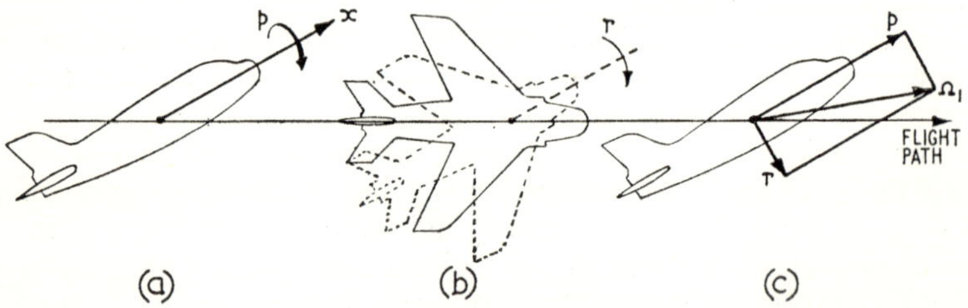

Fig. 16.19. Pitch divergence

Consider the aircraft shown in Fig. 16.19*a* rolling to starboard at a rate p. If the aircraft has an initial incidence of α_0 then, as described in para. 7.2.1*a*, if it were not for the aerodynamic stability effects, the aircraft, after rolling through 90°, would have developed a sideslip to starboard, numerically equal to the initial incidence. This position is shown by the dotted outline in Fig. 16.19*b*. However, as a result of the large weathercock stability that we have assumed, a yawing moment will be developed that will yaw the aircraft back onto the flight path (i.e. nose to starboard), so as to reduce the sideslip angle. The actual position of the aircraft after it has rolled through 90° is thus shown by the full-line outline in Fig. 16.19*b*. As a result of this, the aircraft is rotating about two axes; at a rate p about the principal inertia axis and at a rate r about another axis perpendicular to it. These two rates of rotation can be combined into a single resultant angular velocity $\Omega_1[=\sqrt{(p^2 + r^2)}]$ about an axis inclined at an angle below the principal inertia axis given by $\tan^{-1}(r/p)$ (Fig. 16.19*c*).

If, for simplicity, the inertia distribution of the aircraft is represented by concentrated masses placed at the extremities of the aircraft, then it can be seen that the centrifugal force effects are such that the yawing inertia masses produce a nose-up pitching moment and the rolling inertia masses produce a nose-down pitching moment. If the yawing inertia C is greater than the rolling inertia A, then there will be overall a nose-up pitching moment resulting from the inertia effects. It is shown in Appendix E that this inertia moment is given by

$$M_i = +(C - A)pr \tag{16.6}$$

If the longitudinal stability is small, then the aerodynamic restoring moment in pitch will be too small to balance this inertia moment and the motion will become divergent in the nose-up direction.

The direction of the pitch divergence does not depend on the direction of rotation, but it does depend on the initial inclination of the principal inertia axis. See Table 1.

Table 1

	ROLL TO PORT (i.e. negative roll)	ROLL TO STARBOARD (i.e. positive roll)
Positive incidence	Nose-up (positive) (negative yaw)	Nose-up (positive) (positive yaw)
Negative incidence	Nose-down (negative) (positive yaw)	Nose-down (negative) (negative yaw)

It should be noted that the results shown in Table 1 are consistent with eqn. (16.6) [i.e. a positive roll and a positive yaw produces a positive (i.e. nose-up) pitching moment, etc.].

(*b*) *Yaw Divergence*

When the longitudinal stability is large and the weathercock stability is small, inertia coupling effects result in a divergent motion in yaw. Consider the aircraft rolling to starboard at a rate p. If the aircraft had an initial incidence of α_0, then, if it were not for the aerodynamic stability effects, the aircraft would have developed a sideslip after rolling through 90° and an incidence of $(-\alpha_0)$ after rolling through 180°, a position shown by the dotted outline in Fig. 16.20*a*. However, as a result of the large pitching stability that we have assumed, a pitching moment will be developed that will pitch the nose of the aircraft "up" (actually in the downwards direction when the aircraft is inverted) and the actual position of the aircraft will be shown by the full line outline in Fig. 16.20*a*. Once again, the aircraft is rotating about two axes; at a rate p about the principal inertia axis and at a rate q about another axis perpendicular to it. These two rates can be combined into a single resultant angular velocity $\Omega_2[=\sqrt{(p^2+q^2)}]$ about an axis inclined at an angle to the principal inertia axis given by $\tan^{-1}(q/p)$, Fig. 16.20*b*.

If the inertia distribution of the aircraft is again represented by concentrated masses, then it can be seen that the centrifugal force effects are such that yawing moments are produced, that from the

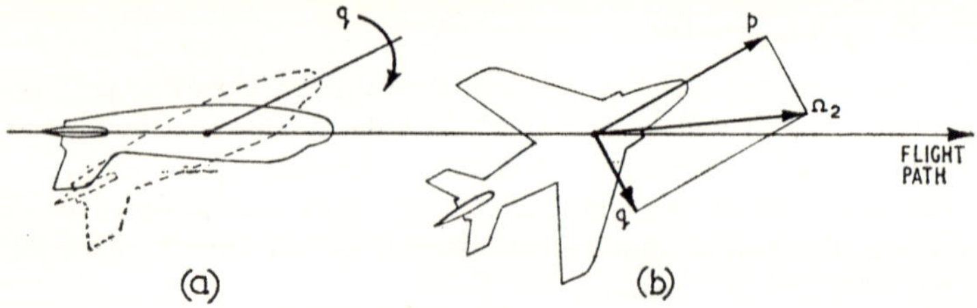

Fig. 16.20. Yaw divergence

pitching inertia weights tending to increase the sideslip, and that from the rolling inertia weights tending to reduce the sideslip. If $B > A$, there will be a resulting inertia yawing moment given (para. 7.3.2) by

$$N_i = -(B - A)pq \tag{16.7}$$

and if the directional stability is small, the motion will become divergent in yaw.

The direction of the yaw divergence depends on the direction of roll and the initial incidence of the principal inertia axis. For $B > A$ see Table 2.

Table 2

	ROLL TO PORT (i.e. negative roll)	ROLL TO STARBOARD (i.e. positive roll)
Positive incidence	Yaw to starboard (positive) (positive pitch)	Yaw to port (negative) (positive pitch)
Negative incidence	Yaw to port (negative) (negative pitch)	Yaw to starboard (positive) (negative pitch)

It should be noted that the results shown in Table 2 above are consistent with eqn. (16.7) [i.e. if $B > A$, a positive roll and a positive pitch produces a negative (i.e. yaw to port) yawing moment, etc.].

Though the explanation that is usually given for inertia coupling considers the effect of the rotation of the aircraft on weights distributed along the fuselage and wings, an alternative explanation can be given in terms of gyroscopic precession. Because the aircraft is both rolling and pitching it tends to yaw, and because it is rolling

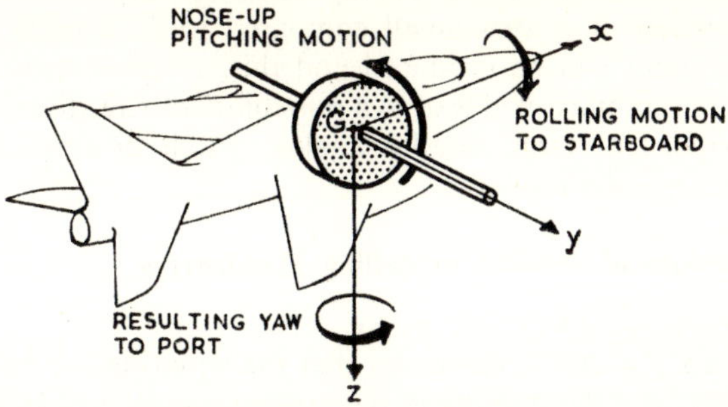

Fig. 16.21. Gyroscopic explanation of inertia coupling—the *Apq* yawing moment term

and yawing it tends to pitch. Let us consider first of all the problem of yaw divergence. The rolling inertia of the aircraft can be represented by a rotor with the axis of rotation in the direction of the longitudinal $0x$ axis (Fig. 16.21). If the aircraft is rolling to starboard and pitching nose-up, then, in accordance with the gyroscopic

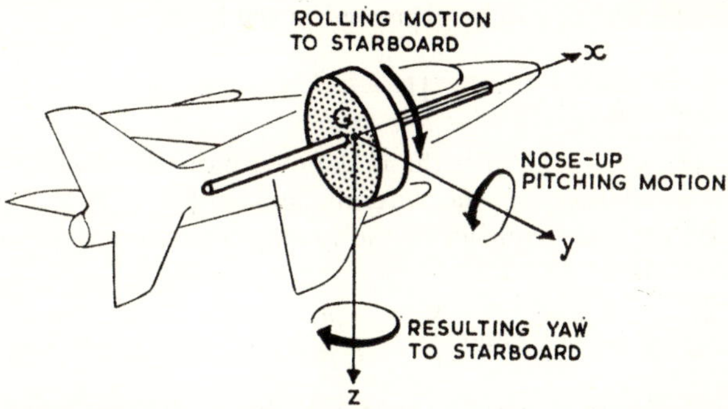

Fig. 16.22. Gyroscopic explanation of inertia coupling—the *Bpq* yawing moment term

rule of precession, it will tend to yaw to starboard. (This explanation is similar to that for the gyroscopic "swing on take-off" experienced by propeller-driven aircraft.) In a similar way, the pitching inertia of the aircraft can be represented by a rotor with the axis of rotation in the direction of the $0y$ axis (Fig. 16.22). If the aircraft is pitching nose-up and rolling to starboard, then it will also tend to yaw to port. Thus there are two inertia effects, acting in opposite directions. As long as the rolling inertia A and the pitching inertia B are nearly equal to one another, the resulting inertia effect will be small and it will be easily counteracted by the directional static stability of the aircraft N_v which opposes the yawing tendency. On modern designs, however, where A is very small compared to B, a large resultant yawing moment will be produced and this, coupled with a loss of aerodynamic stability, will lead to a possibility of yaw divergence and the occurrence of large sideslip angles. Pitch divergence can be considered in a similar way.

7.3. Conditions of Stability in Rolling Manoeuvres

7.3.1. CRITICAL RATES OF ROLL

In Appendix E it has been shown that the equations of motion can be simplified and the stability of the motion represented by a quartic equation. When an aircraft is rolling at a constant rate p_0, the motion of the aircraft usually consists of two oscillatory modes, both of which involve some coupling between the longitudinal and lateral motions. It is possible, however, for certain values of the rate of roll, for the coefficient E_1 to become negative. In these circumstances, the long period oscillatory motion of the aircraft breaks down into two aperiodic modes, one of which is divergent and the other convergent. The aircraft then diverges either in pitch or yaw.

The coefficient E_1 can be expressed in the form

$$E_1 = \left(\frac{C-A}{B} p_0^2 - \omega_\theta^2\right)\left(\frac{B-A}{C} p_0^2 - \omega_\psi^2\right) \qquad (16.8)$$

where $\omega_\theta[=\sqrt{(-M_w/B)}]$ is the undamped natural frequency of the pitching oscillation of the aircraft and $\omega_\psi[=\sqrt{(N_v/C)}]$ is the undamped natural frequency of the yawing oscillation. It should be noted that, since

$$-\frac{M_w}{B} = -\frac{\mu_1 m_w}{i_B \hat{t}^2} \text{ and } \frac{N_v}{C} = \frac{\mu_2 n_v}{i_C \hat{t}^2}$$

the frequencies (and periodic times) of the oscillations can also be expressed in terms of the non-dimensional derivatives m_w and n_v (see Chapters 8 and 9).

The aircraft becomes unstable when E_1 is negative, i.e. when one (but not both) of the factors in eqn. (16.8) is negative. Thus, instability occurs when

$$\frac{B}{C-A}\omega_\theta^2 < p_0^2 < \frac{C}{B-A}\omega_\psi^2 \qquad (16.9)$$

or when

$$\frac{B}{C - A}\,\omega_\theta^2 > p_0^2 > \frac{C}{B - A}\,\omega_\psi^2 \tag{16.10}$$

where p_0 is the steady rate of roll in rads/sec. Eqn. (16.9) corresponds to the occurrence of pitch divergence (i.e. a small longitudinal static stability and a low longitudinal natural frequency), whereas eqn. (16.10) corresponds to the occurrence of yaw divergence (i.e. a small directional static stability and a low directional natural frequency). Since $A + B \simeq C$, $B/(C - A)$ can be approximated to unity and hence eqns. (16.9) and (16.10) can be written

Pitch divergence

$$\omega_\theta^2 < p_0^2 < \frac{C}{B - A}\,\omega_\psi^2 \tag{16.11}$$

Yaw divergence

$$\omega_\theta^2 > p_0^2 > \frac{C}{B - A}\,\omega_\psi^2 \tag{16.12}$$

These stability conditions can be represented graphically (Fig. 16.23). If $(\omega_\theta/p_0)^2$ is plotted vertically and $(\omega_\psi/p_0)^2$ horizontally, then the conditions become

Pitch divergence

$$\left(\frac{\omega_\theta}{p_0}\right)^2 < 1 \text{ and } \left(\frac{\omega_\psi}{p_0}\right)^2 > \frac{B - A}{C}$$

Yaw divergence

$$\left(\frac{\omega_\theta}{p_0}\right)^2 > 1 \text{ and } \left(\frac{\omega_\psi}{p_0}\right)^2 < \frac{B - A}{C}$$

The regions of instability are shown shaded on Fig. 16.23; the critical values are unity on the vertical axis and $(B - A)/C$ on the horizontal axis. It can be seen that when $A = 0$, the latter critical value is also unity, but as A increases, the critical value decreases, until, when $A = B$, the critical value is zero.

These stability criteria can be expressed in terms of the forward speed V_i. From eqn. (16.11), pitch divergence will not occur until a certain critical rate of roll is exceeded, i.e.

$$p_0^2 > \omega_\theta^2$$

But since $\omega_\theta^2 = -\dfrac{M_w}{B} = -\dfrac{m_w \rho_0 V_i^2 S l'}{B}$, this criteria can be expressed

$$p_0^2 > -\frac{m_w S l'}{B}\,\rho_0 V_i^2 \tag{16.13}$$

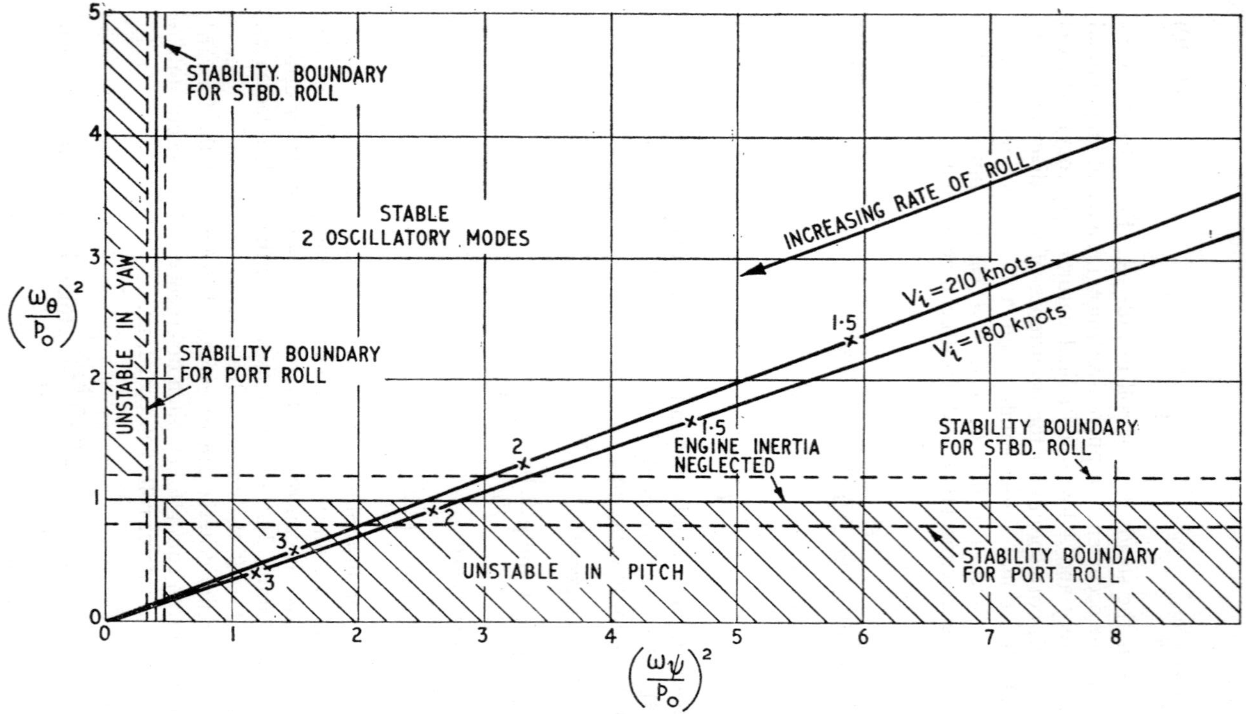

Fig. 16.23. B.P. Delta. Stability of the rolling aeroplane

Note—Numbers adjoining crosses are rates of roll (rad/sec) [From Ref. 47]

Similarly, the criteria for yaw divergence is that

$$p_0^2 > \frac{n_v S s}{B - A} \rho_0 V_i^2 \qquad (16.14)$$

On Fig. 16.23 the characteristics of an aircraft with different steady rates of roll are shown, for two different conditions of flight (V_i = 180 kts and 210 kts). At low steady rates of roll the aircraft is stable and it does not diverge either in pitch or yaw. As the steady rate of roll is increased, a condition is reached at which the motion becomes divergent in pitch. If the value of the steady rate of roll could be further increased, the aircraft would eventually become stable again. It should be noted that the slope of the line representing the characteristics of the aircraft is given by $(\omega_\theta/\omega_\psi)^2$ or $(-m_w/n_v)$

i.e. $\dfrac{\text{the aircraft's longitudinal static stability}}{\text{the aircraft's lateral static stability}}$

7.3.2. COMPLICATING FACTORS

The results of para. 7.3.1 above have to be modified to allow for the effects of various additional factors.

(a) Effect of Damping Terms

If the calculations are performed with all the damping terms retained, then the effect on the stability boundaries is as shown in Fig. 16.24

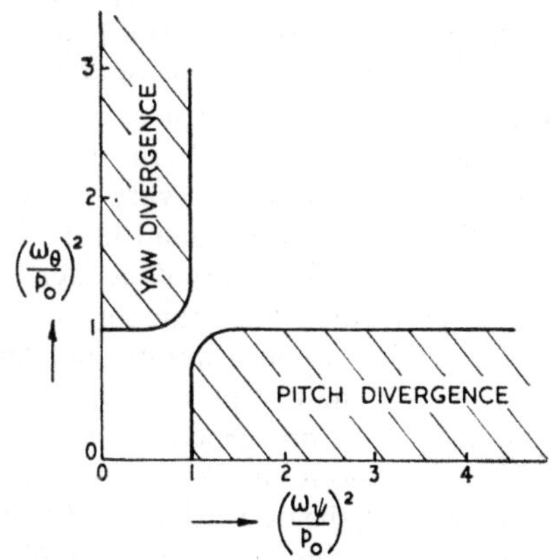

Fig. 16.24. Effect of damping on the stability boundaries ($A/B = 0$)

(for $A/B = 0$). As can be seen, the damping terms have little effect, except in the region of the intersection of the two critical lines, where they have the effect of improving the stability.

(*b*) *Effect of Engine Rotation*

The effect of the momentum of the rotating parts of the engine is shown in Fig. 16.23 for the case where the engine rotation is anti-clockwise viewed from the rear. The effect is asymmetric; when the aircraft is rolling in a sense opposite to the rotation of the engine, divergence occurs at a slightly lower rate of roll than in the case when the aircraft and engine rotate in the same sense. For an assessment of the more severe condition, therefore, rolling should be performed in the sense *opposite* to the rotation of the engine.

7.4. Aircraft Response to Aileron Inputs

In para. 7.3.1 above, rather theoretical conditions have been assumed of an aircraft rotating at different steady rates of roll; nothing has been said about the entry into the roll, the recovery from it, or even about the transition from one steady value of the rate of roll to another. These considerations, however, have been useful, in that they have given us a general appreciation of the problem and a reliable guide as to the necessary criteria for avoiding any divergent motion.

In practical cases, however, the pilot applies aileron, and the rate of roll builds up, possibly approaching or attaining a steady rate of roll. The pilot then applies opposite aileron, and the rate of roll decreases again. To assess the severity of the inertia coupling phenomenon (i.e. the peak loads produced) it is necessary to investigate the response of an aircraft to this type of control input, because it is quite possible for the largest incidences and sideslip angles to be obtained after the controls have been reversed. The actual response of an aircraft can, however, vary quite a lot between different individual flight tests and different individual pilots. This is because the response is very critically dependent on the way the rolling manoeuvre is performed (e.g. rate of control application, the exact phasing between the application of the ailerons, the elevator, and the rudder, etc.). The computer studies are based on an idealized aileron input, and the one generally used is that of the step aileron input. The calculations are based on tunnel and (when available) flight test values of the derivatives, and cover a wide range of aileron angles, speeds, heights, etc. These results are then used to draw up a safe flight test programme and the flight test results obtained from this programme are compared with the original computer estimates. The values of the derivatives are then modified, until the two sets of results agree as closely as possible (Ref. 46). This new set of derivatives is then used on the computer to extend the flight régime, and when this has been cleared on the computer, the flight test programme is also extended into these new conditions of flight. In this way, the computer programme is kept one step ahead of the flight test programme, and the computer estimates are constantly being checked against flight test records to make sure that the values of the derivatives being used are as accurate as possible. Nowadays, the computer studies form an essential part of the flight test clearance programme for any aircraft.

The response characteristics of aircraft to various types of aileron control inputs are discussed in Refs. 41, 42 and 44. In Ref. 41, curves are given showing the way in which the rate of roll, the angle of attack, and the angle of sideslip, vary with time for a standard step aileron input. The maximum changes in the values of the angle of attack $\Delta\alpha_{max}$ and angle of sideslip β_{max} are shown to be dependent on such factors as the aerodynamic and inertia characteristics of the aircraft, the speed and height at which it is flying, the amount of aileron applied, and the angle of bank rolled through before the control is centralized. Fig. 26 of Ref. 41 shows that if the same ratio of inertias is maintained, then increasing all the inertias by any factor greater than unity, up to at least 4, has very little effect on the values of $\Delta\alpha_{max}$ and β_{max}. Multiplication by a factor less than unity, however, shows the expected reduction in the values of incidence and sideslip. The values of $\Delta\alpha_{max}$ and β_{max} are found to be roughly proportional to the inclination of the principal axis. Thus the response of the aircraft will be worse in rolling pull-out manoeuvres.

Figs. 30 and 31 of Ref. 41 show that increasing the damping derivatives m_q and n_r produces a small reduction in $\Delta\alpha_{max}$ and β_{max}. Figs. 24 and 25 of Ref. 41 show, as might be expected, that an increase in m_w reduces the values of $\Delta\alpha_{max}$, without appreciably changing β_{max}, whereas an increase in n_v increases $\Delta\alpha_{max}$, with only a small reduction in β_{max}.

Eqns. (16.13) and (16.14) show that the speed and the height are important factors determining the critical roll rates. Speed has a beneficial effect in increasing the critical rate of roll and reducing the values of $\Delta\alpha_{max}$ and β_{max} (Fig. 22, Ref. 41). Because a constant Mach number corresponds to a lower value of V_i at the higher altitudes, the critical roll rates are decreased, and the corresponding $\Delta\alpha_{max}$ and β_{max} values increased (Fig. 23, Ref. 41). In spite of this, however, the most critical stressing case may well be found at high speeds at low altitudes.

If it is required to perform the rolling manoeuvre while maintaining constant incidence and zero sideslip throughout, then the pitching and yawing moments developed have to be balanced by elevator and rudder deflections. Figs. 37 and 38 of Ref. 41 show the control movements that are necessary to achieve this. The movements are very complicated ones, however, and it would be impossible for the pilot to coordinate the controls in the correct way.

7.5. Auto-rotation

Some of the computer studies in Ref. 41 made for α_0 values of $-5°$ indicated a new type of instability. Step aileron inputs were used and it was found that, on some occasions, when the aileron was removed the rate of roll increased very rapidly (Fig. 16.25). This

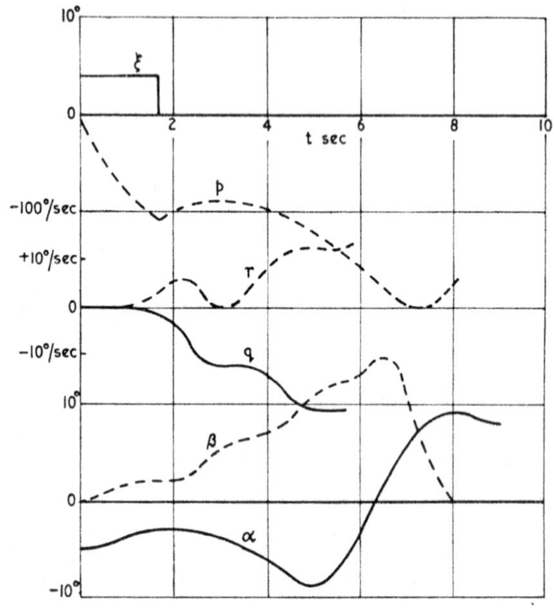

Fig. 16.25. Aircraft response showing auto-rotation [From Ref. 41] 40,000 ft, $M = 0{\cdot}8$, $\alpha_0 = -5°$

phenomenon is called Auto-rotation. (*Note.* This word is also used for the basic motion of the spin; the result is the same but the physical explanation is different.) Calculations have shown that auto-rotation can occur when the principal axis of inertia is inclined at a negative incidence (and even, sometimes, at small positive incidences) to the flight path; it is thus most likely to occur in push-over manoeuvres ($< 1g$ flight).

Though this phenomenon has not yet made any dramatic appearance in flight, it may well be something that will affect aircraft designs of the future. It was quite seriously considered in the investigation following the loss of an aircraft during rolling manoeuvres (Ref. 45). It was found that auto-rotation was a possible cause of the accident, though, in this particular case, an unlikely one, since it would have meant that the pilot had held on aileron in a push-over manoeuvre at about $-\frac{1}{4}g$ for the effects of auto-rotation to have broken up the aircraft.

The physical explanation of auto-rotation is shown in Fig. 16.26. If the aircraft is initially at a negative incidence, then, when it rolls to starboard the inertia coupling effects produce a nose-down pitching moment, as we have seen in para. 7.2.2*a*. When the aircraft pitches nose-down, additional inertia coupling effects yaw the aircraft to increase the sideslip to port, which is developed as a result

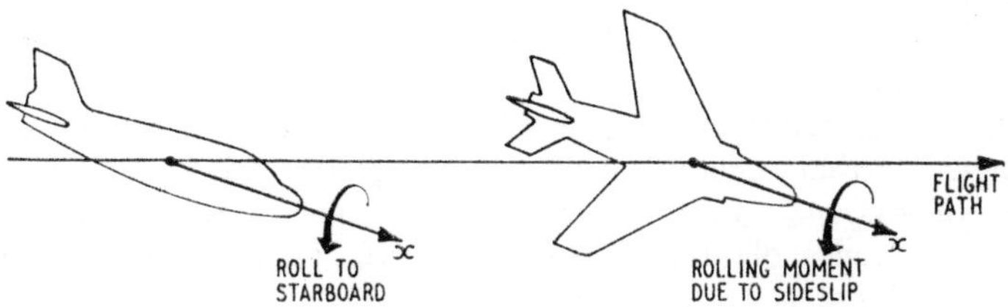

Fig. 16.26. Auto-rotation in roll resulting from inertia cross-coupling

of the interchange of angle of incidence and sideslip (para. 7.2.1*a*). The sideslip to port produces a rolling moment, due to the dihedral effect L_v, assisting the rolling motion of the aircraft to starboard. The rate of roll will build up, until an equilibrium position is reached at which the $L_v v$ moment to starboard just balances the opposing damping moment $L_p p$. This should be compared with what happens when $\alpha_0 > 0$; Fig. 16.17 shows that the sideslip developed is to starboard in a starboard roll and hence the rolling moment from the dihedral effect opposes the rolling motion of the aircraft.

A more detailed mathematical analysis has been carried out in Ref. 41, and auto-rotation has been shown to be possible when α_0 is less than a certain critical value. It can be shown that the critical value of α_0 is usually negative, though for particular values of m_w and n_v it can be positive. Damping in pitch is also an important factor; it moves the instability boundary to more negative values of incidence.

When $\alpha_0 < (\alpha_0)_{critical}$ there are two values of the rate of roll at which equilibrium conditions can be achieved. The smaller value represents a state of neutral equilibrium, however, and if this value, p_1, is exceeded, the aircraft's rate of roll builds up to the other equilibrium value p_2 which is a stable condition. This is equally applicable for positive and negative values of the rate of roll (i.e. rolls to starboard and rolls to port).

In Table 3 of Ref. 41, numerical values are given for p_1 and p_2 on the supersonic fighter design considered in this reference. Values of p_1 vary between 50 and 150°/sec; at $M = 0{\cdot}8$, 40,000 ft for example with $\alpha_0 = 0$, $p_1 = 94°$/sec and $p_2 = 117°$/sec, and with $\alpha_0 = -5°$, $p_1 = 80°$/sec and $p_2 = 125°$/sec. Similar calculations on a supersonic bomber project predicted instability occurring at a rate of roll as low as 20°/sec.

7.6. Flight Test Results

Inertia coupling problems have been encountered on a number of aircraft in the last few years. The first operational aircraft to exhibit a divergence tendency was the F-100; a typical time history of a manoeuvre that occurred at $M = 0{\cdot}7$ at 30,000 ft is shown in Fig. 8.32 of Ref. 48. In the more violent stages of the manoeuvre, negative angles of attack greater than 16° ($-4{\cdot}4g$), and sideslip angles as large as 25° were reached.

Fig. 8.31 of Ref. 48 shows a similar time history for a rolling manoeuvre of the X-3, which occurred at $M = 1{\cdot}05$ at 30,000 ft. The aileron was moved at about $t = 0{\cdot}6$ secs and reversed about 2 secs later, when the aircraft had reached an angle of bank of roughly 270°. Even when the aileron had been reversed, however, the aircraft still continued to roll to port and the sideslip to starboard continued to increase. The rate of roll reversed its direction at $t = 4$ sec, and the sideslip reached its maximum value of 20° a fraction of a second later, by which time the angle of bank had reached 450°. The incidence initially remained constant, but after 2 secs it began to change and after 4 secs it decreased by 14° in $\frac{1}{2}$ sec to $-12°$ ($-6{\cdot}7g$). Following a rearward stick movement, it increased by 30° again within 1 sec, and between $t = 4$ secs and $t = 10$ secs, both β and α varied in an oscillatory manner.

A series of inertia coupling tests carried out on the Boulton-Paul 111A delta wing aircraft ($A = 20 \times 10^6$ lb in², $B = 43{\cdot}5 \times 10^6$ lb in², $C = 58{\cdot}1 \times 10^6$ lb in²) is described in Ref. 47. The graph of $(\omega_\psi/p_0)^2$ against $(\omega_\theta/p_0)^2$ for this aircraft at speeds of 210 and 180 kts (E.A.S.) is shown in Fig. 16.23. As the steady value of the rate of roll is varied at a constant E.A.S., the aircraft comes closer to the divergent boundary, until, eventually, at a steady rate of roll of somewhere about 2 rads/sec ($\simeq 115°$/sec) the pitch divergence boundary is crossed. The divergence experienced by the aircraft with these rate of roll values at these low speeds is thus predominantly in pitch. More accurately determined values of the critical roll rates

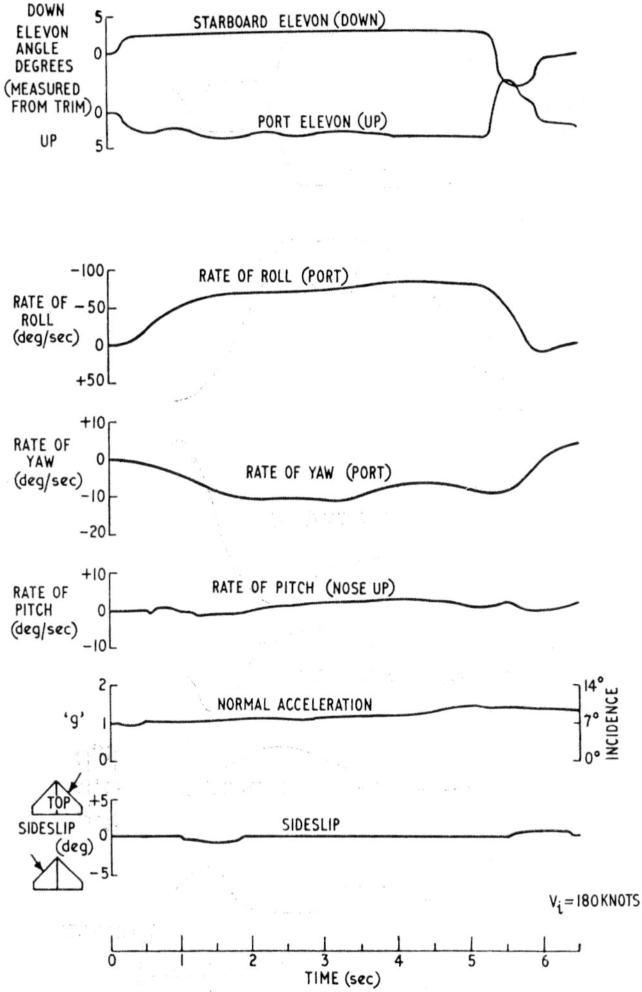

Fig. 16.27. Time history of roll below critical rate
[From Ref. 47]
$V_i = 180$ knots

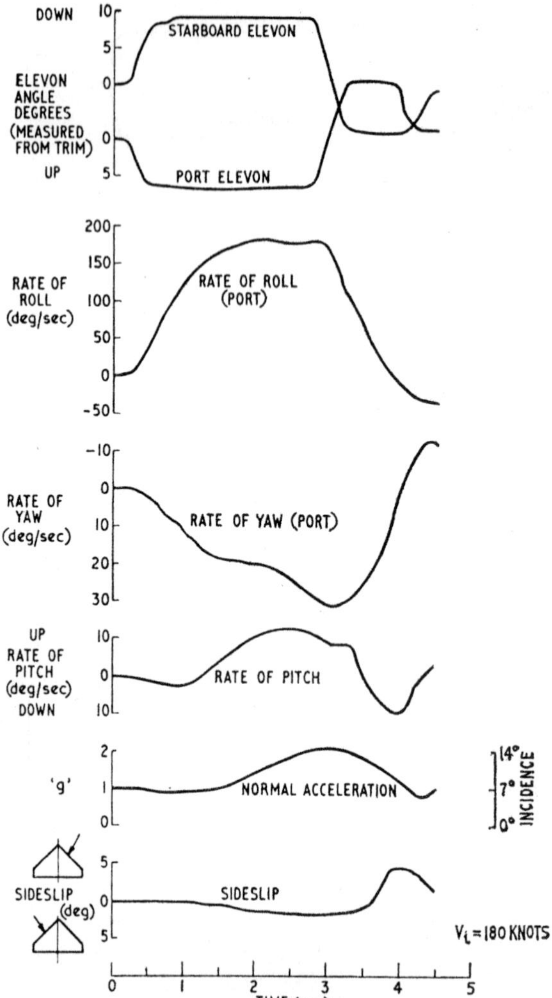

Fig. 16.28. Time history of roll above critical rate
[From Ref. 47]
$V_i = 180$ knots

are shown in Table 3. (*Note.* The engine rotation is anticlockwise viewed from the rear.)

Table 3

	$V_i = 210$ kts	$V_i = 180$ kts
Neglecting engine momentum effects	131°/sec	113°/sec
Including engine momentum effects, starboard roll	125°/sec	107°/sec
Including engine momentum effects, port roll	138°/sec	119°/sec

Figs. 16.27 and 16.28 show two actual trace records obtained during the tests; Fig. 16.27 corresponds to a rate of roll less than the critical, and Fig. 16.28 a rate of roll greater than the critical. In the latter case, the rate of pitch and rate of yaw are seen to increase much more rapidly and the maximum values of incidence (and g) and sideslip are much greater.

During the whole flight programme on the B.P. delta aircraft, pilots had little or no idea that inertia cross coupling effects were present. In particular, they could not appreciate the increase in incidence while rolling, even though the normal g often increased from 1 to more than 2. It was not until the aircraft started to buffet, in fact, that they realized that the g was excessive. However, the pilots did sometimes notice, at the end of a roll, the tendency of the aircraft to yaw.

References

1. Thomas, H. H. B. M.; *State of the Art of Estimation of Derivatives*, Agard Report No. 339, 1961.
2. Margolis, K., Bobbitt, P. J.; *Theoretical Calculations of the Stability Derivatives at Supersonic Speeds for a High-Speed Airplane Configuration*, NACA Research Memorandum RM L53G17, 1953.
3. Babister, A. W.; *Aircraft Stability and Control* (Pergamon Press, 1961).
4. Seckel, E.; *Stability and Control of Airplanes and Helicopters* (Academic Press, 1964).
5. Turner, K. J.; *Measurements of Dynamic Stability from Three Simplified Free-Flight Models of a Supersonic Research Aircraft (Bristol ER134) over the Mach number range* 1·2–2·6, Agard Report No. 378, 1961.
6. Rose, R.; *Flight Measurements of the Dutch Roll Characteristics of a* 60 *degree Delta Wing Aircraft (Fairey F.D.2) at Mach numbers from* 0·4 *to* 1·5 *with Stability Derivatives Extracted by Vector Analysis*, R.A.E. Tech. Note. Aero 2752, C.P. 653, 1961.
7. Jones, A. L.; "The Theoretical Lateral Stability Derivatives for Wings at Supersonic Speeds," *Journal of Aeronautical Sciences*, Vol. 17, page 39, 1950.

8. HEFFERNAN, T. H. J.; *Victor* B Mk. 2 *Handling Trials.* Unpublished M.O.A. Report.
9. YANCEY, R. B., REDIESS, H. A. and ROBINSON, G. H.; *Aerodynamic Derivative Characteristics of the* X-15 *Research Airplane as Determined from Flight Tests for Mach Numbers from* 0·6 to 3·4, NASA Tech. Note D-1060, 1962.
10. STONE, R. W., POLHAMUS, E. C.; *Some Effects of Shed Vortices on the Flow Fields around Stabilizing Tail Surfaces*, Agard Report No. 108, 1957.
11. LETKO, W.; *A Low-Speed Experimental Study of the Directional Characteristics of a Sharp-nosed Fuselage through a Large Angle-of-Attack Range at Zero Angle of Sideslip*, NACA Tech. Note 2911, 1953.
12. HASSELL, J. L., HEWES, D. E.; *Investigation of the Subsonic Stability and Control Characteristics of a* 1/7th *Scale Model of the North American* X-15 *Airplane with and without Fuselage Forebody Strakes*, NASA Tech. Memorandum X-210, 1960.
13. FINCH, T. W., WALKER, J. A.; *Flight Determination of the Static Longitudinal Stability Boundaries of the Bell* X-5 *Research Airplane with* 59° *Sweepback*, NACA Research Memorandum RM L53A 09b, 1953.
14. *Agard Flight Test Manual Vol. II—Stability and Control.*
15. ZIMMERMAN, C. H.; *Recent Stability and Aerodynamic Problems and their Implications as to Load Estimation*, NACA Research Memorandum RM L55E11a, 1955.
16. WILLIAMS, W. C., DRAKE, H. M. and FISCHEL, J.; *Comparison of Flight and Wind-Tunnel Measurements of High-Speed Aircraft Stability and Control Characteristics*, NACA Tech. Note 3859, 1956.
17. BISGOOD, P. L.; *A Review of Recent Handling Qualities Research, and its Application to the Handling Problems of Large Aircraft*, R.A.E. Report Aero 2688, 1964.
18. *Design Requirements for Aircraft for the Royal Air Force and Royal Navy*, Av.P.970, Ministry of Aviation.
19. WHITE, M. D., INNIS, R. C.; *A Flight Investigation of the Low-Speed Handling Qualities of a Tailless Delta-Wing Fighter Aircraft*, NASA Memo. 4-15-59A.
20. MCNAMAR, L. F., ARMSTRONG, W. T.; *F-106A—Case XIV—Case XXIX Wing Stability Evaluation*, AFFTC-TR-59-13 (1959).
21. ETKIN, B.; *Dynamics of Flight* (John Wiley and Sons, 1959).
22. Royal Aeronautical Society Data Sheets, *Performance.*
23. THOMAS, H. H. B. M., NEUMARK, S.; *Stability and Response Characteristics of Supersonic Aircraft*, R.A.E. Tech. Note Aero. 2412, 1955.
24. GATES, O. B., WOODLING, C. H.; *A method for Estimating Variations in the Roots of the Lateral-Stability Quartic due to Changes in Mass and Aerodynamic Parameters of an Airplane*, NACA Tech. Note 3134, 1954.
25. *Lateral Stability of the F*23/49, English Electric Co. Ltd., Aerodynamics Tech. Note Ae 76, 1953.
26. *Canberra PR Mk.*7: *Lateral Characteristics with Reference to the PR Role.* Unpublished M.O.A. Report.
27. *Vulcan B Mk.* 1—*Trials for Initial C.A. Release.* Unpublished M.O.A. Report.
28. PINSKER, W. J. G.; *Features of Large Transport Aircraft Affecting Control During Approach and Landing*, Agard Report No. 421, 1963.
29. SCHADE, R. O., HASSELL, J. L.; *The Effects on Dynamic Lateral Stability and Control of Large Artificial Variations in the Rotary Stability Derivatives*, NACA Report No. 1151, 1953.

30. Gates, O. B.; *A Theoretical Analysis of the Effect of Several Auxiliary Damping Devices on the Lateral Stability and Controllability of a High-Speed Aircraft*, NACA Tech. Note 2565, 1951.
31. White, R. J.; "Investigation of Lateral Dynamic Stability in the XB-47 Airplane," *Journal of the Aeronautical Sciences*, Vol. 17, March 1950.
32. Graham, D., Lathrop, R. C.; "Automatic Feedback and All-Weather Flying," *Aeronautical Engineering Review*, page 70, October 1955.
33. Johns, S. L., Wood, J. W.; *F-104A: Stability and Control with External Stores*, AFFTC-TR-58-14, Addendum 1, 1959.
34. Gates, O. B., Sternfield, L.; *Effect of an Autopilot Sensitive to Yawing Velocity on the Lateral Stability of a Typical High-Speed Airplane*, NACA Tech. Note 2470, 1951.
35. Kolk, W. R.; *Modern Flight Dynamics* (Prentice-Hall International, 1961).
36. Brown, L. W.; *An Investigation of the Dynamic Stability and Control Characteristics for a Transport Cruising at a Mach Number of* 3, NASA Tech. Note D-2483, 1964.
37. Brown, L. W.; *A Theoretical Investigation of the Effect of Cross-control Derivatives on the Stability Characteristics of Airplanes Designed for Flight at High Mach Numbers*, NASA Tech. Note D-1223, 1962.
38. Sherman, W. L.; *A Theoretical Investigation of the Dynamic Lateral Stability of Three Possible Airplane Configurations for Flight at a Mach Number of* 3·0. Unpublished NASA Report.
39. Schy, A. A., Gates, O. B.; *A Theoretical Method of Analysing the Effects of Yaw-damper Dynamics on the Stability of an Aircraft Equipped with a Second-Order Yaw Damper*, NACA Tech. Note 2857, 1952.
40. Sternfield, L., Gates, O. B.; *A Theoretical Analysis of the Effect of Time Lag in an Automatic Stabilization System on the Lateral Oscillatory Stability of an Airplane*, NACA Report 1018, 1951.
41. Pinsker, W. J. G.; *Preliminary Note on the Effect of Inertia Cross-coupling on Aircraft Response in Rolling Manoeuvres*, R.A.E. Tech. Note Aero 2419, 1955 (A.R.C., C.P.435).
42. Pinsker, W. J. G.; *Critical Flight Conditions and Loads Resulting from Inertia Cross-coupling and Aerodynamic Stability Deficiencies*, Agard Rep. No. 107, 1957 (A.R.C., C.P.404).
43. Phillips, W. H.; *Effect of Steady Rolling on Longitudinal and Directional Stability*, NACA Tech. Note 1627, 1948.
44. Pinsker, W. J. G.; *Charts of Peak Amplitudes in Incidence and Sideslip in Manoeuvres due to Inertia Cross-coupling*, R. & M. 3293, 1962.
45. *Investigation of Lightning in Rolling Manoeuvres*, English Electric Aviation Ltd. Aero Tech. Note Ae 128.
46. *Investigations to Improve Computer Estimates of Lightning Response in Rolls* at $M = 1{\cdot}7$ *at* 40,000 *ft*, English Electric Aviation Ltd. Aero Tech. Note Ae 129.
47. Nethaway, J. E., Clark, J.; *Inertia Cross-coupling Effects during Rolling of a Delta Aircraft* (*B.P.* 111*A*), R.A.E. Tech. Note Aero. 2674, 1960.
48. U.S.A.F., *E.F.T.P.S. Pilot's Handbook for Stability and Control Flight Testing*, Vol. II, "Theory."

17 Dynamic Stability Techniques

1. Introduction

In the work of Chapter 8 and 9 we have seen that the motion of an aircraft can be represented by the appropriate equations of motion, which relate the characteristics of the motion (i.e. rate of pitch, angle of sideslip, etc.) to the stability and control derivatives, the aircraft's relative density parameter, and the inertia characteristics. If the values of the derivatives are estimated theoretically, or obtained for wind tunnel or free-flight model tests, then, even before the aircraft flies, the oscillatory characteristics can be determined. This can be done by solving the equations or motion on an analogue computer, or by using the time-vector graphical method, or by using approximate equations derived from the complete equations of motion by making suitable assumptions. These methods can also be used to investigate the effects of varying the basic values of the derivatives on the oscillatory characteristics, and to determine the effects of autostabilization systems.

When the aircraft has been built, flight tests can be carried out to determine the values of the stability and control derivatives. Some of these derivatives (e.g. m_w, n_v, l_v, l_ξ, n_ζ, etc.) can be determined from steady tests (e.g. steady sideslips), but many of the derivatives can only be obtained from dynamic oscillation tests. The oscillatory response of an aircraft can be analysed in various ways, the time-vector and approximate equation methods are again often used. When the values of the derivatives have been obtained, they can be compared with the theoretical estimates and model test values used in the initial calculations. They can then be used in simulators and in analogue computer investigations of particular problems (e.g. inertia coupling).

This chapter will describe briefly the flight test techniques used and the instrumentation required, but it will be mainly concerned with the methods of analysing the results. To perform the analysis, it is necessary to know the inertia characteristics of the aircraft and, possibly, the values of the control surface derivatives. Manufacturer's estimates are often relied on for the inertia characteristics, though experimental methods are available. The control power derivatives can be obtained from an analysis of the transient response of the aircraft, immediately following a control displacement, and the lateral control derivatives can also be obtained from steady sideslip tests.

2. Flight Test Techniques and Instrumentation

2.1. Flight Test Techniques

There are two types of flight test technique available

(i) *Steady-state forced oscillation method* In this method a steady oscillation is forced upon the aircraft by means of a sinusoidal control input applied by an autopilot. When the steady-state oscillation has been established, the amplitude and phase of the aircraft response is measured relative to that of the control surface input. The test is then repeated at a number of different oscillatory frequencies.

(ii) *Transient or free oscillation method* In this method a step or a pulse control input is applied to the aircraft and the corresponding response is recorded. During the oscillation, the controls may be held fixed (i.e. stick-fixed stability), or released (i.e. stick-free stability in the case of a non-powered control system). Measurements are made of the periodic time and the damping of the oscillation, and the relative amplitudes and phase differences of the component motions of the oscillation. Ref. 6, however, makes a distinction between the "free oscillation" and the "transient oscillation" techniques.

These two techniques each have their own advantages and disadvantages which are discussed in detail in Refs. 1 and 6. The free oscillation method is generally used, and the particular technique employed in the tests carried out on the Fairey F.D.2 is discussed in more detail in Refs. 11 and 12.

The results obtained from the flight test experiments can be analysed in a number of different ways. The forced oscillation results are analysed using frequency response techniques (Ref. 1). The transient oscillation results can also be analysed to give frequency response data and it has been shown (Refs. 1 and 2) that this method gives similar results to those obtained from the analysis of the forced oscillation. The free oscillation results are, however, usually analysed by one of the following methods

(i) Method of approximate equations,
(ii) Time-vector method.

These methods are described in Refs. 3, 5 and 7, and used to determine the values of the derivatives of particular aircraft in Refs. 10, 11, 12 and 13.

2.2. Instrumentation

To obtain satisfactory estimates of the values of the derivatives from the analysis of the flight test results, it is essential that the response characteristics of the aircraft (e.g. rates of roll, yaw, and pitch, normal and lateral accelerations, angle of attack and sideslip angles, control surface positions, etc.) should all be measured as accurately as possible. The instrumentation usually used consists of rate gyros, accelerometers, wind vanes, potentiometer pick-offs, etc., and the measurements are recorded on Hussenot paper trace recorders. The general requirements for the instrumentation used in dynamic stability tests are discussed in Chapter 11 of Ref. 1, and in Refs. 5 and 6. In practice the instruments are subject to certain limitations and various errors. The limitations have to be reduced to a minimum and the errors have to be corrected for. Among the corrections that have to be made (Ref. 5) are

(i) *Phase-lag corrections* The instruments themselves have their own phase-lags which depend upon the frequency of the oscillation being measured. The instruments have to be dynamically calibrated, and the appropriate correction made to the measured results.

(ii) *Corrections for position of instruments* The linear accelerometers should be located at the C.G. of the aircraft. This is not always practical, and the necessary corrections have to be applied.

(iii) *Corrections to appropriate system of axes* The instruments are located in the aircraft and measure the response of the aircraft with respect to axes fixed in the body. If the analysis is being performed with respect to the wind axes system, then the measured results must be corrected.

The particular instrumentation used, and the various corrections that had to be applied, in the flight tests on the F.D.2 and Handley-Page H.P.115 are described in Refs. 11, 12 and 13.

3. Methods of Analysis

3.1. Method of Approximate Equations

The motion of an aircraft is represented by the corresponding equations of motion. These equations are rather complicated ones, but, by making suitable assumptions, it is possible to obtain approximate equations relating the oscillatory characteristics of the aircraft to the stability derivatives and the inertia characteristics. These equations are derived in Appendix B. By substituting the measured values of periodic time, damping, etc., in these equations, the values of the derivatives can be determined.

Though only approximate ones, these equations are generally satisfactory for most purposes, and are used quite widely. The relationships are particularly useful and fairly simple for certain of the derivatives. This method has the following advantages

(i) it is simple and easy to use,
(ii) it is very easy to see which derivative is the predominant one determining any particular oscillatory characteristic. Eqn. (B.30), for example, shows that the periodic time of the Dutch Roll oscillation is mainly dependent on the value of n_v, the weathercock stability. Ref. 5 gives some actual numerical values showing the relative importance of the various terms in the equations for one particular high-performance aircraft.

There are, however, a number of disadvantages

(i) the error involved may not be known,
(ii) the equations exist in a large number of different forms, depending on the amount of approximation made, and the parameters used to express the relationships (e.g. if damping is involved in the equations, it can be expressed in terms of δ or $t_{1/2}$ or $t_{1/e}$),
(iii) the form of the equation sometimes depends on the particular type of aircraft being considered (e.g. compare eqns. (B31) and (B35) for the roll/yaw ratio of the Dutch Roll oscillation). The equations discussed in this Chapter and in Appendix B are only those applicable to conventional aircraft designs. In Chapter 21 it will be seen that the approximate equations required for a slender-delta wing configuration are of a different form.

3.2. Time-vector Method

In Chapter 7 it was shown that an oscillating quantity (e.g. a displacement, a force, etc.) can be represented by a vector which rotates with time (i.e. a time-vector). Starting at a suitable position, depending on whether a sine or cosine variation is being represented, the vector rotates (conventionally) anticlockwise with a frequency equal to that of the oscillatory motion (i.e. it completes one rotation in a time T, equal to the periodic time of the oscillation). The value of the quantity at any instant in time is represented by the vertical component of the rotating vector.

During an oscillation the values of the displacements, velocities, accelerations, forces, moments, etc., are all varying with time. Despite the fact that these parameters have different magnitudes (i.e. amplitudes) and that there are phase differences between them

(i) they all have the same frequency and periodic time,
(ii) they maintain the same relative amplitude to one another (i.e. they all have the same damping),
(iii) they maintain the same phase differences relative to one another.

These facts make the time-vector representation a useful one, because a vector diagram which has been drawn representing the oscillatory conditions at any one instant of time, represents the relative amplitudes of, and the phase differences between, the quantities involved during the whole oscillation.

The time-vector method is not new; it has previously been used in electrical engineering and for mechanical vibrations (Ref. 3). It is a method of representing the equations of motion in a pictorial form and as such leads to a much better physical appreciation of the oscillatory motion.

(i) It is very easy to see which terms in the equations are the dominant ones, and which ones are negligible.
(ii) As a consequence of (i), it is possible to check, for the particular case being considered, the validity, or otherwise, of the various assumptions made to derive the approximate formulae (see para. 3.1 and Appendix B).
(iii) By comparing the vector diagrams of different aircraft, it can be seen why the variation of a particular parameter has one effect in one case, and, perhaps, a totally different effect in another.

The time-vector method is described in Ref. 3, and the simple cases discussed below illustrate the technique.

(a) A Mass Oscillating on a Spring with Zero Damping

This problem has already been considered in para. 2.2 of Chapter 7. The displacement of the mass from the equilibrium position can be represented by an equation of the form

$$x = x_0 \cos \omega_n t \tag{17.1}$$

where ω_n rads/sec is the frequency of the undamped oscillation. The corresponding periodic time is given by

$$T = \frac{2\pi}{\omega_n} \text{ secs} \tag{17.2}$$

The oscillatory motion of the mass can be represented by a vector of constant length x_0 (equal to $\frac{1}{2}$ of the total amplitude of the oscillation), rotating at an angular velocity ω_n (Fig. 17.1).

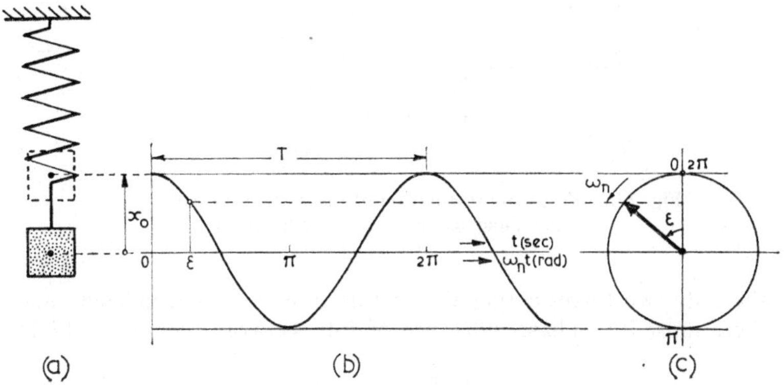

Fig. 17.1. Diagrammatic representation of an undamped oscillation (*a*) Physical system. (*b*) Time history of oscillation. (*c*) Time vector representation

By differentiating eqn. (17.1) with respect to time, the corresponding velocity of the mass (dx/dt or $\dot{x}$) is given by

$$\dot{x} = -\omega_n x_0 \sin \omega_n t = \omega_n x_0 \cos \left(\omega_n t + \frac{\pi}{2}\right) \tag{17.3}$$

and the acceleration by

$$\ddot{x} = -\omega_n^2 x_0 \cos \omega_n t = \omega_n^2 x_0 \cos (\omega_n t + \pi) \tag{17.4}$$

Thus the displacement, velocity and acceleration of the mass can be represented by vectors of length x_0, $\omega_n x_0$ and $\omega_n^2 x_0$, respectively, all rotating with the same frequency ω_n, but with each vector leading the preceding one by an angle of 90° (Fig. 17.2).

The only forces acting in this case are the inertia force and the spring stiffness force. The inertia force is proportional to both the mass and the acceleration and acts in the opposite direction to the acceleration (i.e. $-m\ddot{x}$). The spring force is proportional to the spring stiffness k and the displacement and acts in the opposite direction to the displacement (i.e. $-kx$). These two forces can be represented by rotating vectors, the spring force by a vector of length kx_0 and the inertia force by one of length $m\omega_n{}^2x_0$ and, since these are

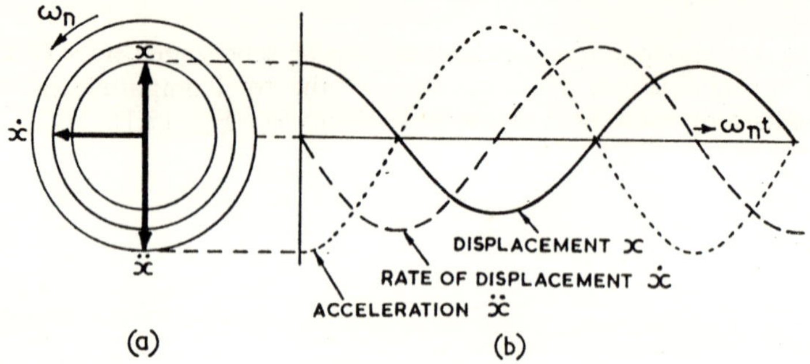

Fig. 17.2. Vector representation of the oscillatory parameters. (*a*) Time vectors. (*b*) Time histories

the only two forces acting, the vectors must be equal in length and there must be a phase difference of 180° between them (Fig. 17.3).

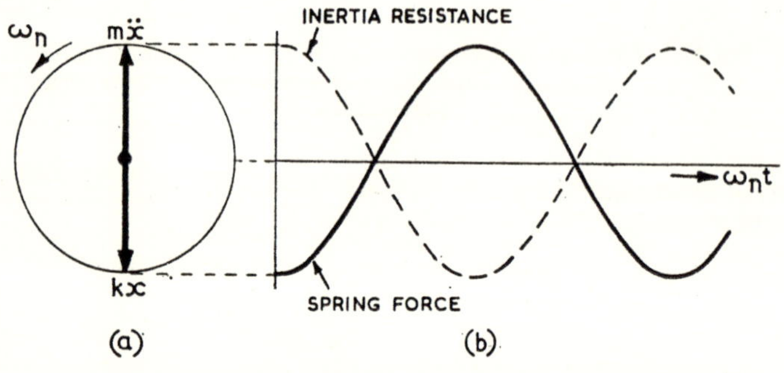

Fig. 17.3. Vector representation of the forces acting in an undamped oscillation
(*a*) Time vectors. (*b*) Time histories

By equating these two quantities it can be seen that

$$\omega_n = \sqrt{\left(\frac{k}{m}\right)} \tag{17.5}$$

and this expression for the undamped natural frequency of the spring-mass system is the same as that derived in Chapter 7 (eqn. 7.5).

(b) A Mass Oscillating on a Spring with Damping

This problem was also considered in para. 2.3 of Chapter 7. The displacement of the mass from the equilibrium position is in this case represented by

$$x = x_0 e^{-\mu t} \cos \omega t \tag{17.6}$$

where ω rads/sec is the frequency of the damped oscillation and μ is the damping factor. By differentiating eqn. (17.6), the velocity of the mass is given by

$$\dot{x} = -\omega x_0 e^{-\mu t} \sin \omega t - \mu x_0 e^{-\mu t} \cos \omega t$$

Rearranging this equation

$$\dot{x} = -\omega_n x_0 e^{-\mu t} \left(\frac{\omega}{\omega_n} \sin \omega t + \frac{\mu}{\omega_n} \cos \omega t \right)$$

and this can be rewritten in the form

$$\dot{x} = -\omega_n x_0 e^{-\mu t} [\sin(\omega t + \varepsilon_D)] \tag{17.7}$$

where

$$\left.\begin{aligned} \cos \varepsilon_D &= \frac{\omega}{\omega_n} \\ \sin \varepsilon_D &= \frac{\mu}{\omega_n} \end{aligned}\right\} \tag{17.8}$$

Equation (17.7) can also be expressed in the form

$$\dot{x} = \omega_n x_0 e^{-\mu t} \cos \left[\omega t + \left(\frac{\pi}{2} + \varepsilon_D \right) \right] \tag{17.9}$$

In a similar way the acceleration can be shown to be

$$\ddot{x} = \omega_n^2 x_0 e^{-\mu t} \cos [\omega t + (\pi + 2\varepsilon_D)] \tag{17.10}$$

By comparing eqns. (17.6), (17.9) and (17.10), it can be seen that the displacement, velocity, and acceleration can again be represented by vectors of length x_0, $\omega_n x_0$ and $\omega_n^2 x_0$ (note that it is ω_n, the undamped frequency, and not ω, the actual frequency, in these expressions) rotating at a constant frequency ω, with each vector leading the preceding one by an angle of $(90 + \varepsilon_D)°$. The decay factor $e^{-\mu t}$ occurs in all the equations; thus as the pattern rotates it keeps its shape, but shrinks exponentially with time, the ends of the vectors sweeping out a spiral path (Fig. 17.4).

The angle ε_D is known as the Damping Angle. It increases with an increase of the damping in the system (see eqn. 17.17) and, as we have already seen, in the undamped case $\varepsilon_D = 0$. It will be useful to derive here some relationships between the various oscillatory characteristics (e.g. ε_D, ζ, $t_{1/2}$, T, etc.) that are employed in dynamic stability work. From Chapter 6 the frequency of oscillation can be expressed in the form

$$\omega = \frac{2\pi}{T} = \sqrt{\left[\frac{k}{m} - \left(\frac{b}{2m}\right)^2\right]} \tag{17.11}$$

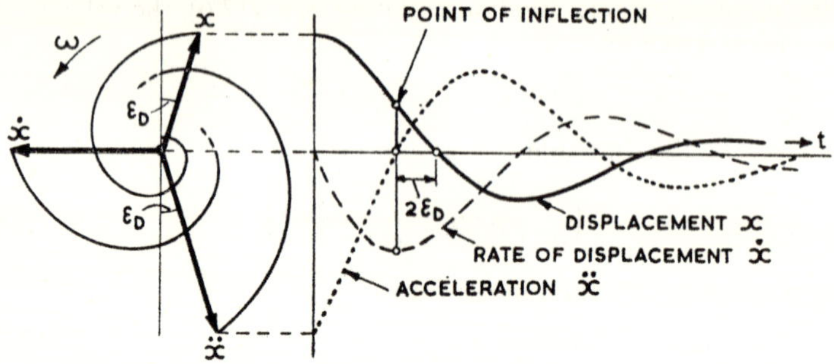

Fig. 17.4. Vector representation of the oscillatory parameters in a damped oscillation
(*a*) Time vectors. (*b*) Time histories

Substituting $\omega_n = \sqrt{\left(\frac{k}{m}\right)}$, eqn. (17.11) becomes

$$\omega = \omega_n \sqrt{\left(1 - \frac{b^2}{4mk}\right)} \tag{17.12}$$

$$\underline{\underline{\omega = \omega_n \sqrt{(1 - \zeta^2)}}} \tag{17.13}$$

where $\zeta = \dfrac{b}{2\sqrt{(mk)}}$ is the relative damping. By comparing eqns. (17.8) and (17.13) it can be seen that

$$\underline{\underline{\cos \varepsilon_D = \frac{\omega}{\omega_n} = \sqrt{(1 - \zeta^2)}}} \tag{17.14}$$

Alternatively, since $\zeta\omega_n = \dfrac{b}{2m} = \dfrac{1}{t_{1/e}} = \dfrac{0{\cdot}693}{t_{1/2}}$, eqn. (17.13) can be expressed either in the form

$$\omega = \sqrt{\left[\omega_n^2 - \left(\frac{1}{t_{1/e}}\right)^2\right]}$$

i.e.

$$\omega_n^2 = \omega^2 + \left(\frac{1}{t_{1/e}}\right)^2 \tag{17.15}$$

or as

$$\omega_n^2 = \left(\frac{2\pi}{T}\right)^2 + \left(\frac{0{\cdot}693}{t_{1/2}}\right)^2 \tag{17.16}$$

The damping factor μ is equal to $\zeta\omega_n$, hence eqn. (17.8) for $\sin \varepsilon_D$ becomes

$$\sin \varepsilon_D = \zeta \tag{17.17}$$

Dividing eqn. (17.17) by (17.14) gives

$$\tan \varepsilon_D = \frac{\zeta\omega_n}{\omega} = \frac{\zeta}{\sqrt{(1-\zeta^2)}}$$

and substituting $\omega = \dfrac{2\pi}{T}$ and $\zeta\omega_n = \dfrac{1}{t_{1/e}}$, this becomes

$$\tan \varepsilon_D = \frac{T}{2\pi t_{1/e}} \tag{17.18}$$

or in terms of $t_{1/2}$

$$\tan \varepsilon_D = \frac{0{\cdot}693T}{2\pi t_{1/2}} \tag{17.19}$$

Rearranging

$$\varepsilon_D = \tan^{-1}\left(\frac{0{\cdot}693T}{2\pi t_{1/2}}\right) = \tan^{-1}\left[0{\cdot}1103\,\frac{T}{t_{1/2}}\right] \tag{17.20}$$

In this case there are three forces acting, the inertia force ($-m\ddot{x}$), the spring stiffness force ($-kx$) and the damping force ($-b\dot{x}$) These forces act in the opposite directions to $\ddot{x}$, x and $\dot{x}$, respectively (Fig. 17.5*a*). Since, at any instant of time, the forces acting in the oscillation add up to zero, the vectors representing the forces must form a closed isosceles triangle (Fig. 17.5*b*). The inertia force and the spring force are equal in magnitude, and there is an angle of $2\varepsilon_D$

between them. The side of the triangle opposite the angle $2\varepsilon_D$ represents the damping force $b\dot{x}$.

(*c*) *An Oscillatory System with More than One Degree of Freedom*
In the more general case, a separate vector diagram is required for each degree of freedom (i.e. one diagram to represent each equation of motion). Because there are usually more than the three basic terms of "inertia," "damping" and "spring stiffness" involved in these equations (e.g. cross-coupling terms, etc.), the simple isosceles triangle is replaced by a polygon of forces (or moments), although

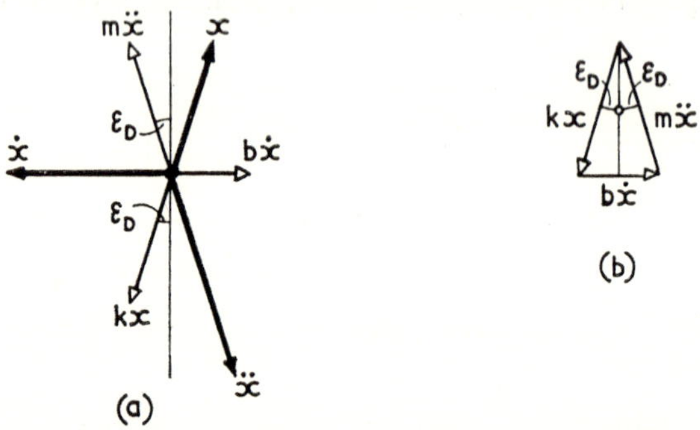

Fig. 17.5. Vector representation of the forces acting in a damped oscillation
(*a*) Time vectors. (*b*) Polygon of forces

it is often possible to determine the "equivalent" isosceles triangle, and to measure the damping angle ε_D. It is important to realize that the vector polygons representing a particular motion are closely related to one another. They are all rotating with the same frequency and decaying at the same rate, and all the sides in the different polygons which are related to a particular motion variable (e.g. all the forces and moments which depend on the rate of roll, etc.) always remain parallel to one another.

This method may be used to represent the forces and moments involved during the longitudinal (see para. 4) or lateral (see para. 5) oscillatory motions of an aircraft; in each case three vector polygons are required. From flight tests the relative amplitude of, and the phase differences between, the various components of the motion can be measured and these relationships form the basis of the vector polygons. In order to construct the polygons it is necessary to know beforehand the inertia characteristics of the aircraft (see para. 6.2) and one or two of the derivatives. By closing the polygons (this is equivalent to solving the corresponding equation of motion), the values of the unknown derivatives can be determined. The method is an extremely useful one, but it does suffer from certain disadvantages

(i) Only a limited number of derivatives can be determined in this way. In order to construct the polygons, the values of some of the derivatives must already be known. These values may be obtained from theoretical estimations, from model tests, or from previous flight tests (e.g. l_v and n_v from steady sideslip tests).

(ii) Iterative procedures may be necessary to obtain the required solution. This could be time consuming and to make the work easier, tables are provided with the necessary steps clearly laid out (see Ref. 3 for the phugoid oscillation and Ref. 14 for the Dutch Roll).

(iii) The method cannot be used for heavily damped oscillatory motions.

4. Longitudinal Dynamic Stability

4.1. Method of Approximate Equations

The longitudinal motion of an aircraft is represented by eqns. B1, 2 and 3 of Appendix B. These equations can be simplified to give approximate solutions for the short period oscillatory characteristics in terms of the longitudinal derivatives (eqns. B7–15). Alternatively, the derivatives can be expressed in terms of the oscillatory characteristics (eqns. B16–21).

The application of these approximate relationships has already been discussed in Chapter 8. The simplest forms of the equations for the periodic time and the time to half amplitude of the longitudinal short period oscillation are given by

$$T = 2\pi \sqrt{\left(\frac{i_B}{-m_w \mu_1}\right)}\, \hat{t} \text{ secs} \tag{B11}$$

and

$$t_{1/2} = -1{\cdot}386\, \hat{t} \Big/ \left(\frac{m_{\dot{\theta}}}{i_B} + z_w\right) \text{ secs} \tag{B13}$$

Rearranging these equations we get

$$m_w = -4\pi^2 \frac{i_B}{\mu_1} \left(\frac{\hat{t}}{T}\right)^2 \tag{B18}$$

and

$$m_{\dot{\theta}} = -i_B \left(\frac{1{\cdot}386\hat{t}}{t_{1/2}} + z_w\right) \tag{B19}$$

Other relationships are given in Appendix B. Alternative forms of the relationships are given in Ref. 4, and these equations are used in Ref. 12 to determine the longitudinal derivatives of the F.D.2 aircraft.

4.2. Time-vector Method

4.2.1. PHUGOID OSCILLATION

The application of the time-vector method to the phugoid oscillation is discussed in Ref. 3, and an example given of the iterative procedure that is necessary for determining the oscillatory characteristics of the aircraft from given values of the derivatives. For a rough approximation it is possible to neglect changes of incidence and pitching moment, but for more accurate results it is necessary to retain all three degrees of freedom. The corresponding vector diagrams are illustrated in Ref. 3, and the report also discusses the effect of varying the derivatives on the oscillatory characteristics of the aircraft and the different methods of automatic control available for improving the damping of the phugoid motion.

4.2.2. SHORT PERIOD OSCILLATION

In the case of the short period oscillation, changes in forward speed (and hence the *X*-force equation) can usually be neglected. The remaining equations for *Z* and *M* can be simplified to

$$(D - z_w)\bar{w} - \left(1 + \frac{z_q}{\mu_1}\right)\bar{q} = 0 \tag{17.21}$$

and

$$-\left(m_w + \frac{m_{\dot{w}}}{\mu_1} D\right)\bar{w} + \left(\frac{i_B}{\mu_1} D - \frac{m_q}{\mu_1}\right)\bar{q} = 0 \tag{17.22}$$

Sometimes (Refs. 4 and 6), these equations are written in terms of the so-called concise stability derivatives (e.g. $\chi = -m_{\dot{w}}/i_B$, $\nu = -m_q/i_B$, $\omega = -(\mu_1 m_w)/i_B$), eqn. (17.22) then becomes

$$(\omega + \chi D)\bar{w} + (D + \nu)\bar{q} = 0 \tag{17.23}$$

A typical pitching moment vector diagram is shown in Fig. 1 of Ref. 6, others are given in Ref. 3. The *Z*-force diagram can be used to check the validity of the assumption made in para. 1 of Appendix B, that the z_q term is small and can be neglected.

5. Lateral Dynamic Stability

5.1. Method of Approximate Equations

The lateral motion of an aircraft is represented by eqns. B22, 23 and 24 of Appendix B. These equations can also be simplified to give approximate solutions for the Dutch Roll oscillatory characteristics in terms of the lateral derivatives (eqns. B28–39). Alternatively, the derivatives can be expressed in terms of the oscillatory characteristics (eqns. B40–47).

The application of these equations has already been discussed in Chapter 9. Some of the more important formulae taken from Appendix B are

$$T = 2\pi \sqrt{\left(\frac{i_c}{\mu_2 n_v}\right)}\, \hat{t} \text{ secs} \tag{B30}$$

and neglecting the n_p term in eqn. (B39)

$$\delta = \pi\left[\sqrt{\left(\frac{i_c}{\mu_2 n_v}\right)}\right]\left(-\frac{n_r}{i_c} - y_v\right)$$

Alternatively, rearranging these equations, we have

$$\underline{\underline{n_v = 4\pi^2 \frac{i_c}{\mu_2}\left(\frac{\hat{t}}{T}\right)^2}} \tag{B40}$$

and

$$n_r = -i_c\left[\frac{\delta}{\pi}\sqrt{\left(\frac{\mu_2 n_v}{i_c}\right)} + y_v\right] \tag{B45}$$

or

$$\underline{\underline{n_r = -i_c\left(\frac{2\delta\hat{t}}{T} + y_v\right)}} \tag{B46}$$

The value of the derivative n_v, the weathercock stability, is a very important derivative on modern high-performance aircraft. A loss of n_v leads to directional instability and inertia coupling problems, and eqn. (B40) is often used to determine the value of this derivative. In Refs. 10 and 11 the values obtained for the oscillatory characteristics from these approximate equations are compared with those obtained by other methods.

5.2. Time-vector Method

The time-vector method is widely used in the analysis of the Dutch Roll oscillation. Ref. 14 describes how the time-vector method is used to calculate the oscillatory characteristics from known values of the derivatives and gives a table laying down the steps that are necessary in the iterative process. This method is applied in Ref. 15 to determine the variation of the values of the periodic time, the roll/yaw ratio and the log. dec., for a particular aircraft at various speeds and heights.

In Refs. 10 and 11 the time-vector method is used to determine the values of the lateral derivatives. The method of analysis is described in detail in Ref. 11. In order to construct the vector polygons in this particular case it was necessary to know the values of the inertia characteristics, the rudder derivatives (because some rudder motion was allowed for) and one rolling and one yawing derivative. Since the cross-derivatives n_p and l_r only play a small part in the oscillatory motion, their values were estimated theoretically.

A typical set of vector diagrams, taken from Ref. 10, is shown in Fig. 17.6. From the rolling moment diagram it can be seen that, in this case, the i_E and l_r terms are relatively small and this confirms the simplifying assumptions made in para. 3 of Appendix B. In a similar way, the n_p term in the yawing moment diagram can be seen to be relatively unimportant.

Other Dutch Roll vector diagrams can be found in Refs. 5, 11 and 13. It is extremely interesting and instructive to compare the vector diagrams of two aircraft with different configurations. It will be seen that different terms of the equation often have a different relative importance. This will explain why one particular approximate equation will not be satisfactory for a number of different aircraft, and why different forms may exist. This is particularly applicable to the equations for the roll/yaw ratio (para. 3, Appendix B).

6. Aircraft Inertia Characteristics and Control Surface Derivatives

In order to analyse the oscillatory response of an aircraft, it is necessary to know the inertia characteristics of the aircraft, and, if control movements are involved, the values of the control derivatives.

6.1. Control Surface Derivatives

There are two techniques available for measuring the values of the derivatives

(i) steady sideslip tests with asymmetric wing weights and wing tip parachutes,
(ii) transient response tests.

(i) *Steady Sideslip Tests*

These methods are discussed in Refs. 10, 13, 16 and 17. When an aircraft is symmetrically loaded, the rolling moment equation in a steady straight sideslip is given by

$$l_\beta \beta + l_\xi \xi + l_\zeta \zeta = 0 \qquad (17.24)$$

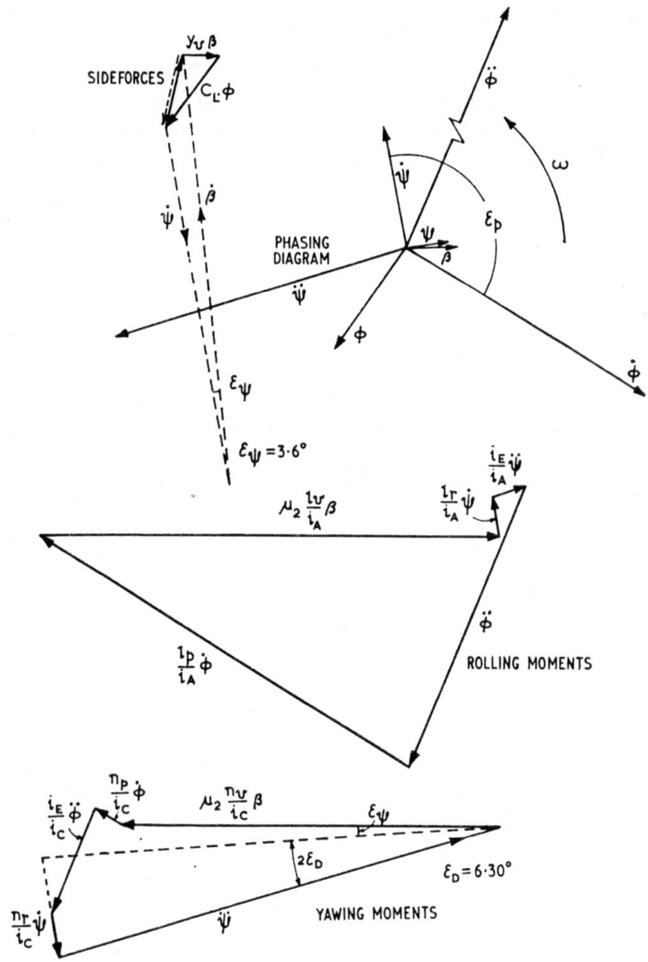

Fig. 17.6. Vector diagram of the lateral oscillation of an aircraft $C_L = 0{\cdot}40$; 10,000 ft [From Ref. 10]

With ballast weights mounted at one of the wing tips, eqn. (17.24) becomes, for the same sideslip angle

$$l_\beta \beta + l_\xi \xi' + l_\zeta \zeta' + \frac{L_W}{\rho V^2 S s} = 0 \tag{17.25}$$

where ξ' and ζ' are the new trim positions of the aileron and rudder, and L_W is the rolling moment resulting from the asymmetric wing weights (divided by $\rho V^2 S s$ to make it non-dimensional). Subtracting eqn. (17.24) from eqn. (17.25), we have

$$l_\xi(\xi' - \xi) + l_\zeta(\zeta' - \zeta) + \frac{L_W}{\rho V^2 S s} = 0 \tag{17.26}$$

or replacing $(\xi' - \xi)$ by $\Delta\xi$ and $(\zeta' - \zeta)$ by $\Delta\zeta$

$$l_\xi \Delta\xi + l_\zeta \Delta\zeta + \frac{L_W}{\rho V^2 S s} = 0 \tag{17.27}$$

The change in rudder angle required is due to the yawing moment produced by the aileron movement. If the effect of rudder deflection on the rolling moment equation can be neglected, then the aileron rolling power can be expressed in the form

$$l_\xi = -\left(\frac{L_W}{\rho V^2 S s}\right)\frac{1}{\Delta\xi} \tag{17.28}$$

In Ref. 16, various asymmetric weights are used and the non-dimensional weight rolling moment is plotted against the corresponding values of $\Delta\xi$ at a number of selected speeds. The slopes of these lines give the corresponding values of l_ξ.

The rudder power is determined by attaching a small parachute to one wing tip of the aircraft to produce an asymmetric yawing moment. Without the parachute, the yawing moment equation in a steady sideslip is given by

$$n_\beta \beta + n_\xi \xi + n_\zeta \zeta = 0 \tag{17.29}$$

With the parachute, the equation becomes, for the same sideslip angle

$$n_\beta \beta + n_\xi \xi' + n_\zeta \zeta' + \frac{N_P}{\rho V^2 S s} = 0 \tag{17.30}$$

where ξ' and ζ' are the new trim positions of the aileron and the rudder and N_P is the yawing moment produced by the parachute. By subtracting eqn. (17.29) from eqn. (17.30) we have

$$n_\xi \Delta\xi + n_\zeta \Delta\zeta + \frac{N_P}{\rho V^2 S s} = 0 \tag{17.31}$$

Hence

$$n_\zeta = -\left(\frac{N_P}{\rho V^2 Ss}\right)\frac{1}{\Delta\zeta} - n_\xi \frac{\Delta\xi}{\Delta\zeta} \tag{17.32}$$

(ii) *Transient Response Techniques*

When a control surface is deflected, the initial response is one of acceleration, though this is soon followed by a slower build up of velocity and displacement. If it is assumed that the peak on the acceleration curve occurs before the velocity and displacement have appreciably altered, then the corresponding equation of motion can be considerably simplified to contain only the control deflection and the direct inertia terms (though in the case of the lateral control powers it may be necessary to include the product of inertia terms). The rudder power derivative, for example, is given in dimensional form by

$$N_\zeta = \frac{C\dot{r} - E\dot{p}}{\zeta} \tag{17.33}$$

This method is discussed in Refs. 1 and 5 and is employed in Ref. 13, in preference to the wing tip parachute method, to determine the rudder characteristics of the H.P.115.

6.2. Aircraft Inertia Characteristics

The inertia characteristics of the aircraft need to be known fairly accurately, though manufacturer's estimates are usually considered adequate for most purposes (Refs. 5 and 8). However, experimental methods are available for determining the values of the moments of inertia A, B and C, the product of inertia E and the inclination of the principal inertia axis. The methods are discussed in Refs. 1, 5, 8 and 9. The technique employed is that of oscillating the aircraft about a fixed datum point, with a restoring force provided by restraining springs. From the measured values of the periodic time, the inertia characteristics can be determined.

References

1. *Agard Flight Test Manual*, Vol. II, "Stability and Control," Chapters 10 and 11.
2. ETKIN, B.; *Dynamics of Flight*, page 348 (J. Wiley and Sons, 1959).
3. DOETSCH, K. H.; *The Time-vector Method for Stability Investigations*, R. & M. No. 2945, 1957.
4. NEUMARK, S.; *Analysis of Short-period Longitudinal Oscillations of an Aircraft*, R.A.E. Rep. Aero. 2479, 1952.
5. WOLOWICZ, C. H. and HOLLEMAN, E. C.; *Stability-derivative Determination from Flight Data*, Agard Report No. 224, 1958.

6. Zbrozek, J. K.; *On the Extraction of Stability Derivatives from Full-scale Flight Data*, Agard Report No. 190, 1958.
7. Huff, W. W.; *Application of Dynamic Testing Procedures to Stability and Control Flight Test Programmes*, Agard Report No. 191, 1958.
8. Wener, N. L.; *Measurement of Aircraft Moments of Inertia*, Agard Report No. 248, 1959.
9. Bradley, W. G.; *Measurements of the Moments of Inertia of the Morane-Saulnier M.S.* 760 *"Paris" Aircraft*, College of Aeronautics Note No. Aero. 160, 1963.
10. Perry, D. H., Morrall, J. C. and Port, W. G. A.; *Low Speed Flight Tests on a Tailless Delta Aircraft with Wings Swept Back* 44·5° (*Avro* 707*B*). Unpublished M.O.A. Report.
11. Rose, R.; *Flight Measurements of the Dutch-roll Characteristics of a* 60° *Delta Wing Aircraft* (*Fairey Delta* 2) *at Mach Numbers from* 0·4 *to* 1·5 *with Stability Derivatives Extracted by Vector Analysis*, R.A.E. Tech. Note Aero 2752, CP 653, 1961.
12. Andrews, D. R.; *Measurements in Flight of the Longitudinal Derivatives of a* 60° *Delta Wing Aircraft* (*F.D.*2), R.A.E. Rep. Aero. 2617 (C.P.639) 1959.
13. Bisgood, P. L. and O'Leary, C. O.; *Interim Report on Low-speed Flight Tests of a Slender-wing Research Aircraft* (*Handley-Page H.P.*115), R.A.E. Tech. Note Aero. 2921, 1963.
14. English Electric Co., Ltd., *Aero T.N. Ae* 58.
15. *Lateral Stability of the* F/23/49 English Electric Co., Ltd., *Aero. T.N. Ae* 76.
16. Dee, F. W.; *Flight Measurements at Subsonic Speeds of the Aileron Rolling Power and Lateral Stability Derivatives* l_v *and* y_v *on a* 60° *Delta Wing Aircraft* (*F.D.*2) R.A.E. Tech. Note Aero. 2897, 1963.
17. Dee, F. W., *Proving Tests of a Wing Tip Parachute Installation on a Venom Aircraft, with some Measurements of Directional Stability and Rudder Power,* R.A.E. Tech. Note Aero. 2824, 1962.

18 The Use of Analogue Computers in Aeronautics

1. Introduction

As we have seen in previous chapters, the motion of an aircraft can be represented in terms of the corresponding equations of motion. The differential equations representing the aircraft system are basically the same as those representing a simple spring-mass system or an electrical system containing capacitances, resistances, etc. (see Chapters 7 and 19). Thus these alternative mechanical and electrical systems can be used as an analogue representation of the actual aircraft system.

Analogue computers can be used to obtain the solutions of differential equations by representing the variables of a given problem by convenient physical quantities (e.g. the rotation of a shaft, or a continuously varying voltage). These physical quantities are made to obey mathematical relationships analogous to those of the problem being considered and the records of the variation of the physical quantities (e.g. on voltmeters or cathode ray tubes) constitute solutions of the given mathematical problem.

Analogue computers can be classified in various ways, e.g. according to the nature of the physical quantities used to represent the mathematical variables (i.e. mechanical or electronic analogue computers), or according to the way in which they are employed (i.e. differential analysers, simulators, etc.). An ordinary slide rule is an example of a mechanical analogue computer, in which a length is used to represent a number, the addition of two lengths a multiplication, and so on. The majority of analogue computers are D.C. analogue computers, in which the mathematical variables are represented by D.C. voltages varying with time. D.C. analogue computers contain electrical circuits employing amplifiers, resistances, capacitances, etc., arranged to perform the mathematical processes of addition, subtraction, multiplication, integration, and so on. They are particularly useful for solving differential equations and, when used in this way, they are often called Differential Analysers. The analogue computer is a very important tool in the early development stages of an aircraft. In order to determine an aircraft's stability and control characteristics it is necessary to know the values of the stability derivatives and the moments of inertia of the aircraft and to feed these values into the computer. In the early design stages, the values of the derivatives may be obtained from theoretical estimates or from wind tunnel model tests. These values are checked later on with those obtained from flight tests, and the values of the derivatives are modified as necessary for any further calculations on the computer. Analogue computers are used to assist in the clearance programme on an aircraft, and are particularly valuable in determining the handling characteristics of an aircraft in those flight areas which are potentially dangerous (e.g. because of inertia coupling problems, etc.).

Simulators are basically analogue computers, with a human pilot contained in the closed-loop system. In the case of ground-based simulators it is essential to provide the pilot with a feedback of the correct visual and motion cues. Within recent years, in-flight simulators have been used in the U.S.A.; these are often referred to as Variable-stability Aircraft.

The basic principles and methods of operation of analogue computers and their various advantages and disadvantages are discussed in more detail in Refs. 1 to 4.

2. Basic Principles of D.C. Analogue Computers

2.1. Components of a D.C. Analogue Computer

A D.C. analogue computer will, in general, consist of the following components

(*a*) linear electronic computing elements for performing the operations of addition, subtraction, multiplication by a constant, differentiation [though this is normally avoided because of the noise in the input signal (Ref. 4, page 26)] and integration.

(*b*) Non-linear electronic computing elements for performing the operations of multiplication and division of variables, generation of mathematical functions, etc.

(*c*) Potentiometers for setting in the values of the constant coefficients (e.g. the values of the aircraft's stability derivatives).

(*d*) Equipment for observing and recording the varying D.C. voltages representing the solution of the problem (e.g. voltmeters, cathode ray oscilloscopes, pen-recorder oscillographs).

(*e*) Wiring, control circuits and power supplies.

2.2. Computing Elements of a D.C. Analogue Computer

At first it may seem strange that electronic circuits consisting of various arrays of resistances, capacitances and so on can be used to perform the various mathematical processes of multiplication, integration, etc. However, if one remembers the basic laws governing the relationships between the voltages and currents in simple electrical circuits, it will be seen that these processes, are, in fact, possible. Two very simple examples of this can be considered

(i) if a voltage is placed across a number of resistances in series, the total voltage is the sum of the separate voltages across the individual resistances.

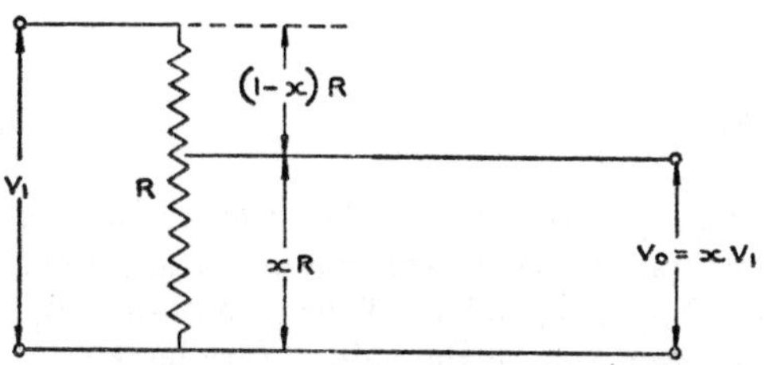

Fig. 18.1. Multiplication of a variable by a number less than 1

(ii) if a voltage V_1 is placed across a resistance R, then a movable tapping can be used to take off a proportion of this voltage, i.e.

$$V_0 = xV_1 \tag{18.1}$$

where the value of the constant coefficient x depends on the position of the movable tapping (Fig. 18.1). In practice, the potentiometer usually consists of a number of coils of wire with a rotary contact and the numerical value of the coefficient x can be set on a calibrated dial.

This process is equivalent to the multiplication of a variable (represented by the voltage V_1) by a constant factor of less than unity (i.e. $0 \leqslant x \leqslant 1$). The system can be modified to represent the multiplication of two variable quantities, if it is arranged so that the setting on the potentiometer is made dependent on the value of the second variable. The output voltage will then be proportional to the product of the two variables.

2.2.1. D.C. AMPLIFIERS

In practice, however, most of the mathematical processes are performed using high-gain directly-coupled amplifiers (sometimes called "operational amplifiers" because they are used to perform mathematical operations). A D.C. amplifier consists of a number of separate stages of amplification, the output of each stage being directly coupled through a resistance to the input of the following stage. Some typical characteristics of operational amplifiers (e.g. total D.C. gain = 2×10^7) are given on page 24 of Ref. 4.

The way that the D.C. amplifier is used to perform the various mathematical operations is shown below.

(*a*) *Multiplication of a variable by a factor greater than unity*
In the arrangement shown in Fig. 18.2, the output voltage V_0 is given by

$$V_0 = \frac{-R_F}{R_1} V_1 \tag{18.2}$$

where V_1 is the input voltage, R_1 the input resistance, and R_F the "feedback" resistance. There are two important points to note

(i) Eqn. (18.2) is independent of the gain of the amplifier. This is true, however, only if the gain is very large.

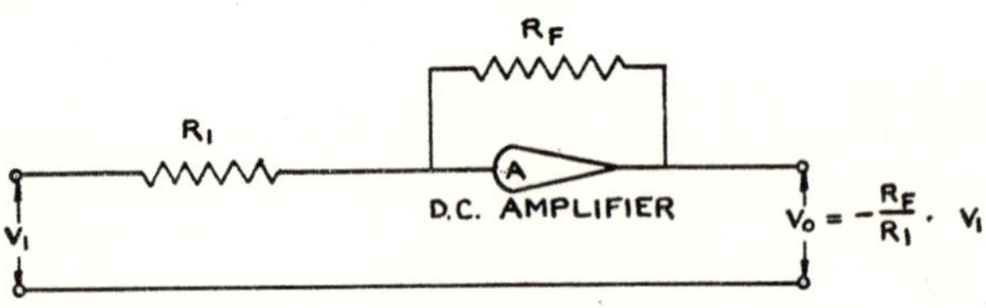

Fig. 18.2. Multiplication of a variable by a number which can be greater than 1

(ii) The practical arrangement of the circuit is such that there is a reversal of sign between the output and input. Because the amplifier is a high-gain one, the voltage at the input of the amplifier must be relatively small and since the current flow into the input of the amplifier is assumed to be zero, the same current must flow through R_F as through R_1. This can only occur if there is a drop in voltage across R_F and thus the output voltage must be a negative one (Ref. 4, page 24; Ref. 5, page 37).

(*b*) *Addition*
If a number of resistances are placed in parallel at the input to the amplifier, then the arrangement can be used to represent the process

of addition. The output voltage in Fig. 18.3 is given by

$$V_0 = -\left(\frac{R_F}{R_1} V_1 + \frac{R_F}{R_2} V_2 + \frac{R_F}{R_3} V_3\right) \tag{18.3}$$

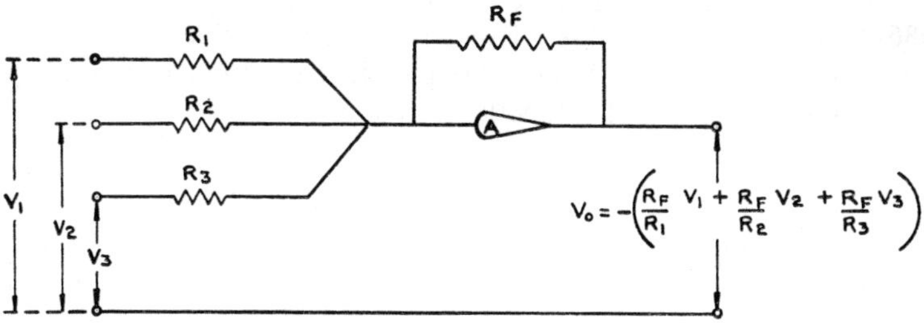

Fig. 18.3. A summing amplifier

(*c*) *Integration*
If the feedback resistance is replaced by a capacitor C, then the resulting circuit (Fig. 18.4) can be used for integration, the output

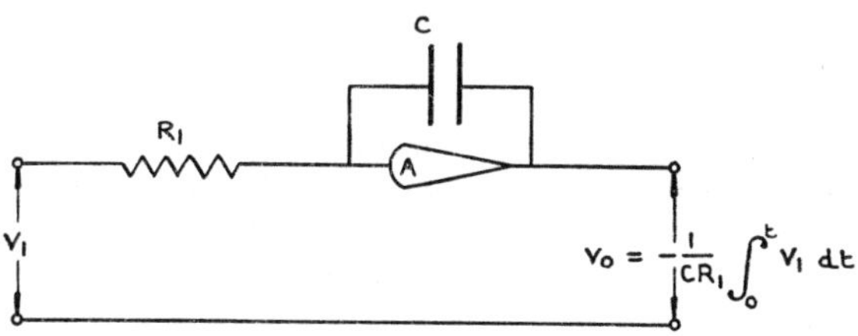

Fig. 18.4. An integrator

voltage being given by

$$V_0 = -\frac{1}{CR_1}\int_0^t V_1\,dt \tag{18.4}$$

The product CR_1 is sometimes referred to as the time constant of the circuit.

The integrating amplifier may, like the summing amplifier, have any number of inputs, and the output is then the sum of the integrals of the various inputs (with the signs reversed).

2.2.2. FUNCTION GENERATORS
In addition to the simple mathematical processes described above, it is also necessary to be able to generate voltages which are representative of many varied functions of the input voltage (i.e. $V_0 = f(V_1)$, where f may represent a sine, a cosine, an exponential or any

other required variation). In aeronautical work these function generators are required to produce a voltage variation which is representative of such features as backlash and break-out forces in a control system.

Function generators may be mechanical ones (Ref. 1, page 128–137) or electronic ones (Ref. 1, page 137–154). Tables 5 and 6 on pages 424–428 of Ref. 2 show the circuit diagrams corresponding to a number of particular functional relationships.

3. Simple Differential Analysers

When an analogue computer is used to solve differential equations, it is usually referred to as a Differential Analyser. In order to understand the more complicated problems, it is extremely valuable to consider first of all the solutions of some simple examples.

(a) *Variation with Time of the Position of a Moving Body*

The equation relating the distance s travelled in a time t by a body moving at a velocity v is given by

$$s = \int_0^t v \, dt \tag{18.5}$$

Eqn. (18.5) can be compared with eqn. (18.4), and it can be seen that, if the input voltage is made proportional to the velocity of motion, then the output voltage will be proportional to the distance travelled by the body.

(b) *Mass and Spring System*

The particular problem of a mass and spring system is considered in Ref. 1 (page 20). Consider a mass m with a spring attached resting

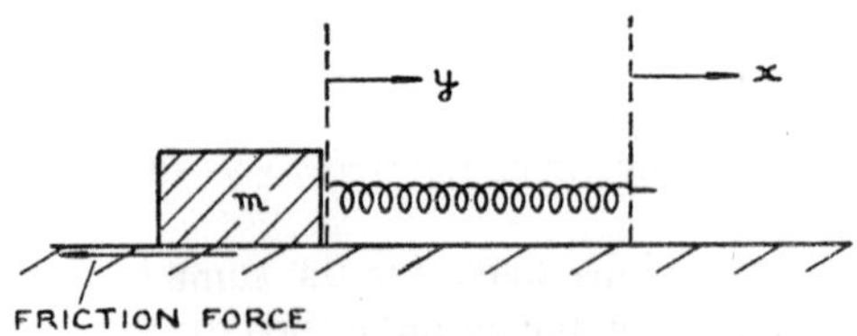

Fig. 18.5. Mass and spring system

on a horizontal table (Fig. 18.5). If the free end of the spring is displaced by an amount x (which can be any desired function of time), then the mass will be displaced through a corresponding distance y (which will also be a function of time) and the solution of this problem is concerned with finding the way in which y varies for a given variation in x. The equation of motion for this particular system is given by

$$m\frac{d^2y}{dt^2} = k(x - y) - m\mu\frac{dy}{dt} \tag{18.6}$$

where k is the spring stiffness coefficient and μ the coefficient of friction (i.e. frictional force per unit velocity per unit mass). Using the dot notation, eqn. (18.6) becomes

$$m\ddot{y} = k(x - y) - m\mu\,\dot{y} \tag{18.7}$$

On the L.H.S. of this equation the product of the mass and the acceleration of the mass represents the inertia force and the terms

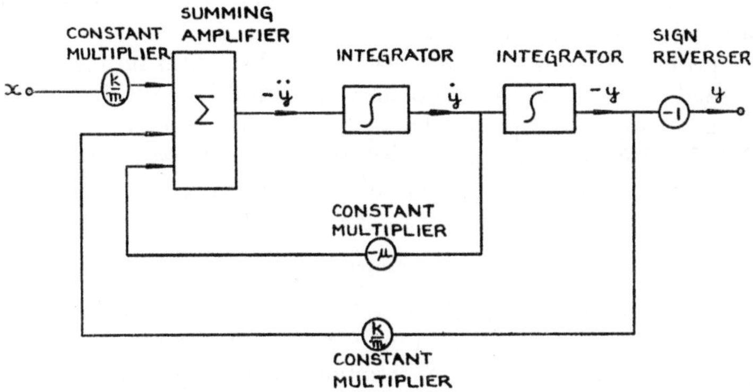

Fig. 18.6. Differential analyser for mass and spring system

on the R.H.S. represent the force resulting from the extension of the spring and the frictional force acting on the mass.

Eqn. (18.7) can be rewritten in the form

$$\ddot{y} = \frac{k}{m}x - \frac{k}{m}y - \mu\dot{y} \tag{18.8}$$

and Fig. 18.6 shows the arrangement of the differential analyser that is required for the solution of this problem. It forms a closed-loop system and its method of operation can be explained by considering the signal representing the quantity $-\ddot{y}$. This voltage is fed to an operational amplifier which is being used as an integrator (represented by a rectangular block containing an integration sign), and the output $\dot{y}$ is fed to a further integrator. The resulting output represents $-y$ and by reversing the sign of this quantity the variation of y can be obtained. Eqn. (18.8) shows that $\ddot{y}$ is made up of three terms, one of which is related to the input into the system x and the others are related to the displacement of the mass y and the velocity of the mass $\dot{y}$. Thus the $\dot{y}$ signal has to be multiplied by the constant $-\mu$, and the $-y$ signal has to be multiplied by the constant k/m, and both these signals have to be fed back to the input of a summing amplifier (represented by a rectangular block containing a summation sign). The voltage variation representing the input into the system is obtained by multiplying a signal proportional to x, the displacement of the mass, by the constant k/m, and this is also fed into the input of the summing amplifier. Thus, allowing for the sign reversal, the resulting signal coming out of the summing amplifier represents the quantity $-\ddot{y}$ which we started with.

It may seem rather strange that you cannot obtain the total input required for the first (summing) amplifier before you have the answers coming out of the second and third amplifiers. However, this problem is commonly met with in mathematical problems and the method of successive approximations is widely used. In the case of an analogue computer, the signals are transmitted so quickly (Ref. 3, page 97), that, in practice, the correct answer is produced immediately, without any appreciable time delay.

4. Application of Analogue Computers to the Motion of Aircraft

As we have seen in our work on stability and control, the motion of an aircraft can be expressed in the form of a number of differential equations, one equation representing each degree of freedom considered. The technique of using analogue computers to determine the response of an aircraft to control inputs and atmospheric disturbances under different conditions of flight is an extremely valuable one, and it is used quite widely. It has a number of advantages

(i) The values of the aerodynamic derivatives and the inertias representing the aircraft can be changed merely by changing the setting on the dial of a potentiometer. This technique enables the tests to be carried out quickly over a wide range of simulated heights, speeds and aircraft configurations. This saves a tremendous amount of time compared with flight test methods, when it would be quite impractical to be continually changing the shape of the wing or the size of the vertical tail.

(ii) The technique has the advantage over flight test methods of not getting the pilot into dangerous conditions of flight. On the computer it is possible to extend the "flight" régime into probable danger areas (e.g. to investigate roll-coupling problems) and, if the results are satisfactory, the aircraft can then actually be flown under these particular conditions in order to confirm the computer predictions.

The longitudinal motion of an aircraft following a small elevator deflection η is obtained by solving the three equations of motion—those for X, Z and M. There are various ways in which these equations can be written, depending on the notation used and the assumptions which are made. These equations have been given as

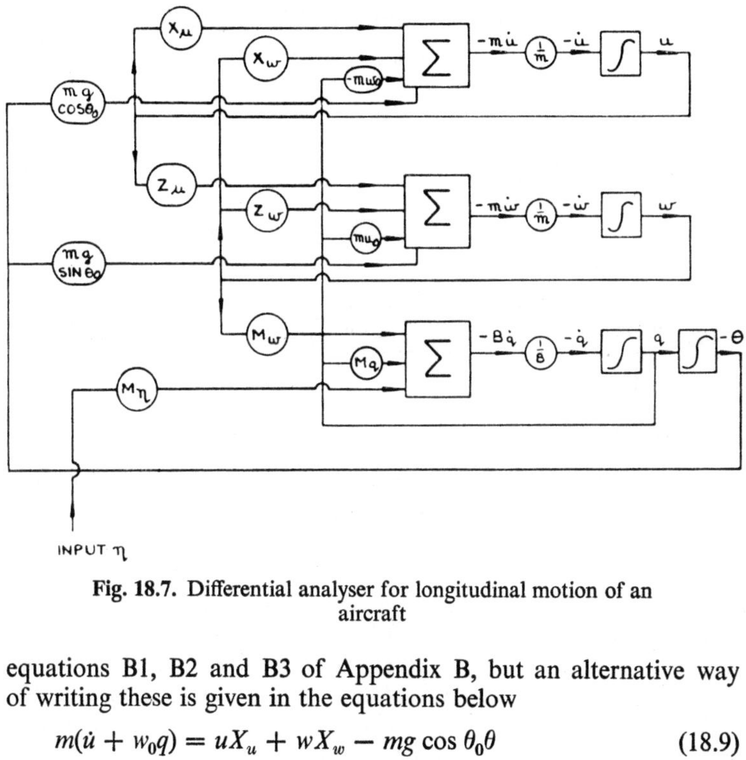

Fig. 18.7. Differential analyser for longitudinal motion of an aircraft

equations B1, B2 and B3 of Appendix B, but an alternative way of writing these is given in the equations below

$$m(\dot{u} + w_0 q) = uX_u + wX_w - mg\cos\theta_0\theta \tag{18.9}$$

$$m(\dot{w} - u_0 q) = uZ_u + wZ_w - mg\sin\theta_0\theta \tag{18.10}$$

$$B\dot{q} = wM_w + qM_q + \eta M_\eta \tag{18.11}$$

For an aircraft initially in level flight $\theta_0 = 0$, and for wind axes $w_0 = 0$.

The layout of the analogue computer required for the solution of these equations of motion is shown in Fig. 18.7. It is first of all necessary to set up the constant coefficients $mg\cos\theta_0$, $mg\sin\theta_0$, X_u, X_w, Z_u, Z_w, M_η, M_w, M_q, $-mw_0$ and mu_0. The variable η is then represented by a voltage which is fed directly into the computer, while the signals representing the other variables (i.e. u, w, q and θ) have to be fed back from the various outputs to the appropriate inputs (since they are not independent variables). For example, it is necessary to feed the values of wX_w, wZ_w, and wM_w into the X, Z and M summing amplifiers respectively.

The various outputs of the computer (i.e. u, w, q and θ) can then be recorded by pen recorders or used to operate a display.

5. Ground-based "Flight" Simulators

The term "simulator" is sometimes used for an analogue computer which is arranged in a particular way to represent the various separate parts of the dynamic system under consideration, but it is

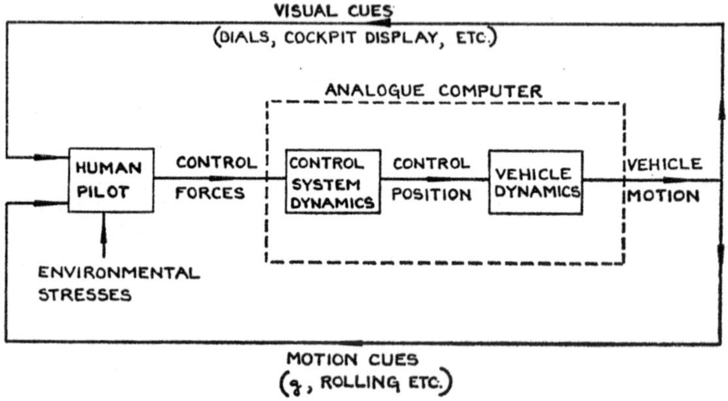

Fig. 18.8. Block diagram of pilot-controlled simulator

usually reserved for computers which have a human pilot in the control loop. The various types of simulators that are available are described in Refs. 6 to 11. Ref. 7 discusses the various ways in which simulators can be classified and contains a large bibliography of 212 reports dealing with simulators and a list of the simulator facilities that are available in the U.S.A.

Simulators are used both for training purposes and for aeronautical research and development. Refs. 9, 10 and 12 discuss the application of simulator techniques to the solution of particular problems (e.g. speed stability on the approach, pitch-up, inertia coupling, etc.). They can be used before the aircraft is built to give the pilot an idea of the possible handling characteristics that he will experience in flight, and during the development work on an aircraft they may save an appreciable amount of actual flying time.

A block diagram of a pilot-controlled simulator is shown in Fig. 18.8. To obtain satisfactory results, it is necessary to give the pilot many of the same sensations that he would experience when actually flying the aircraft. This is done by sitting him in a representative cockpit mock-up and by feeding back to him appropriate visual cues (e.g. cockpit instrumentation, outside views, etc.) and motion cues (e.g. rate of roll, *g*, etc.). It is, however, very difficult to simulate all the sensations that may be important to a pilot when he is flying an aircraft and a selection has to be made depending on the particular problem being investigated (Ref. 9). Simulators vary in cost and complexity from a very simple type of device, to extremely complicated ones allowing the pilot to have six degrees of freedom.

The pilot assesses the handling characteristics of the simulated aircraft on the basis of one of the available rating scales (e.g. the Cooper Rating Scale), and the results are used to prepare the corresponding pilot-opinion diagrams, on which the handling quality requirements of aircraft can then be based. The results obtained on the simulator can later be compared with the variable-stability aircraft and full-scale flight test results.

6. In-flight Simulators (Variable Stability Aircraft)

Though the pilot in the ground-based simulator can be presented with a number of visual and motion cues, better results can be obtained by carrying out the tests in the air. These in-flight simulators are usually referred to as variable-stability aircraft and they have been in use in the United States since about 1948. Various aircraft (e.g. the F94A jet fighter, B.26 and T33 aircraft) have been modified in this way for research purposes. The theory and application of in-flight simulation is discussed in Ref. 13 and Ref. 7 gives a list of about 26 variable-stability aircraft that have been used in the U.S.A. The list includes four variable-stability helicopters which have been used to simulate VTOL aircraft designs. The use of variable-stability helicopters in simulating the handling characteristics of V/STOL aircraft is discussed in Refs. 14 to 17. Recent work has also included simulation of the handling characteristics of possible supersonic transport designs (Ref. 18).

A variable stability aircraft is fitted with a complicated auto-stabilization system which allows all the various derivatives of the aircraft to be varied to simulate the required aircraft. This is achieved by arranging for the measured response signals of the aircraft to move the control surfaces in the way required to produce the appropriate change in the particular derivative being considered. These automatic control movements are not fed back to the pilot, who flies the aircraft in the usual way by superimposing his control inputs on top of the automatic control deflections. To simulate another aircraft, it is necessary to arrange for the moments acting on the aircraft and the corresponding moment of inertia to be in the correct proportion to one another (Ref. 13). To change the effective value of the damping-in-pitch derivative, for example, having made suitable allowance for the moment of inertia in pitch characteristics, the system is arranged to act like a pitch autostabilizer, the elevator control being deflected through an amount $(\Delta\eta)_q$ which is dependent on the measured rate of pitch q of the aircraft. Thus if M_q represents the basic pitch damping of the aircraft, the effective value of M_q is given by

$$M_{q_{eff}} = \frac{M_q q + M_\eta(\Delta\eta)_q}{q} \tag{18.12}$$

or

$$M_{q_{eff}} = M_q + \frac{M_\eta(\Delta\eta)_q}{q} \tag{18.13}$$

Since the automatic control movements are not fed back to the pilot, the control movements and forces which he experiences are those that would occur on an aircraft with stability characteristics equal to the "effective" values.

These variable-stability aircraft are very useful research tools (Refs. 19 and 20). They are extremely useful for showing how the characteristics vary with a change in the values of the derivatives, since each derivative can be varied separately over a wide range. From the results obtained, pilot-opinion plots can be produced and handling quality requirements can be formulated (Refs. 21 and 22). In this way, it can be determined how much instability may be acceptable on the more advanced designs of the future.

References

1. Wass, C. A. A.; *Introduction to Electronic Analogue Computers* (McGraw-Hill Co., Ltd., New York, 1955).
2. Korn, G. A. and Korn, T. M.; *Electronic Analogue Computers* (McGraw-Hill Co. Ltd., New York, Second Edition 1956).
3. Hersee, E. W. E.; *A Simple Approach to Electronic Computers* (Blackie and Sons, Ltd., London, 1959).
4. Harris, D. J.; *Analogue and Digital Computer Methods* (Temple Press, London, 1964).
5. Karplus, W. J. and Soroka, W. W.; *Analog Methods* (McGraw-Hill Co., Ltd., New York, 1959).
6. Rathert, G. A., Creer, B. Y. and Sadoff, M.; *The Use of Piloted Flight Simulators in General Research*, Agard Rep. No. 365, 1961.
7. Westbrook, C. B.; *Simulation in Modern Aero-space Vehicle Design*, Agard Rep. No. 366, 1961.

8. PINSKER, W. J. G.; *The Flight Simulator in Aircraft Control Research and Design*, R.A.E. T.N. Aero 2467, 1956.
9. *The Proceedings of a Symposium on the Use of Simulators in Aeronautical Research*, Symposium held at the Empire Test Pilots' School, Farnborough, 23rd October, 1964, unpublished.
10. PERRY, D. H.; *Flight Simulators and the Study of Aircraft Handling Characteristics*, Symposium on Take-off and Landing Problems held at the R.A.E., December 1962.
11. FOODY, J. J. and PAUL, R. J. A.; "Simulation Techniques in Aeronautics," *Journal of the Royal Aeronautical Society*, Vol. 62, page 878, December 1958.
12. WEIL, J.; *Application of Analytical Techniques to Flight Evaluations in Critical Control Areas*, Agard Rep. No. 369, 1961.
13. KIDD, E. A., BULL, G. and HARPER, R. P.; *In-flight Simulation—Theory and Application*, Agard Report No. 368, 1961.
14. DAW, D. and MCGREGOR, D. M.; *Development of a Model-Controlled V/STOL Airborne Simulator*, Canadian NRC-AR-LR-352.
15. MCGREGOR, D. M.; *An Investigation of the Effects of Lateral-Directional Control Cross-coupling on Flying Qualities using a V/STOL Airborne Simulator*, Canadian NRC-AR-LR-390, 1963.
16. GOULD, D. G.; *The Model-Controlled Method for the Development of Variable-Stability Aircraft*, Agard Report 402, July, 1962.
17. GARREN, J. F. and KELLY, J. R.; *Description of an Analog Computer Approach to V/STOL Simulation Employing a Variable-stability Helicopter*, NASA Tech. Note D1970, 1964.
18. BREUHAUS, W. O.; *Airborne Simulation for the S.S.T.*, Paper presented at the Aviation and Space, Hydraulic, Gas Turbine Conference and Products Show, Los Angeles, California, March 3–7, 1963.
19. BREUHAUS, W. O.; "Flight Research Utilising Variable Stability Aircraft," *Aeronautical Engineering Review*, November, 1955, page 49.
20. KAUFFMAN, W. M. and DRINKWATER, F. J.; "Variable-Stability Airplanes in Lateral Stability Research," *Aeronautical Engineering Review*, August, 1955, page 29.
21. BULL, G.; *Minimum Flyable Handling Qualities of Airplanes*, Paper presented at the S.A.E. National Aeronautical Meeting, New York, April 5–8th, 1960.
22. NEWELL, F. D.; *Criteria for Acceptable Representation of Airplane Dynamic Responses in Simulators used for Pilot Training*, Cornell Aeronautical Laboratory, CAL Report No. BM-1642-F-1, September, 1962.

19 Automatic Control Systems for Aircraft

1. Introduction

Automatic control systems are dependent on the principle of feedback, and feedback control systems are often termed servomechanisms. Although the greatest development of automatic control systems has taken place within the last thirty years, they have been in existence for very much longer. One of the earliest examples was the steam-engine governor devised by James Watt in about 1784. In about 1870 ships of the Royal Navy were fitted with hydraulic steering systems incorporating feedback control and the earliest Sperry autopilots were used in aircraft as early as 1914. Since 1930 automatic control systems have been widely used in industry, but the greatest impetus was given by the various military applications during the Second World War. Automatic control is now very widely used in aeronautics in various ways, but, before discussing particular aircraft control systems, we will discuss some of the basic principles of automatic control.

2. Basic Principles of Feedback Control Systems

2.1. General

The term *servo-system* is generally used to refer to a feedback control system in which the difference between the desired and actual value of the controlled variable is amplified, and fed to the control elements which operate on the controlled system so as to reduce the error signal between the input and output to zero. The term *servomechanism* is also used to refer to this type of system, though a more restricted definition of a servomechanism is that of a feedback control system in which the controlled variable is one of mechanical position. The essential feature of a servo system is that it uses a low-power error signal to control the supply of power to the control elements which are used to effect the desired control (i.e. it can be described as an "error-actuated" control system).

There are two main types of control systems

(i) open-loop control systems,
(ii) closed-loop control systems.

Fig. 19.1 shows a block diagram of a closed-loop control system

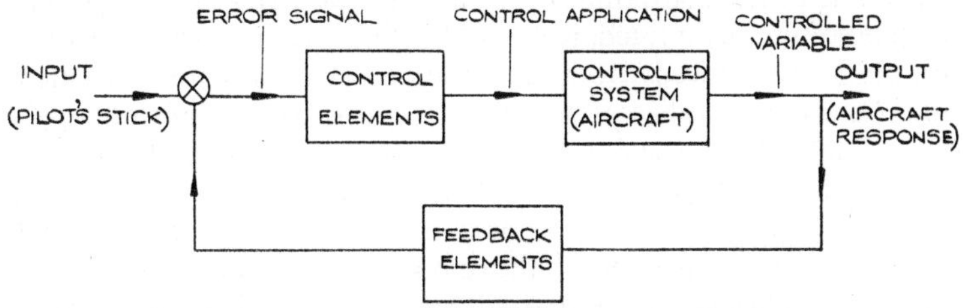

Fig. 19.1. Closed-loop feedback control system

in which the output of the system is fed back to the input, and the error signal used, as described above, to operate the controls automatically as required. This should be compared to the open-loop system shown in Fig. 19.2. The output is once again controlled by

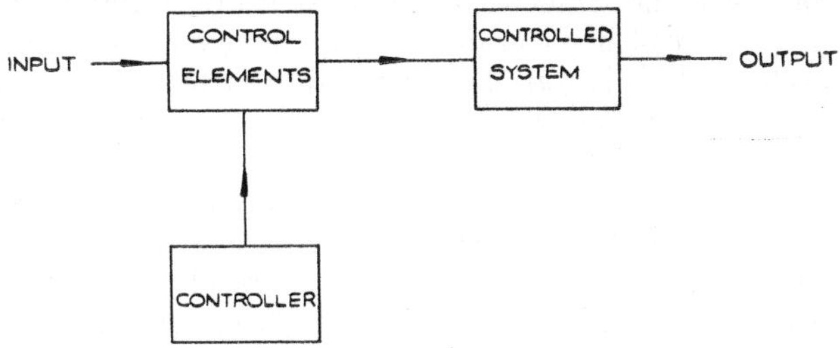

Fig. 19.2. Open-loop control system

means of the control elements, but in this case they have to be adjusted by a "controller" which may consist of a pre-determined programme or a human operator. However, if a human operator is used, then the system, in effect, becomes a closed-loop one, with the human operator closing the loop and feeding back the output signals. (Even the simple process of threading a needle can be analysed in this way, with the human operator closing the loop by means of the visual error signal between the desired and actual position of the thread with reference to the eye of the needle.) In general, closed-loop systems are superior to open-loop ones, but they do suffer from certain problems which will be discussed later (para. 6). The open-loop system may be the most economical method for certain processes, but it suffers from the following disadvantages

(*a*) It is necessary to have a known programme for the controller.
(*b*) The system is not able to deal with any disturbing factors which have not been included in the programme.

A simple practical example of the use of these two possible systems, which is commonly quoted, is that of heating a room. The controlled variable is the desired room temperature, and the control element could be an electric heater supplied from the electric power supply of the house. If the outside air temperature is known, and the doors and windows of the room are kept closed, then a programme could be worked out (e.g. a constant heater setting, or one which varies with time throughout the day) which would maintain the room at approximately the desired temperature. This system would not, however, work satisfactorily if the outside temperature changed unexpectedly, or if a door or a window was left open. A closed-loop system on the other hand would use a thermostat device to control the amount of heat produced by the electric heater, so that the temperature remained approximately constant. (A simpler automatic system would be an "on-off" device, that switched the electric heater on when the temperature dropped below a certain pre-set value, and switched it off again when the temperature exceeded a certain value.) In this way the temperature of the room could be maintained approximately constant and the latter system would compensate automatically for any internal or external disturbances.

Automatic control systems may be of many different types (e.g. mechanical, electrical, hydraulic, etc.). The mathematical analysis is, however, common to them all, though most of the notation and terminology that is used has been taken over from electrical engineering. In this chapter, we are particularly interested in the application of automatic control systems to aircraft. It is impossible to deal at any length with the general principles of automatic control systems and the reader should refer to the various standard textbooks that are available (Refs. 1 to 8).

2.2. Basic Elements of Control Systems

Because of the general application of the results obtained by a mathematical analysis to the many different types of control systems, it is very convenient and useful to employ the block diagram technique of representation. Each block then represents a particular part (or a number of parts) of a system which can be either electrical, hydraulic or mechanical. Unfortunately, as a result of this, one tends to lose a practical appreciation of the systems being considered, and before considering the mathematical analysis of control systems, it is useful to describe briefly the different basic elements of which these systems may be composed (Refs. 7 and 8). In addition to the particular element which is being controlled (e.g. the aircraft), the complete system will also contain the feedback elements, the error detector, the amplifier, the power source, and the control elements.

(*a*) *Feedback Elements* The feedback elements are required to measure the value of the controlled variable at the output of the system and to feed it back to the error detector. The output signal must be measured in a way which does not affect the value of the output. The type of element used (often referred to as a "transducer") depends upon the quantity being measured; it may be necessary to measure displacements (using potentiometers, position gyros, strain gauges, etc.), velocities and accelerations (rate gyros, accelerometers, tachometers, etc.), forces, torques and pressures (strain gauges), temperatures (resistance thermometers) or volume and mass flows (flowmeters).

(*b*) *Error Detector* The output signal is fed back and subtracted from the input signal and the resulting error signal is used to operate the controls. The error signal can be produced by mechanical methods (e.g. differential gears (Ref. 7, page 95)), electrical methods (e.g. potentiometer, Wheatstone bridge, summing amplifiers) or electromechanical methods (e.g. synchros (Ref. 8; Ref. 7, page 49)).

(*c*) *Amplifiers* The error signal is generally very small, and it is usually amplified before it is delivered to the control elements. Depending on the type of system being considered, the amplifiers that can be used are the normal electronic amplifiers (Ref. 7, page 104), magnetic amplifiers or "mag-amps" (Ref. 7, pages 126 and 227) and hydraulic amplifiers (Ref. 7, page 162).

(*d*) *Control Elements* The amplified error signals are fed to the control elements which operate so as to alter the controlled variable in the desired way. These control elements usually consist of pneumatic, hydraulic or electrical servomotors (Ref. 7, pages 132–168; Ref. 8, Pt. 2), and they consume power from the power source provided. In aircraft, power-operated control systems usually incorporate hydraulically powered systems (Chapter 12).

2.3. Control System Requirements

When a given input is applied to a control system, it is desirable that the output should follow the input as soon as possible, that any transient oscillations should be quickly damped out and that the output should attain the steady value corresponding to the initial input. Thus the main requirements of a control system are

(*a*) *Transient response characteristics* The system should respond quickly to an input demand, the output should not overshoot the final steady value by too large an amount and the oscillations that develop in the output should rapidly die away. If the output continues to oscillate about a particular value, this is described as "hunting."

(*b*) *Steady-state response characteristics* The transient effects should die away quickly and the output should become equal to the input demand, i.e. the steady-state error should be zero.

Unfortunately these various requirements are not compatible and the final overall design has to be a compromise between the different requirements. To obtain a fast response it is necessary to increase the gain (or amplification) of the system, but this unfortunately leads to a large overshoot of the output signal and to possible instability in the system. Various methods of modifying the characteristics of control systems will be discussed in para. 6.

3. Control System Analysis

3.1. General

The performance of a particular system can be determined by the appropriate mathematical analysis and this analysis can be used to predict the transient and steady-state errors and the stability characteristics of the system. In an earlier chapter (Chapter 7) we have already considered the analysis of the transient (i.e. free-oscillation) response of a simple spring-mass-viscous damper system and, as we shall see, this is an example of what is known as a second-order system. In Chapters 8 and 9 these results were applied to determine the longitudinal and lateral dynamic stability characteristics of an aircraft. In this chapter, the analysis will be extended to cover more general systems and alternative methods of solving the equations involved and presenting the results will be described.

The behaviour of any system can be expressed in the form of a differential equation. These equations may be fairly simple ones (e.g. first-order and second-order differential equations with constant coefficients), or they may be extremely complicated ones. The simpler equations are amenable to solution by the usual methods, but any complicated differential equations have to be solved by the use of more advanced techniques (e.g. Laplace transform methods, etc.).

The general solution of a differential equation consists of two parts

(*a*) the complementary function,
(*b*) the particular integral.

The complementary function corresponds to the transient (i.e. free-oscillation) characteristics of the system, whereas the particular integral depends on the input to the system and often represents the final steady-state output. The response of any system to a given input can be determined from a knowledge of the response of the system to particular forms of standard inputs (e.g. a step input, a sinusoidal input, etc.). If the response of an open-loop system is known, then the response characteristics of the corresponding closed-loop system can also be determined.

The type of analysis discussed is strictly applicable only to *linear* systems represented by differential equations with constant coefficients (i.e. ones in which a sine wave output is produced from a sine wave input, though the amplitude and phase may be altered). In practice,

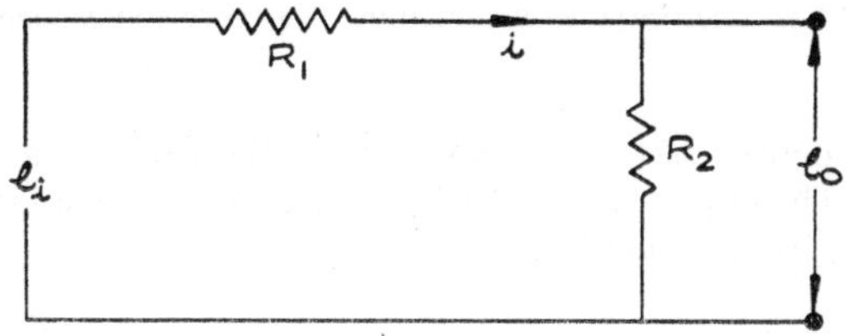

Fig. 19.3. Electrical circuit containing pure resistances

however, various unavoidable effects give rise to non-linearities (e.g. Coulomb friction [i.e. friction independent of speed], backlash in mechanical drives, overloading of amplifiers, saturation of motors, etc.). Fortunately most systems can be regarded as approximately linear, and they behave roughly in a manner predicted by the linear theory. Some types of control systems, on the other hand, are nowhere near linear (e.g. an on/off or bang/bang control system) and a general description of the behaviour of these systems is given in Chapter 5, of Ref. 5, and Chapter 9 of Ref. 4.

3.2. Transfer Functions

The behaviour of a particular system can be represented in terms of the transfer function, which, for our purposes, can be defined simply as the ratio of the output signal to the input signal of the system.

3.2.1. TRANSFER FUNCTIONS OF SOME SIMPLE SYSTEMS

A very simple electrical circuit containing only pure resistances is shown in Fig. 19.3. It is shown in Appendix D that the transfer function of this system is given by

$$\frac{e_0}{e_i} = \frac{R_2}{R_1 + R_2} \tag{19.1}$$

where e_i is the input voltage and e_0 the output voltage.

Most electrical circuits, however, contain inductances and/or capacitances and the relationship between the input and output

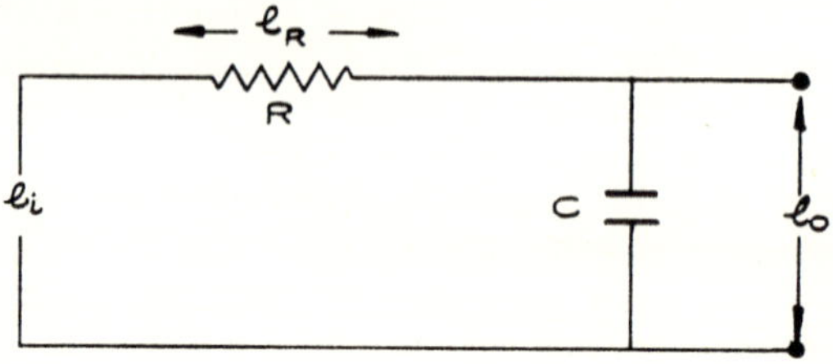

Fig. 19.4. Electrical circuit containing resistance and capacitance

voltages generally has to be expressed in the form of a differential equation. The circuit in Fig. 19.4 can be shown (Appendix D) to be represented by the equation

$$e_i = CR\frac{de_0}{dt} + e_0 \tag{19.2}$$

This equation can be written in operational form by replacing d/dt by the symbol or "operator" D

$$e_i = CRDe_0 + e_0$$

Under certain conditions, the operator D can be replaced by the Laplace operator p or s (Ref. 4, page 41), and the transfer function of this system can then be expressed in the form

$$\frac{e_0}{e_i} = \frac{1}{1 + pCR} \tag{19.3}$$

Fig. 19.5 shows a circuit containing resistance, capacitance and inductance, and the corresponding transfer function is given (Appendix D) by

$$\frac{e_0}{e_i} = \frac{1}{CLp^2 + CRp + 1} \tag{19.4}$$

3.2.2. SYSTEMS ANALOGY

Though the results of para. 3.2.1 above have been obtained from an analysis of electrical systems, the results can also be applied to mechanical and hydraulic systems (Ref. 4, paras. 2.2 and 2.7). Eqn. (19.4) represents the transfer function of the second-order electrical system. The mechanical system of the spring-mass-viscous

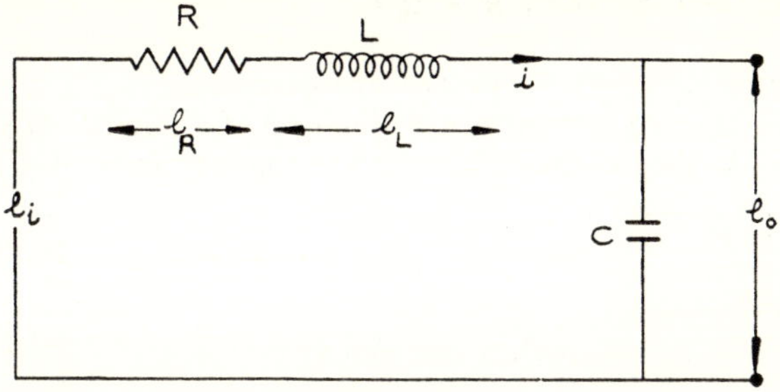

Fig. 19.5. Electrical circuit containing resistance, inductance and capacitance

damper, shown on page 42 of Ref. 4, is also a second-order system and its characteristics can be expressed in a similar form

$$\frac{\theta_0}{\theta_i} = \frac{k}{mp^2 + bp + k} \tag{19.5}$$

where θ_0 represents the output of the system, θ_i the input, m the mass, b the damping coefficient and k the spring stiffness coefficient.

Mechanical		Electrical
LINEAR	ROTATIONAL	
Mass	Moment of Inertia	Inductance
Spring Stiffness	Torsional Stiffness	(Capacitance)$^{-1}$
Viscous Damping	Frictional Torque	Resistance
Linear Displacement	Angular Displacement	Charge
Linear Velocity	Angular Velocity	Current

Fig. 19.6. Systems analogy

Fig. 19.6 shows the analogy between the various system elements. The second-order transfer function can be expressed in the more general form

$$\frac{\theta_0}{\theta_i} = \frac{1}{\dfrac{p^2}{\omega_n^2} + 2\zeta\dfrac{p}{\omega_n} + 1} \tag{19.6}$$

where ω_n is the natural undamped frequency of the system and ζ the relative damping. The characteristics of a second-order system are determined once the values of ω_n and ζ are known. By comparing eqns. (19.4) and (19.5) with eqn. (19.6) it can be seen that in the mechanical system $m/k = 1/\omega_n^2$ and $b/k = 2\zeta/\omega_n$ (compare the equations in para. 2.4 of Chapter 7), whereas in the electrical system $CL = 1/\omega_n^2$ and $CR = 2\zeta/\omega_n$.

3.2.3. MORE GENERAL TRANSFER FUNCTIONS

More complicated systems can be built up of a number of separate elements. As long as these elements are non-interacting, the transfer

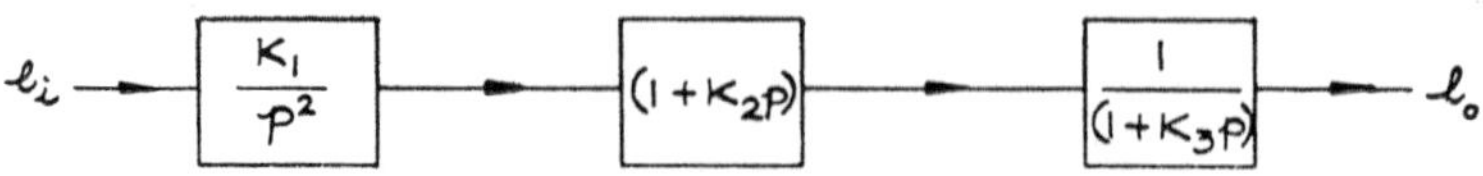

Fig. 19.7. System with separate units

functions of the separate elements can be used to determine the overall transfer function of the complete system. If the system shown in Fig. 19.7 is made up of three elements, with individual transfer functions of K_1/p^2, $(1 + K_2p)$ and $1/(1 + K_3p)$, then the overall transfer function is given by

$$\frac{e_0}{e_i} = \frac{K_1(1 + K_2p)}{p^2(1 + K_3p)} \qquad (19.7)$$

The overall transfer function of a system can be expressed generally in the form

$$\frac{e_0}{e_i} = KF(p) \qquad (19.8)$$

where K is the steady-state gain of the system and the function $F(p)$ represents the frequency dependent part of the gain. If a constant input signal is applied, say $e_i = E$, then the final steady-state output is given by $e_0 = KE$. The initial transient response of the system is determined by the function $F(p)$.

3.2.4. OPEN-LOOP AND CLOSED-LOOP TRANSFER FUNCTIONS

In the open-loop system shown in Fig. 19.8, the transfer function

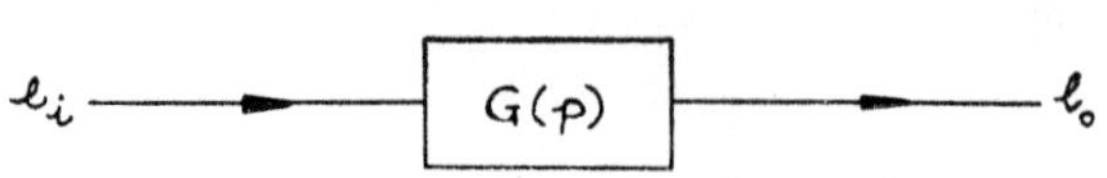

Fig. 19.8. Diagrammatic representation of open-loop system

is represented for convenience by the symbol G [more strictly $G(p)$ to show that it will usually be a function of p]

$$\frac{e_0}{e_i} = G \tag{19.9}$$

Fig. 19.9 shows the corresponding closed-loop system with negative feedback and an element in the feedback loop with a transfer

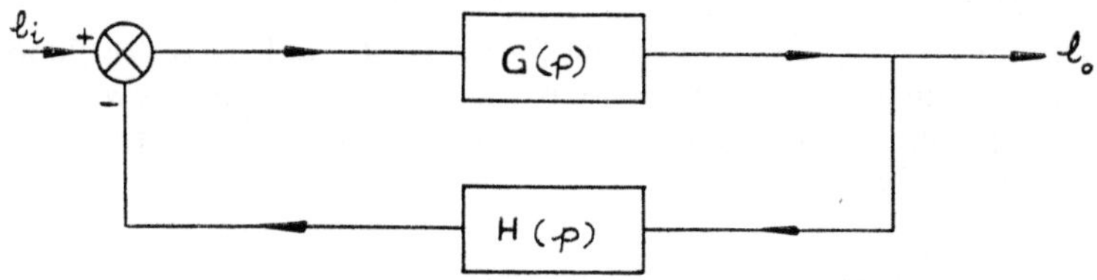

Fig. 19.9. Diagrammatic representation of closed-loop system

function H [i.e. $H(p)$]. Various transfer functions can be defined which are applicable to this closed-loop system

(i) Forward-path transfer function, $e_0/e = G$,
(ii) Loop (or Open-loop) transfer function, $He_0/e = GH$,
(iii) Closed-loop (or Overall) transfer function, e_0/e_i,

where $e(= e_i - He_0)$ is the error signal. From the above relationships, the closed-loop transfer function is given by

$$\frac{e_0}{e_i} = \frac{e_0}{e}\frac{e}{e_i} = G\left(1 - H\frac{e_0}{e_i}\right) \tag{19.10}$$

and solving for e_0/e_i

$$\frac{e_0}{e_i} = \frac{G}{1 + GH} \tag{19.11}$$

The denominator of the closed-loop transfer function is thus $1 + GH$, where GH is the loop transfer function. The equation $1 + GH = 0$ is often referred to as the characteristic equation of the closed-loop system.

When the forward path gain G is very high, then, since $1/G \simeq 0$, eqn. (19.11) reduces to

$$\frac{e_0}{e_i} = \frac{1}{H} \tag{19.12}$$

In this case, the characteristics of the system are independent of the transfer function of the element in the forward loop and are determined entirely by the properties of the feedback network. This is the basic principle of the high-gain adaptive control system (para. 7.4.2*a*).

The output signal is often fed back directly, and subtracted from the input signal. In this case $H = 1$ and the error signal is represented by $(e_i - e_0)$. Eqn. (19.11) then becomes

$$\frac{e_0}{e_i} = \frac{G}{1 + G} \tag{19.13}$$

Thus G now represents both the forward-path and the (open-)loop transfer functions and if the loop transfer function is known, the corresponding closed-loop transfer function can easily be determined from eqn. (19.13).

4. Response of Simple Systems to Step Inputs

The general response of any system can be determined if the response of the system to various standard inputs (e.g. step, ramp, pulse and

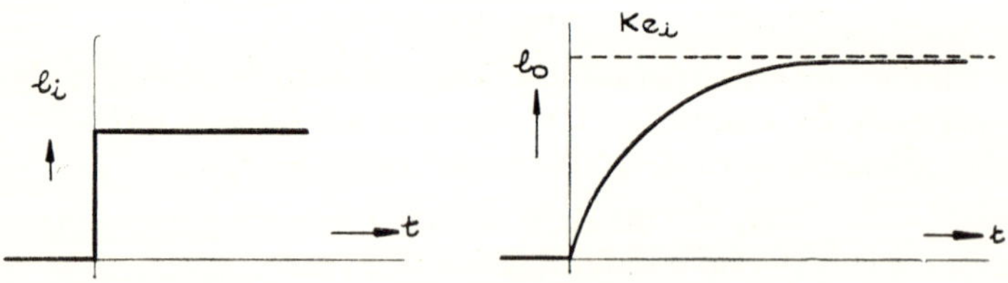

Fig. 19.10. Response of first-order system to a step input

sinusoidal inputs) is known. These response characteristics are dealt with in detail in the standard textbooks; in this paragraph only the response of first and second order systems to step inputs will be considered.

4.1. Response of a First-order System to a Step Input

The transfer function of a first-order system can be expressed in its most general form as

$$\frac{e_0}{e_i} = \frac{K}{1 + p\tau} \tag{19.14}$$

where τ is the time constant of the system. (By a comparison with eqn. (19.3), it can be seen that the time constant of the first order electrical system is equal to CR.)

The general response of this type of system to a step input is shown in Fig. 19.10; if the step input is applied at time $t = 0$, the output increases exponentially up to the steady state value of Ke_i. The response can be represented by an equation of the form

$$e_0 = Ke_i(1 - \mathrm{e}^{-t/\tau}) \tag{19.15}$$

(It should be noted that e_0 and e_i in this equation represent voltages, whereas $\mathrm{e} = 2{\cdot}7818\ldots$ is the exponential index.) From this equation it can be seen that

(i) When $t = \tau$

$$e_0 = Ke_i\left(1 - \frac{1}{\mathrm{e}}\right) = 0{\cdot}632Ke_i \tag{19.16}$$

and at this time the output has increased to 63·2 per cent of its final steady value.

(ii) as $t \to \infty$, $e_0 \to Ke_i$.

Because of its characteristics, this first-order system is often referred to as an "exponential lag network."

4.2. Response of a Second-order System to a Step Input

The various possible responses of a second-order system to a step input are shown in Fig. 19.11. The form of the response depends on

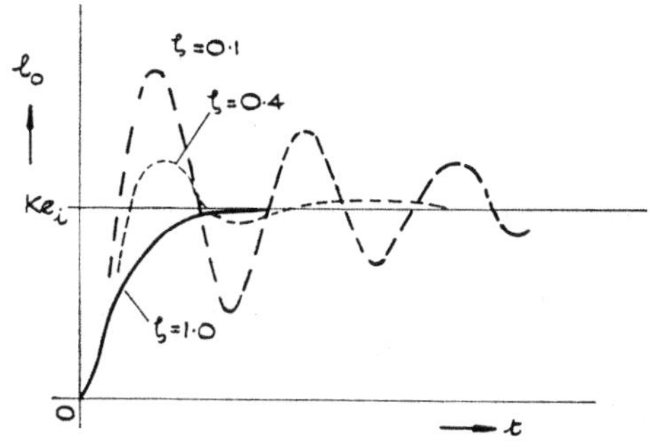

Fig. 19.11. Response of second-order system to a step input

the amount of damping in the system. For low values of the relative damping ζ the output initially overshoots the final steady-state value and then oscillates about the steady-state value before settling down. As ζ increases, the size of the oscillations decrease, and when $\zeta = 1{\cdot}0$ (i.e. critical damping) no oscillations occur and the motion is termed a "dead-beat" one. With further increase in ζ, the motion remains dead-beat, but the time response becomes more sluggish. The second-order system is often referred to as a "complex lag network."

5. Response of Systems to Sinusoidal Inputs (Frequency Response)

5.1. Definition of Frequency Response

The characteristics of a system can be determined by measuring the response of the system to sinusoidal inputs. The steady-state frequency (or harmonic) response of a system is defined as the variation with frequency of the steady-state amplitude and phase difference of the output, corresponding to a forced sinusoidal input of fixed amplitude.

It should be noted that the frequency response and the transient response characteristics of a system both depend on the same factors and when one of the responses is known, the other can be determined. Both methods are used in flight test work to determine the dynamic stability characteristics of aircraft, though, in this particular case, the transient technique is generally employed.

If the amplitude of the sinusoidal input is represented by θ_i and the amplitude of the corresponding sinusoidal output by θ_0, then the

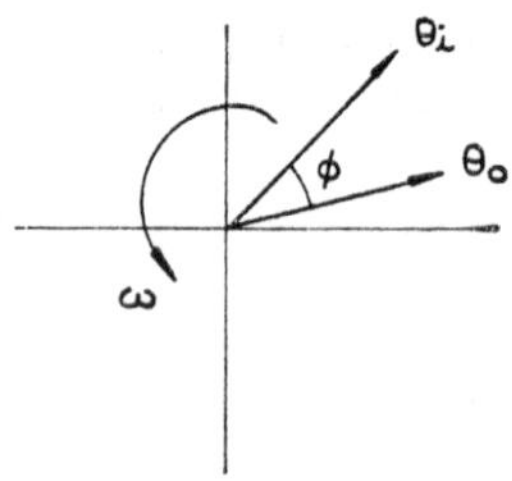

Fig. 19.12. Vector representation of input and output

relationship between the output and the input can be expressed in the form

$$\frac{\theta_0}{\theta_i} = A\,e^{-i\phi} \qquad (19.17)$$

This equation states that the amplitude of the output is A times the amplitude of the input (i.e. the system has an amplification factor or gain of A), but that the output lags behind the input by an angle ϕ (i.e. the system has a phase lag of ϕ). The input and output are both sinusoidal oscillations of frequency, say, ω rads/sec, and they can be represented by two vectors of length θ_i and θ_0, respectively, rotating at a rate of ω rads/sec, with the output vector θ_0 lagging behind the input vector θ_i by an angle ϕ (Fig. 19.12). We have seen in our discussion of the time-vector method of analysis of aircraft dynamic stability in Chapter 17, that this rotating vector representation of a sinusoidal oscillation is an extremely useful one.

5.2. Graphical Representations of the Frequency Response of a System

The characteristics of a system, or of an element of a system, are determined by the way in which the amplification or gain A and the phase lag ϕ vary with the frequency ω of the input signal. There are a number of ways of presenting this information, each of the various methods having its own advantages and disadvantages.

If the loop or open-loop response of a system is known, the corresponding closed-loop response can easily be determined. For this reason, most of the graphical presentations for closed-loop systems plot the open-loop characteristics and not the actual closed-loop ones. In this way, it is possible to check the stability characteristics of a proposed closed-loop system, before it is actually connected up as a closed-loop system.

5.2.1. FREQUENCY DIAGRAMS

One of the most straightforward methods of presenting the necessary information is a graphical presentation of the values of A and ϕ plotted against the frequency ω. This representation is unsatisfactory, however, and the same information can be plotted more usefully in the form of $\log_{10}A$ against $\log_{10}\omega$ and ϕ against $\log_{10}\omega$. One advantage of these logarithmic plots is that the information corresponding to very low frequencies (a few cycles per second) and to very high frequencies (hundreds of thousands of cycles per second) can all be conveniently presented on the one graph.

In this logarithmic presentation, the vertical scale of $\log_{10}A$ (i.e. $\log_{10} \theta_0/\theta_i$) is usually calibrated in terms of a unit called a decibel. The gain in decibels is defined as

$$\text{Gain} = 20 \log_{10} \left(\frac{\theta_0}{\theta_i}\right) \text{ decibels} \tag{19.18}$$

For example, when $\theta_0/\theta_i = 100$, the gain is 40 decibels and when $\theta_0/\theta_i = 1000$, the gain is 60 decibels (i.e. an increase of gain by a factor of 10 increases the number of decibels by 20).

More detailed information on the use of logarithmic plots is given in Chapter 10 of Ref. 6 and Chapter 7 of Ref. 4.

(*Note.* The decibel is also used as the unit for the measurement of noise intensity. It is defined by $20 \log_{10}(p/p_0)$, where $p =$ the measured pressure and $p_0 =$ a reference pressure. The value of p_0

is often taken as 0·0002 dynes/cm². Thus a pressure of $p = 15$ lb/ft² is equivalent to 151 dB and one of 150 lb/ft² to 171 dB.)

5.2.2. FREQUENCY RESPONSE LOCUS PLOTS (TRANSFER LOCUS PLOTS OR NYQUIST DIAGRAMS)

Another convenient representation of the frequency response characteristics of a system is obtained by combining together the gain and phase characteristics into a single curve, often called the Nyquist diagram of the system. The relationship between θ_0 and θ_i given by equation (19.17), i.e.

$$\frac{\theta_0}{\theta_i} = A\,\mathrm{e}^{-i\phi}$$

can be represented by a vector on what is known as an Argand

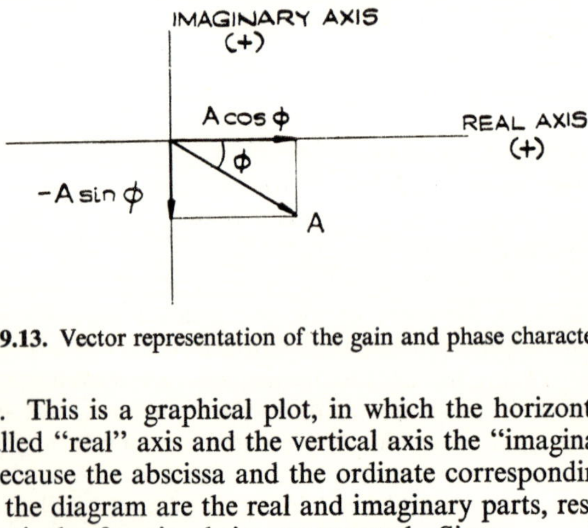

Fig. 19.13. Vector representation of the gain and phase characteristics

Diagram. This is a graphical plot, in which the horizontal axis is the so-called "real" axis and the vertical axis the "imaginary" axis. This is because the abscissa and the ordinate corresponding to any point on the diagram are the real and imaginary parts, respectively, of the particular function being represented. Since

$$\mathrm{e}^{-i\phi} = \cos\phi - i\sin\phi,$$

we have

$$\frac{\theta_0}{\theta_i} = A\,\mathrm{e}^{-i\phi} = A\cos\phi - iA\sin\phi \tag{19.19}$$

In Fig. 19.13 the θ_0/θ_i relationship is represented by a vector of length proportional to the gain A, at an angle ϕ below the positive real axis. The abscissa is seen to be $A\cos\phi$ (i.e. the real part of the R.H.S. of eqn. (19.19)), and the ordinate is $-A\sin\phi$ (i.e. the imaginary part of the R.H.S. of eqn. (19.19)).

It should be noted that, whenever the vector lies in the lower half of the Argand Diagram, the output *lags behind* the input; whenever it lies in the upper half of the diagram, the output *leads* the input.

As the frequency ω of the input is varied from $\omega = 0$ to $\omega = \infty$, the values of A (i.e. the length of the vector) and ϕ (i.e. the angle of the vector to the real axis) vary and the end of the vector sweeps out a curve or a locus plot. This curve is the Nyquist diagram of the system; it can have the appropriate values of ω or ω/ω_n noted alongside it, as is indicated in Fig. 19.16*a*.

5.3. Response of Simple Systems to Sinusoidal Inputs

5.3.1. RESPONSE OF FIRST-ORDER SYSTEM TO A SINUSOIDAL INPUT

When a sinusoidal input is fed to a first-order system, a sinusoidal

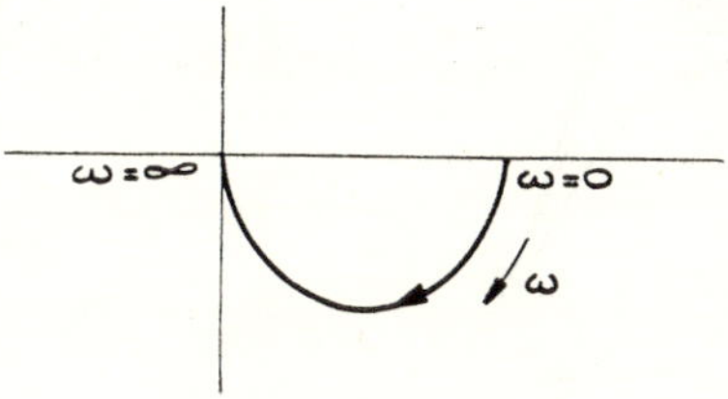

Fig. 19.14. Nyquist diagram for the response of a first-order system to a sinusoidal input

output is produced. At low frequencies the signals are transmitted with little change of amplitude (unless the system has a steady-state gain represented by K, in which case the amplitude is approximately K times the input) or phase. At the higher frequencies the amplitude is reduced and the output lags behind the input and as $\omega \to \infty$, $A \to 0$ and $\phi \to -90°$. The Nyquist diagram for this system is shown in Fig. 19.14.

5.3.2. RESPONSE OF SECOND-ORDER SYSTEM TO A SINUSOIDAL INPUT

The response of a second-order system to sinusoidal oscillations of varying frequency is shown in the frequency response curves of Fig. 19.15. The gain and phase lag of the system depend on both the frequency of the input and the relative damping of the system. For low values of damping, as the input frequency is increased the gain increases up to a maximum at a frequency just less than the natural frequency ω_n of the system. This peak amplification frequency is given by $\omega_n\sqrt{(1 - 2\zeta^2)}$. For values of $\zeta > 1/\sqrt{2}$ no peak exists in the amplification curve and at the higher values of ζ, the amplification decreases continuously with increase of frequency. The corresponding phase lag increases with increasing frequency, being higher for the higher values of ζ up to $\omega = \omega_n$ and smaller at the higher values of $\omega(>\omega_n)$. Irrespective of the amount of damping in the system, the phase difference is always $-90°$ when $\omega = \omega_n$ (i.e. when the applied frequency is equal to the natural frequency of the system).

The corresponding Nyquist diagram of the system is shown in Fig. 19.16*a*, and the effect of varying ζ is shown in Fig. 19.16*b*.

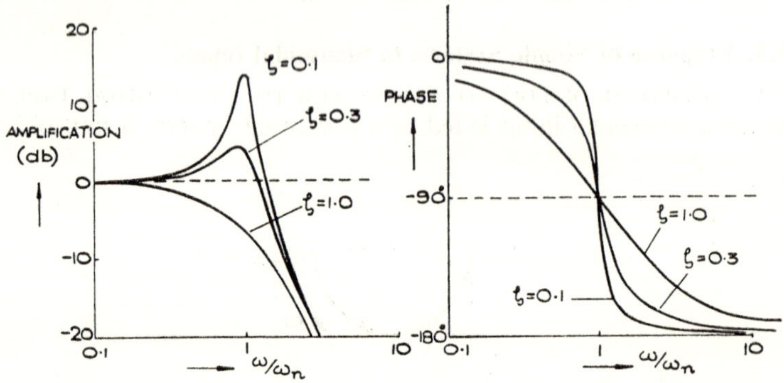

Fig. 19.15. Response of a second-order system to a sinusoidal input

When $\omega = 0$ the gain of the system is K (i.e. the steady-state gain) and the phase is zero. As the frequency increases the gain and phase vary, until at $\omega = \omega_n$, $\phi = -90°$ and the gain $= 1/2\zeta$. With further increase in ω, the phase increases from $-90°$ to $-180°$, and the gain decreases steadily until, when $\phi = -180°$, the gain is zero.

6. Stability Characteristics of Closed-loop Control Systems

Ideally, the output of a control system should follow exactly any variations of the input signal. In practice this is not possible, but, as we have seen in para. 2.3, it is desirable that the system should respond quickly, that steady-state errors should be eliminated, and that the system should be a stable one. The various errors involved depend upon the type of system and the type of input being considered. Various feedback signals can be used to improve the performance of a particular system, e.g. signals proportional to the error, to the derivative of the output ("velocity feedback"), to the derivative of the error, to the integral of the output, to the integral of the error, etc., or any desired combination of these. A compromise is always necessary in the design of any system; a high gain produces good response, but may lead to problems of instability, whereas a large amount of damping improves the stability but increases the response time and the magnitude of the steady-state errors.

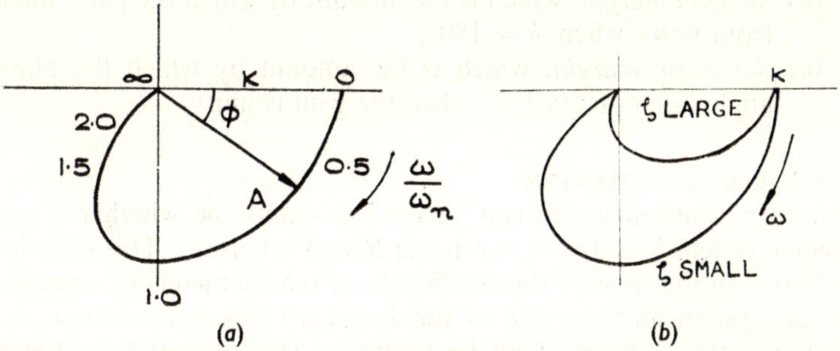

Fig. 19.16. (*a*) and (*b*) Nyquist diagrams for the response of a second-order system

6.1. Stability Criteria

The closed-loop transfer function is given by eqn. (19.11)

$$\frac{\theta_0}{\theta_i} = \frac{G}{1 + GH}$$

where G is the forward-path transfer function and GH the loop transfer function. From this equation it can be seen that, when the denominator is zero, the output would be theoretically infinite and the system would be unstable. Thus the condition of instability can be stated in the form

$$GH = -1 \qquad (19.20)$$

i.e. the system is unstable when the loop gain becomes equal to -1 (i.e. when the loop gain has an amplitude of unity and a phase lag of 180°).

6.1.1. NYQUIST CRITERION

The critical condition can be represented on the Nyquist diagram by the $(-1, 0)$ point and if the locus of the loop transfer function passes through, or to the left of this point, the corresponding closed-loop system is an unstable one. More generally, the Nyquist criterion can be stated by saying that the locus, when traversed in the direction from $\omega = 0$ to $\omega = \infty$, must pass the $(-1, 0)$ point in such a way that the point lies to the left of the locus.

The Nyquist diagram of the loop transfer function can also be used to determine the amount of stability that would be possessed by the corresponding closed loop system. This stability is expressed in terms of

(*a*) *the gain margin*, which is the amount by which the gain differs from unity when $\phi = 180°$,
(*b*) *the phase margin*, which is the amount by which the phase angle differs from 180° when the gain is unity.

6.1.2. ROUTH CRITERION

Another criterion which can be used to determine whether a particular system is stable or not is the Routh criterion. This criterion is based on the signs of the coefficients of the characteristic equation of the system and the sign of the Routhian Discriminant and, for stability, the signs must all be positive. The method has already been used in Chapters 8 and 9 to determine the longitudinal and lateral dynamic stability characteristics of aircraft. Although it indicates whether a system is a stable or an unstable one, it does not, unfortunately, give an exact measure of the amount of stability possessed by the system. The characteristic equation of a system is obtained from the expression for the closed loop transfer function (eqn. 19.11) by equating the denominator (i.e. $1 + GH$) to zero. The solution of the equation

$$(1 + GH)\theta_0 = 0 \tag{19.21}$$

represents the transient (i.e. free) response of the system and this is indicative of the overall stability of the system. In order that this equation should have no solutions involving real roots (i.e. no divergent motions), the coefficients of the equation must satisfy the Routh criterion.

6.1.3. ROOT-LOCUS PLOTS

A more exact measure of the stability of the system can be obtained by determining the actual roots of the characteristic equation (i.e. by actually solving eqn. (19.21), rather than just checking that the coefficients obey certain rules). These roots may be either real or complex and they can then be represented graphically on an Argand diagram, with the "real" axis representing the damping and the "imaginary" axis representing the frequency of the particular mode being considered. The roots of the equation will depend on the value of the loop transfer function and the root locus plot can be used to show how the values of the roots vary for different values of gain in the system.

The characteristic equation of the closed-loop system is given by

$$1 + GH = 0 \tag{19.22}$$

As has been mentioned in para. 3.2.3, a transfer function can be represented in its most general form as $KF(p)$ where K is the steady-state gain and $F(p)$ represents the frequency-dependent part of the gain. Replacing the loop gain GH in eqn. (19.22) by $KF(p)$ we have the characteristic equation expressed in the form

$$1 + KF(p) = 0 \tag{19.23}$$

The frequency-dependent part of the gain usually consists of a numerator and a denominator, both of which are, in general, functions of p. Thus eqn. (19.23) can be written

$$1 + K\frac{F_N(p)}{F_D(p)} = 0 \tag{19.24}$$

or alternatively

$$K\frac{F_N(p)}{F_D(p)} = -1 \tag{19.25}$$

The starting point of the root locus plot can be considered to be the condition of zero steady-state loop gain (i.e. zero feedback). If $K = 0$, eqn. (19.25) can only be satisfied if

$$F_D(p) = 0 \tag{19.26}$$

The values of p which satisfy this equation represent the basic dynamic stability characteristics of the system; they are usually referred to as the *poles* of the root locus plot. When feedback is introduced the values of the roots of the characteristic equation are changed and as the value of K is increased the roots sweep out the root locus plot. The ends of the root locus plot correspond to $K = \infty$. In order to satisfy eqn. (19.25)

$$F_N(p) = 0 \tag{19.27}$$

(i.e. the ends of the root locus plot correspond to the solution of the equation with the numerator of the loop transfer function equated to zero). These points are referred to as the *zeros* of the root locus plot. The system is a stable one if the roots lie in the left hand half of the Argand diagram (i.e. if the real parts of the roots have negative values), and any points in the right hand half indicate instability.

The root locus plot technique is particularly useful for representing the effects on the dynamic characteristics of an aircraft of moving the C.G. (i.e. changing the static stability), or of changing the gain in an autostabilization system.

6.2. Methods of Improving the Stability of a System

The stability characteristics of a system can be improved by modifying the basic characteristics of the system, by the addition of extra feedback signals, or by incorporating additional electronic circuits.

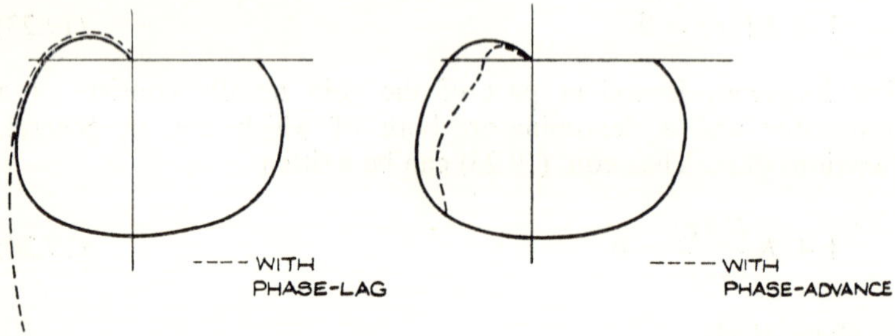

Fig. 19.17. Phase-lag and phase-advance circuits

The stability can be improved by reducing the gain of the system, but this has the disadvantage of reducing the performance as well. The damping can be increased by using "velocity feedback," i.e. feedback of a signal proportional to the derivative of the output. This has the disadvantage of producing a lag in some systems, and this method can be improved by using "transient velocity feedback" so that a feedback signal is produced only when the output is changing.

It is desirable to have a high gain in the system, as long as it does not result in instability. To achieve this it is possible to add electrical circuits which either increase the gain at low frequencies leaving the high frequency gain unchanged, or reduce the gain at high frequencies leaving the low frequency gain unchanged. These phase-lag and phase-advance circuits are made up of different combinations of resistances and capacitances (Ref. 6, page 163) and their effects are illustrated in Fig. 19.17.

7. Aircraft Automatic Control Systems

7.1. General*

Until fairly recently, aircraft control systems were usually fairly simple, consisting of an arrangement of cables and rods connecting the pilot's stick and rudder pedals to the control surfaces. The dynamic response of the aircraft to control inputs was determined by the design of the aircraft and the particular flight conditions. However, as a result of the improvement in the speed and height performance of aircraft and the consequent changes in aircraft design, it is no longer possible for the aerodynamicist to provide the necessary basic stability and control requirements over the whole flight régime. Power controls were introduced to handle the large aerodynamic control hinge moments encountered, and stability augmentation systems are used to provide artificial damping of the aircraft's dynamic response. Auto-Mach trimmers have also been incorporated to give satisfactory handling characteristics in the transonic speed range. As the performance of aircraft has continued to improve, these simple stability augmentation systems were not sufficient to produce an acceptable aircraft response, due to the large variation in the dynamic characteristics over the whole flight range. Command (or Manoeuvre Demand) Flight Control Systems were developed, such that any control input by the pilot produces a corresponding change in the response of the aircraft (rather than, as in the case of conventional controls, a corresponding change in the control surface position, the response then being determined by the dynamics of the aircraft). More recently, Adaptive Flight Control Systems have been used on some aircraft (e.g. X-15) and missiles. Initially these systems were designed such that the performance was adjusted to suit the measured values of airspeed, Mach number, altitude, etc., but they suffer from the disadvantage that some pre-knowledge of the aircraft's characteristics is necessary. Modern "Self-adaptive" Systems are a further improvement, in that they continually adjust themselves to produce the desired response from the aircraft.

The reasons for the introduction of automatic control systems into present-day aircraft are discussed more fully in Ref. 9 (pages 353–357) and Ref. 11. Doetsch (Ref. 11) considers that the rightful place of a servomechanism is where the pilot's natural limitations

* A recent book (Ref. 30) has described in more detail the application of automatic control systems to aircraft and missiles. The basic longitudinal and lateral dynamic characteristics are analysed, and the effects of various types of automatic control systems are discussed. The work is illustrated with a number of analogue computer traces showing the response characteristics of aircraft to different control inputs under particular conditions of flight. These computer traces are very useful in giving a more physical appreciation of the variations that occur in the speed, height, angle of attack, sideslip, and other flight characteristics of an aircraft during the short period, phugoid, Dutch roll, and spiral motions.

make the task of controlling the aircraft more difficult. He classifies these limitations in various ways, e.g.

(*a*) *Absolute limitations* These are limitations which cannot be avoided (e.g. maximum possible control forces, pilot's reaction time, etc.).

(*b*) *Deficiencies that can be alleviated by training and experience* These are deficiencies that can be alleviated in various ways (e.g. co-ordination of information, overall time lag, etc.).

7.2. Autostabilization, Autopilot and Flight Control Systems

7.2.1. GENERAL

As a result of the wide range of speeds and heights at which aircraft are required to operate and the changes that have taken place in the design configurations of modern high-performance aircraft, there has been a general deterioration in the aircraft's overall handling characteristics. In order to improve these handling characteristics and to ease the task of the pilot, various forms of automatic control systems have been incorporated. Many aircraft control systems are completely integrated systems which can be operated in various modes, as selected by the pilot (e.g. in the autostabilizer mode or in the full autopilot mode). For convenience, it is desirable to consider autostabilizers and autopilots separately, though the basic principles and many of the elements used are common to both systems. They can, however, be differentiated as follows.

(*a*) *Autostabilization Systems* These are provided to improve the basic stability of the aircraft and to help the pilot in flying the aircraft more accurately. This may be very important in improving the operational efficiency of the aircraft when being used for bombing, weapon firing, photography, etc. The pilot still flies the aircraft and he does not see or feel the control surface inputs which are introduced by the autostabilization system.

(*b*) *Autopilot Systems* Though the autostabilization system improves the stability of the aircraft about a basic datum, it does not provide any absolute (or long-term) datum hold. An autopilot, on the other hand, provides basic attitude stabilization and a long term datum hold in pitch, roll and yaw. In addition to the basic attitude hold about each of the three axes, the system can also be extended to include height locks, speed locks, etc. When the autopilot is switched in, the pilot is no longer flying the aircraft. In this case, the signals produced by the autopilot system are usually introduced into the control system in such a way that they are fed back to the pilot's stick, though the requirements have recently been changed to allow systems with no feedback to the stick.

The signals of an automatic system can be introduced into the control circuit in one of two ways

(i) series system,
(ii) parallel system.

In the series system the actuator of the automatic system is placed in the main control circuit in such a way that it moves the control surface without moving the pilot's stick. In this case, it is necessary to ensure that the forces involved in moving that part of the control circuit "downstream" of the actuator are less than those required to move the "upstream" part, otherwise the pilot's stick would move. Because of this, series systems are usually associated with powered controls, where the "downstream" forces required are small (i.e. those required to operate the pilot valve of the hydraulic jack). In the parallel system the actuator of the automatic system is placed in the control circuit so that both the control surface and the pilot's stick move in response to an input signal. Because of the different requirements mentioned above, autopilot systems usually incorporate a full-authority parallel-system actuator, whereas an autostabilizer uses a limited-authority series-system actuator.

Though the overall handling characteristics of an aircraft can be improved in this way, it is still important that the flying qualities of the basic aircraft design should be reasonably satisfactory, in order that the aircraft can still be flown in the event of a failure of the automatic systems. In many cases a failure would seriously affect the operational efficiency of the aircraft and it is very important that all the automatic systems should be safe and reliable. The required reliability is usually achieved by duplicating or multiplexing the various parts of the system. The methods employed for testing automatic systems for malfunctioning (e.g. straight runaways and oscillatory malfunctions) are discussed in Ref. 17.

The signals produced by autostabilization and autopilot systems are generally electrical ones. In a conventional mechanical control system these electrical signals are fed to an electromechanical actuator, and the resulting mechanical signal is fed, together with the mechanical signal from the pilot's input, to the power control unit. A complete electrical signalling control system would be very much more convenient, since it would allow the automatic signals and the pilot's input signals to be added together directly (see Chapter 12).

Autopilots and autostabilizers are described in more detail in Refs. 16 (Chapter 16), and 17. The effects of various types of autostabilizer systems on the longitudinal and lateral oscillatory characteristics of an aircraft are discussed in Chapters 15 and 16.

7.2.2. AUTOPILOT SYSTEMS

As has been mentioned, automatic pilots are concerned with the long term motion of an aircraft; they are used to maintain the aircraft at a constant attitude, height, etc., and to improve the damping of the long-period longitudinal phugoid and lateral spiral modes. The autopilot relieves the pilot of the mental and physical fatigue of controlling the aircraft for long periods, and allows him to concentrate on other tasks. In addition, facilities are provided to enable the pilot to carry out accurate and automatically co-ordinated flight manoeuvres (e.g. automatic turns, automatic I.L.S. approaches, etc.).

To keep the aircraft on a selected flight path, the behaviour of the aircraft is measured and appropriate signals fed back into the control system in such a way as to correct the deviations of the aircraft from the flight path. The signals may be obtained from attitude or rate gyroscopes, or from accelerometers, depending on the type of signals required. Gyroscopes are usually used; they are fitted with potentiometer pick-offs, and the electrical signals produced can be modified as necessary (i.e. displacement signals can be differentiated electrically to give rate signals, etc.) mixed and amplified, and used to operate the control surfaces through hydraulic or pneumatic servos. Negative feedback signals are used to ensure that the resulting control deflections are proportional to the input signals.

The control law of an autopilot system can be of many different types. The control deflections can be made proportional to the angular deviation of the aircraft from its desired datum, to the rate of change of angular deviation, to the angular acceleration, to the integral of the deviation, or to any desired combination of these various parameters. The effect of various control laws on the stability and control characteristics of an aircraft are described in Refs. 10, 14, 15 and 16.

The simplest type of longitudinal autopilot system is one that deflects the elevator control by an amount proportional to the angular deviation of the aircraft in pitch, i.e.

$$\eta = G_\theta \theta \tag{19.28}$$

where η is the elevator deflection, θ the deviation in pitch from the required value of pitch angle and G_θ (measured in units of, say, degrees of elevator deflection per degree of pitch deviation) is a constant, representing the gearing of the system. The effects of this particular system on the longitudinal characteristics of an aircraft are shown in Fig. 21.3 of Part I of Ref. 10 or in Fig. 141 of Ref. 16. The basic aircraft considered (i.e. $G_\theta = 0$) has a heavily damped short period oscillation and a poorly damped phugoid. An increase in the value of the gearing G_θ increases the damping of the phugoid oscillation, but the short period damping decreases. Eventually, the increase in G_θ results in a reduction of the phugoid frequency and the oscillation breaks down into two convergent aperiodic modes.

The addition of a rate term to the control law, i.e.

$$\eta = G_\theta \theta + G_{\dot{\theta}} \dot{\theta} \tag{19.29}$$

(where the gearing $G_{\dot{\theta}}$ is expressed possibly in degrees of elevator deflection per degree/sec) is beneficial because it increases the overall damping of the aircraft. An increase in the value of the gearing $G_{\dot{\theta}}$ increases the damping of the short period oscillation. A rate term may also be beneficial in counteracting the various lag effects that occur in the autopilot system.

Various other types of control laws, involving height, u, $\dot{u}$, etc., terms, are also discussed in Refs. 10 and 16. Ref. 20 shows how various control laws can be used to improve the phugoid damping, by incorporating the appropriate terms in the vector diagrams representing the phugoid oscillation.

In order to control the lateral motion of the aircraft, the deflections of the aileron and rudder control surfaces can be related to the various lateral motions, e.g. sideslip velocity v, angle of bank ϕ, rate of roll $\dot{\phi}$ or p, angle of yaw ψ, rate of yaw $\dot{\psi}$ or r, etc. The simplest type of autopilot system is one in which the aileron deflection is proportional to the angle of bank

$$\xi = G_\phi \phi \tag{19.30}$$

or one in which the rudder deflection is proportional to the angle of yaw

$$\zeta = G_\psi \psi \tag{19.31}$$

The effects of these control laws are discussed in Chapter 15 of Pt. II of Ref. 10 and in Ref. 16. The effect of the aileron law in eqn. (19.30) is shown in Fig. 144 of Ref. 16; the basic aircraft considered has a heavily damped rolling mode, a moderately damped Dutch Roll, and a slightly divergent spiral mode. An increase in the value of G_ϕ decreases the damping of the roll mode and increases the damping of the spiral mode, but the damping of the oscillation only increases slowly. The effect of the rudder law in eqn. (19.31) is shown in Fig. 143 of Ref. 16; an increase in G_ψ has no effect on the roll and oscillatory modes, but the spiral stability decreases and an unstable oscillation is soon produced.

Other lateral control laws are discussed in Ref. 16.

7.2.3. AUTOSTABILIZATION SYSTEMS

Autostabilizers are more concerned with the short period motion of the aircraft. They are used to increase the damping of the short period oscillations about the various axes, in order to improve the handling characteristics of the aircraft and to assist the pilot in achieving more precise control (e.g. for photography, weapon aiming, etc.). The pilot still flies the aircraft and the automatic system is usually limited to a fairly small amount of authority (between 5 and 10 per cent).

The effect on the damping of the oscillatory modes of moving the control surfaces has already been discussed in Chapters 8 and 9. It was shown that the control movement could result in either an increase or a decrease in the oscillatory damping, depending on the phase relationship between the motion of the aircraft and the motion of the control surface. In order to increase the damping, the control surface has to be deflected by the maximum amount, in the appropriate direction to oppose the motion of the aircraft, when the aircraft is rotating at its maximum rate and not when it has its maximum displacement. An automatic stabilization system is designed to produce the necessary phase relationship between the control surface and the aircraft.

A block diagram of a typical basic autostabilization system is shown in Fig. 16.12. The motion of the aircraft is measured by means of a rate gyro and the electrical signal produced is amplified and fed to an electro-mechanical actuator (which consists of a torque motor, a valve and a hydraulic jack) in order to convert the electrical signal into a mechanical one, which can be fed to the power control unit. In order to design a stable system which will produce a control deflection proportional to the measured rate signal, the system incorporates both position feedback and rate feedback signals which are fed back to the input of the amplifier. The position feedback is obtained from an electrical pick-off on the rod to the control surface and the rate feedback is obtained from a signal generated by a pick-off on the torque motor, since the displacement of the valve is a measure of the jack velocity.

The autostabilization system is designed to oppose any aircraft motion, and thus it will also oppose any steady-state pilot demands. The system must incorporate a device that will allow the pilot to manoeuvre the aircraft, while damping out any transient oscillations. This may be done in various ways

(*a*) by incorporating a band-pass filter (i.e. an arrangement of resistances and capacitances) that will prevent direct current and low-frequency signals from being transmitted, but will allow high-frequency signals to pass,
(*b*) by incorporating a stick pick-off and feeding the signal produced to the input of the amplifier where it is added in the appropriate way to the other inputs.

In addition to the normal rate term, a pitch autostabilizer may also include an attitude term. The required attitude signal can be obtained either from an attitude gyro, or by integrating the signal from a rate gyro. This pseudo-attitude term is often used to improve the stability of VTOL aircraft and helicopters (Chapter 22, para. 9).

The amount of damping produced by an automatic system depends on the control surface gearing (expressed in degrees of control surface movement per degree or radian per second of rate of rotation) and the speed and height at which the aircraft is flying. Since the dynamic characteristics of an aircraft may vary considerably through its speed and height range, it is usually desirable to vary the control surface gearing. This is usually done automatically in accordance with a pre-determined height/Mach number schedule. Typical values of the rudder gearing of a yaw damper, for example, may be 0·5° per deg/sec at high speeds, and 1° per deg/sec at low speeds.

Because of the various time lags involved in any practical system, the resulting control surface motion may not have the correct phasing that is required to increase the damping of the oscillatory motion. In this case, a phase advance electrical circuit can be introduced to produce the appropriate phase relationship.

7.3. Manoeuvre Demand or Command Flight Control Systems

In this type of control system, the pilot's input, which may be a stick movement or a stick force, commands a proportional change in some aircraft response characteristics (e.g. normal acceleration, rate of roll, etc.), rather than a proportional change in the deflection of the control surface. These systems are designed to make the response of the aircraft to a control input the same for all conditions of flight (i.e. independent of the actual aerodynamic characteristics, which will themselves vary with speed and height). In practice, however, it may not be possible to make the response characteristics completely invariant with the flight conditions, owing to various limitations in the control system (e.g. the limit to the amount of elevator power available, or the limit to the possible rate of elevator movement).

Ref. 21 describes a series of flight tests carried out on a Grumman F9F-2 with different types of longitudinal command control systems

(*a*) *A pitch-attitude system,* in which the pitch attitude angle of the aircraft is made proportional to the pilot's control input.

(*b*) *A pitch-rate system,* in which the angular rate of pitch is made proportional to the pilot's input.

(*c*) *A normal acceleration system,* in which the normal acceleration is made proportional to the input.

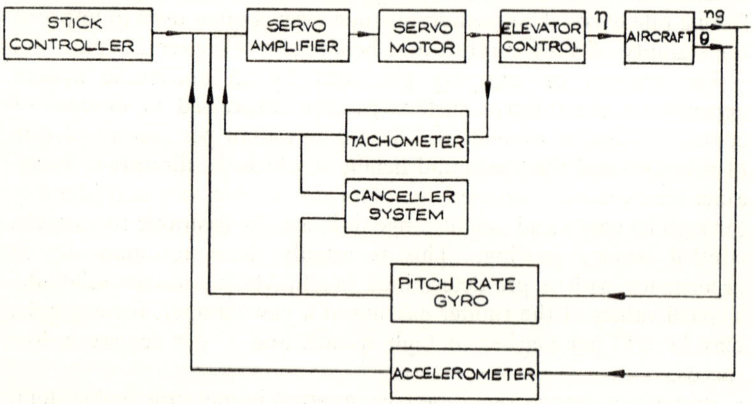

Fig. 19.18. Manoeuvre demand control system

Time histories are given showing the response characteristics of these three systems. The pitch-attitude system was not found to be very satisfactory, but the handling qualities associated with the other two systems were good. The pilot's input could be in the form of either a stick movement or a stick force and both centre-position control columns and miniature side-position sticks were used during the tests. A block diagram of the normal acceleration control system used is shown in Fig. 19.18. The pilot's input generates an electrical signal, which is amplified and fed to the servo-motor. This controls the position of the elevator and the aircraft responds, changing the pitch attitude, the pitch rate and the normal acceleration. The motion of the aircraft is sensed by a pitch rate gyro and an accelerometer and electrical signals are fed back to oppose the pilot's input signal. An electrical signal is also fed back to the input from the servo-motor output; this signal and the pitch rate signal provide stability for the system. Steady pitch rate and servo signals are eliminated by the canceller system, so that, for steady state conditions, the only feedback signal is that from the accelerometer. Thus, when the feedback signal from the accelerometer is equal to the pilot's input signal, a constant value of normal acceleration is obtained, independent of the speed and height at which the aircraft is flying, and the position of the C.G. The stick force (or stick movement) per g is then a constant, irrespective of the particular flight conditions. In the absence of any pilot input, the system acts like a conventional autopilot; the rate gyro and accelerometer detect any aircraft motion and elevator movements are produced to maintain the aircraft at a constant acceleration.

A theoretical analysis of a longitudinal control system for a supersonic interceptor again using either a pitch-rate or a normal acceleration demand system is given in Ref. 22. A similar theoretical analysis of a pitch-rate system for use on an aircraft fitted with electric signalling controls (the Avro 707C) is given in Ref. 23. Fig. 1 of the reference shows a block diagram of this particular control system. The pitch-rate demand system is achieved by feeding the input signal from the pilot's control to a rate gyro, which is measuring the rate of pitch response of the aircraft and arranging for the elevator control surface to move in proportion to the difference between these two signals. The pilot's input signal demands a certain rate of pitch (q_D) and the elevator control law can be expressed in the form

$$\eta = G_q(q - q_D) \tag{19.32}$$

where G_q is the gearing between the elevator and the signal from the rate gyro. The analysis shows that when G_q is high (of the order of 2·5 degrees of elevator deflection per degree per second of pitch rate) the response becomes independent of the flight conditions and the rate of pitch attains the required steady-state value in a small fraction of a second. This high value of gearing G_q, however, leads to instability in the servo system and the report discusses the various ways in which the stability characteristics of the system can be improved (e.g. by velocity feedback; phase advance of the gyro signal, an increase in the undamped natural frequency of the servo, etc.). Though the rate of pitch response of the aircraft is almost invariant with different conditions of flight, the elevator movements required and the normal acceleration response of the aircraft will vary depending on the height and speed at which the aircraft is being flown. The g-response is shown to build up exponentially with time, the time constant being inversely proportional to the value of the lift curve slope. The resulting elevator movement is not directly related to the stick movement applied by the pilot, and it varies as necessary to give the required g-response.

Ref. 11 shows the response curves for a hypothetical aircraft with different types of control systems at various speeds and heights. ($M = 0{\cdot}26$ at sea-level, $M = 0{\cdot}9$ at sea-level and $M = 1{\cdot}2$ at 50,000 ft). The basic aircraft has a very sluggish response to an elevator input, particularly in the build-up of normal acceleration and at high speeds the response is oscillatory. The final steady-state values of q and g vary considerably for different conditions of flight. A simple autostabilizer damps the oscillatory response, but does not equalize the steady-state responses. A g-manoeuvre demand system equalizes the steady-state value of g corresponding to a given stick movement so that, about 2 secs after initiating the demand, it is independent of the conditions of flight. The initial transient g response, the q response and the resulting variation of elevator deflection, however, all vary considerably for different conditions of flight (see Fig. 6, Ref. 11).

Ref. 24 discusses the characteristics of a roll rate demand control system installed on the Avro 707C. The report describes the theoretical analysis of the system, and the associated simulator and flight test work. In this type of system, a given lateral stick movement demands a certain value of the rate of roll, which is independent of the actual conditions of flight. The control law of the system can be expressed in the form

$$\xi = G_p(p - p_D) \tag{19.33}$$

where the gearing G_p is given in terms of degrees of aileron deflection per degree per second of roll rate. The analysis shows that if G_p is made large, the steady-state response is equal to the demanded response and independent of the flight conditions. Because of servo lag occurring in the system, however, a high value of G_p leads to problems associated with the stability of the system. Various methods of improving the stability, similar to those mentioned above for the pitch-rate system of the same aircraft, are discussed and Figs. 2, 3 and 4 of Ref. 24 show the behaviour of the system for various values of servo damping and control gearing and for different simulated conditions of flight. With velocity feedback incorporated, the value of the gearing G_p can be increased to 0·75 degrees per degree per second, while still maintaining satisfactory stability characteristics, and this value has been used in the flight test work on the aircraft.

7.4. Adaptive Control Systems

The latest developments in the field of automatic flight control systems have been with what are known as adaptive control systems. These systems can modify their gains with reference to the measured conditions of flight or to the actual response of the aircraft, in this way achieving satisfactory dynamic characteristics over the complete range of speeds and heights. The case for adaptive control systems is presented in Ref. 25; the reference describes the historical development of adaptive controls since the early 1950s and discusses some particular types of adaptive systems. Adaptive control systems are also discussed in Refs. 3, 12, 26 and 29.

These systems can be divided into two types, "semi-adaptive" control systems, which require an accurate knowledge of the aircraft characteristics over the flight range, and "self-adaptive" systems, which are able to adjust themselves and do not require any knowledge of the aircraft's characteristics.

7.4.1. SEMI-ADAPTIVE CONTROL SYSTEMS

A block diagram of a semi-adaptive system is shown in Fig. 19.19.

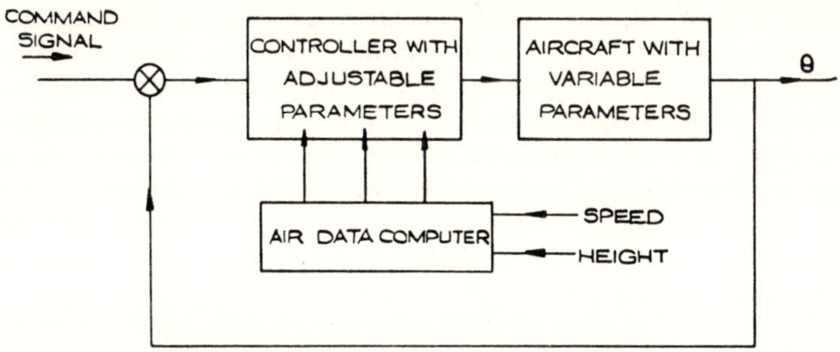

Fig. 19.19. Open-loop adaptive control system

It is an open-loop system in which the gain adjustments in the system are made in accordance with a pre-determined programme depending on the measured values of Mach number and height. The system adjusts the gain to the value that will give satisfactory aircraft response characteristics at the particular condition of flight. The semi-adaptive system has, however, a number of disadvantages (Ref. 25)

(*a*) The system is an open-loop one. It does not measure its own performance and a pre-determined programme has to be fed into it. For this reason the system cannot make any allowance for any errors in the programme, any deterioration in the system, etc.

(*b*) To determine the programme for the system, it is necessary to know the aircraft's response characteristics (i.e. the aircraft's transfer function) under all possible conditions of flight. Again, any incorrect data or a change in the configuration of the aircraft will result in an incorrect value of the gain in the system.

(*c*) It is necessary to measure the value of height, speed, etc., for all conditions of flight.

(*d*) It is necessary to tailor the system to the particular aircraft being considered. Each new aircraft requires a different system to be developed and this may occupy a large amount of development time.

Another block diagram of a possible type of semi-adaptive control system is shown in Fig. 7 of Ref. 25.

7.4.2. SELF-ADAPTIVE CONTROL SYSTEMS

To overcome the disadvantages of the semi-adaptive systems, it is necessary to have a closed-loop control system which has the capability of adjusting itself in order to produce the required response. Thus a self-adaptive control system has been defined as one which is capable of changing its parameters through an internal process of measurement, evaluation, and adjustment, to adapt to a changing external or internal environment. These systems can be designed without any accurate knowledge of the aircraft transfer functions, they can be applied fairly easily to different aircraft and they do not require a lot of flight development work to be carried out.

There are a number of different types of self-adaptive systems; the two most commonly used on aircraft can be described as the high loop gain system and the model-reference system.

(*a*) *High Loop Gain Self-adaptive Systems*

It is shown in Ref. 25 (pages 3 to 6) that if there is a high gain in the forward loop of a closed-loop aircraft control system, then the overall response of the system is independent of the particular characteristics (i.e. the transfer function) of the actual aircraft. The limit to the amount of gain that is permissible is determined by the stability of the system and hence it is necessary for the system to adjust the gain to keep within the stability limits.

A block diagram of a high loop gain control system is shown in Figs. 8 and 11 of Ref. 25. The pilot's command signal is initially fed to a "model" (i.e. the pilot flies the model), which consists of an electrical analogue system with the desired characteristics (e.g. a second-order system with the desired values of the undamped natural frequency ω_n and relative damping ζ). The model response is compared with the actual measured response of the aircraft and the error signal is then amplified and fed to the servo and used to actuate the controls. The servo response is used to adjust the gain of the system, the gain being increased until the system becomes neutrally stable. A small amplitude oscillation develops in the servo output which is carefully controlled by the system (Ref. 25), and for this reason the system is often referred to as a "limit-cycle high-gain" adaptive system. This system has been employed in the control systems of a number of research aircraft (e.g. F101 and X15).

The series model type of adaptive system has also been used to develop a variable-stability helicopter. This development is discussed in Ref. 27, and graphs are given showing the actual and desired pitch rate responses of the system. Ref. 28 describes the application of this particular technique to the roll and pitch autopilot channels of a ballistic missile.

A modification of this system is discussed in Ref. 12. In this system the "model" is moved to the feedback loop, and it is designed to be the inverse of the desired response. The actual aircraft response is fed back through the model and compared with the pilot's demand signal. The error signal is then amplified and used to actuate the controls. The system again adjusts its gain as high as possible, while maintaining the system stable. Block diagrams representing this system are shown in Figs. 9 and 10 of Ref. 12.

(*b*) *Model-reference Self-adaptive Systems*

These systems differ from those discussed in para. *a* above in having the electrical "model" in parallel with the aircraft flight control system. A block diagram of this type of system is shown in Fig. 19.20. The pilot's command signal is fed to both the model and the control system. The actual response of the aircraft and the response of the model (i.e. the "desired" response) are compared and the gain adjusted to minimize a particular selected "performance criterion." The response of the aircraft will thus be made the same as that of the analogue model. Since it is only possible to compare certain selected response characteristics, the overall response of the aircraft is not identical with that of the model in all respects, only the selected parameters are made the same.

Adaptive systems are designed to give the same response under all conditions of flight. As a result of the large changes in aircraft dynamics, however, it may not be possible to achieve this uniform high performance for all flight conditions. In this case it is better to have a system that will give the best possible performance for each

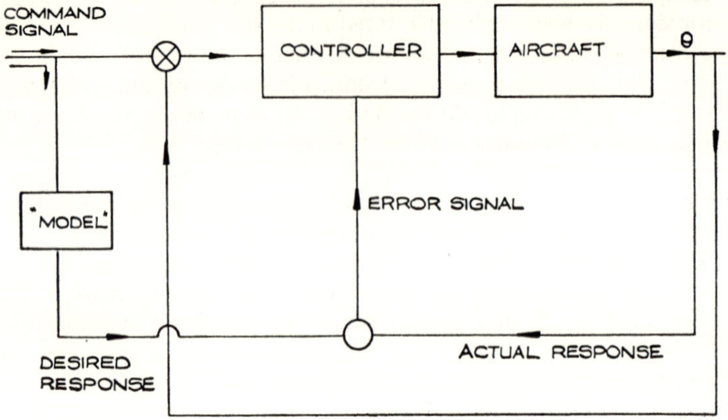

Fig. 19.20. Model-reference adaptive control system

given set of conditions. These optimalizing systems are discussed on page 370 of Ref. 3.

References

1. MACMILLAN, R. H.; *Automation—Friend or Foe* (Cambridge University Press, 1956).
2. HAMMOND, P. H.; *Feedback Theory and its Applications* (English Universities Press, Ltd., 1958).
3. DOEBLIN, E. O.; *Dynamic Analysis and Feedback Control* (McGraw-Hill Co., Ltd. New York, 1962).
4. WEBB, C. R.; *Automatic Control* (McGraw-Hill Co., Ltd., New York, 1964).
5. DE BARR, A. E.; *Automatic Control.* Monograph of the Institute of Physics and the Physical Society (Chapman and Hall, Ltd., London, 1962).
6. WESTWATER, F. L. and WADDELL, W. A.; *An Introduction to Servomechanisms* (The English Universities Press Ltd., London, 1961).
7. ZEINES, B.; *Servomechanism Fundamentals* (McGraw-Hill Co., New York, 1959).
8. *Basic Synchros and Servomechanisms Pt.* 1 *and* 2 (The Technical Press Ltd., New York, British and Commonwealth Edition, 1960).
9. ETKIN, B.; *Dynamics of Flight* (John Wiley and Sons, 1959).
10. BABISTER, A. W.; *Aircraft Stability and Control* (Pergamon Press, Oxford, 1961).
11. DOETSCH, K. H.; *The Place of Servo-Mechanisms in the Design of Aircraft with Good Flight Characteristics*, Agard Report No. 360, 1961.
12. GAYNOR, F. A.; *Effect of Servomechanism Characteristics on Aircraft Stability and Control*, Agard Report No. 361, 1961.
13. DUBLIN, M.; *An Approach to the Control of Statically Unstable Manned Flight Vehicles*, Agard Report No. 364, 1961.

14. HOPKIN, H. R. and DUNN, R. W.; *Theory and Development of Automatic Pilots*, 1937–47, R.A.E. Report No. I.A.P. 1459, 1947.
15. NEUMARK, S.; *The Disturbed Longitudinal Motion of an Uncontrolled Aircraft and of an Aircraft with Automatic Control*, R. & M. No. 2078, 1943.
16. COULTHARD, W. H.; *Aircraft Instrument Design* (Sir Isaac Pitman & Sons, London, 1952).
17. TURNER, H. W.; *Flight Testing of Automatic Stabilization and Control Systems in Manned Aircraft*, Boscombe Down A. & A.E.E., Unpublished Note.
18. BOLLAY, W.; "Aerodynamic Stability and Automatic Control," *Journal of Aeronautical Sciences*, Vol. 18, September 1951, page 569–623.
19. SEAMANS, R. C., BARNES, F. A., GARBER, T. B. and HOWARD, V. W.; "Recent Developments in Aircraft Control," *Journal of Aeronautical Sciences*, Vol. 22, March 1955, page 145–164.
20. DOETSCH, K. H.; *The Time Vector Method for Stability Investigations*, R. & M. No. 2945, 1957.
21. SJOBERG, S. A.; *Flying Qualities Associated with Several Types of Command Flight Control Systems*, I.A.S. Paper 59–134.
22. WOODLING, C. H. and GATES, O. B.; *Theoretical Analysis of the Longitudinal Behaviour of an Automatically Controlled Supersonic Interceptor During the Attack Phase Against Manoeuvring and Non-Manoeuvring Targets*, NASA Tech. Note D-454, 1960.
23. HOWELL, G. C.; *A Pitch Rate Demand System and its Effect on the Short Period Longitudinal Stability of an Aircraft*, R.A.E. Tech. Note I.A.P. 1115, 1960.
24. FORREST, J. M.; *The Effect of Servo Characteristics on the Stability of an Aircraft (Avro 707C) Under Roll Rate Demand Control*, R.A.E. Tech. Note I.E.E. No. 40, 1964.
25. OSTGAARD, M. A., STEAR, E. B. and GREGORY, P. C.; *The Case for Adaptive Controls*, Agard Report No. 406, 1962.
26. OSTGAARD, M. A. and BUTSCH, L. M.; "Adaptive and Self-Organising Flight Control Systems," *Aerospace Engineering*, September 1962, page 80.
27. GOULD, D. G.; *The Model-Controlled Method for the Development of Variable-Stability Aircraft*, Agard Report No. 402, 1962.
28. GATES, O. B. and KASTE, O. C.; *Synthesis and Flight Test of a Ballistic Missile Adaptive Flight Control System*, I.A.S. Paper No. 59–124.
29. *Proceedings of the First International Symposium on Optimizing and Adaptive Control*, Held in Rome April 26–28th, 1962. (Published by the Instrument Society of America.)
30. BLAKELOCK, J. H.; *Automatic Control of Aircraft and Missiles* (John Wiley and Sons Inc., New York, 1965).

20 Tail-less, Tail-first, and Variable-sweep Aircraft

The plates relating to this chapter are positioned between pages 538 and 539

1. Introduction

Most of the work of the previous chapters has been concerned with the conventional aircraft layout of the straight wing–fuselage–tailplane combination. Though this configuration was initially used almost exclusively, a number of aircraft with different configurations have been built and flown successfully. The problems associated with delta planforms have already been discussed in Chapters 15 and 16. Though some of the other possible configurations are, as yet, less familiar, they will probably become much more common during the next few years. This chapter discusses the stability and control problems associated with tail-less and tail-first aircraft and with aircraft employing variable sweep. It combines two articles which were originally published in the Royal Air Force Magazine, "Air Clues"; the article on tail-less aircraft was written for the September, 1964, edition, and the one on variable-sweep for the March, 1965, edition of this magazine. They are reproduced here with only minor modifications.

2. Tail-less and Tail-first Aircraft

2.1. General

One of the greatest dilemmas facing aircraft designers at the present time is whether or not to have a tailplane and, more important still, where to position it on the aircraft if one is required. Looking back over the last 60 years of flight it can be seen that, although many aspects of aircraft design have changed considerably, the basic layout of a wing-fuselage combination, stabilized and trimmed by a horizontal tailplane and elevator at the rear of the aircraft, has remained unchanged. When one considers the way in which the rear-tail configuration satisfies the stability and control requirements of an aircraft this is perhaps not surprising, and alternative layouts might not appear to have any advantages. The first aircraft to fly successfully was, however, of the tail-first (i.e. Canard) configuration, and since the time of the Wright brothers many tail-less and tail-first aircraft have been built and flown successfully.

It was not until the advent of the jet engine and the consequent increase of flight speeds into the transonic and supersonic régimes, that the conventional layout was seriously challenged. Since the end of the Second World War the tail-less delta configuration has been used extensively on many fighter and bomber aircraft all over the world. The rear-tail layout has been retained on the present-day subsonic civil transport aircraft, but the supersonic airliner of the future will have a different configuration. The BAC/Sud Concorde (Plate 1) is a tail-less delta design and it is quite likely that the American supersonic airliner will have either a tail-less delta or a canard configuration.

2.2. Historical Review

The main object of this chapter is to discuss the longitudinal stability and control characteristics of the various possible layouts, not to describe all the different tail-less and canard configurations that have been built and flown during the past 60 years. It is, however, of interest to mention briefly a few of the more successful developments that have taken place. Many of the early tail-less and canard aircraft were built only for experimental purposes, often in the form of a glider, and for various reasons few of them reached an advanced stage of development.

(*a*) *Canard Aircraft*

The first aircraft to fly successfully was that of the Wright brothers. It had a canard configuration with the horizontal tailplane at the front of the aircraft. Between the years 1903 and 1909 a number of aircraft were built with a similar layout, though in some cases a rear tailplane was also fitted. The forward tailplane was eventually discarded altogether and the rear tailplane configuration remained almost unchallenged for the next 35 years.

During the Second World War experimental fighter-type aircraft with canard configurations were built and flown in Italy, Japan and the United States. The first of these aircraft to fly was the Italian S.A.I. S.S.4 in May, 1939, but although the results were promising the development of this aircraft was abandoned in 1941. The Japanese Kyushu Shinden was initially tested in the form of a glider and the prototype did not fly until August 1945. This appeared to have the potential of being a very successful aircraft and plans had been made for large-scale production. The Kyushu Shinden was similar in appearance to the American Curtiss XP-55 Ascender (Plate 2), both of them having sweptback wings with the engine and pusher airscrew at the rear of the fuselage. Development of this aircraft began in 1939, and the prototype first flew in July, 1943. The stalling characteristics were unsatisfactory, however, and further development was abandoned.

After the war a number of experimental canard aircraft were designed and built in this country by Miles Aircraft Ltd., though it would be more appropriate to describe these as tandem-wing designs, since the forward horizontal surface was almost as large as the rear one.

It was not until about 1957 that the canard layout was seriously considered again, though it had meanwhile been used on the French Nord Griffon experimental aircraft. In that year preliminary design work was started on the North American B-70 Valkyrie (Plate 3). This aircraft was originally designed as an intercontinental bomber to fly at around Mach 3 at 70–75,000 ft for a range of 9,000 miles. It has a trimming foreplane with trailing-edge flaps set high on the fuselage in front of a delta wing fitted with elevon control surfaces. Though this was cancelled as a bomber project, the aircraft has flown as a research vehicle, and it is expected to provide valuable design data for the American supersonic airliner.

Three possible designs were initially put forward for this supersonic aircraft, but the final selection has still to be made. One design was for a rear-tail layout incorporating a wing with variable sweep, another was a tail-less delta planform and the third was a canard layout with a delta wing. This latter design was the NAC-60 of the North American Company who also built the B-70.

(*b*) *Tail-less Aircraft*

The development of the tail-less aircraft began in England with J. W. Dunne, who, between the years of 1907 and 1912, built a number of successful tail-less biplanes and monoplanes. These aircraft obtained their longitudinal stability from the sweepback and wash-out on the wings.

The next important series of experiments were those carried out by G. R. T. Hill. Between the years 1926 and 1932, he built a number of tail-less aircraft, culminating in the Pterodactyl Mk. V. This aircraft was designed as a two-seat fighter. It had a high sweptback wing, a much smaller lower unswept wing and the engine and propeller were at the front of the fuselage. It is interesting to note that the Pterodactyl Mk. IV, which was a 3-seat cabin monoplane, incorporated in-flight variable sweep for trimming purposes.

Similar work was being carried out in America by the Northrop

Aircraft Company, and by the end of the Second World War a large 300,000 lb multi-engined experimental long-range bomber, the Northrop XB-35, had been designed and built. It first flew in June 1946. A turbo-jet version of this aircraft, the YB 49, first flew in October 1947.

Some of the most important development work was, however, carried out in Germany, in particular by Alexander Lippisch. After some initial work on tail-less gliders, he developed the delta planform and in the period 1931 to 1937 a whole series of delta aircraft were built and flown. During the war his work resulted in the development of the rocket-powered Me163 (Plate 4).

In the years following the Second World War, the results of the wartime German research were used, both in this country and in America, to develop the swept wing and delta configurations. In England the de Havilland Company began work on the DH108, an experimental swept-wing tail-less jet aircraft built to investigate the stability and control problems associated with this particular layout. It was the first British aircraft to exceed the speed of sound in a dive, and in 1948 it set up the world closed circuit speed record of 605·23 m.p.h. At the same time Avros were developing the Avro 707 series of delta research aircraft, and the first of these flew on the 4th September, 1949. The results of this research programme were used in the development of the Vulcan bomber.

Two other important research aircraft with delta planforms are the Fairey F.D.2 (Plate 5) and the Handley Page H.P.115 (Plate 6). The F.D.2 first flew in 1954, and since that time has been engaged on high-speed research. One of these F.D.2 aircraft is still flying at the Royal Aircraft Establishment at Bedford, while another one, renamed the BAC Type 221 (Plate 7), has been modified to test an ogee-wing planform similar to the one designed for the BAC/Sud Concorde. The BAC 221 flew for the first time on May 1st, 1964. The H.P. 115 on the other hand, was built in order to test the low-speed handling characteristics of the slender-delta planform.

Within the last few years, many delta wing aircraft have reached squadron service in several countries of the world. In the U.S.A. the Convair Aircraft Company has built a series of delta wing fighter aircraft (e.g. the F-102A Delta Dagger and F-106A Delta Dart) and the B-58 Hustler bomber. Similar development has occurred in France, resulting in the Mirage series of aircraft built by the Dassault Company.

2.3. Stability and Control Requirements

Any aircraft design must satisfy certain basic requirements—those of trim, static and dynamic stability, and control.

(i) *Trim* It is essential that the pilot should be able to trim out the various pitching moments acting on the aircraft for all possible configurations throughout the speed range. This trimming should be achieved, as far as possible, without any performance penalties in the form of loss of lift or increase of drag.

(ii) *Stability* The aircraft should possess satisfactory static and dynamic stability characteristics under all conditions of flight.

(iii) *Control* The pilot must have sufficient control over the aircraft to enable him to manoeuvre it as required to fulfil the particular role of the aircraft. The stalling characteristics and, possibly, the spinning characteristics should also be satisfactory.

These requirements can be satisfied by either a rear-tail, a tail-first or a tail-less configuration. Whether or not a separate tailplane is

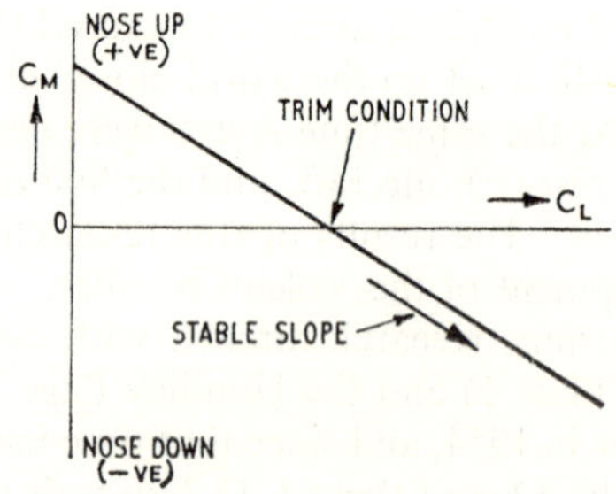

Fig. 20.1. Typical C_M/C_L characteristics of a stable aircraft

required is largely dependent on the various features incorporated in the design of the aircraft, and on its operational requirements.

Before discussing the advantages and disadvantages of the different possible configurations, the various methods of satisfying the trim and stability requirements will be described. In previous chapters it has been shown that these characteristics can be represented, for any one selected position of the longitudinal control, by a graph of the pitching moment coefficient measured about the C.G. of the aircraft plotted against the lift coefficient (Fig. 20.1).

The aircraft is in trim (i.e. the pitching moments all add up to zero) at the point where the graph crosses the horizontal axis, and, due to the sign convention used, this condition is a stable one when the line has a negative slope. In order to satisfy both the conditions the curve must intersect the vertical axis at a positive value of C_M, i.e. there must be a nose-up pitching moment at the condition of zero-lift. (It should perhaps be pointed out that, although the intercept is a very useful reference point to use in this work, it does not correspond to a condition that can be achieved in flight.)

An aircraft design consisting of an unswept wing planform with a normal cambered sectional shape has unsatisfactory characteristics,

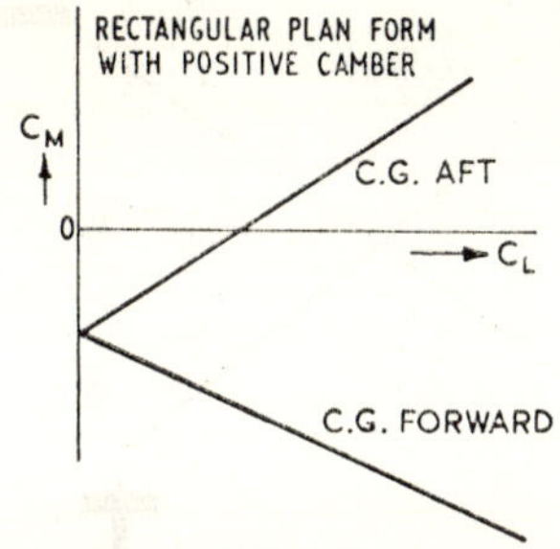

Fig. 20.2. Effect of C.G. position

since the curve of C_M against C_L intersects the vertical axis at a negative value of C_M, and, for normal C.G. positions, the line has a positive slope (Fig. 20.2). This results in an aircraft which, although in trim at one particular speed, is statically unstable (i.e. an increase

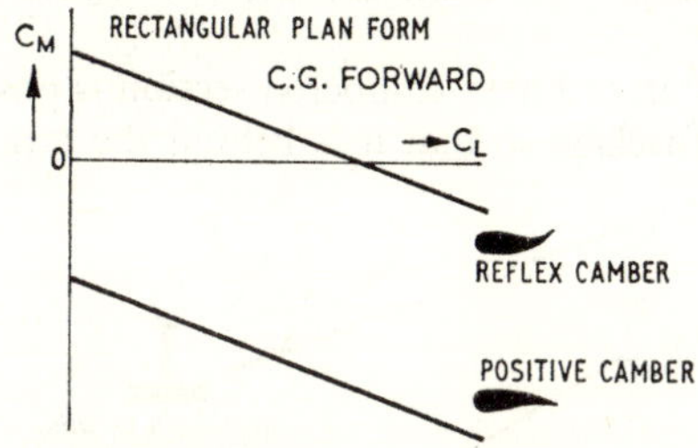

Fig. 20.3. Effect of reflex camber

in the angle of attack produces a nose-up pitching moment which further increases the angle of attack).

The C.G. can be moved forward to produce a C_M/C_L curve with a stable slope, but then the aircraft is not in trim at any speed. The aircraft can be trimmed at any desired C_L value, while still retaining the same amount of static stability, by adding a constant nose-up pitching moment coefficient. This can be achieved by deflecting a trailing edge control surface upwards, or by designing the basic wing section to have a reflex cambered shape (Fig. 20.3). Tail-less aircraft using unswept wing planforms with this type of sectional shape have been built and flown.

When the C.G. is farther back the aircraft is statically unstable and, to satisfy the requirements, it is necessary to add the nose-up pitching moment mentioned above and to make the aircraft stable. This can be done by means of a horizontal tailplane at the rear of the aircraft

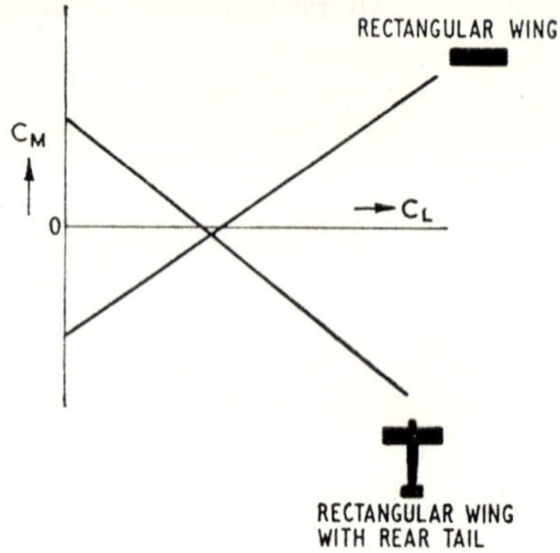

Fig. 20.4. Effect of a horizontal tailplane

(Fig. 20.4). In order to produce the required nose-up pitching moment, the tailplane is fixed onto the fuselage at a negative angle relative to the wing. This angle is sometimes called the longitudinal dihedral angle.

If a wing with an ordinary cambered section is positioned towards the rear of the fuselage so that it is behind the C.G. of the aircraft,

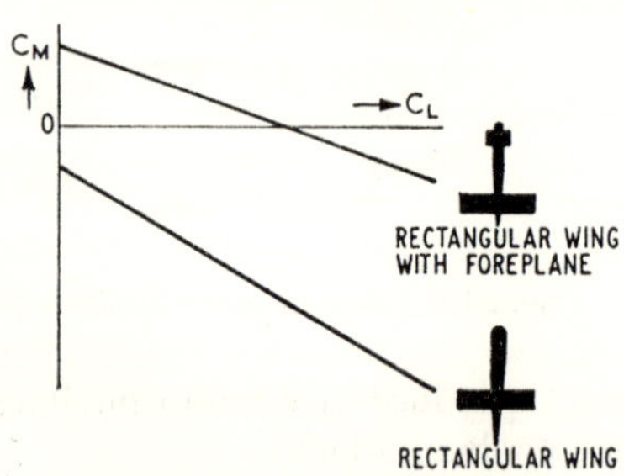

Fig. 20.5. Effect of a foreplane

then, as we have seen already, an aircraft is produced which, although it possesses stable characteristics, cannot be trimmed. The nose-up pitching moment required to trim the aircraft can also be produced by means of a foreplane fixed onto the fuselage at a positive angle relative to the wing. In this tail-first or "canard" layout, the wing is stabilizing and the tailplane destabilizing, but, as always, the design has to be such that the complete aircraft is statically stable (Fig. 20.5).

So far only unswept wing planforms have been mentioned. Another way of designing a wing to have satisfactory trim and stability characteristics is by a suitable combination of sweep and twist. Sweepback is generally used and this requires "wash-out" on the wings to produce the nose-up zero-lift pitching moment. Wash-out reduces the angles of attack of the wing tip sections relative to those at the wing root, so that, when the total lift on the wing is zero, there is an upload on the centre part of the wing and a down-load at the tips (Fig. 20.6). This produces a nose-up pitching

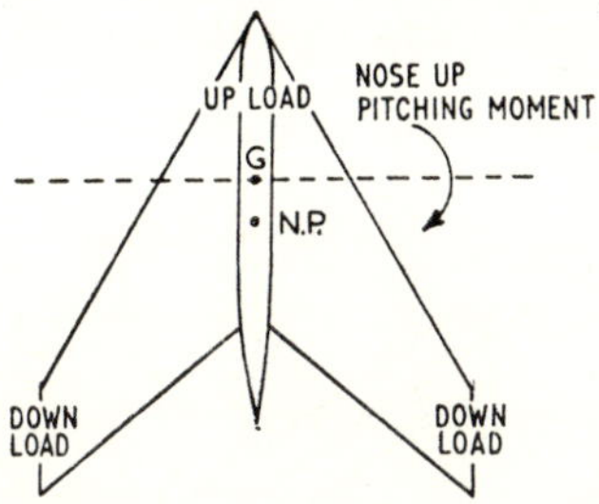

Fig. 20.6. Effect of sweepback and washout

moment in a similar way to the longitudinal dihedral on a tailplane at the rear of the aircraft.

The stability requirements are also satisfied by this design, because the sweepback has the effect of moving the Neutral Point rearwards. The effect of sweeping the wings on the position of the C.G. is, on the other hand, much less, because the C.G. position is dependent on the distribution of the weight and this is usually mainly concentrated in the fuselage. In this way, the Neutral Point can be brought behind the C.G. and the aircraft designed to have the required amount of static stability.

It is thus possible to design various configurations for an aircraft which will enable it to fly in a trimmed and stable condition at one particular speed. It is necessary, however, for the pilot to be able to vary the speed at which the aircraft is in trim, while still retaining the static stability. This can be achieved by means of movable control surfaces; these can consist of trailing edge flap controls on the tailplane, the wing, or the foreplane, or of all-moving tailplane or foreplane surfaces.

2.4. Advantages and Disadvantages of Different Configurations

Although all these different configurations can be designed to satisfy the stability and control requirements, they each have their advantages and disadvantages. The configuration finally chosen is a compromise, depending on the particular type of aircraft being considered.

2.4.1. REAR-TAIL LAYOUT

This configuration has the obvious disadvantages of the extra drag and structure weight that are associated with any tailplane, but it also has some additional disadvantages.

(i) *Trim Drag*

On an aircraft fitted with a rear tailplane and elevator controls, down elevator is required to trim the aircraft at a high forward

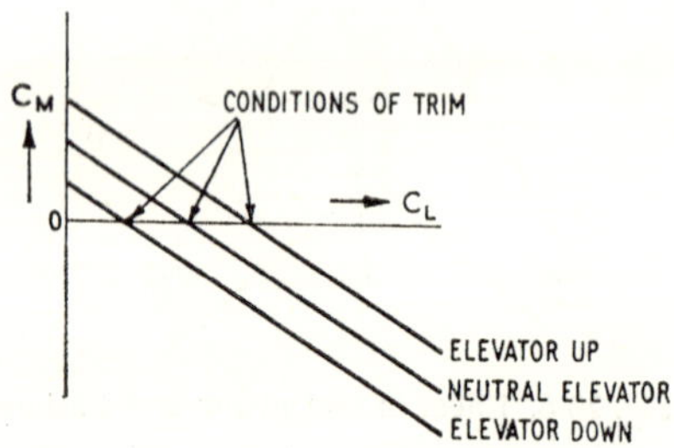

Fig. 20.7. Effect of elevator deflection

speed (i.e. at a low value of C_L) and up elevator to trim at a low forward speed (i.e. at a high value of C_L). The corresponding C_M/C_L curves intersect the horizontal axis at different points, but the slope (i.e. the static stability) remains the same (Fig. 20.7).

The load on the tailplane is up at low speeds and down at high speeds. This means that at a high forward speed the wing has to produce a lift force greater than the weight of the aircraft. To achieve this extra lift the angle of attack of the wing has to be increased and both the profile and induced drags are correspondingly increased. This extra drag is sometimes called "trim drag"; it can be a large and important part of the total drag of an aircraft at high speeds.

(ii) *Pitch-up, Superstall and Spinning Problems*

With the advent of sweptback wings for the purpose of delaying the onset of compressibility effects, a form of longitudinal instability known as "pitch-up" became a serious problem. This phenomenon is the result of the particular stalling characteristics of these wing planforms. The stall begins at the wing tips and spreads inboard as the angle of attack is further increased.

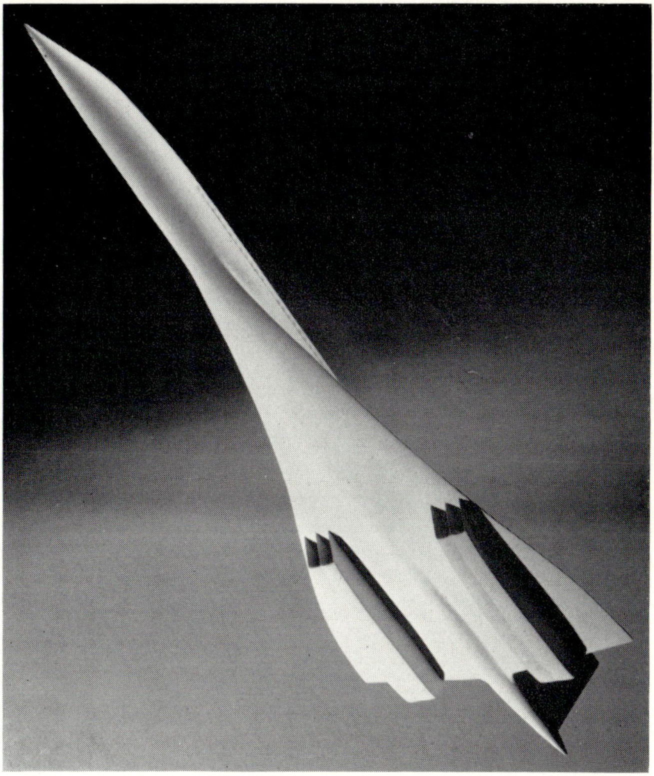

Plate 1. BAC/Sud Concorde

Plate 2. Curtiss XP-55 Ascender

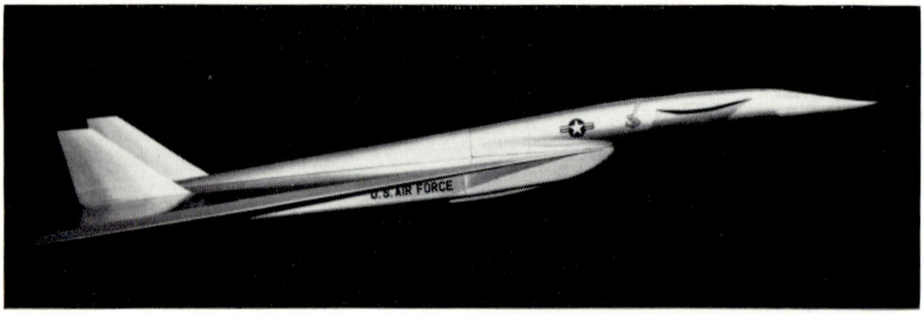

Plate 3. B-70 Valkyrie

Plate 4. Me 163

Plate 5. F.D.2

Plate 6. H.P. 115

Plate 8. Model of Boeing 733 with wings in the low speed configuration

Plate 9. Model of Boeing 733 with wings swept back for supersonic flight

Plate 7 (*opposite*). BAC Type 221

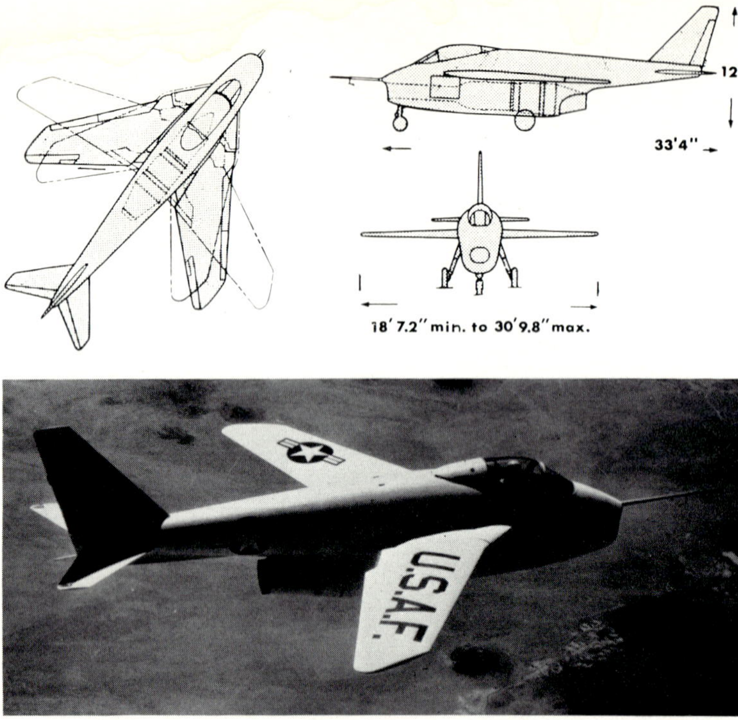

Plate 10. Bell X-5

Plate 11. Grumman XF 10F-1

Plate 12. Short SB5

Plate 13. F-111A in flight

Plate 14. Model of F-111A showing variable wing positions

The concentration of the lift on the middle part of the wing causes the centre of pressure to move forward and for the downwash from this part of the wing to increase. The increased downwash striking the tail-plane and the forward movement of the centre of pressure both have the effect of producing a nose-up change in pitching moment. This "pitch-up" may cause large g values to be developed on an aircraft during a manoeuvre, and it may limit the aircraft's manoeuvring performance.

Various modifications can be made to the design of the wing planform to reduce or eliminate this particular problem. With the rear-tail layout the "pitch-up" effect can also be reduced by moving the tailplane out of the worst of the downwash region. This can be done by placing the tailplane high up on the fin or low down on the fuselage; it may be necessary to incorporate some anhedral or dihedral on the tailplane to obtain satisfactory results.

If a low-set tailplane can be designed so that it is out of the way of the jet efflux, and it does not restrict the landing attitude of the aircraft, then this is usually the best position for the elimination of pitch-up, though the configuration may have other disadvantages at transonic and supersonic speeds. The tailplane is often placed on top of the vertical fin, but this position has the associated disadvantages of extra structure weight and problems of flutter, and also the possibility of the pitch-up causing a condition known as the "super-stall" to develop.

Recovery from this condition may be extremely difficult, since, at the high angles of attack developed, the tailplane is in a region of intense turbulence produced by the wing and the effectiveness of the longitudinal control may be considerably reduced.

2.4.2. TAIL-LESS LAYOUT

If the tailplane can be eliminated altogether, there is an immediate and obvious reduction in structure weight and drag. One of the most satisfactory planforms for a tail-less layout is the delta wing. Though, as we have seen, it is possible to satisfy the trim and static stability requirements, the aircraft may suffer from a number of disadvantages.

(i) *Trim Drag and Reduction in C_{Lmax}*

When the elevator controls form part of the trailing edge of the wing, upward deflection of the control may produce an appreciable loss of lift, which has to be balanced by an increase in the wing angle of attack. This effect is increased if high-lift flaps producing a large nose-down pitching moment are used, because more up elevator will be required for trim. In these cases a separate tailplane is often necessary to give the amount of trimming power required.

Trim drag will also be important on a delta wing aircraft at high speeds. When the forward speed is increased from a subsonic to a supersonic value the centre of lift on the wing moves back. This produces a nose-down change in pitching moment that has to be balanced out by up elevator movement and there is a corresponding increase in the drag. On the BAC/Sud Concorde it is planned to use a system of fuel transfer to move the C.G. of the aircraft in order to reduce the trim change and the associated drag.

(ii) *Reduction in Manoeuvrability*

A tail-less aircraft may also suffer from a reduction in manoeuvrability and may have poor spin recovery characteristics. To improve these characteristics it is often necessary to fit a separate tailplane.

2.4.3. TAIL-FIRST (CANARD) LAYOUT

On an aircraft configuration consisting of a long slender fuselage with the engines at the tail and a C.G. position well to the rear of the aircraft, the canard layout has the obvious geometric advantage of the longer moment arm. This enables the stability and trim requirements to be satisfied by a foreplane of smaller area. The trim drag problem will also be reduced, because at high speeds an upload will be required on the foreplane to trim the aircraft. The lift on the wing will consequently be less than the weight of the aircraft, and this will tend to reduce the total drag.

The layout does, however, suffer from certain disadvantages.

(i) *Stalling Problems*

On the rear-tail configuration the wing stalls before the tailplane, and longitudinal control and stability are maintained at the stall. On a canard layout if the wing stalls first stability is lost, but if the foreplane stalls first then control is lost and the maximum value of C_L is reduced. Various methods of alleviating this problem have been considered. One possible approach is to use a canard surface and a wing trailing edge flap in combination, with one surface acting as a trimming device and the other as a control. An alternative suggestion is to use an auxiliary horizontal tail-plane at the rear of the aircraft for trim and control at low speeds.

(ii) *Interference Problems*

On a rear-tail configuration the airflow from the wing produces interference problems with the tail unit. In a similar way on a canard layout the airflow from the foreplane interferes with the flow round the main wing and the vertical fin. This interference causes a reduction of lift on the main wing, and may also result in stability problems.

The interference effects between a canard and a single centrally-placed vertical fin often result in the directional static stability becoming marginal at high angles of attack. This loss of stability is due to the effect of the wing tip vortices from the canard and the stability can be improved by replacing the single fin by twin vertical fins as on the B-70 Valkyrie. A canard configuration was initially considered for the BAC/Sud Concorde, but it was eventually rejected because of the difficulty of obtaining a satisfactory solution to this interference problem.

3. Variable-sweep Aircraft

3.1. General

The final design of any aircraft is always a compromise between the various conflicting requirements. Most modern aircraft are designed to fly at high subsonic or supersonic speeds, and to enable them to achieve the desired performance they need to have thin symmetrical wing sections with highly swept or delta planforms. Unfortunately, these types of wings are very inefficient for flight at low forward speeds. For this flight régime the optimum wings have unswept planforms and cambered sectional shapes. An aircraft which is going to fly at supersonic speeds most of the time will be designed primarily for this condition of flight, and the poor low speed characteristics will have to be accepted, though various high-lift devices are available for reducing the landing and take-off speeds and improving the low speed handling qualities. In the case of an aircraft which is required to operate efficiently in both the subsonic and supersonic speed régimes, variable wing sweep is a desirable feature to incorporate in the design. The wings can then be swept back when the aircraft is being flown at high speeds and swept forward again when flying at low speeds.

Variable geometry is not new. From the very earliest days of flight, aircraft have been designed with various forms of variable geometry to provide methods of controlling the aircraft, and for improving the performance capabilities and the handling characteristics. Wing warping, movable flaps, and all-moving surfaces have been used to control the aircraft, and leading-edge and trailing-edge flaps and leading-edge slots to reduce the landing and take-off speeds. The top speed performance of aircraft was increased by the invention of retractable undercarriages, and the pilot's view on the approach has been improved more recently by the use of variable incidence wings and drooping fuselage noses. The XB-70A Valkyrie has wing tips that can be raised or lowered; in low speed flight they are raised to form part of the wing, and at high speeds they are lowered to act as additional fins. The proposed designs for the British and American supersonic transport aircraft also incorporate many variable geometry features; the shape of the fuselage nose can be varied to reduce the drag at high speeds and to improve the view at low speeds, and the engine intakes and exhaust nozzles are adjustable to suit the particular conditions of flight.

Variable wing sweep is the next logical step in the continuing story of variable geometry. But even variable sweep is not new—one of the early Hill Pterodactyl aircraft is supposed to have used variable sweep for in-flight trimming. However, the more recent development work on variable sweep was started, as was that on many other important aerodynamic design features, in Germany during the Second World War. After the war, the work was continued both in this country and in the United States, and a small number of research aircraft were built and flown.

It is in the United States that the first operational variable sweep aircraft have been designed and built. The General Dynamics-Grumman F-111 strike aircraft being built for the USAF and the USN incorporate variable sweep, and this design feature is also included in the Boeing 733 (Plates 8 and 9), one of the three project design studies that were the early proposals for the American supersonic transport.

3.2. Historical Development

To support the research on variable sweep being carried out in Germany during the Second World War, the Messerschmitt P1101 was designed. This was a fighter-type aircraft designed so that it would have been possible to vary the sweep of the wings between flights. By the end of the war, however, this project had only reached the mock-up stage. After the war, the research on variable sweep was continued in the United States, and two in-flight variable sweep aircraft were built, the Bell X-5 (Plate 10) and the Grumman XF10F-1 (Plate 11). In Great Britain swept-wing research was carried on using the Short SB5 (Plate 12), though this aircraft was similar to the P1101 in that the sweep angle could only be adjusted on the ground.

Bell X-5

The Bell X-5 was a jet fighter-type aircraft designed to fly at high subsonic speeds. It flew for the first time on 20th June, 1951. On this aircraft the whole wing rotated, except for a very small leading-edge fillet, and the sweep angle could be varied in flight from 20° to

59°. The wing rotated about a pivot point 27·2 in. outboard from the line of symmetry of the aircraft, and 38 per cent back along the wing root chord. In flight it took about 10 sec to change the sweep angle by 10°, and as the wing rotated through the full range it moved fore and aft a distance of about 2 ft. This translation of the wing was allowed for in order to reduce the trim change associated with sweeping the wing.

Quite an extensive flight test programme was carried out on this aircraft, and the sweep angle was often varied in flight over the whole range, without experiencing any mechanical problems. Unfortunately, the performance of the aircraft was rather poor, and it suffered from a number of deficiencies in its longitudinal and lateral stability and control characteristics. This tended to discourage any further development at the time (1955).

Grumman XF10F-1 Jaguar

In the late 1940s the Grumman Company put forward a design for a long-range fighter for the United States Navy. In an attempt to keep the landing and take-off speeds down, they incorporated a wing which could be swept from $13\frac{1}{2}°$ to $42\frac{1}{2}°$.

This aircraft first flew on 19th May, 1952, and within the next year made over 200 flights. The design was such that the whole wing moved, and as on the Bell X-5, it was arranged to translate as it rotated in order to reduce the associated trim change. The aircraft suffered, unfortunately, from rather severe pitch-up and lateral stability problems, and consequently the development was abandoned. Though this aircraft did not go into production, it did at least show that variable sweep was structurally practical and operationally feasible.

Short SB5

The Short SB5 was built for carrying out research into the low speed handling characteristics of swept wing aircraft. Provision was made for adjusting the angle of sweep on the ground to 50°, 60° or 69°, and for positioning the horizontal tailplane either low down underneath the fuselage or high up on top of the fin.

This aircraft first flew in December, 1952, and flight tests were carried out on it with a large number of different configurations. It was used for all the low speed research necessary for the development of the English Electric Lightning.

3.3. Aerodynamics of Variable Sweep

The design of any aircraft depends on the particular role it has to play. Variable sweep is desirable on an aircraft that needs to be versatile, and requiring efficient flight in both the subsonic and supersonic speed ranges. Though the performance of an aircraft can be improved in this way, there are some obvious aerodynamic, structural and operational problems associated with the variable sweep concept. The pivoting mechanism must work under the dynamic loads acting in flight, and the system must be as reliable as the basic control system of the aircraft. The design limits the possible positions for mounting the engines, and restricts the carriage of external weapons and stores. It also involves extra structural complexity, weight and cost. To justify the use of variable sweep in any design it is essential that all these disadvantages should be more than compensated for by the improved performance and the saving in weight of the fuel used.

3.3.1. AIRCRAFT PERFORMANCE

The top speed and range flying performance of an aircraft depends upon its drag characteristics. The overall drag is made up of a number of separate parts, the relative importance of which varies throughout the speed range of the aircraft. At low speeds induced drag is the predominant form of drag and this can be kept small by employing a high aspect ratio unswept wing planform; at supersonic speeds wave drag is more important and this is reduced by using short span highly-swept or delta planforms, and by applying the principle of the area rule.

The range performance of an aircraft is expressed by the following formula

$$R \propto \frac{M}{S}\frac{L}{D}\log\left(\frac{W_i}{W_f}\right)$$

where M is the Mach number, S the engine specific fuel consumption, L/D the lift-drag ratio and W_i/W_f the ratio of the initial to the final all-up weight of the aircraft. This equation shows that the range is dependent on the product of M, the Mach number and L/D. Fig. 20.8 shows typical values of L/D for various design configurations; the use of variable sweep enables the best L/D ratio to be achieved at all speeds, though the values are much less at supersonic speeds than under subsonic conditions. The corresponding variation of range (i.e. $M \times L/D$) is shown in Fig. 20.9, the range decreases at low supersonic speeds due to the large reduction in L/D, but increases again at higher speeds due to the increase in M. Between $M = 2$ and $M = 3$ the curve tends to level off again, and this is one of the reasons that a Mach number of 2·2 has been decided upon as the optimum for the design of the BAC/Sud Concorde.

In order to achieve a good airfield performance, the maximum

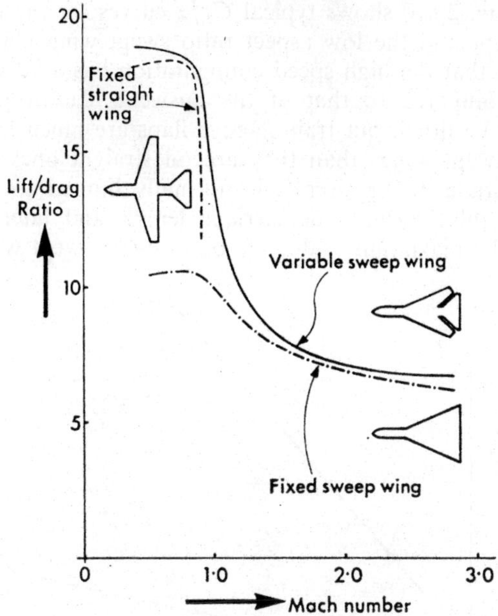

Fig. 20.8. Variation of the L/D ratio with Mach number for different configurations

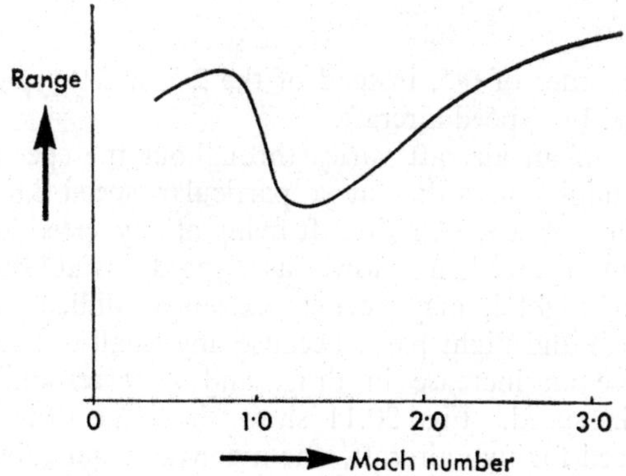

Fig. 20.9. Variation of range with Mach number

usable lift coefficient available from the wing should be as large as possible. Fig. 20.10 shows typical C_L/α curves for the high aspect ratio unswept and the low aspect ratio swept wing planforms. It can be seen that the high speed configuration has a very low value of C_{Lmax} compared to that of the unswept planform. High-lift devices can be fitted, but trailing-edge flaps are much less effective on highly swept wings than they are on straight ones. Since the landing attitude of an aircraft is normally limited to about 15°, because of pilot view, undercarriage length and lateral stability problems, the maximum usable C_L on a highly swept wing may be

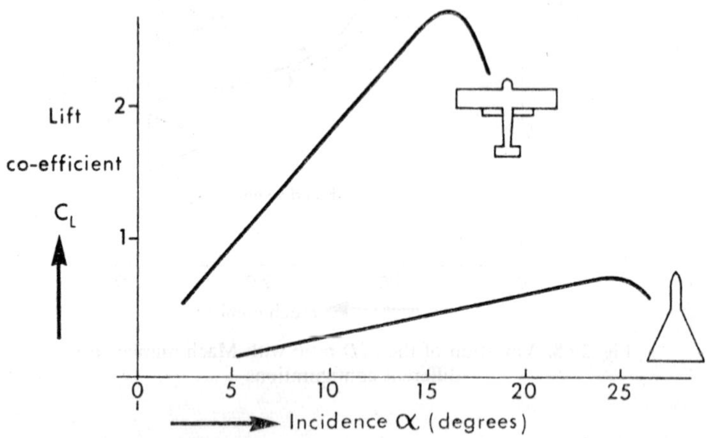

Fig. 20.10. Effect of aspect ratio on the C_L/α curve

only of the order of 0·5, instead of the 2·5 or 3·0 typical of many conventional low speed aircraft.

The drag of an aircraft varies throughout the speed range and reaches a minimum value at a particular speed known as the minimum drag speed. An aircraft flying at any speed less than this suffers from a problem known as "speed instability." In this condition of flight it may become extremely difficult to maintain a given speed and flight path, because any small decrease in speed will produce an increase in drag, and a corresponding further reduction in speed. Fig. 20.11 shows how the drag varies with forward speed for two aircraft, one a conventional subsonic design and the other a supersonic one. The minimum drag speed can be seen to be much higher on the supersonic aircraft than on the subsonic one and it is obvious that the problem of speed stability will become more severe on aircraft designed for high speed flight.

Mild instability (i.e. a time to double the amplitude of any disturbance of more than 20 sec) may be acceptable, but it should not be

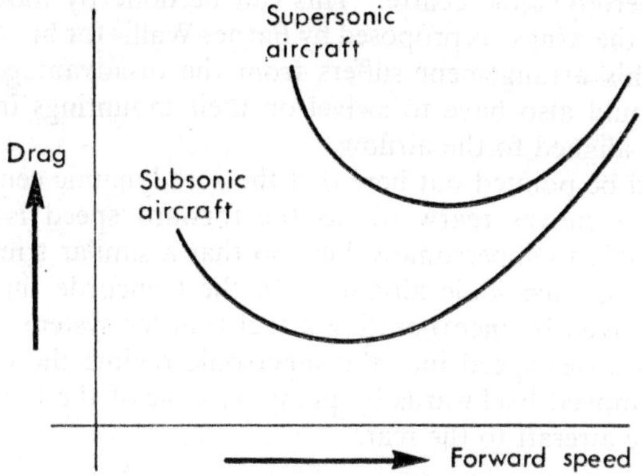

Fig. 20.11. Variation of drag with forward speed

worse than this, and it may be necessary to fit an automatic throttle in order to improve the handling characteristics.

3.3.2. STABILITY AND CONTROL PROBLEMS

Though the concept of variable sweep enables a definite gain to be achieved in the performance capabilities of an aircraft, a number of stability and control problems may arise associated with the sweeping of the wings. When the wing of an aircraft is swept backwards the aerodynamic centre (i.e. the "centre of lift") moves rearwards. The C.G. of the aircraft also moves rearwards at the same time, but since most of the weight of an aircraft is concentrated in the fuselage, the movement is less than that of the aerodynamic centre. The rearward movement of the aerodynamic centre produces a nose-down change in trim, and an increase in the longitudinal static stability of the aircraft. Additional up-elevator is required to trim the aircraft, and this results in an additional drag called "trim drag." This extra drag can be a relatively large part of the total drag of an aircraft at supersonic speeds, and it is essential that it should be kept as small as possible. Various design methods are available for reducing or eliminating the trim changes produced by sweeping the wings.

(a) Wing Translation

The aerodynamic centre can be moved forwards again by translating the wing forwards as it rotates. This method was used on the Bell X-5 and the Grumman Jaguar, but it obviously involves extra weight and structural complication.

(b) C.G. Movement

The aircraft can be designed so that the C.G. moves rearwards in step with the aerodynamic centre. This can be done by mounting the engines on the wings, as proposed by Barnes Wallis for his "Swallow" design. This arrangement suffers from the disadvantage that the engines would also have to swivel on their mountings in order to keep them aligned to the airflow.

It should be pointed out here that the aerodynamic centre on the wing always moves rearwards as the forward speed is increased from subsonic to supersonic values, so that a similar trim problem occurs on all supersonic aircraft. On the Concorde this problem has been solved by incorporating a fuel transfer system; when the aircraft increases speed into the supersonic régime the C.G. of the aircraft is moved backwards by pumping some of the fuel from the front of the aircraft to the rear.

(c) Leading Edge Fillet and Pivot Position

The aerodynamic centre A.C. movement can be reduced by a suitable choice of wing planform, pivot position and ratio of the areas of the fixed to the moving part of the wing. Research has shown that aspect ratio has little effect on the A.C. movement, but that taper ratio is an important parameter. To reduce the movement of the A.C. a high taper ratio is desirable, but, because this would make it difficult to satisfy the area rule requirements, some compromise is necessary. A pivot point inside the fuselage is desirable from structural considerations, but this would result in a large A.C. movement. The need for fore and aft translation of the wing can be eliminated by positioning the pivot point outboard of the fuselage inside a fixed leading edge fillet. The optimum pivot position depends on the wing planform, but it is usually about 20 per cent out along the span of the wing and about half way back along the chord. As the sweep is increased the lift on the moving part of the wing decreases and to maintain the same total lift the overall incidence of the wing has to be increased. More load is then carried by the fixed part of the wing and the corresponding A.C. movement is reduced. Fig. 20.12 compares the A.C. movement on a wing with an outboard pivot position and fixed fillet to that on a wing with an inboard pivot position and no fillet.

In this way the aircraft can be designed with a simple pivot point for the wing, no fore and aft translation being required. This is obviously preferable, and it is the method used on all recent variable sweep designs. The fixed fillet does, however, have some disadvantages associated with it. It is highly swept, and in low speed flight the flow may separate at the kink on the leading edge of the wing causing possible pitch-up problems. Various devices can be used to improve the flow over the wing (e.g. leading-edge flaps and slots); on the Boeing 733 a leading-edge saw cut and a boundary layer fence have been placed at the outboard end of the fixed fillet.

While on this particular subject it is also worthwhile mentioning that the large increase in longitudinal static stability resulting from the rearward movement of the aerodynamic centre can be reduced

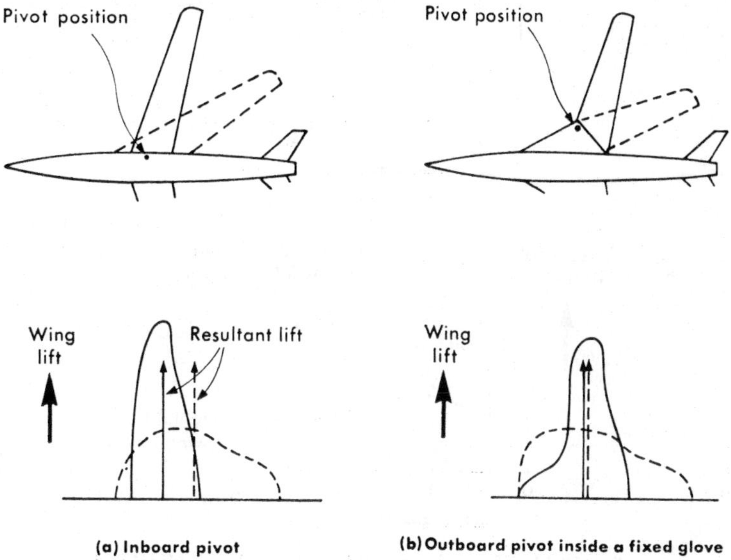

Fig. 20.12. Effect of wing sweep on the movement of the aerodynamic centre

by incorporating folding tailplanes or wing-tips in the design of the aircraft. At supersonic speeds these surfaces would fold down, thus reducing the longitudinal stability and increasing the directional stability (Fig. 20.13). This design feature serves two useful purposes in that it reduces the excessive longitudinal stability and increases the directional stability which is usually low at these speeds. Folding wing tips are used on the B-70 Valkyrie.

One other important design feature that has to be considered when designing the planform of the wing of a variable sweep aircraft is that of the shape of the wing tips. If the wing tips are shaped so that they are in the streamwise direction during the cruise when the wing is highly swept, then they will be swept back when the wing is in the low speed configuration. This may result in an early tip stall, with consequent pitch-up problems and loss of aileron power. As always, a compromise is necessary between the requirements for high and low speed flight. On the Bell X-5 the wing tips were streamwise with the wing in the 50° sweep position, they were thus

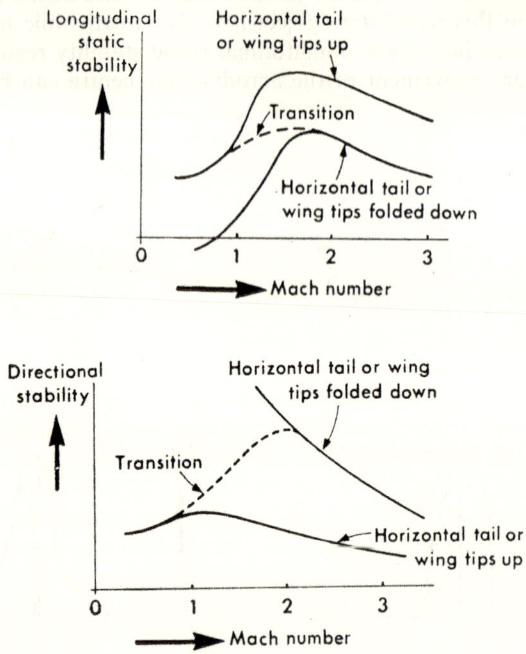

Fig. 20.13. Effect of wing folding the longitudinal and directional static stabilities

sweptback at low speeds and swept forward at high speeds. The TFX design, on the other hand, has the tips in the streamwise position when they are moved to the fully forward position.

3.3.3. GUST RESPONSE

The response of an aircraft to atmospheric turbulence is very important from pilot fatigue and structural loading considerations, particularly in the case of a high speed low level strike aircraft. If v is the vertical gust velocity and V the forward speed of the aircraft, then, to a first approximation, the change in the angle of attack is given by v/V (in radians) and the corresponding change in lift is proportional to

$$\frac{v}{V} \times \frac{dC_L}{d\alpha}$$

where $dC_L/d\alpha$ is the lift curve slope of the wing. The change in g on the aircraft can be shown to be proportional to

$$\frac{vV}{(W/S)}\frac{dC_L}{d\alpha}$$

where W/S is the wing loading. The response to gusts is thus reduced by having a small value of $dC_L/d\alpha$ and a large value of W/S. Variable sweep is an advantage in this respect, because with the wing in the fully swept position the configuration has a much smaller value of $dC_L/d\alpha$ than it has when the wing is in the low speed position.

3.4. Recent Developments

Until a few years ago, the variable sweep aircraft that had flown had all been research aircraft, though it was originally intended that the Grumman Jaguar should go into operational service. Most of the work in this country on developing a practical engineering solution for variable sweep has been done by Dr. Barnes Wallis of Vickers-Armstrongs (Aircraft) Ltd. This pioneering work has been useful to the Americans in the development of their two variable sweep projects, the TFX strike aircraft and the Boeing 733 supersonic transport.

Boeing 733

The initial Boeing design study for the American supersonic transport was for a variable sweep aircraft designed to cruise at $M2{\cdot}7$ at around 65,000 ft. The all-up weight of the basic design would be about 430,000 lb and the aircraft would carry 150 passengers over a range of 4,000 miles. The front portion of the inner 20 per cent of the wing is fixed, and the outer part rotates about a large 3 ft diameter pivot. The four engines are mounted in pods under the fixed part of the wing.

The fixed part of the wing has a sweep of 74°, and at its outboard end there is a saw-cut on the leading edge of the wing and a boundary layer fence on the upper surface. The leading edge sweep of the outer part of the wing is designed to vary from 20° for take-off and landing, to 30–35° for the subsonic climb, to 50–60° for flight at transonic speeds and finally to 74° for the supersonic cruise régime. The span of the wing varies from 173 ft 4 ins. in the fully forward position to 86 ft 4 ins. in the supersonic configuration. The wing tips are shaped so that they are in the streamwise direction when the wing is in the mid-sweep position.

On this aircraft the wing sweep would be under the control of the pilot and it would take about 4 min for the wing to sweep through the full range of 54°. In any variable sweep design it is essential that the operating mechanism should be extremely reliable, but in an emergency this aircraft could still be landed on a long runway with the wing in the fully swept position. The design is such that the aerodynamic centre movement is very small over the full sweep and Mach number range of the aircraft; no fuel transfer, autostabilization, or wing fold systems are required.

The rudder and elevator controls are conventional. Ailerons would be used for lateral control at subsonic speeds, but spoilers would be used with the wing swept at supersonic speeds. Double-slotted trailing edge flaps are fitted, but these could only be lowered when the wing is in the unswept position.

TFX (*F*-111*A and F*-111*B*)

Two versions of this strike fighter aircraft have been designed and built, the F-111A for the USAF (Plates 13 and 14) and the F-111B for the USN. The two aircraft are basically the same, though there are small differences in configuration and all-up weight. Early information indicated that the AUW of the first versions of the F-111A would be about 77,000 lb, that of the F-111B about 69,000 lb. These aircraft have been designed for $M1{\cdot}2$ at low level, with a high altitude dash capability exceeding $M2{\cdot}5$ and with good range and endurance characteristics and good airfield performance. The fixed portion of the wing has a leading edge sweep of 69°, and the sweep of the outer part varies from 16° to $72\frac{1}{2}$°. The span of the F-111A is about 63 ft with the wing in the fully forward position. The F-111B has a slightly larger span of 70 ft, the higher aspect ratio being to improve the low speed performance. The wing tips are in the streamwise direction in the low speed flight configuration.

To ensure good airfield performance conventional full-span trailing-edge flaps and leading-edge slots are used. Lateral control is by means of a differential tailplane (i.e. "tailerons," as used on the TSR-2) though this is augmented by the use of spoilers when the wings are in the swept forward position. The aircraft is powered by two turbojet engines, each producing a thrust of 19,000 lb with afterburning.

The first F-111A was rolled out on 15th October, 1964, and the prototype of the first variable sweep aircraft to go into operational service made its first flight on Monday, 21st December, 1964.

References

1. SPENCER, B. and SLEEMAN, W. C.; *Low-speed Longitudinal Characteristics of an Airplane Configuration Including Effects of Canard and Wing Trailing-edge Flap Controls in Combination*, NASA Tech. Note D-1397, 1962.
2. HALL, C. F. and BOYD, J. W.; *Effects of Canards on Airplane Performance and Stability*, NACA RM A58D24, 1958.

3. Johnson, J. L. and Paulson, J. W.; *Free-Flight-Tunnel Investigation of the Low-Speed Stability and Control Characteristics of a Canard Airplane Model*, NACA RM L53I11, 1953.
4. Bishop, W.; "The Development of Tailless Aircraft and Flying Wings," *Journal of the Royal Aeronautical Society*, December, 1961, page 799.
5. Northrop, J. K.; "The Development of All-Wing Aircraft," *Journal of the Royal Aeronautical Society*, 1947, Vol. LI, page 481.
6. Weyl, A. R.; "Tailless Aircraft and Flying Wings," *Aircraft Engineering*, 1944–45.
7. Meyer, C. H.; *Flight Testing a Variable Sweep Aircraft*, Agard Report No. 439, 1963.
8. "German Military Aircraft of World War II," *Aireview* (Kantosha Co. Ltd., Japan), English Edition, 1959.
9. *Jane's All the World's Aircraft* (Sampson Low, London).
10. "Variable Sweep," *Aircraft Engineering*, Oct. 1965, pages 294–308.

21 Slender Delta Wing Aircraft

1. Introduction

The optimum design of a particular aircraft is very dependent of the role it has to fulfil. A fighter-type aircraft must have satisfactory performance and manoeuvring capabilities, and though the handling characteristics must be acceptable and the economy reasonable, these requirements are usually subordinate to the main ones. A transport aircraft, on the other hand, has got to be economical with a good range performance, and it must have good handling characteristics at all speeds, though automatic stabilization systems can be incorporated to improve the basic characteristics of the aircraft.

An analysis of the drag characteristics of an aircraft at supersonic speeds shows that the optimum configuration for a cruising speed of about $M = 2{\cdot}0$ consists of a slender delta wing planform. Ideally, the fuselage would be integrated into the wing, but, in practice, there are a number of advantages in having a non-integrated fuselage.

The flow pattern round a slender delta wing is rather unusual, large spiral vortices forming over the top surface of the wing under all conditions of flight. Associated with this are non-linear lift characteristics and possible pitch-up problems. The slender delta planform suffers from a number of trim and stability problems, but, by suitable design, the majority of these can be reduced or eliminated. Many of these problems occur at the low speed end of the speed range. At one time it was thought that the instability of the Dutch Roll oscillation would severely limit the amount of usable incidence, but so far, in practice, this has not proved a major handling problem for the pilot.

This particular configuration is being used in the design of the BAC/Sud Concorde aircraft being built at the present time and which is due to make its first flight in 1968.

2. Flow Patterns of Slender Delta Wings

On aircraft with unswept high aspect ratio wing planforms, the most noticeable features of the flow pattern round the aircraft are the vortices which originate at the wing tips. On swept-back and delta planforms of moderate aspect ratio, however, the vortices usually originate from a point on the leading edge of the wing and these vortices move inboard as the angle of attack is increased.

On the highly swept leading edges of a slender delta wing, separation is unavoidable and these wings are designed with sharp leading

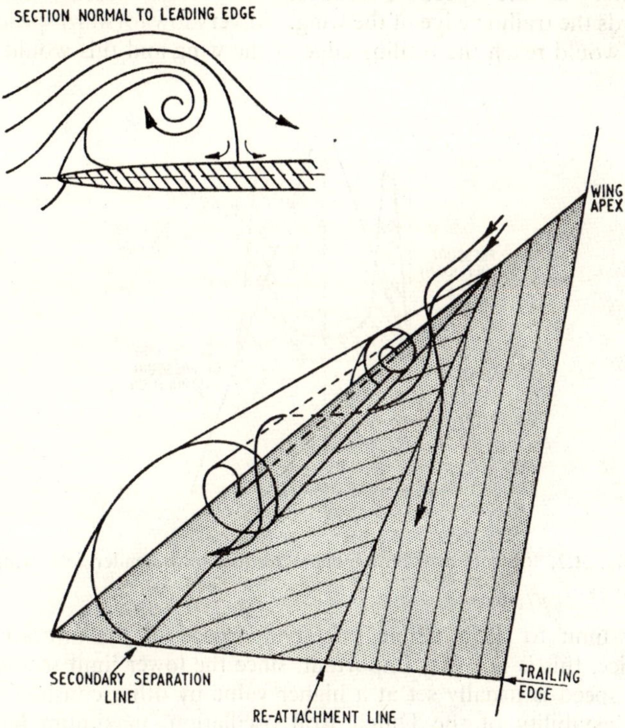

Fig. 21.1. Diagrammatic representation of the vortex flow over a slender delta wing

edges so that separation occurs along the whole length of the leading edge, even at very low incidences (Ref. 3). This produces a very stabilized flow pattern which is basically the same through the complete speed and incidence range of the aircraft. The flow separates from the leading edges of the wing and the vortex sheets roll up into two large spiral vortices which lie above and inboard of the leading edges and dominate the flow round the wing (Fig. 21.1). These vortices produce strong spanwise flow components at the surface of the wing underneath them. The flow which is not entrained in the vortices passes over the top, and re-attaches to the surface of the wing farther downstream. In the centre part of the wing, the flow is almost parallel to the centre line of the aircraft. The vortices continue behind the aircraft, but eventually the flow becomes less steady and breaks down (the vortices are said to "burst"). At high forward speeds the vortex breakdown is a long way behind the aircraft; as the speed is reduced the "burst" moves forward towards the trailing edge of the wing. At very low forward speeds the burst would reach the trailing edge of the wing and this would set a

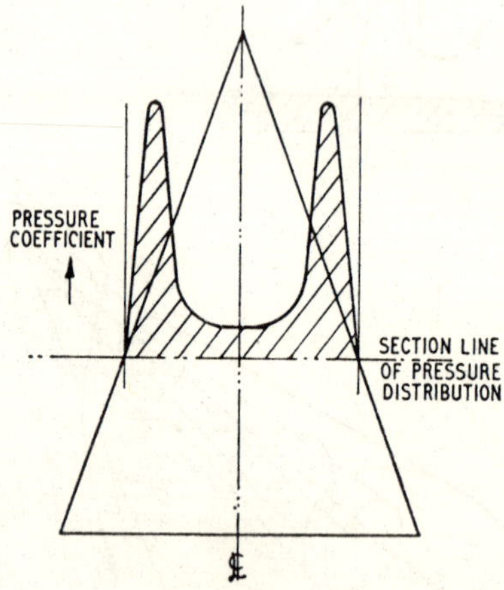

Fig. 21.2. Pressure distribution characteristics on a slender delta wing

lower limit to the possible safe flying speed of the aircraft. In practice, this is not very important, since the lower limit to the forward speed is usually set at a higher value by other considerations (e.g. instability of the Dutch Roll oscillation, maximum landing attitude, etc.). At supersonic speeds, shock waves occur in the flow pattern and the vortices are smaller and lie closer to the surface of the wing. The presence of the vortices causes large pressure reductions over the outer parts of the upper surface of the wing (Fig. 21.2), with corresponding effects on the various aerodynamic characteristics.

This particular flow pattern is found to be almost independent of Reynolds Number, thus a reliable guide to its characteristics can be obtained from wind tunnel tests carried out at low values of *Re*. Various methods are available for investigating the flow pattern round a wing in wind tunnel and free flight tests. Ref. 6 discusses the results of some wind tunnel tests using wool tufts, smoke, and oil flow methods. Ref. 5 discusses in detail the results obtained from flight tests on a delta wing research aircraft (though one with a relatively "rounded" leading edge) and compares the results with those obtained from wind tunnel tests. The tests were made over the speed range of $M = 0{\cdot}4$ to $1{\cdot}29$ and wool tufts were used to show up the flow patterns. The corresponding pressure distributions were also measured. Ref. 12 discusses some of the wind tunnel and free flight test results obtained on the Handley-Page H.P.115, a low-speed slender-delta research aircraft; kaolin and red ink were used to show up the surface flow patterns and trailing smoke to show the development of the vortex motions behind the wing.

3. Lift, Drag and Pitching Moment Characteristics

3.1. Lift and Pitching Moment Characteristics

The slender delta wing has rather unusual lift characteristics. On a large aspect ratio rectangular wing, the lift increases linearly with

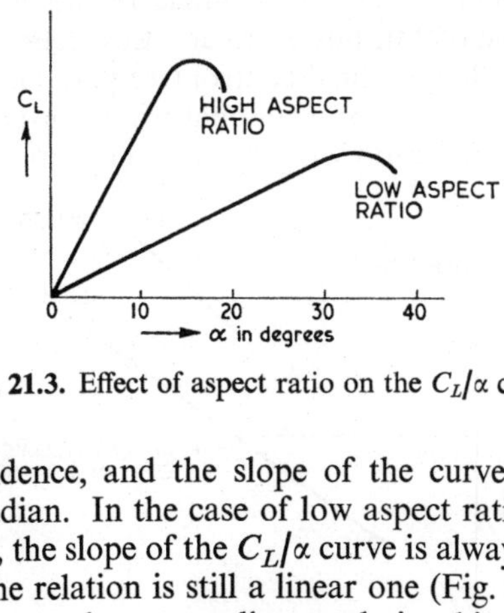

Fig. 21.3. Effect of aspect ratio on the C_L/α curve

increase of incidence, and the slope of the curve (i.e. $dC_L/d\alpha$) is about 2π per radian. In the case of low aspect ratio swept or delta wing planforms, the slope of the C_L/α curve is always very much less than this, but the relation is still a linear one (Fig. 21.3). A slender delta wing, however, has a non-linear relationship between lift and incidence, the value of C_L increasing with incidence at an increasing rate (Fig. 21.4). For aspect ratios of about unity, the non-linear lift is of the same order as the linear part of the lift. The non-linear lift is associated with the reduced pressure regions on the top surface of the wing, which are produced by the large vortices. These vortices increase in size as the angle of attack of the wing is increased, and this results in the non-linear lift variation. The C_L/α relationship can be expressed in the form

$$C_L = k_1\alpha + k_2\alpha^n \tag{21.1}$$

where k_1, k_2, and n have different numerical values depending on the particular condition of flight. The index n may have the value of 2 or 5/3 and the value of k_2 is sometimes quoted as 4.

The low value of lift-curve slope at low incidences is an advantage, in that it results in an aircraft which is relatively insensitive to

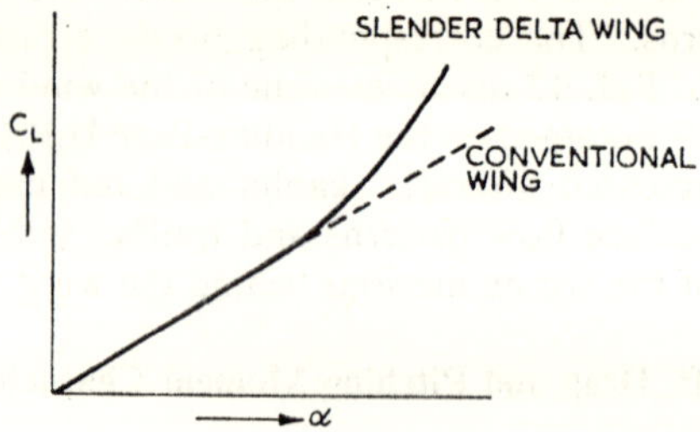

Fig. 21.4. Non-linear lift characteristics of a slender delta wing

vertical gusts. Large changes of angle of attack are required to produce a change of lift, however, and this does make the aircraft less responsive to longitudinal control (see para. 6.1).

The stalling angle of a conventional wing is usually round about 15°. On a low aspect ratio wing the lift will continue to increase up

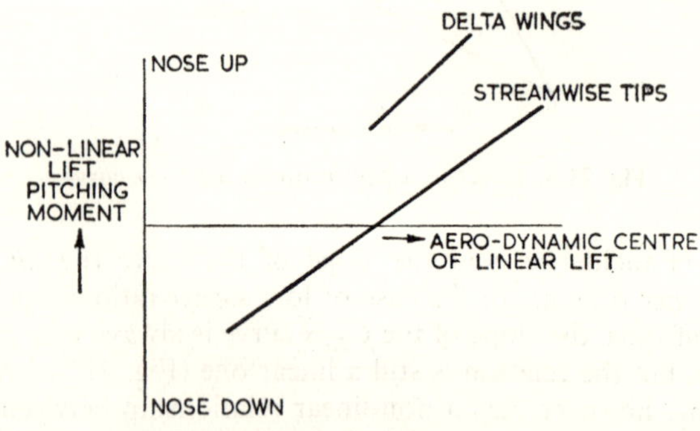

Fig. 21.5. Effect of wing planform shape on the pitching moment characteristics

to angles of attack of 35° or 40° (Fig. 21.3). The loss of C_L and the buffeting, which denote the "stall" on a slender delta wing, are associated with the vortex breakdown occurring near the trailing edge of the wing. As we have seen in para. 2, however, the usable angle of attack is usually limited to about 15° or so by other considerations. The non-linear lift characteristic is very beneficial at the higher angles of attack, because with it, the slender delta wing is able to produce a reasonably high value of C_L that will result in good take-off performance and low approach speeds.

The point at which the non-linear lift acts is a very important factor in determining the pitching moment characteristics of the aircraft (see para. 4). In the case of a pure delta planform, the centre of the non-linear lift occurs well ahead of the centre of the linear lift, thus producing a nose-up change in pitching moment with increase of incidence. The centre of the non-linear lift can be moved rearwards, nearer to the centre of linear lift, by incorporating streamwise wing tips and this reduces the pitch-up tendency (Fig. 21.5).

The theoretical aspects of the non-linear lift and pitching moment characteristics of slender delta wings are discussed in more detail in Ref. 4.

3.2. Drag Characteristics

The cruising efficiency of an aircraft is dependent upon its drag characteristics and the resulting optimum configuration depends upon the selected design speed. At low forward speeds, the main drag components are the skin friction drag, the form drag and the induced (i.e. vortex) drag. The optimum configuration for these conditions of flight must have a wing of fairly high aspect ratio, in order to keep the vortex drag reasonably low. At supersonic speeds two additional drag components are present, the wave drag due to lift and the wave drag due to volume. To obtain the optimum design for a supersonic aircraft, it is necessary to minimize the sum of all these individual drag components. The problem is discussed by Küchemann in Refs. 1 and 2, and the slender delta configuration is shown to be the optimum one for a cruising speed of $M = 2{\cdot}2$. This was chosen as the design cruising speed for the BAC/Sud Concorde supersonic transport because it represents the upper limiting speed for which conventional aluminium alloys can still be used.

The vortex drag component is given (Ref. 1, Appendix 2) by the equation

$$D_v = \frac{K_v L^2}{\pi b^2 q} = \frac{K_v L^2}{4\pi s^2 q} \tag{21.2}$$

where L is the wing lift, b ($= 2s$) is the span of the wing, $q = \frac{1}{2}\rho V^2$ and K_v is a factor which depends on the way in which the lift is distributed across the span of the wing. K_v has its theoretical minimum value of unity when the spanwise lift distribution is made elliptical. The wave drag due to lift is also given in Appendix 2 of Ref. 1 by

$$D_{WL} = \frac{K_W(M^2 - 1)L^2}{2\pi l^2 q} \tag{21.3}$$

where l is the length of the wing and K_W is a factor which depends on the way the lift is distributed along the length of the aircraft. K_W has a value between 1 and 2, and it has its minimum value when the chordwise lift distribution is also elliptical. The wave drag due to volume is given by

$$D_{WV} = \frac{128}{\pi} K_o \frac{q(Vol)^2}{l^4} \tag{21.4}$$

where *Vol* (to distinguish from V, the velocity) is the volume of the wing and K_o another factor, which this time depends on the way the volume is distributed. The value of K_o is unity for an ideal body of revolution; for slender delta wings, however, the value may be less than 1·0.

The skin friction drag can be kept small by reducing the surface area of the wing. The above equations show that the vortex drag can be reduced by keeping the span large, while the wave drags due to lift and volume are reduced by increasing the length of the wing. In addition, the volume must be kept small and it must be distributed in a way which can be utilized, and which will give satisfactory C.G. positions, with the aircraft both loaded and unloaded, for longitudinal stability considerations (see para. 4). The resulting optimum design is a compromise between these various conflicting requirements. For a $M = 2{\cdot}2$ transport aircraft the optimum design turns out to be a slender delta wing aircraft, with the following approximate dimensions—

Ratio of planform area to area of the enclosing rectangle $\simeq 0{\cdot}5$.
Ratio of span to length $\simeq 0{\cdot}4$ to $0{\cdot}5$.

The leading edges of the wing are highly swept so that they lie behind the Mach cone (i.e. leading edges are "subsonic" ones), whereas the trailing edge is straight or only slightly swept (i.e. a "supersonic trailing edge").

4. Longitudinal Trim and Stability Characteristics

The longitudinal stability characteristics of a slender delta wing aircraft are basically similar to those of more conventional aircraft, though differences will occur as a result of the changes occurring in the configuration and in the inertia characteristics of the aircraft. On a conventional aircraft, important longitudinal stabilizing components are obtained from the tailplane; in the case of a slender delta design all the stability has to be obtained from the wing. The elimination of the tailplane usually results in tailless delta wing designs suffering from a lack of pitch damping, particularly at transonic speeds (Chapter 15, para. 3.1(*d*)). There are three particular problems which must be discussed, however, that are associated with high speed aircraft designs,

(i) trim drag at high speeds,
(ii) speed instability at low speeds, below V_{imd},
(iii) pitch-up.

(i) *Trim Drag*
Trim drag has already been discussed in Chapter 20; it can be defined as the additional component of the drag that arises as a result of having to satisfy the condition of trim on an aircraft. This problem is made worse at high forward speeds by the rearward movement of the aerodynamic centre of the wing, since an additional download is required on the tailplane (or, in the case of a tailless aircraft, on the rear part of the wing) to balance the resulting nose-down pitching moment.

The trim drag may be as much as 10 per cent of the total drag of an aircraft at supersonic speeds. In the case of a supersonic transport aircraft the design is very critical and the overall drag must be reduced to a minimum. The payload may be only of the order of 7 to 10 per cent of the all-up weight, and a relatively small reduction in drag will produce a saving in fuel, which will result in an appreciable improvement in the percentage payload.

The trim drag problem can be eliminated by suitably designing the wing. In addition to satisfying the condition of trim, an aircraft must also be statically stable and this requires the aerodynamic centre of the slender delta wing to be behind the C.G. of the aircraft. On conventional wings, the A.C. moves rearwards from about the $\frac{1}{4}\bar{c}$ point at subsonic speeds, to about the $\frac{1}{2}\bar{c}$ point at supersonic speeds. On slender wing designs the total rearward movement is reduced to about 7 or 8 per cent of the wing length. However, this does mean that, if the aircraft is statically stable at subsonic speeds, the stability will be excessive at supersonic speeds. If the C.G. is made to coincide with the A.C. position at subsonic speeds, the aircraft will possess neutral static stability. At supersonic speeds, the A.C. will then be 7 or 8 per cent of the wing length behind the C.G. and the aircraft will be statically stable. To trim the aircraft out, the overall centre of pressure must be made to coincide with the C.G. If this is achieved by deflecting the elevator control, then a substantial trim drag is produced. This increase in drag can be eliminated by suitably designing the wing of the aircraft. The centre of pressure of the wing can be made to coincide with the C.G. of the aircraft by designing the wing with the appropriate amount of camber and twist. In this condition of flight, no download will be required to trim the aircraft and the trim drag will be zero. To achieve this, the wing has to be designed with reverse or negative chordwise camber (i.e. the incidence of the leading edge of the wing has to be increased, relative to that of the trailing edge). To ensure that the flow separation occurs all along the leading edge at the same time, the leading edges of the wing have to be bent down until they point into the local wind direction when the aircraft is at a particular selected incidence, known as the "attachment" incidence.

The trim drag problem can also be eliminated by other methods (e.g. a fuel transfer system), which are discussed in Ref. 9.

(ii) *Speed Instability*

The problem of speed instability has also been discussed in Chapter 20. Though some instability can be tolerated, a severe amount of instability (e.g. a time to double amplitude <10 secs) cannot be accepted. Automatic throttle control can be used to eliminate this problem and Ref. 8 discusses the results of some flight tests carried out on the Avro 707A delta-wing research aircraft, with the automatic system arranged to give various levels of speed stability and instability. Speed instability does not, however, present any handling problem to the pilot in the case of the H.P.115 (Ref. 12).

(iii) *Pitch-up*

Swept and delta wing planforms often suffer from pitch-up problems at high angles of attack (Chapter 15, para. 4.4), though a reduction

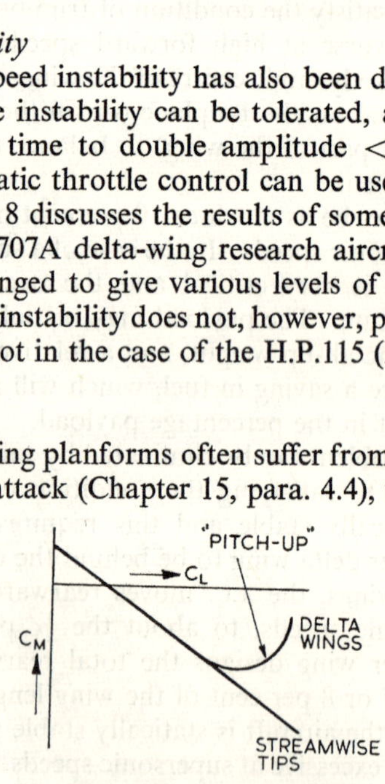

Fig. 21.6. Pitch-up characteristics

in aspect ratio is usually beneficial in eliminating this problem. On slender delta wings, the aerodynamic centre usually moves forward with increasing angle of attack, thus producing a pitch-up tendency. As we have seen (para. 3.1), this can be reduced by designing the wing planform with streamwise tips (Fig. 21.6).

5. Lateral Stability Characteristics

5.1. General

The lateral dynamic stability characteristics of a slender-delta aircraft can be analysed in a similar way to those of a more conventional aircraft. The lateral motion of this type of aircraft usually consists of the three classical modes

(i) the roll subsidence mode,
(ii) the spiral mode,
(iii) the Dutch Roll oscillation,

though there is the possibility at very high incidences that the roll subsidence and the spiral modes may combine together to produce a long period, highly damped oscillation, which is sometimes referred to as the "lateral phugoid" (Ref. 7).

Though the basic lateral characteristics are similar to those of conventional aircraft, differences occur due to the aerodynamic and inertia changes involved in the slender wing design. The main aerodynamic effects occur in the derivatives l_v (the dihedral effect) and l_p (the roll-damping). Because of the high sweepback of slender delta wings, the numerical value of l_v becomes very large at high incidences. The damping-in-roll derivative, on the other hand, is fairly small due to the short span of the wing. The main difference, however, occurs in the inertia characteristics of these designs. Table 1 (taken from Ref. 7) compares the roll, pitch, and yaw inertias of three different aircraft (the roll inertia being made equal to unity in each case).

Table 1

	CONVENTIONAL AIRCRAFT	BIG JET	SLENDER AIRCRAFT
Roll inertia A	1	1	1
Pitch inertia B	1	3·5	7
Yaw inertia C	2	4·5	8

In the case of the slender delta aircraft, the inertia in yaw is 8 times the inertia in roll, whereas on a conventional aircraft it is only twice as large.

The slender delta aircraft does not suffer from any lateral stability and control problems at high speeds, except that the roll damping is fairly low. The important problems occur at low forward speeds (i.e. at high C_L values). As a result of the particular inertia characteristics of this configuration, when the aircraft rolls, it tends to roll around its principal axis of inertia, instead of, as in the case of more conventional aircraft, about the direction of flight. This behaviour is the basis of the inertia coupling problem, which is discussed in more detail in Chapter 16. As a result of this tendency to roll about the inertia axis of the aircraft, angle of incidence and sideslip angle are continually interchanged [i.e. what is initially angle of attack, becomes, 90° later, an angle of sideslip, and, 90° later still, becomes an angle of attack again (Fig. 16.17 of Chapter 16)]. At low speeds and high incidences this effect becomes particularly noticeable, because of the large angle between the principal inertia axis and the direction of flight.

One other aerodynamic effect may be present at very high incidences, which is associated with the "bursting" of the wing vortices. At these incidences, the vortices may become unstable and erratic, and the directional stability n_v may well become negative for small sideslip angles (Ref. 7).

5.2. Roll Subsidence Mode

Though this mode is part of the lateral dynamic motion of an aircraft, it is usually so rapidly damped that it plays little part in the overall motion. It is very important, however, in determining the roll response characteristics of the aircraft to aileron deflection and for this reason is discussed in para. 6.2.

5.3. Spiral Mode

In Chapter 9, the spiral stability criterion was expressed in the form

$$(l_v n_r - l_r n_v) > 0 \tag{21.5}$$

and the spiral stability characteristics of an aircraft will vary through the speed range due to the changes occurring in the stability derivatives involved. In the case of a slender delta aircraft, both $(-l_v)$ and $(-n_r)$ increase as the forward speed is reduced, and this tends to increase the spiral stability. Any increase in l_r and n_v tends to reduce the stability, though, as has been mentioned, n_v may possibly decrease at the higher angles of attack. The amount of dihedral effect can be reduced to a suitable value at the higher incidences by using anhedral on the wings, but this may have the result of making l_v change sign at the lower angles of attack.

Spiral stability is relatively unimportant in assessing the handling characteristics of an aircraft. Pilots will accept a spirally unstable aircraft with a time to double amplitude of about 20 secs and even a time to double amplitude of only 4 secs can be controlled in an emergency. If the amount of instability becomes unacceptable, automatic stabilization systems can be used to improve the handling characteristics.

5.4. Dutch Roll Oscillation

In the case of a conventional aircraft, the lateral oscillation involves both yawing and rolling motions, the predominant effect being the yawing. The Dutch Roll oscillation of a slender delta aircraft, however, is found to consist mainly of a rolling motion about the fuselage axis of the aircraft (Refs. 10 and 14).

The development of the lateral oscillatory characteristics of aircraft, from a pure directional oscillation, through the "classical" Dutch Roll oscillation, to the pure rolling oscillation, is described in detail in Ref. 10. This report is extremely valuable in giving a physical appreciation of the motions involved. Pinsker shows that the pure rolling oscillation associated with slender delta wing designs can be more readily understood if the analysis is carried out using the principal inertia system of axes, instead of the more usual stability axes system. Approximate relationships can be determined for the oscillatory characteristics; the periodic time T and the logarithmic decrement δ can be expressed in the form

$$T = \frac{16{\cdot}08}{V_i} \sqrt{\left(-\frac{b}{\sin\alpha_o}\,\frac{(W/S)i_{Ao}}{l_{vo}}\right)} \tag{21.6}$$

and

$$\delta = -0{\cdot}614\,\frac{l_{po}}{i_{Ao}} \sqrt{\left(\frac{b\sigma i_{Ao}}{(W/S)l_{vo}\sin\alpha_o}\right)} \tag{21.7}$$

where b is the wing span, σ the relative density, α_o the angle of incidence of the principal inertia axis and the suffix o indicates that the quantities are measured with respect to the principal inertia system of axes.

Though these equations are rather complicated ones, it is worth while writing them down, in order that they can be compared to the corresponding equations for the conventional Dutch Roll oscillation (eqns. (9.29) and (9.30)). From eqn. (21.6) it can be seen that the periodic time of the oscillation of the slender delta aircraft is related to l_{vo}, the dihedral stability, instead of to n_v, the weathercock stability. Similarly, eqn. (21.7) shows that the damping of the oscillation is determined by the value of l_{po}, the damping in roll derivative, rather than by n_r, the damping in yaw. The physical explanation for the dependence of the periodic time of the motion on the dihedral effect is that, when the aircraft rolls about its inertia axis, sideslip is developed, and the dihedral effect produces the static restoring moment (i.e. the "spring" term) acting during the oscillation.

The roll/yaw ratio for the slender-delta aircraft is given approximately by

$$\left|\frac{\phi}{\beta}\right| = \frac{1}{\sin \alpha_o} \tag{21.8}$$

and thus it decreases with increase of incidence (i.e. decrease of speed) and is independent of height. This particular oscillatory characteristic is often expressed in terms of

$$\left|\frac{\phi}{v_e}\right|, \text{ where } \left|\frac{\phi}{v_e}\right| = \frac{57\cdot3}{V\sqrt{\sigma}}\left|\frac{\phi}{\beta}\right| \quad \text{(see Chapter 9).}$$

A change in forward speed changes the characteristics of the oscillation. Forward speed occurs directly in the equation for the periodic time (eqn. 21.6), but also occurs indirectly in all the equations, as a result of the associated changes in α_o, l_{vo} and l_{po}. A reduction in forward speed (i.e. an increase in C_L) results in a reduction in the values of periodic time, the roll/yaw ratio, and the log. dec. δ. The reduction in the value of the damping is mainly due to the large increase in the numerical value of l_{vo} that is associated with sweptback wing aircraft at high values of C_L, though the associated changes in the values of some of the other derivatives may also have important effects (Ref. 14). Eventually, an incidence is reached at which the Dutch Roll oscillation becomes unstable.

Some flight test results for the Handley-Page 115 slender-delta research aircraft are given in Ref. 11, and the low-speed handling characteristics of the aircraft are discussed in Ref. 12. To the pilot the oscillation is mainly one of rolling about the body axis. At the high speed end, the oscillation damps out quickly, but as the speed is reduced the damping becomes first neutral, and eventually negative. Under these conditions, the oscillation diverges, but stabilizes itself with the angle of bank varying over the range of something like $\pm 30°$, depending on the position of the C.G. At one time, the results of some early model tests seemed to indicate that the problem of Dutch Roll instability would severely limit the amount of angle of attack that could be used on slender delta wings. In practice, however, it has turned out that the H.P.115 can be flown quite comfortably at speeds well below that at which the Dutch Roll becomes divergent. This is due to the highly effective lateral controls of the aircraft, which can be applied to stop the oscillation very quickly.

Because the Dutch Roll oscillation does not contain much yawing motion, a yaw damper is consequently not very effective in stabilizing the oscillation. From eqn. (21.7) it can be seen that the motion can best be stabilized by fitting a roll damper (Ref. 10).

6. Aircraft Response to Controls and Atmospheric Gusts

6.1. Response to Longitudinal Control

The longitudinal pitching control of a slender delta aircraft can be provided either by trailing-edge elevators, or by a separate canard control surface (i.e. a foreplane). The effective arm of trailing-edge elevons is relatively small, and thus large area control surfaces have to be used in order to develop sufficient change in wing lift to produce the required amount of control power (i.e. m_η). Up elevator deflection causes the aircraft to rotate in the nose-up direction and, as a result of the increase of incidence of the wing, the overall lift increases. However, due to the large pitching inertia of this particular configuration and the small value of $dC_L/d\alpha$ associated with it, the response of the aircraft may be fairly sluggish, and the initial effect of an up-elevator movement may be a loss in the overall lift. The adverse effect of a large numerical value of z_η may be particularly important when considering the response of an aircraft to control of height on the approach. On conventional aircraft the response to elevator control is almost immediate ($<3/4$ sec), but on the slender delta configuration there may be a time lag of 2 or 3 sec before the height starts to increase significantly.

This adverse characteristic has not been noticed in the flight tests discussed in Ref. 12, due to the crisp and immediate pitch response on this particular aircraft. It may be a more important problem on the slender delta transport, however, due to the much greater inertia in pitch of the larger aircraft.

6.2. Response to Lateral Control

The lateral controls are required to trim the aircraft in steady sideslips or in cross-wind landing conditions and also to manoeuvre the aircraft. In steady flight conditions, the amount of aileron and rudder deflections required depends upon the values of the derivatives l_v and n_v (see Chapter 11). On slender delta aircraft at low speeds and, hence, high angles of attack, the numerical values of l_v and n_v may be large and correspondingly large control deflections will be required. In addition to their direct effects, however, the controls also produce important cross-control effects, the rudder producing a rolling moment, and the ailerons a yawing moment. On conventional configurations the yawing moment produced by the ailerons is generally described as an "adverse" one, though in this particular condition of flight it is, if anything, a favourable one, since it tends to reduce the amount of rudder deflection required. In the case of a slender delta aircraft, the yawing moment produced by the aileron deflection may act in the opposite direction, i.e. n_ξ may become negative (see Chapter 6) and more rudder deflection would be required for the cross-wind landing condition of flight. This particular problem may determine the design area of both the rudder and aileron control surfaces.

Since large area aileron controls are required for handling at low speeds, only small deflections are required for rolling control at high

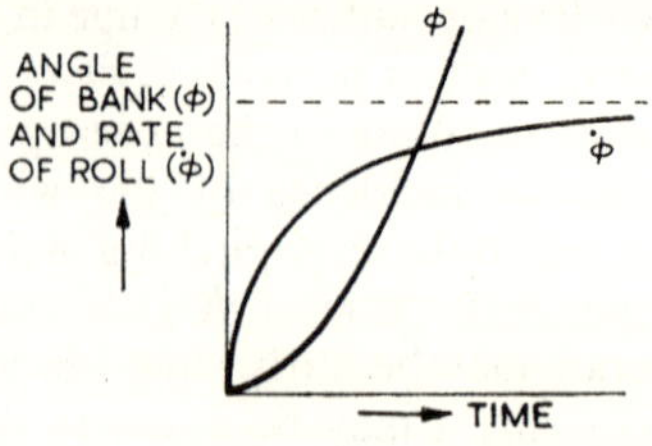

Fig. 21.7. Rolling response of a conventional aircraft

forward speeds. Because the aircraft tends to roll around its principal inertia axis, however, the aileron response characteristics may be very different from those of a more conventional aircraft. The normal time response of an aircraft to aileron deflection (see Chapter 11) is shown in Fig. 21.7. When a slender delta aircraft rolls around

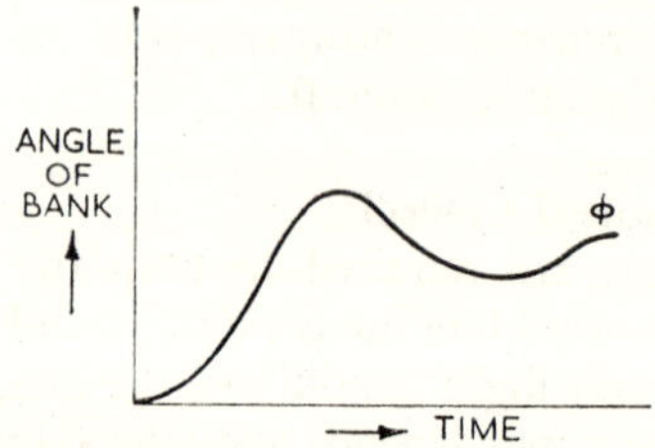

Fig. 21.8. Rolling response of a slender delta wing aircraft

its principal axis of inertia, sideslip is developed which has an "adverse" effect, because the rolling moment produced by the dihedral (i.e. l_v) effect opposes the rolling moment of the ailerons. The angle of bank of the aircraft may be limited to the value at which the rolling moments produced by the ailerons and the dihedral effect are exactly equal and opposite. Since an aileron input has, in this case, produced a change in the angular position of the aircraft, the controls can be described as "displacement" controls. This response should be compared to the normal response, when an aileron deflection produces a constant rate of roll and the controls can then be described as "rate" controls. In practice, the angle of bank may oscillate about a particular value (Fig. 21.8) and, eventually, if the controls are held on, the bank may continue to increase. One method of eliminating the lag in the roll response is by gearing the rudder to the ailerons in such a way that the rudder moves to reduce the initial sideslip.

Ref. 12 discusses the roll response and cross-wind landing characteristics of the Handley Page 115; at the lower speeds the oscillatory nature of the roll response is very noticeable, but the effects can be eliminated by the instinctive use of coordinating rudder. Though, under some conditions of flight, this aircraft has a divergent Dutch Roll oscillation, the lateral controls are so effective that the pilot can prevent the oscillation developing and it does not present any handling difficulties. This is a good example of an aircraft which, despite its poor inherent stability, has satisfactory handling characteristics, due to its effective controls.

6.3. Response to Atmospheric Gusts

The low value of $dC_L/d\alpha$ associated with a slender delta wing makes it less sensitive to vertical gusts. At high altitudes and high Mach numbers, the low damping in pitch may result in a poorly damped short period oscillation, though this can be easily prevented by incorporating a pitch damper (Ref. 7).

The response of an aircraft to a lateral gust is similar to that produced by aileron deflection. On a slender delta configuration the initial rolling acceleration is high, but the aircraft soon becomes stabilized at a particular value of the angle of bank. This equilibrium condition will be reached when the sideslip produced as a result of the rolling motion (i.e. $\beta = \alpha_o \sin\phi \simeq \alpha_o\phi$) is equal to that produced by the lateral gust (i.e. $\beta = v_g/V$). Thus, the equilibrium angle of bank (ϕ_e) is given for small angles by

$$\phi_e = \frac{v_g}{\alpha_o V} \tag{21.9}$$

where α_o is the incidence of the principal inertia axis, v_g the gust velocity and V the forward speed of the aircraft. At low forward speeds, the angle of bank may reach a value of about 45° and if this occurred very quickly it would be a very serious problem. On a supersonic transport aircraft near the ground, angles of bank of more than 5° cannot be tolerated. The pilot may be able to react quickly enough to prevent large angles of bank being developed (Ref. 12), otherwise some form of automatic stabilization will be required.

References

1. KÜCHEMANN, D.; *Some Considerations of Aircraft Shapes and their Aerodynamics for Flight at Supersonic Speeds*, Unpublished M.O.A. Report.
2. KÜCHEMANN, D.; "Aircraft Shapes and their Aerodynamics for Flight at Supersonic Speeds," *Advances in Aeronautical Sciences*, Vol. 3, page 221–252 (Pergamon Press, 1961).
3. MASKELL, E. C. and WEBER, J.; "On the Aerodynamic Design of Slender Wings," *Journal of the Royal Aeronautical Society*, December, 1959.
4. GERSTEN, K.; *Calculation of Non-Linear Aerodynamic Stability Derivatives of Aeroplanes*, Agard Report No. 342, 1961.
5. DEE, F. W. and NICHOLAS, O. P.; *Flight Determination of Wing Flow Patterns and Buffet Boundaries for the Fairey Delta 2 Aircraft at Mach Numbers Between* 0·4 *and* 1·3, *and Comparison with Wind Tunnel Results*, R.A.E. Tech. Report No. 64012, 1964.
6. LEE, G. H.; *Note on the Flow Around Delta Wings with Sharp Leading Edges*, R. & M. No. 3070, 1958.
7. GOLDSMITH, H. A.; *Stability and Control of Slender Aircraft*, Symposium on Recent Developments in Slender Wings for Supersonic Aircraft, Hatfield College of Technology, May 1963.
8. SPENCE, A. and LEAN, D.; *Some Low-speed Problems of High-speed Aircraft*, Agard Report No. 357, 1961.
9. DUBLIN, M.; *An Approach to the Control of Statically Unstable Manned Flight Vehicles*, Agard Report No. 364, 1961.
10. PINSKER, W. J. G.; *The Lateral Motion of Aircraft and in Particular of Inertially Slender Configurations*, R.A.E. Aero. Report 2656, R. & M. 3334, 1961.
11. BISGOOD, P. L. and O'LEARY, C. O.; *Interim Report on Low-speed Flight Tests of a Slender-Wing Research Aircraft* (*Handley-Page H.P.*115), R.A.E. Aero. Tech. Note 2921, 1963.
12. HENDERSON, J. M.; "Low Speed Handling of a Slender Delta (H.P.115)", *Journal of the Royal Aeronautical Society*, May 1965.
13. HENSHAW, J. T. (Ed.); *Supersonic Engineering* (Heinemann, 1962).
14. ROSS, A. J.; *The Lateral Oscillation of Slender Aircraft*, R.A.E. Aero. Rep. No. 2666, 1962.

22 Helicopters and Other Types of VTOL Aircraft

1. Introduction

The considerations of the stability and control characteristics of helicopters and other types of VTOL aircraft are basically similar to those of fixed-wing aircraft. It is necessary to consider the motion with respect to a set of three mutually perpendicular axes and to determine the amount of static and dynamic stability possessed by the aircraft about these axes. The aircraft must be provided with a control system that will produce satisfactory handling characteristics over the complete range of speeds and heights. There are, however, very many different configurations for helicopters and VTOL aircraft and the type of control system used depends on the particular configuration being considered. The different possible design configurations are discussed in Refs. 1, 5, 7 and 8. Most present-day helicopters are of the shaft-driven fully-articulated single main rotor type, with a small tail rotor for balancing the torque, and the following treatment of helicopter stability and control is mainly applicable to this particular design configuration.

The notation and sign convention and many of the definitions used are the same as those for fixed-wing work and many aspects of the work are identical. The analysis of helicopter stability and control can be done dimensionally, or, alternatively, by using non-dimensional parameters as is usual in fixed-wing aircraft analysis. This chapter contains little of the mathematical analysis, and has been written to give a more physical appreciation of the stability and control characteristics of these types of aircraft.

The essential differences that have to be considered are

(i) In addition to normal forward flight, helicopters and VTOL aircraft have the ability to hover, to fly forwards at very low speeds, and even to fly sideways and backwards. It is essential to consider the stability and control characteristics of these aircraft in all possible conditions of flight. Since the aerodynamic forces involved are negligible at low forward speeds, aerodynamic damping is small, and the normal fixed-wing aerodynamic controls are ineffective. To improve the stability characteristics of the aircraft under these conditions of flight, it is usually necessary to incorporate an automatic stabilization system. To achieve satisfactory low-speed control on VTOL aircraft, it is necessary to provide an alternative type of control; this usually consists of reaction air jets at the nose and tail and at the wing tips of the aircraft.

(ii) Though the equations of motion can be expressed in the form of non-dimensional derivatives, these derivatives do not in general have the same physical interpretation as on a fixed-wing aircraft. The overall value of any derivative is always the sum of a number of separate parts. In the case of a helicopter, the main contributions come from the main rotor and from the tail rotor. Smaller contributions are produced by the fuselage of the aircraft, by the fixed vertical fin and by the horizontal tailplane (if fitted), and these particular contributions are analysed in a similar way to the corresponding ones in fixed-wing work.

(iii) To avoid the difficulty of the non-dimensional derivatives (based on forward speed) going to infinity as the forward speed goes to zero at the hover, a different standard speed (or standard dynamic pressure q) has to be used in the definition of the derivatives for a VTOL aircraft. In the case of the helicopter, the rotor tip speed is often used; in the case of other VTOL aircraft the slipstream dynamic pressure is sometimes used instead of the free stream dynamic pressure. Alternatively, the derivatives are often left in their dimensional form, and the mass or the appropriate inertia is included in the definition (see Appendix F).

Helicopters and VTOL aircraft must satisfy the same sort of requirements as fixed-wing aircraft and these design requirements are specified in Refs. 9 and 10. The specifications are checked by carrying out various flight test exercises which are almost identical with those carried out on fixed-wing aircraft. During these tests, the appropriate physical quantities are measured and the pilot makes a qualitative assessment (based, for example, on the Cooper Scale) of the handling qualities of the aircraft. It is desirable that helicopters and VTOL aircraft should possess both positive static and dynamic stability about each of the three axes, though a small amount of instability may be acceptable as long as it does not result in any adverse handling characteristics. Helicopters and VTOL aircraft are, however, inherently less stable than conventional fixed-wing aircraft and it is generally necessary to use some form of autostabilization to improve the stability and control characteristics of the aircraft.

Helicopters

2. Introduction to Helicopter Aerodynamics

The stability of any aircraft is concerned with the forces and moments produced on it in response to various linear speed, angular displacement, and rate of rotation disturbances. Many of the major contributions to the overall stability of the helicopter are provided by the main rotor and thus it is necessary to investigate the way in which the rotor responds to these various disturbances. The work of this paragraph is based mainly on Ref. 1.

2.1. Blade Motion

In the case of a fully articulated rotor, the blades have freedom to rotate about their flapping, lagging (or dragging), and feathering

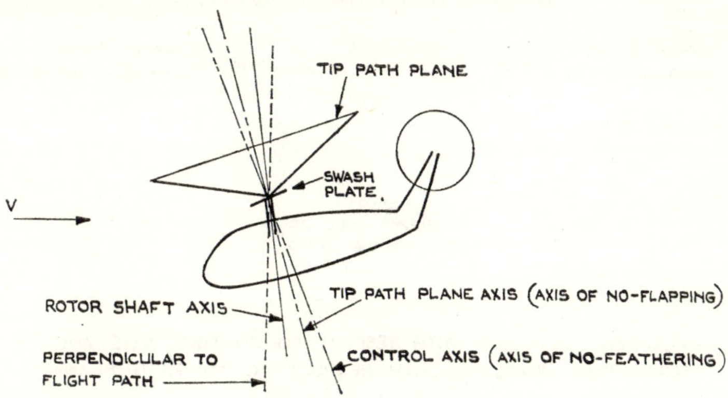

Fig. 22.1. Axes systems

hinges. When the blades are rotating, they move about the hinges until they achieve a condition of equilibrium under the action of the various aerodynamic and inertia forces acting on them. The flapping and feathering motions of the blades can be expressed mathematically in various ways, depending on the particular set of reference axes chosen. There are three possible systems of axes (Fig. 22.1).

(a) Tip-path Plane Axis (or Axis of No-flapping)

With respect to a system based on the axis perpendicular to the tip-path plane, the motion of the blades can be regarded as a pure feathering motion.

(b) Control Axis (or Axis of No-feathering)
With respect to this axis system, the motion of the blades can be regarded as a pure flapping motion. For convenience, the control axis can be assumed to be perpendicular to the swash plate, and its position is determined by the position of the pilot's longitudinal cyclic control.

(c) Shaft Axis
With respect to the rotor shaft of the helicopter, the overall motion of the blades can be considered to consist of both flapping and feathering motions.

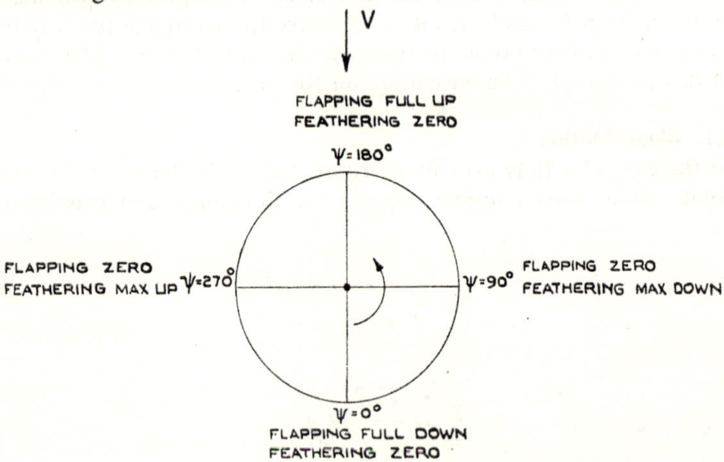

Fig. 22.2. Variation of blade position round the rotor disc

The relative positions of these axes depend on the particular conditions of flight. In hovering flight they may all coincide with one another, but in forward flight they take up different positions. Most analyses discuss the motion of the blades with respect to the control axis (i.e. as a pure "flapping" motion) and, if only the first harmonic terms are included (Ref. 1, page 191), the motion can be expressed in the form

$$\beta = a_0 - a_1 \cos \psi - b_1 \sin \psi \tag{22.1}$$

where β is the blade flapping angle and ψ is the azimuth angle, which gives the position of the blade around the rotor disc measured from the downstream position (Fig. 22.2). The coefficients a_0, a_1 and b_1

can be shown to represent the blade coning angle, the longitudinal rotor disc tilt, and the lateral rotor disc tilt, respectively, and the signs in eqn. (22.1) have been chosen so that a_0, a_1 and b_1 are positive for normal conditions of forward flight.

(*a*) *Coning angle* a_0 The coefficient a_0 represents that part of the flapping motion which is entirely independent of the azimuth angle

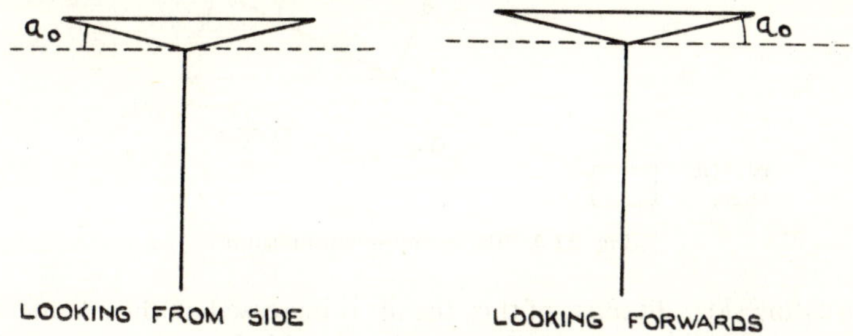

Fig. 22.3. Pure coning motion

$$\beta = a_0$$

ψ. Thus it is the pure coning motion and it is represented by Fig. 22.3.

(*b*) *Longitudinal tilting* a_1 The cosine term has its maximum positive (i.e. upwards) value at $\psi = 180°$ and its maximum negative (i.e. downwards) value at $\psi = 0°$. Thus, as long as a_1 is positive, this term represents the rearwards flapping of the rotor disc.

(*c*) *Lateral tilting* b_1 The sine term has its maximum positive (i.e. upwards) value at $\psi = 270°$, and its maximum negative (i.e. downwards) value at $\psi = 90°$. Thus, as long as b_1 is positive, this term represents the tilt of the rotor disc towards the side of the advancing blades.

Typical numerical values of these flapping angles are referred to on page 153 of Ref. 1, and on pages 127–129 of Ref. 3. The values of a_0, a_1 and b_1 vary with forward speed. In hovering flight, only blade coning occurs. Forward motion modifies the value of a_0 and causes a_1 and b_1 flapping motions to develop.

The flapping motion of the blades is determined by the effects of the various aerodynamic and inertia forces acting on them. The blades will cone until the angle a_0 is reached at which the resultant force acts along the blade, i.e. passes through the hinge and therefore produces no moment (Fig. 22.4). The weight of the blade itself should also be included in this equilibrium diagram, but since it is small compared to the other forces acting, it is usually neglected. The rearward flapping of the rotor disc is the result of the higher relative airflow velocity on the advancing side of the disc than on the

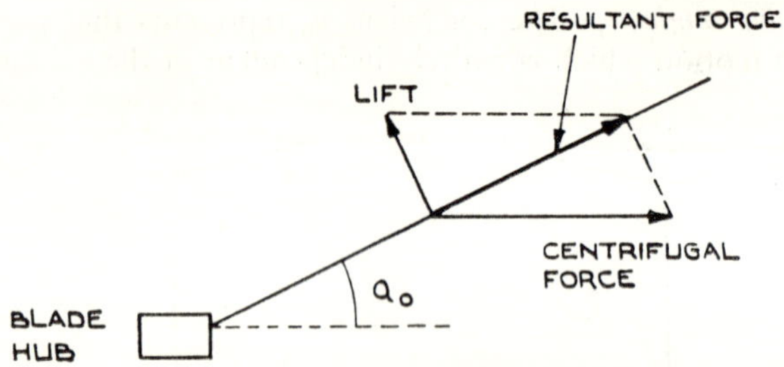

Fig. 22.4. Blade coning equilibrium

retreating side. Because of this, the lift is increased on the advancing side and reduced on the retreating side. The blades will consequently achieve their maximum flapping velocities at these points, but, due to the 90° phase difference that occurs between maximum velocity and maximum displacement, the maximum upward displacement occurs at $\psi = 180°$ and the maximum downward displacement at

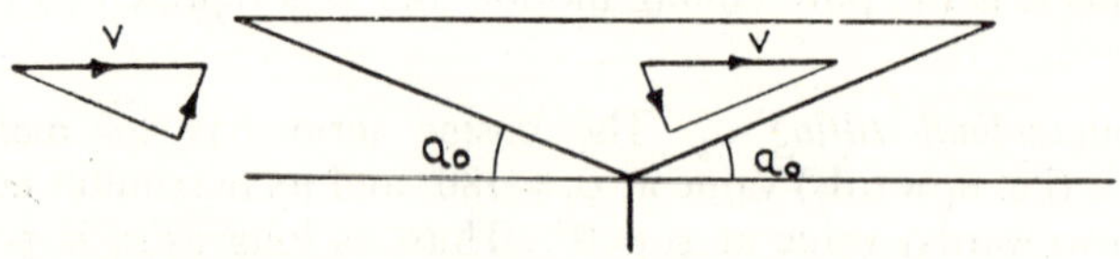

Fig. 22.5. Change in blade angle of attack due to coning

$\psi = 0°$. The sideways flapping of the rotor disc is the result of the blade coning motion and the consequent change in the angle of attack of the rotor blades round the disc. On the front half of the disc the effective angle of attack is increased and on the rear half decreased (Fig. 22.5). Once again, however, due to the 90° phase lag, the maximum upward displacement occurs at $\psi = 270°$ and the maximum downward displacement at $\psi = 90°$.

2.2. Effect of Incidence (or Angle of Attack) Disturbances

In hovering flight the control axis usually coincides with the shaft axis, and the tip-path plane is perpendicular to them. When the helicopter is given an angular displacement, the rotor shaft moves, but, since the blades are free to flap and no moments can be transmitted from the hub to the blades, the blades will not immediately align themselves with the shaft. However, due to the changes occurring in the angle of attack of the blades and the corresponding changes in the aerodynamic lift forces, the blades flap up and down and the overall result is that the tip-path plane moves until it again

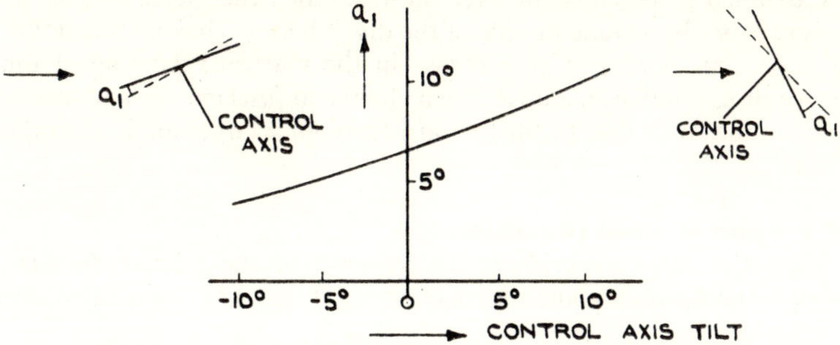

Fig. 22.6. Variation of tip-path plane tilt with control axis tilt

becomes perpendicular to the rotor shaft (Ref. 1, page 272). In practice, the time delay involved in this motion is only a very small fraction of a second.

In forward flight an additional effect is present, which results in the tip-path plane rotating through a larger angle than the rotor shaft. This is due to the variation of the fore-and-aft tilt a_1 with shaft axis or control axis position, which occurs because of the variation of inflow velocity (Fig. 22.6). If the helicopter is given a nose-up

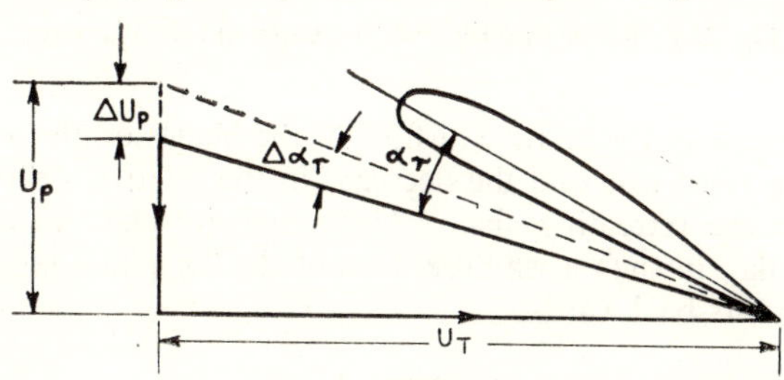

Fig. 22.7. Change in blade angle of attack due to change in fuselage angle of attack

disturbance, the the tip-path plane tilts rearwards. As a result of this, the downflow through the rotor is decreased by an amount ΔU_P and this increases the angle of attack of a typical blade element by $\Delta\alpha_r$ (Fig. 22.7). If this angle is assumed to be small (Ref. 1, page 280), then

$$\Delta\alpha_r = \frac{\Delta U_P}{U_T} \tag{22.2}$$

where U_T is the velocity component tangential to the rotor disc. The corresponding change in the lift produced by the blade section is approximately proportional to $\Delta\alpha_r U_T^2$, i.e. to $\Delta U_P U_T$. If ΔU_P is assumed to be constant over the disc, then the increase in lift is greatest on the advancing side of the disc where U_T has its maximum value. This results in an increase in the rearward flapping of the rotor disc. Similarly, a nose-down change in fuselage angle of attack causes the rotor disc to tilt forward through a larger angle than the fuselage.

2.3. Effect of Speed Disturbances

When the rotor experiences an increase in the relative forward speed, the tip-path plane tilts backwards (Fig. 22.8). This tilt is due

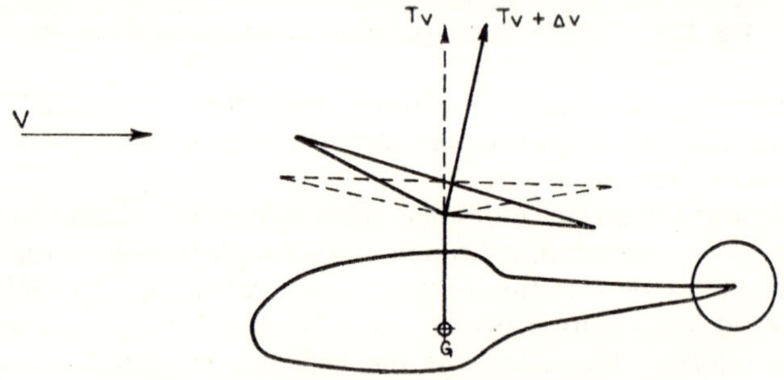

Fig. 22.8. Rotor disc tilt due to change in forward speed

to the increase in the relative airflow of the blades on the advancing side of the rotor disc and the decrease in the relative airflow of the blades on the retreating side. These velocity changes cause the blades to flap in such a way (because of the 90° phase lag) that the rotor disc tilts backwards.

2.4. Effect of Rate of Rotation Disturbances

In para. 2.2 we have seen that when the rotor shaft of a hovering helicopter is displaced through an angle, the tip-path plane follows the shaft, but with a small time delay. If, instead of being given an abrupt step displacement, the rotor shaft is rotated at a constant angular rate ω, then, because of the time delay, the tip-path plane will follow the shaft at the same angular rate, but will lag behind it by an amount δ (Fig. 22.9) given by

$$\delta = \frac{16}{\gamma\Omega}\omega \qquad (22.3)$$

where ω rads/sec is the angular rate of rotation of the aircraft, Ω rads/sec the rotational speed of the rotor and $\gamma(= c\rho a R^4/I_1)$ the non-dimensional Lock number. The Lock number can be regarded as being proportional to the ratio of the aerodynamic and the inertia forces acting on a rotor blade. In eqn. (22.3), the quantity $16/\gamma\Omega$ sec represents the time delay between the rotor shaft reaching a certain position and the thrust vector reaching the same position.

Eqn. (22.3) shows that the angle of lag is inversely proportional to the speed of the rotor and inversely proportional to the Lock

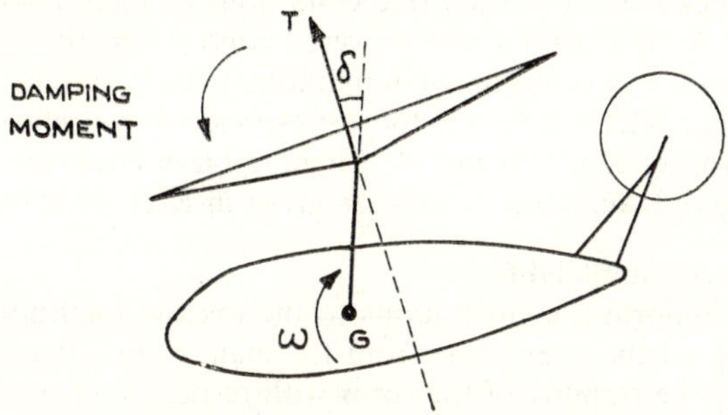

Fig. 22.9. Rotor damping due to pitching

number (and, hence, directly proportional to the blade moment of inertia about the flapping hinge I_1). Thus the lag can be increased by using lower rotor speeds and by increasing the inertia of the blades. It can also be increased by using off-set hinges.

3. Longitudinal Static Stability

Static stability is usually concerned with the initial motion of a body immediately any disturbance is removed. Since this represents only a small part of the complete motion of the helicopter that results following a disturbance, the amount of static stability possessed by an aircraft can be determined by solving the complete equations of motion. However, as in fixed-wing work, it is convenient to separate out the longitudinal and lateral motions and also to discuss the static stability and dynamic stability characteristics separately.

The static stability of a helicopter can be expressed in terms of a Static Margin, though this has much less significance than on a fixed-wing aircraft because C.G. movement is not such an important factor. The mathematical expressions for static stability can be written in terms of the stability derivatives. These derivatives are made up of a number of separate contributions, and, in the analysis of the longitudinal motion, the main contributions come from the main rotor, though the fuselage and the horizontal tailplane (if one is fitted) may have important effects.

The stick-fixed static stability is an important factor, because it can be regarded as determining the magnitude of the longitudinal cyclic control movements that are required to trim the helicopter at different forward speeds. The control forces required to trim are also important, and their variation is often taken as giving a measure of the stick-free static stability of the helicopter, though, since most helicopters have power controls, there is not the same significance between stick-fixed and stick-free conditions as there is on a fixed-wing aircraft fitted with a non-powered control system.

A qualitative description of longitudinal static stability is given in Ref. 1 (pages 279–283), and the corresponding theoretical analysis in Refs. 3 (pages 250–253) and 14. Flight test techniques are discussed in Ref. 16, and some test results are given in Refs. 16, 17 and 18.

3.1. Rotor Static Stability

The most important contributions to the overall longitudinal static stability of a helicopter come from the main rotor. It is necessary to consider the stability of the rotor with respect to changes in both angle of attack and forward speed.

(*a*) *Rotor Stability with Respect to Forward Speed Disturbances*

As we have seen in para. 2.3, when a helicopter experiences a disturbance in the forward speed, the tip-path plane tilts, an increase in speed causing the tip-path plane to tilt backwards. It is usually assumed that the rotor thrust acts normal to the tip-path plane and thus a rearward tilt produces a rearward thrust component and a nose-up pitching moment, which tend to reduce the forward speed back to its original value. The disturbance may also produce a change in the rotor thrust and if the thrust increases when the rotor disc tilts backwards, then the nose-up moment is increased (Fig. 22.8).

Thus the main rotor possesses positive static stability with respect to changes in forward speed and, using the normal sign convention, this can be expressed by saying that the derivative M_V (or, more strictly, M_u) is positive. In later work we shall see that static stability with respect to forward speed may be an undesirable feature when considering the longitudinal dynamic stability characteristics of the helicopter.

(*b*) *Rotor Stability with Respect to Angle of Attack Disturbances*

In para. 2.2 it has been shown that, in hovering flight, when the rotor shaft is rotated through an angle, the tip-path plane moves through exactly the same angle. Thus the rotor thrust continues to act along the shaft axis (assuming that it was acting in this direction in the trimmed condition) and no change of pitching moment is produced. Thus a hovering helicopter rotor possesses neutral stability with respect to angle of attack disturbances.

In forward flight, however, the tip-path plane does not rotate through the same angle, but through a larger one, and consequently changes of pitching moment are produced. The tilting of the tip-path plane following an angle of attack disturbance is such that the pitching moments act in the direction to increase the initial disturbance. Thus the main rotor is unstable with respect to angle of attack disturbances.

In addition to the changes occurring in the disc tilt, there will also be corresponding changes in the total thrust produced by the rotor.

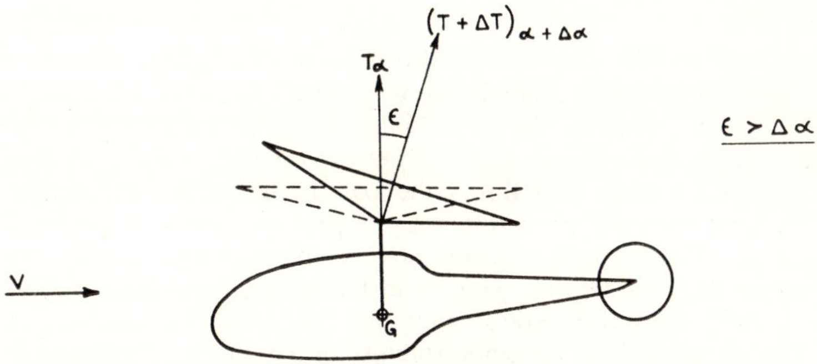

Fig. 22.10. Rotor disc tilt and change of thrust due to change in fuselage angle of attack

An increase of angle of attack will result in an increase of thrust (Fig. 22.10), whereas a decrease of angle of attack will decrease the thrust. Thus the instability of the rotor will be greater for nose-up disturbances than for nose-down ones (Ref. 1, page 281). This instability can be expressed by saying that the derivative M_w (or M_α) has a positive value. As we shall see later, the lack of angle of attack stability is one of the main deficiencies of the basic single-rotor helicopter configuration.

3.2. Static Stability of the Complete Helicopter

Additional contributions to the overall static stability are obtained from the fuselage and the horizontal tailplane. The conventional helicopter fuselage is usually destabilizing with respect to both speed and angle of attack disturbances. A horizontal tailplane, on the other hand, always produces a stabilizing contribution with respect to angle of attack disturbances. If the tailplane has a download acting on it in the trimmed condition of flight, then it will also have a stabilizing contribution with respect to speed changes.

The overall static stability of the helicopter is expressed in Ref. 14 in terms of dC_M/dC_T, and since it is made up of both speed and angle of attack contributions, it can be written

$$\frac{dC_M}{dC_T} = \frac{\partial C_M}{\partial \alpha}\frac{d\alpha}{dC_T} + \frac{\partial C_M}{\partial V}\frac{dV}{dC_T} \tag{22.4}$$

where C_M is the pitching moment about the C.G. of the aircraft and C_T (which replaces C_L in fixed-wing work) is the rotor thrust coefficient [$C_T = T/(\frac{1}{2}\rho V^2 S)$ where $S = \pi R^2$ the rotor disc area]. Alternatively, since

$$\frac{dC_M}{dV} = -\frac{2C_T}{V}\frac{dC_M}{dC_T}$$

the static stability can be expressed in terms of dC_M/dV.

The mathematical analysis can be carried out as for a fixed-wing aircraft by expressing the overall pitching moment acting about the C.G. of the helicopter in terms of the various rotor thrust and drag forces, the fuselage drag forces, the load on the horizontal tailplane, etc., and then differentiating with respect to speed and angle of attack. The separate contributions can be expressed as the derivatives M_α and M_V (i.e. M_u), or in non-dimensional form m_α and m_u, and for positive static stability dC_M/dV should be positive.

In the analysis of the static stability of a helicopter, it is usually assumed that the rotor speed remains constant during the disturbance. Any variation in rotor speed will affect the static stability of the aircraft and thus an auto-rotating helicopter will have different stability characteristics from one in power-on flight.

Other factors which affect the value of the static stability are

(a) Thrust Axis Offset from the C.G.

Ref. 1, pages 294–5, shows that offsetting the rotor thrust vector behind the C.G. produces a stabilizing contribution with respect to angle of attack changes, that will have the effect of counteracting the basic angle of attack instability of the rotor. An offset thrust vector may result from a need to cancel out pitching moments produced by the fuselage, or by the horizontal tailplane of the helicopter.

(b) Movement of the C.G.

Movement of the C.G. does not have the direct effect on the static stability that it has on a fixed-wing aircraft. If there are no fuselage or tailplane pitching moments which vary with angle of attack and no off-set flapping hinges, then movement of the C.G. will not have any effect on the static stability (Ref. 1, page 298).

3.3. Longitudinal Cyclic Control Required to Trim

In order to trim the aircraft at different forward speeds, the controls of the helicopter must be applied to balance the stabilizing moments that would occur with the controls held fixed. Thus the control movements required to trim can be directly related to the stick-fixed static stability characteristics of the aircraft. The control changes may involve movements in both the longitudinal cyclic and the collective controls and there are thus two methods of determining the static stability of a helicopter from flight tests.

(i) By measuring the longitudinal cyclic control movement required to trim the aircraft at different forward speeds in level flight, varying the collective pitch as necessary. This is usually referred to as the "apparent" static stability; it is important to the pilot, however, because it represents the overall static stability of the aircraft in level flight.

(ii) By measuring the longitudinal cyclic control movement required to trim the aircraft at different forward speeds, maintaining the collective pitch constant. In this test, the helicopter is trimmed at a number of selected speeds and then the speed is varied by ± 5 or 10 knots on either side of the trimmed value, using only the longitudinal cyclic control. This is often regarded as a measure of the "true" static stability of the aircraft and the handling qualities of the helicopter are usually assessed under constant collective pitch conditions.

Fig. 22.11, taken from Ref. 16, shows results obtained using these two techniques. It is desirable, as on fixed-wing aircraft, that forward movement of the stick should be required to trim the aircraft out at high forward speeds, and rearward movement to trim at low speeds. The slope of the trim curves (dB_1/dV) is a measure of the respective stabilities, and the values determined from Fig. 22.11 are shown in Fig. 22.12.

4. Longitudinal Manoeuvring Stability

The manoeuvre stability of an aircraft is related to its behaviour in response to a disturbance involving both angle of attack and rate of pitch changes (Chapter 3). It can be expressed in terms of the Manoeuvre Margin, though, once again, this particular parameter has much less significance on a helicopter than it has on a fixed-wing aircraft. The main contribution to the manoeuvre stability is obtained from the main rotor, though the fuselage and horizontal

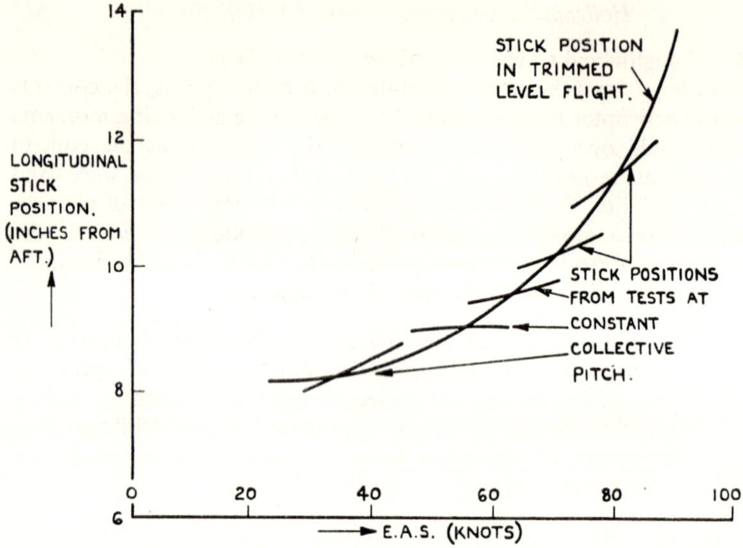

Fig. 22.11. Measurements of longitudinal stick positions on a Dragonfly Mk. III [From Ref. 16]

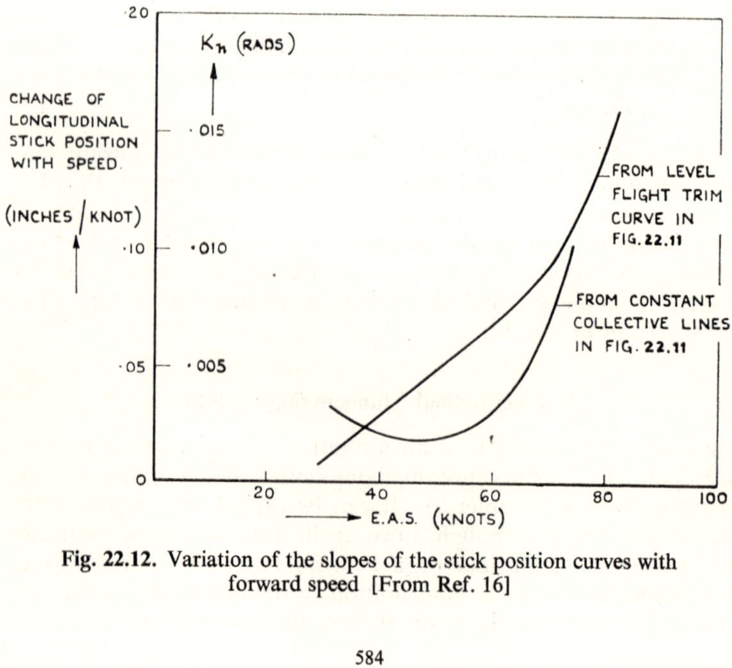

Fig. 22.12. Variation of the slopes of the stick position curves with forward speed [From Ref. 16]

tailplane may have important effects. The incidence contribution of the main rotor to the manoeuvre stability is basically the same as that considered during the analysis of static stability and to this it is necessary to add the contribution arising from the angular rate of rotation of the aircraft.

The stick-fixed manoeuvre stability can be related to the stick movements required to manoeuvre the aircraft and the stability characteristics can be measured by using similar flight test techniques to those used on fixed-wing aircraft (e.g. pull-outs and steady turns).

4.1. Rotor Manoeuvre Stability

The most important contribution to the overall longitudinal manoeuvring stability of a helicopter comes from the main rotor. The angle of attack contribution is destabilizing and has already been discussed in para. 3.1*b*. In para. 2.4 it was shown that when a helicopter is pitching at a constant angular rate ω rads/sec, the rotor disc rotates at the same rate, but lags behind the motion of the helicopter by a constant amount given by $\delta = (16/\gamma\Omega)\omega$ rads. If the thrust vector is assumed to act normal to the tip-path plane, then this angular lag results in a pitching moment acting about the C.G. in the direction to oppose the motion. Thus the effect is a stabilizing one; the pitch damping can be expressed in the form of the derivative M_q and it will usually be negative.

4.2. Manoeuvre Stability of the Complete Helicopter

Additional contributions to the overall manoeuvre stability are obtained from the fuselage and from the horizontal tailplane. These contributions are similar to those produced on a fixed-wing aircraft. The tailplane produces an important stabilizing contribution with respect to rate of pitch disturbances.

The mathematical analysis is similar to that already mentioned for static stability, except that the differentiation is performed with respect to angle of attack and rate of pitch disturbances, rather than angle of attack and forward speed disturbances (Ref. 14). The overall manoeuvre stability can be expressed in terms of dC_M/dC_T, and this can again be separated into its two parts

$$\frac{dC_M}{dC_T} = \frac{\partial C_M}{\partial \alpha}\frac{d\alpha}{dC_T} + \frac{\partial C_M}{\partial q}\frac{dq}{dC_T} \tag{22.5}$$

The two contributions to manoeuvre stability are illustrated diagrammatically in Fig. 22.13.

4.3. Longitudinal Cyclic Control Required to Manoeuvre

In order to hold the aircraft in a steady manoeuvre, the controls must be applied to balance the stabilizing moments produced as a result of the angle of attack and rate of pitch changes. The control movements may involve changes in both the longitudinal cyclic and

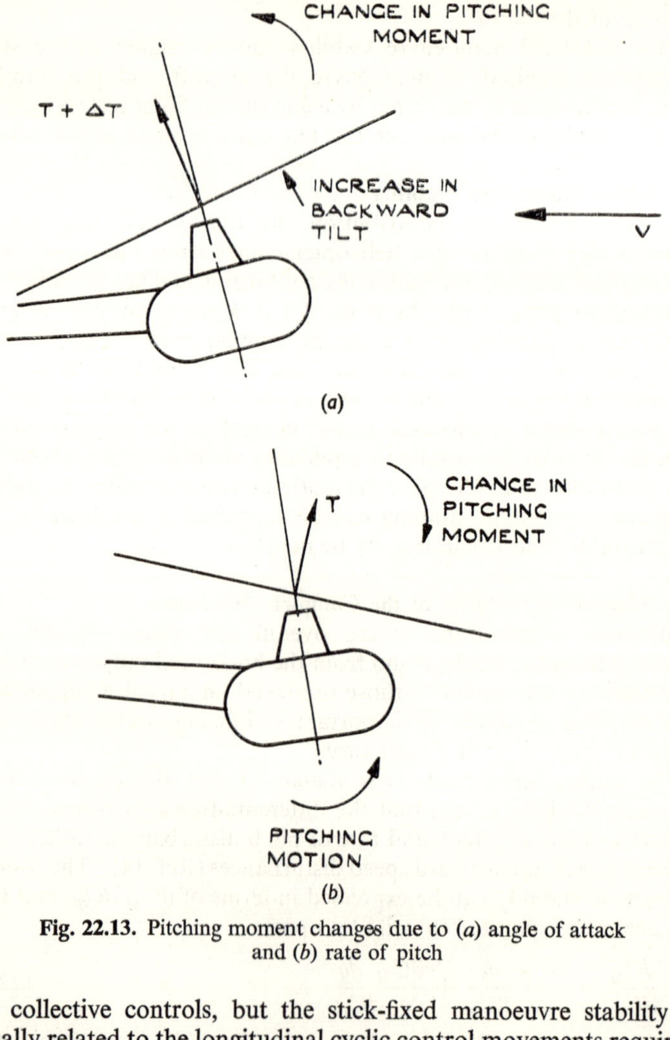

Fig. 22.13. Pitching moment changes due to (*a*) angle of attack and (*b*) rate of pitch

the collective controls, but the stick-fixed manoeuvre stability is usually related to the longitudinal cyclic control movements required for constant collective pitch conditions.

Steady pull-out manoeuvres and steady turns can both be analysed, in a similar way to that used in fixed-wing work, and the mathematical relationships between the manoeuvre stability and dB_1/dn, the slope of the curve of the longitudinal cyclic control position plotted against applied *g*, can be derived. The two techniques can both be used in flight test work to determine the manoeuvre stability of a

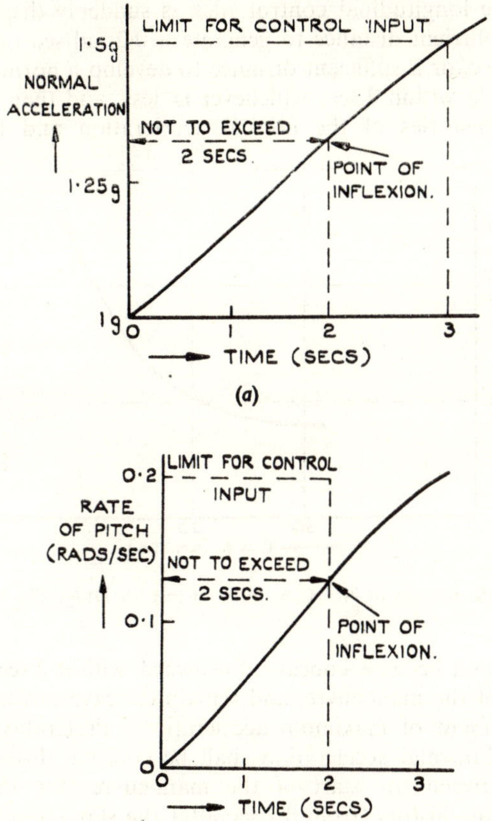

Fig. 22.14. Typical normal acceleration curves at (*a*) and rate of pitch response curves at (*b*)

helicopter. The tests are carried out at various selected forward speeds, using constant collective pitch and constant rotor R.P.M.

These techniques are, however, less satisfactory in the case of a helicopter than on fixed-wing aircraft and it is more usual to estimate the amount of manoeuvre stability possessed by a helicopter by measuring the transient response of the aircraft to a longitudinal cyclic control input. Ref. 14 shows that these different methods are related to one another and the helicopter is said to possess a satisfactory amount of manoeuvre stability if it satisfies the N.A.C.A. Criterion, which is usually stated in the form—

> "After the longitudinal control stick is suddenly displaced rearward a sufficient distance to generate a 0·2 rad/sec pitching rate within 2 sec, or a sufficient distance to develop a normal acceleration of 1·5*g* within 3 sec, whichever is less, and then held fixed, the time-histories of the normal acceleration and the angular

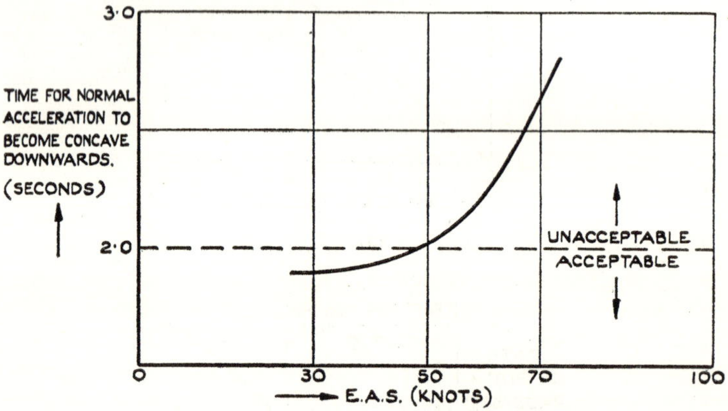

Fig. 22.15. Results of N.A.C.A. pull-out test for Dragonfly Mk. III

> velocity shall become concave downward within 2 sec following the start of the manoeuvre, and remain concave downward until the attainment of maximum acceleration. Preferably, the time history of normal acceleration shall be concave throughout the period between the start of the manoeuvre (for the angular velocity time history, from 0·2 sec after the start) and the attainment of maximum acceleration."

This test technique is much easier and quicker for a pilot to carry out, though it may be necessary for him to take rapid recovery action from the marked nose-up attitude that may develop following the control input, particularly at the lower airspeeds. Some typical response curves are shown in Fig. 22.14, for a case in which the control input was limited by the normal acceleration response. By measuring the time taken for the curves to become concave downwards and plotting the variation against forward speed (Fig. 22.15), the longitudinal manoeuvre stability characteristics of the helicopter can be assessed.

5. Lateral-directional Static Stability

The lateral-directional static stability of a helicopter can be considered in a similar way to that of a fixed-wing aircraft. The major differences are that the wing contribution to the dihedral stability is replaced by the contribution of the main rotor and that the chief contribution to the weathercock stability is provided by the tail rotor. The contributions from the fuselage and the vertical fin can be added as necessary.

The control movements required in steady sideslipping flight can be related to the corresponding static stabilities; the lateral cyclic control movements to the lateral static (i.e. dihedral) stability and the rudder pedal movements to the directional static (i.e. weathercock) stability. In order that these control movements should be in the conventional directions, it is necessary for both the lateral and directional static stabilities to be positive.

5.1. Stick-fixed Lateral Static (Dihedral) Stability

As we have already seen, a hovering helicopter is neutrally stable with respect to angular disturbances. When the helicopter rolls through an angle (and is prevented from any translational motion), the thrust vector remains aligned to the rotor shaft and thus no restoring rolling moments are produced about the C.G. of the helicopter. However, if the rolling disturbance also produces a sideways translational motion, then, since the rotor is always stable with respect to speed disturbances (para. 2.3), it tilts sideways and produces a rolling moment acting in the direction to correct the initial disturbance. Thus, using the normal sign convention, the derivative L_v has a negative value and the helicopter possesses positive dihedral stability. Additional rolling moments are produced from the sideforces, which result from the sideslipping motion, acting on the fuselage and the other side areas. It is interesting to note that these results are the same as those for a fixed-wing aircraft; both types of aircraft possess neutral stability with respect to rolling angular disturbances, and obtain their lateral static stability from the sideslip velocity.

When the helicopter is sideslipping, an additional component to the overall lateral static stability is obtained from the tail rotor. The rotor acts like a vertical fin and, if it is above the C.G. of the aircraft, it produces a positive dihedral effect, irrespective of the direction in which the aircraft is sideslipping. The restoring rolling moment is produced as a result of the sideways velocity component which acts normal to the rotor disc. Consider a helicopter, in which the tail rotor thrust acts to starboard, sideslipping to starboard. The normal velocity component $V \sin \beta$ causes a reduction in the blade angle of attack of the tail rotor (Fig. 22.16) and a corresponding reduction in thrust. This is equivalent to a force acting to port and, if the tail rotor is above the C.G. of the aircraft, this will produce a restoring

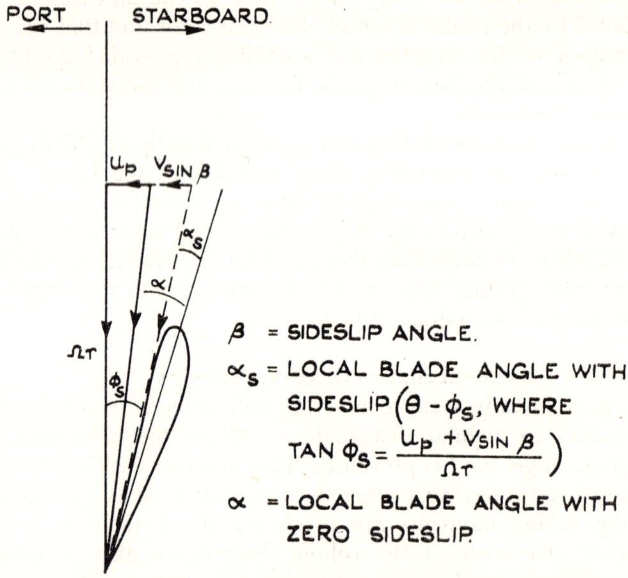

Fig. 22.16. Effect of sideslip on tail rotor blade angle of attack

moment acting to port. Similarly, if the helicopter sideslips to port, a rolling moment acting to starboard will be produced.

5.2. Stick-fixed Directional Static (Weathercock) Stability

The helicopter has neutral directional static stability in hovering flight, if translational motion is again prevented. In forward flight the aircraft possesses positive weathercock stability. The contribution of the main rotor to directional stability is small and the major contributions are provided by the fuselage, the vertical fin and the tail rotor. The contributions from the fuselage and the vertical fin are similar to those on fixed wing aircraft; the fuselage effect is usually destabilizing and the vertical fin stabilizing. The contribution of the tail rotor is usually a stabilizing one. When the helicopter is sideslipping, the velocity component normal to the tail rotor ($V \sin \beta$) produces a change of thrust which has been discussed in para. 5.1. In addition to producing a rolling moment, this change in the tail rotor thrust force acts in such a way as to produce a stabilizing yawing moment. Thus a helicopter normally possesses positive directional stability (i.e. N_v is positive).

5.3. Control Positions to Trim in Steady Sideslips

The lateral control positions required to trim the helicopter in steady sideslips are related to the stick-fixed dihedral and weathercock stabilities, the lateral cyclic stick position (A_1) to the dihedral

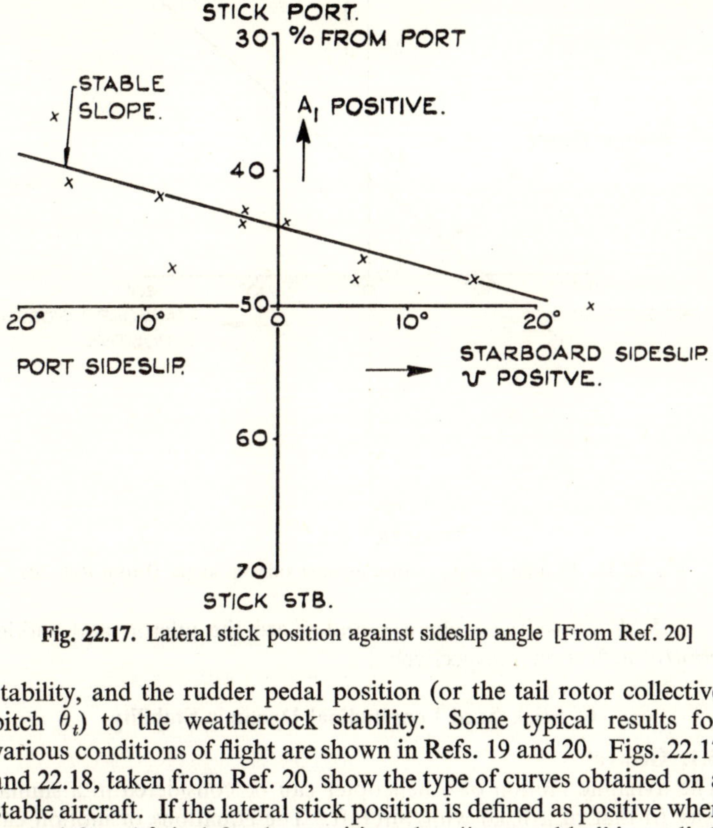

Fig. 22.17. Lateral stick position against sideslip angle [From Ref. 20]

stability, and the rudder pedal position (or the tail rotor collective pitch θ_t) to the weathercock stability. Some typical results for various conditions of flight are shown in Refs. 19 and 20. Figs. 22.17 and 22.18, taken from Ref. 20, show the type of curves obtained on a stable aircraft. If the lateral stick position is defined as positive when to the left and θ_t is defined as positive when "port rudder" is applied, then the condition for positive dihedral stability is that

$$\frac{dA_1}{dv} \text{ should be negative}$$

and the condition for positive weathercock stability is that

$$\frac{d\theta_t}{dv} \text{ should be positive}$$

(It should be noted that the sign convention used is the same as that on fixed-wing aircraft and that the conditions of stability are also the same. On a stable aircraft in steady sideslips $d\xi/dv$ is negative

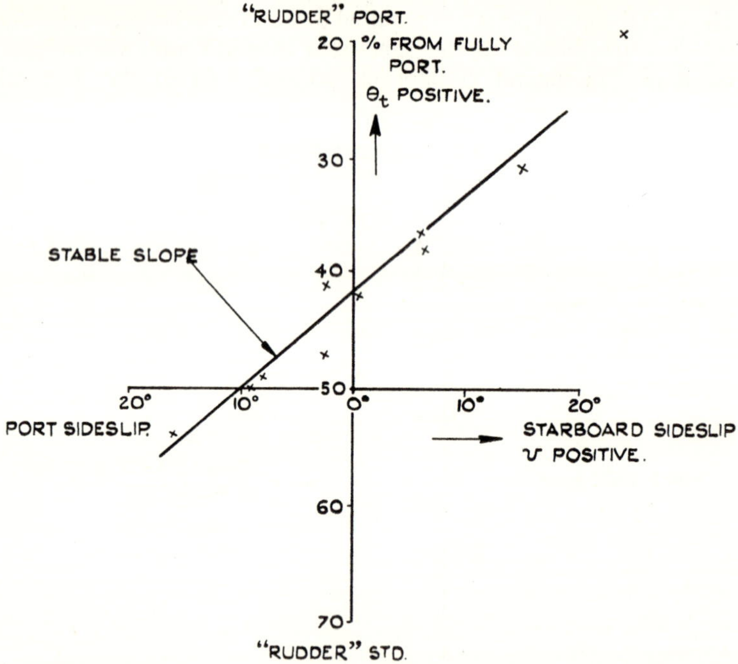

Fig. 22.18. Rudder pedal position against sideslip angle [From Ref. 20]

and $d\zeta/dv$ is positive, where ξ and ζ are the aileron and rudder control deflections, respectively.)

6. Stick-fixed Longitudinal Dynamic Stability

6.1. General

The dynamic motion of a helicopter can be considered in a similar way to that of a fixed-wing aircraft. The equations of motion can be expressed in terms of the stability and control derivatives and assuming that the longitudinal and lateral motions do not couple together, the equations can be divided into their two groups and analysed separately. Strictly, in the case of a helicopter, it is no longer possible to ignore the coupling between the longitudinal and lateral motions, but an estimate of the magnitude of the coupling effects usually shows them to be small, so that the errors involved in the assumption are not large (Ref. 3, page 276).

If the disturbances are assumed to be exponential in form, then the problem of stick-fixed longitudinal dynamic stability is reduced to the solution of the stability quartic equation of the form

$$A\lambda^4 + B\lambda^3 + C\lambda^2 + D\lambda + E = 0 \tag{22.6}$$

where A is equal to unity, and B, C, D and E are functions of the stability derivatives and the aircraft inertias. From this quartic equation the values of λ can be determined and these values determine the dynamic (and static) stability characteristics of the helicopter.

The mathematical analysis of longitudinal dynamic stability is described in Ref. 15, whereas Ref. 1 gives clearer physical descriptions of the motions involved. During the longitudinal motion, disturbances occur in forward speed, angle of attack and rate of pitch, and these disturbances result in corresponding changes in the X and Z forces and in the M pitching moment. Thus the non-dimensional derivatives involved are x_u, z_w, m_u, m_q, etc., as in fixed-wing work. Though the basic equations are the same, the physical interpretation and the relative importance of the various derivatives will be different.

The pitching moment derivatives m_u, m_α and m_q have already been discussed and it has been seen how the contributions from the main rotor can be expressed in terms of the way in which the rotor thrust force and the rotor tilt vary with disturbances in forward speed, angle of attack and rate of pitch, respectively. Additional contributions have to be added to allow for the effects of the fuselage and the horizontal tailplane. Similar physical explanations can be given for the other derivatives. The derivative x_u, for example, is a measure of the change occurring in the X-force due to a change in forward speed and the contribution of the main rotor is made up of three main parts

(i) the change in the horizontal thrust component produced as a result of the longitudinal disc tilt with change in forward speed,
(ii) the change in the horizontal thrust component produced as a result of the change in the thrust of the rotor with change in forward speed,
(iii) the change in the horizontal force component due to the change in the rotor H-force with change in forward speed.

In addition, the change in the drag characteristics of the fuselage will also produce an additional contribution to x_u.

Ref. 15 analyses the various derivatives involved and expresses them in the form of mathematical relationships. The report uses these theoretical equations to derive the numerical values of the derivatives for a particular helicopter design (with and without a horizontal tailplane) and the results are presented graphically showing the variation of the derivatives against forward speed for different conditions of flight.

The mathematical equations for the longitudinal derivatives of a helicopter are discussed in detail in Chapter XXI of Ref. 5.

6.2. Longitudinal Dynamic Stability in Hovering Flight

6.2.1. TYPICAL DYNAMIC STABILITY CHARACTERISTICS

A description of the longitudinal dynamic motion of a helicopter, following a disturbance from trimmed hovering flight conditions, is given in pages 283–291 of Ref. 1. Consider the helicopter to be disturbed nose-up in pitch, then, in the absence of rotor damping, the tip-path plane re-aligns itself perpendicular to the rotor shaft (i.e. it has neutral angle of attack stability). The thrust from the rotor then has a rearward component, which tends to move the helicopter backwards. As soon as the helicopter moves backwards, due to the speed stability of the rotor, the tip-path plane tilts forward, producing a nose-down pitching moment. The consequent nose-down rotation of the helicopter rotates the tip-path plane with it and a component of the rotor thrust now acts in the forward direction. This tends to reverse the motion of the helicopter, making it move forward. This forward motion, in its turn, makes the tip-path plane tilt backwards again and thus an oscillatory motion develops. This oscillation is usually a divergent one, that is the motion is dynamically unstable. Fig. 11.10 of Ref. 1 illustrates the conditions during one half of a cycle of the pitching oscillation.

In the above considerations, the effect of rotor damping has been neglected. When damping is allowed for, the rotor tends to lag behind the pitching motion of the helicopter. Though the damping modifies the motion and, consequently, the values of the periodic time and the damping of the oscillation, it does not change the basis of the above description.

6.2.2. THEORETICAL ANALYSIS

As has been mentioned above, the dynamic motion of a helicopter can be expressed in terms of the stability quartic equation. Ref. 15 (page 7) shows that, for hovering flight conditions, this equation can be simplified in a way that implies that the vertical motion of the aircraft can be considered separately from that of the fore-and-aft and pitching motions. This vertical motion is always damped, because the stability derivative which determines the damping (i.e.

z_w) is invariably negative (Ref. 2, page 269). This simplification reduces the quartic equation to a cubic one, and the three roots of this equation generally represent a well-damped subsidence and an undamped or divergent oscillation. The convergent mode is not important from the point of view of the handling characteristics of the helicopter, and the pilot is chiefly interested in the characteristics of the phugoid oscillation.

The mathematical analysis shows that the dynamic instability is the inevitable result for an aircraft which possesses neutral stability with respect to angle of attack disturbances and positive stability with respect to speed disturbances. It can be shown (Ref. 3, page 265) that the basic helicopter cannot be both statically and dynamically stable with the stick fixed, since the two conditions are mutually exclusive.

Expressions can be determined for the periodic time, and the damping (or the time to double amplitude) of the oscillatory mode. Ref. 1 (page 289) gives the periodic time as

$$T = 2\pi \sqrt{\left(\frac{1}{g}\right)} \sqrt{\left(\frac{-M_q}{M_u}\right)} \tag{22.7}$$

whereas Ref. 15 (page 8) give it in the non-dimensional form

$$T = 2\pi\hat{t} \sqrt{\left(\frac{-m_q}{\mu_2 m_u t_c}\right)} \tag{22.8}$$

An expression for the damping coefficient is given on page 269 of Ref. 2, and one for the time to double amplitude on page 274 of Ref. 3.

Eqn. (22.8) shows that the two main factors affecting the periodic time are the speed stability derivative m_u and the pitch damping derivative m_q. An increase in speed stability can be seen to result in a decrease in periodic time, whereas an increase in pitching stability results in an increased periodic time. This is in agreement with the simple physical interpretation of the oscillation. Eqn. (22.8) is, however, only an approximation. It assumes that the fuselage pitching inertia of the aircraft is zero; in practice, an increase in inertia will increase the periodic time (Ref. 1, page 290).

The expressions for the damping of the oscillation show that speed stability has an adverse affect on the damping, whereas an increase in the damping in pitch m_q is beneficial in reducing the amount of instability. However, an increase in m_q can only reduce the amount of instability and, at best, the helicopter will only approach neutral stability (Ref. 15, page 8).

6.2.3. MODIFICATION OF HOVERING DYNAMIC STABILITY CHARACTERISTICS

The single-rotor helicopter configuration is inherently dynamically unstable in hovering flight. The rate of divergence can be reduced by modifying certain design features. One of the most important factors affecting the stability is the blade moment of inertia; an increase in blade inertia (i.e. a decrease in the value of the Lock number γ) will increase the time to double amplitude. Ref. 2 shows on pages 270–271 that the time to double amplitude can also be increased by

(i) the use of offset flapping hinges,
(ii) increasing the vertical height of the rotor above the C.G.,
(iii) decreasing the moment of inertia of the fuselage.

Finally, artificial autostabilization systems can be used to provide some angle of attack stability and to improve the pitch damping, and in this way the oscillation can be made convergent (para. 9).

6.3. Longitudinal Dynamic Stability in Forward Flight

6.3.1. TYPICAL DYNAMIC STABILITY CHARACTERISTICS

A description of the longitudinal dynamic motion of a helicopter, following a disturbance from trimmed level flight conditions, is given in pages 299–304 of Ref. 1. This description, though it is only an approximate one since secondary effects are neglected, does show up the important stability derivatives. Consider a nose-down disturbance from trim on a helicopter which has neutral angle of attack stability (i.e. $m_\alpha = 0$). The weight component along the flight path causes the helicopter to accelerate and, due to the rotor speed stability, this causes a backward tilt of the rotor plane, a nose-up pitching moment, and consequently, a nose-up angular acceleration. These effects cause the helicopter to climb again and it will pass through the original trim position. It will continue to climb, the forward speed will decrease and the helicopter will begin to pitch nose down again. The cycle is described in detail and shown diagrammatically in Ref. 1 (Fig. 11.20). The motion is usually dynamically unstable, the amplitude of the oscillations increasing with time.

6.3.2. THEORETICAL ANALYSIS

An analysis of the dynamic motion of the aircraft in forward flight can again be made from the basic equations of motion, though the simplifications that were made for hovering flight cannot now be used. Ref. 15 shows how the values of the derivatives and the values of the coefficients of the stability quartic vary with forward speed. The report also compares the theoretically derived values of the periodic time and damping of the phugoid oscillation of a particular helicopter with some measured values.

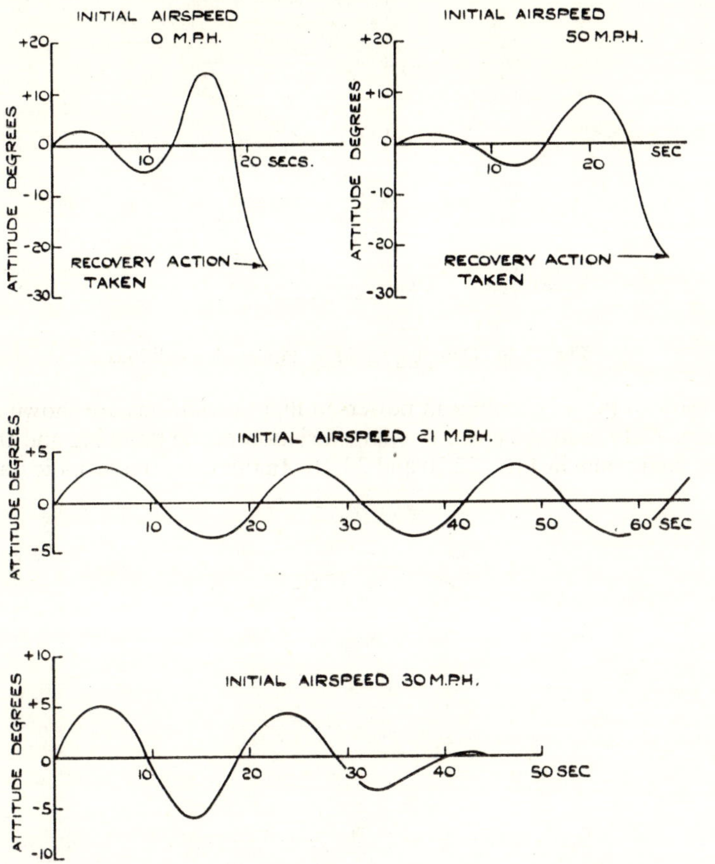

Fig. 22.19. Typical attitude-time histories. Power-on conditions

The mathematical expressions derived for the periodic time and the damping are generalized forms of the corresponding hovering equations. The periodic time is given on page 300 of Ref. 1 by

$$T = 2\pi \sqrt{\left(\frac{-M_q - \dfrac{WV}{g}\dfrac{M_\alpha}{T_\alpha}}{M_u g}\right)} \text{ sec} \tag{22.9}$$

where M_q is in ft lb/rad/sec, M_α in ft lb/deg, M_u in ft lb/ft/sec, and T_α in lb/deg.

Eqn. (22.9) shows that in forward flight, the angle of attack stability derivative is also an important factor. Some typical time

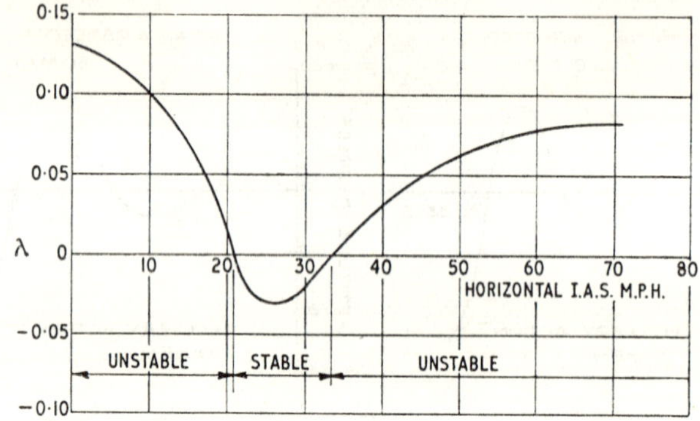

Fig. 22.20. Damping factor λ. Power-on conditions

histories for a helicopter in power-on flight conditions are shown in Fig. 22.19, and the corresponding variation of the damping and the periodic time in Figs. 22.20 and 22.21. In this case, the periodic time

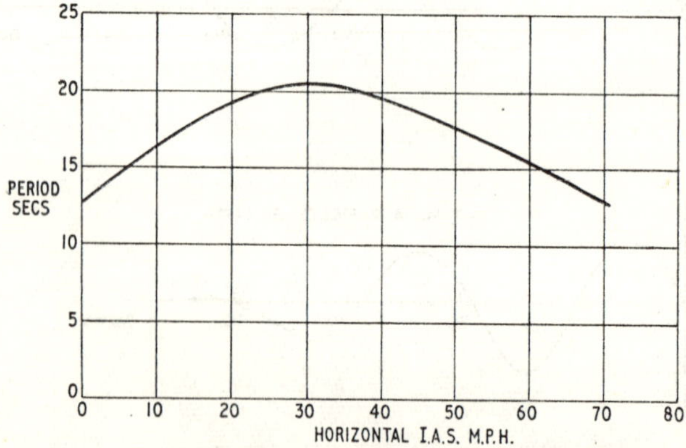

Fig. 22.21. Periodic time of oscillation. Power-on conditions

of the oscillation varies little with forward speed, though there is a tendency for the period to increase slightly in the middle part of the speed range. The helicopter is dynamically unstable at both high and low speeds and the motion just becomes stable in the middle part of the speed range. This aircraft has similar characteristics under auto-rotational conditions, except that the dynamic stability does not become positive.

6.3.3. MODIFICATION OF THE DYNAMIC STABILITY CHARACTERISTICS IN FORWARD FLIGHT

The main factors affecting the dynamic characteristics are the derivatives M_q, M_u and M_α. If the helicopter is unstable with respect to angle of attack disturbances, then the damping of the oscillation can be increased by decreasing the speed stability M_u, and by increasing the damping in pitch (i.e. by increasing $-M_q$).

A horizontal tailplane can be fitted to make the helicopter stable with respect to angle of attack disturbances (i.e. to make M_α negative). This improves the dynamic characteristics and the oscillation will become convergent at the higher forward speeds. In this case, the effects of the derivatives on the dynamic stability are reversed (Ref. 2, page 273).

The effect of the derivatives on the periodic time is obvious from eqn. (22.9). An increase in speed stability M_u reduces the periodic time, whereas an increase of pitch damping increases it. A positive value of M_α reduces the periodic time and the addition of a tailplane increases it.

An increase of fuselage inertia has the effect of increasing the periodic time and reducing the damping of the oscillation.

7. Stick-fixed Lateral Dynamic Stability

7.1. General

When an aircraft is disturbed laterally, it banks, yaws and sideslips and an oscillation develops. This lateral oscillation is usually referred to as a Dutch Roll, and its characteristics are dependent on the appropriate speed stability and angular damping derivatives (i.e. on the derivatives l_v, n_r, l_p, etc.). The overall values of the derivatives are made up of the contributions of the main and tail rotors, the fuselage and other parts of the aircraft. The mathematical expressions for the derivatives are given in Refs. 5 (Chapter XXIV) and 20. A physical interpretation of the l_v and n_v derivatives has already been given in paras. 5.1 and 5.2. The main rotor contribution to the damping in roll is produced as a result of the rotor lag which develops when the helicopter has an angular rate of roll (para. 2.4); a rolling moment is produced which acts to stabilize the aircraft (i.e. the roll damping derivative l_p has a negative value). The main contribution to the damping in yaw is produced by the tail rotor, which acts like a vertical fin to produce a negative value of n_r. The physical explanation for this derivative is similar to that for the n_v derivative given in para. 5.2. Ref. 20 shows graphically how the lateral derivatives of a particular helicopter vary with forward speed.

The physical description of hovering dynamic stability given on pages 283–291 of Ref. 1 is equally applicable to longitudinal and lateral motions, as is the equation for the periodic time. The lateral dynamic stability characteristics of a helicopter are considered in more detail in Refs. 20 and 24.

7.2. Theoretical Analysis

The analysis of the stick-fixed lateral-directional dynamic motion of a helicopter results in a stability quartic and the values of the coefficients are dependent on the lateral derivatives and the rolling

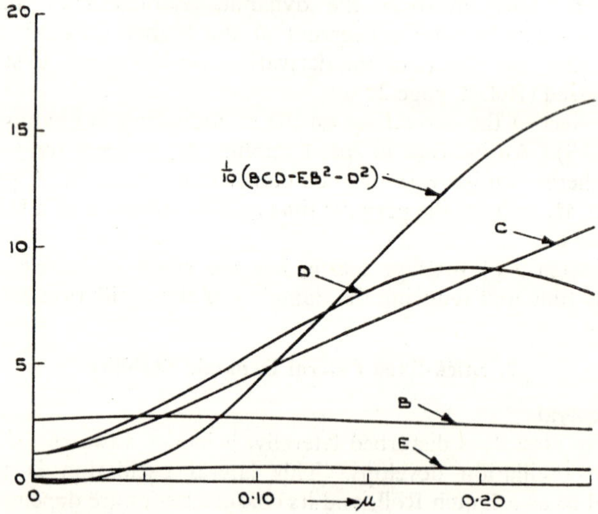

Fig. 22.22. Coefficients of non-dimensional stability quartic and Routh's discriminant for a Sycamore [From Ref. 20]

and yawing inertias of the aircraft. The solution of the quartic usually consists of two real negative roots and a complex root, with a negative real part. These roots indicate that the motion consists of three modes

(i) a very heavily damped rolling motion,
(ii) a damped spiral motion,
(iii) an oscillation, which is usually damped.

These modes are, in fact, similar to those found on ordinary fixed-wing aircraft.

The dynamic characteristics of any particular aircraft are dependent upon the values of the derivatives and the speed and height at which the aircraft is flying. Some typical results are shown in Figs. 22.22 and 22.23; Fig. 22.22 shows the way in which the coefficients of the stability quartic vary with forward speed (or μ the tip speed ratio) and Fig. 22.23 shows the corresponding variation of the roots of the equation. The spiral mode is only slightly damped and the damping decreases with speed. The heavily damped roll mode also becomes less damped at high forward speeds. The oscillatory mode, on the other hand, is damped at high forward speeds, but becomes slightly unstable at low forward speeds and at the hover. The corresponding periodic times and damping coefficients are shown in Fig. 22.24.

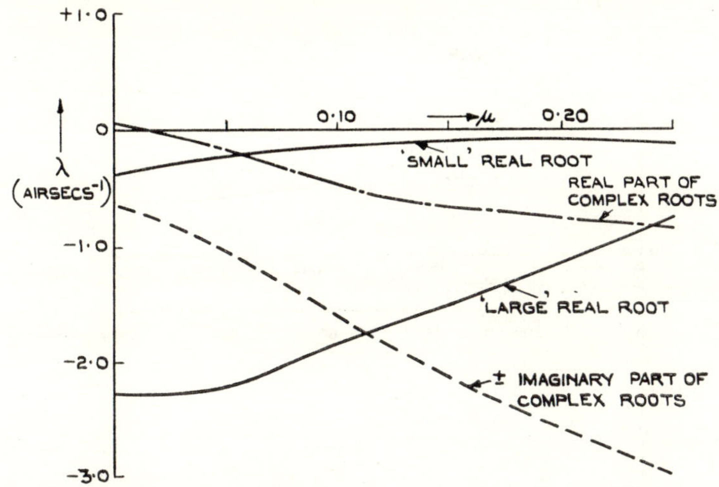

Fig. 22.23. Roots of non-dimensional stability quartic for a Sycamore [From Ref. 20]

The dynamic characteristics of the helicopter can be assessed in the usual way from the signs of the coefficients of the stability quartic and of the Routhian Discriminant ($R = BCD - B^2E - D^2$). If these are all positive, the aircraft is a stable one. The coefficient E can be expressed (Ref. 20) in the form

$$E = \frac{t_c\mu_2}{i_A i_C}(l_v n_r - n_v l_r) \tag{22.10}$$

and if this is positive [i.e. if $(l_v n_r - n_v l_r) > 0$] then the helicopter possesses spiral stability. When the Routhian Discriminant R goes negative, the aircraft is dynamically unstable and the oscillatory mode becomes divergent. The conditons of spiral and oscillatory stability can be represented on an l_v/n_v diagram, as in the case of a fixed-wing aircraft and Fig. 22.25 shows one of a series of graphs, taken from Ref. 20, representing one particular condition of flight. The oscillatory mode is damped in this case, but the spiral mode is only just stable.

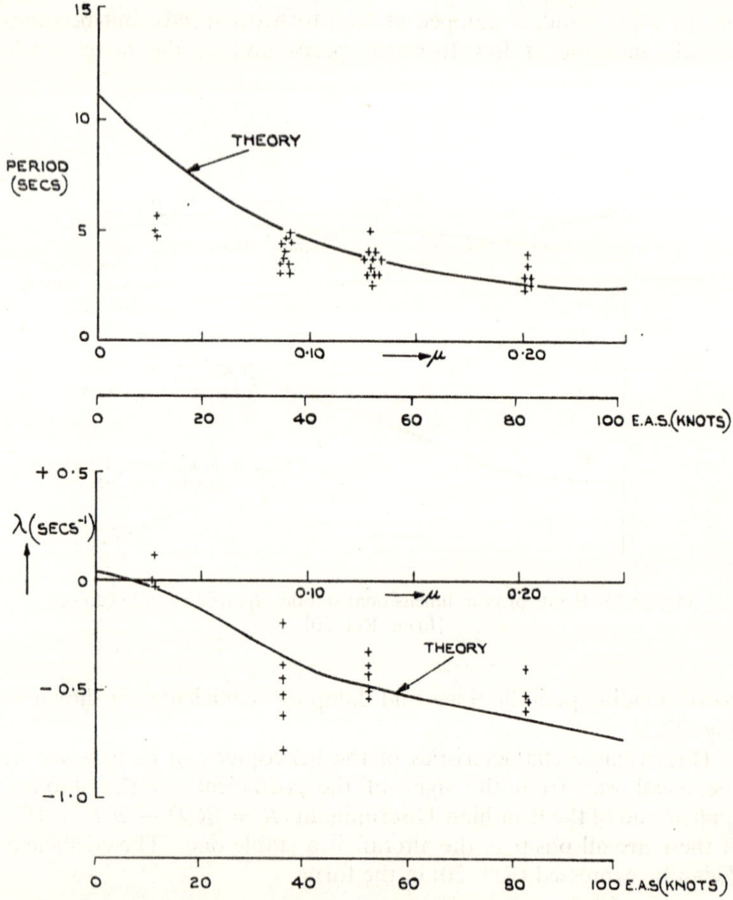

Fig. 22.24. Period and damping of lateral-directional oscillation—comparison with experimental results (Sycamore at 2,000 ft ICAN) [From Ref. 20]

The value of the damping in roll derivative $-l_p$ decreases with increase of forward speed and at high forward speeds the motion of the helicopter may consist of two oscillatory modes, one of which may be unstable (Ref. 20, para. 6.2).

7.3. Modification of the Lateral Dynamic Stability Characteristics

The lateral characteristics can be modified by varying the values of the derivatives or by changing the rolling or yawing inertias. The values of the derivatives are dependent on such physical features of the helicopter as the horizontal and vertical arms of the tail rotor, the moment of inertia of the main rotor blades and offset hinges on the main rotor (Refs. 20 and 24). The damping of the oscillatory mode can be increased by increasing the values of the damping in roll ($-l_p$) and the damping in yaw ($-n_r$) derivatives. Offset hinges

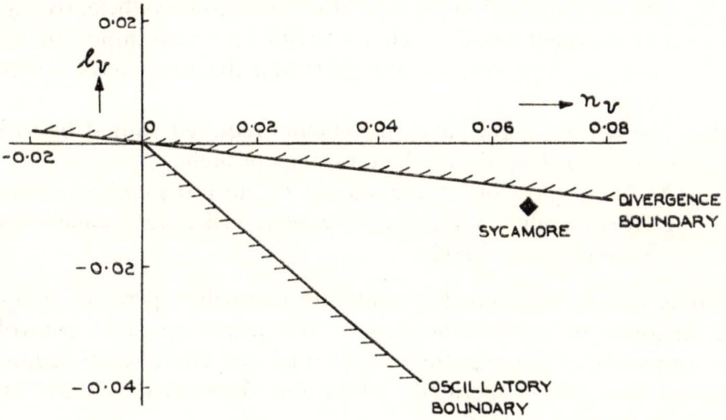

Fig. 22.25. The stability boundaries for a Sycamore at $\mu = 0{\cdot}15$ [From Ref. 20]

can be used to increase the value of $-l_p$. An increase in the blade inertia increases the damping of the oscillation, due to the increased "lag" of the rotor disc when the aircraft is rotating, but the periodic time remains unchanged.

8. Helicopter Control Systems

8.1. General

The helicopter is provided with a set of controls in order that the pilot can manoeuvre the aircraft about each of the three axes. The actual methods of achieving control are dependent on the particular configuration which is being used, but, as in the rest of this work, we are interested in the single main rotor configuration with an anti-torque vertical tail rotor. In this case, the pilot achieves longitudinal (pitching) and lateral (rolling) control by moving the stick which operates the cyclic pitch of the main rotor blades. Directional (yawing) control is achieved by the pilot moving the rudder pedals which operate the collective pitch control of the tail rotor. In addition, the helicopter pilot can control the vertical motion by operating his collective pitch control.

It is desirable that any control system should possess certain characteristics

(*a*) The overall response of the aircraft should be satisfactory. To achieve this, a deflection of the control should develop enough moment about the appropriate axis of the aircraft to produce a sufficient amount of angular acceleration. It is also necessary for the final steady rate of rotation to be sufficiently high. The transient response characteristics are also important and these are dependent on the particular dynamic characteristics of the aircraft.

(*b*) The control forces and movements required should be satisfactory to the pilot in all conditions of flight.

(*c*) As far as possible, there should be no cross-control effects. Unfortunately, in the the case of a helicopter, some cross-effects are unavoidable.

Some helicopter control systems are manually operated, though the majority of modern helicopters use power-operated controls. On a number of aircraft, the longitudinal and lateral cyclic control systems are power-operated, while the directional control still retains a manual system.

8.2. Control Response Characteristics

The transient response characteristics of a helicopter depend on the dynamic stability characteristics of the aircraft. If the aircraft is an unstable one, it tends to diverge, but if the aircraft is stable the response is an oscillatory one, the amount of oscillation depending on the damping. Figs. 22.26 and 22.27 show possible aircraft responses to different types of control inputs.

The initial and steady-state response characteristics of a control can be expressed in terms of the control power and the control sensitivity, which are defined in Ref. 1 as follows

(*a*) *Control Power* This is the moment produced about the C.G. of the helicopter by a unit displacement of the stick, expressed in lb ft per foot or lb ft per inch. By dividing the value of control power by the fuselage moment of inertia, the corresponding maximum angular acceleration can be determined (rads/sec^2/ft or rads/sec^2/in.).

(*b*) *Control Sensitivity* This can be defined as the final steady-state value of the angular velocity of the aircraft produced by a unit

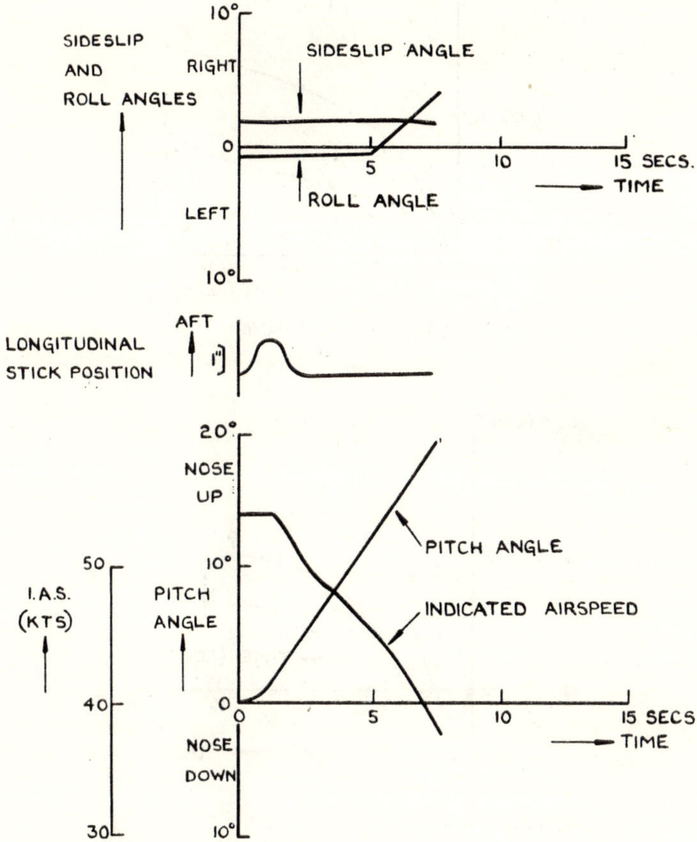

Fig. 22.26. Response of the Belvedere to a longitudinal pulse at 50 kts at 2,000 ft

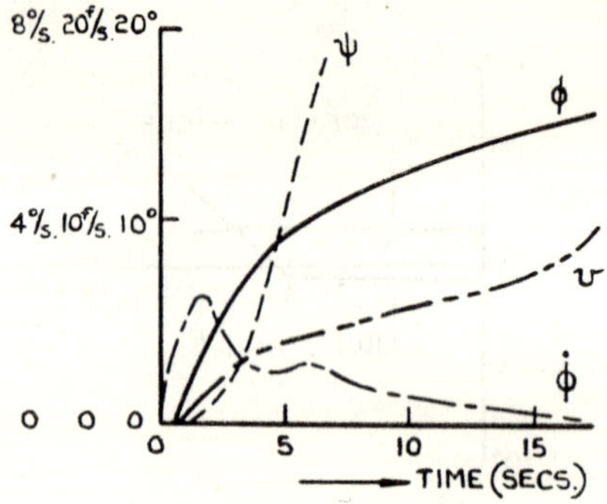

30 KTS. E.A.S. ($\mu = 0{\cdot}076$)

(*a*)

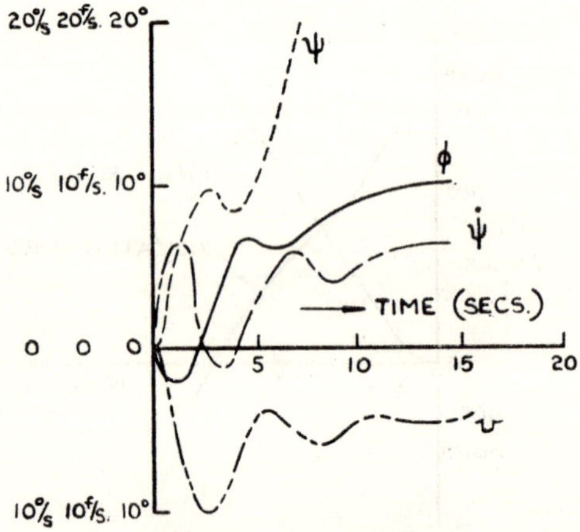

30 KTS. E.A.S. ($\mu = 0{\cdot}076$)

(*b*)

Fig. 22.27. Response to lateral inputs

(*a*) $\frac{1}{4}$° lateral cyclic pitch change
(*b*) 1° change in tailrotor pitch

displacement of the stick (expressed in rads/sec/ft or rads/sec/inch) i.e.

$$\text{Control sensitivity} = \frac{\text{maximum equilibrium angular velocity}}{\text{stick displacement}}$$

When the helicopter achieves the equilibrium condition, the moment produced by the control is exactly balanced by the damping moment produced by the rate of rotation. Thus an equivalent definition of control sensitivity is given by

$$\text{Control sensitivity} = \frac{\text{control power}}{\text{rotational damping}}$$

The optimum values of control power and sensitivity depend on which of the three axes is being considered. The results of various flight tests are reported in Refs. 17 and 19 and Appendix I of Ref. 6 compares the longitudinal, lateral and directional rate and acceleration responses of a number of U.S. helicopters.*

8.3. Control Movements and Forces Required for Trim and Manoeuvring Flight

The control movements required are dependent on the values of the stick-fixed static and manoeuvring stabilities. These have already been discussed in paras. 3.3, 4.3 and 5.3.

The control forces experienced by a pilot are also extremely important on a helicopter as on fixed-wing aircraft and they have a considerable influence on the pilot's impressions of the handling characteristics of the aircraft. In the case of a helicopter with manual controls, both steady and oscillatory aerodynamic forces are produced in the control system. The characteristics of these systems are usually rather unsatisfactory and as a result most helicopters employ power operated control systems with artificial spring feel.

9. Helicopter Autostabilization Systems

As we have seen, a helicopter often possesses poor inherent stability characteristics, particularly at the hover and at low forward speeds. In order to improve the handling characteristics of the aircraft,

* It should be noted that different definitions are sometimes used for control power and control sensitivity. In VTOL work they are often defined as the angular acceleration for full control displacement and the angular acceleration per unit control displacement, respectively. This definition of control sensitivity is the same as that of control power in Ref. 1.

autostabilization systems are usually incorporated. There are two different types of systems used

(i) mechanical autostabilization systems,
(ii) electrical autostabilization systems.

The most familiar mechanical systems that are used are the Bell bar and the Hiller paddle systems. The Bell bar is mounted on the shaft below the rotor. It is pivoted so that it is free to move, but it is connected to the shaft through a viscous damper. When the helicopter is disturbed, then, due to the gyroscopic effects, there is a time delay (of the order of 2 secs) between the movement of the shaft and the movement of the bar. The bar is linked to the rotor and the system introduces a cyclic control input which is proportional to both the attitude and the rate of change of attitude of the aircraft. In this way the attitude stability and the damping of the helicopter and, hence, the overall dynamic stability characteristics, can be improved and positive dynamic stability can be achieved. The Hiller bar operates basically in a similar way and the two systems are, in fact, just two particular examples of a more general system. A more detailed analysis of the control input involved (Ref. 5), shows that these systems produce a "lagged rate" signal rather than an "attitude plus rate" signal.

Most modern helicopters employ autostabilization systems similar to those used on fixed-wing aircraft. The basic principles and characteristics of these systems have been discussed in Chapter 19. Electrical signals from the rate and/or attitude gyros which measure the response of the helicopter are fed through various electrical circuits, amplified, and used to operate the various controls. The autostabilizer usually forms part of the fully-integrated autopilot system and usually consists of a limited-authority series system. The systems used on the pitch, roll and yaw channels of a particular aircraft are discussed in Ref. 25, and Figs. 19 and 20 of Ref. 13 show typical longitudinal and lateral response curves of a Wessex I with, and without, the autostabilizer acting.

The pitch channel of a helicopter autostabilizer usually includes an attitude θ and a rate $\dot{\theta}$ term. The attitude term can be produced by obtaining a signal from an attitude gyro, or by integrating the signal from a rate gyro. The attitude term is not, however, required to provide a long-term datum (this is the task of the autopilot) but only a short-term datum. Hence, if the attitude signal is produced from the rate signal, the circuit is modified to produce what is known as a "leaky-integrator," which allows the attitude signal to leak slowly away. In order that the system will not oppose the pilot in controlling the aircraft, a stick pick-off is used and a signal proportional to the stick displacement is introduced into the system to cancel the stabilization signals that would otherwise be produced.

The effect of different control laws on the characteristics of the helicopter can be represented graphically in the various ways discussed in Chapter 19. The effects of rate damping, lagged rate damping and attitude stability and rate damping, on the hovering dynamic stability characteristics of a helicopter are shown in the root locus plots on pages 387 of Ref. 5.

Other VTOL Aircraft

10. General

The helicopter is only one of the many different types of aircraft now in existence that have the ability of taking off and landing vertically and of hovering, as well as being capable of conventional forward flight. These VTOL aircraft range from the tilt-rotor, tilt-propeller and tilt-wing configurations, which at the hover are similar to a helicopter, through the tilt-duct and deflected slipstream designs, to the fan-in-wing, deflected-thrust, and separate lift engine configurations, which are more like conventional aircraft (Refs. 5 and 7). In England, most work has been done on the Short SC1, an aircraft with separate lift and propulsion engines, and the Hawker P 1127 (Kestrel), an aircraft with a single jet engine employing the principle of deflected thrust. In the U.S.A. the main research has been carried out using the tilt-propeller (Curtiss-Wright X-19), the tilt-wing (Ling-Temco-Vought XC-142, Vertol VZ-2), the tilt-duct (Doak VZ 4, Bell X22-A), and the fan-in-wing (Ryan XV-5) configurations. In Germany the tilt-engine design has been used (VJ 101C), whereas France has used the dual-propulsion concept (Dassault Mirage III V), and Canada the tilt-wing configuration (Canadair CL-84).

Any VTOL aircraft is required to have satisfactory handling characteristics in all conditions of flight, from the hover up to the top speed of the aircraft. At the lower speeds the requirements are similar to those of helicopters, and at the higher forward speeds similar to those of conventional aircraft. At the present time the hover and low-speed requirements of VTOL aircraft are based on the corresponding requirements of U.S. military helicopters (Ref. 10), though a lot of work is being done in simulators, on variable-stability helicopters (Ref. 36), and in flight tests on VTOL aircraft, to substantiate these specifications.

In spite of the numerous possible configurations, the stability and control characteristics of VTOL aircraft (including helicopters) are basically similar to one another, and the analysis is performed in very much the same way. Each particular design does, however, have its own special features, and appropriate modifications to the various equations are necessary. For this reason, it is not possible to generalize all the results obtained. The problems associated with VTOL aircraft are discussed in detail in Refs. 7, 11, and 27 to 44.

11. Dynamic Stability Analysis

11.1. Conventional Analysis

When analysing the dynamic stability of VTOL aircraft, the problem involved is not one of deriving the equations of motion, but rather one of deciding which assumptions and simplifications may be applied to these equations to make their solution easier. The equations of motion are well known, and are derived at length in Ref. 41. Before any simplifying assumptions are made, the equations represent a six-degree of freedom set of non-linear differential equations, written with respect to a set of axes fixed in the body (Ref. 41, p. 116). In the stability and control analysis of conventional aircraft, these equations are then linearized by using the small perturbation theory (Chapter 1, para. 4.4), and by ignoring various coupling terms. They are then further simplified by writing the equations with respect to a special set of body axes, referred to as wind or stability axes (i.e. $W_0 = 0$), and by neglecting many of the smaller stability derivatives (Ref. 41, pp. 123–24). When this has been done, it is possible to separate the six equations of motion into two distinct groups, the longitudinal equations (those for X, Z, and M) and the lateral equations (those for Y, L, and N). The equations are then usually non-dimensionalized before being written in their final form (Ref. 41, p. 133, Appendix F).

The analysis of the stability and control of VTOL aircraft in conventional flight can be carried out using these equations of motion, and they can also be specialized for the hover condition of flight by making appropriate assumptions ($U_0 = 0$, etc.). The hovering equations are given in equations (F7) and (F8) of Appendix F.

Even for conventional aircraft, however, this approach may not always be satisfactory. When the spinning motion (Chapter 13) or the inertia coupling problem (Chapter 16 and Appendix E) of an aircraft is being discussed, it is essential to leave the terms in the equations which couple together the longitudinal and lateral motions. In this analysis the equations are usually written with respect to a different set of body axes, called the principal inertia axes ($W_0 \neq 0$; I_{xz} (or E) $= 0$). The equations are often left in a dimensional form, and they are basically non-linear (though, as shown in Appendix E, it may still be possible to linearize them in order to obtain analytical solutions). The solution of the complete non-linear equations can be obtained using an analogue computer.

11.2. VTOL Analysis

The analysis of the stability and control characteristics of VTOL aircraft is carried out in a similar way to that for conventional aircraft. The equations of motion can be written with respect to general body axes ($W_0 \neq 0$, $I_{xz} \neq 0$), and then, if required, specialized to wind axes ($W_0 = 0$) or principal inertia axes ($I_{xz} = 0$). The equations can also be linearized and non-dimensionalized, if the approximations made are sufficiently accurate for the problem under consideration (see para. 1). If more accuracy is required, as, for example, for use in a VTOL pilot simulator, then the more complete non-linear equations of motion must be used (Ref. 43). Ref. 44 derives the complete dimensional, non-linear equations of motion with respect to general body axes for five categories of V/STOL aircraft, the tilt-wing, the tilt-duct, the tilt-propeller, the fan-in-wing, and the rotating thrust configurations. However, in VTOL analysis, it is usual to make the same simplifications to the equations of motion as are made in conventional aircraft work:

(i) The equations are linearized,
(ii) The longitudinal and lateral equations are uncoupled,
(iii) Various terms in the equations are neglected.

The resulting equations of motion are shown in Appendix F. Unfortunately, the problems of non-linear and coupling terms in the equation are more important in the case of VTOL aircraft than they are in conventional aircraft analysis, and these approximations are not always so acceptable. However, if the analysis is only required to give an approximate idea of the aircraft's characteristics (e.g. periodic times and dampings of the oscillatory modes, etc.), then the linearized, decoupled, simplified equations of motion usually give results of sufficient accuracy, and they are often employed in place of the more complete equations (Ref. 39, p. 28). Most of the reports mentioned in the references use this linearized approach in their stability and control analysis.

The problem of non-linearities in the equations of motion is discussed in Refs. 30, 31, and 34. In Ref. 30 the justification for neglecting the non-linearities is discussed on page 19; though non-linearities are present, the report concludes that it still appears reasonable to neglect them. In the analysis in Ref. 31, which is concerned with a tilt-wing VTOL design, the six-degree of freedom, non-linear equations of motion are initially derived, and are then linearized by conventional small-perturbation techniques. The linearization is then validated by comparing the actual response of the non-linear equations obtained on an analogue computer, with the response obtained from the linearized approximate equations (see Figures 32 and 33 of Ref. 31). Though non-linearities are considered to be more predominant than in conventional aircraft, the results indicate that the linearized equations are sufficiently accurate. However, in Ref. 34, it is stated that most V/STOL aircraft have strong and dominant non-linear characteristics at low speeds, which cannot be neglected because they significantly affect the pilot. The linearized equations can then only be applied to small disturbance problems.

To uncouple the longitudinal and lateral equations of motion it is necessary to neglect the aerodynamic and inertia coupling terms, and also any gyroscopic coupling defects. Helicopters and other VTOL aircraft generally have more important coupling terms than conventional aircraft. Gyroscopic coupling terms are included in the complete non-linear equations of Ref. 43, and are also discussed on page 25 of Ref. 44. Coupling between the longitudinal pitching motion and the lateral rolling motion of a VTOL aircraft is discussed in Ref. 32, and the particular aspects of gyroscopic coupling in Refs. 31 (p. 4–14) and 33. The uncoupling of the equations of motion for a helicopter has already been discussed in para 6.1. The gyroscopic coupling effects may be important at the hover and at low forward speeds, and they are often eliminated by rotating different parts of the aircraft in opposite directions. On the Hawker P 1127 the low-pressure and high-pressure compressors of the jet engine rotate in opposite directions (Ref. 28); in the case of the Canadair CL-84 the main propellers on the wing counter-rotate, and the tail rotor is made up of two, two-bladed, counter-rotating propellers (Ref. 31, p. 4–14). In Ref. 31 (p. 7–1) the longitudinal equations are separated from the lateral equations. However, the lateral dynamics of the particular tilt-wing design being studied in this report are found to depend on the longitudinal acceleration, and it is suggested that the lateral-directional analysis will, in this particular case, reduce to a minimum of five degrees of freedom (only the pitching moment equation can be neglected).

As we have mentioned, the equations of motion can usually be further simplified by neglecting certain of the terms. In the analysis of conventional aircraft this may go so far as eliminating some of the equations altogether, and using two-degree of freedom approximations to represent the particular motion of interest (as in the analysis of the longitudinal short-period and phugoid motions in Appendix B and F). When dealing with helicopters and other VTOL aircraft, the same simplifying assumptions may not be possible. As an example of this, we can consider the speed stability derivative M_u. In the low-speed analysis of conventional aircraft this derivative is usually neglected; on a helicopter, however, even at the lowest forward speeds, M_u is a very important derivative (para. 3). It is, in fact, the positive value of the M_u derivative that results in oscillatory pitching instability at the hover condition. The actual approximations that can be made depend on the configuration being discussed, and the particular condition of flight. The various terms in the equations must be analysed for the magnitude of their contributions, before final simplification can be made (Ref. 44, p. 1). Ref. 30 discusses the form of the approximate equations (and their validity) that can be used for four particular aircraft designs, a single-rotor helicopter, a tandem-rotor helicopter, a tilt-duct and a tilt-wing configuration. In this report it was found that, above a forward speed of about 50 ft/sec, the approximate equations applicable to the longitudinal motion of conventional aircraft are also applicable to the tilt-duct and single-rotor helicopter designs. However, because of a reduction in the numerical value, and even, possibly, a change in the sign, of M_w, the approximate equations for the tilt-wing VTOL aircraft and for the tandem-rotor helicopter are different from those of conventional aircraft (see Tables VI and VII of Ref. 30). When lateral motions are considered, Ref. 30 shows that the approximate equations of a single-rotor helicopter are also the same as those for a conventional aircraft at forward speeds greater than about 50 ft/sec. The conventional equations have to be slightly modified before they can be applied to the tilt-duct configuration, and they are not applicable to the tandem-rotor design. In the latter case, due to the small N_v of this aircraft, the lateral characteristics in forward flight remain similar to those at the hover (Ref. 30, p. 45). A review of the best approximations then (October 1965) currently available for various types of VTOL aircraft, in both low and high speed flight, is given in Appendix A of Ref. 38.

Though the three-degree of freedom equations of motion of a VTOL aircraft may be similar to those of a conventional aircraft, it is not generally valid to approximate these by two-degree of freedom equations, in order to represent the longitudinal short-period and phugoid modes, an approximation which is frequently used in conventional analysis. The reason for this is discussed on page 5–10 of Ref. 31.

12. Hovering Flight

When a VTOL aircraft using separate lift engines, deflected thrust, deflected slipstream or tilt-thrust is in hovering flight, the aerodynamic effects which are present are the result of the air-flow induced round the aircraft by the lifting engines. These induced effects may result in aerodynamic pitching, rolling, and yawing moments acting upon the aircraft. If the aircraft is hovering close to the ground, the flow pattern and the corresponding aerodynamic forces and moments may be considerably modified by the ground effect.

When a hovering aircraft is disturbed, the resulting aerodynamic forces and moments are very small, and the aircraft possesses very little inherent static stability or damping. At low forward speeds conventional aerodynamic controls are ineffective, and some other form of control (e.g. compressed air jets at the wing tips, nose and tail, or modulated thrust) has to be incorporated. It is desirable that the pilot should be able to control the aircraft precisely during the hover. The lack of stability makes this difficult, but in some cases it is more than compensated for by the particular control system used. Though some VTOL aircraft can be hovered without an autostabilization system, it is generally considered necessary to incorporate some stability augmentation in order to improve the handling characteristics of the aircraft in this condition of flight. The autostabilization system used on the Short SC1 is described in detail in Ref. 29.

12.1. Longitudinal Characteristics at the Hover

The theoretical analysis of the longitudinal stability of any VTOL aircraft is similar to that discussed previously for a helicopter. The simplified equations of motion for an aircraft at the hover have been derived in equation (F7) of Appendix F. If the stick-fixed stability of the aircraft is being considered, the terms on the right-hand side of the equations are zero, and the characteristic equation can be written in the form (Ref. 30, pp. 9 and 10)

$$(s - Z_w)\,[s^3 - (X_u + M_q)s^2 + X_u M_q s + gM_u] = 0 \qquad (22.11)$$

The first factor on the left-hand side of this equation shows that the vertical or "plunging" motion of the aircraft can be considered entirely separately from the motion involving forward speed and pitching motion (as for a helicopter in para. 6.2.2). The second factor on the left-hand side of equation (22.11) is referred to as the longitudinal hovering cubic. The solution of this equation usually results in a negative real root (representing a well-damped subsidence motion, which is not very important), and a complex pair of roots with a positive real part (representing an undamped oscillatory mode). The subsidence mode is sometimes referred to as the short-period mode, because, as we shall see, as forward speed is increased, it combines with the real root of the plunging mode to produce the conventional longitudinal short-period oscillation (Ref. 38).

Ref. 39 is an analytical study of the stability of a tilt-wing aircraft, and Fig. 29 of the reference shows the analogue computer traces of the longitudinal motion of the aircraft following a disturbance at the hover. Traces are given of u, w, and θ, and the aircraft can be seen to be very unstable in u and θ, whereas w remains zero.

The method of the root locus plot (Chapters 8 and 19) is an extremely valuable visual aid in appreciating the aperiodic and oscillatory characteristics of an aircraft, and how they vary with change in forward speed, variations in the values of the stability derivatives, the addition of an autostabilization system, etc. Ref. 5 makes use of these diagrams to show the effects of various factors on the stability and control characteristics of conventional aircraft and helicopters. In the root locus plot, the real part of the root (represented by μ or σ, the damping factor of the mode, see Chapter 7) is plotted along the horizontal axis, and the imaginary part (represented by ω, the frequency of the oscillation) along the vertical axis. If a root of the stability quartic lies on the real axis, it represents an aperiodic or non-oscillatory mode (convergent if σ is negative, and divergent if σ is positive). In general, a root (represented by λ) will contain both a real and an imaginary part; it will then consist of a complex pair of roots and represent an oscillatory mode (Chapter 7):

$$\lambda = \sigma \pm i\omega$$

where σ replaces $-\mu$ used in Chapter 7. These two roots are always symmetrically positioned about the horizontal axis, and, for this reason, sometimes only the top half of the diagram is drawn. The damping of an oscillation is often represented in terms of ζ ($\sigma = -\zeta\omega_n$), referred to as the relative damping or the damping ratio (Chapter 7). It can be shown that straight lines drawn through the origin of the root locus plot represent constant values of the relative damping ζ, varying from $\zeta = 0$ along the vertical axis (i.e. $\sigma = 0$) to $\zeta \geqslant 1$ along the horizontal axis (i.e. $\omega = 0$) (Fig. 22.28).

The root locus plot in Fig. 22.28 shows how the roots of the characteristic equation may vary with forward speed from the hover through transition to conventional forward flight. The hover roots are seen to correspond to one large negative real root ($s = Z_w$), another smaller negative real root, and one complex pair of roots with a positive real part (representing the undamped oscillation). Similar root locus plots showing the hover roots for the XC-142 tilt-wing aircraft and the X-22 A tilt-duct aircraft are shown in Figs. 26 and 27, respectively, of Ref. 37. Refs. 37 and 30 (Figure 7) show that the dynamic instability at the hover is caused by the positive value of *Mu* associated with these designs. This result is similar to that obtained for the conventional helicopter (Ref. 5, p. 357, Fig. 3).

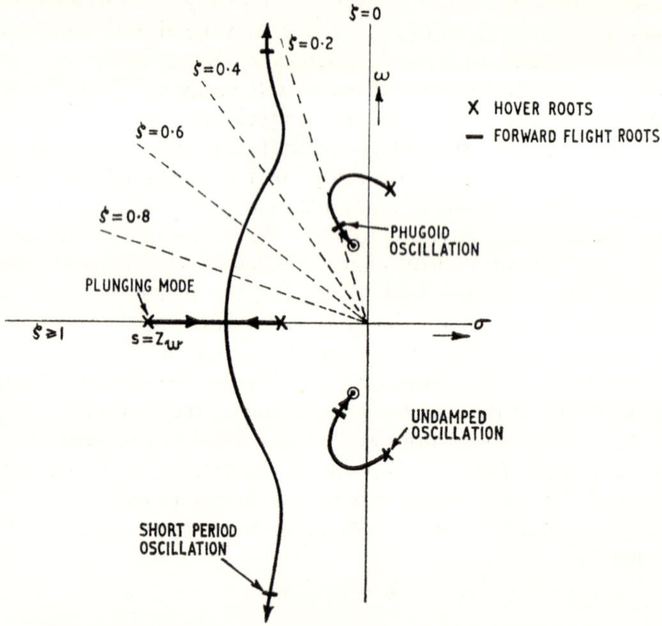

Fig. 22.28. Effect of forward speed on the longitudinal roots

12.2. Lateral Characteristics at the Hover

The corresponding simplified equations for the lateral motion at the hover have been given in equation (F8) of Appendix F. For the stick-fixed condition, the characteristic equation can be written in the form (Ref. 30, p. 14)

$$(s - N_r)\,[s^3 - (Y_v + L_p)s^2 + Y_vL_ps - gL_v] = 0 \qquad (22.12)$$

This equation is similar to the corresponding one (eqn. 22.11) for the longitudinal motion. The first term on the left-hand side of eqn. (22.12) represents an aperiodic convergence involving only *r* perturbations. It should be noted, however, that in the longitudinal case the corresponding "plunging" mode was a linear motion, whereas in the lateral case it is an angular one. The second term on the left-hand side of eqn. (22.12) is referred to as the lateral hovering cubic. The solution of this cubic equation represents one damped aperiodic mode, and an undamped oscillatory mode involving roll and sideslip. The four roots are thus similar (i.e. two real negative roots and one complex pair of roots with a positive real part) to those obtained in the longitudinal analysis. Typical lateral hovering roots are illustrated by the root locus plot in Fig. 22.29. The root locus plots given in Figs. 46 and 47 of Ref. 37 show the lateral hovering roots of the XC-142 and the X-22A, respectively.

The above result is applicable to tilt-duct VTOL and tandem-rotor helicopter configurations, but not to a single-rotor helicopter

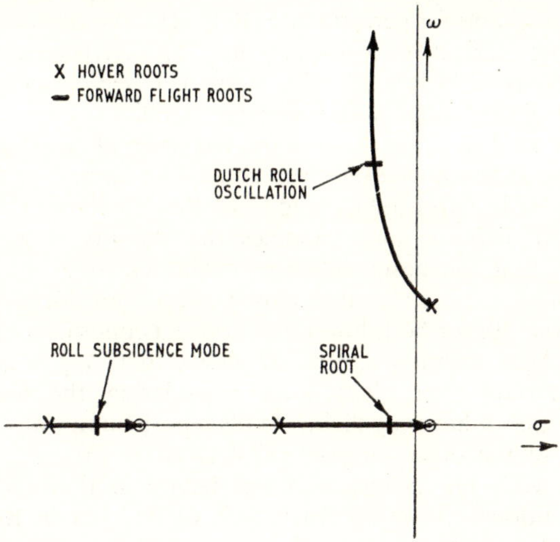

Fig. 22.29. Effect of forward speed on the lateral roots

design (Ref. 30, p. 43). In the latter case, the hover modes are similar to those of a conventional aircraft in forward flight, consisting of a heavily damped rolling mode, a damped spiral motion, and an oscillatory Dutch Roll mode (para. 7.2). Ref. 30 (p. 16) shows that the hover lateral cubic does not usually apply to the single-rotor helicopter, because the various assumptions used in obtaining the cubic equation become invalid.

Fig. 31 of Ref. 39, which analyses the stability of a tilt-wing aircraft, shows the analogue computer traces of the lateral motion of an aircraft disturbed at the hover. Curves are given for v, ψ, and ϕ; those for v and ϕ show a divergent oscillatory motion, whereas the aircraft's heading is only slightly affected. The corresponding mathematical analysis, on p. 101 of the reference, gives two aperiodic modes with a time to half amplitude ($t_{1/2}$ or $T_{1/2}$) of 1·08 sec and 9·9 sec, and an oscillation with a periodic time of 14·6 sec and a time to double amplitude (t_2 or T_2) of 3·6 sec.

12.3. Hover Control

An iso-opinion plot showing the desirable yaw control characteristics (with N_{δ_r} plotted against $-1/N_r$) in the hover condition is given in Fig. 47 of Ref. 30. Other iso-opinion plots showing the pitch and yaw control requirements at the hover are given in Ref. 36. It is very important to minimize the control requirements at the hover, because, since both the thrust and the control moments are obtained from the propulsion system, the control power requirements reduce the thrust available for performance (Ref. 34). The effects of changes in the various derivatives on the handling characteristics at the hover are also discussed in Ref. 34; this report is based on an analysis of the X-22A aircraft fitted with a variable stability system.

Ref. 38 analyses the closed-loop response of a pilot-airframe system, and shows how a pilot can control an aircraft at the hover, when it is basically unstable. Fig. 3 of Ref. 38 shows that closure of the pilot-airframe loop modifies the stability roots, and, in particular, that the basic unstable oscillatory roots may become stable. Similar pilot-airframe closed loop root locus plots are given in Ref. 30, for both hover and cruising conditions of flight.

As has been mentioned, Ref. 37 shows that M_u is one of the most important derivatives when considering the longitudinal characteristics of hovering flight. In his report Rampy also discusses the effect of the other longitudinal derivatives (M_q, Z_w, M_w, X_u), and he extends his analysis to cover lateral motion and forward flight conditions. A similar analysis is carried out in Ref. 38 for various types of VTOL configurations.

The problem of the longitudinal hovering instability associated with aircraft supported by lifting propellers or rotors is also discussed in Ref. 35. A physical explanation of the instability is given, an analysis is made of methods of improving the stability, and the actual hardware of some automatic stabilization systems is described.

13. Low Speed Forward Flight

The transition regime covers the range of speeds from the hover up to the forward speed at which the VTOL aircraft achieves conventional wing-borne flight. The stability and control characteristics must remain satisfactory during transition; the trim changes should be small, the "lift" should be transferred smoothly from the lift jets or the deflected thrust to the wing of the aircraft, and there should be no appreciable change in the characteristics of the control system as the change over is made to the conventional aerodynamic flap-type controls. Some stability augmentation may again be necessary to ensure satisfactory handling qualities in this speed range.

13.1. Longitudinal Characteristics at Low Speeds

The root locus plot method is extremely useful in illustrating how the roots of the characteristic equation, and, hence, the characteristics of the modes of the aircraft, change as the forward speed increases. Fig. 22.28, already referred to in para. 12.1, shows that the longitudinal roots change as the forward speed is increased from the two real roots and one complex pair of roots obtained at the hover (representing two aperiodic convergent modes and one undamped oscillation), to the two complex pairs of roots obtained in conventional flight (representing two damped oscillatory modes, i.e. the long-period phugoid and the short-period oscillation). The unstable oscillation at the hover becomes the damped phugoid in conventional flight, and the two negative real roots combine to produce the damped short period oscillation. In forward flight, the characteristic equation no longer breaks down as shown in eqn. (22.11), but it splits up instead into two quadratic equations, both of which have complex roots similar to those of conventional aircraft (Chap. 8, para. 8.1).

A similar root locus plot is shown in Fig. 26 of Ref. 37 for the longitudinal motion of the XC-142; the two aperiodic modes again combine together to produce the short-period oscillation, and the undamped oscillation at the hover changes to a damped phugoid as forward speed is increased. The corresponding plot of the X-22A (Fig. 27, Ref. 37) shows very different results, however, and, up to the maximum speed used in the analysis, the modes remain similar to those obtained at the hover. The oscillatory motion, in fact, becomes even more undamped.

Ref. 39 presents results at various stages through the transition range. At a forward speed of 25 ft/sec the longitudinal modes are similar to those in the hover condition, the unstable oscillation having a periodic time (P or T) of 14·9 sec, and a time-to-double amplitude (t_2 or T_2) of 2·9 sec (Fig. 32, Ref. 39). At 70 ft/sec the conventional damped short period and phugoid modes are present, having characteristics of $P = 6{\cdot}8$ sec, $T_{1/2} = 2{\cdot}7$ sec, and $P = 25$ sec, $T_{1/2} = 3{\cdot}2$ sec, respectively (Fig. 34, Ref. 39). At a forward speed of 140 ft/sec, the characteristics of the short period oscillation have become $P = 3{\cdot}15$ sec and $T_{1/2} = 1{\cdot}12$ sec, while the corresponding phugoid values are $P = 44$ sec and $T_{1/2} = 6$ sec.

Similar results are shown for the transition of a particular tilt-wing configuration in the root locus plots of Fig. 12 in Ref. 40, and the results are compared to those of typical conventional aircraft and helicopters. Regions showing acceptable handling qualities are also marked. In this case, a small increase in forward speed improves the damping of the initially undamped mode. With further increase in forward speed, the original oscillatory mode becomes damped, and the two real roots combine together into an oscillatory mode (in Fig. 12 of Ref. 40 this mode is represented eventually at the high forward speed as the phugoid mode, but see the sketch on page 72 of the reference, and the associated discussion), whereas the initial oscillation transforms into the level flight short-period oscillation.

Ref. 5 (Fig. 5, p. 361) gives typical root locus plots for a helicopter, showing the effect of forward speed on the longitudinal dynamic characteristics.

Ref. 34 describes the results of an analysis carried out on the X-22A tilt-duct aircraft fitted with a variable stability system, designed to vary the static and dynamic characteristics of the aircraft and the pilot's controls over wide ranges. The report discusses the handling characteristics of the aircraft in hovering, translation, and conventional flight conditions, illustrating the results by means of root locus plots.

13.2. Lateral Characteristics at Low Speeds

A change also takes place in the lateral roots during the transition. Fig. 22.29 shows how the roots may change from the two negative real roots and one complex pair of roots with a positive real part obtained at the hover (representing two convergent aperiodic modes and one undamped oscillation), to the one negative real root, one real root which may be either positive or negative, and one complex pair of roots with a negative real part typical of conventional flight (representing the roll subsidence mode, the divergent or convergent spiral mode, and the damped Dutch roll oscillation. Similar results are shown for the XC-142 aircraft in Fig. 46 of Ref 37; the Dutch roll oscillation becomes damped when the forward speed exceeds about 70 ft/sec. The corresponding plot for the X-22A (Fig. 47, Ref. 37) shows different characteristics; though the roll and spiral modes remain stable, the Dutch roll is unstable up to the highest speed used in the analysis (i.e. 101·5 ft/sec).

Fig. 6 on page 331 of Ref. 27 shows the Dutch roll characteristics of the Curtiss-Wright X-19 tilt-propeller tandem-wing V/STOL aircraft during transition, and compares these characteristics with various specified handling requirements.

Other root locus plots presented in Ref. 30 show the effect of variation in the numerical values of particular derivatives on the roots obtained in hovering and conventional forward flight. Fig. 2

on page 381 of Ref 5 gives root locus plots showing the effect of forward speed on the lateral characteristics of a helicopter.

Ref. 34 shows that the lateral and directional characteristics of some VTOL aircraft in conventional flight may differ from those usually obtained. One of these differences is that the roll and spiral roots may couple together to form an oscillatory pair of roots, representing what can be called the roll-spiral mode. This coupling can be achieved by a reduction in roll damping, a reduction in I_Z, and by letting $I_{xz} = 0$. As a result of this coupling of the roots into an oscillatory mode, there is a deterioration in the handling qualities of the aircraft. Fig. 10 of Ref. 34 is a root locus plot showing how the roll-spiral oscillation arises. The coupling of the roll and spiral roots, into an oscillation of higher frequency than the Dutch roll, is also discussed in Ref. 42. It is shown that this could occur on aircraft designed for high speed, high altitude flight, with low roll damping and low directional static stability. In extreme cases the oscillation could become divergent. This roll-spiral oscillation is sometimes referred to as the lateral phugoid. Ref. 5 (pp. 265–70) discusses the effect of varying the stability derivatives and the product of inertia on the lateral modes of conventional aircraft, and the root locus plots given show that the roll-spiral coupling could occur.

13.3. Control at Low Speeds

Iso-opinion plots showing the desirable pitch and yaw characteristics during a low speed approach are given in Refs. 30 (Fig. 50) and 36. An analysis of the handling qualities of a tilt-duct and a tilt-wing aircraft through the speed range from the hover to conventional flight is given in Ref. 38.

Fig. 5 on page 330 of Ref. 27 shows the various control requirements for the Curtiss-Wright X-19 during transition flight from 0 to 160 knots, and illustrates the proportion of each control that is obtained from the aerodynamic surfaces and from the propellers. Similar curves showing the transition characteristics of the German VJ101C aircraft can be found on page 473 of Ref. 27. On the VJ101C, the coupling between the roll and yaw controls during transition is automatically compensated for by a mechanical sine-cosine linkage in the control system (Ref. 27, p. 458). The control system is described in more detail on pages 457–58 of this reference.

References

1. Gessow, A. and Myers, G. C.; *Aerodynamics of the Helicopter* (Macmillan Co., New York, 1952).
2. Shapiro, J.; *Principles of Helicopter Engineering* (Temple Press Ltd., London, 1955).

3. PAYNE, P. R.; *Helicopter Dynamics and Aerodynamics* (Sir Isaac Pitman & Sons, Ltd., London, 1959).
4. NIKOLSKY, A. A.; *Helicopter Analysis* (John Wiley and Sons, New York, 1951).
5. SECKEL, E.; *Stability and Control of Airplanes and Helicopters* (Academic Press, New York, 1964).
6. SWOPE, W. A.; *Introduction to Rotary Wing Stability and Control,* U.S. Army Aviation Test Office.
7. CAMPBELL, J. P.; *Vertical Take-off and Landing Aircraft* (Macmillan Co., New York, 1962).
8. SWANBOROUGH, F. G.; *Vertical Flight Aircraft of the World* (Temple Press Books, London, 1964).
9. *Design Requirements for Aircraft for the Royal Air Force and Royal Navy*, Av.P.970 Vol. 3, Ministry of Aviation.
10. *Recommendations for V/STOL Handling Qualities*, Agard Report No. 408A, 1964.
11. *Symposium on Vertical and Short Take-off and Landing Aircraft*, Agardograph No. 46, Parts 1 and 2, 1960.
12. *The Proceedings of a Symposium on V/STOL Aircraft*, Held at the Empire Test Pilots' School, Farnborough, 7th July, 1964.
13. POOLE, J.; "Rotorcraft Work at the Aeroplane and Armament Experimental Establishment," *Journal of the Royal Aeronautical Society*, August 1963, page 501.
14. O'HARA, F.; *An Analysis of the Longitudinal Stability and Control of a Single Rotor Helicopter*, R. & M. No. 2958, 1957.
15. BRAMWELL, A. R. S.; *Longitudinal Stability and Control of the Single-Rotor Helicopter*, R. & M. No. 3104, 1959.
16. GREGORY, J. D. L., LANGDON, G. F. and COOMBES, R. W.; *An Account of Some Longitudinal Stability Tests on a Dragonfly Helicopter*, Boscombe Down, Rep. No. A. & A.E.E./Res/293, revised.
17. FERRELL, K. R. and BALFE, P. J.; *Limited Stability and Control Evaluation of the S-62A Helicopter* A.R.D.C. Rep. AFFTC-TR-61-8, 1961.
18. *Saunders-Roe Skeeter Mk. 6—Brief Performance and Handling Trials*, Boscombe Down, A. & A.E.E./897/2/1st Part.
19. KOELLE, E. A. and HODGSON, W. J.; *ARDC, YH41 Stability and Control Evaluation*, AFFTC-TR-60-7, 1960.
20. LANGDON, G. F. and NEALE, M. C.; *An Analysis of the Lateral-Directional Stability and Control of the Single-Rotor Helicopter*, Boscombe Down Rep. No. AAEE/Res/299, 1958.
21. ZBROZEK, J. K.; *Introduction to Dynamic Longitudinal Stability of Single-Rotor Helicopter with Hinged Blades*, R.A.E. Rep. Aero 2248, 1948.
22. STEWART, W.; *Dynamic Longitudinal Stability Measurements on a Single-Rotor Helicopter* (*Hoverfly Mk.* 1), R. & M. No. 2505, 1948.
23. BURLE, M. F. and STEWART, W.; *Dynamic Longitudinal Stability Measurements on a Single-Rotor Helicopter* (*Sikorsky S*-51), R.A.E. Tech. Note 2021, 1949.
24. ZBROZEK, J. K.; *Investigation of Lateral and Directional Behaviour of a Single Rotor Helicopter* (*Hoverfly Mk.* 1), R. & M. No. 2509, 1948.
25. CIASTULA, T. L.; "The Development of the P.531," *Journal of the Royal Aeronautical Society*, June 1964, page 399.
26. SULLIVAN, N. and SIMPSON, J. A.; "Integrated Flight Systems for Multi-Engined Transport Rotorcraft," *Journal of the Royal Aeronautical Society*, February 1962, page 103.
27. *V/STOL Aircraft*, Agardograph No. 89, Part 1, Sept. 1964.

28. BEDFORD, A. W.; "The Hawker 1127 V/STOL Strike Fighter," *Journal of the Royal Aeronautical Society*, December 1962, page 743.
29. LASBREY, I. K. B. and CONWAY, H. G.; "The Autostabilizer System for the S.C.1 Aircraft," *The Royal Aeronautical Society and the Institution of Electrical Engineers Joint Conference on the Importance of Electricity in the Control of Aircraft*, 26–28th February, 1962.
30. WOLKOVITCH, J. and WALTON, R. P.; *VTOL and Helicopter Approximate Transfer Functions and Closed-loop Handling Qualities*, Systems Technology Inc. (U.S.A.), S.T.I. Report No. 128–1, June 1965.
31. *Optimal Blending of VTOL Transition Controls*, Attachment No. 1: CL-84 Aerodynamics Notebook; Sperry Phoenix Co., Phoenix, U.S.A., Tech. Memo. No. LJ–1264–0490, June 1965.
32. GARREN, J. F.; *Effects of Coupling between Pitch and Roll Control Inputs on the Handling Qualities of VTOL Aircraft*, NASA Tech. Note D1233, March 1962.
33. GARREN, J. F.; *Effects of Gyroscopic Cross Coupling between Pitch and Roll on the Handling Qualities of VTOL Aircraft*, NASA Tech. Note D812, April 1961.
34. SCHULER, J. M., KROLL, J., and KEY, D. L.; "VTOL Flying Qualities and the Flight Research Use of VTOL Variable Stability Airplanes," *Canadian Aeronautics and Space Journal*, Vol. 12, No. 6, June 1966.
35. ELLIOTT, D. A. and SCHREIBER, C. R.; "Improving the Stability of Hovering Aircraft," *Canadian Aeronautics and Space Journal*, Vol. 11, No. 3, March 1965.
36. GOULD, D. G.; *Results of Recent Investigations of V/STOL Control and Response Requirements Using Variable Stability Helicopters*, Agard Report No. 519, October 1965.
37. RAMPY, J. M.; *Important V/STOL Aircraft Stability Derivatives in Hover and Transition*, University of Tennessee Report, Dec. 1965. (Also Edwards Air Force Base, U.S.A., Report FTC-TR-66-29, Oct. 66.)
38. STAPLEFORD, R. L., WOLKOVITCH, J., *et al.*; *An Analytical Study of V/STOL Handling Qualities in Hover and Transition*, Systems Technology Inc., S.T.I. Report TR-140-1; Wright-Patterson A.F.B., U.S.A., AFFDL-TR-65-73, October 1965.
39. CROMWELL, C. H. and PAYNE, H. E.; *A Stability Analysis of Tilt-Wing Aircraft (Analytical)*, Princeton University, Department of Aeronautical Engineering, Report No. 477, May 1960.
40. HARGRAVES, C. R.; *An Analytical Study of the Longitudinal Dynamics of a Tilt-wing VTOL*, Princeton University, Department of Aeronautical Engineering, Report No. 561, June 1961.
41. ETKIN, B.; *Dynamics of Flight* (John Wiley and Sons, New York, 1959).
42. NEWELL, F. D.; *Ground Simulator Evaluations of Coupled Roll-Spiral Mode Effects on Aircraft Handling Qualities*, Cornell Aeronautical Laboratory Technical Report TC-1921-F-1 and Wright-Patterson A.F.B., Ohio, U.S.A., AFFDL-TR-65-39, March 1965.
43. BALSINK, E. B. and SOVINE, D. M.; *Analog Computer Mechanization of a Tilt-wing VTOL Aircraft*, Wright-Patterson A.F.B., Ohio, U.S.A., WADD-TN-59-344.
44. MCINTYRE, W.; *Simulation of Helicopter and V/STOL Aircraft, Vol. II V/STOL Analysis Report*, U.S.N. NAVTRADEVCEN TR-1205-2.

Appendix A

Expressions for the Coefficients of the Stability Quartic Equations

1. Longitudinal Stability Quartic

Using the definitions of Ref. 1, the longitudinal stick-fixed motion of an aircraft in horizontal flight can be represented by the quartic equation

$$A_1\lambda^4 + B_1\lambda^3 + C_1\lambda^2 + D_1\lambda + E_1 = 0 \tag{A1}$$

where

$$A_1 = 1$$

$$B_1 = -(x_u + z_w) - \frac{m_q}{i_B} - \left(1 + \frac{z_q}{\mu_1}\right)\frac{m_{\dot{w}}}{i_B}$$

$$C_1 = (x_u z_w - x_w z_u) + \frac{m_q}{i_B}(x_u + z_w) + \frac{m_{\dot{w}}}{i_B}\left[x_u\left(1 + \frac{z_q}{\mu_1}\right) - \frac{x_q z_u}{\mu_1}\right] - \frac{m_w}{i_B}(\mu_1 + z_q) - \frac{x_q m_u}{i_B}$$

$$D_1 = -\frac{m_q}{i_B}(x_u z_w - x_w z_u) + \frac{m_{\dot{w}}}{i_B}\frac{C_L}{2}z_u + \frac{m_w}{i_B}[x_u(\mu_1 + z_q) - x_q z_u] + \frac{m_u}{i_B}\left[\mu_1\frac{C_L}{2} - x_w(\mu_1 + z_q) + x_q z_w\right]$$

$$E_1 = \frac{\mu_1}{2i_B}C_L(m_w z_u - m_u z_w)$$

2. Lateral Stability Quartic

In a similar way, the lateral stick-fixed motion can be represented by the equation

$$A_2\lambda^4 + B_2\lambda^3 + C_2\lambda^2 + D_2\lambda + E_2 = 0 \qquad \text{(A2)}$$

where

$$A_2 = 1 - \frac{i_E{}^2}{i_A i_C} \simeq 1$$

$$B_2 = -y_v\left(1 - \frac{i_E{}^2}{i_A i_C}\right) - \frac{l_p}{i_A} - \frac{n_r}{i_C} - \frac{i_E}{i_C}\frac{l_r}{i_A} - \frac{i_E}{i_A}\frac{n_p}{i_C}$$

$$C_2 = y_v\left(\frac{l_p}{i_A} + \frac{n_r}{i_C} + \frac{i_E}{i_A}\frac{n_p}{i_C} + \frac{i_E}{i_C}\frac{l_r}{i_A}\right) + \left(\frac{l_p}{i_A}\frac{n_r}{i_C} - \frac{l_r}{i_A}\frac{n_p}{i_C}\right) + \frac{\mu_2 l_v}{i_A}\left[\frac{i_E}{i_C}\left(1 - \frac{y_r}{\mu_2}\right) - \frac{y_p}{\mu_2}\right] + \frac{\mu_2 n_v}{i_C}\left[1 - \frac{y_r}{\mu_2} - \frac{i_E}{i_A}\frac{y_p}{\mu_2}\right]$$

$$D_2 = -y_v\left(\frac{l_p}{i_A}\frac{n_r}{i_C} - \frac{l_r}{i_A}\frac{n_p}{i_C}\right) + \frac{\mu_2 l_v}{i_A}\left[\frac{n_p}{i_C}\left(1 - \frac{y_r}{\mu_2}\right) + \frac{n_r}{i_C}\frac{y_p}{\mu_2} - \frac{C_L}{2}\right] + \frac{\mu_2 n_v}{i_C}\left[-\frac{l_p}{i_A}\left(1 - \frac{y_r}{\mu_2}\right) - \frac{l_r}{i_A}\frac{y_p}{\mu_2} - \frac{i_E}{i_A}\frac{C_L}{2}\right]$$

$$E_2 = \frac{\mu_2}{2 i_A i_C} C_L\,(l_v n_r - n_v l_r)$$

Reference

1. DUNCAN, W. J.; *Control and Stability of Aircraft* (Cambridge University Press, 1952, pages 129, 157).

Appendix B

The Determination of Approximate Relationships from the Equations of Motion

1. Approximate Equations for the Determination of the Characteristics of the Longitudinal Short Period Oscillation from the Longitudinal Stability Derivatives

The characteristics of the longitudinal oscillatory motion are determined by the equations for X, Z and M. These equations are given on page 128 of Ref. 1; if the controls are fixed, and the aircraft is in level flight and x_q is assumed to be zero, the equations reduce to

(*a*) *X-force Equation*

$$(D - x_u)\bar{u} - x_w\bar{w} + \tfrac{1}{2}C_L\theta = 0 \tag{B1}$$

(*b*) *Z-force Equation*

$$-z_u\bar{u} + (D - z_w)\bar{w} - \left(1 + \frac{z_q}{\mu_1}\right) D\theta = 0 \tag{B2}$$

(*c*) *M-moment Equation*

$$-m_u\bar{u} - \left(m_w + \frac{m_{\dot{w}}}{\mu_1} D\right)\bar{w} + \left(\frac{i_B}{\mu_1} D^2 - \frac{m_q}{\mu_1} D\right)\theta = 0 \tag{B3}$$

where $\bar{u} = \dfrac{u}{V}$, $\bar{w} = \dfrac{w}{V}$, $D = \dfrac{d}{d\tau}$ and $\tau = \dfrac{t}{\hat{t}}$.

For the short period motion, changes in forward speed can be neglected; hence, neglecting the z_u and m_u terms and also the z_q term, we have, replacing $D\theta$ by $\bar{q}$ $\left(\text{i.e. } \bar{q} = \dfrac{d\theta}{d\tau} = q\hat{t}\right)$

Z-force Equation

$$(D - z_w)\bar{w} - \bar{q} = 0 \tag{B4}$$

M-moment Equation

$$-\left(m_w + \frac{m_{\dot{w}}}{\mu_1} D\right)\bar{w} + \left(\frac{i_B}{\mu_1} D - \frac{m_q}{\mu_1}\right)\bar{q} = 0 \tag{B5}$$

From eqn. (B4)

$$\bar{q} = (D - z_w)\bar{w}$$

and substituting this value in eqn. (B5)

$$-\left(m_w + \frac{m_{\dot{w}}}{\mu_1} D\right)\bar{w} + \left(\frac{i_B}{\mu_1} D - \frac{m_q}{\mu_1}\right)(D - z_w)\bar{w} = 0$$

Rearranging this equation, we have

$$\left[\frac{i_B}{\mu_1} D^2 + \left(\frac{-m_{\dot{w}}}{\mu_1} - \frac{m_q}{\mu_1} - \frac{i_B\, z_w}{\mu_1}\right) D + \left(-m_w + \frac{m_q\, z_w}{\mu_1}\right)\right]\bar{w} = 0 \tag{B6}$$

Let this equation be represented by

$$[D^2 + \alpha D + \beta]\,\bar{w} = 0$$

where

$$\alpha = -\frac{\left(\dfrac{m_{\dot{w}}}{\mu_1} + \dfrac{m_q}{\mu_1} + \dfrac{i_B\, z_w}{\mu_1}\right)}{(i_B/\mu_1)}$$

$$\beta = \frac{\left(-m_w + \dfrac{m_q\, z_w}{\mu_1}\right)}{(i_B/\mu_1)}$$

Then, if the solution is exponential in form (i.e. $\bar{w} = \bar{w}_0\, e^{\lambda t}$, where $\lambda = -\mu \pm i\omega$), we have that

$$\lambda^2 + \alpha\lambda + \beta = 0$$

The solution of this quadratic equation is given by

$$\lambda = \tfrac{1}{2}[-\alpha \pm \sqrt{(\alpha^2 - 4\beta)}]$$

or alternatively

$$\lambda = \tfrac{1}{2}[-\alpha \pm i\sqrt{(4\beta - \alpha^2)}]$$

Thus

$$\mu = \tfrac{1}{2}\alpha = \frac{-(m_{\dot{w}} + m_q + i_B z_w)}{2i_B} \tag{B7}$$

and

$$\omega = \tfrac{1}{2}\sqrt{(4\beta - \alpha^2)} = \sqrt{[\beta - (\tfrac{1}{2}\alpha)^2]}$$

$$\omega = \sqrt{\left[\frac{-m_w \mu_1 + m_q z_w}{i_B} - \left(\frac{m_{\dot{w}} + m_q + i_B z_w}{2i_B}\right)^2\right]} \tag{B8}$$

This formula should be compared to the equation for ω given in eqn. 7.7 of Chapter 7, i.e.

$$\omega = \sqrt{\left[\frac{k}{m} - \left(\frac{b}{2m}\right)^2\right]}$$

If the effect of the damping on the frequency, and also the m_q term are neglected, eqn. (B8) reduces to

$$\omega = \sqrt{\left(\frac{-m_w \mu_1}{i_B}\right)} \tag{B9}$$

From the values of μ and ω, the period, the time to half amplitude and the loc. dec. can be determined from the formulae given in Chapter 7. The period is given by

$$T = \frac{2\pi}{\omega}\hat{t} \text{ secs} \tag{B10}$$

and substituting from eqn. (B9) we get

$$T = 2\pi\sqrt{\left(\frac{i_B}{-m_w \mu_1}\right)}\hat{t} \text{ secs} \tag{B11}$$

The time to half amplitude is given by

$$t_{1/2} = \frac{0{\cdot}693}{\mu}\hat{t} \text{ secs} \tag{B12}$$

and substituting from eqn. (B7) we get

$$t_{1/2} = \frac{1{\cdot}386 i_B}{-(m_{\dot{w}} + m_q + i_B z_w)}\hat{t} \text{ secs} = \frac{1{\cdot}386 t}{-\left(\dfrac{m_{\dot{\theta}}}{i_B} + z_w\right)} \text{ secs} \tag{B13}$$

where $m_{\dot{\theta}} = m_{\dot{w}} + m_q$.

The log. dec. δ is given by

$$\delta = \frac{0{\cdot}693T}{t_{1/2}} = \frac{\mu T}{\hat{t}} \tag{B14}$$

Eqn. (B8) can be written in terms of δ as

$$\omega = \sqrt{\left[\frac{(-m_w\mu_1 + m_q\, z_w)}{i_B} - \left(\frac{\delta\hat{t}}{T}\right)^2\right]} \tag{B15}$$

If the values of the longitudinal derivatives are known, then μ and ω, and hence T, $t_{1/2}$ and δ, can be calculated.

2. Approximate Equations for the Determination of the Longitudinal Stability Derivatives from the Characteristics of the Longitudinal Short Period Oscillation

In para. 1 the approximate equations for the period T, the time to half amplitude $t_{1/2}$, and the log. dec. δ, are derived in terms of the longitudinal derivatives. These equations are now used to derive expressions for the derivatives in terms of T, $t_{1/2}$ and δ.

(*a*) m_w

From eqns. (B10) and (B15)

$$\omega^2 = 4\pi^2\left(\frac{\hat{t}}{T}\right)^2 = \frac{(-m_w\mu_1 + m_q\, z_w)}{i_B} - \left(\frac{\delta\hat{t}}{T}\right)^2$$

Rearranging for the m_w term

$$\frac{-m_w\,\mu_1}{i_B} = 4\pi^2\left(\frac{\hat{t}}{T}\right)^2 + \delta^2\left(\frac{\hat{t}}{T}\right)^2 - \frac{m_q\, z_w}{i_B}$$

$$= [4\pi^2 + \delta^2]\left(\frac{\hat{t}}{T}\right)^2 - \frac{m_q\, z_w}{i_B}$$

$$m_w = -\frac{i_B}{\mu_1}[4\pi^2 + \delta^2]\left(\frac{\hat{t}}{T}\right)^2 + \frac{m_q\, z_w}{\mu_1} \tag{B16}$$

The term $\dfrac{m_q\, z_w}{\mu_1}$ is often neglected, since it may be only 3 to 5 per cent of the total (Ref. 2, para. 5.2.2) and hence a further approximation for m_w is given by

$$m_w = -\frac{i_B}{\mu_1}[4\pi^2 + \delta^2]\left(\frac{\hat{t}}{T}\right)^2 \tag{B17}$$

This equation can be written in different forms by substituting for

δ any of the alternatives from eqn. (B14). If the δ^2 term can be neglected, we have the approximation

$$m_w = -4\pi^2 \frac{i_B}{\mu_1}\left(\frac{\hat{t}}{T}\right)^2 = \frac{-i_B}{\mu_1\omega^2} \tag{B18}$$

Note that this is the same as equation (B11) transposed to read for m_w. It neglects the effect of damping and the m_q term on the frequency and period of the oscillation.

(*b*) $m_{\dot\theta} = m_q + m_{\dot w}$

From equation (B7) we have

$$m_q + m_{\dot w} = -\,i_B(2\mu + z_w) = -\,i_B\left(\frac{1{\cdot}386\hat{t}}{t_{1/2}} + z_w\right) \tag{B19}$$

where

$$z_w = -\tfrac{1}{2}\frac{dC_L}{d\alpha} = -\tfrac{1}{2}a$$

(*c*) *Manoeuvre margin* (H_m)

The manoeuvre margin is given by

$$H_m = \frac{m_w}{z_w} + \frac{m_q}{\mu_1} \tag{B20}$$

Substituting from eqn. (B18) we have

$$H_m = -4\pi^2 \frac{i_B}{\mu_1 z_w}\left(\frac{\hat{t}}{T}\right)^2 + \frac{m_q}{\mu_1}$$

and since $\omega^2 = 4\pi^2 (\hat{t}/T)^2$, this can be written

$$H_m = -\frac{i_B}{\mu_1}\frac{1}{z_w}\,\omega^2 + \frac{m_q}{\mu_1}$$

and substituting $-\tfrac{1}{2}a$ for z_w,

$$H_m = \frac{2i_B}{\mu_1 a}\,\omega^2 + \frac{m_q}{\mu_1} \tag{B21}$$

If the value of m_q is known, then the value of H_m can be determined.

3. Approximate Equations for the Determination of the Characteristics of the Dutch Roll Oscillation from the Lateral Stability Derivatives

The characteristics of the lateral oscillatory motion are determined by the equations for Y, L and N. These equations are given on page 156 of Ref. 1; if the controls are fixed and the aircraft is in level flight, the equations reduce to

(*a*) *Y-force Equation*

$$(D - y_v)\bar{v} - \left(\frac{y_p}{\mu_2} D + \frac{C_L}{2}\right)\phi + \left(1 - \frac{y_r}{\mu_2}\right) D\psi = 0 \qquad \text{(B22)}$$

(*b*) *L-moment Equation*

$$-l_v\bar{v} + \left(\frac{i_A}{\mu_2} D^2 - \frac{l_p}{\mu_2} D\right)\phi - \left(\frac{i_E}{\mu_2} D^2 + \frac{l_r}{\mu_2} D\right)\psi = 0 \qquad \text{(B23)}$$

(*c*) *N-moment Equation*

$$-n_v\bar{v} - \left(\frac{i_E}{\mu_2} D^2 + \frac{n_p}{\mu_2} D\right)\phi + \left(\frac{i_C}{\mu_2} D^2 - \frac{n_r}{\mu_2} D\right)\psi = 0 \qquad \text{(B24)}$$

where $\bar{v} = \dfrac{v}{V}$ and $D = \dfrac{d}{d\tau}$.

In Reference 3 the approximate equations are obtained by simplifying the original equations of motion. The Time-vector representation is used, and from the vector diagram corresponding to the equations of motion the appropriate approximations can be determined. The yawing moment equation can be approximated by

$$-n_v\bar{v} - \frac{i_E}{\mu_2} D^2\phi + \frac{i_C}{\mu_2} D^2\psi = 0$$

or since $\bar{v} = \dfrac{v}{V} = \beta$ rads

$$\ddot{\psi} = \frac{\mu_2 n_v}{i_C}\beta + \frac{i_E}{i_C}\ddot{\phi} \qquad \text{(B25)}$$

and the rolling moment equation by

$$-l_v\bar{v} + \frac{i_A}{\mu_2} D^2\phi = 0$$

or

$$\ddot{\phi} = \frac{\mu_2 l_v}{i_A}\beta \qquad \text{(B26)}$$

The last approximation neglects the small contribution of $(i_E/i_A)\ddot{\psi}$ to the rolling moment; this is permissible since eqn. (B26) is used to substitute for $\ddot{\phi}$ in eqn. (B25). In fact, the term $(i_E/i_C)\ddot{\phi}$ in eqn. (B25) could be neglected and eqn. (B30) derived much more easily.

Substituting from eqn. (B26) in eqn. (B25) we have

$$\ddot{\psi} = \frac{\mu_2 n_v}{i_C}\beta + \frac{i_E \mu_2 l_v}{i_A i_C}\beta$$

From para. 3 of Chapter 17

$$\ddot{\psi} = -\omega_n^2 \psi$$

and since $-\psi = \beta$, we have

$$\omega_n^2 = \frac{\mu_2 n_v}{i_C} + \frac{i_E \mu_2 l_v}{i_A i_C}$$

$$\therefore \quad \omega_n = \sqrt{\left[\frac{\mu_2 n_v}{i_C}\left(1 + \frac{i_E l_v}{i_A n_v}\right)\right]} \qquad \text{(B27)}$$

where ω_n is the non-dimensional natural (undamped) frequency parameter [i.e. $\omega_n^2 = \omega^2 + \mu^2$ where ω is the actual (damped) non-dimensional frequency and μ the damping parameter]. The corresponding undamped periodic time is given by

$$T = \frac{2\pi}{\omega_n}\hat{t}$$

and hence

$$T = \frac{2\pi\hat{t}}{\sqrt{\left[\frac{\mu_2 n_v}{i_C}\left(1 + \frac{i_E l_v}{i_A n_v}\right)\right]}} \text{ secs} \qquad \text{(B28)}$$

or, since $\hat{t} = \mu_2 s/V$, where s is the standard length for the lateral equations of motion

$$T = \frac{2\pi s}{V}\sqrt{\left[\frac{\mu_2 i_C/n_v}{(1 + i_E l_v/i_A n_v)}\right]} \text{ secs} \qquad \text{(B29)}$$

When i_E/i_A is small, eqn. (B28) can be approximated further to

$$T = 2\pi\sqrt{\left(\frac{i_C}{\mu_2 n_v}\right)}\,\hat{t} \text{ secs} \qquad \text{(B30)}$$

As noted earlier, this equation could be obtained more directly if the $(i_E/i_C)\ddot{\phi}$ term is neglected initially in eqn. (B25).

To determine the roll/yaw ratio (p/r) we have that, since $\ddot{\psi} = -\omega_n^2\psi$, $\dot{\psi} = i\omega_n\psi$, $\ddot{\phi} = -\omega_n^2\phi$ and $\dot{\phi} = i\omega_n\phi$,

$$\frac{p}{r} = \frac{\dot{\phi}}{\dot{\psi}} = \frac{\phi}{\psi} = \frac{\ddot{\phi}}{\ddot{\psi}}$$

and substituting from eqns. (B25) and (B26) we have

$$\frac{p}{r} = \left(\frac{\mu_2 l_v}{i_A}\beta\right)\bigg/\left(\frac{\mu_2 n_v}{i_C}\beta + \frac{i_E\,\mu_2 l_v}{i_C i_A}\beta\right)$$

$$\therefore\ \frac{p}{r} = \left(\frac{i_C l_v}{i_A n_v}\right)\bigg/\left(1 + \frac{i_E l_v}{i_A n_v}\right) \tag{B31}$$

or alternatively,

$$\frac{p}{r} = \left(\frac{i_C}{i_E}\right)\bigg/\left(1 + \frac{i_A n_v}{i_E l_v}\right) \tag{B32}$$

In Ref. 4, the roll/yaw ratio is given by

$$\frac{p}{r} = \left(\frac{i_C l_v}{i_A n_v} + \frac{i_E}{i_A}\right)\bigg/\left(1 + \frac{i_E l_v}{i_A n_v}\right) \tag{B33}$$

The presence of the i_E/i_A term in the numerator of eqn. (B33) results from the $(i_E/i_A)\ddot{\psi}$ term in the rolling moment equation, which is neglected in eqn. (B26).

An alternative equation for (p/r) is given in Ref. 5; from the vector diagram for the rolling moment equation for the particular aircraft concerned (Avro 707B), it is found that a better approximation than eqn. (B26) is given by

$$\left.\begin{aligned} -l_v\bar{v} - \frac{l_p}{\mu_2}D\phi = 0 \\ \text{or} \qquad \\ \dot{\phi} = -\frac{\mu_2 l_v}{l_p}\beta \end{aligned}\right\} \tag{B34}$$

From this the roll/yaw ratio can be determined as

$$\frac{p}{r} = \frac{l_v}{l_p}\sqrt{\left(\frac{\mu_2 i_C}{n_v}\right)} \tag{B35}$$

The damping of the oscillation is related to the damping angle ε_D. In Ref. 3 it is shown that, for low values of C_L and i_E/i_A,

$$2\varepsilon_D = \left[\sqrt{\left(\frac{i_C}{\mu_2 n_v}\right)}\right]\left(-\frac{n_r}{i_C} - \frac{n_p}{i_C}\frac{p}{r} - y_v\right) \tag{B36}$$

When i_E/i_A is small, we have from eqn. (B31)

$$\frac{p}{r} = \frac{i_C l_v}{i_A n_v} \tag{B37}$$

and substituting for p/r in eqn. (B36) gives

$$2\varepsilon_D = \left[\sqrt{\left(\frac{i_C}{\mu_2 n_v}\right)}\right]\left(-\frac{n_r}{i_C} - \frac{n_p l_v}{i_A n_v} - y_v\right) \tag{B38}$$

The log. dec. δ is given by

$$\delta = 2\pi \tan \varepsilon_D \simeq 2\pi\varepsilon_D$$

$$\therefore \quad \delta = \pi\left[\sqrt{\left(\frac{i_C}{\mu_2 n_v}\right)}\right]\left(-\frac{n_r}{i_C} - \frac{n_p l_v}{i_A n_v} - y_v\right) \tag{B39}$$

The approximate equation given for δ in Ref. 5, is identical with eqn. (B39), except that the $\frac{n_p l_v}{i_A n_v}$ term is omitted.

4. Approximate Equations for the Determination of the Lateral Stability Derivatives from the Characteristics of the Dutch Roll Oscillation

In para. 3 the approximate equations for the period, the log. dec. and the roll/yaw ratio, are derived in terms of the lateral derivatives. These equations are now used to derive expressions for the derivatives in terms of T, δ and p/r.

(*a*) n_v

From eqn. (B30) we have, transposing for n_v

$$n_v = 4\pi^2 \frac{i_C}{\mu_2}\left(\frac{\hat{t}}{T}\right)^2 \tag{B40}$$

A more accurate equation can be derived from eqn. (B28), but the value of the derivative l_v needs to be known

$$n_v = 4\pi^2 \frac{i_C}{\mu_2}\left(\frac{\hat{t}}{T}\right)^2 - \frac{i_E}{i_A} l_v \tag{B41}$$

Eqn. (B40) can be written as

$$n_v = \frac{i_C}{\mu_2}\,\omega_n^2 \tag{B42}$$

where ω_n is the undamped non-dimensional frequency. In terms of ω, the damped non-dimensional frequency, we can write eqn. (B42) as

$$n_v = \frac{i_C}{\mu_2}(\omega^2 + \mu^2) \tag{B43}$$

Since $\omega = \dfrac{2\pi t}{T}$ and $\mu = \dfrac{0{\cdot}693\hat{t}}{t_{1/2}} = \dfrac{\delta\hat{t}}{T}$, eqn. (B43) can be expressed in the form

$$n_v = \frac{i_C}{\mu_2}\left[4\pi^2\left(\frac{\hat{t}}{T}\right)^2 + \left(\frac{0{\cdot}693\hat{t}}{t_{1/2}}\right)^2\right]$$

or alternatively as

$$n_v = \frac{i_C}{\mu_2}[4\pi^2 + \delta^2]\left(\frac{\hat{t}}{T}\right)^2 \qquad \text{(B44)}$$

(*b*) n_r

From eqn. (B39) we have, neglecting the $\dfrac{n_p l_v}{i_A n_v}$ term and transposing for n_r

$$n_r = -i_C\left[\frac{\delta}{\pi}\sqrt{\left(\frac{\mu_2 n_v}{i_C}\right)} + y_v\right] \qquad \text{(B45)}$$

Substituting for n_v from eqn. (B40) we have

$$n_r = -i_C\left[\frac{2\delta t}{T} + y_v\right]$$

or

$$n_r = -i_C[2\mu + y_v] \qquad \text{(B46)}$$

(*c*) l_v

In Ref. 6 the approximate equation for l_v is given as

$$l_v = -\frac{i_A}{\mu_2}\omega^2\frac{\left(\dfrac{p}{r} - \dfrac{i_E}{i_A}\right)}{1 + \dfrac{i_E i_E}{i_A i_C}} \qquad \text{(B47)}$$

References

1. Duncan, W. J.; *Control and Stability of Aircraft* (Cambridge University Press, 1952).
2. Wolowicz, C. H. and Hollemann, E. C.; *Stability-Derivative Determination from Flight Data*, Agard Report No. 224, 1958.

3. *Lateral Stability of the F/23/49*, English Electric Aero T.N. Ae. 76.
4. Thomas, H. H. B. M. and Neumark, S.; *Interim Note on the Stability and Response Characteristics of Supersonic Aircraft*, R.A.E. T.N. Aero 2412, 1955.
5. Perry, D. H., Morrall, J. C. and Port, W. G. A.; *Low Speed Flight Tests on a Tailless Delta Aircraft with Wings Swept Back* 44·5° (*Avro* 707*B*), R.A.E. Rep. Aero 2638, 1960.
6. Rose, R.; *Flight Measurements of the Dutch-Roll Characteristics of a* 60° *Delta Wing Aircraft* (*Fairey Delta* 2) *at Mach Numbers from* 0·4 *to* 1·5 *with Stability Derivatives Extracted by Vector Analysis*, R.A.E. T.N. Aero 2752, C.P. 653, 1961.

Appendix C
Typical Solutions of Longitudinal and Lateral Stability Quartic Equations

1. Longitudinal Stability Quartic

If the longitudinal motion of an aircraft is represented by the quartic equation of the form

$$\lambda^4 + 5\lambda^3 + 7\lambda^2 + 0{\cdot}5\lambda + 0{\cdot}2 = 0$$

it can be resolved approximately into two quadratic equations each of which represents an oscillatory motion

$$(\lambda^2 + 4{\cdot}948\lambda + 6{\cdot}7078)(\lambda^2 + 0{\cdot}0526\lambda + 0{\cdot}0299) = 0$$

Solving each of these quadratics in turn, we have

(i) $\underline{\lambda^2 + 4{\cdot}948\lambda + 6{\cdot}7078 = 0}$

The solution is

$$\lambda = -2{\cdot}474 \pm 0{\cdot}766i$$

and it represents an oscillation with a periodic time given by

$$T = \frac{2\pi}{\omega}\hat{t} = \frac{2\pi}{0{\cdot}766}\hat{t} \text{ secs}$$

and a time to half amplitude of

$$t_{1/2} = \frac{0{\cdot}693}{\mu}\hat{t} = \frac{0{\cdot}693}{2{\cdot}474}\hat{t} \text{ secs}$$

Assuming a value of 2 secs for $\hat{t}$, these values become

$$T = 16{\cdot}4 \text{ secs and } t_{1/2} = 0{\cdot}56 \text{ secs}$$

Thus the oscillation (i.e. the short period oscillation) is a very heavily damped one with a fairly short periodic time; it is almost critically damped, the relative damping ζ being about 0·995.

(ii) $\lambda^2 + 0{\cdot}0526\lambda + 0{\cdot}0299 = 0$

The solution is

$$\lambda = -0{\cdot}0263 \pm 0{\cdot}171i$$

The periodic time is given by

$$T = \frac{2\pi}{0{\cdot}171}\,\hat{t} \text{ secs}$$

and the time to half amplitude by

$$t_{1/2} = \frac{0{\cdot}693}{0{\cdot}0263}\,\hat{t} \text{ secs}$$

Assuming $\hat{t} = 2$ secs

$$T = 73{\cdot}5 \text{ secs and } t_{1/2} = 52{\cdot}7 \text{ secs}$$

Thus the oscillation (i.e. the Phugoid) is one with a relatively long periodic time and a smaller amount of damping ($\zeta = 0{\cdot}152$).

2. Lateral Stability Quartic

If the lateral quartic is of the form

$$\lambda^4 + 5{\cdot}63\lambda^3 + 21{\cdot}4\lambda^2 + 87{\cdot}35\lambda + 0{\cdot}113 = 0$$

it can be resolved approximately into the following factors

$$(\lambda + 4{\cdot}90)(\lambda + 0{\cdot}00129)(\lambda^2 + 0{\cdot}73\lambda + 17{\cdot}82) = 0$$

The two factors with real roots represent convergent motions and the quadratic represents an oscillatory mode—

(i) $\lambda + 4{\cdot}90 = 0$

The solution is

$$\lambda = -4{\cdot}9$$

and it represents a very heavily convergent mode (i.e. the roll subsidence mode) with a time to half amplitude given (assuming $\hat{t} = 2$ secs) by

$$t_{1/2} = \frac{0{\cdot}693 \times 2}{4{\cdot}9} = 0{\cdot}28 \text{ secs}$$

(ii) $\underline{\lambda + 0{\cdot}00129 = 0}$

The solution is

$$\lambda = -0{\cdot}00129$$

and it represents a mode which is just stable (i.e. the spiral mode) with a time to half amplitude of

$$t_{1/2} = \frac{0{\cdot}693 \times 2}{0{\cdot}00129} = 1{,}074 \text{ secs}$$

(iii) $\underline{\lambda^2 + 0{\cdot}73\lambda + 17{\cdot}82 = 0}$

The solution is

$$\lambda = -0{\cdot}365 \pm 4{\cdot}205i$$

The periodic time of the Dutch Roll oscillation is given by

$$T = \frac{2\pi}{4{\cdot}205}\,\hat{t} = 2{\cdot}99 \text{ secs}$$

and the time to half amplitude by

$$t_{1/2} = \frac{0{\cdot}693}{0{\cdot}365}\,\hat{t} = 3{\cdot}80 \text{ secs}$$

Appendix D

Determination of the Transfer Functions of Some Simple Electrical Circuits

1. Circuit Containing Pure Resistances

In an electrical circuit containing pure resistances R_1 and R_2 (Fig. 19.3), the current i, corresponding to an input voltage e_i, is given by

$$i = \frac{e_i}{R_1 + R_2} \tag{D1}$$

The output voltage e_o across the resistance R_2 is then given by

$$e_o = i \times R_2 = \frac{e_i R_2}{R_1 + R_2} \tag{D2}$$

and this is the equation relating the input and output voltages of this circuit.

The transfer function of the circuit can then be expressed in the form of

$$\frac{e_o}{e_i} = \frac{R_2}{R_1 + R_2} \tag{D3}$$

This, however, is a very simple circuit, and the eqn. (D2) relating e_o and e_i does not contain any terms involving differentiation.

2. Circuit Containing Resistance and Capacitance

Fig. 19.4 shows a circuit containing both resistance and capacitance. The current i in the circuit is given in terms of the resistance R by

$$i = \frac{e_R}{R} \tag{D4}$$

and in terms of the capacitance C by

$$i = C\frac{de_o}{dt} \tag{D5}$$

where de_o/dt is the rate at which the voltage across the capacitance varies with time.

Considering the voltages round the circuit we have

$$e_i = e_R + e_o \tag{D6}$$

From eqn. (D4) $e_R = iR$, and from eqn. (D5) $\frac{de_o}{dt} = \frac{i}{C}$, hence

$$e_R = CR\frac{de_o}{dt} \tag{D7}$$

Substituting from eqn. (D7) into (D6)

$$\underline{\underline{e_i = CR\frac{de_o}{dt} + e_o}} \tag{D8}$$

and this differential equation represents the relation between the input and output voltages.

If de_o/dt is represented by pe_o, then eqn. (D8) can be written

$$e_i = CRpe_o + e_o \tag{D9}$$

The transfer function of the circuit is then given by

$$\underline{\underline{\frac{e_o}{e_i} = \frac{1}{1 + CRp}}} \tag{D10}$$

3. Circuit Containing Resistance, Capacitance and Inductance

In the circuit shown in Fig. 19.5, the capacitance is represented by C, the inductance by L, and the resistance by R.

The current in the circuit can be represented in terms of the capacitance C by the equation

$$i = C\frac{de_o}{dt} \tag{D11}$$

where de_o/dt is the rate at which the voltage across the capacitance varies with time. The voltage across the resistance is given by

$$e_R = iR = CR\frac{de_o}{dt} \tag{D12}$$

and across the inductance by

$$e_L = L\frac{di}{dt} = CL\frac{d^2e_o}{dt^2} \tag{D13}$$

But

$$e_i = e_o + e_L + e_R \tag{D14}$$

hence

$$e_i = e_o + CL\frac{d^2e_o}{dt^2} + CR\frac{de_o}{dt} \tag{D15}$$

and rearranging

$$e_i = e_o(1 + CRp + CLp^2) \tag{D16}$$

The transfer function of this circuit is thus given by

$$\frac{e_o}{e_i} = \frac{1}{1 + CRp + CLp^2} \tag{D17}$$

Appendix E

A More Detailed Theory of the Problem of Inertia Coupling

1. Introduction

A theoretical analysis of inertia coupling is an extremely complicated problem. However, by making a number of assumptions, the equations involved can be simplified and the criterion for stability in rolling manoeuvres established.

The equations of motion of an aircraft can be written in a number of different forms, depending on the particular axes system selected for the analysis. Most stability work is carried out with reference to the "wind axes"—a set of axes fixed in the body, with the $0x$ axis pointing in the direction of the relative airflow. For the inertia coupling analysis it is more convenient to use the principal inertia axes system. In this case, all the terms containing the products of inertia $D\ (=\int yz\,dm)$, $E(=\int xz\,dm)$ and $F\ (=\int xy\,dm)$ are eliminated, since $D = E = F = 0$.

2. Equations of Motion with Respect to Principal Inertia Axes System

The motion of an aircraft can be represented by six equations of motion, one each for the three forces X, Y and Z and one each for the three moments L, M and N. These equations can be expressed in the form (Ref. 1)

$$X = mV\left(\frac{\dot{u}}{V} - r\beta + q\alpha\right) \tag{E1}$$

$$Y = mV(\dot{\beta} - p\alpha + r) \tag{E2}$$

$$Z = mV(\dot{\alpha} - q + p\beta) \tag{E3}$$

$$A\dot{p} = L + (B - C)qr \tag{E4}$$

$$B\dot{q} = M + (C - A)rp \tag{E5}$$

$$C\dot{r} = N + (A - B)pq \tag{E6}$$

where V is the forward speed of the aircraft. The forces X, Y and Z on the left hand side of eqns. (E1) to (E3), and the moments L, M and N on the right hand side of eqns. (E4) to (E6), consists of aerodynamic components (including the effect of any control deflections), gravity components and engine gyroscopic couple components. Initially, only the aerodynamic effects will be considered; gravity effects can generally be neglected in fast rolling manoeuvres and the engine effects can be allowed for separately (see Chapter 16, para 7.3.2).

The aerodynamic forces and moments can be expressed in terms of the stability and control derivatives (N.B. measured with respect to the principal inertia system of axes). Strictly, it is necessary to include the effects of the various cross derivatives (e.g. N_p), but these derivatives would introduce non-linear terms in the equations and are usually neglected (see below). The aerodynamic forces Y and Z and the moments L, M and N can be expressed in the form

$$Y = Y_v v + Y_p p + Y_r r \tag{E7}$$

$$Z = Z_w w + Z_q q \tag{E8}$$

$$L = L_v v + L_p p + L_r r + L_\xi \xi + L_\zeta \zeta \tag{E9}$$

$$M = M_w w + M_{\dot{w}} \dot{w} + M_q q + M_\eta \eta \tag{E10}$$

$$N = N_v v + N_p p + N_r r + N_\xi \xi + N_\zeta \zeta \tag{E11}$$

In eqns. (E1) to (E6), a number of terms occur which contain the product of two disturbances. In stability theory, these disturbances (e.g. v, p, r, etc.) are usually assumed to be small and any products are neglected. This allows the set of six simultaneous equations to be separated into two independent (uncoupled) groups, each containing three equations of motion. One group of equations (that of the X, Y and M equations) represents the longitudinal motion, and the other group (that of the Z, L and N equations) represents the lateral motion. In order to study the possible effects of coupling which may occur at high rates of roll, it is necessary to retain all those terms containing p. Thus it is necessary to retain the terms containing $p\alpha$, $p\beta$, pr and pq, but those involving $r\beta$, $q\alpha$ and qr can still be neglected.

It is obvious that the term $(C - A)rp$, which is retained in the pitching moment M equation, is the one that leads to pitch divergence, whereas the $(A - B)pq$ term in the yawing moment N

equation leads to yaw divergence. The physical interpretation of these terms is discussed in Chapter 16, para. 7.2.2. These inertia moments will be large if $(C - A)$ and $(A - B)$ [or $(B - A)$] are large (i.e. if the aircraft has a large pitching and yawing inertia and a small rolling inertia).

In order to study the elevator and rudder fixed condition of flight, the control terms $L_\zeta\zeta$, $M_\eta\eta$ and $N_\zeta\zeta$ in eqns. (E9) to (E11) are put equal to zero. Substituting eqns. (E7) to (E11) (with $\zeta = \eta = 0$) into eqns. (E1) to (E6) we get

$$X = m\dot{u} \tag{E12}$$

$$Y_v v + Y_p p + Y_r r = mV(\dot{\beta} - p\alpha + r) \tag{E13}$$

$$Z_w w + Z_q q = mV(\dot{\alpha} - q + p\beta) \tag{E14}$$

$$A\dot{p} = L_v v + L_p p + L_r r + L_\xi \xi \tag{E15}$$

$$B\dot{q} = M_w w + M_{\dot{w}} \dot{w} + M_q q + (C - A)rp \tag{E16}$$

$$C\dot{r} = N_v v + N_p p + N_r r + N_\xi \xi + (A - B)pq \tag{E17}$$

The non-linear equations can be further simplified and linearized by

(i) Assuming that the forward speed V is constant. Since $\dot{u}$ is then zero, this eliminates the X-force equation.
(ii) Assuming that many of the derivatives (e.g. Y_p, Y_r, Z_q, $M_{\dot{w}}$, N_ξ and N_p) can be neglected.
(iii) Assuming a steady value p_o for the rate of roll. Since $\dot{p} = 0$, eqn. (E15) is redundant for our analysis, though it can be used to determine the necessary amount of aileron deflection, if required.

Using these assumptions, the equations of motion can be written

$$Y_v v = mV(\dot{\beta} - p_o\alpha + r) \tag{E18}$$

$$Z_w w = mV(\dot{\alpha} - q + p_o\beta) \tag{E19}$$

$$B\dot{q} = M_w w + M_q q + (C - A)rp_o \tag{E20}$$

$$C\dot{r} = N_v v + N_r r + (A - B)p_o q \tag{E21}$$

where $\alpha = \alpha_o + \Delta\alpha$, α_o being the initial trimmed incidence of the principal inertia axis, and $\Delta\alpha$ the corresponding disturbance.

These four equations constitute a set of linear differential equations with constant coefficients. The control-fixed stability of the aircraft can be considered in the usual way by assuming small exponential disturbances of the form $q = q_o\,e^{\lambda t}$, $r = r_o\,e^{\lambda t}$, etc. The resulting characteristic equation has the form of a quartic equation in λ given by

$$\lambda^4 + B_1\lambda^3 + C_1\lambda^2 + D_1\lambda + E_1 = 0 \quad \text{(E22)}$$

where the coefficients B_1, C_1, D_1 and E_1 are quantities which are dependent on the values of the stability derivatives and the aircraft inertias involved in eqns. (E18) to (E21). This quartic equation represents two oscillatory modes; as p_o approaches to zero, these modes approach to the normal uncoupled longitudinal short period and lateral oscillations (Ref. 2, page 552).

In this Appendix, the equations of motion have been retained in dimensional form. They can, if required, be made non-dimensional (Ref. 2, page 552).

References

1. PINSKER, W. J. G.; *Preliminary Note on the Effect of Inertia Cross-coupling on Aircraft Response in Rolling Manoeuvres*, R.A.E. Tech. Note Aero. 2419, 1955 (A.R.C., C.P.435).
2. BABISTER, A. W.; *Aircraft Stability and Control* (Pergamon Press, Oxford, 1961).

Appendix F

Alternative Forms of the General and Simplified Equations of Motion

1. General Equations of Motion

The equations of motion applicable to an aircraft under different conditions of flight can be expressed in various ways, depending on the notation and definitions used, and the simplifying assumptions made. The longitudinal and lateral motions are usually considered separately, and the equations are then simplified further in the analysis of particular aircraft modes. The equations are usually written in terms of the non-dimensional stability derivatives, and they can be found in this form in Appendix B (also on p. 133 of Ref. 1). By applying the Laplace transform and the transfer function concept the equations can be written in the alternative form shown on page 136 of Ref. 1.

Using the Laplace transform notation the general longitudinal equations of motion can be written (making the usual permissible assumptions) in a dimensional form as

$$\left.\begin{aligned} (s - X_u)u(s) - X_w w(s) \qquad\qquad + (g - X_q s)\theta(s) &= X_\delta \delta(s) \\ - Z_u u(s) + (s - Z_w)w(s) \qquad - (Z_q + U_0)s\theta(s) &= Z_\delta \delta(s) \\ - M_u u(s) - (M_{\dot{w}} s + M_w)w(s) + (s^2 - M_q s)\theta(s) &= M_\delta \delta(s) \end{aligned}\right\} \quad \text{(F1)}$$

The equations can be found in this form on page 7 of Ref. 2; those on page 32 of Ref. 3 are similar but non-dimensional derivatives and a different notation are employed, and no control terms are included. Further simplifications can often be made, and in Ref. 4 (pg. III-7) equations (F1) are written without the X_q, X_δ and Z_q terms (i.e. it is assumed that $X_q = X_\delta = Z_q = 0$).

Care must be taken over the definitions of the quantities involved in eqns. (F1). The dimensional derivatives are often defined (Ref. 7,

Notation, p. 408) to include the mass of the aircraft or the appropriate inertia, e.g.

$$\left.\begin{aligned} X_u &= \frac{1}{m}\cdot\frac{\partial X}{\partial u} \\ N_v &= \frac{1}{I_z}\cdot\frac{\partial N}{\partial v} \\ M_q &= \frac{1}{I_y}\cdot\frac{\partial M}{\partial q}\ \text{etc.} \end{aligned}\right\} \tag{F2}$$

Because these dimensional derivatives are defined in this way they will have units, X_u, for instance, will have to be expressed in units of $\sec^{-1}$, N_v in $(\text{ft.sec})^{-1}$, and M_q in $\sec^{-1}$ and so on (Ref. 7, p. 461 gives some actual numerical values of the derivatives for the Sikorsky S-58 helicopter). The terms on the right-hand side of eqns. (F1) allow for the input from a control or an atmospheric gust. In the case of a conventional aircraft it is assumed that the aileron and rudder controls do not produce any effect in the longitudinal equations, and these terms will represent the effect of elevator deflection. In the case of helicopters and other VTOL aircraft, the longitudinal control may not be the conventional elevator, and these terms will have to include the effects of rotors, air jets, etc., depending on the method of pitch control used.

The corresponding general lateral equations can be found on page 13 of Ref. 2. Making various permissible assumptions they can be written

$$\left.\begin{aligned} &(s - Y_v)v(s) - (Y_p s + g)\phi(s) && + (U_0 - Y_r)r && = Y_\delta\delta(s) \\ &- L_v v(s) + (s^2 - L_p s)\phi(s) && - \left(\frac{I_{xz}}{I_x} s + L_r\right) r && = L_\delta\delta(s) \\ &- N_v v(s) - \left(\frac{I_{xz}}{I_z} s + N_p\right) s\phi(s) && + (s - N_r)r && = N_\delta\delta(s) \end{aligned}\right\} \tag{F3}$$

I_{xz} (usually E in British notation) is the product of inertia. If this is assumed small and neglected, or if the definitions of the dimensional derivatives are modified (Ref. 2, p. 17), then the terms containing I_{xz} in the lateral equations are eliminated. The lateral equations are often further simplified by assuming that $Y_p = Y_r = Y_\delta = 0$. The equations thus reduce to

$$\left.\begin{aligned} &(s - Y_v)v(s) - g\phi(s) && + U_0 r && = 0 \\ &- L_v v(s) + (s^2 - L_p s)\phi(s) - L_r r && && = L_\delta\delta(s) \\ &- N_v v(s) - N_p s\phi(s) && + (s - N_r)r && = N_\delta\delta(s) \end{aligned}\right\} \tag{F4}$$

2. Simplified Equations of Motion for Conventional Aircraft

As discussed in Appendix B, it is possible to simplify the general equations of motion in the discussion of certain problems, by assuming that the motion involves only one or two of the u, w and θ (or v, ϕ, and r) degrees of freedom instead of three.

(a) Equations for the Longitudinal Short Period Motion
By letting $u = 0$ and eliminating the first of equations (F1) we obtain (assuming $Z_q = 0$)

$$\left.\begin{aligned} (s - Z_w)w(s) \qquad - sU_0\theta(s) &= Z_\delta\delta(s) \\ -(M_{\dot{w}}s + M_w)w(s) + (s^2 - M_q s)\theta(s) &= M_\delta\delta(s) \end{aligned}\right\} \tag{F5}$$

Similar equations can be found on page 40 of Ref. 3. These different forms of the approximate short period equations should be compared with those derived in Appendix B.

(b) Equations for the Longitudinal Phugoid Motion
By letting $w = 0$ in eqns. (F1) we obtain the phugoid approximation (assuming $X_\delta = X_q = Z_q = 0$),

$$\left.\begin{aligned} (s - X_u)u(s) + g\theta(s) \qquad &= 0 \\ -Z_u u(s) \quad - sU_0\theta(s) \qquad &= Z_\delta\delta(s) \\ -M_u u(s) + (s^2 - M_q s)\theta(s) &= M_\delta\delta(s) \end{aligned}\right\} \tag{F6}$$

Similar equations can be found on page 44 of Ref. 3, though in this case the pitching moment equation is even further simplified by neglecting both the M_u and the inertia terms.

(c) Longitudinal Equations for a VTOL Aircraft at the Hover
In the case of an aircraft at the hover, the values of M_w, $M_{\dot{w}}$ and X_w are usually negligible due to considerations of symmetry, and, retaining the X_δ term, the equations for longitudinal motion become (still assuming that $X_q = Z_q = 0$)

$$\left.\begin{aligned} (s - X_u)u(s) + g\theta(s) \qquad &= X_\delta\delta(s) \\ -Z_u u(s) \ + (s - Z_w)w(s) \quad &= Z_\delta\delta(s) \\ -M_u u(s) + (s^2 - sM_q)\theta(s) &= M_\delta\delta(s) \end{aligned}\right\} \tag{F7}$$

(d) Lateral Equations for a VTOL Aircraft at the Hover
The corresponding lateral hover equations can be written, assuming that I_{xz}, N_p, Y_r, Y_p and L_r are all negligible in this condition of flight, and retaining the Y_δ term (Ref. 2, p. 14),

$$\left.\begin{aligned} (s - Y_v)v(s) - g\phi(s) \qquad &= Y_\delta\delta(s) \\ -L_v v(s) \ + (s^2 - L_p s)\phi(s) &= L_\delta\delta(s) \\ -N_v v(s) \ + (s - N_r)r(s) \quad &= N_\delta\delta(s) \end{aligned}\right\} \tag{F8}$$

3. Aircraft Transfer Functions

3.1. Transfer Functions

Transfer functions were discussed in para. 3.2 of Chapter 19, where they were expressed in terms of the operator $p(=\frac{d}{dt})$. As we have seen in this appendix, the equations of motion of an aircraft are often written using the Laplace transform notation. The general longitudinal motion of an aircraft, for example, is represented by eqn. (F1).

The transfer function concept is widely used in present-day stability and control analyses (Refs. 1–4, 6). These analyses may be concerned with either

(*a*) The open-loop response of the aircraft to a given (control or atmospheric) disturbance, or

(*b*) The closed-loop response of the aircraft to a given disturbance. The closure of the loop may be achieved either by means of a human pilot or by an automatic pilot or stabilization system.

The characteristics of open-loop and closed-loop systems have been discussed in paras. 2–6 of Chapter 19, and aircraft automatic control systems in para. 7 of Chapter 19.

The equations of motion are often written in determinant notation; eqn. (F1), for example, can be expressed in the form

$$\begin{bmatrix} s - X_u & -X_w & g - X_q s \\ -Z_u & s - Z_w & -(Z_q + U_0)s \\ -M_u & -(M_{\dot{w}}s + M_w) & s^2 - M_q s \end{bmatrix} \begin{bmatrix} u \\ w \\ \theta \end{bmatrix} = \begin{bmatrix} X_\delta \\ Z_\delta \\ M_\delta \end{bmatrix} \delta \tag{F9}$$

These particular equations represent the variation in the longitudinal characteristics of the motion of the aircraft (u, w, and θ) in response to a control deflection (δ). The individual transfer functions can be written

$$\left.\begin{aligned} G_{u\delta} &= \frac{u(s)}{\delta(s)} \\ G_{w\delta} &= \frac{w(s)}{\delta(s)} \\ G_{\theta\delta} &= \frac{\theta(s)}{\delta(s)} \end{aligned}\right\} \tag{F10}$$

and these transfer functions represent the speed u response, the speed w response, and the angle θ response, respectively, to a control deflection δ.

3.2. Aircraft Open-loop Response

The basic longitudinal dynamic stability characteristics of an aircraft can be obtained from eqn. (F9) by putting all the control input terms equal to zero. If the determinant on the left hand side is expanded and equated to zero, the characteristic equation of the motion is obtained:

$$A_1 s^4 + B_1 s^3 + C_1 s^2 + D_1 s + E_1 = 0 \tag{F11}$$

where $A_1 = 1{\cdot}0$.

This process is similar to that carried out in conventional stability and control analysis (Ref. 1, p. 193), except that s, the Laplace transform variable or Laplace operator, replaces λ (compare eqn. (8.1) of Chapter 8). Expressions can be obtained, as in Ref. 1, for the coefficients B_1, C_1, D_1, and E_1 in terms of the aerodynamic derivatives. The solutions of the characteristic equation (F11) represent the modes of the open-loop response of the aircraft (Chapter 8, para. 2.1).

The open-loop response of an aircraft to a control input can be expressed in terms of the transfer functions given in eqn. (F10). In determinant notation these are

$$G_{u\delta} = \frac{u(s)}{\delta(s)} = \begin{vmatrix} X_\delta & -X_w & g - X_q s \\ Z_\delta & s - Z_w & -(Z_q + U_0)s \\ M_\delta & -(M_{\dot{w}}s + M_w) & s^2 - M_q s \end{vmatrix} \div \Delta \tag{F12}$$

$$G_{w\delta} = \frac{w(s)}{\delta(s)} = \begin{vmatrix} s - X_u & X_\delta & g - X_q s \\ -Z_u & Z_\delta & -(Z_q + U_0)s \\ -M_u & M_\delta & s^2 - M_q s \end{vmatrix} \div \Delta \tag{F13}$$

$$G_{\theta\delta} = \frac{\theta(s)}{\delta(s)} = \begin{vmatrix} s - X_u & -X_w & X_\delta \\ -Z_u & s - Z_w & Z_\delta \\ -M_u & -(M_{\dot{w}}s + M_w) & M_\delta \end{vmatrix} \div \Delta \tag{F14}$$

where the symbol Δ, the common denominator on the right-hand side of these three equations, represents the left-hand side of the characteristic equation (F11), i.e.

$$\Delta = s^4 + B_1 s^3 + C_1 s^2 + D_1 s + E \tag{F15}$$

The determinants in the numerators of these three equations can also be expanded, and the response characteristics of the system can be represented using Bode plots (Ref. 4, III-3) or root-locus plots.

All the other various forms of the equations of motion (e.g. eqns. (F3–8)) can be expressed in determinant form, and the corresponding transfer functions obtained. The approximation for the phugoid motion (eqn. (F6)), for example, can be further simplified (Ref. 4, p. III–11) and written

$$\begin{bmatrix} s - X_u & g \\ -Z_u & -sU_0 \end{bmatrix} \begin{bmatrix} u \\ \theta \end{bmatrix} = \begin{bmatrix} 0 \\ Z_\delta \end{bmatrix} \delta \tag{F16}$$

The transfer function $u(s)/\delta(s)$ becomes in this case

$$G_{u\delta} = \frac{u(s)}{\delta(s)} = \frac{\begin{vmatrix} 0 & g \\ Z_\delta & -sU_0 \end{vmatrix}}{\begin{vmatrix} s - X_u & g \\ -Z_u & -sU_0 \end{vmatrix}} \tag{F17}$$

and there is a corresponding expression for $\theta(s)/\delta(s)$. Expanding the determinants in eqn. (F17) we obtain:

$$G_{u\delta} = \frac{-gZ_\delta}{-s^2U_0 + X_uU_0s + gZ_u} = \frac{gZ_\delta}{U_0\left(s^2 - X_us - \dfrac{gZ_u}{U_0}\right)} \tag{F18}$$

Typical response characteristics of an aircraft based on this two-degree of freedom approximation of the phugoid mode are shown in Ref. 4 (Chapter III, Section 5).

3.3. Aircraft Closed-loop Response

The transfer functions in para. 3.2 represent the basic stability characteristics of the aircraft, and thus would be associated with the block marked "Controlled System (Aircraft)" in Figs. 19.1 and 19.2 of Chapter 19. However, as shown in Fig. 19.1, feedback is often used to complete the loop, and the overall transfer function of the complete system is a combination of the transfer function of the aircraft, that of the feedback elements, and that of the control elements (see para. 3.2.4 of Chapter 19, and eqn. 19.11). As already mentioned, the closure of the loop may be by means of a human pilot or by means of an automatic control system.

Ref. 2 discusses the closed-loop response characteristics of helicopters and VTOL aircraft, with special emphasis on human rather than on automatic pilots. In order to carry out the mathematical analysis, the human pilot is given a transfer function (Ref. 2, p. 59; Ref. 6, p. 5) of the form

$$Y_p = \frac{K_p e^{-\tau s}(T_L s + 1)}{(T_I s + 1)(T_N s + 1)} \tag{F19}$$

where K_p = pilot gain
τ = reaction time delay
T_L = lead time constant
T_I = lag time constant
T_N = neuromuscular lag time constant.

The closed loop response characteristics for various conditions of flight are illustrated in Ref. 2 by Bode and root locus plots. Similar results from the analysis of the closed-loop response of V/STOL aircraft can be found in Ref. 6.

The response characteristics of aircraft using automatic systems to close the loop can be analysed in a similar way, as long as the transfer function of the feedback loop is known. This topic is discussed briefly in para. 7 of Chapter 15, and para. 6 of Chapter 16, but more details can be found in the appropriate references of Chapters 15 and 16, and in Ref. 3. Ref. 8 discusses the stability problem of VTOL aircraft at the hover, and shows how different feedback transfer functions can be used to improve the stability. The corresponding closed-loop transfer functions are given in Fig. 4 of the reference, and root locus plots and analogue traces are given to show how the stability is improved.

References

1. Etkin, B.; *Dynamics of Flight*, (John Wiley and Sons, New York, 1959).
2. Wolkovitch, J. and Walton, R. P.; *VTOL and Helicopter Approximate Transfer Functions and Closed-Loop Handling Qualities*, Systems Technology Inc. U.S.A., S.T.I. Technical Report No. 128–1, June 1965.
3. Blakelock, J. H.; *Automatic Control of Aircraft and Missiles* (John Wiley and Sons, New York, 1965).
4. *Dynamics of the Airframe*, Bureau of Aeronautics, U.S. Navy, BU AER Report AE-61-4 II, September 1952.
5. Woodcock, R. J. and Drake, D. E.; *Estimation of Flying Qualities of Piloted Airplanes*, Wright-Patterson Air Force Base, Ohio, U.S.A., Report AFFDL-TR-65-218.
6. Stapleford, R. L., Wolkovitch, J. *et al.*; *An Analytical Study of V/STOL Handling Qualities in Hover and Transition.* Wright-Patterson Air Force Base, Ohio, U.S.A., Report AFFDL-TR-65-73.
7. Seckel, E.; *Stability and Control of Airplanes and Helicopters* (Academic Press, New York, 1964).
8. Elliott, D. A. and Schreiber, C. R.; "Improving the Stability of Hovering Aircraft," *Canadian Aeronautics and Space Journal*, Vol. II, No. 3, March 1965, p. 79.

Index

Stick shakers and pushers, 96, 107